FIRST EDIT

ELEMENTARY LINEAR ALGEBRA

W. KEITH NICHOLSON
UNIVERSITY OF CALGARY

Toronto Montréal Boston Burr Ridge, IL Dubuque, IA Madison, WI New York San Francisco
St. Louis Bangkok Bogotá Caracas Kuala Lumpur Lisbon London Madrid Mexico City Milan
New Delhi Santiago Seoul Singapore Sydney Taipei

*McGraw-Hill
Ryerson Limited*
A Subsidiary of The **McGraw·Hill** Companies

**Elementary Linear Algebra
First Edition
W. Keith Nicholson**

Copyright © 2001 by McGraw-Hill Ryerson Limited, a Subsidiary of The McGraw-Hill Companies. All rights reserved. No part of this publication may be reproduced or transmitted in any form or by any means, or stored in a data base or retrieval system, without the prior written permission of McGraw-Hill Ryerson Limited, or in the case of photocopying or other reprographic copying, a licence from CANCOPY (the Canadian Copyright Licensing Agency), 6 Adelaide Street East, Suite 900, Toronto, Ontario, M5C 1H6.

ISBN: 0-07-089229-6

2 3 4 5 6 7 8 9 10 TCG 0 9 8 7 6 5 4 3 2 1

Printed and bound in Canada.

Care has been taken to trace ownership of copyright material contained in this text; however, the publisher will welcome any information that enables them to rectify any reference or credit for subsequent editions.

Vice-President, Editorial Director: Pat Ferrier
Senior Sponsoring Editor: Cathy Koop
Marketing Manager: Rhondda McNabb
Developmental Editor: Darren Hick
Supervising Editor: Alissa Messner
Production Coordinator: Nicla Dattolico
Cover Design: Sharon Lucas
Cover Image: © Jeff Smith/Imagebank
Printer: Transcontinental Printing Group

To Kathleen, Jason and Mark

Contents

PREFACE   xi

ACKNOWLEDGEMENTS   xvii

ILAW: Interactive Linear Algebra on the Web   xix

1 LINEAR EQUATIONS AND MATRICES   1
- 1.1 MATRICES 1
  - 1.1.1 Matrices 2
  - 1.1.2 Matrix Addition 3
  - 1.1.3 Scalar Multiplication 6
  - 1.1.4 Transposition 8
  - Exercises 11
- 1.2 LINEAR EQUATIONS 14
  - 1.2.1 Linear Equations 14
  - 1.2.2 Systems of Linear Equations 15
  - 1.2.3 Gaussian Elimination 19
  - 1.2.4 Rank of a Matrix 24
  - Exercises 26
- 1.3 HOMOGENEOUS SYSTEMS 29
  - 1.3.1 Homogeneous Systems 30
  - 1.3.2 Basic Solutions 31
  - Exercises 33
- 1.4 MATRIX MULTIPLICATION 35
  - 1.4.1 Matrix Multiplication 36
  - 1.4.2 Properties of Matrix Multiplication 39
  - 1.4.3 Matrix Multiplication and Linear Equations 41
  - 1.4.4 Block Multiplication 43
  - 1.4.5 Directed Graphs 45
  - 1.4.6 Proof of Theorem 1 47
  - Exercises 48
- 1.5 MATRIX INVERSES 52
  - 1.5.1 Matrix Inverses 53
  - 1.5.2 Inverses and Systems of Linear Equations 56

		1.5.3	The Matrix Inversion Algorithm	57
		1.5.4	Properties of Inverses	60
		1.5.5	Conditions for Invertibility	63
		Exercises		65
	1.6	ELEMENTARY MATRICES		69
		1.6.1	Elementary Matrices	69
		1.6.2	Applications to Inverses and Rank	71
		1.6.3	Uniqueness of Reduced Row-echelon Matrices	74
		Exercises		75
	1.7	LU-FACTORIZATION		78
		1.7.1	Triangular Matrices	78
		1.7.2	LU-Factorization	79
		1.7.3	Dealing with Row Interchanges	82
		Exercises		84
	1.8	APPLICATION TO MARKOV CHAINS		86
		1.8.1	An Example	86
		1.8.2	Markov Chains	89
		1.8.3	Proof of (1) of Theorem 1	92
		Exercises		92

2 DETERMINANTS AND DIAGONALIZATION — 97

	2.1	THE LAPLACE EXPANSION		97
		2.1.1	Laplace Expansion	98
		2.1.2	Elementary Operations and Determinants	101
		Exercises		106
	2.2	DETERMINANTS AND INVERSES		109
		2.2.1	The Product Theorem	109
		2.2.2	Adjoint of a Matrix	111
		2.2.3	Cramer's Rule	114
		2.2.4	Proof of the Product Theorem	115
		2.2.5	Proof of the Adjoint Formula	116
		Exercises		117
	2.3	DIAGONALIZATION AND EIGENVALUES		119
		2.3.1	Diagonalization	119
		2.3.2	Eigenvalues and Eigenvectors	121
		2.3.3	Diagonalization	124
		2.3.4	Similar Matrices	128
		Exercises		131
	2.4	LINEAR DYNAMICAL SYSTEMS		133
		2.4.1	Linear Dynamical Systems	134
		2.4.2	Dominant Eigenvalues	136
		2.4.3	A Predator-Prey Model	139
		2.4.4	Graphical Description of Solutions	142
		Exercises		145
	2.5	COMPLEX EIGENVALUES		146
		2.5.1	Complex Numbers	147

	2.5.2	Quadratics	150
	2.5.3	Complex Linear Algebra	150
	2.5.4	Roots of Polynomials	152
	2.5.5	Symmetric Matrices	153
	2.5.6	Polar Form	153
	2.5.7	Proof of Theorem 3	158
	Exercises	....	159
2.6	LINEAR RECURRENCES		162
	2.6.1	Linear Recurrences	162
	Exercises	....	166
2.7	POLYNOMIAL INTERPOLATION		168
	2.7.1	Polynomial Interpolation	168
	Exercises	....	171
2.8	SYSTEMS OF DIFFERENTIAL EQUATIONS		173
	2.8.1	The Diagonalization Method	173
	Exercises	....	177

3 VECTOR GEOMETRY — 179

3.1	GEOMETRIC VECTORS		179
	3.1.1	Coordinate Systems	179
	3.1.2	Vectors and Scalars	180
	3.1.3	Matrix Form	181
	3.1.4	Geometric Descriptions	182
	3.1.5	Vector Addition	184
	3.1.6	Scalar Multiplication	188
	3.1.7	A Proof of Pythagoras' Theorem	190
	Exercises	....	191
3.2	DOT PRODUCT AND PROJECTIONS		194
	3.2.1	The Dot Product	194
	3.2.2	Angles	196
	3.2.3	Projections	199
	Exercises	....	202
3.3	LINES AND PLANES		205
	3.3.1	Position Vectors	205
	3.3.2	Lines	206
	3.3.3	Planes	211
	Exercises	....	214
3.4	THE CROSS PRODUCT		217
	3.4.1	The Cross Product	217
	3.4.2	Other Properties of the Cross Product	221
	3.4.3	Geometrical Meaning of the Cross Product	224
	Exercises	....	225
3.5	MATRIX TRANSFORMATIONS OF $\mathbb{R}^2$		228
	3.5.1	Transformations	228
	3.5.2	Linear Transformations	231
	3.5.3	Effect on the Unit Square	234

		3.5.4	Composition of Matrix Transformations	237

        3.5.4   Composition of Matrix Transformations 237
        3.5.5   Inverse of a Matrix Transformation 239
        3.5.6   Computer Graphics 241
        3.5.7   Isometries 242
        Exercises 243

4 THE VECTOR SPACE $\mathbb{R}^n$ — 249

4.1 SUBSPACES AND SPANNING — 249
- 4.1.1 An Example — 249
- 4.1.2 Euclidean n-space — 250
- 4.1.3 Subspaces of $\mathbb{R}^n$ — 251
- 4.1.4 Spanning Sets — 253
- Exercises — 257

4.2 LINEAR INDEPENDENCE — 260
- 4.2.1 Independent Sets of Vectors — 260
- 4.2.2 Invertibility of Matrices — 263
- 4.2.3 Linear Dependence — 264
- Exercises — 266

4.3 DIMENSION — 268
- 4.3.1 Fundamental Theorem — 269
- 4.3.2 Existence of Bases — 271
- 4.3.3 Proof of the Fundamental Theorem — 275
- Exercises — 276

4.4 RANK — 279
- 4.4.1 Row and Column Spaces — 280
- 4.4.2 The Rank Theorem — 281
- 4.4.3 Null Space and Image — 286
- Exercises — 288

4.5 ORTHOGONALITY — 291
- 4.5.1 Dot Product, Length and Distance — 291
- 4.5.2 Orthogonal Sets and the Expansion Theorem — 295
- 4.5.3 The Gram-Schmidt Algorithm — 297
- 4.5.4 QR-Factorization — 300
- 4.5.5 Proof of Uniqueness in the QR-Factorization — 303
- Exercises — 304

4.6 PROJECTIONS AND APPROXIMATION — 306
- 4.6.1 Orthogonal Complements — 307
- 4.6.2 Projections — 308
- 4.6.3 Approximation — 311
- 4.6.4 Inconsistent Systems — 312
- 4.6.5 Least Squares Approximation — 315
- Exercises — 319

4.7 ORTHOGONAL DIAGONALIZATION — 322
- 4.7.1 Diagonalization Revisited — 322
- 4.7.2 Orthogonal Matrices — 326
- 4.7.3 The Principal Axis Theorem — 328

		4.7.4 Triangulation	331
		Exercises	333
	4.8	QUADRATIC FORMS	336
		4.8.1 Quadratic Forms	336
		4.8.2 Positive Definite Matrices	341
		4.8.3 Constrained Optimization	345
		4.8.4 Statistical Principal Component Analysis	348
		Exercises	350
	4.9	LINEAR TRANSFORMATIONS	353
		4.9.1 Transformations $\mathbb{R}^n \to \mathbb{R}^m$	353
		4.9.2 Changing Coordinates	359
		4.9.3 The $\mathcal{F}$-matrix of a Linear Operator	362
		4.9.4 Similarity	366
		4.9.5 Isometries	368
		Exercises	373
	4.10	COMPLEX MATRICES	376
		4.10.1 Complex Inner Products	376
		4.10.2 Complex Matrices	378
		4.10.3 Hermitian and Unitary Matrices	379
		4.10.4 Unitary Diagonalization	382
		Exercises	385
	4.11	SINGULAR VALUE DECOMPOSITION	387
		4.11.1 The Singular Value Decomposition	388
		4.11.2 The Fundamental Subspaces	392
		4.11.3 The Polar Decomposition	393
		Exercises	396

5 VECTOR SPACES — 399

	5.1	EXAMPLES AND BASIC PROPERTIES	399
		5.1.1 Vector Spaces	399
		5.1.2 Subspaces	405
		Exercises	408
	5.2	INDEPENDENCE AND DIMENSION	412
		5.2.1 Independence and the Fundamental Theorem	412
		Exercises	420
	5.3	LINEAR TRANSFORMATIONS	425
		5.3.1 Linear Transformations	425
		5.3.2 Kernel and Image	430
		5.3.3 The Dimension Theorem	433
		Exercises	435
	5.4	ISOMORPHISMS AND MATRICES	439
		5.4.1 Isomorphisms	439
		5.4.2 Composition	442
		5.4.3 Coordinates	445
		5.4.4 The Matrix of a Linear Transformation	449
		Exercises	452

5.5 LINEAR OPERATORS AND SIMILARITY ... 456
- 5.5.1 The $\mathcal{B}$-Matrix of an Operator ... 456
- 5.5.2 Change of Basis ... 459
- 5.5.3 Diagonalization ... 461
- Exercises ... 466

5.6 INVARIANT SUBSPACES ... 468
- 5.6.1 Invariant Subspaces ... 469
- 5.6.2 Direct Sums ... 472
- 5.6.3 Reducible Operators ... 476
- 5.6.4 The Cayley-Hamilton Theorem ... 479
- Exercises ... 482

5.7 GENERAL INNER PRODUCTS ... 485
- 5.7.1 Inner Products ... 485
- 5.7.2 Norms and Orthogonality ... 488
- 5.7.3 Projections and Approximation ... 493
- 5.7.4 Fourier Approximation ... 496
- 5.7.5 Legendre Polynomials ... 499
- Exercises ... 501

APPENDIX 505
- A.1 Basic Trigonometry ... 505
- A.2 Induction ... 509
- A.3 Polynomials ... 511

SELECTED HINTS AND SOLUTIONS 515

INDEX 581

PREFACE

This book springs from three sources: The needs of the users of linear algebra as a service course, the trend away from the abstract view of the subject toward a more matrix approach, and the desire to utilize technology to help with instruction and computation. Let me explain.

In the late 1990's discussions with Engineering and Science revealed a desire for early introduction of diagonalization and for more emphasis on linear transformations. Consequently eigenvalues and diagonalization are introduced in Chapter 2 (using only determinants and matrix algebra), motivated by examples of dynamical systems that the students can relate to; and linear transformations of the plane are discussed in Chapter 3, giving geometrical interpretations of multiplication, determinants and inverses of 2×2 matrices, and providing a solid introduction to the treatment in Chapter 4 of Euclidean n-space.

The desire for a more matrix oriented linear algebra course expressed by the engineers and scientists echos a more general trend in the subject away from the abstract approach. The second half of the book (Chapters 4 and 5) continues this evolution. Every teacher of linear algebra knows that the students "hit the wall" when the notion of an abstract vector space is introduced. One reason for this is that they are coping simultaneously with *two* new ideas: the concept of an abstract structure, and mastering difficult notions like spanning, independence and linear transformations. This double jeopardy is difficult to deal with for students, even the most talented ones. Consequently, in Chapter 4 topics like independence are discussed first in the context of $\mathbb{R}^n$, so the students can deal with them without having to cope with the abstract baggage of an n-dimensional space. Then the general setup is introduced in Chapter 5, with $\mathbb{R}^n$ as the principal motivating example. This has been tested in the classroom, and it works very well pedagogically.

The third theme is to involve technology in a way that is more than merely a "book-in-the-computer", and is more accessible than a computational package. This has been achieved in the creation of ILAW—Interactive Linear Algebra on the Web. This is an (optional) internet based computer tutorial that provides audio-enhanced lessons covering the entire text, interactive explorations allowing the student to "play" with the central concepts or perform the more difficult computations, and self-assessment in the form of labs with feedback to help the student discover where and why he or she made an error. The system has been class tested, and is available (at a nominal price) to users of this book.[1]

On the other hand, this book serves as a stand-alone text for the traditional linear algebra syllabus (see the Table of Contents). It can be used for a two-semester course for beginners with a working knowledge of High School Algebra, or for a single semester course for more advanced students. For students in many application areas, Chapters 1 through 4 may be all the linear algebra they need. While the text is designed with Science and Engineering students in mind, it will also serve the needs of students in Management and the Social Sciences. Calculus is not a prerequisite; the few places where it is needed are clearly marked and can be omitted with no loss of continuity.

Wherever possible, concepts are introduced with real-world examples that are meaningful to the students, and theorems are stated simply with short proofs often preceded by an example. (Longer, or more difficult proofs are deferred to the end of the section and can be omitted.) Nearly 350 solved examples are included to aid understanding, introduce techniques, illustrate algorithms and motivate theorems. These examples are reinforced by the exercises (about 800 in all) at the end of each section, many with answers or solutions at the end of the book.

Features

- *Early eigenvalues.* Eigenvalues and diagonalization are introduced in Chapter 2 using only determinants and matrix inverses, and not requiring notions like dimension and independence. This allows a number of applications (for example dynamical systems) that are seen as relevant by the students.

- *Early introduction of linear transformations.* In response to a request from the Engineers, rotations, reflections, shears, etc. in $\mathbb{R}^2$ are described using matrix multiplication in Section 3.5. The idea of a linear transformation is introduced in this context, and its relation to matrix multiplication is made explicit. This leads to geometrical interpretations of matrix multiplication, inverses and determinants, and provides a basis for the study of linear transformations in $\mathbb{R}^n$ in Chapter 4.

- *Matrix approach.* In keeping with a trend over the past few years, more emphasis is placed on matrix computations. For example, the following topics are discussed: Positive definite matrices, LU-factorization, QR-factorization, Triangular form, Complex matrices, Singular value decomposition. This has the effect of reducing the overall level of abstraction in the book.

---

[1]More information about ILAW can be found on page xix.

- *Vector space concepts introduced first in $\mathbb{R}^n$.* Independence, spanning, subspaces and dimension are introduced first in $\mathbb{R}^n$ (in Chapter 4), before the student has to cope with abstract vector spaces. The general notions then come easily in the abstract context (Chapter 5), allowing the student to focus on the new conceptual framework.

- *Applications.* These occur both as separate sub-sections or as examples in the text. Examples: Markov chains, systems of differential equations, directed graphs, linear dynamical systems, computer graphics, statistical principal component analysis, constrained optimization.

- *Motivating examples.* Wherever possible, concepts are introduced with examples that are not only understandable to the students but are seen as relevant to the real world. A good illustration is the motivation (in Section 2.3) of diagonalization by looking at the possible extinction of a species of birds.

- *Solved examples.* Nearly 350 solved examples are given to help the student to understand the techniques of the subject, and to motivate theorems and proofs.

- *Exercises.* Nearly 800 exercises are included, keyed to the examples. They start with routine, computational problems and progress to more theoretical exercises. Every exercise has a hint at the end of the book, and about half have answers or solutions.

- *Difficult proofs.* These are deferred to the end of the section.

- *Approximation.* Best possible solutions to inconsistent systems are discussed, leading to the normal equations and to the method of least squares.

- *Complex matrices.* Optional treatment including Hermitian, unitary and normal matrices, the spectral theorem, and Schur's theorem.

- *Singular value decomposition.* Optional treatment including the fundamental subspaces and the polar decomposition.

- *Appendices.* Brief reviews are included of trigonometry, induction and polynomials.

Supplements

- *Text website:* A website containing information about the text and supplements can be found at http://www.mcgrawhill.ca/college/nicholson.

- *Instructor solution manual.* Includes an answer or solution to every exercise in the text.

- *Internet tutorial and assessment.* A student tutorial is available over the internet at a reduced cost to any student using this book. It includes several features to aid student understanding (see description on page xix), and the labs can be used by instructors for secure testing.

Chapter Summaries

Chapter 1. Linear Equations and Matrices. After a brief introduction to matrix addition, scalar multiplication and transposition, the Gaussian algorithm for solving systems of linear equations is given. Then matrix multiplication is defined and applied to linear systems, and the main features of matrix inverses are presented. General block multiplication is mentioned, but the emphasis is on three basic cases that are used throughout the book. Elementary matrices are included (but may be omitted with no loss of continuity), and optional sections on LU-factorization and Markov chains are presented.

Chapter 2. Determinants and Diagonalization. Determinants are defined inductively, the Laplace expansion is stated (motivated by the 2×2 case), and the usual rules deduced. Hence the student can immediately begin computing determinants using row and column operations. A major innovation is that diagonalization is introduced at this level using only determinants and matrix arithmetic, with no need for vector space concepts like basis and dimension. This leads to linear dynamical systems which provide many motivating examples that are meaningful for the students. Because of the possibility of non-real eigenvalues, this leads naturally into a discussion of complex numbers. The chapter closes with optional applications to linear recurrences and polynomial interpolation.

Chapter 3. Vector Geometry. Motivated by examples, vectors are introduced in $\mathbb{R}^2$ and $\mathbb{R}^3$ as matrices, and the matrix operations are interpreted geometrically. Then angles, orthogonality and projections are brought in via the dot product, and used to describe lines and planes in $\mathbb{R}^3$. This leads to a discussion of the cross product. The chapter concludes with a discussion of matrix transformations in $\mathbb{R}^2$ viewed as linear transformations, leading to geometric interpretations of matrix multiplication, determinants, and inverses.

Chapter 4. The Vector Space $\mathbb{R}^n$. The basic concepts of subspace, spanning, independence, basis and dimension are all introduced for subspaces of $\mathbb{R}^n$. This allows the student to learn these new ideas in a familiar context, without the extra complication of having to deal with abstract spaces. After discussing rank, the dot product is introduced, the expansion theorem is proved, and orthogonal bases are introduced (including the Gram-Schmidt algorithm). Then orthogonal complements are defined, the projection theorem is proved, and the approximation theorem is given (leading to best possible solutions of inconsistent systems, and to least squares). Next, orthogonal diagonalization is developed and the principal axis theorem for symmetric matrices is proved. This leads to quadratic forms, positive definite matrices and the Cholesky decomposition (and to applications to constrained optimization and statistical principal component analysis). Then linear transformations are discussed for $\mathbb{R}^n$, the effect of a basis change is noted and related to similarity, and the isometries of $\mathbb{R}^3$ are determined. The chapter concludes with optional sections on complex matrices and the singular value decomposition.

Chapter 5. Vector Spaces. General vector spaces are introduced, with spaces of matrices, polynomials and functions as the main new examples, and their basic properties are deduced (with $\mathbb{R}^n$ as the motivating example). Concepts like independence and basis are now routine extensions of their $\mathbb{R}^n$ counterparts, and the student can focus on becoming accustomed to working in the abstract setting. General linear transformations (and their matrices) are introduced, one-to-one and onto mappings are discussed, and isomorphisms are characterized. Then the effect of a basis change on operators is described, with its relation to similarity. This leads to a discussion of invariant subspaces and, as an (optional) illustration, to the Cayley-Hamilton theorem. The chapter (and the book) ends with a discussion of general inner products, with applications to numerical integration, Fourier approximation and orthogonal polynomials.

Chapter Dependencies

The diagram suggests how the material in each chapter depends on earlier chapters. The dependence of Chapter 4 on Chapter 3 is primarily for motivation and examples. A dotted line in the diagram indicates a minor dpendency.

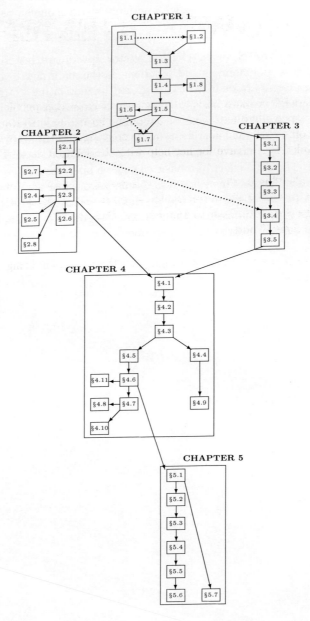

Visit the text website at http://www.mcgrawhill.ca/college/nicholson

ACKNOWLEDGEMENTS

It is a pleasure to recognize the contributions of many other people to this book. Thanks are especially due to Claude Laflamme for his enthusiastic support and invaluable comments, and for his drive in creating the ILAW tutorial. I also want to thank Claire Sauvé for her help with the solution manual.

Thanks are also due to the Editorial and Production staff at McGraw-Hill Ryerson. First to Cathy Koop for her enthusiasm and determination to get the project under way, and also to Pat Ferrier, Darren Hick, Jason Stanley, Jeff MacLean, and Bill Todd.

I also want to thank the group in Calgary for their efforts in bringing the ILAW project to life.

ILAW: Interactive Linear Algebra on the Web
http://www.mcgrawhill.ca/college/nicholson

Lyryx Learning Inc.
lyryx.com

ILAW is a self paced, interactive tutorial and assessment system available on the internet at greatly reduced cost to users of this book. It provides a college-level introduction to the concepts of linear algebra. Our objective has been to create appropriate tools to support students and instructors in all aspects of the course.

Features for Students:

- *Lessons*. These are animated, audio enhanced tutorials, that reinforce concepts in much the way a patient human tutor would do it. Students find the lessons useful for study or review, and they are available any time of the day or night, whenever you are in the mood to study.

- *Explorations*. These provide hands-on student interaction with the concepts of the course to enhance understanding. We believe that exploring the concepts, and so developing your understanding through concrete manipulation, is an essential part of learning and *doing* mathematics.

- *Labs*. These allow you to gauge your understanding of the material by working problems, with feed-back explaining how and why a submitted answer was incorrect, and guidance for improvement utilizing the Lessons and Explorations. The lab problems are randomly generated but similar each time, and so can be taken as often as desired.

- *Computing Tool*. This has an easy to use, cut-and-paste capability that allows you to do all the calculations in the Labs. Moreover, it utilizes the same standard notation used in the text, rather than the artificial programming format of calculators and computing packages, and so helps overcome another barrier to using computing tools for learning mathematics. In particular, matrix objects *look* like a matrix and *behave* as you would expect (when entering data directly, for example).

- *Message Board.* This is a feature that allows you to communicate with other students (and with an ILAW monitor) and exchange views about the course.

Features for Instructors:

- *Secure Testing.* The lab grades are held securely on our server, and instructors can access them at any time, along with labs themselves.

- *Economical.* Using ILAW for your course has little or no cost to your college. We have found that most students have their own computer, so very little expensive infrastructure is required.

- *Inexpensive for your students.* If you adopt this book, your students can obtain ILAW for less than the cost of couple of hours of private tutoring.

Chapter 1

LINEAR EQUATIONS AND MATRICES

1.1 MATRICES

In mathematics and its applications, numbers frequently appear naturally in ordered rectangular arrays. Here are a few examples.

Example 1. The coordinates of a point in the plane are usually written as an ordered pair (x, y). Here *ordered* means that the order of the coordinates x and y is important. For example, $(2, 3)$ and $(3, 2)$ represent *different* points. Similarly the coordinates of a point in space are written as an ordered triple (x, y, z). □

Example 2. The system of linear equations
$$\begin{aligned} 3x - 5y &= 2 \\ 4x + 7y &= 1 \end{aligned}$$
is completely described by the 2×3 array of numbers
$$\begin{bmatrix} 3 & -5 & 2 \\ 4 & 7 & 1 \end{bmatrix}.$$
In fact this is how the system of equations is represented in a computer. We will return to this in Section 1.2. □

Example 3. Statistical data are often displayed in arrays called tables. For example, the number of male, female, undergraduate and graduate students at a small college is given in the following table

	undergraduate	graduate
women	3013	256
men	3155	511

Such a table gives a clear, graphic description of the data, and has other uses as we shall see. □

Example 4. Directed graphical networks like the one shown occur frequently in the scheduling of large projects. They are described by the adjacency table

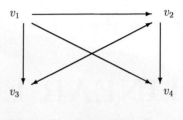

	v_1	v_2	v_3	v_4
v_1	0	1	1	1
v_2	0	0	1	1
v_3	0	1	0	0
v_4	0	0	0	0

where a 1 (or a 0) in row v_i and column v_j indicates an arrow (or none) from vertex v_i to vertex v_j. For example, the vertices could represent cities, and the arrows available flights. By eliminating the arrowheads and allowing more than one edge between two vertices, this gives a model for chemical bonding. □

1.1.1 Matrices

Motivated by these examples, a rectangular array of numbers is called a **matrix**[1], and the numbers themselves are called the **entries**[2] of the matrix. Matrices are usually denoted by upper case letters A, B, C, etc.

Example 5. The following are matrices:

$$A = \begin{bmatrix} 1 & 0 & -3 \\ 5 & 3 & 8 \end{bmatrix} \quad B = \begin{bmatrix} 2 & -5 \end{bmatrix} \quad C = \begin{bmatrix} 2 \\ 0 \\ -3 \end{bmatrix} \quad D = \begin{bmatrix} 3 & 1 & -1 \\ 2 & 0 & 5 \\ -2 & -2 & 1 \end{bmatrix}$$

Matrices come in various (rectangular) shapes depending on the number of **rows** and **columns**. For example, the matrix A in Example 5 has 2 rows and 3 columns. In general a matrix with m rows and n columns is called an **m × n matrix**, or is said to have **size m × n**. Thus the matrices A, B, C and D in Example 5 have size 2×3, 1×2, 3×1 and 3×3 respectively. Not surprisingly, a matrix of size $1 \times n$ is called a **row matrix** or an n**-row**, and one of size $m \times 1$ is called a **column matrix** or an m**-column**. A matrix with equal numbers of rows and columns is called a **square matrix**. Thus each square matrix has size $n \times n$ for some n.

Each entry of a matrix is located by the row and column in which it lies. The rows of the matrix are numbered from the top down, and the columns are

---
[1] While the term *matrix* was first used in 1848 by James Joseph Sylvester (1814-1897), it was Arthur Cayley (1821-1895) who first considered matrices as single entities in a paper in 1858 entitled "A memoir on the theory of matrices".

[2] Definitions of new terms appear in bold face throughout the text.

1.1. MATRICES

numbered from left to right. Then the entry in row i and column j is called the **(i,j)-entry** of the matrix. For example:

The (1,2)-entry of $\begin{bmatrix} 1 & 0 & -3 \\ 5 & 3 & 8 \end{bmatrix}$ is 0, while the (2,3)-entry is 8.

A special notation has been devised for the entries of a matrix A. If A is of size $m \times n$, and if the (i,j)-entry of A is denoted by a_{ij}, then A is displayed as follows:

$$A = \begin{bmatrix} a_{11} & a_{12} & a_{13} & \cdots & a_{1n} \\ a_{21} & a_{22} & a_{23} & \cdots & a_{2n} \\ \vdots & \vdots & \vdots & & \vdots \\ a_{m1} & a_{m2} & a_{m3} & \cdots & a_{mn} \end{bmatrix}.$$

This is usually written simply as $A = [a_{ij}]$, and a_{ij} is the entry in row i and column j. For example, using this notation the general 2×3 matrix is written

$$A = \begin{bmatrix} a_{11} & a_{12} & a_{13} \\ a_{21} & a_{22} & a_{23} \end{bmatrix}.$$

Note that, as for ordered pairs, we regard two matrices A and B as **equal** (written $A = B$) if they have the same number of rows and columns and if corresponding entries are equal. More precisely:

$A = B$ if and only if $\begin{cases} 1. \ A \text{ and } B \text{ have the same size} \\ 2. \text{ Corresponding entries are equal} \end{cases}$

If $A = [a_{ij}]$ and $B = [b_{ij}]$ are both $m \times n$, this takes the form

$$[a_{ij}] = [b_{ij}] \quad \text{if and only if} \quad a_{ij} = b_{ij} \text{ for all } i \text{ and } j.$$

Example 6. Given $A = \begin{bmatrix} 1 & -3 & 0 \\ 2 & 5 & 0 \end{bmatrix}$, $B = \begin{bmatrix} 1 & -3 \\ 2 & 5 \end{bmatrix}$ $C = \begin{bmatrix} x & y \\ z & w \end{bmatrix}$, discuss the possibility that $A = B$, $A = C$ and $B = C$.

Solution. $A = B$ is impossible because A and B are of different sizes: A is 2×3 while B is 2×2. Thus A and B are *not* equal (written $A \neq B$). Similarly $A \neq C$. However $B = C$ is possible provided that corresponding entries are equal: $\begin{bmatrix} 1 & -3 \\ 2 & 5 \end{bmatrix} = \begin{bmatrix} x & y \\ z & w \end{bmatrix}$ can only happen if $x = 1$, $y = -3$, $z = 2$ and $w = 5$. □

1.1.2 Matrix Addition

Matrices can be added in much the same way as numbers. The following example shows how this arises in practise.

Example 7. The yearly government tax revenue from Corporation 1 (in millions) is $175m in corporation tax, $35m in income tax and $17m in sales tax. This data is summarized in the revenue matrix

$$T_1 = \begin{bmatrix} 175 & 35 & 17 \end{bmatrix}.$$

Similarly, Corporation 2 has revenue matrix

$$T_2 = \begin{bmatrix} 190 & 41 & 22 \end{bmatrix}.$$

The combined revenue matrix for both corporations is thus

$$T = \begin{bmatrix} 175+190 & 35+41 & 17+22 \end{bmatrix} = \begin{bmatrix} 365 & 76 & 39 \end{bmatrix}.$$

This new matrix is obtained by adding corresponding entries in T_1 and T_2. □

In general, if A and B are two matrices of the same size, their **sum** $A+B$ is defined to be the matrix of the same size formed by adding corresponding entries. If $A = [a_{ij}]$ and $B = [b_{ij}]$ this takes the form

$$[a_{ij}] + [b_{ij}] = [a_{ij} + b_{ij}].$$

Similarly, the **difference** $A - B$ of A and B is defined by subtracting corresponding entries:

$$[a_{ij}] - [b_{ij}] = [a_{ij} - b_{ij}].$$

Note that addition and subtraction are not defined for matrices of different sizes. Hence:

When we write $A + B$ or $A - B$, it is assumed that A and B are the same size

We will use this convention without comment below.

Example 8. If $A = \begin{bmatrix} 2 & -1 & 7 \\ -3 & 5 & 0 \end{bmatrix}$ and $B = \begin{bmatrix} 4 & 6 & -2 \\ 8 & 1 & 9 \end{bmatrix}$ then

$$A + B = \begin{bmatrix} 2+4 & -1+6 & 7-2 \\ -3+8 & 5+1 & 0+9 \end{bmatrix} = \begin{bmatrix} 6 & 5 & 5 \\ 5 & 6 & 9 \end{bmatrix}$$

$$A - B = \begin{bmatrix} 2-4 & -1-6 & 7-(-2) \\ -3-8 & 5-1 & 0-9 \end{bmatrix} = \begin{bmatrix} -2 & -7 & 9 \\ -11 & 4 & -9 \end{bmatrix}$$

□

Many general properties of numerical addition also hold for matrix addition. If A, B and C are matrices of the same size, then:

$$A + B = B + A \qquad \text{(Commutative Law)}$$
$$A + (B + C) = (A + B) + C \qquad \text{(Associative Law)}$$

1.1. MATRICES

In fact if $A = [a_{ij}]$ and $B = [b_{ij}]$ then the (i,j)-entries of $A + B$ and $B + A$ are $a_{ij} + b_{ij}$ and $b_{ij} + a_{ij}$ respectively. Since these are equal for all i and j, we obtain

$$A + B = [a_{ij} + b_{ij}] = [b_{ij} + a_{ij}] = B + A.$$

This proves the commutative law, and the associative law is similarly verified.

The $m \times n$ matrix in which every entry is zero is called the **zero matrix** of that size, and is denoted by 0 (or 0_{mn} if the size must be noted).[3] It is clear that

$$0 + A = A \quad \text{for every } m \times n \text{ matrix } A.$$

The **negative** of the $m \times n$ matrix A (denoted $-A$) is the $m \times n$ matrix obtained by negating every entry of A. If $A = [a_{ij}]$ this can be written

$$-[a_{ij}] = [-a_{ij}].$$

In other words, the (i,j)-entry of $-A$ is $-a_{ij}$. Since matrix addition is carried out component-wise, we have

$$A + (-A) = 0 \quad \text{for every } m \times n \text{ matrix } A$$

where 0 denotes the $m \times n$ zero matrix. The following theorem collects these four basic properties of matrix addition for reference later.

Theorem 1. *If A, B and C denote arbitrary $m \times n$ matrices, then*

(1) $A + B = B + A$.

(2) $A + (B + C) = (A + B) + C$.

(3) $0 + A = A$ where 0 is the $m \times n$ zero matrix.

(4) $A + (-A) = 0$.

Example 9. If $A = \begin{bmatrix} 3 & 1 \\ 6 & 4 \end{bmatrix}$ and $B = \begin{bmatrix} 1 & -2 \\ 0 & 4 \end{bmatrix}$, find a matrix X such that $X + B = A$.

Solution. To solve a numerical equation $x + b = a$, one simply subtracts b from both sides to obtain $x = a - b$. Similarly, the matrix equation $X + B = A$ leads to

$$X = A - B = \begin{bmatrix} 3 - 1 & 1 - (-2) \\ 6 - 0 & 4 - 4 \end{bmatrix} = \begin{bmatrix} 2 & 3 \\ 6 & 0 \end{bmatrix}.$$

---

[3]The same symbol 0 is commonly used for a zero matrix and for the number zero. This causes very little confusion in practice since the meaning is nearly always clear from the context.

The reader can check that the matrix X does indeed satisfy the equation $X+B = A$. □

The properties in Theorem 1 have several useful consequences. For example, property (2) asserts that the sum $A + (B + C) = (A + B) + C$ is the same no matter how it is formed, and so it is written simply as $A+B+C$. Similarly, the sum $A + B + C + D$ is independent of how it is formed. For example, it equals both $(A + B) + (C + D)$ and $A + [B + (C + D)]$. Furthermore, property (1) in Theorem 1 ensures that, for example, $B + D + A + C = A + B + C + D$. In other words, the *order* in which these four matrices are added does not matter. Similar remarks apply to sums of five or more matrices.

1.1.3 Scalar Multiplication

Example 10. Suppose a corporation has a tax revenue matrix $T = [350 \ 65 \ 40]$ by which (as in Example 7) we mean that the government tax revenue from the corporation is \$350m in corporate tax, \$65m in income tax and \$40m in sales tax. If the government decided to increase all forms of tax by half, the new tax revenue matrix would be

$$T_1 = \begin{bmatrix} \tfrac{3}{2} \cdot 350 & \tfrac{3}{2} \cdot 65 & \tfrac{3}{2} \cdot 40 \end{bmatrix} = \begin{bmatrix} 525 & 97.5 & 60 \end{bmatrix}.$$

This new matrix is computed from T by multiplying each entry by $\tfrac{3}{2}$. □

In general, if A is a matrix and c is a number, the **scalar product**[4] cA is the matrix formed from A by multiplying each entry of A by the number c. If $A = [a_{ij}]$ this takes the form

$$cA = [c \, a_{ij}],$$

that is the (i,j)-entry of cA is $c \, a_{ij}$. Note that the matrix cA is the same size as A.

Example 11. If $A = \begin{bmatrix} 1 & 0 & -3 \\ 2 & 7 & 3 \end{bmatrix}$ and $B = \begin{bmatrix} 5 & -6 & 1 \\ 8 & 8 & 9 \end{bmatrix}$, we have

$$3A = \begin{bmatrix} 3 & 0 & -9 \\ 6 & 21 & 9 \end{bmatrix} \qquad \tfrac{1}{2}B = \begin{bmatrix} \tfrac{5}{2} & -3 & \tfrac{1}{2} \\ 4 & 4 & \tfrac{9}{2} \end{bmatrix}$$

$$5A - 2B = \begin{bmatrix} 5 & 0 & -15 \\ 10 & 35 & 15 \end{bmatrix} - \begin{bmatrix} 10 & -12 & 2 \\ 16 & 16 & 18 \end{bmatrix} = \begin{bmatrix} -5 & 12 & -17 \\ -6 & 19 & -3 \end{bmatrix}.$$

□

If A is any matrix, the following facts are clear:

$$1A = A \quad \text{and} \quad (-1)A = -A.$$

Furthermore, for any scalar c and matrix A

---
[4]In linear algebra numbers are often referred to as *scalars*; hence the name.

1.1. MATRICES

$$c0 = 0 \quad \text{and} \quad 0A = 0.$$

Note that the symbol 0 is playing two roles in the equation $0A = 0$: It represents the number zero on the left side, and it represents the zero matrix on the right side. This ambiguity is harmless as it is always clear from the context which meaning is correct.

There are four basic properties of scalar multiplication which we record for reference.

Theorem 2. *Let A and B denote matrices and let c and d denote numbers. Then* :

(1) $c(A + B) = cA + cB$.

(2) $(c + d)A = cA + dA$.

(3) $c(dA) = (cd)A$.

(4) $1A = A$.

Proof. We have already mentioned (4). To verify (1), write $A = [a_{ij}]$ and $B = [b_{ij}]$. Then $A + B = [a_{ij} + b_{ij}]$ as before, so the (i,j)-entry of the matrix $c(A + B)$ is

$$c(a_{ij} + b_{ij}) = ca_{ij} + cb_{ij}.$$

But ca_{ij} and cb_{ij} are the (i,j)-entries of cA and cB respectively, so $ca_{ij} + cb_{ij}$ is the (i,j)-entry of $cA + cB$. It follows that $c(A + B) = cA + cB$, proving (1). The verifications of (2) and (3) are similar and are left as exercises for the reader. □

Properties (1) and (2) in Theorem 2 extend to sums of more than two matrices. For example

$$\begin{aligned} c(A + B + C) &= cA + cB + cC \\ (c + d + e)A &= cA + dA + eA \end{aligned}$$

Similar expressions hold for more than two summands. These facts, together with the properties in Theorem 1, enable us to simplify matrix expressions by collecting like terms, expanding, and taking out common factors, in exactly the same way that algebraic expressions are simplified. Here are some examples.

Example 12. *Simplify the matrix expression $3(8A - 5B) + 4(4B - 6A)$.*

Solution. The procedure is the same as if A and B were numbers:

$$3(8A - 5B) + 4(4B - 6A) = 24A - 15B + 16B - 24A = B.$$

□

Example 13. *Find A if* $\frac{1}{5}\left\{4A - \begin{bmatrix} 1 \\ -2 \end{bmatrix}\right\} = \begin{bmatrix} 0 \\ 5 \end{bmatrix} - 2A.$

Solution. Multiplying both sides by 5 yields

$$4A - \begin{bmatrix} 1 \\ -2 \end{bmatrix} = 5\left\{\begin{bmatrix} 0 \\ 5 \end{bmatrix} - 2A\right\}$$

$$4A - \begin{bmatrix} 1 \\ -2 \end{bmatrix} = \begin{bmatrix} 0 \\ 25 \end{bmatrix} - 10A.$$

Adding $10A$ to both sides gives $14A - \begin{bmatrix} 1 \\ -2 \end{bmatrix} = \begin{bmatrix} 0 \\ 25 \end{bmatrix}$; then adding $\begin{bmatrix} 1 \\ -2 \end{bmatrix}$ to both sides gives

$$14A = \begin{bmatrix} 0 \\ 25 \end{bmatrix} + \begin{bmatrix} 1 \\ -2 \end{bmatrix} = \begin{bmatrix} 1 \\ 23 \end{bmatrix}.$$

Finally, dividing by 14 yields $A = \frac{1}{14}\begin{bmatrix} 1 \\ 23 \end{bmatrix} = \begin{bmatrix} \frac{1}{14} \\ \frac{23}{14} \end{bmatrix}.$ □

We have already observed that a scalar product cA is the zero matrix if either $c = 0$ is the zero number or $A = 0$ is the zero matrix. The next useful result shows that the converse holds.

Theorem 3. *If $cA = 0$ then either $c = 0$ or $A = 0$.*

Proof. If $c = 0$ there is nothing to prove; if $c \neq 0$ we show that $A = 0$. Write $A = [a_{ij}]$ where a_{ij} denotes the (i, j)-entry of A. Since $cA = 0$ we have $c\,a_{ij} = 0$ for each i and j ($c\,a_{ij}$ is the (i, j)-entry of cA). Since $c \neq 0$ this gives $a_{ij} = 0$ for each i and j; that is $A = 0$. □

1.1.4 Transposition

Writing column matrices is cumbersome and takes up text space. The idea of the transpose of a matrix helps with this problem by turning columns into rows (it has many other uses as well as we shall see). Before defining it, we must discuss another important notion.

If $A = [a_{ij}]$ is an $m \times n$ matrix, the **main diagonal** of A consists of the entries

$$a_{11}, a_{22}, a_{33}, \ldots.$$

Thus the main diagonal entries lie on the diagonal line starting from the upper left corner of A. For example, in each of the following matrices the main diagonal consists of the entries a and b:

$$\begin{bmatrix} a & x \\ y & b \end{bmatrix} \quad \begin{bmatrix} a & x & p \\ y & b & q \end{bmatrix} \quad \begin{bmatrix} a & x \\ y & b \\ r & s \end{bmatrix}$$

1.1. MATRICES

Note that the main diagonal of a square matrix extends from the upper left corner to the lower right corner.

We now informally define the **transpose** of an $m \times n$ matrix A to be the $n \times m$ matrix A^T obtained when A is "flipped" over its main diagonal, that is A is rotated $180°$ about the line containing the main diagonal entries.

Example 14. *Find the transpose of each of the following matrices:*

$$A = \begin{bmatrix} 3 \\ 5 \end{bmatrix}, B = \begin{bmatrix} 0 & -1 \\ 2 & 11 \end{bmatrix}, C = \begin{bmatrix} 5 & -2 & 3 \end{bmatrix}, D = \begin{bmatrix} 1 & 6 & 9 \\ 3 & 2 & 7 \end{bmatrix}.$$

Solution.

$$A^T = \begin{bmatrix} 3 & 5 \end{bmatrix}, B^T = \begin{bmatrix} 0 & 2 \\ -1 & 11 \end{bmatrix}, C^T = \begin{bmatrix} 5 \\ -2 \\ 3 \end{bmatrix}, D^T = \begin{bmatrix} 1 & 3 \\ 6 & 2 \\ 9 & 7 \end{bmatrix}.$$

□

Clearly the transpose of any row matrix is a column matrix, and the transpose of any column matrix is a row matrix. This gives another way to think of the transpose of a matrix A:

The columns of A^T are just the transposes of the rows of A, in the same order.

Similarly:

The rows of A^T are just the transposes of the columns of A, in the same order.

This is illustrated by the matrices in Example 14, and leads to a more formal description of A^T:

If $A = [a_{ij}]$ then $A^T = [b_{ij}]$ where $b_{ij} = a_{ji}$ for all i and j.

The following theorem collects four basic properties of transposition which will be used below repeatedly and without comment.

Theorem 4. *Let A and B denote matrices of the same size, and let c denote a scalar.*

(1) *If A is an $m \times n$ matrix, then A^T is an $n \times m$ matrix.*

(2) $(A^T)^T = A.$

(3) $(cA)^T = c A^T.$

(4) $(A + B)^T = A^T + B^T.$

Proof. (1). This is part of the definition of A^T.

(2). Flipping A about the main diagonal gives A^T, and then flipping again gives $(A^T)^T$. But flipping A twice gets you back to A, so $(A^T)^T = A$.

(3). $(cA)^T$ is the result of multiplying each entry of A by c and then flipping. This is the same as the result cA^T of first flipping A and then multiplying each entry by c. This proves (3).

(4). $(A+B)^T$ is the result of adding corresponding entries of A and B, and then flipping about the main diagonal. This is the same as first flipping A and B separately, and then adding corresponding entries. Hence (4) holds. $\square$

Example 15. *Find the matrix A if* $\left\{ A^T + 2 \begin{bmatrix} 1 & -2 \\ 5 & 0 \end{bmatrix} \right\}^T = 5 \begin{bmatrix} 3 & -1 \\ 1 & 7 \end{bmatrix}.$

Solution. Using Theorem 4 several times we get

$$5 \begin{bmatrix} 3 & -1 \\ 1 & 7 \end{bmatrix} = (A^T)^T + \left\{ 2 \begin{bmatrix} 1 & -2 \\ 5 & 0 \end{bmatrix} \right\}^T$$

$$= A + 2 \begin{bmatrix} 1 & -2 \\ 5 & 0 \end{bmatrix}^T$$

$$= A + 2 \begin{bmatrix} 1 & 5 \\ -2 & 0 \end{bmatrix}.$$

Hence $A = 5 \begin{bmatrix} 3 & -1 \\ 1 & 7 \end{bmatrix} - 2 \begin{bmatrix} 1 & 5 \\ -2 & 0 \end{bmatrix} = \begin{bmatrix} 13 & -15 \\ 9 & 35 \end{bmatrix}.$ $\square$

A matrix A is called **symmetric** if $A^T = A$. Such matrices are necessarily square (if A is $m \times n$ then A^T is $n \times m$, so $A^T = A$ forces $m = n$). The name comes from the fact that the symmetric matrices exhibit a symmetry about the main diagonal. Indeed, the condition $A^T = A$ means that flipping A about the main diagonal does not change it, and so requires that entries directly across the main diagonal from each other are equal. Hence the general 2×2 and 3×3 symmetric matrices have the form

$$\begin{bmatrix} a & x \\ x & b \end{bmatrix} \quad \text{and} \quad \begin{bmatrix} a & x & y \\ x & b & z \\ y & z & c \end{bmatrix}$$

respectively. Clearly every 1×1 matrix is symmetric.

1.1. MATRICES

Example 16. An undirected graph consists of vertices some of which are connected by edges with no direction assigned (as distinguished from a directed graph). The adjacency matrix of such a graph has a 1 in the (i,j)-position if v_i and v_j are connected, and so is a symmetric matrix, as is illustrated below for the given graph.

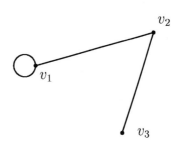

$$\begin{bmatrix} 1 & 1 & 0 \\ 1 & 0 & 1 \\ 0 & 1 & 0 \end{bmatrix}$$

Note the contrast with Example 4 in the directed graph situation. □

The elementary properties of symmetric matrices come from Theorem 4. Here is an example of how they are used.

Example 17. *If A and B are symmetric, show that $A + B$ is also symmetric.*

Solution. We are given that $A^T = A$ and $B^T = B$. Hence Theorem 4 gives

$$(A + B)^T = A^T + B^T = A + B$$

This shows that $A + B$ is symmetric. □

Exercises

1. A store sells road and mountain bikes. The manager has the following two data tables available, the first giving sales volumes of each kind for each quarter of 1998, and the second for each quarter of 1999:

1998	Road Bikes	Mountain Bikes
1st Quarter	32	27
2nd Quarter	65	78
3rd Quarter	53	74
4th Quarter	35	46

1999	Road Bikes	Mountain Bikes
1st Quarter	38	43
2nd Quarter	72	97
3rd Quarter	76	102
4th Quarter	43	65

(a) Express each of the tables as a matrix.

(b) Find a matrix operation that will produce the total sales volumes for these two years combined, and perform the calculations.

(c) The data for 1997 was given in a slightly different format as follows:

1997	1st Quarter	2nd Quarter	3rd Quarter	4th Quarter
Road Bikes	24	29	32	21
Mountain Bikes	23	31	37	29

Express this data also as a matrix, find a matrix operation that will produce the total sales volumes for these three years combined, and perform the calculations.

(d) A second store sold exactly three times as many of each kind of bike as the first store for each quarter in 1997, exactly twice as many for 1998, and exactly four times as many for 1999. Find a matrix operation that will produce the total sales volumes for this second store for these three years combined, and perform the calculations.

2. (a) If A is an $m \times n$ matrix, what do the m and n represent?

   (b) What are the requirements needed to add two matrices A and B?

3. How could matrix addition, scalar multiplication and transposition be helpful in practice?

4. Compute the following:

   (a)
   $$3\left(\begin{bmatrix} 1 & -3 & 2 \\ 0 & 4 & -9 \\ 2 & -3 & 1 \end{bmatrix} - 21\begin{bmatrix} 2 & 0 & 12 \\ -15 & -3 & 2 \\ 12 & -8 & -32 \end{bmatrix}\right) + 5\begin{bmatrix} 1 & -5 & -43 \\ 0 & 7 & 0 \\ 21 & 45 & -34 \end{bmatrix}^T$$

   (b)
   $$\begin{bmatrix} 21 & 20 & -21 \\ 56 & 11 & 34 \\ 45 & 83 & -34 \end{bmatrix} - 9\left(2\begin{bmatrix} 90 & -45 & 20 \\ -23 & 32 & 72 \\ 65 & 23 & 89 \end{bmatrix} - 13\begin{bmatrix} 19 & -25 & 87 \\ 47 & 40 & 81 \\ 85 & 35 & 24 \end{bmatrix}^T\right)$$

1.1. MATRICES

5. (a) Describe when a matrix is the same as its transpose. Such matrices are called *symmetric* matrices.

   (b) Describe when a matrix equal to the negative of its transpose. Such matrices are called *skew-symmetric* matrices.

6. In each case either show that the statement is true or give an example showing that it is false. Assume throughout that A, B and C denote matrices.

   (a) If $A + B = A + C$ then B and C have the same size.

   (b) If $A + B = 0$ then $B = 0$.

   (c) If the $(2,3)$-entry of A is 7, then the $(3,2)$-entry of A^T is -7.

   (d) If $A = -A$, then $A = 0$.

   (e) A and A^T have the same main diagonal for every matrix A.

7. Find a, b, c and d if:

   (a) $\begin{bmatrix} a & b \\ c & d \end{bmatrix} = \begin{bmatrix} b-c & c \\ d & 1 \end{bmatrix}$.

   (b) $3 \begin{bmatrix} a \\ b \end{bmatrix} - 2 \begin{bmatrix} b \\ 0 \end{bmatrix} = \begin{bmatrix} 1 \\ 4 \end{bmatrix}$.

8. In each case find the matrix A:

   (a) $2A - \begin{bmatrix} 1 & 0 & -2 \\ 4 & 7 & 3 \end{bmatrix}^T = \begin{bmatrix} 2 & 0 \\ -3 & 4 \\ 0 & 8 \end{bmatrix}$.

   (b) $\left(3A^T - 2 \begin{bmatrix} 0 & 2 \\ -5 & 2 \\ 1 & 3 \end{bmatrix}\right)^T = \begin{bmatrix} 7 & 0 & -1 \\ 2 & 1 & 5 \end{bmatrix}$.

9. (a) Simplify $3(5A - 3B) + 6(B - 4A) + 3(2A + B)$.

   (b) Simplify $2(3A - 5B) + 4(B - 2A) + 2(3B - A)$.

10. A matrix A is called *diagonal* if it is square and every entry off the main diagonal is zero. If A and B are both diagonal, show that each of the following matrices is diagonal:

    (a) $A + B$. (b) cA for any scalar c. (c) A^T.

11. (a) If A is any matrix, show that $(c + d)A = cA + dA$ for all scalars c and d. [Part (2) of Theorem 2]

    (b) If A is any matrix, show that $(cd)A = c(dA)$ for all scalars c and d. [Part (3) of Theorem 2]

1.2 LINEAR EQUATIONS

One of the historical motivations for the development of mathematics has been to find a way to analyze and solve practical problems. Apart from calculus, the most common method has been to reduce a problem to the solution of a set of linear equations. This has proved to be effective in science, engineering and social science. Here is a simple example.

1.2.1 Linear Equations

Example 1. *A charity wishes to endow a fund that will provide $50,000 per year for cancer research. The charity has $480,000 and, to reduce risk, wants to invest in two banks paying 10% and 11% respectively. How much should be invested in each bank?*

Solution. If x dollars are invested at 10% and y dollars are invested at 11%, then $x + y = 480,000$ and the yearly interest is $\frac{10}{100}x + \frac{11}{100}y$. Hence x and y must satisfy the equations

$$\begin{aligned} x + y &= 480,000 \\ \tfrac{10}{100}x + \tfrac{11}{100}y &= 50,000 \end{aligned}$$

If the first equation is multiplied by 10, and the second by 100, the resulting equations are

$$\begin{aligned} 10x + 10y &= 4,800,000 \\ 10x + 11y &= 5,000,000 \end{aligned}$$

Subtracting the first equation from the second gives $y = 200,000$, whence $x = 280,000$. In other words, the charity should invest $280,000 at 10% and $200,000 at 11%. □

An equation of the form $ax + by = c$ is called a linear equation[5] in the variables x and y. If more than two variables are present, they will usually be denoted by $x_1, x_2, \cdots, x_n$. Then an equation of the form

$$a_1 x_1 + a_2 x_2 + \cdots + a_n x_n = b \qquad (*)$$

is called a **linear equation** in the **variables** $x_1, x_2, \cdots, x_n$. Here $a_1, a_2, \cdots, a_n$ denote real numbers, called the **coefficients** of the variables $x_1, x_2, \cdots, x_n$, and the number b is called the **constant term** of the equation. The column matrix

$$X = \begin{bmatrix} x_1 \\ x_2 \\ \vdots \\ x_n \end{bmatrix} = \begin{bmatrix} x_1 & x_2 & \cdots & x_n \end{bmatrix}^T$$

---

[5] The graph of the equation $ax + by = c$ is a straight line if a and b are not both zero; hence the name.

1.2. LINEAR EQUATIONS

is called the **matrix of variables**. A set of numbers $s_1, s_2, \cdots, s_n$ is called a **solution** to the linear equation (*) if

$$a_1 s_1 + a_2 s_2 + \cdots + a_n s_n = b,$$

that is if the equation is satisfied when the substitutions $x_1 = s_1, x_2 = s_2, \cdots, x_n = s_n$ are made. We express this by saying that $X = \begin{bmatrix} s_1 & s_2 & \cdots & s_n \end{bmatrix}^T$ is a solution[6] to equation (*).

Example 2. Show that $X = \begin{bmatrix} 1 & -2 \end{bmatrix}^T$ is a solution to the equation $2x_1 - 3x_2 = 8$, but that $Y = \begin{bmatrix} 1 & 1 \end{bmatrix}^T$ is not a solution.

Solution. $\begin{bmatrix} 1 & -2 \end{bmatrix}^T$ is a solution because $x_1 = 1$ and $x_2 = -2$ satisfy the equation: $2(1) - 3(-2) = 8$. But $x_1 = 1$ and $x_2 = 1$ do not satisfy the equation since $2(1) - 3(1) \neq 8$, so $\begin{bmatrix} 1 & 1 \end{bmatrix}^T$ is not a solution. $\square$

1.2.2 Systems of Linear Equations

A finite collection of linear equations is called a **system of linear equations**, or simply a **system**. A solution to *every* equation in the system is called a **solution to the system**. The heart of linear algebra is a routine procedure for finding all solutions to any system of linear equations. But first some examples.

Example 3. In Example 1, the system of two linear equations

$$\begin{array}{rcl} x + y &=& 480,000 \\ \tfrac{10}{100}x + \tfrac{11}{100}y &=& 50,000 \end{array} \quad \text{has solution } X = \begin{bmatrix} x \\ y \end{bmatrix} = \begin{bmatrix} 280,000 \\ 200,000 \end{bmatrix}.$$

Note that this system has a *unique* solution. $\square$

Example 4. A system of equations need not have a solution. For example, the system

$$\begin{array}{rcl} x + y &=& 1 \\ x - z &=& 2 \\ y + z &=& 1 \end{array}$$

has *no* solution. Indeed adding the last two equations gives $x + y = 3$, contrary to the first equation. $\square$

A system of linear equations is called **inconsistent** if it has no solutions, and the system is called **consistent** if it has one or more solutions.

Example 5. Verify that $X = \begin{bmatrix} 1+t-s & 2+t+s & s & t \end{bmatrix}^T$ is a solution to the system

$$\begin{array}{rcl} x_1 - 2x_2 + 3x_3 + x_4 &=& -3 \\ 2x_1 - x_2 + 3x_3 - x_4 &=& 0 \end{array}$$

---

[6] The reason for writing solutions (or variables) as *columns* will be clear later. We will frequently use the transpose notation $\begin{bmatrix} s_1 & s_2 & \cdots & s_n \end{bmatrix}^T$ to simplify printing.

for all values of the numbers s and t, called **parameters** *in this context.*

Solution. Simply substitute $x_1 = 1 + t - s$, $x_2 = 2 + t + s$, $x_3 = s$ and $x_4 = t$ into each equation:

$$\begin{aligned} x_1 - 2x_2 + 3x_3 + x_4 &= (1+t-s) - 2(2+t+s) + 3s + t &= -3 \\ 2x_1 - x_2 + 3x_3 - x_4 &= 2(1+t-s) - (2+t+s) + 3s - t &= 0 \end{aligned}$$

Because both equations are satisfied, X is a solution for all s and t. Note that this system has *infinitely many* solutions since there are infinitely many choices for the parameters s and t. □

In fact, *every* solution of the system in Example 5 arises as shown for some parameters s and t. To see why this is so, and to see how one arrives at sets of solutions like those in Example 5, we develop a general procedure for finding solutions in this form. To simplify the computations we introduce a matrix notation for describing systems of linear equations. Given the system

$$\begin{aligned} x_1 - 2x_2 + 3x_3 + x_4 &= -3 \\ 2x_1 - x_2 + 3x_3 - x_4 &= 0 \end{aligned}$$

of 2 equations in 4 variables, the coefficients of the variables form a 2×4 matrix

$$\begin{bmatrix} 1 & -2 & 3 & 1 \\ 2 & -1 & 3 & -1 \end{bmatrix}$$

called the **coefficient matrix** for the system. The 2×5 matrix

$$\begin{bmatrix} 1 & -2 & 3 & 1 & -3 \\ 2 & -1 & 3 & -1 & 0 \end{bmatrix}$$

is called the **augmented matrix** for the system (it is just the coefficient matrix *augmented* by the column of constants). It is evident that the system is completely described by the augmented matrix, [7] so it is not surprising that by manipulating this matrix we can find all solutions to the system.

To see how this is done, it is convenient to call two systems of linear equations **equivalent** if they have the same set of solutions. Starting with a given system of linear equations, we solve it by writing a series of systems, one after the other, each of which is equivalent to the previous one. Since all these systems have the same solutions, the aim is to find one that is easy to solve. Surprisingly enough, there is a simple, routine method for doing this. The following example provides an illustration.

Example 6. *Find all solutions to the system in Example 5:*

$$\begin{cases} x_1 - 2x_2 + 3x_3 + x_4 &= -3 \\ 2x_1 - x_2 + 3x_3 - x_4 &= 0 \end{cases}$$

---
[7] When the system is solved in a computer, it is the augmented matrix that is stored.

1.2. LINEAR EQUATIONS

Solution. The system is written below together with its augmented matrix:

$$\begin{array}{rcl} x_1 - 2x_2 + 3x_3 + x_4 &=& -3 \\ 2x_1 - x_2 + 3x_3 - x_4 &=& 0 \end{array} \qquad \begin{bmatrix} 1 & -2 & 3 & 1 & -3 \\ 2 & -1 & 3 & -1 & 0 \end{bmatrix}$$

We first eliminate x_1 from equation 2 by subtracting twice the first equation from the second. The result is the following system (with its augmented matrix).

$$\begin{array}{rcl} x_1 - 2x_2 + 3x_3 + x_4 &=& -3 \\ 3x_2 - 3x_3 - 3x_4 &=& 6 \end{array} \qquad \begin{bmatrix} 1 & -2 & 3 & 1 & -3 \\ 0 & 3 & -3 & -3 & 6 \end{bmatrix}$$

This new system is equivalent to the original system (see Theorem 1 below). Note that the new augmented matrix can be obtained directly from the original one by subtracting twice the first row from the second row.

We now multiply the second equation by $\frac{1}{3}$ to obtain another equivalent system

$$\begin{array}{rcl} x_1 - 2x_2 + 3x_3 + x_4 &=& -3 \\ x_2 - x_3 - x_4 &=& 2 \end{array} \qquad \begin{bmatrix} 1 & -2 & 3 & 1 & -3 \\ 0 & 1 & -1 & -1 & 2 \end{bmatrix}$$

Again the new augmented matrix comes from the preceding one by multiplying the second row by $\frac{1}{3}$.

Finally, we eliminate x_2 from equation 1 by adding twice the second equation to the first. This results in the (equivalent) system

$$\begin{array}{rcl} x_1 + x_3 - x_4 &=& 1 \\ x_2 - x_3 - x_4 &=& 2 \end{array} \qquad \begin{bmatrix} 1 & 0 & 1 & -1 & 1 \\ 0 & 1 & -1 & -1 & 2 \end{bmatrix}$$

This system is *easy* to solve. Indeed, if the numbers x_3 and x_4 are chosen arbitrarily, then x_1 and x_2 can be found so that the equations are satisfied. More precisely, if we set $x_3 = s$ and $x_4 = t$ where s and t are arbitrary parameters, the equations become $x_1 + s - t = 1$ and $x_2 - s - t = 2$, whence

$$x_1 = 1 + t - s \qquad \text{and} \qquad x_2 = 2 + t + s$$

This gives the solutions exhibited in Example 5 and, because all the systems in the series are equivalent (as will be proved in Theorem 1), we have obtained *all* the solutions to the original system. □

Observe that at each stage in the above procedure a certain operation is performed on the system (and thus on the augmented matrix) to produce an equivalent system. The following operations, called **elementary operations**, can routinely be performed on systems to produce equivalent systems:

    I   *Interchange two equations.*
    II   *Multiply an equation by a nonzero number.*
    III   *Add a multiple of one equation to a different equation.*

We only needed operations of type II and III in the computation in Example 6, but type I operations are sometimes useful as well. The next theorem is crucial for the method we are developing.

Theorem 1. *If an elementary operation is performed on a system of linear equations, the resulting system is equivalent to the original system.*

Proof. We prove it for type III operations; similar arguments work for operations of type I or II.

Suppose we modify the original system by replacing equation p by a new equation, formed by adding a multiple of a different equation q to equation p. Then any solution to the original system will satisfy this new equation (it satisfies both equations p and q), and so will be a solution to the new system. On the other hand, the new system contains equation q (because p and q are *different* equations). This means that the new system can be transformed *back* to the original system by *subtracting* the same multiple of equation q from the new equation. Hence the same argument shows that every solution to the new system is a solution to the original system. Thus the two systems have the same set of solutions, so operations of type III produce equivalent systems. □

Theorem 1 has profound consequences for linear algebra. In particular it allows us to use the procedure in Example 6 on *any* system of equations. The idea is to apply a series of elementary operations to the system with the goal of finding a system that is easy to solve. Since all the systems created in this way are equivalent (by Theorem 1), the solutions to the easy system are the solutions to the original system.

As in Example 6, elementary operations performed on a system of equations produce corresponding manipulations of the *rows* of the augmented matrix (regarded as row matrices). In hand calculations (and in computer programs) it is easier to manipulate rows than equations. For this reason we restate these elementary operations for rows:

    I   *Interchange two rows.*
    II   *Multiply a row by a nonzero number.*
    III   *Add a multiple of one row to a different row.*

These are called **elementary row operations**. Here is another example of our method where the entire calculation is carried out by manipulating the augmented matrix.

Example 7. *Find all solutions to the following system of linear equations:*

$$\begin{aligned} x_1 + x_2 - 3x_3 &= 3 \\ -2x_1 - x_2 &= -4 \\ 4x_1 + 2x_2 + 3x_3 &= 7 \end{aligned}$$

Proof: The augmented matrix of the original system is

$$\begin{bmatrix} 1 & 1 & -3 & 3 \\ -2 & -1 & 0 & -4 \\ 4 & 2 & 3 & 7 \end{bmatrix}.$$

1.2. LINEAR EQUATIONS

We begin by using the 1 in the upper left corner to "clean up" column 1, that is to obtain 0's in the other locations (this corresponds to eliminating x_1 from equations 2 and 3). More precisely, we add twice row 1 to row 2, and we also subtract 4 times row 1 from row 3. The result is

$$\begin{bmatrix} 1 & 1 & -3 & 3 \\ 0 & 1 & -6 & 2 \\ 0 & -2 & 15 & -5 \end{bmatrix}.$$

This completes the work on column 1.

We now use the 1 in the second position of row 2 to clean up column 2, that is to obtain 0's in the top and bottom positions (this corresponds to eliminating x_2 from equations 1 and 3). We do this by subtracting row 2 from row 1, and adding twice row 2 to row 3. This gives

$$\begin{bmatrix} 1 & 0 & 3 & 1 \\ 0 & 1 & -6 & 2 \\ 0 & 0 & 3 & -1 \end{bmatrix}.$$

Note that these row operations have not disturbed column 1 because the first entry of row 2 is a *zero*.

We next divide row 3 by 3 to obtain a 1 in the third position:

$$\begin{bmatrix} 1 & 0 & 3 & 1 \\ 0 & 1 & -6 & 2 \\ 0 & 0 & 1 & -\frac{1}{3} \end{bmatrix}.$$

Finally, we clean up column 3 by subtracting 3 times row 3 from row 1, and adding 6 times row 3 to row 2:

$$\begin{bmatrix} 1 & 0 & 0 & 2 \\ 0 & 1 & 0 & 0 \\ 0 & 0 & 1 & -\frac{1}{3} \end{bmatrix}.$$

The corresponding system of equations is

$$\begin{aligned} x_1 & & & = & 2 \\ & x_2 & & = & 0 \\ & & x_3 & = & -\frac{1}{3} \end{aligned}$$

and the (unique) solution $X = [\ 2 \ \ 0 \ \ -\frac{1}{3}\ ]^T$ is apparent. Since this system of equations is equivalent to the original system, this is the solution to the original system. □

1.2.3 Gaussian[8] Elimination

In the computations in Examples 6 and 7, elementary row operations (on the augmented matrix) led to matrices of the form

---

[8]Carl Friedrich Gauss (1777-1855) was one of the greatest mathematicians of all time. He made ground-breaking discoveries in every part of mathematics, and also made important contributions to astronomy and physics.

$$\begin{bmatrix} 1 & 0 & * & * & * \\ 0 & 1 & * & * & * \end{bmatrix} \quad \text{and} \quad \begin{bmatrix} 1 & 0 & 0 & * \\ 0 & 1 & 0 & * \\ 0 & 0 & 1 & * \end{bmatrix}$$

respectively, where each * indicates a number. In both cases the solution was easily obtained from the corresponding system of equations. The matrices that arise in general are described as follows.

A matrix is said to be in **row-echelon form** (and will be called a **row-echelon matrix**) if the following conditions are satisfied:

1. *All zero rows are at the bottom.*

2. *The first nonzero entry from the left in each nonzero row is a 1, called the **leading 1** for that row.*

3. *Each leading 1 is to the right of all leading 1's in the rows above it.*

A row-echelon matrix is said to be in **reduced row-echelon form** if, in addition, it satisfies

4. *Each leading 1 is the only nonzero entry in its column.*

Thus the row-echelon matrices have a "staircase" form as indicated below (as before the asterisks indicate arbitrary numbers):

$$\begin{bmatrix} 0 & 1 & * & * & * & * & * \\ 0 & 0 & 0 & 1 & * & * & * \\ 0 & 0 & 0 & 0 & 1 & * & * \\ 0 & 0 & 0 & 0 & 0 & 0 & 1 \\ 0 & 0 & 0 & 0 & 0 & 0 & 0 \end{bmatrix}.$$

The leading 1's proceed down and to the right through the matrix, and every entry below and to the left of a leading 1 is a 0. In a *reduced* row-echelon matrix the additional requirement is that all entries *above* a leading 1 must also be zero. Any row-echelon matrix can be carried to reduced form using a few more row operations (clean up above the leading 1's one after the other).

Example 8. The first matrix below is in row-echelon form, and the second matrix is the reduced row-echelon matrix to which it can be carried by row operations:

$$\begin{bmatrix} 1 & * & * & * & * \\ 0 & 0 & 1 & * & * \\ 0 & 0 & 0 & 0 & 1 \end{bmatrix} \rightarrow \begin{bmatrix} 1 & * & 0 & * & 0 \\ 0 & 0 & 1 & * & 0 \\ 0 & 0 & 0 & 0 & 1 \end{bmatrix}$$

In general we use an arrow $\rightarrow$ to indicate that row operations have been performed. □

1.2. LINEAR EQUATIONS

Here is a procedure by which any matrix can be carried to row-echelon form (and hence to reduced form if desired) using nothing but elementary row operations.

Gaussian Algorithm.[9] *Every matrix can be carried to row-echelon form as follows*:

Step 1. *If the matrix consists entirely of zeros, stop: it is already in row-echelon form.*

Step 2. *Otherwise, find the first column from the left containing a nonzero entry k, and move the row containing that entry to the top of the matrix.*

Step 3. *Multiply the top row by $\frac{1}{k}$ to create the first leading 1.*

Step 4. *Make each entry below the leading 1 zero by subtracting multiples of its row from lower rows.*

This completes the first row; all further row operations are carried out on the remaining rows.

Step 5. *Repeat steps 1–4 on the matrix consisting of the remaining rows.*

Note that the Gaussian Algorithm is recursive in the sense that, after the first leading 1 has been created, the whole procedure is repeated on the remaining rows. This makes it easy to use on a computer. Note further that at step 4 we can also make every entry *above* the leading 1 zero. Then the algorithm carries the matrix to *reduced* row-echelon form (as in Examples 6 and 7). The reason for distinguishing two row-echelon forms will be discussed later.

The Gaussian Algorithm certainly proves the following important theorem.

Theorem 2. *Every matrix can be carried to (reduced) row-echelon form by a sequence of elementary row operations.*

Example 9. *Find all solutions to the following system of linear equations*:

$$\begin{aligned} x_1 - 2x_2 - x_3 + 3x_4 &= 1 \\ 2x_1 - 4x_2 + x_3 &= 5 \\ x_1 - 2x_2 + 2x_3 - 3x_4 &= 4 \end{aligned}$$

Solution. The augmented matrix is given below. The first leading 1 is in place so we clean up column 1:

$$\begin{bmatrix} 1 & -2 & -1 & 3 & 1 \\ 2 & -4 & 1 & 0 & 5 \\ 1 & -2 & 2 & -3 & 4 \end{bmatrix} \to \begin{bmatrix} 1 & -2 & -1 & 3 & 1 \\ 0 & 0 & 3 & -6 & 3 \\ 0 & 0 & 3 & -6 & 3 \end{bmatrix}$$

---
[9] While Gauss did use this procedure, the method has been attributed to the Chinese several centuries earlier.

Now subtract the second row from the third, and then[10] multiply the second row by $\frac{1}{3}$, to get the matrix below (now in row-echelon form):

$$\rightarrow \begin{bmatrix} 1 & -2 & -1 & 3 & 1 \\ 0 & 0 & 1 & -2 & 1 \\ 0 & 0 & 0 & 0 & 0 \end{bmatrix}$$

Next use the second leading 1 (in column 3) to clean up column 3 and so achieve reduced row-echelon form:

$$\rightarrow \begin{bmatrix} 1 & -2 & 0 & 1 & 2 \\ 0 & 0 & 1 & -2 & 1 \\ 0 & 0 & 0 & 0 & 0 \end{bmatrix}$$

This is as far as the Gaussian Algorithm will take us. The corresponding system of equations is

$$\begin{aligned} x_1 - 2x_2 \quad\quad + \quad x_4 &= 2 \\ x_3 - 2x_4 &= 1 \\ 0 &= 0 \end{aligned}$$

The leading 1's are in columns 1 and 3 here, and the corresponding variables x_1 and x_3 are called the leading variables. To solve the system the non-leading variables are assigned arbitrary values (called parameters), and then the two equations are used to determine the leading variables in terms of the parameters. More precisely, we set $x_2 = s$, and $x_4 = t$ where s and t are arbitrary parameters, so the equations become $x_1 - 2s + t = 2$ and $x_3 - 2t = 1$. Solving these gives $x_1 = 2 + 2s - t$ and $x_3 = 1 + 2t$. Hence the solutions are given by

$$X = \begin{bmatrix} 2+2s-t & s & 1+2t & t \end{bmatrix}^T = \begin{bmatrix} 2+2s-t \\ s \\ 1+2t \\ t \end{bmatrix}.$$

$\square$

The solution $X = \begin{bmatrix} 2+2s-t & s & 1+2t & t \end{bmatrix}^T$ in Example 9 is called the **general solution** of the system because *every* solution has this form for some values of the parameters s and t.

When the augmented matrix of a linear system has been carried to reduced row-echelon form, the variables corresponding to the leading 1's are called the **leading variables**. Then the method of solution in Example 9 provides a way of writing down the solutions of any linear system (assuming that there *are* solutions).

Gaussian Elimination. *Assume that a system of linear equations has at least one solution. Then the general solution can be found in parametric form as follows:*

---

[10] These steps are not strictly in the order specified by the algorithm. However, the point is to carry the matrix to reduced form using *some* sequence of row operations. The sequence in the algorithm will always work, but it may not be the most efficient.

1.2. LINEAR EQUATIONS

Step 1. *Carry the augmented matrix of the system to reduced*[11] *row-echelon form.*

Step 2. *Assign the non-leading variables as parameters.*

Step 3. *Use the equations corresponding to the reduced row-echelon form to solve for the leading variables in terms of the parameters.*

This procedure solves any system of linear equations which *has* a solution. The following example shows how the method reveals that a system has *no* solution.

Example 10. In Example 4 it was shown directly that the system

$$\begin{aligned} x + y & = 1 \\ x \phantom{{}+y} - z & = 2 \\ \phantom{x+{}} y + z & = 1 \end{aligned}$$

has no solution. The reduction of the augmented matrix to row-echelon form is as follows:

$$\begin{bmatrix} 1 & 1 & 0 & 1 \\ 1 & 0 & -1 & 2 \\ 0 & 1 & 1 & 1 \end{bmatrix} \rightarrow \begin{bmatrix} 1 & 1 & 0 & 1 \\ 0 & -1 & -1 & 1 \\ 0 & 1 & 1 & 1 \end{bmatrix}$$

$$\rightarrow \begin{bmatrix} 1 & 1 & 0 & 1 \\ 0 & 1 & 1 & -1 \\ 0 & 0 & 0 & 2 \end{bmatrix}$$

$$\rightarrow \begin{bmatrix} 1 & 1 & 0 & 1 \\ 0 & 1 & 1 & -1 \\ 0 & 0 & 0 & 1 \end{bmatrix}$$

This last matrix corresponds to a system in which the last equation is

$$0x + 0y + 0z = 1.$$

It is clear that *no* choice of x, y and z will satisfy *this* equation, so this last system (and hence the original system) has no solution. This is typical of what happens when there is no solution. □

---

[11] When solving a large system, it is more efficient to carry the augmented matrix only to row-echelon form, assign parameters to the non-leading variables, and then solve for the leading variables using **back-substitution**: Use the last equation to find the last leading variable in terms of the parameters, then substitute that value in the second last equation to solve for the second last leading variable, and so on.

1.2.4 Rank of a Matrix

It will be proved in Section 1.6 (Theorem 4) that:

The reduced row-echelon form of a matrix A *is uniquely determined by* A

That is, no matter which series of row operations is used to carry A to a reduced row-echelon matrix, the result will always be the same. By contrast, the same matrix can be taken to *different* row-echelon matrices. For example, if $A = \begin{bmatrix} 1 & 3 & 4 \\ 2 & 7 & 9 \end{bmatrix}$, then

$$A \to \begin{bmatrix} 1 & 3 & 4 \\ 0 & 1 & 1 \end{bmatrix} \quad \text{and} \quad A \to \begin{bmatrix} 1 & 3 & 4 \\ 0 & 1 & 1 \end{bmatrix} \to \begin{bmatrix} 1 & 2 & 3 \\ 0 & 1 & 1 \end{bmatrix}.$$

Hence A can be carried to two different row-echelon matrices $\begin{bmatrix} 1 & 3 & 4 \\ 0 & 1 & 1 \end{bmatrix}$ and $\begin{bmatrix} 1 & 2 & 3 \\ 0 & 1 & 1 \end{bmatrix}$. However, we will show in Section 4.4 that:

The number of leading 1's must be the same no matter how A is carried to row-echelon form.

This number of leading 1's is called the **rank** of the matrix A and is denoted $rank A$.

Example 11. *Compute the rank of* $A = \begin{bmatrix} 1 & 2 & -1 & 3 \\ 2 & 1 & 1 & 5 \\ -1 & 4 & -5 & -1 \end{bmatrix}$.

Solution. The matrix A is carried to row-echelon form as follows:

$$A = \begin{bmatrix} 1 & 2 & -1 & 3 \\ 2 & 1 & 1 & 5 \\ -1 & 4 & -5 & -1 \end{bmatrix} \to \begin{bmatrix} 1 & 2 & -1 & 3 \\ 0 & -3 & 3 & -1 \\ 0 & 6 & -6 & 2 \end{bmatrix} \to \begin{bmatrix} 1 & 2 & -1 & 3 \\ 0 & 1 & -1 & \frac{1}{3} \\ 0 & 0 & 0 & 0 \end{bmatrix}$$

Because there are two leading 1's, we have $rank A = 2$. □

The relationship between rank and systems of equations is given in the following theorem.

Theorem 3. *Suppose a system of m equations in n variables has at least one solution. If the rank of the augmented matrix is r, the set of solutions has exactly $n - r$ parameters.*

Proof. Carry the augmented matrix to a reduced row-echelon matrix R. Then R has r leading 1's (since the rank is r), so there are exactly r leading variables. Hence there are $n - r$ non-leading variables, and each of these is assigned a parameter. □

1.2. LINEAR EQUATIONS

Theorem 3 has a surprising number of consequences and will be referred to several times below. We have seen examples of systems with no solution, one solution or infinitely many solutions; the first application of Theorem 3 is to show that these are the only possibilities.

Theorem 4. *For any system of linear equations there are exactly three possibilities:*

1. *No solution.*

2. *A unique solution.*

3. *Infinitely many solutions.*

Proof. If there is a solution, then either every variable is a leading variable (unique solution) or there is at least one non-leading variable (infinitely many solutions because a parameter is involved). □

A system of linear equations is called **consistent** if it has at least one solution, and the system is called **inconsistent** if it has no solution. Thus a consistent system has either a unique solution or infinitely many solutions; it cannot have (say) exactly two solutions.

The truth of Theorem 4 can be seen graphically for a system of two equations in two variables x and y. Recall that the graph of an equation $ax + by = c$ is a straight line if a and b are not both zero, and that $\begin{bmatrix} s \\ t \end{bmatrix}$ is a solution of the equation exactly when the point $P(s,t)$ with coordinates (s,t) lies on the line. Now consider a system

$$\begin{aligned} a_1 x + b_1 y &= c_1 \\ a_2 x + b_2 y &= c_2 \end{aligned} \qquad (**)$$

The graphs of these equations are two straight lines, L_1 and L_2 assuming that a_1 and b_1 are not both zero, and that a_2 and b_2 are not both zero. Geometrically there are three possibilities for these two lines (illustrated in Figure 1):

(1) Lines are parallel and distinct — System (**) has no solution because there is no point on both lines

(2) Lines are not parallel — System (**) has a unique solution corresponding to the point of intersection of the lines

(3) Lines are identical — System (**) has infinitely many solutions, one for each point on the (common) line

Clearly these three possibilities correspond to those in Theorem 4.

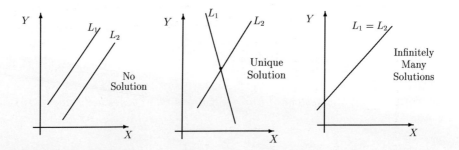

Figure 1

The graph of an equation $ax + by + cz = d$ is a plane in space if a, b, and c are not all zero (this is discussed in detail in Section 3.3). Thus a system of 2 equations in 3 variables must either have no solutions (the planes are parallel) or have infinitely many solutions (the planes coincide or intersect in a line). A unique solution is not possible here. This illustrates Theorem 3 since there are $n = 3$ variables and the augmented matrix has rank $r \leq 2$ (because there are 2 equations) so there are $n - r \geq 3 - 2 = 1$ parameters.

A similar graphical argument can be given that a system of 3 equations in 3 variables must have zero, one or infinitely many solutions as in Theorem 4. However, this geometrical argument fails for systems with more than three variables, and we must rely on Theorem 4.

Exercises

1. In each case show that the column matrix X is a solution to the given system of linear equations; if it involves parameters s and t, show that X is a solution for all possible values of these parameters.

    (a) $x_1 - 2x_2 + 3x_3 + x_4 = -3 \quad X = [1\ 2\ 0\ 0]^T$
    $ 2x_1 - x_2 + 3x_3 - x_4 = 0 \quad X = [1\ 4\ 1\ 1]^T$
    $ X = [-s+t+1\ \ s+t+2\ \ s\ \ t]^T$

    (b) $\begin{array}{llllll} x_1 & -2x_2 & +3x_3 & -3x_4 & = 2 & X = [-2\ -2\ 1\ 1]^T \\ 2x_1 & -x_2 & +3x_3 & & = 1 & X = [-3\ -1\ 2\ 1]^T \\ -x_1 & -4x_2 & +3x_3 & -9x_4 & = 4 & X = [-s-t\ \ -1+s-2t\ \ s\ \ t]^T \\ 2x_1 & -7x_2 & +9x_3 & -12x_4 & = 7 & \end{array}$

2. Describe each of the following in your own words: A linear equation, a solution to a linear equation, a system of linear equations, and a solution to a system of linear equations.

1.2. LINEAR EQUATIONS

3. In each case carry the given matrix to reduced row-echelon form. For a methodical approach, follow the Gaussian Algorithm.

    (a) $\begin{bmatrix} 3 & -1 & 2 & 1 & 2 & 1 \\ -4 & 1 & -2 & 2 & 7 & 2 \\ 2 & -2 & 4 & 3 & 7 & 1 \\ 0 & 3 & -6 & 1 & 6 & 4 \end{bmatrix}$

    (b) $\begin{bmatrix} 30 & -1 & 3 & 1 & 3 & 2 & 1 \\ -34 & -2 & 6 & 1 & -5 & 0 & -1 \\ 2 & 3 & -9 & 2 & 4 & 1 & -1 \\ 1 & 1 & -3 & -1 & 3 & 0 & 1 \end{bmatrix}$

4. For each of the following systems of linear equations, form the augmented matrix, perform elementary row operations to transform it to reduced row-echelon form, and solve the given system this way. For a methodical approach to reducing to row-echelon form, follow the Gaussian Algorithm.

    (a) $\begin{aligned} 3x_1 - x_2 &= 4 \\ 2x_1 - \tfrac{1}{2}x_2 &= 1 \end{aligned}$

    (b) $\begin{aligned} 2x_1 - 3x_2 &= 4 \\ x_1 - 3x_2 &= 1 \end{aligned}$

    (c) $\begin{aligned} x_1 + x_2 - x_3 &= 1 \\ 3x_1 - x_2 + x_3 &= 0 \\ x_1 - 3x_2 + 3x_3 &= -2 \end{aligned}$

    (d) $\begin{aligned} 2x_1 + 2x_2 - 3x_3 &= 1 \\ x_1 + x_3 &= 5 \\ 3x_1 + 4x_2 - 7x_3 &= -3 \end{aligned}$

    (e) $\begin{aligned} x_1 - 2x_2 + 2x_3 &= 4 \\ -2x_1 + x_2 + x_3 &= 1 \\ x_1 - 5x_2 + 7x_3 &= -1 \end{aligned}$

    (f) $\begin{aligned} 4x_1 + x_2 - 8x_3 &= 1 \\ 3x_1 - 2x_2 + 3x_3 &= 5 \\ -x_1 + 8x_2 - 25x_3 &= -3 \end{aligned}$

    (g) $\begin{aligned} x_1 - 3x_2 + x_3 + x_4 - x_5 &= 8 \\ -2x_1 + 6x_2 + x_3 - 2x_4 - 4x_5 &= -1 \\ 3x_1 - 9x_2 + 8x_3 + 4x_4 - 13x_5 &= 49 \end{aligned}$

    (h) $\begin{aligned} x_1 - 2x_2 + x_3 + 3x_4 - x_5 &= 1 \\ -3x_1 + 6x_2 - 4x_3 - 9x_4 + 3x_5 &= -1 \\ -x_1 + 2x_2 - 2x_3 - 4x_4 - 3x_5 &= 3 \\ x_1 - 2x_2 + 2x_3 + 2x_4 - 5x_5 &= 1 \end{aligned}$

5. In your own words, can you describe an algorithm that will solve a system of linear equations for you?
    [ILAW: Use Exploration 1.2.2 to test your algorithm.]

6. In each of the systems of linear equations given in Exercise 4, calculate the rank of the corresponding augmented matrix, and find a connection with the number of variables, the number of solutions and the number of parameters in the solution.

7. If a system of 5 equations in 7 variables has a solution, explain why there is more than one solution.

8. Consider the system $ax = b$ of one equation in one variable. Give conditions on a and b such that the system has no solution, a unique solution, or infinitely many solutions.

9. In each case find (if possible) conditions on the numbers a, b and c that the given system has no solution, a unique solution, or infinitely many solutions.

   (a) $\begin{aligned} 2x_1 - 3x_2 - 3x_3 &= a \\ -x_1 + x_2 + 2x_3 &= b \\ x_1 - 3x_2 &= c \end{aligned}$

   (b) $\begin{aligned} x_1 - 2x_2 + 2x_3 &= a \\ -2x_1 + x_2 + x_3 &= b \\ x_1 - 5x_2 + 7x_3 &= c \end{aligned}$

   (c) $\begin{aligned} x_1 + ax_2 &= 1 \\ bx_1 + 2x_2 &= 5 \end{aligned}$

   (d) $\begin{aligned} ax_1 + x_2 &= -1 \\ 2x_1 + x_2 &= b \end{aligned}$

10. Let $A = \begin{bmatrix} 1 & -2 & 0 & 3 \\ -4 & 1 & 5 & -2 \\ -1 & -5 & 5 & 7 \end{bmatrix}$. Show that A and A^T have the same rank.

    [*Remark*: We will see in Section 4.4 that this is true for *every* matrix A.]

11. In your own words, can you describe an algorithm that will compute the rank of a matrix?

12. Four one-way streets feed into a traffic circle as in the diagram, where the traffic flows are measured in cars per minute.

    In order to design the roads, the engineers must determine the flows f_1, f_2, f_3, and f_4 in the circle. Use the fact that the traffic flow into any intersection must equal the traffic flow out of that intersection to obtain four equations relating these flows. Solve the equations and express f_1, f_2, and f_3 in terms of f_4. Which part of the circle will carry the most traffic?

    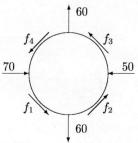

13. A man must take 5 units of vitamin A, 13 units of vitamin B and 23 units of vitamin C each day. Three brands of vitamin pills are available and the number of units of each vitamin per pill are given in the following table.

		VITAMIN		
		A	B	C
BRAND	1	1	2	4
	2	1	1	3
	3	0	1	1

(a) Find all combinations of pills that provide the exact daily requirement (no partial pills).

(b) If brands 1, 2 and 3 cost $0.90, $0.60, and $1.50 per pill respectively, find the least expensive treatment.

14. In each case either show that the statement is true or give an example showing that it is false. Assume that a system of equations is given with augmented matrix A, coefficient matrix C and reduced row-echelon form R.

(a) A and R are the same size.

(b) If R has a row of zeros, there are infinitely many solutions.

(c) If there is more than one solution, R must have a row of zeros.

(d) If there exists a solution, there are infinitely many solutions.

(e) If there are more variables than equations, there are infinitely many solutions.

(f) If every row of R has a leading 1, the system has at least one solution.

(g) If the system has a solution, $rank(A) = rank(C)$.

(h) $rank(C) \leq rank(A)$.

(i) $rank(A) \leq 1 + rank(C)$.

(j) If A is $m \times n$ and $rank A = m$, the system is consistent.

1.3 HOMOGENEOUS SYSTEMS

In this section, we concentrate on a particular class of systems of linear equations, that is those systems where the constant matrix is 0.

1.3.1 Homogeneous Systems

A system of linear equations is called **homogeneous** if all the constant terms are zero. Thus a typical homogeneous linear equation in the n variables $x_1, x_2, \cdots, x_n$ has the form

$$a_1x_1 + a_2x_2 + \cdots + a_nx_n = 0.$$

Because the constants are all zero, any homogeneous system always has the **trivial** solution

$$x_1 = 0, \ x_2 = 0, \cdots, x_n = 0$$

in which every variable is zero. Many practical problems come down to discovering whether or not some homogeneous system has a **nontrivial** solution, that is a solution where at least one of the variables is nonzero. The next theorem gives an important situation where this *must* happen.

Theorem 1. *If a homogeneous system of linear equations has more variables than equations then it has nontrivial solutions.*

Proof. Suppose there are m equations in n variables, so our assumption is that $n > m$. If r is the rank of the augmented matrix, then $r \leq m$ because the number r of leading 1's cannot exceed the number m of equations. Hence $r \leq m < n$, whence $r < n$. By Theorem 3 §1.2, this means that the number $n - r$ of parameters is not zero, so there are (infinitely many) nontrivial solutions. $\square$

The existence of a nontrivial is often the desired outcome for a homogeneous system. The following example provides an illustration of how Theorem 1 can be used in geometry.

Example 1. *The graph of an equation $ax^2 + bxy + cy^2 + dx + ey + f = 0$ is called a **conic** if a, b and c are not all zero. (Circles, ellipses, hyperbolas and parabolas are all examples of conics.) Show that there is at least one conic through any five points in the plane that are not all on a line.*

Solution. Suppose that the coordinates of the five points are (p_1, q_1), (p_2, q_2), (p_3, q_3), (p_4, q_4), and (p_5, q_5). The graph of the equation $ax^2 + bxy + cy^2 + dx + ey + f = 0$ passes through the point (p_i, q_i) if

$$ap_i^2 + bp_iq_i + cq_i^2 + dp_i + eq_i + f = 0$$

Since there are five points, this gives five homogeneous equations which are linear in the six variables a, b, c, d, e, and f. Hence there is a nontrivial solution by Theorem 1. If $a = b = c = 0$ in this solution, then the five points all lie on the line with equation $dx + ey + f = 0$, contrary to our assumption. Hence one of a, b and c is nonzero and we have a conic. $\square$

1.3.2 Basic Solutions

Of course Gaussian elimination also works for homogeneous systems. In fact it provides a way to write the solutions in a convenient form which will be needed later. This is illustrated in the following example.

Example 2. *Solve the following homogeneous system*

$$\begin{aligned} x_1 - 2x_2 + x_3 + x_4 &= 0 \\ -x_1 + 2x_2 + x_4 &= 0 \\ 2x_1 - 4x_2 + x_3 &= 0 \end{aligned}$$

and express the solutions as sums of scalar multiples of specific solutions.

Solution. The augmented matrix is reduced as follows:

$$\begin{bmatrix} 1 & -2 & 1 & 1 & 0 \\ -1 & 2 & 0 & 1 & 0 \\ 2 & -4 & 1 & 0 & 0 \end{bmatrix} \to \begin{bmatrix} 1 & -2 & 1 & 1 & 0 \\ 0 & 0 & 1 & 2 & 0 \\ 0 & 0 & -1 & -2 & 0 \end{bmatrix}$$

$$\to \begin{bmatrix} 1 & -2 & 0 & -1 & 0 \\ 0 & 0 & 1 & 2 & 0 \\ 0 & 0 & 0 & 0 & 0 \end{bmatrix}$$

Hence the leading variables are x_1 and x_3, and the non-leading variables x_2 and x_4 become parameters: $x_2 = s$ and $x_4 = t$. Then the equations in the final system determine the leading variables in terms of the parameters:

$$x_1 = 2s + t \quad \text{and} \quad x_3 = -2t$$

This means that the general solution is $X = [2s+t \ \ s \ \ -2t \ \ t]^T$. We separate this into parts involving only s or t:

$$X = \begin{bmatrix} 2s+t \\ s \\ -2t \\ t \end{bmatrix} = \begin{bmatrix} 2s \\ s \\ 0 \\ 0 \end{bmatrix} + \begin{bmatrix} t \\ 0 \\ -2t \\ t \end{bmatrix} = s \begin{bmatrix} 2 \\ 1 \\ 0 \\ 0 \end{bmatrix} + t \begin{bmatrix} 1 \\ 0 \\ -2 \\ 1 \end{bmatrix}$$

Hence $X_1 = \begin{bmatrix} 2 \\ 1 \\ 0 \\ 0 \end{bmatrix}$ and $X_2 = \begin{bmatrix} 1 \\ 0 \\ -2 \\ 1 \end{bmatrix}$ x are specific solutions, and the general solution X has the form $X = sX_1 + tX_2$. □

The specific solutions X_1 and X_2 in Example 2 are called the basic solutions of the homogeneous system. They have the property that *every* solution X has the form

$$X = sX_1 + tX_2 \quad \text{where } s \text{ and } t \text{ are arbitrary parameters.}$$

This happens for every homogeneous system.

To describe the general situation, the following terminology is useful: If X_1, $X_2, \cdots, X_k$ are columns, expressions of the form

$$s_1 X_1 + s_2 X_2 + \cdots + s_n X_k \quad \text{where } s_1,\ s_2, \cdots, s_k \text{ are arbitrary parameters}$$

are called **linear combinations** of the columns X_1, $X_2, \cdots, X_k$. Given *any* homogeneous system of linear equations, a computation like that in Example 2 expresses every solution to the system as a linear combination of certain particular solutions. These particular solutions are called the **basic solutions** produced by the Gaussian Algorithm. (Of course the system may have only the trivial solution; in this case we say that the system has **no basic solutions**.)

Theorem 2. *Suppose a system of homogeneous linear equations in n variables is given. Assume that the rank of the augmented matrix is r. Then:*

(1) *The Gaussian Algorithm produces exactly $n - r$ basic solutions.*

(2) *Every solution is a linear combination of these basic solutions.*

Proof. We have already observed that every solution is a linear combination of the basic ones produced by the Gaussian Algorithm. By Theorem 3 §1.2 there are exactly $n - r$ parameters in the general solution, and so there are exactly $n - r$ basic solutions (one for each parameter). □

Example 3. *Find the basic solutions for the following homogeneous system, and illustrate Theorem 2.*

$$\begin{array}{rcrcrcrcrcrcl}
x_1 & - & 2x_2 & + & 4x_3 & - & x_4 & & & + & 5x_6 & = & 0 \\
-2x_1 & + & 4x_2 & - & 7x_3 & + & x_4 & + & 2x_5 & - & 8x_6 & = & 0 \\
3x_1 & - & 6x_2 & + & 12x_3 & - & 3x_4 & + & x_5 & + & 15x_6 & = & 0 \\
2x_1 & - & 4x_2 & + & 9x_3 & - & 3x_4 & + & 3x_5 & + & 12x_6 & = & 0
\end{array}$$

Solution. The reduction of the augmented matrix to reduced form is as follows:

$$\begin{bmatrix} 1 & -2 & 4 & -1 & 0 & 5 & 0 \\ -2 & 4 & -7 & 1 & 2 & -8 & 0 \\ 3 & -6 & 12 & -3 & 1 & 15 & 0 \\ 2 & -4 & 9 & -3 & 3 & 12 & 0 \end{bmatrix} \rightarrow \begin{bmatrix} 1 & -2 & 4 & -1 & 0 & 5 & 0 \\ 0 & 0 & 1 & -1 & 2 & 2 & 0 \\ 0 & 0 & 0 & 0 & 1 & 0 & 0 \\ 0 & 0 & 1 & -1 & 3 & 2 & 0 \end{bmatrix}$$

$$\rightarrow \begin{bmatrix} 1 & -2 & 0 & 3 & 0 & -3 & 0 \\ 0 & 0 & 1 & -1 & 0 & 2 & 0 \\ 0 & 0 & 0 & 0 & 1 & 0 & 0 \\ 0 & 0 & 0 & 0 & 0 & 0 & 0 \end{bmatrix}$$

Hence the non-leading variables are assigned as parameters: $x_2 = s$, $x_4 = t$ and $x_6 = u$. Then the equations give the leading variables: $x_1 = 2s - 3t + 3u$,

1.3. HOMOGENEOUS SYSTEMS

$x_3 = t - 2u$, and $x_5 = 0$. Thus the general solution X is

$$X = \begin{bmatrix} x_1 \\ x_2 \\ x_3 \\ x_4 \\ x_5 \\ x_6 \end{bmatrix} = \begin{bmatrix} 2s - 3t + 3u \\ s \\ t - 2u \\ t \\ 0 \\ u \end{bmatrix} = s \begin{bmatrix} 2 \\ 1 \\ 0 \\ 0 \\ 0 \\ 0 \end{bmatrix} + t \begin{bmatrix} -3 \\ 0 \\ 1 \\ 1 \\ 0 \\ 0 \end{bmatrix} + u \begin{bmatrix} 3 \\ 0 \\ -2 \\ 0 \\ 0 \\ 1 \end{bmatrix}$$

so the basic solutions are $X_1 = \begin{bmatrix} 2 \\ 1 \\ 0 \\ 0 \\ 0 \\ 0 \end{bmatrix}$, $X_2 = \begin{bmatrix} -3 \\ 0 \\ 1 \\ 1 \\ 0 \\ 0 \end{bmatrix}$ and $X_3 = \begin{bmatrix} 3 \\ 0 \\ -2 \\ 0 \\ 0 \\ 1 \end{bmatrix}$.

Thus there are 3 basic solutions, which agrees with Theorem 2 because the rank of the augmented matrix is $r = 3$ in this case and there are $n = 6$ variables. □

Exercises

1. In each case, consider the system of homogeneous equations with the given matrix as coefficient matrix, find the basic solutions to the system, and express the general solution as a linear combination of these basic solutions.

    (a) $\begin{bmatrix} 1 & 2 & 3 \\ -1 & 3 & 0 \\ -1 & 8 & 3 \end{bmatrix}$
    (b) $\begin{bmatrix} 1 & 2 & 3 \\ -1 & 3 & 0 \\ 1 & 8 & 3 \end{bmatrix}$

    (c) $\begin{bmatrix} 1 & 3 & -2 & -1 \\ -2 & 1 & 0 & 4 \\ 1 & 10 & -6 & 1 \end{bmatrix}$
    (d) $\begin{bmatrix} 1 & -2 & 0 & 3 \\ -3 & 6 & 1 & -4 \\ 2 & -4 & -1 & 1 \end{bmatrix}$

    (e) $\begin{bmatrix} 1 & 2 & 0 \\ 2 & 3 & -1 \\ 3 & 4 & -2 \\ 1 & 3 & 1 \\ 6 & 9 & -3 \end{bmatrix}$
    (f) $\begin{bmatrix} 2 & -1 & 6 & 0 & -3 & 7 \\ -3 & 1 & 3 & -5 & 1 & 4 \end{bmatrix}$

    (g) $\begin{bmatrix} 1 & 2 & 1 & -1 & 3 \\ 1 & 2 & 2 & 1 & 2 \\ 2 & 4 & 2 & -1 & 7 \end{bmatrix}$
    (h) $\begin{bmatrix} 1 & 1 & -2 & 3 & 2 \\ 2 & -1 & 3 & 4 & 1 \\ -1 & -2 & 3 & 0 & 1 \\ 3 & 0 & 1 & 7 & 3 \end{bmatrix}$

2. A chemical equation represents the rearrangement of atoms to form new chemical units, where reactants are transformed into products. A chemical

reaction is considered to be balanced when every atom of every different element which appears on one side of the reaction also appears on the other side of the reaction. For example, the equation $2Na + Cl_2 \to 2NaCl$ can be used to represent the reaction of two molecules of sodium plus one molecule of chlorine into molecules of sodium chloride.

For example, to balance the following reaction,

Methane (CH_4) + Oxygen (O_2) $\to$ Carbon Dioxide (CO_2) + Water (H_2O),

let

$$\begin{aligned} x &= \text{the number of methane molecules present,} \\ y &= \text{the number of oxygen molecules present,} \\ w &= \text{the number of carbon dioxide molecules present,} \\ z &= \text{the number of water molecules present.} \end{aligned}$$

The reaction can now be represented by

$$xCH_4 + yO_2 \to wCO_2 + zH_2O$$

The number of atoms of each element on both side of the equation must be the same, giving the system of linear equations:

$$\begin{aligned} x &= w & \text{(To balance C)} \\ 4x &= 2z & \text{(To balance H)} \\ 2y &= 2w + z & \text{(To balance O)}. \end{aligned}$$

This can be written as a linear system

$$\begin{aligned} x \phantom{{}+2y} \phantom{{}-2w} -w \phantom{{}-2z} &= 0 \\ 4x \phantom{{}-w} \phantom{{}-2w} -2z &= 0 \\ 2y -2w -z &= 0 \end{aligned}$$

In this case, the general solution is $X = [s \; 2s \; s \; 2s]^T$ where s is arbitrary. Since we want positive integers for a chemical reaction, a particular solution is $x = 1 = w$, and $y = 2 = z$, giving

$$CH_4 + 2O_2 \to CO_2 + 2H_2O$$

Following this method, find a balanced equation for the following chemical reactions:

(a) Ammonia (NH_3) + Copper Oxide (CuO) $\to$ Nitrogen (N_2) + Copper (Cu) + Water (H_2O).

(b) Octane (C_8H_{18}) + Oxygen (O_2) $\to$ Carbon Dioxide (CO_2) + Water (H_2O).

3. Explain the difference between general and homogeneous systems of linear equations, and what the implications are for the solutions.

4. Suppose a homogeneous system has 4 equations and 6 variables, and let A denote the augmented matrix.

    (a) Can the system have a unique solution? No solution? Support your answers.

    (b) How many parameters are possible? Support your answer.

    (c) How many parameters are possible if a row of A is a multiple of a different row? Support your answer.

    (d) How many parameters are possible if $rank A = 4$? Support your answer.

    (e) How many parameters are possible if $rank A = 2$? Support your answer.

5. Show that the converse of Theorem 1 is not true. That is, show that if a homogeneous system has a nontrivial solution it may *not* have more variables than equations.

6. If a system has more equations than variables, can it have a unique solution? Justify your answer.

7. In each case either show that the statement is true or give an example showing that it is false. Assume that a system of equations is given with augmented matrix A and reduced row-echelon form R.

    (a) If the system is homogeneous, every solution is trivial.

    (b) If the system has a nontrivial solution, it cannot be homogeneous.

    (c) If the system has a trivial solution, it must be homogeneous.

    (d) If the system is homogeneous and has a nontrivial solution, R has a row of zeros.

    (e) If the system is homogeneous and has a nontrivial solution, then it has infinitely many nontrivial solutions.

    (f) If the system is homogeneous and R has a row of zeros, there are nontrivial solutions.

    (g) If A is $m \times n$ and $rank A = m$, the system has only the trivial solution.

8. For a homogencous system with coefficient matrix C and augmented matrix A, show that $rank A = rank C$.

1.4 MATRIX MULTIPLICATION

In addition to the operations in Section 1.1, there is another operation by which certain matrices can be multiplied together to yield another matrix. This is useful for systems of linear equations, and turns out to have many other applications. We begin with an example of how matrix multiplication arises in a commercial context.

1.4.1 Matrix Multiplication

Example 1. A store sells two brands of toasters, brand X and brand Y. The manager has the following two data matrices available, the first giving sales volumes for three months, and the second giving the retail price and the dealer cost (in dollars):

$$\begin{array}{c} \\ \text{Jan} \\ \text{Feb} \\ \text{Mar} \end{array} \begin{array}{c} X \quad Y \\ \left[\begin{array}{cc} 11 & 9 \\ 10 & 7 \\ 14 & 8 \end{array} \right] \end{array} \text{ and } \begin{array}{c} \\ X \\ Y \end{array} \begin{array}{c} \text{Retail} \quad \text{Cost} \\ \left[\begin{array}{cc} 29 & 22 \\ 31 & 26 \end{array} \right] \end{array}$$

The manager wants a systematic way to calculate the new matrix

$$\begin{array}{c} \\ \text{Jan} \\ \text{Feb} \\ \text{Mar} \end{array} \begin{array}{c} \text{Retail} \quad \text{Cost} \\ \left[\begin{array}{cc} 598 & 476 \\ 507 & 402 \\ 654 & 516 \end{array} \right] \end{array}$$

of total sales and costs for the three months. For example, for January, the total retail sales were

$$\$598 = 11 \cdot 29 + 9 \cdot 31$$

because there were 11 brand-X toasters sold at \$29 each, and 9 brand-Y toasters sold at \$31 each. This gives the (1,1)-entry 598 of the new matrix. Observe that this entry can be calculated directly from the first two matrices by going along the first row $\begin{bmatrix} 11 & 9 \end{bmatrix}$ of the left matrix and the first column $\begin{bmatrix} 29 \\ 31 \end{bmatrix}$ of the right matrix, multiplying corresponding entries, and adding the results. This is called taking the dot product of this row and column. Similarly, the (3,2)-entry, 516, of the new matrix is obtained by taking the dot product of row 3 of the left matrix and column 2 of the right matrix:

$$\$516 = 14 \cdot 22 + 8 \cdot 26$$

Every entry in the new matrix is calculated in the same way: The (i,j)-entry of the new matrix is the dot product of row i of the left matrix and column j of the right matrix. $\square$

The calculation in Example 1 is a special case of a general matrix multiplication. To describe it we use the following terminology: If $R = [r_1 \; r_2 \; \cdots \; r_n]$ is a row matrix and $C = [c_1 \; c_2 \; \cdots \; c_n]^T$ is a column matrix, each with n entries, we define their **dot product** to be the number

1.4. MATRIX MULTIPLICATION

$$r_1c_1 + r_2c_2 + \cdots + r_nc_n.$$

In other words, the dot product of R and C is the number formed by multiplying corresponding entries and adding the results. As in Example 1, this leads to a way to multiply two matrices:

> The **product** of an $m \times n$ matrix A and an $n \times p$ matrix B is the $m \times p$ matrix AB whose (i,j)-entry is the dot product of row i of A and column j of B.

Thus computing the (i,j)-entry of the product AB means going *across* row i of A and *down* column j of B, multiplying corresponding entries, and adding the results. This is illustrated in Figure 1.

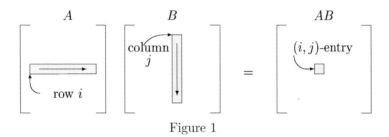

Figure 1

Example 2. Here are some typical matrix products.

$$\begin{bmatrix} 5 & 1 \\ 7 & 6 \\ -4 & 9 \end{bmatrix} \begin{bmatrix} 2 & -1 \\ 0 & 3 \end{bmatrix} = \begin{bmatrix} 5 \cdot 2 + 1 \cdot 0 & 5(-1) + 1 \cdot 3 \\ 7 \cdot 2 + 6 \cdot 0 & 7(-1) + 6 \cdot 3 \\ (-4)2 + 9 \cdot 0 & (-4)(-1) + 9 \cdot 3 \end{bmatrix} = \begin{bmatrix} 10 & -2 \\ 14 & 11 \\ -8 & 31 \end{bmatrix}$$

$$\begin{bmatrix} p & q & r \end{bmatrix} \begin{bmatrix} 5 & 7 \\ -3 & 9 \\ 2 & -5 \end{bmatrix} = \begin{bmatrix} 5p - 3q + 2r & 7p + 9q - 5r \end{bmatrix}$$

$$\begin{bmatrix} 2 & 7 & -5 \\ 3 & 0 & 1 \end{bmatrix} \begin{bmatrix} x \\ y \\ z \end{bmatrix} = \begin{bmatrix} 2x + 7y - 5z \\ 3x + z \end{bmatrix} \qquad \square$$

Note that forming the product AB is not always possible; it requires that the rows of A have the same length as the columns of B (so the dot products can be computed). The following rule is a useful way to remember when AB can be formed and what the size of AB is when A is $m \times n$ and B is $n' \times p$:

$$\begin{array}{cc} A & B \\ m \times \boxed{n \quad n'} \times p \end{array}$$

The product AB is defined if and only if $n = n'$, and in this case the matrix AB has size $m \times p$.

When these conditions are met we say that A and B are **compatible for multiplication**. Whenever we write a matrix product AB, it is tacitly assumed that this is the case.

The next example shows that matrix multiplication has properties not encountered in numerical arithmetic. If A is a square matrix, we write the **powers** of A as for numbers:
$$\begin{aligned} A^2 &= AA \\ A^3 &= AAA \quad \text{etc.} \end{aligned}$$

Example 3. *Show that the following can happen for 2×2 matrices A and B.*

(1) $A^2 = 0$ *even though* $A \neq 0$.

(2) $AB \neq BA$.

Solution. Let $A = \begin{bmatrix} 0 & 1 \\ 0 & 0 \end{bmatrix}$ and $B = \begin{bmatrix} 1 & 0 \\ 0 & 0 \end{bmatrix}$. Then $A^2 = \begin{bmatrix} 0 & 1 \\ 0 & 0 \end{bmatrix} \begin{bmatrix} 0 & 1 \\ 0 & 0 \end{bmatrix} = \begin{bmatrix} 0 & 0 \\ 0 & 0 \end{bmatrix} = 0$, even though $A \neq 0$. Moreover $AB = \begin{bmatrix} 0 & 1 \\ 0 & 0 \end{bmatrix} \begin{bmatrix} 1 & 0 \\ 0 & 0 \end{bmatrix} = 0$ while $BA = \begin{bmatrix} 1 & 0 \\ 0 & 0 \end{bmatrix} \begin{bmatrix} 0 & 1 \\ 0 & 0 \end{bmatrix} = A$. Hence $AB \neq BA$. □

Two matrices A and B are said to **commute** if $AB = BA$. Thus Example 3 exhibits two matrices that do *not* commute. In fact commuting is relatively rare: Given two matrices, the chances are they will not commute.

Example 4. *If $B = \begin{bmatrix} 1 & 1 \\ 0 & 0 \end{bmatrix}$, find all 2×2 matrices A that commute with B.*

Solution. The condition on A is that $AB = BA$. If we write $A = \begin{bmatrix} x & y \\ z & w \end{bmatrix}$ in terms of its entries, this condition becomes
$$\begin{bmatrix} x & y \\ z & w \end{bmatrix} \begin{bmatrix} 1 & 1 \\ 0 & 0 \end{bmatrix} = \begin{bmatrix} 1 & 1 \\ 0 & 0 \end{bmatrix} \begin{bmatrix} x & y \\ z & w \end{bmatrix},$$
that is
$$\begin{bmatrix} x & x \\ z & z \end{bmatrix} = \begin{bmatrix} x+z & y+w \\ 0 & 0 \end{bmatrix}.$$

Equating entries gives $z = 0$ and $x = y+w$, so A has the form $A = \begin{bmatrix} y+w & y \\ 0 & w \end{bmatrix}$ where y and w are arbitrary numbers. □

1.4.2 Properties of Matrix Multiplication

Example 3 shows that matrix arithmetic differs in basic ways from ordinary numerical arithmetic. On the other hand, many similarities do occur (as seen for matrix addition in Theorem 1, §1.1).

The number 1 has the property that $1 \cdot a = a = a \cdot 1$ for every number a. An analogous role is played for matrix multiplication by square matrices of the following type:

$$I_2 = \begin{bmatrix} 1 & 0 \\ 0 & 1 \end{bmatrix} \quad I_3 = \begin{bmatrix} 1 & 0 & 0 \\ 0 & 1 & 0 \\ 0 & 0 & 1 \end{bmatrix} \quad I_4 = \begin{bmatrix} 1 & 0 & 0 & 0 \\ 0 & 1 & 0 & 0 \\ 0 & 0 & 1 & 0 \\ 0 & 0 & 0 & 1 \end{bmatrix} \quad \text{etc.}$$

In general the $n \times n$ **identity matrix** I_n is the square $n \times n$ matrix with 1's on the main diagonal and 0's elsewhere. It is often written simply as I when no confusion can result. It is routine to verify that identity matrices enjoy the following properties

$$IA = A \quad \text{and} \quad BI = B$$

whenever the products are defined. These are the first of the following important facts about matrix multiplication.

Theorem 1. *Let c be a scalar and let A, B and C denote matrices of sizes such that the products below can be performed.*

(1) $IA = A$ and $BI = B$.

(2) $A(BC) = (AB)C$.

(3) $A(B + C) = AB + AC$ $\quad$ and $\quad$ $A(B - C) = AB - AC$.

(4) $(B + C)A = BA + CA$ $\quad$ and $\quad$ $(B - C)A = BA - CA$.

(5) $c(AB) = (cA)B = A(cB)$.

(6) $(AB)^T = B^T A^T$.

The proofs of these properties are routine and are given at the end of this section.

The property that $A(BC) = (AB)C$ is called the **associative law** for matrix multiplication. It asserts that the product of three matrices A, B and C (in that order) is the same no matter how it is formed and so is denoted simply as ABC. Similarly, the product $ABCD$ of four matrices can be formed in several ways (examples: $A[B(CD)]$, $(AB)(CD)$, $[A(BC)]D$, etc.). However the associative law shows that they are all equal, and so are written simply as $ABCD$. Similar remarks apply to products of five or more matrices.

However, a note of caution is needed here. The fact that AB and BA may not be equal (see Example 3) means that the *order* of the factors in a matrix product *is* important. For example, the products $ADCB$ and $ABCD$ may *not* be equal.

Warning. *If the order of the factors in a product of matrices is changed, the product matrix may change (or may not exist.)*

Ignoring this is the source of many errors by linear algebra students! For example, when multiplying a matrix equation

$$B = C$$

by a matrix A care must be taken to multiply B and C on the *same side* by A. Hence we speak of *left-multiplying* by A to get $AB = AC$, and *right-multiplying* by A to get $BA = CA$. The next example illustrates how this is used.

Example 5. *If A, B and C are matrices such that $AB = I$ and $CA = I$, show that $B = C$.*

Solution. Left multiply the given equation $I = AB$ by C, and use (1) and (2) of Theorem 1, to get

$$C = CI = C(AB) = (CA)B = IB = B. \qquad \square$$

Properties (3) and (4) of Theorem 1 are called the **distributive laws**. They extend to sums of more than two terms, and they combine with property (5) to allow algebraic manipulations very much like those in numerical arithmetic. For example:

$$\begin{aligned} A(2B - C + 3D) &= 2AB - AC + 3AD \\ (A + 2C - 3E + 5D)B &= AB + 2CB - 3EB + 5DB \end{aligned}$$

Note that the warning is still in effect: the order of each of the products here must not be changed.

Example 6. *Simplify the expression* $A(BC - 2CD) + A(2C - B)D + (BA - AB)C$.

Solution.

$$\begin{aligned} &A(BC - 2CD) + A(2C - B)D + (BA - AB)C \\ &= ABC - 2ACD + 2ACD - ABD + BAC - ABC \\ &= BAC - ABD. \qquad \square \end{aligned}$$

Recall that two matrices A and B are said to commute if $AB = BA$. Many formulas that are valid for numbers will fail for non-commuting matrices. For example, the well known "difference of squares" formula $a^2 - b^2 = (a-b)(a+b)$ of arithmetic may not hold for matrices; the whole thing depends upon whether the matrices commute.

Example 7. *Let A and B denote $n \times n$ matrices.*

1.4. MATRIX MULTIPLICATION

(a) If $AB = BA$ show that $A^2 - B^2 = (A - B)(A + B)$.

(b) If $A^2 - B^2 = (A - B)(A + B)$, show that necessarily $AB = BA$.

Solution. The distributive laws give
$$(A - B)(A + B) = A(A + B) - B(A + B) = A^2 + AB - BA - B^2. \quad (*)$$
This holds whether or not $AB = BA$, and so is helpful in both (a) and (b).

(a) If $AB = BA$ the right side of equation (*) equals $A^2 - B^2$, giving (a).

(b) If $A^2 - B^2 = (A - B)(A + B)$, equation (*) becomes $A^2 - B^2 = A^2 + AB - BA - B^2$. Subtracting $A^2 - B^2$ from both sides gives $0 = AB - BA$, whence $AB = BA$. This gives (b). □

Recall that a matrix A is called symmetric if $A^T = A$ where A^T is the transpose of A.

Example 8. If A and B are symmetric matrices such that $AB = BA$, show that AB is also symmetric.

Solution. We are given $A^T = A$ and $B^T = B$. Compute the transpose of AB using (6) of Theorem 1:
$$(AB)^T = B^T A^T = BA$$
Since we are assuming that $AB = BA$, this gives $(AB)^T = AB$; that is AB is symmetric. □

1.4.3 Matrix Multiplication and Linear Equations

One of the most important motivations for matrix multiplication is in the study of systems of linear equations begun in Section 1.2. Consider the following system of 3 linear equations in 4 variables:

$$\begin{array}{rcrcrcrcl} x_1 & - & 2x_2 & + & 2x_3 & - & x_4 & = & 1 \\ 2x_1 & - & 4x_2 & + & 3x_3 & + & x_4 & = & 3 \\ 3x_1 & - & 6x_2 & + & 5x_3 & & & = & 4 \end{array}$$

This system can be written as a single matrix equation

$$\begin{bmatrix} x_1 - 2x_2 + 2x_3 - x_4 \\ 2x_1 - 4x_2 + 3x_3 + x_4 \\ 3x_1 - 6x_2 + 5x_3 \end{bmatrix} = \begin{bmatrix} 1 \\ 3 \\ 4 \end{bmatrix}.$$

The column matrix on the left can be factored as a matrix product

$$\begin{bmatrix} 1 & -2 & 2 & -1 \\ 2 & -4 & 3 & 1 \\ 3 & -6 & 5 & 0 \end{bmatrix} \begin{bmatrix} x_1 \\ x_2 \\ x_3 \\ x_4 \end{bmatrix} = \begin{bmatrix} 1 \\ 3 \\ 4 \end{bmatrix}.$$

Thus the *system* of linear equations becomes a *single* matrix equation

$$AX = B$$

where $A = \begin{bmatrix} 1 & -2 & 2 & -1 \\ 2 & -4 & 3 & 1 \\ 3 & -6 & 5 & 0 \end{bmatrix}$, $X = \begin{bmatrix} x_1 \\ x_2 \\ x_3 \\ x_4 \end{bmatrix}$ and $B = \begin{bmatrix} 1 \\ 3 \\ 4 \end{bmatrix}$. These are called the **coefficient matrix**, the **matrix of variables** and the **matrix of constants** respectively. Clearly every system of linear equations can be written in this way.

Given a system $AX = B$ written in matrix form, the new system

$$AX = 0$$

is called the **associated homogeneous system**. It differs from $AX = B$ only in that all the constants have been set equal to zero. The solutions of the two systems are closely related as the next theorem demonstrates.

Theorem 2. *Suppose X_0 is any particular solution to the system $AX = B$ of linear equations.*

(1) *If X' is any solution to the associated homogeneous system $AX = 0$, then $X = X_0 + X'$ is a solution to the system $AX = B$.*

(2) *Every solution X to $AX = B$ has the form $X = X_0 + X'$ for some solution X' to $AX = 0$.*

Proof. We are assuming that $AX_0 = B$.
 (1). If $AX' = 0$ we have $AX = A(X_0 + X') = AX_0 + AX' = B + 0 = B$.
 (2). If X is *any* solution to the system $AX = B$, write $X' = X - X_0$. Clearly $X = X_0 + X'$, and we have $AX' = A(X - X_0) = AX - AX_0 = B - B = 0$. Thus X' is indeed a solution to the associated homogeneous system $AX = 0$.□

Example 9. If $A = \begin{bmatrix} 1 & -2 & 2 & -1 \\ 2 & -4 & 3 & 1 \\ 3 & -6 & 5 & 0 \end{bmatrix}$ and $B = \begin{bmatrix} 1 \\ 3 \\ 4 \end{bmatrix}$ as above, use Gaussian elimination to express every solution to the system $AX = B$ as the sum of a particular solution and the general solution to the associated homogeneous system $AX = 0$.

Solution. The reduction of the augmented matrix to reduced form is as follows:

$$\begin{bmatrix} 1 & -2 & 2 & -1 & 1 \\ 2 & -4 & 3 & 1 & 3 \\ 3 & -6 & 5 & 0 & 4 \end{bmatrix} \to \begin{bmatrix} 1 & -2 & 0 & 5 & 3 \\ 0 & 0 & 1 & -3 & -1 \\ 0 & 0 & 0 & 0 & 0 \end{bmatrix}$$

Hence the leading variables are x_1 and x_3, and the nonleading variables become parameters: $x_2 = s$ and $x_4 = t$. Thus the reduced form gives the general solution

$$X = \begin{bmatrix} x_1 \\ x_2 \\ x_3 \\ x_4 \end{bmatrix} = \begin{bmatrix} 3+2s-5t \\ s \\ -1+3t \\ t \end{bmatrix} = \begin{bmatrix} 3 \\ 0 \\ -1 \\ 0 \end{bmatrix} + s \begin{bmatrix} 2 \\ 1 \\ 0 \\ 0 \end{bmatrix} + t \begin{bmatrix} -5 \\ 0 \\ 3 \\ 1 \end{bmatrix}$$
$$= X_0 + sX_1 + tX_2.$$

Here $X_0 = \begin{bmatrix} 3 & 0 & -1 & 0 \end{bmatrix}^T$ is a particular solution to $AX = B$ (corresponding to $s = t = 0$) and $X' = sX_1 + tX_1$ is the general solution to the associated homogeneous system $AX = 0$ (where $X_1 = \begin{bmatrix} 2 & 1 & 0 & 0 \end{bmatrix}^T$ and $X_2 = \begin{bmatrix} -5 & 0 & 3 & 1 \end{bmatrix}^T$ are the basic solutions). (This can also be seen directly from the algorithm by replacing the last column in the augmented matrix by zeros.) □

A version of Theorem 2 holds in other contexts as well. For example, it splits the problem of solving a system of linear differential equations into two problems: (1) Finding a particular solution; and (2) Solving the associated homogeneous system. In this situation the two problems are attacked using quite different methods.

1.4.4 Block Multiplication

When working with very large matrices which exceed computer memory capacity, it is important to regard a matrix as being made up of smaller matrices (called blocks), and to be able to carry out matrix multiplication by multiplying one pair of blocks at a time. This is only one reason for the importance of block multiplication.

Recall that the augmented matrix of a system $AX = B$ is the matrix $[A\ B]$ consisting of the coefficient matrix A and the (column) matrix B of constants. This as an example of a **block partitioning** of the augmented matrix, and we say that the augmented matrix $[A\ B]$ is written in **block form**.

Another common example is to write an $m \times n$ matrix A as a row of columns. If the columns are denoted $C_1, C_2, \cdots, C_n$, we write A as follows:

$$A = [C_1\ C_2\ \cdots\ C_n]$$

Then the definition of a matrix product BA can be expressed as follows:

$$BA = B[C_1\ C_2\ \cdots\ C_n] = [BC_1\ BC_2\ \cdots\ BC_n]$$

In other words, each column of BA is formed by multiplying B times the corresponding column of A.

Partitioning a matrix A in this way with its columns as blocks will be used several times below. Here is another way in which it arises.

Example 10. Write the 2×3 matrix A as $A = [C_1 \ C_2 \ C_3]$ where C_1, C_2 and C_3 denote the columns of A. If $X = [x_1 \ x_2 \ x_3]^T$, show that

$$AX = [C_1 \ C_2 \ C_3] \begin{bmatrix} x_1 \\ x_2 \\ x_3 \end{bmatrix} = x_1 C_1 + x_2 C_2 + x_3 C_3.$$

Solution. Write $A = \begin{bmatrix} a_1 & a_2 & a_3 \\ b_1 & b_2 & b_3 \end{bmatrix}$ so that $C_1 = \begin{bmatrix} a_1 \\ b_1 \end{bmatrix}$, $C_2 = \begin{bmatrix} a_2 \\ b_2 \end{bmatrix}$ and $C_3 = \begin{bmatrix} a_3 \\ b_3 \end{bmatrix}$. Then

$$\begin{aligned} AX &= \begin{bmatrix} a_1 & a_2 & a_3 \\ b_1 & b_2 & b_3 \end{bmatrix} \begin{bmatrix} x_1 \\ x_2 \\ x_3 \end{bmatrix} = \begin{bmatrix} a_1 x_1 + a_2 x_2 + a_3 x_3 \\ b_1 x_1 + b_2 x_2 + b_3 x_3 \end{bmatrix} \\ &= x_1 \begin{bmatrix} a_1 \\ b_1 \end{bmatrix} + x_2 \begin{bmatrix} a_2 \\ b_2 \end{bmatrix} + x_3 \begin{bmatrix} a_3 \\ b_3 \end{bmatrix} \\ &= x_1 C_1 + x_2 C_2 + x_3 C_3 \end{aligned}$$

as required. $\square$

The calculation in Example 10 generalizes, and is given as part of the following theorem.

Theorem 3. Let $A = [C_1 \ C_2 \ \cdots \ C_n]$ be an $m \times n$ matrix with columns C_1, $C_2, \cdots, C_n$.

(1) If B is a $k \times m$ matrix the matrix product BA can be written as

$$BA = B[C_1 \ C_2 \ \cdots \ C_n] = [BC_1 \ BC_2 \ \cdots \ BC_n]$$

(2) If $X = [x_1 \ x_2 \ \cdots \ x_n]^T$ is any column, then

$$AX = [C_1 \ C_2 \ \cdots \ C_n] \begin{bmatrix} x_1 \\ x_2 \\ \vdots \\ x_n \end{bmatrix} = x_1 C_1 + x_2 C_2 + \cdots + x_n C_n$$

Theorem 3 will be used frequently later.

Observe that the products in Theorem 3 are computed using ordinary matrix multiplication, regarding the columns C_j as entries. This observation holds much more generally as the next example illustrates.

Example 11. Consider $A = \begin{bmatrix} 1 & -1 & | & 7 & 8 & 0 \\ 3 & 2 & | & 5 & 6 & 7 \\ \hline 0 & 0 & | & -1 & -2 & -3 \end{bmatrix} = \begin{bmatrix} P & X \\ 0 & Q \end{bmatrix}$

1.4. MATRIX MULTIPLICATION

and $B = \begin{bmatrix} -2 & 0 \\ 5 & 8 \\ \hline 1 & 1 \\ 2 & 0 \\ -4 & 6 \end{bmatrix} = \begin{bmatrix} U \\ V \end{bmatrix}$ partitioned as shown. If we use the blocks as entries, the product is

$$AB = \begin{bmatrix} P & X \\ 0 & Q \end{bmatrix}\begin{bmatrix} U \\ V \end{bmatrix} = \begin{bmatrix} PU + XV \\ 0U + QV \end{bmatrix} = \begin{bmatrix} PU + XV \\ QV \end{bmatrix} = \begin{bmatrix} 16 & -1 \\ -7 & 63 \\ \hline 7 & -19 \end{bmatrix}$$

since $PU + XV = \begin{bmatrix} 16 & -1 \\ -7 & 63 \end{bmatrix}$ and $QV = \begin{bmatrix} 7 & -19 \end{bmatrix}$. This agrees with ordinary matrix multiplication as the reader can verify. □

Block multiplication can be used to compute *any* matrix product AB, subject only to the requirement that the partitioning of A and B are **compatible for multiplication**: The block sizes must be such that every product of blocks that occurs make sense. Then the product can be computed by matrix multiplication using blocks as entries.

Block Multiplication. *If matrices A and B are partitioned compatibly into blocks, the product AB can be computed by matrix multiplication using the blocks as entries.*

A useful occurrence of block multiplication is given in the following Theorem.

Theorem 4. *Let* $A = \begin{bmatrix} B & X \\ 0 & C \end{bmatrix}$ *and* $A_1 = \begin{bmatrix} B_1 & X_1 \\ 0 & C_1 \end{bmatrix}$, *in block form, where B and B_1 are square matrices of the same size, and C and C_1 are square of the same size. These are compatible partitionings, and block multiplication gives*

$$AA_1 = \begin{bmatrix} B & X \\ 0 & C \end{bmatrix}\begin{bmatrix} B_1 & X_1 \\ 0 & C_1 \end{bmatrix} = \begin{bmatrix} BB_1 & BX_1 + XC_1 \\ 0 & CC_1 \end{bmatrix}.$$

There is a similar theorem for block matrices of the form $\begin{bmatrix} B & O \\ X & C \end{bmatrix}$ where B and C are square.

1.4.5 Directed Graphs

Matrix multiplication arises in a variety of situations other than the study of linear equations. In Example 1 it is used to describe how to systematically compute the total sales of retail items. We now describe how it is used to compute the number of paths in a directed graph from one vertex to another.

A **directed graph** is a set of points (called **vertices**) together with arrows (called **edges**) joining various pairs of vertices. Sometimes a double arrow indicates an edge in both directions. For example the vertices could represent cities with arrows indicating direct flights between them. If the graph has n vertices, the **adjacency matrix** $A = [a_{ij}]$ is the $n \times n$ matrix whose (i,j)-entry a_{ij} is 1 if there is an edge from vertex v_i to vertex v_j, and 0 otherwise. For example, the adjacency matrix of the directed graph below is $A = \begin{bmatrix} 1 & 1 & 1 \\ 1 & 0 & 0 \\ 0 & 1 & 0 \end{bmatrix}$. Observe that the (i,j)-entry a_{ij} of the matrix A is the *number* of edges (0 or 1) from vertex v_i to vertex v_j.

The next result gives an important extension of this observation. A **path of length** r from vertex v_i to vertex v_j is a sequence of r edges beginning at v_i and leading one after the other to v_j. Thus, in the graph,

$$v_1 \to v_2 \to v_1 \to v_1 \to v_3$$

is a path of length 4 from v_1 to v_3. The powers of A determine the number of such paths.

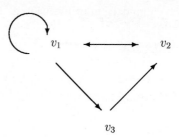

Theorem 5. *If A is the adjacency matrix of a directed graph, the (i,j)-entry of A^r is the number of paths of length r in the graph from vertex v_i to vertex v_j.*

For example, consider the adjacency matrix A of the graph above. Then

$$A = \begin{bmatrix} 1 & 1 & 1 \\ 1 & 0 & 0 \\ 0 & 1 & 0 \end{bmatrix}, \quad A^2 = \begin{bmatrix} 2 & 2 & 1 \\ 1 & 1 & 1 \\ 1 & 0 & 0 \end{bmatrix} \quad \text{and} \quad A^3 = \begin{bmatrix} 4 & 3 & 2 \\ 2 & 2 & 1 \\ 1 & 1 & 1 \end{bmatrix}$$

Since the $(1,2)$-entry of A^2 is 2, there must be exactly 2 paths of length 2 from vertex v_1 to v_2. This can be verified directly in this case: In fact the paths are $v_1 \to v_1 \to v_2$ and $v_1 \to v_3 \to v_2$. The other entries of A^2 can also be checked. Similarly, the fact that A^3 has no zero entries shows that it is possible to go from *any* vertex to *any* other vertex in exactly three steps.

Proof of Theorem 5. Let b_{ij} denote the number of paths of length r from v_i to v_j, and write $B_r = [b_{ij}]$. We must show that $B_r = A^r$ for each r. This is clear if $r = 1$ because $B_1 = A$. In general we show that, if $B_r = A^r$ holds for some $r \geq 1$, then it follows necessarily that $B_{r+1} = A^{r+1}$. Then $B_1 = A = A^1$ gives $B_2 = A^2$; this in turn gives $B_3 = A^3$; from which we get $B_4 = A^4$, and so on.

1.4. MATRIX MULTIPLICATION

So assume that $B_r = A^r$ holds for some $r \geq 1$. Every path of length $r+1$ from v_i to v_j must pass first through *some* vertex v_k :

$$v_i \xrightarrow{\text{length } 1} v_k \xrightarrow{\text{length } r} v_j$$

But the number of such paths is $a_{ik}b_{kj}$ because there are a_{ik} paths of length 1 from v_i to v_k, and b_{kj} paths of length r from v_k to v_j. Hence the *total* number of paths of length $r+1$ from v_i to v_j is the sum

$$a_{i1}b_{1j} + a_{i2}b_{2j} + \cdots + a_{in}b_{nj} \tag{*}$$

In other words, (*) is the (i,j)-entry of B_{r+1}. But this is what we wanted because, by the definition of matrix multiplication, (*) is just the (i,j)-entry of the matrix $AB_r = AA^r = A^{r+1}$. (Indeed, row i of A is $\begin{bmatrix} a_{i1} & a_{i2} & \cdots & a_{in} \end{bmatrix}$ and column j of B is $\begin{bmatrix} b_{1j} & b_{2j} & \cdots & b_{nj} \end{bmatrix}^T$, and (*) is their dot product.) □

1.4.6 Proof of Theorem 1

Theorem 1 is a list of the most basic facts about matrix multiplication. Hence, before looking at the proof, we must formalize the definition of matrix multiplication. If $A = [a_{ij}]$ is $m \times n$ and $B = [b_{ij}]$ is $n \times p$, then row i of A and column j of B are, respectively

$$[a_{i1}\ a_{i2}\ \cdots\ a_{in}] \quad \text{and} \quad \begin{bmatrix} b_{1j} \\ b_{2j} \\ \vdots \\ b_{nj} \end{bmatrix}$$

Hence the (i,j)-entry of AB is the dot product

$$a_{i1}b_{1j} + a_{i2}b_{2j} + \cdots + a_{in}b_{nj} = \Sigma_{k=1}^n a_{ik}b_{kj}$$

where we are using **summation notation**[12] for convenience. With this we prove parts 2, 3 and 6 of Theorem 1; the rest is similar and is left as an exercise.

Proof of (2): $A(BC) = (AB)C$. Assume C is of size $p \times q$ and write $C = [c_{ij}]$ and $BC = [d_{ij}]$. Then the (i,j)-entry of $A(BC)$ is $\Sigma_{k=1}^n a_{ik}d_{kj}$. Now $d_{kj} = \Sigma_{l=1}^p b_{kl}c_{lj}$, so the (i,j)-entry of of $A(BC)$ is

$$\Sigma_{k=1}^n a_{ik}(\Sigma_{l=1}^p b_{kl}c_{lj}) = \Sigma_{k=1}^n \Sigma_{l=1}^p a_{ik}b_{kl}c_{lj}$$

A similar argunment shows that this is also the (i,j)-entry of $(AB)C$.

Proof of (3): $A(B + C) = AB + AC$. Now assume C is of size $n \times p$, and write $C = [c_{ij}]$ so that $B + C = [b_{ij} + c_{ij}]$. Hence the (i,j)-entry of $A(B+C)$ is

$$\Sigma_{k=1}^n a_{ik}(b_{kj} + c_{kj}) = \Sigma_{k=1}^n (a_{ik}b_{kj} + a_{ik}c_{kj}) = \Sigma_{k=1}^n a_{ik}b_{kj} + \Sigma_{k=1}^n a_{ik}c_{kj}$$

---

[12]Summation notation is a convenient shorthand way to write sums. For example: $a_1 + a_2 + a_3 + a_4 = \Sigma_{k=1}^4 a_k$, and $a_3b_3 + a_4b_4 + a_5b_5 = \Sigma_{k=3}^5 a_k b_k$.

This is the (i,j)-entry of $AB + BC$ because the sums on the right are the (i,j)-entries of AB and BC respectively.

Proof of (6): $(AB)^T = B^T A^T$. Write $A^T = [a'_{ij}]$ and $B^T = [b'_{ij}]$ where $a'_{ij} = a_{ji}$ and $b'_{ij} = b_{ji}$ for all i and j. Then the (i,j)-entry of $B^T A^T$ is

$$\Sigma_{k=1}^n b'_{ik} a'_{kj} = \Sigma_{k=1}^n b_{ki} a_{jk} = \Sigma_{k=1}^n a_{jk} b_{ki}$$

But this is the (j,i)-entry of AB, that is the (i,j)-entry of $(AB)^T$. Hence $B^T A^T = (AB)^T$.

Exercises

1. In each case compute the indicated matrix products.

    (a) $\begin{bmatrix} 2 & 0 & -3 \\ -1 & 6 & 5 \end{bmatrix} \begin{bmatrix} 0 & 3 \\ -4 & 7 \\ 2 & -5 \end{bmatrix}$ (b) $\begin{bmatrix} 5 & 0 & -3 \\ 1 & 2 & 1 \end{bmatrix} \begin{bmatrix} 8 \\ -5 \\ 2 \end{bmatrix}$

    (c) $\begin{bmatrix} 1 & 3 & -6 \end{bmatrix} \begin{bmatrix} 2 & -3 \\ 0 & 1 \\ -4 & 5 \end{bmatrix}$ (d) $\begin{bmatrix} 2 & 7 & -4 \end{bmatrix} \begin{bmatrix} -1 \\ 2 \\ 8 \end{bmatrix}$

    (e) $\begin{bmatrix} 3 \\ -5 \\ 2 \end{bmatrix} \begin{bmatrix} 5 & 9 & -3 \end{bmatrix}$ (f) $\begin{bmatrix} 4 & 2 \\ -2 & -5 \\ 1 & 5 \end{bmatrix} \begin{bmatrix} 1 & 2 & 3 \\ 0 & -3 & 5 \end{bmatrix}$

2. Let $A = \begin{bmatrix} -2 & 5 & 1 \\ -3 & -6 & 5 \end{bmatrix}$, $B = \begin{bmatrix} -2 & -5 \\ 0 & 3 \\ 9 & -4 \end{bmatrix}$, $C = \begin{bmatrix} 1 & 3 \\ -2 & 5 \end{bmatrix}$ and $D = \begin{bmatrix} -2 \\ 3 \end{bmatrix}$. In each case either compute the indicated product or explain why it is not defined.

    (a) AB (b) A^2 (c) CD (d) DC (e) BC (f) AC (g) C^2 (h) AD

3. In your own words, describe an algorithm to check whether two given matrices can be multiplied, and an algorithm to perform the multiplication when they are compatible.

4. If A and B are matrices and AB is square, explain why BA is also square.

5. If AB and BA can both be formed, describe the sizes of A and B.

6. If $AC = CB$ explain why A and B are both square.

1.4. MATRIX MULTIPLICATION

7. Simplify the following expressions where A, B and C represent matrices.
   (a) $A(3B - C) + A^3 C$
   (b) $A(3B - C) + A^2 B + 3A(C - 2B)$
   (c) $(A - B)(A - B) - A^2 + B^2$

8. (a) If $AB = BA$ show that $(A + B)^2 = A^2 + 2AB + B^2$.
   (b) If $AB \neq BA$ explain why $(A + B)^2 \neq A^2 + 2AB + B^2$.

9. If $AB = A$ and $BA = B$, show that $A^2 = A$ and $B^2 = B$.

10. In each case either show that the statement is true or give an example showing that it is false. Let A and B denote matrices.
    (a) If AB is defined then BA is defined.
    (b) If AB is defined and is a square matrix then BA is defined.
    (c) If $AB = BA$ then A and B are both square and the same size.
    (d) If A^2 can be formed then A must be square.
    (e) If A has a row of zeros then AB has a row of zeros.
    (f) If A has a column of zeros then AB has a column of zeros.
    (g) If $AB = 0$ then $A = 0$ or $B = 0$.
    (h) $(AB)^2 = A^2 B^2$ always holds.
    (i) $(A + B)^2 = A^2 + 2AB + B^2$ always holds.
    (j) If $AJ = A$ then J is an identity matrix.
    (k) If $A^2 = A$ then $A = 0$ or $A = I$.
    (l) If A is $m \times n$ and $m > n$ then the system $AX = 0$ has infinitely many solutions.
    (m) If A is $m \times n$ and $m < n$ then the system $AX = B$ has a solution for any column B.

11. If A is any matrix show that AA^T and $A^T A$ are both symmetric matrices. [*Hint*: Theorem 1(6).]

12. Suppose that A and B are both symmetric matrices.
    (a) If AB is symmetric, show that $AB = BA$. Explain how this relates to Example 8.
    (b) Show conversely that if $AB = BA$, then AB is symmetric.

13. In each case write the system in matrix form $AX = B$ and express the general solution as a sum of a particular solution plus the general solution of the associated homogeneous system $AX = 0$ (See Example 9).
    (a) $\begin{aligned} x_1 + x_2 + x_3 &= 2 \\ 2x_1 + x_2 &= 3 \\ x_1 - x_2 - 3x_3 &= 0 \end{aligned}$

(b) $\begin{aligned} x_1 - x_2 - 4x_3 &= -4 \\ x_1 + 2x_2 + 5x_3 &= 2 \\ x_1 + x_2 + 2x_3 &= 0 \end{aligned}$

(c) $\begin{aligned} x_1 + x_2 - x_3 - 5x_5 &= 2 \\ x_2 + x_3 - 4x_5 &= -1 \\ x_2 + x_3 + x_4 + x_5 &= -1 \\ 2x_1 - 4x_3 + x_4 + x_5 &= 6 \end{aligned}$

(h) $\begin{aligned} 2x_1 + x_2 - x_3 - x_4 &= -1 \\ 3x_1 + x_2 + x_3 - 2x_4 &= -2 \\ -x_1 - x_2 + 2x_3 + x_4 &= 2 \\ -2x_1 - x_2 + 2x_4 &= 3 \end{aligned}$

14. Find an example of an inconsistent system of 4 equations in 4 variables such that the associated homogeneous system has a 2-parameter family of solutions.

15. Suppose A is a 3×3 matrix and $AX = 0$ has a unique solution. Is $AX = B$ consistent for each column B? If so is the solution unique? Support your answer.

16. Let A be a 4×6 matrix with $rank A = 4$. Explain why $AX = B$ has a solution for any 4×1 matrix B.

17. If A is a matrix and the system $AX = 0$ has a nontrivial solution, show that there is no matrix B such that $BA = I$.

18. Let $D = \begin{bmatrix} a_1 & 0 & 0 \\ 0 & a_2 & 0 \\ 0 & 0 & a_3 \end{bmatrix}$. A matrix of this type is called a **diagonal matrix**.

    (a) Calculate the products D^2, D^3, D^4, and make a guess at what D^n looks like.

    (b) If A is a 3×3 matrix, describe the products AD and DA in terms of the columns, respectively rows, of A.

19. Let $A = \begin{bmatrix} 3 & -1 & 7 \\ 0 & 2 & 0 \\ 0 & 0 & -1 \end{bmatrix}$. Matrices of this form are called *upper triangular*.

    (a) Compute A^2, A^3, A^4, and make a conjecture about A^n for $n \geq 1$.

    (b) Make a conjecture about the product AB of any two 3×3 upper triangular matrices.

20. Let $A = \begin{bmatrix} 1 & 0 & | & 2 \\ 1 & 0 & | & 0 \\ \hline 0 & 0 & | & -1 \end{bmatrix}$. Compute A^2, A^3, A^4, A^5. Make a conjecture about A^n for $n \geq 1$.

1.4. MATRIX MULTIPLICATION

21. Verify that $A^2 - 5A + 4I = 0$ if: (a) $A = \begin{bmatrix} 1 & 7 \\ 0 & 4 \end{bmatrix}$ (b) $A = \begin{bmatrix} -5 & 6 \\ -9 & 10 \end{bmatrix}$

22. (a) If $A = \begin{bmatrix} 0 & 1 \\ 0 & 0 \end{bmatrix}$ show that $A^2 = 0$.

    (b) If A is a symmetric 2×2 matrix with $A^2 = 0$, show that $A = 0$.

    (c) If $A^2 = 0$ show that $(A^T)^2 = 0$, and that $(cA)^2 = 0$ for all constants c.

23. Compute the matrix product AB using block multiplication where

$$A = \left[\begin{array}{cc|ccc} -2 & 1 & 6 & -2 & 0 \\ 0 & 3 & 1 & 3 & 7 \\ \hline 0 & 0 & -1 & 2 & 3 \end{array} \right]$$

and $B = \left[\begin{array}{cc|cc} 1 & -3 & 0 & 0 \\ 5 & 9 & 0 & 0 \\ \hline -1 & 5 & -1 & -2 \\ 7 & 0 & 3 & 3 \\ 2 & -3 & -1 & 0 \end{array} \right]$

24. In each case compute all powers A, A^2, A^3 and A^4 using the preceding exercise. Then find an expression for A^n for any $n \geq 1$.

    (a) $A = \left[\begin{array}{c|cc} 1 & 0 & 0 \\ \hline 1 & 1 & -1 \\ 1 & -1 & 1 \end{array} \right]$

    (b) $A = \left[\begin{array}{cc|cc} 1 & -1 & 2 & -1 \\ 0 & 1 & 0 & 0 \\ \hline 0 & 0 & -1 & 1 \\ 0 & 0 & 0 & 1 \end{array} \right]$

25. Compute the following matrix products using block multiplication. Assume that the partitionings are compatible for multiplication.

    (a) $\begin{bmatrix} I & X \\ -Y & I \end{bmatrix} \begin{bmatrix} I & 0 \\ Y & I \end{bmatrix}$ (b) $\begin{bmatrix} I & X \\ 0 & I \end{bmatrix} \begin{bmatrix} I & -X \\ 0 & I \end{bmatrix}$

    (c) $\begin{bmatrix} I & X \end{bmatrix} \begin{bmatrix} I \\ Y \end{bmatrix}$ (d) $\begin{bmatrix} X & I \end{bmatrix} \begin{bmatrix} I \\ -X \end{bmatrix}$

26. Let $A = \begin{bmatrix} 0 & I \\ B & 0 \end{bmatrix}$ where each block is square of the same size. Compute A, A^2, A^3, and A^4. What is A^n if n is even? What if n is odd? Support your answer.

27. In each of the directed graphs, determine the number of paths of length 4 and 10 from any vertex to any other vertex, and see if you can check the graph itself to verify your findings.

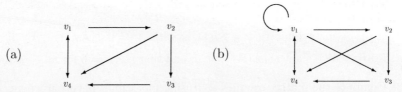

28. If B is $n \times n$ and $BC = C$ for all columns C, show that $B = I_n$ is the $n \times n$ identity matrix.

29. (a). If the 2×2 matrix A commutes with every 2×2 matrix, show that $A = aI_2$ for some number a.

    (b). If a is any number, show that aI_2 commutes with *every* 2×2 matrix.

    [*Remark*: The only $n \times n$ matrices that commute with every $n \times n$ matrix are the scalar multiples aI_n of the $n \times n$ identity matrix. You might like to show this as in (a).]

30. If $A = \begin{bmatrix} a & b & c \\ a' & b' & c' \end{bmatrix}$ and $AA^T = 0$, show that $A = 0$. [*Remark*: More generally, if A is *any* matrix such that $AA^T = 0$, then necessarily $A = 0$.]

31. A square matrix E is called an **idempotent** if $E^2 = E$.

    (a) Show that $0, I, \begin{bmatrix} 1 & a \\ 0 & 0 \end{bmatrix}$, and $\frac{1}{\sqrt{2}}\begin{bmatrix} 1 & 1 \\ 1 & 1 \end{bmatrix}$ are all idempotents.

    (b) If E is an idempotent, show that $I - E$ and E^T are both idempotents.

    (c) If $AB = I$ for some matrices A and B, show that $E = BA$ is an idempotent.

32. If $A = \begin{bmatrix} a & b \\ c & d \end{bmatrix}$ where $a \neq 0$, show that A can be factored in the form
    $A = \begin{bmatrix} 1 & 0 \\ x & 1 \end{bmatrix}\begin{bmatrix} y & z \\ 0 & w \end{bmatrix}.$

33. If $A = \begin{bmatrix} 3 & 2 \\ 2 & 3 \end{bmatrix}$ compute A^2, A^3, and A^4. Describe the pattern. Explain the pattern by examining $\begin{bmatrix} a+1 & a \\ a & a+1 \end{bmatrix}^2$.

1.5 MATRIX INVERSES

It is frequently important to "reverse" the effect of multiplying by a matrix. Here is an example.

1.5. MATRIX INVERSES

Example 1. *A spy plane flies over enemy territory and radios its position* $X = \begin{bmatrix} x \\ y \end{bmatrix}$ *back to headquarters (here x and y denote the longitude and latitude respectively). These transmissions will likely be intercepted, so they must be encoded to keep the exact location secret. The method chosen is to multiply the position coordinates by the matrix* $A = \begin{bmatrix} 3 & -4 \\ 2 & 7 \end{bmatrix}$ *to obtain encoded coordinates*

$$\begin{bmatrix} x' \\ y' \end{bmatrix} = A \begin{bmatrix} x \\ y \end{bmatrix} = \begin{bmatrix} 3x - 4y \\ 2x + 7y \end{bmatrix}$$

which are sent to headquarters. Devise a method for headquarters to decode these coordinates.

Solution. Write the encoded coordinates as $X' = \begin{bmatrix} x' \\ y' \end{bmatrix}$. Headquarters receives x' and y', and must recover the true coordinates x and y. They know that $X' = AX$, that is

$$\begin{array}{rcl} 3x - 4y & = & x' \\ 2x + 7y & = & y' \end{array} \quad (*)$$

and so they solve this system for x and y. The solution is

$$\begin{array}{rcl} x & = & \frac{7}{29}x' + \frac{4}{29}y' \\ y & = & -\frac{2}{29}x' + \frac{3}{29}y' \end{array}$$

In matrix form this is $X = CX'$ where $C = \begin{bmatrix} \frac{7}{29} & \frac{4}{29} \\ -\frac{2}{29} & \frac{3}{29} \end{bmatrix} = \frac{1}{29}\begin{bmatrix} 7 & 4 \\ -2 & 3 \end{bmatrix}$.
Hence headquarters can decode the received coordinates X' by multiplying by C. □

1.5.1 Matrix Inverses

The coding matrix A in Example 1 transforms the true coordinates X into coded coordinates $X' = AX$. Then the decoding matrix C transforms X' back to $X = CX'$. If we combine these equations we get $X = CX' = C(AX) = (CA)X$. In fact the reader can verify that

$$CA = I \quad \text{and also} \quad AC = I.$$

These equations reveal the relationship between the matrices C and A. Moreover, C is obtained from A by solving a system of equations, so we are led once again to examine how systems of equations are solved.

As we have seen, every system of linear equations can be written in matrix form

$$AX = B$$

where the column X is to be determined. This suggests that we first investigate how we solve *numerical* equations of the form $ax = b$. If $a = 0$ there is no solution (unless $b = 0$). But if $a \neq 0$ we can multiply both sides by the inverse a^{-1} to get $a^{-1}ax = a^{-1}b$, that is $x = a^{-1}b$. The property of a^{-1} that makes this work is $a^{-1}a = 1$. This condition can be stated for matrices when we remember that the role of 1 in arithmetic is played by the identity matrix I in matrix algebra.

With this in mind, we make the following definition: If A is a square matrix, a matrix C is called an **inverse** of A if

$$AC = I \quad \text{and} \quad CA = I.$$

A square matrix that has an inverse is called an **invertible matrix**[13]. Since matrix multiplication is not commutative, we insist that *both AC and CA equal I*. Note that the inverse matrix C must be the same size as A.

Example 2. *Show that* $A = \begin{bmatrix} -1 & 1 \\ 1 & 0 \end{bmatrix}$ *has inverse* $C = \begin{bmatrix} 0 & 1 \\ 1 & 1 \end{bmatrix}$.

Solution. We verify that $AC = I$ and $CA = I$:

$$AC = \begin{bmatrix} -1 & 1 \\ 1 & 0 \end{bmatrix} \begin{bmatrix} 0 & 1 \\ 1 & 1 \end{bmatrix} = \begin{bmatrix} 1 & 0 \\ 0 & 1 \end{bmatrix} = I$$

and

$$CA = \begin{bmatrix} 0 & 1 \\ 1 & 1 \end{bmatrix} \begin{bmatrix} -1 & 1 \\ 1 & 0 \end{bmatrix} = \begin{bmatrix} 1 & 0 \\ 0 & 1 \end{bmatrix} = I \qquad \square$$

If a square matrix A has a row (or a column) of zeros then A is not invertible. The next example illustrates why.

Example 3. *Show that* $A = \begin{bmatrix} 2 & -3 \\ 0 & 0 \end{bmatrix}$ *has no inverse.*

Solution. If $C = \begin{bmatrix} a & b \\ c & d \end{bmatrix}$ is *any* matrix, then $AC = \begin{bmatrix} 2 & -3 \\ 0 & 0 \end{bmatrix} \begin{bmatrix} a & b \\ c & d \end{bmatrix} = \begin{bmatrix} 2a - 3c & 2b - 3d \\ 0 & 0 \end{bmatrix}$. Since the (2,2)-entry of AC is not 1, AC can never equal I for any choice of the matrix C. $\qquad \square$

Example 3 shows that, contrary to the situation for numbers, *it is possible for a nonzero square matrix to have no inverse*. On the other hand, if a matrix A has an inverse, that inverse is uniquely determined by A. Indeed, if C and D are both inverses of A then certainly $AC = I$ and $DA = I$, so

$$D = DI = D(AC) = (DA)C = IC = C.$$

---

[13] We will see later that only square matrices can be invertible.

1.5. MATRIX INVERSES

Because of this, we speak of *the* inverse of A and denote it by A^{-1} (when it exists). Hence the inverse matrix A^{-1} is determined by the following conditions:

$$AA^{-1} = I \quad \text{and} \quad A^{-1}A = I.$$

The above discussion shows that these equations characterize A^{-1} in the following sense.

Theorem 1. *If a matrix C can be found such that*

$$AC = I = CA$$

then C is the inverse of A, that is $C = A^{-1}$. If no such matrix C can be found, then A has no inverse.

Example 4. *If $A = \begin{bmatrix} -1 & -1 \\ 1 & 0 \end{bmatrix}$, compute A^3 and so find A^{-1}.*

Solution. $A^2 = \begin{bmatrix} -1 & -1 \\ 1 & 0 \end{bmatrix} \begin{bmatrix} -1 & -1 \\ 1 & 0 \end{bmatrix} = \begin{bmatrix} 0 & 1 \\ -1 & -1 \end{bmatrix}$, so

$$A^3 = A^2 A = \begin{bmatrix} 0 & 1 \\ -1 & -1 \end{bmatrix} \begin{bmatrix} -1 & -1 \\ 1 & 0 \end{bmatrix} = \begin{bmatrix} 1 & 0 \\ 0 & 1 \end{bmatrix} = I$$

Hence $A \cdot A^2 = I$ and $A^2 \cdot A = I$, so A^2 is the inverse of A by Theorem 1. In other words $A^{-1} = A^2 = \begin{bmatrix} 0 & 1 \\ -1 & -1 \end{bmatrix}$. □

The inverse of A is usually *not* a power of A as in Example 4. If A is 2×2, there is a simple formula for the inverse (when it exists) which will be used frequently below.

Example 5. *If $A = \begin{bmatrix} a & b \\ c & d \end{bmatrix}$ and $ad - bc \neq 0$, show that*

$$A^{-1} = \frac{1}{ad - bc} \begin{bmatrix} d & -b \\ -c & a \end{bmatrix}.$$

Solution. If we write $C = \frac{1}{ad-bc} \begin{bmatrix} d & -b \\ -c & a \end{bmatrix}$, then

$$AC = \frac{1}{ad-bc} \begin{bmatrix} a & b \\ c & d \end{bmatrix} \begin{bmatrix} d & -b \\ -c & a \end{bmatrix} = \frac{1}{ad-bc} \begin{bmatrix} ad-bc & 0 \\ 0 & ad-bc \end{bmatrix} = I.$$

The reader can verify that also $CA = I$, so C is indeed the inverse of A. □

The number $ad - bc$ in Example 5 is called the **determinant** of $A = \begin{bmatrix} a & b \\ c & d \end{bmatrix}$, and $\begin{bmatrix} d & -b \\ -c & a \end{bmatrix}$ is called the **adjoint** of A. In Chapter 2 we will discuss the determinant of an arbitrary $n \times n$ matrix, and use it to give a formula for the inverse (when it exists).

Example 6. *If $YA = 0$ for some nonzero matrix Y, show that A has no inverse.*

Solution. Suppose A^{-1} does exist. Then right-multiply the equation $YA = 0$ by A^{-1} to obtain $YAA^{-1} = 0A^{-1}$. This becomes $YI = 0$, that is $Y = 0$, contrary to the assumption that Y is nonzero. So A^{-1} does not exist. □

1.5.2 Inverses and Systems of Linear Equations

With the notion of the inverse of a matrix in hand, we return to the study of a system of linear equations, written as a single matrix equation

$$AX = B.$$

Here A is the coefficient matrix, B is the column of constants, and the column X is to be determined. If A is (square and) invertible we can solve the system by left-multiplying each side by A^{-1} to get

$$\begin{aligned} A^{-1}AX &= A^{-1}B \\ IX &= A^{-1}B \\ X &= A^{-1}B. \end{aligned}$$

Thus $X = A^{-1}B$ is a solution to the system. [Check: $AX = A(A^{-1}B) = IB = B$.] Moreover the argument shows that if X is *any* solution to the system, then necessarily $X = A^{-1}B$, so the solution is unique. This discussion is summarized in the following theorem.

Theorem 2. *Suppose a system of n equations in n variables is written in matrix form as*

$$AX = B.$$

If the $n \times n$ coefficient matrix A is invertible, the system has the unique solution

$$X = A^{-1}B.$$

Example 7. *Solve the system of equations* $\begin{cases} 5x - 3y &= -2 \\ 2x + 7y &= 3 \end{cases}$ *by finding the inverse of the coefficient matrix and then using Theorem 2.*

Solution. In matrix form the system is $AX = B$ where $A = \begin{bmatrix} 5 & -3 \\ 2 & 7 \end{bmatrix}$, $X = \begin{bmatrix} x \\ y \end{bmatrix}$, and $B = \begin{bmatrix} -2 \\ 3 \end{bmatrix}$. The determinant of A is $5 \cdot 7 - (-3) \cdot 2 = 41$.

1.5. MATRIX INVERSES

Since this is not zero, Example 5 gives $A^{-1} = \frac{1}{41}\begin{bmatrix} 7 & 3 \\ -2 & 5 \end{bmatrix}$. Hence the solution of the system of equations is

$$X = A^{-1}B = \frac{1}{41}\begin{bmatrix} 7 & 3 \\ -2 & 5 \end{bmatrix}\begin{bmatrix} -2 \\ 3 \end{bmatrix} = \frac{1}{41}\begin{bmatrix} -5 \\ 19 \end{bmatrix}$$

by Theorem 2. In other words, $x = \frac{-5}{41}$ and $y = \frac{19}{41}$. □

1.5.3 The Matrix Inversion Algorithm

Except for the formula in Example 5 for the inverse of a 2×2 matrix, we have not given a systematic way to find the inverse of a square matrix. Fortunately an easy method exists which is based on Gaussian elimination. To illustrate the technique, consider the invertible matrix $A = \begin{bmatrix} 1 & 2 \\ 3 & 7 \end{bmatrix}$. If we write $A^{-1} = \begin{bmatrix} x_1 & x_2 \\ y_1 & y_2 \end{bmatrix}$, the condition $AA^{-1} = I$ becomes

$$A\begin{bmatrix} x_1 & x_2 \\ y_1 & y_2 \end{bmatrix} = \begin{bmatrix} 1 & 0 \\ 0 & 1 \end{bmatrix}.$$

Equating columns, this becomes two systems of linear equations with A as coefficient matrix:

$$A\begin{bmatrix} x_1 \\ y_1 \end{bmatrix} = \begin{bmatrix} 1 \\ 0 \end{bmatrix} \quad \text{and} \quad A\begin{bmatrix} x_2 \\ y_2 \end{bmatrix} = \begin{bmatrix} 0 \\ 1 \end{bmatrix}.$$

Since A is invertible each of these systems has a unique solution (by Theorem 2). Because A is square, this means that the reduced row-echelon form R of A cannot have a row of zeros (otherwise there would be at least one non-leading variable, and hence at least one parameter). But then R is the identity matrix (it is square and reduced row-echelon), so there exists a sequence of row operations carrying A to the identity matrix. This sequence carries the augmented matrices of both systems to reduced row-echelon form as follows:

$$\begin{bmatrix} A & 1 \\ & 0 \end{bmatrix} = \begin{bmatrix} 1 & 2 & 1 \\ 3 & 7 & 0 \end{bmatrix} \to \begin{bmatrix} 1 & 2 & 1 \\ 0 & 1 & -3 \end{bmatrix} \to \begin{bmatrix} 1 & 0 & 7 \\ 0 & 1 & -3 \end{bmatrix},$$

$$\begin{bmatrix} A & 0 \\ & 1 \end{bmatrix} = \begin{bmatrix} 1 & 2 & 0 \\ 3 & 7 & 1 \end{bmatrix} \to \begin{bmatrix} 1 & 2 & 0 \\ 0 & 1 & 1 \end{bmatrix} \to \begin{bmatrix} 1 & 0 & -2 \\ 0 & 1 & 1 \end{bmatrix}.$$

Hence the solutions are $\begin{bmatrix} x_1 \\ y_1 \end{bmatrix} = \begin{bmatrix} 7 \\ -3 \end{bmatrix}$ and $\begin{bmatrix} x_2 \\ y_2 \end{bmatrix} = \begin{bmatrix} -2 \\ 1 \end{bmatrix}$ respectively, so $A^{-1} = \begin{bmatrix} 7 & -2 \\ -3 & 1 \end{bmatrix}$. Now observe that the two row reductions can be carried out *simultaneously* because the row operations used are the same (the ones carrying $A \to I$). The result is

$$\left[\begin{array}{cc|cc} A & \begin{matrix}1 & 0\\ 0 & 1\end{matrix} \end{array}\right] = \left[\begin{array}{cc|cc} 1 & 2 & 1 & 0 \\ 3 & 7 & 0 & 1 \end{array}\right]$$

$$\to \left[\begin{array}{cc|cc} 1 & 2 & 1 & 0 \\ 0 & 1 & -3 & 1 \end{array}\right]$$

$$\to \left[\begin{array}{cc|cc} 1 & 0 & 7 & -2 \\ 0 & 1 & -3 & 1 \end{array}\right].$$

This last reduction can be written more compactly as

$$[\, A \quad I \,] \to [\, I \quad A^{-1} \,].$$

In other words, the set of row operations that carry $A \to I$ also carry $I \to A^{-1}$.

This procedure works for any invertible square matrix A. In fact all that is needed is that A can be carried to I by row operations (this is equivalent to invertibility as will be seen in Theorem 5). Hence we get the desired algorithm.

The Matrix Inversion Algorithm. *Suppose A is a square matrix and there exists a sequence of elementary row operations that carry $A \to I$. Then A is invertible and this same sequence carries $I \to A^{-1}$. Thus*

$$[\, A \quad I \,] \to [\, I \quad A^{-1} \,]$$

where the row operations on A and I are carried out simultaneously.

Example 8. *Find the inverse of the matrix* $A = \begin{bmatrix} 1 & 4 & -1 \\ 2 & 7 & 1 \\ 1 & 3 & 0 \end{bmatrix}.$

Solution. The reduction $[\, A \quad I \,] \to [\, I \quad A^{-1} \,]$ is as follows:

1.5. MATRIX INVERSES

$$\begin{bmatrix} 1 & 4 & -1 & 1 & 0 & 0 \\ 2 & 7 & 1 & 0 & 1 & 0 \\ 1 & 3 & 0 & 0 & 0 & 1 \end{bmatrix}$$

$$\rightarrow \begin{bmatrix} 1 & 4 & -1 & 1 & 0 & 0 \\ 0 & -1 & 3 & -2 & 1 & 0 \\ 0 & -1 & 1 & -1 & 0 & 1 \end{bmatrix}$$

$$\rightarrow \begin{bmatrix} 1 & 0 & 11 & -7 & 4 & 0 \\ 0 & 1 & -3 & 2 & -1 & 0 \\ 0 & 0 & -2 & 1 & -1 & 1 \end{bmatrix}$$

$$\rightarrow \begin{bmatrix} 1 & 0 & 11 & -7 & 4 & 0 \\ 0 & 1 & -3 & 2 & -1 & 0 \\ 0 & 0 & 1 & -\frac{1}{2} & \frac{1}{2} & -\frac{1}{2} \end{bmatrix}$$

$$\rightarrow \begin{bmatrix} 1 & 0 & 0 & -\frac{3}{2} & -\frac{3}{2} & \frac{11}{2} \\ 0 & 1 & 0 & \frac{1}{2} & \frac{1}{2} & -\frac{3}{2} \\ 0 & 0 & 1 & -\frac{1}{2} & \frac{1}{2} & -\frac{1}{2} \end{bmatrix}$$

Hence $A^{-1} = \begin{bmatrix} -\frac{3}{2} & -\frac{3}{2} & \frac{11}{2} \\ \frac{1}{2} & \frac{1}{2} & -\frac{3}{2} \\ -\frac{1}{2} & \frac{1}{2} & -\frac{1}{2} \end{bmatrix} = \frac{1}{2} \begin{bmatrix} -3 & -3 & 11 \\ 1 & 1 & -3 \\ -1 & 1 & -1 \end{bmatrix}$ by the matrix inversion algorithm. Of course this can be checked by verifying that $AA^{-1} = I$ and $A^{-1}A = I$. □

Remark. If A is not invertible then no sequence of row operations will carry $A \to I$ as we shall see in Theorem 5 below. Hence the algorithm breaks down because a row of zeros is encountered. Here is an example.

Example 9. Show that the matrix $A = \begin{bmatrix} 1 & 3 & -2 \\ 1 & 2 & 0 \\ 2 & 8 & -8 \end{bmatrix}$ is not invertible.

Solution. We try the algorithm on A to get:

$$[A \ I] = \begin{bmatrix} 1 & 3 & -2 & 1 & 0 & 0 \\ 1 & 2 & 0 & 0 & 1 & 0 \\ 2 & 8 & -8 & 0 & 0 & 1 \end{bmatrix}$$

$$\rightarrow \begin{bmatrix} 1 & 3 & -2 & 1 & 0 & 0 \\ 0 & -1 & 2 & -1 & 1 & 0 \\ 0 & 2 & -4 & -2 & 0 & 1 \end{bmatrix}$$

$$\rightarrow \begin{bmatrix} 1 & 3 & -2 & 1 & 0 & 0 \\ 0 & -1 & 2 & -1 & 1 & 0 \\ 0 & 0 & 0 & -4 & 2 & 1 \end{bmatrix}$$

Since A will never be transformed to the identity matrix by elementary row operations, A is not invertible. $\square$

1.5.4 Properties of Inverses

For reference later, the following theorem collects several properties of inverses which are used throughout linear algebra.

Theorem 3. *The following properties hold where the matrices are all square.*

(1) *The identity matrix I is invertible, and $I^{-1} = I$.*

(2) *If A is invertible then A^{-1} is also invertible, and $(A^{-1})^{-1} = A$.*

(3) *If A and B are invertible, then AB is invertible, and $(AB)^{-1} = B^{-1}A^{-1}$.*

(4) *If $A_1, A_2, \cdots, A_{k-1}, A_k$ are all invertible, their product $A_1 A_2 \cdots A_{k-1} A_k$ is also invertible, and $(A_1 A_2 \cdots A_{k-1} A_k)^{-1} = A_k^{-1} A_{k-1}^{-1} \cdots A_2^{-1} A_1^{-1}$.*

(5) *If A is invertible, then A^k is invertible for $k \geq 1$, and $(A^k)^{-1} = (A^{-1})^k$.*

(6) *If A is invertible, then A^T is also invertible, and $(A^T)^{-1} = (A^{-1})^T$.*

(7) *If A is invertible and $c \neq 0$ is a number, then cA is also invertible, and $(cA)^{-1} = \frac{1}{c} A^{-1}$.*

Proof. (1). Clearly $I^2 = I$ so this follows from Theorem 1.

(2). We have $AA^{-1} = I = A^{-1}A$. By Theorem 1, these equations show that A is the inverse of A^{-1}, proving (2).

(3). We must find a matrix C such that $(AB)C = I$ and $C(AB) = I$. The proposed candidate is $C = B^{-1}A^{-1}$ (which exists because A^{-1} and B^{-1} are assumed to exist). We test it as follows:

$$(AB)C = (AB)(B^{-1}A^{-1}) = A(BB^{-1})A^{-1} = AIA^{-1} = AA^{-1} = I.$$

Similarly, $C(AB) = I$ as the reader can verify. So $C = (AB)^{-1}$ by Theorem 1, proving (3).

(4). *Claim.* If (4) holds for some integer $k \geq 1$, then it holds for $k+1$.
Proof. If $(A_1 A_2 \cdots A_k)^{-1} = A_k^{-1} \cdots A_2^{-1} A_1^{-1}$ holds for an integer k, then (3) gives

$$\begin{aligned}(A_1 A_2 \cdots A_k A_{k+1})^{-1} &= [(A_1 A_2 \cdots A_k)(A_{k+1})]^{-1} \\ &= A_{k+1}^{-1}(A_1 A_2 \cdots A_k)^{-1} \\ &= A_{k+1}^{-1}(A_k^{-1} \cdots A_2^{-1} A_1^{-1}).\end{aligned}$$

This proves the Claim.

1.5. MATRIX INVERSES

Now (4) clearly holds if $k = 1$, and it holds for $k = 2$ by (3). Then the Claim shows that (4) holds for $k = 3$; then the Claim again shows it holds for $k = 4$, etc. It follows that (4) holds for every $k \geq 0$. [14]

(5). This restates (4) in the special case that $A_1 = A_2 = \cdots = A_k = A$.

(6). We must find a matrix C such that $A^T C = C A^T$, and a candidate $C = (A^{-1})^T$ is suggested. Using Theorem 1 §1.4, we have

$$A^T C = A^T (A^{-1})^T = (A^{-1} A)^T = I^T = I.$$

Similarly $C A^T = I$, so Theorem 1 shows that C is the inverse of A^T, that is $(A^T)^{-1} = C$. This proves (6).

(7). This is left to the reader (Exercise 30). □

Part (6) of Theorem 3 yields a useful criterion that a square matrix is invertible.

Corollary. *If A is a square matrix, then A is invertible if and only if A^T is invertible.*

Proof. If A is invertible, then A^T is invertible by (6) of Theorem 3. Conversely, if A^T is invertible, then $A = (A^T)^T$ is invertible, again by (6) of Theorem 3 (applied to A^T). □

We now give four examples showing how the properties in Theorem 3 are used.

Example 10. *Find A if $(A^{-1} - 3I)^T = 2 \begin{bmatrix} -1 & 2 \\ 5 & 4 \end{bmatrix}$.*

Solution. We use Theorem 3 several times. First transpose both sides to get

$$\{(A^{-1} - 3I)^T\}^T = \left\{ 2 \begin{bmatrix} -1 & 2 \\ 5 & 4 \end{bmatrix} \right\}^T,$$

that is

$$A^{-1} - 3I = 2 \begin{bmatrix} -1 & 5 \\ 2 & 4 \end{bmatrix}.$$

Hence $A^{-1} = 3I + 2 \begin{bmatrix} -1 & 5 \\ 2 & 4 \end{bmatrix} = \begin{bmatrix} 1 & 10 \\ 4 & 11 \end{bmatrix}$, and so the fact that $[A^{-1}]^{-1} = A$ gives

$$A = [A^{-1}]^{-1} = \begin{bmatrix} 1 & 10 \\ 4 & 11 \end{bmatrix}^{-1} = \frac{1}{-29} \begin{bmatrix} 11 & -10 \\ -4 & 1 \end{bmatrix}$$

by Example 5, as required. □

---

[14]This method of proof is called **mathematical induction**: Suppose that $p_1, p_2, \cdots$ are statements such that p_1 is true; and p_{k+1} is true whenever the preceding statement p_k is true. Then p_k is true for every $k \geq 1$. The method is discussed in Appendix A.2.

Example 11. *If A, B and C are invertible, simplify $C^T B(AB)^{-1} \left[C^{-1}A^T\right]^T$.*

Solution.
$$\begin{aligned} C^T B(AB)^{-1} \left[C^{-1}A^T\right]^T &= C^T B(B^{-1}A^{-1}) \left[(A^T)^T (C^{-1})^T\right] \\ &= C^T I A^{-1} \left[A(C^T)^{-1}\right] \\ &= C^T I (C^T)^{-1} \\ &= I. \end{aligned}$$
$\square$

Example 12. *If A and AB are both invertible, show that B is also invertible.*

Solution. We have $B = A^{-1}(AB)$. Now A^{-1} is invertible by Theorem 3, because A is invertible. Since AB is invertible by assumption, it follows that $A^{-1}(AB) = B$ is invertible by (3) of Theorem 3. $\square$

Example 13. *A matrix E is called an idempotent if $E^2 = E$. Show that I is the only invertible idempotent.*

Solution. I is invertible (by Theorem 3) and $I^2 = I$, so I is indeed an invertible idempotent. Conversely, assume that $E^2 = E$ and that E^{-1} exists. Left multiply the equation $E^2 = E$ by E^{-1} to get $E^{-1}E^2 = E^{-1}E$, that is $E = I$. $\square$

A square matrix is called **upper triangular** if every entry below the main diagonal is zero. Similarly a matrix A is **lower triangular** if every entry above the main diagonal is zero, equivalently if A^T is upper triangular. A matrix is called **triangular** if it is either upper or lower triangular. One of the reasons for the importance of these matrices is that it is easy to tell if they are invertible.

Theorem 4. *Let A denote a square triangular matrix.*

(1) *A is invertible if and only if no entry on the main diagonal is zero.*

(2) *If A is upper (lower) triangular then A^{-1} is also upper (lower) triangular.*

Proof. (1). If A is upper triangular and every main diagonal entry is nonzero, it is clear that $A \to I$ by row operations, so A^{-1} exists by the matrix inversion algorithm. Conversely, if some main diagonal entry of A is zero then $A \to I$ is impossible because the row containing the lowest diagonal zero can be carried to a row of zeros by row operations (verify). Hence the inverse does not exist by Theorem 5 below. Thus (1) holds for upper triangular matrices. But then it holds for a lower triangular matrix B because B is invertible if and only B^T is invertible (Corollary of Theorem 3), and B^T is upper triangular with the same main diagonal as A.

(2). If A is upper triangular and invertible, then the matrix inversion algorithm shows that A^{-1} will also be upper triangular because no multiple of any row is added to a lower row. If B is lower triangular then B^T is upper

1.5. MATRIX INVERSES

triangular, whence $(B^{-1})^T = (B^T)^{-1}$ is upper triangular (by the above). Thus $B^{-1} = [(B^{-1})^T]^T$ is lower triangular. □

The next example illustrates Theorem 4. The verification (using Example 5) is left to the reader.

Example 14. *If* $A = \begin{bmatrix} a & b \\ 0 & c \end{bmatrix}$ *where* $a \neq 0$ *and* $c \neq 0$, *show that the inverse is given by* $A^{-1} = \begin{bmatrix} 1/a & -b/ac \\ 0 & 1/c \end{bmatrix}$.

1.5.5 Conditions for Invertibility

The following theorem collects a number of conditions on a matrix all equivalent[15] to invertibility. It will be referred to frequently below.

Theorem 5. *The following conditions are equivalent for a square matrix* A:

(1) A *is invertible.*

(2) *The homogeneous system* $AX = 0$ *has only the trivial solution* $X = 0$.

(3) A *can be carried to* I *by elementary row operations.*

(4) *The system* $AX = B$ *has a solution* X *for every choice of column* B.

(5) *There exists a matrix* C *such that* $AC = I$.

Proof. We show that each of these conditions implies the next, and that (5) implies (1).

(1)⇒(2). If A^{-1} exists then $AX = 0$ gives $X = IX = A^{-1}AX = A^{-1}0 = 0$. This is (2).

(2)⇒(3). Assume that (2) is true. Certainly A can be carried to a reduced, row-echelon matrix R; we show that $R = I$. Suppose on the contrary that $R \neq I$. Then R has a row of zeros (being square), so the augmented matrix $\begin{bmatrix} A & 0 \end{bmatrix}$ of the system $AX = 0$ also has a row of zeros when carried to row-echelon form. Since A is square, this means that there is at least one non-leading variable, and hence at least one parameter. Thus $AX = 0$ has infinitely many solutions, contrary to (2). So $R = I$ after all.

(3)⇒(4). Consider the augmented matrix $\begin{bmatrix} A & B \end{bmatrix}$ of the system $AX = B$. Using (3) let $A \to I$ by a sequence of row operations. Then these same operations carry $\begin{bmatrix} A & B \end{bmatrix} \to \begin{bmatrix} I & C \end{bmatrix}$ for some column C. Hence, the system $AX = B$ has a solution (in fact unique) by Gaussian Elimination.

(4)⇒(5). By (4), assume that the system $AX = B$ has a solution for *any* choice of the column B. If we take $B = E_j$ where E_j is column j of

---

[15] If p and q are statements, we say that p **implies** q (written $p \Rightarrow q$) if q is true whenever p is true. The statements are called **equivalent** if both $p \Rightarrow q$ and $q \Rightarrow p$ (written $p \Leftrightarrow q$, spoken "p if and only if q").

the identity matrix I_n, this gives a column C_j such that $AC_j = E_j$. Now let $C = [C_1 \ C_2 \ \cdots \ C_n]$ be the $n \times n$ matrix with these matrices C_j as its columns. Then the definition of matrix multiplication gives

$$AC = A \ [C_1 \ C_2 \ \cdots \ C_n] = [AC_1 \ AC_2 \ \cdots \ AC_n] = [E_1 \ E_2 \ \cdots \ E_n] = I_n.$$

This proves that (5) is true.

(5)$\Rightarrow$(1). Assume that (5) is true so that $AC = I$ for some matrix C.

Claim. If $CX = 0$ then $X = 0$.

Proof. If $0 = CX$, left multiply by A to get $A0 = ACX0$. Thus $0 = IX = X$, which proves the Claim.

Thus condition (2) holds for the matrix C rather than A. Hence the argument above that (2)$\Rightarrow$(3)$\Rightarrow$(4)$\Rightarrow$(5) (with A replaced by C) shows that a matrix C' exists such that $CC' = I$. But then

$$A = AI = A(CC') = (AC)C' = IC' = C'.$$

Thus $CA = CC' = I$ which, together with $AC = I$, shows that C is the inverse of A. This proves (1). $\square$

Theorem 5 has some wonderful consequences. First, to verify that a square matrix A is invertible it is enough to find some matrix C such that either $AC = I$ or $CA = I$. Recall that the definition of invertibility requires that *both* conditions are satisfied.

Corollary 1. *Let A and C denote square matrices, and assume that $AC = I$. Then also $CA = I$, and so A and C are both invertible, $C = A^{-1}$, and $A = C^{-1}$.*

Proof. If $AC = I$, Theorem 5 shows that A is invertible. But then left multiplication by A^{-1} gives

$$A^{-1} = A^{-1}I = A^{-1}(AC) = (A^{-1}A)C = IC = C$$

Thus C is invertible (by Theorem 3) and $C^{-1} = (A^{-1})^{-1} = A$. This proves Corollary 1. $\square$

Observe that Corollary 1 is false if A and C are not square matrices. For example, we have

$$\begin{bmatrix} 1 & 2 & 1 \\ 1 & 1 & 1 \end{bmatrix} \begin{bmatrix} -1 & 1 \\ 1 & -1 \\ 0 & 1 \end{bmatrix} = I_2 \quad \text{but} \quad \begin{bmatrix} -1 & 1 \\ 1 & -1 \\ 0 & 1 \end{bmatrix} \begin{bmatrix} 1 & 2 & 1 \\ 1 & 1 & 1 \end{bmatrix} \neq I_3.$$

In fact, we can say more. It is quite plausible that $AC = I_m$ and $CA = I_n$ could happen for some $m \times n$ matrix A and some $n \times m$ matrix C. We claim

1.5. MATRIX INVERSES

that these requirements force $m = n$. Suppose, for example, that $m < n$. Then the system of equations

$$AX = 0$$

has a nontrivial solution $X \neq 0$ by Theorem 1 §1.3. But then $X = I_n X = CAX = C0 = 0$, a contradiction. Thus $m < n$ cannot be true. Similarly we show that $n < m$ cannot happen by considering the system $CX = 0$. So the only possibility is $m = n$. This proves another consequence of Theorem 5:

Corollary 2. *The only invertible matrices are square. That is, if A is an $m \times n$ matrix, and if $AC = I_m$ and $CA = I_n$ hold for some $n \times m$ matrix C, then $m = n$.*

Exercises

1. If $A = \begin{bmatrix} -5 & 2 \\ 2 & -1 \end{bmatrix}$, then show that $A^{-1} = \begin{bmatrix} -1 & -2 \\ -2 & -5 \end{bmatrix}$ by verifying that $AA^{-1} = I = A^{-1}A$. What is $(A^{-1})^{-1}$?

2. In each case, use the matrix inversion algorithm to find the inverse of the given matrix.

   (a) $\begin{bmatrix} 3 & -7 \\ -2 & 5 \end{bmatrix}$
   (b) $\begin{bmatrix} 3 & 5 \\ 1 & 2 \end{bmatrix}$

   (c) $\begin{bmatrix} 1 & 0 & 1 \\ 2 & 2 & 3 \\ 0 & 3 & 1 \end{bmatrix}$
   (d) $\begin{bmatrix} 1 & 3 & -1 \\ -1 & 0 & 0 \\ 0 & 2 & -1 \end{bmatrix}$

   (e) $\begin{bmatrix} 3 & 5 & 1 \\ 1 & 2 & -1 \\ -1 & 0 & +1 \end{bmatrix}$
   (f) $\begin{bmatrix} 2 & 1 & 2 \\ -3 & -1 & -1 \\ 5 & 2 & 1 \end{bmatrix}$

   (g) $\begin{bmatrix} 1 & 1 & 0 & 1 \\ -1 & 0 & 1 & -1 \\ 5 & 7 & 3 & 5 \\ 2 & 5 & 6 & 1 \end{bmatrix}$
   (h) $\begin{bmatrix} 4 & 0 & 2 & -1 \\ -1 & 0 & 1 & 0 \\ 5 & 0 & -2 & -1 \\ 2 & -1 & 0 & 0 \end{bmatrix}$

3. Write your own version of the matrix inverse algorithm.

4. If $A = \begin{bmatrix} 3 & -7 & 2 \\ -2 & 5 & -3 \\ 1 & -2 & 0 \end{bmatrix}$, find A^{-1} and $(A^{-1})^{-1}$. Explain your result.

5. In each case find the matrix A.

(a) $(3A)^{-1} = \begin{bmatrix} 1 & 0 \\ -5 & -1 \end{bmatrix}$ (b) $(5A)^T = \begin{bmatrix} 2 & -3 \\ -1 & 4 \end{bmatrix}^{-1}$

(c) $[2A^T - 3I]^{-1} = \begin{bmatrix} 3 & 2 \\ 1 & 1 \end{bmatrix}$ (d) $(A^{-1} - 3I)^T = 5\begin{bmatrix} 1 & 2 \\ 3 & 4 \end{bmatrix}$

(e) $\left[A^T - 3\begin{bmatrix} 1 & 0 \\ 2 & -1 \end{bmatrix}\right]^{-1} = \begin{bmatrix} 3 & 1 \\ 1 & 1 \end{bmatrix}$

(f) $\left[2\begin{bmatrix} 1 & 1 \\ -2 & 3 \end{bmatrix} - 5A^{-1}\right]^T = (4A^T)^{-1}$

6. Write a strategy to solve a system of linear equations $AX = B$ if A is invertible.

7. In each case, solve the system by finding the inverse of the coefficient matrix if possible.

   (a) $\begin{aligned} 2x + 3y &= -2 \\ x - y &= 5 \end{aligned}$ (b) $\begin{aligned} 3x + 4y &= 3 \\ 2x + 2y &= -1 \end{aligned}$

   (c) $\begin{aligned} x + y + z &= 0 \\ x + y + 2z &= -2 \\ 2x + y + 4z &= 3 \end{aligned}$ (d) $\begin{aligned} 3x + 5y &= -5 \\ x + 2y + z &= 1 \\ 3x + 7y + z &= 0 \end{aligned}$

8. What conclusion can you draw about the number of solutions for a system of linear equations $AX = B$ if A is invertible? Does it depend on B?

9. Assume that you are given that $A^{-1} = \begin{bmatrix} 1 & 0 & 2 \\ 1 & 2 & 1 \\ 3 & 5 & 3 \end{bmatrix}$.

   (a) Find X such that $AX = \begin{bmatrix} 2 & -1 \\ 1 & 0 \\ 0 & -3 \end{bmatrix}$

   (b) Find X such that $XA = \begin{bmatrix} 2 & 3 & -1 \\ -1 & 0 & 5 \end{bmatrix}$

10. If A is an invertible $n \times n$ matrix, show that $AX = B$ has a unique solution for any $n \times k$ matrix B. [*Hint*: Theorem 2.]

11. Suppose that $CA = I_n$ where C is $n \times m$ and A is $m \times n$.

    (a) Show that the system $AX = B$ has a solution for every $n \times 1$ column B.

    (b) Show that the solution in (a) is uniquely determined by B.

12. Show that $A = \begin{bmatrix} a & 0 \\ b & 0 \end{bmatrix}$ is not invertible for any choice of a and b.

13. In each case either show that the statement is true or give an example showing that it is false. Assume throughout that A and B are square matrices.

1.5. MATRIX INVERSES

(a) If $A \neq 0$ then A is invertible.

(b) If A is invertible then $A \neq 0$.

(c) If $A^3 = 3I$ then A is invertible.

(d) If $A^2 = A$ and $A \neq 0$ then A is invertible.

(e) If A and B are both invertible then $A + B$ is invertible.

(f) If $AB = 0$ and $A \neq 0$ then $B = 0$.

(g) If A is invertible and $AC = I$ then $C = A^{-1}$.

(h) If $AX = B$ has no solution for some column B, then $AX = 0$ has no solution.

(i) If $AX = 0$ has only the trivial solution, then $AX = B$ has a unique solution for every column B.

(j) If A^2 is invertible, then A is invertible.

14. If $U = \begin{bmatrix} 3 & -4 \\ 7 & 5 \end{bmatrix}$ and $AU = 0$ for some matrix A, show that necessarily $A = 0$.

15. Let $A = \begin{bmatrix} 5 & -3 \\ 6 & -4 \end{bmatrix}$. (a) Show that $A^2 - A - 2I = 0$. (b) Use (a) to obtain a formula for A^{-1} in terms of A.

16. (a) Simplify the matrix product $(I - A)(I + A)$. If $A^2 = 0$, show that $I - A$ is invertible and that $(I - A)^{-1} = I + A$.

    (b) If $A^3 = 0$, show that $I - A$ is invertible and that $(I - A)^{-1} = I + A + A^2$.

    (c) Generalize.

17. Consider the matrix $A = \begin{bmatrix} 2 & 1 & -1 \\ -3 & -3 & 2 \\ 1 & 3 & -1 \end{bmatrix}$.

    (a) Find the inverse of A using the matrix inversion algorithm.

    (b) Now that you know that the matrix A is invertible, how could you convince someone that a system of linear equations $AX = B$ has a unique solution for any B? If you fix some B and transform the augmented matrix $\begin{bmatrix} A & B \end{bmatrix}$ to reduced row-echelon form, what connection do you see with the computations in part (a)?

    (c) If $AX = B$ has a unique solution for any B, what about the homogeneous system $AX = 0$?

    (d) Solve $AX = 0$ by transforming the augmented matrix $\begin{bmatrix} A & 0 \end{bmatrix}$ to reduced row-echelon form; what happens to A?

    (e) Now since A can be carried to I by elementary row operations, what does that say about solving a system $AX = B$? Use this to find a matrix C such that $AC = I$ by successively finding each column of C.

(f) Now that $AC = I$, verify that $CA = I$ as well. Will this happen all the time?

(g) What happens to all these conditions for $A = \begin{bmatrix} 1 & 1 & 2 \\ -2 & -3 & -5 \\ -1 & 1 & 0 \end{bmatrix}$?

18. If A is a square matrix, a matrix of the form $P^{-1}AP$ where P is invertible is called a **conjugate** of A.

    (a) Is $P^{-1}AP = A$? Explain.

    (b) If some conjugate of A is invertible, show that A is also invertible.

    (c) Show that $(P^{-1}AP)^2 = P^{-1}A^2P$ and $(P^{-1}AP)^3 = P^{-1}A^3P$. Generalize.

19. (a) If P is invertible, solve the equation $P^{-1}XP = B$ for X.

    (b) If P and Q are invertible, solve the equation $P(A+X)Q = B$ for X.

20. (a) If A is invertible and A commutes with C, show that A^{-1} also commutes with C.

    (b) If A and B are invertible and commute, show that A^{-1} and B^{-1} also commute.

21. (a) If the first row of a matrix A consists of zeros, show that A^{-1} does not exist.

    (b) If the first column of a matrix A consists of zeros, show that A^{-1} does not exist.

22. Find the inverse of $A = \begin{bmatrix} sin\theta & cos\theta \\ -cos\theta & sin\theta \end{bmatrix}$ for any angle θ.

23. A matrix $A = \begin{bmatrix} B & X \\ 0 & C \end{bmatrix}$ in block form is called a **block upper triangular matrix** if B and C are both square matrices.

    (a) If B and C are invertible show that A is invertible and
    $$A^{-1} = \begin{bmatrix} B^{-1} & -B^{-1}XC^{-1} \\ 0 & C^{-1} \end{bmatrix}$$
    in block form. [Hint: Theorem 4 §1.4.]

    (b) If A is invertible, show that B and C must also be invertible.

24. If $AB = AC$ where A is an invertible matrix, show that $B = C$.

25. (a) If $AB = BA$ show that $(AB)^2 = A^2B^2$.

    (b) If A and B are both invertible and $(AB)^2 = A^2B^2$, show that $AB = BA$.

1.6. ELEMENTARY MATRICES

26. Suppose that A and B are nonzero square matrices such that $AB = 0$. Show that neither A nor B has an inverse.

27. What is wrong with the following solution to Example 12? Support your answer.

    "*Solution*": Since AB is invertible we have $(AB)^{-1} = B^{-1}A^{-1}$ by Theorem 3, so $B^{-1} = (AB)^{-1}A$.

28. (a) If A and B are square matrices and BA is invertible, show that A and B are invertible.

    (b) If A, B, and C are square matrices and ABC is invertible, show that A, B, and C are invertible.

29. A matrix A is called **selfinverse** if $A^{-1} = A$. Recall that E is an idempotent if $E^2 = E$.

    (a) Show that A is selfinverse if and only if $A^2 = I$.

    (b) If A is selfinverse, show that $E = \frac{1}{2}(I - A)$ is an idempotent.

    (c) If E is an idempotent, show that $A = I - 2E$ is selfinverse.

30. Prove (7) of Theorem 3 by verifying directly that $\frac{1}{c}A^{-1}$ is the inverse of A and using Theorem 1.

1.6 ELEMENTARY MATRICES[16]

By now the reader will be convinced of the importance of elementary row operations. It turns out that these operations can be performed by left multiplication by certain invertible matrices. This gives a new way of looking at matrix inverses, and has other uses as well.

1.6.1 Elementary Matrices

A square matrix E that is obtained by doing a single elementary row operation to an identity matrix is called an **elementary matrix**. We say that E is of Type I, II or III when the corresponding row operation is of type I, II or III. Here are some examples, one of each type.

Example 1. $E_1 = \begin{bmatrix} 1 & 0 & 0 \\ 0 & 0 & 1 \\ 0 & 1 & 0 \end{bmatrix}$, $E_2 = \begin{bmatrix} 1 & 0 & 0 \\ 0 & 7 & 0 \\ 0 & 0 & 1 \end{bmatrix}$ and $E_3 = \begin{bmatrix} 1 & 0 & -3 \\ 0 & 1 & 0 \\ 0 & 0 & 1 \end{bmatrix}$

are elementary matrices of type I, II and III respectively, obtained respectively by performing the following elementary row operations on the 3×3 identity matrix: (1) Interchange rows 2 and 3; (2) Multiplying row 2 by 7; and (3) Subtract 3 times row 3 from row 1. □

---
[16] This section can be ommitted at a first reading as it is used sparingly in the sequel.

Consider the 2×2 elementary matrices $E_1 = \begin{bmatrix} 0 & 1 \\ 1 & 0 \end{bmatrix}$, $E_2 = \begin{bmatrix} 1 & 0 \\ 0 & 5 \end{bmatrix}$, and $E_3 = \begin{bmatrix} 1 & 3 \\ 0 & 1 \end{bmatrix}$. If a 2×3 matrix $A = \begin{bmatrix} a & b & c \\ x & y & z \end{bmatrix}$ is left multiplied by these elementary matrices, the result is:

$$E_1 A = \begin{bmatrix} 0 & 1 \\ 1 & 0 \end{bmatrix} \begin{bmatrix} a & b & c \\ x & y & z \end{bmatrix} = \begin{bmatrix} x & y & z \\ a & b & c \end{bmatrix}$$

$$E_2 A = \begin{bmatrix} 1 & 0 \\ 0 & 5 \end{bmatrix} \begin{bmatrix} a & b & c \\ x & y & z \end{bmatrix} = \begin{bmatrix} a & b & c \\ 5x & 5y & 5z \end{bmatrix}$$

$$E_3 A = \begin{bmatrix} 1 & 3 \\ 0 & 1 \end{bmatrix} \begin{bmatrix} a & b & c \\ x & y & z \end{bmatrix} = \begin{bmatrix} a + 3x & b + 3y & c + 3z \\ x & y & z \end{bmatrix}$$

Observe that in each case $E_i A$ is the matrix obtained from A by performing the elementary operation that created E_i from I_2. This works in general.

Lemma 1[17]. *If an elementary row operation is performed on an $m \times n$ matrix A, the result is the matrix EA where E is the $m \times m$ elementary matrix created by performing the same row operation on I_m.*

We omit the routine proof.

Any elementary row operation can be reversed by another elementary row operation (called its **inverse**). For example:

The inverse of multiplying row 3 by 7 is to multiply row 3 by $\frac{1}{7}$.

The inverse of adding 3 times row 2 to row 1 is to *subtract* 3 times row 2 from row 1.

The inverse of interchanging rows 2 and 3 is the *same* operation; interchange them again.

This observation translates into the fact that every elementary matrix E is invertible; moreover, it gives a simple way to find the inverse matrix. In fact, since E corresponds to some row operation, we let F be the elementary matrix corresponding to the *inverse* operation. Since E is created by doing a row operation to I, then doing the inverse operation will carry E back to I. But Lemma 1 shows that E is carried to FE by the inverse operation, whence $FE = I$. This shows that F is the inverse of the matrix E, and so proves

Lemma 2. *Every elementary matrix E is invertible, and E^{-1} is the elementary matrix (of the same type) obtained from I by the inverse of the operation that produced E from I.*

---

[17]The term "lemma" means an auxiliary proposition used in the demonstration of another proposition.

1.6. ELEMENTARY MATRICES

Example 2. *Given elementary matrices*

$$E_1 = \begin{bmatrix} 1 & 0 & 5 \\ 0 & 1 & 0 \\ 0 & 0 & 1 \end{bmatrix}, \quad E_2 = \begin{bmatrix} 1 & 0 & 0 \\ 0 & 3 & 0 \\ 0 & 1 & 1 \end{bmatrix} \quad \text{and} \quad E_3 = \begin{bmatrix} 1 & 0 & 0 \\ 0 & 0 & 1 \\ 0 & 1 & 0 \end{bmatrix},$$

write down the inverses.

Solution. The elementary matrix E_1 corresponds to adding 5 times row 3 of I_3 to row 1, so the inverse operation is *subtracting* 5 times row 3 to row 1. Hence $E_1^{-1} = \begin{bmatrix} 1 & 0 & -5 \\ 0 & 1 & 0 \\ 0 & 0 & 1 \end{bmatrix}$. Similarly $E_2^{-1} = \begin{bmatrix} 1 & 0 & 0 \\ 0 & \frac{1}{3} & 0 \\ 0 & 0 & 1 \end{bmatrix}$ because the inverse of multiplying row 2 by 3 is to multiply row 2 by $\frac{1}{3}$. Finally, $E_3^{-1} = E_3$ because the corresponding operation (interchanging rows 2 and 3) is self-inverse. □

We note in passing that a similar argument shows that if an elementary column operation is performed on a matrix A, the result is the matrix AF where F is an elementary matrix, and that F can be obtained by doing the same column operation to the identity matrix.

1.6.2 Applications to Inverses and Rank

As in Gaussian elimination, suppose that a series of k elementary row operations is applied to an arbitrary matrix A, and let E_i denote the elementary matrix corresponding to the ith row operation. Then Lemma 1 asserts that step i of the row-reduction is given by left-multiplication by E_i, so the reduction becomes

$$A \to E_1 A \to E_2 E_1 A \to E_3 E_2 E_1 A \to \cdots \to E_k \cdots E_2 E_1 A.$$

Thus
$$A \to UA \quad \text{where} \quad U = E_k \cdots E_2 E_1.$$

This matrix U is invertible because each E_i is invertible (by Lemma 2). Moreover U can be easily constructed without finding the E_i. Indeed, if we apply the same sequence of row operations to the identity matrix I (in place of A), the result is $I \to UI = U$. Thus these operations carry the "double matrix" $[A \ I]$ to $[UA \ U]$. If we denote this as $[A \ I] \to [UA \ U]$, we get part (2) of the following theorem; the rest of the theorem follows from the above discussion.

Theorem 1. *Suppose a series of row operations carries $A \to B$. Then:*

(1) $B = UA$ *for some invertible matrix U.*

(2) U *can be constructed by performing the same row operations on the double matrix* $[A \ I]$:
$$[A \ I] \to [B \ U]$$

CHAPTER 1. LINEAR EQUATIONS AND MATRICES

(3) $U = E_k \cdots E_2 E_1$ where $E_1, E_2, \cdots, E_k$ are the elementary matrices corresponding in order to the row operations carrying A to B.

Example 3. Find the reduced row-echelon form R of $A = \begin{bmatrix} 3 & -2 & 5 \\ 1 & -1 & 0 \end{bmatrix}$, and express it as $R = UA$ where U is invertible.

Solution. The reduction of the double matrix $[A \ \ I]$ is as follows:

$$\begin{bmatrix} 3 & -2 & 5 & 1 & 0 \\ 1 & -1 & 0 & 0 & 1 \end{bmatrix} \rightarrow \begin{bmatrix} 1 & -1 & 0 & 0 & 1 \\ 3 & -2 & 5 & 1 & 0 \end{bmatrix}$$

$$\rightarrow \begin{bmatrix} 1 & -1 & 0 & 0 & 1 \\ 0 & 1 & 5 & 1 & -3 \end{bmatrix}$$

$$\rightarrow \begin{bmatrix} 1 & 0 & 5 & 1 & -2 \\ 0 & 1 & 5 & 1 & -3 \end{bmatrix}$$

Hence $R = \begin{bmatrix} 1 & 0 & 5 \\ 0 & 1 & 5 \end{bmatrix}$ and $U = \begin{bmatrix} 1 & -2 \\ 1 & -3 \end{bmatrix}$. The reader can verify that $UA = R$. □

Theorem 1 is a generalization of the matrix inversion algorithm given in Section 1.5. Indeed, suppose that A is invertible so that $A \rightarrow I$ by row operations by Theorem 5 §1.5. If we take $B = I$, the reduction in Theorem 1 is $[A \ \ I] \rightarrow [I \ \ U]$. But part (1) of Theorem 1 asserts that $UA = B = I$, so $U = A^{-1}$. Thus the reduction becomes

$$[A \ \ I] \rightarrow [I \ \ A^{-1}]$$

This is the matrix inversion algorithm.

However we can say more. We have $U = E_k \cdots E_2 E_1$ by part (3) of Theorem 1 where $E_1, E_2, \cdots, E_k$ are the elementary matrices corresponding (in order) to the row operations carrying A to I. Hence

$$A = (A^{-1})^{-1} = U^{-1} = (E_k \cdots E_2 E_1)^{-1} = E_1^{-1} E_2^{-1} \cdots E_k^{-1}$$

Since each E_i^{-1} is an elementary matrix by Lemma 2, this shows that A is a product of elementary matrices. On the other hand, every product of elementary matrices is invertible (by Lemma 2), so this proves

Theorem 2. *A square matrix A is invertible if and only if A is a product of elementary matrices.*

Example 4. *Express the invertible matrix $A = \begin{bmatrix} 2 & 1 \\ 1 & 1 \end{bmatrix}$ as a product of elementary matrices.*

1.6. ELEMENTARY MATRICES

Solution. Because A is invertible, its reduced row-echelon form is I. The reduction is

$$\begin{bmatrix} 2 & 1 \\ 1 & -1 \end{bmatrix} \xrightarrow{E_1} \begin{bmatrix} 1 & -1 \\ 2 & 1 \end{bmatrix} \xrightarrow{E_2} \begin{bmatrix} 1 & -1 \\ 0 & 3 \end{bmatrix} \xrightarrow{E_3} \begin{bmatrix} 1 & -1 \\ 0 & 1 \end{bmatrix} \xrightarrow{E_4} \begin{bmatrix} 1 & 0 \\ 0 & 1 \end{bmatrix}$$

where the corresponding elementary matrices are

$$E_1 = \begin{bmatrix} 0 & 1 \\ 1 & 0 \end{bmatrix} \quad E_2 = \begin{bmatrix} 1 & 0 \\ -2 & 1 \end{bmatrix} \quad E_3 = \begin{bmatrix} 1 & 0 \\ 0 & \frac{1}{3} \end{bmatrix} \quad E_4 = \begin{bmatrix} 1 & 1 \\ 0 & 1 \end{bmatrix}.$$

Thus the reduction is $A \to E_1 A \to E_2 E_1 A \to E_3 E_2 E_1 A \to E_4 E_3 E_2 E_1 A = I$, so

$$A = (E_4 E_3 E_2 E_1)^{-1} = E_1^{-1} E_2^{-1} E_3^{-1} E_4^{-1}$$

$$= \begin{bmatrix} 0 & 1 \\ 1 & 0 \end{bmatrix} \begin{bmatrix} 1 & 0 \\ 2 & 1 \end{bmatrix} \begin{bmatrix} 1 & 0 \\ 0 & 3 \end{bmatrix} \begin{bmatrix} 1 & -1 \\ 0 & 1 \end{bmatrix}.$$

This is the desired factorization. □

Theorem 1 has another application. Suppose that an $m \times n$ matrix A is given, and let $A \to R$ by row operations where R is a reduced row-echelon matrix. Then Theorem 1 gives $R = UA$ where the invertible matrix U can be found by

$$[A \quad I_m] \to [R \quad U].$$

If $\operatorname{rank} A = r$ then R has r nonzero rows and, because R is reduced, the top r rows contain each column of the $r \times r$ identity matrix I_r. Hence elementary *column* operations will carry R to the block $m \times n$ matrix $\begin{bmatrix} I_r & 0 \\ 0 & 0 \end{bmatrix}$. Furthermore, these column operations can be accomplished by doing the corresponding *row* operations to R^T to obtain its $n \times m$ row-echelon form: $R^T \to U_1 R^T = \begin{bmatrix} I_r & 0 \\ 0 & 0 \end{bmatrix}$ where U_1 is $n \times n$ and invertible. If we write $V = U_1^T$, we get

$$UAV = RV = RU_1^T = (U_1 R^T)^T = \begin{bmatrix} I_r & 0 \\ 0 & 0 \end{bmatrix}^T = \begin{bmatrix} I_r & 0 \\ 0 & 0 \end{bmatrix}.$$

By Theorem 1, the matrix $U_1 = V^T$ can be obtained by

$$[R^T \quad I_n] \to \left[\begin{bmatrix} I_r & 0 \\ 0 & 0 \end{bmatrix} \quad V^T \right].$$

Hence we get

Theorem 3. *Let A be an $m \times n$ matrix of rank r. Then there exist invertible matrices U and V (of sizes $m \times m$ and $n \times n$ respectively) such that*

$$UAV = \begin{bmatrix} I_r & 0 \\ 0 & 0 \end{bmatrix}.$$

Moreover, U and V can be computed in two steps:

Step 1. $[A \quad I_m] \to [R \quad U]$ where R is a reduced row-echelon matrix.

Step 2. $[R^T \quad I_n] \to \left[\begin{bmatrix} I_r & 0 \\ 0 & 0 \end{bmatrix} \quad V^T\right].$

The procedure in Theorem 3 to compute U and V is easily adapted for a computer.

Example 5. If $A = \begin{bmatrix} 1 & -2 & 3 & 1 \\ -1 & 2 & -1 & 1 \\ 2 & -4 & 5 & 1 \end{bmatrix}$, show that $\operatorname{rank} A = 2$, and find invertible matrices U and V such that $UAV = \begin{bmatrix} I_2 & 0 \\ 0 & 0 \end{bmatrix}.$

Solution. The reduction in Step 1 of Theorem 3 is $[A \quad I_3] \to [R \quad U]$ where, as the reader can verify,

$$R = \begin{bmatrix} 1 & -2 & 0 & -2 \\ 0 & 0 & 1 & 1 \\ 0 & 0 & 0 & 0 \end{bmatrix} \quad \text{and} \quad U = \begin{bmatrix} -5 & 0 & 3 \\ 2 & 0 & -1 \\ -3 & 1 & 2 \end{bmatrix}$$

Thus $\operatorname{rank} A = 2$, so Step 2 of Theorem 3 is the reduction

$$[R^T \quad I_4] \to \left[\begin{bmatrix} I_2 & 0 \\ 0 & 0 \end{bmatrix} \quad V^T\right]$$

where
$V = \begin{bmatrix} 1 & 0 & 2 & 2 \\ 0 & 0 & 1 & 0 \\ 0 & 1 & 0 & -1 \\ 0 & 0 & 0 & 1 \end{bmatrix}$. The reader can verify that $UAV = \begin{bmatrix} I_2 & 0 \\ 0 & 0 \end{bmatrix}.$ □

1.6.3 Uniqueness of Reduced Row-echelon Matrices [18]

We stated earlier that the reduced row-echelon form R of an $m \times n$ matrix A is uniquely determined by A; that is if $A \to R$ and $A \to S$ by row operations where R and S are both reduced row-echelon matrices, then $R = S$. We conclude this section with a proof of this.

By Theorem 1 there are invertible matrices P and Q such that $R = PA$ and $S = QA$. Then $U = QP^{-1}$ is an invertible matrix with the property that $UR = S$. Hence we prove the following statement:

Theorem 4. Suppose $UR = S$ where R and S are $m \times n$ reduced row-echelon matrices and U is invertible. Then $R = S$.

---

[18]This section is not used in the sequel, and so may be omitted.

1.6. ELEMENTARY MATRICES

The proof is by induction on m. The case $m = 1$ is left to the reader. Let R_j and S_j denote column j of R and S respectively. Then matrix multiplication gives

$$UR_j = S_j \quad \text{for each } j. \tag{*}$$

Since U is invertible, it follows that $R_j = 0$ if and only if $S_j = 0$; that is R and S have the same zero columns. Hence, by passing to the matrices obtained by deleting all zero columns from R and S, we may assume that there are no zero columns.

But then the first column from the left in R and S is column 1 of I_m (both are row-echelon). Applying (*) to this column shows that column 1 of U equals column 1 of I_m. Hence write U, R and S in block form as follows:

$$U = \begin{bmatrix} 1 & X \\ 0 & V \end{bmatrix}, \quad R = \begin{bmatrix} 1 & Y \\ 0 & R' \end{bmatrix} \quad \text{and} \quad S = \begin{bmatrix} 1 & Z \\ 0 & S' \end{bmatrix}$$

Since $UR = S$, block multiplication shows that $VR' = S'$. Hence $R' = S'$ by induction because V is invertible (U is invertible).

It follows that R and S have leading 1's in the same columns, say k of them. Applying (*) to these columns then yields that the first k columns of U are the first k columns of I_m. Hence we can write U, R and S in block form as follows:

$$U = \begin{bmatrix} I_k & M \\ 0 & W \end{bmatrix} \quad R = \begin{bmatrix} R_1 & R_2 \\ 0 & 0 \end{bmatrix} \quad S = \begin{bmatrix} S_1 & S_2 \\ 0 & 0 \end{bmatrix}$$

where R_1 and S_1 are $k \times k$. But then $UR = R$ by block multiplication that is $S = R$. This completes the proof. □

Exercises

1. For each of the following elementary matrices, describe the corresponding row operation and write the inverse.

   (a) $E = \begin{bmatrix} 0 & 1 & 0 \\ 1 & 0 & 0 \\ 0 & 0 & 1 \end{bmatrix}$ (b) $E = \begin{bmatrix} 1 & 0 & 0 \\ 5 & 1 & 0 \\ 0 & 0 & 1 \end{bmatrix}$

   (c) $E = \begin{bmatrix} 2 & 0 & 0 \\ 0 & 1 & 0 \\ 0 & 0 & 1 \end{bmatrix}$ (d) $E = \begin{bmatrix} 1 & 0 & -3 \\ 0 & 1 & 0 \\ 0 & 0 & 1 \end{bmatrix}$

2. In each case find elementary matrices E and F such that $B = EA$ and $A = FB$. What is the relationship between E and F?

(a) $A = \begin{bmatrix} 2 & -1 \\ 5 & 3 \end{bmatrix}$ $B = \begin{bmatrix} 2 & -1 \\ 1 & 5 \end{bmatrix}$

(b) $A = \begin{bmatrix} 3 & 2 \\ -5 & 7 \end{bmatrix}$ $B = \begin{bmatrix} -2 & 9 \\ -5 & 7 \end{bmatrix}$

(c) $A = \begin{bmatrix} 3 & -2 & 0 \\ 1 & 5 & 6 \\ -2 & 1 & 3 \end{bmatrix}$ $B = \begin{bmatrix} 3 & -2 & 0 \\ -3 & 7 & 12 \\ -2 & 1 & 3 \end{bmatrix}$

(d) $A = \begin{bmatrix} 1 & 3 & -7 \\ -1 & 2 & -5 \\ -3 & 0 & 6 \end{bmatrix}$ $B = \begin{bmatrix} 1 & 3 & -7 \\ -1 & 2 & -5 \\ -1 & 6 & -8 \end{bmatrix}$

3. In each case find elementary matrices E_1 and E_2 such that $B = E_2 E_1 A$.

(a) $A = \begin{bmatrix} 3 & -5 \\ 2 & 4 \end{bmatrix}$ $B = \begin{bmatrix} 0 & -11 \\ 1 & 2 \end{bmatrix}$

(b) $A = \begin{bmatrix} 5 & -3 \\ 2 & 1 \end{bmatrix}$ $B = \begin{bmatrix} 2 & 1 \\ 1 & -5 \end{bmatrix}$

4. Describe the connection between elementary row operations and elementary matrices.

5. In each case express the invertible matrix A as a product of elementary matrices.

(a) $A = \begin{bmatrix} 3 & -1 \\ 1 & 0 \end{bmatrix}$ (b) $A = \begin{bmatrix} 2 & -1 \\ 1 & -1 \end{bmatrix}$

6. In each case find an invertible matrix U such that $UA = R$ is the reduced row echelon form of A.

(a) $A = \begin{bmatrix} 3 & -1 \\ -2 & 0 \end{bmatrix}$ (b) $A = \begin{bmatrix} 3 & -1 \\ 1 & -3 \end{bmatrix}$

(c) $A = \begin{bmatrix} 3 & -1 & 4 \\ -2 & 0 & -3 \\ -1 & 3 & 0 \end{bmatrix}$ (d) $A = \begin{bmatrix} 1 & -3 & -1 \\ 1 & -3 & 5 \\ 3 & -9 & -1 \end{bmatrix}$

7. Explain why any matrix has many row-echelon forms.

8. In each case find an invertible matrix U such that $UA = B$.

(a) $A = \begin{bmatrix} 3 & 1 \\ 2 & 5 \end{bmatrix}$ $B = \begin{bmatrix} 1 & -4 \\ 4 & -3 \end{bmatrix}$

(b) $A = \begin{bmatrix} 2 & -3 \\ 5 & 1 \end{bmatrix}$ $B = \begin{bmatrix} 1 & 7 \\ 2 & -3 \end{bmatrix}$

9. What is the connection between invertible matrices and elementary matrices?

1.6. ELEMENTARY MATRICES

10. Find an elementary matrix F such that $B = AF$, where $A = \begin{bmatrix} 2 & -3 \\ 5 & 7 \end{bmatrix}$ and $B = \begin{bmatrix} 8 & -3 \\ -9 & 7 \end{bmatrix}$.

11. Let E be an elementary matrix. Show that E^T is also elementary of the same type as E.

12. In each case either show that the statement is true or give an example showing that it is false.

    (a) If $B = EA$ where E is elementary, then $A = FB$ for some elementary F.

    (b) The product of two elementary matrices is again elementary.

    (c) The transpose of an elementary matrix is again elementary.

    (d) If $A \to R$ and $B \to R$ where R is a reduced row-echelon matrix, then $A = B$.

13. In each case determine $\operatorname{rank} A$ and find invertible matrices U and V such that $UAV = \begin{bmatrix} I_r & 0 \\ 0 & 0 \end{bmatrix}$.

    (a) $A = \begin{bmatrix} 1 & -1 & 2 & 1 \\ 2 & -1 & 0 & 3 \\ 0 & 1 & -4 & 1 \end{bmatrix}$ (b) $A = \begin{bmatrix} 1 & 1 & 0 & -1 \\ 3 & 2 & 1 & 1 \\ 1 & 0 & 1 & 3 \end{bmatrix}$

14. While trying to invert A, the double matrix $[A \ \ I]$ is carried to $[P \ \ Q]$. Show that $P = QA$.

15. Use Theorem 1 to show that the only matrix A that can be carried to 0 by row operations is $A = 0$.

16. If $B = UA$ where U is invertible, show that $A \to B$ by row operations.

17. (a) Show that every $m \times n$ matrix A can be factored as $A = UR$ where U is invertible and R is in reduced row-echelon form.

    (b) Show that R in the factorization in (a) is unique, that is, if $A = U_1 R_1$ as in (a), then $R_1 = R$.

    (c) Show that the factorization in (a) may not be unique in general.

18. Suppose that $A \to B$ by column operations, and we apply these same column operations to the double matrix $[A \ \ I]$ get $[A \ \ I] \to [B \ \ V]$, explain why V is invertible and $AV = B$.

19. Two matrices A and B are called **row-equivalent** (written $A \stackrel{r}{\sim} B$) if A can be carried to B by row operations.

    (a) Show that $A \stackrel{r}{\sim} B$ if and only if $B = UA$ for some invertible matrix U.

    (b) If $A \stackrel{r}{\sim} B$, show that also $B \stackrel{r}{\sim} A$.

(c) If $A \stackrel{r}{\sim} B$ and $B \stackrel{r}{\sim} C$, show that $A \stackrel{r}{\sim} C$.

(d) If A and B are both row-equivalent to some third matrix, show that $A \stackrel{r}{\sim} B$.

(e) If $A \stackrel{r}{\sim} B$, show that $rank A = rank B$.

(f) If $A \stackrel{r}{\sim} B$ and A is invertible, show that B is also invertible.

1.7 LU-FACTORIZATION

The solution of a system of linear equations $AX = B$ can be computed much more quickly if the matrix A can be factored in the form $A = LU$ where L and U are matrices of a particularly nice form. In this section we show that Gaussian elimination can be used to find such factorizations.

1.7.1 Triangular Matrices

As for square matrices, an $m \times n$ matrix A is called **upper triangular** if each entry below and to the left of the main diagonal is zero. For example, every row-echelon matrix is upper triangular, as is each of the following matrices:

$$\begin{bmatrix} 2 & -3 & 0 & 1 \\ 0 & 1 & 3 & 0 \\ 0 & 0 & 0 & 2 \end{bmatrix} \begin{bmatrix} 0 & -1 & 0 & 5 & 2 \\ 0 & 0 & 0 & 5 & -2 \\ 0 & 0 & 1 & 0 & 3 \end{bmatrix} \begin{bmatrix} 1 & 2 & 3 \\ 0 & -2 & 5 \\ 0 & 0 & 0 \\ 0 & 0 & 0 \end{bmatrix}$$

Similarly, a matrix A is called **lower triangular** if every entry above and to the right of the main diagonal is zero, equivalently if A^T is upper triangular. A matrix is called **triangular** if it is either upper or lower triangular. The following result will be needed, and follows from the definition of matrix multiplication.

Lemma 1. *The product of two lower (upper) triangular matrices is again lower (upper) triangular.*

One reason for the importance of triangular matrices is the ease with which a linear system $AX = B$ can be solved when the coefficient matrix A is triangular.

Example 1. *Solve the system*

$$\begin{aligned} -x_1 - 2x_2 + x_3 + 3x_4 + 2x_5 &= 2 \\ x_3 + 5x_4 - 3x_5 &= -1 \\ 2x_5 &= 3 \end{aligned}$$

where the coefficient matrix is upper triangular.

1.7. LU-FACTORIZATION

Solution. As for a row-echelon matrix, let $x_2 = s$ and $x_4 = t$ be parameters. We solve for x_5, x_3 and x_1 in that order. The third equation gives

$$x_5 = \tfrac{3}{2}$$

Then substitution into the second equation gives

$$x_3 = \tfrac{7}{2} - 5t$$

as the reader can verify. Finally substitute these into the first equation to get

$$x_1 = \tfrac{9}{2} - 2s - 2t.$$

This gives the general solution. □

The method used in Example 1 is called **back substitution** because later variables are substituted into earlier equations. It works because the coefficient matrix is upper triangular. Similarly, if the matrix of coefficients is lower triangular the system can be solved by **forward substitution** where the earlier variables are substituted in later equations.

Now consider a system $AX = B$ where A is an arbitrary matrix. If A can be factored as $A = LU$ where L is an invertible lower triangular and U is upper triangular, the system $AX = B$ can be solved in two stages as follows:

1. First solve $LY = B$ for Y by forward substitution
2. Then solve $UX = Y$ for X by back substitution

Then X is the solution to the system $AX = B$. Indeed, since $A = LU$ we have $AX = LUX = LY = B$ as required. The method adapts easily for use in computer programs.

1.7.2 LU-Factorization

Hence if A is any $m \times n$ matrix, we try to factor A as $A = LU$ where L is lower triangular and U is upper triangular. The idea is to carry A by row operations to a row-echelon matrix U (which is upper triangular), and look for ways to find L. Suppose that k row operations are needed, and that the corresponding elementary matrices are $E_1, E_2, \cdots, E_k$. Then the reduction takes the form

$$A \to E_1 A \to E_2 E_1 A \to \cdots \to E_k \cdots E_2 E_1 A = U.$$

Thus $PA = U$ where $P = E_k \cdots E_2 E_1$ is invertible because each E_i is invertible.

If we do not insist that U is reduced, the row-reduction can be carried out by adding multiples of a row to rows below it, and possibly by row interchanges. The point is that no row of A need ever be added to a row *above* it. Thus, apart from row-interchanges, the only row operations needed are those for which the corresponding elementary matrix E_i is *lower* triangular. In particular, if no row interchanges are needed, the matrix P is lower triangular by Lemma 1. Since

$PA = U$ we obtain $A = P^{-1}U$, and P^{-1} is also lower triangular by Theorem 4 §1.5. If we write $L = P^{-1}$, this proves the following theorem.

For convenience we say that a matrix A can be **lower reduced** if it can be carried to a row-echelon matrix using no row interchanges (and adding no multiple of a row to a row above it).

Theorem 1. *Suppose that a matrix A can be lower reduced to a row-echelon matrix U. Then*
$$A = LU$$
where L is lower triangular and invertible, and U is upper triangular and row-echelon.

A factorization $A = LU$ as in Theorem 1 is called an **LU-factorization** of A. Such a factorization may not exist (Exercise 6) because A cannot be carried to row-echelon form without using at least one row interchange. A procedure for dealing with this situation will be outlined later.

If an LU-factorization $A = LU$ does exist the above discussion shows that U can be any row-echelon form for A (obtained by lower reduction), and L is the inverse of the product of elementary matrices needed to carry $A \to U$. However, there is a simpler way to obtain L whereby the columns of L are obtained one by one. The following examples illustrate the technique. For convenience, the first nonzero column from the left of a matrix A is called the **leading column** of A.

Example 2. *Find an LU-factorization for* $A = \begin{bmatrix} 0 & 2 & -6 & -2 & 4 \\ 0 & -1 & 3 & 3 & 2 \\ 0 & -1 & 3 & 7 & 10 \end{bmatrix}$.

Solution. We lower reduce A to row echelon form as follows:

$$A = \begin{bmatrix} 0 & \boxed{2} & -6 & -2 & 4 \\ 0 & -1 & 3 & 3 & 2 \\ 0 & -1 & 3 & 7 & 10 \end{bmatrix}$$

$$\to \begin{bmatrix} 0 & 1 & -3 & -1 & 2 \\ 0 & 0 & 0 & \boxed{2} & 4 \\ 0 & 0 & 0 & 6 & 12 \end{bmatrix}$$

$$\to \begin{bmatrix} 0 & 1 & -3 & -1 & 2 \\ 0 & 0 & 0 & 1 & 2 \\ 0 & 0 & 0 & 0 & 0 \end{bmatrix} = U$$

The boxed columns are determined as follows: The first is the leading column of A, and it is used to create the first leading 1 (using lower reduction). At this stage, we are finished with row 1, and we repeat the procedure on the matrix consisting of the remaining rows. Thus the second boxed column is the leading

1.7. LU-FACTORIZATION

column of this smaller matrix, and we create the second leading 1. As the remaining row is zero in this case, we are finished. Then $A = LU$ where

$$L = \begin{bmatrix} 2 & 0 & 0 \\ -1 & 2 & 0 \\ -1 & 6 & 1 \end{bmatrix}$$

This matrix L is obtained from the identity matrix by replacing the bottom of columns 1 and 2 by the leading columns (in order) that were boxed in the lower reduction. Note that $rank A = 2$ here, and this is the number of boxed columns. □

The procedure works for any $m \times n$ matrix A that can be lower reduced to row-echelon form. The procedure can be stated formally as follows:

LU-Algorithm. *Suppose $A \to U$ by lower reduction where U is a row-echelon matrix. Then $A = LU$ where the lower triangular, invertible matrix L is created as follows:*

1. *If $A = 0$ take $L = I_m$ and $U = 0$.*

2. *If $A \neq 0$ let C_1 denote the leading column of A. Then use C_1 to create the first leading 1 and make the rest of the entries in that column zero (using lower reduction). When this is done, let A_2 denote the resulting matrix with the first row deleted.*

3. *If $A_2 \neq 0$ let C_2 denote its leading column and repeat Step 2 on A_2 to create A_3.*

4. *Continue in this way until U is reached where all rows below the last leading 1 consist of zeros.*

5. *Create L by placing $C_1, C_2, C_3, \cdots, C_r$ at the bottom of the first r columns of I_m, where $r = rank A$.*

The proof that $LU = A$ in the LU-algorithm involves induction and block multiplication; we omit the details.

Example 3. *Find an LU-factorization for* $A = \begin{bmatrix} 5 & -5 & 10 & 0 & 5 \\ -3 & 3 & 2 & 2 & 1 \\ -2 & 2 & 0 & -1 & 0 \\ 1 & -1 & 10 & 2 & 5 \end{bmatrix}$.

Solution. The lower reduction to row-echelon form is as follows:

$$\begin{bmatrix} 5 & -5 & 10 & 0 & 5 \\ -3 & 3 & 2 & 2 & 1 \\ -2 & 2 & 0 & -1 & 0 \\ 1 & -1 & 10 & 2 & 5 \end{bmatrix} \rightarrow \begin{bmatrix} 1 & -1 & 2 & 0 & 1 \\ 0 & 0 & 8 & 2 & 4 \\ 0 & 0 & 4 & -1 & 2 \\ 0 & 0 & 8 & 2 & 4 \end{bmatrix}$$

$$\rightarrow \begin{bmatrix} 1 & -1 & 2 & 0 & 1 \\ 0 & 0 & 1 & \frac{1}{4} & \frac{1}{2} \\ 0 & 0 & 0 & -2 & 0 \\ 0 & 0 & 0 & 0 & 0 \end{bmatrix}$$

$$\rightarrow \begin{bmatrix} 1 & -1 & 2 & 0 & 1 \\ 0 & 0 & 1 & \frac{1}{4} & \frac{1}{2} \\ 0 & 0 & 0 & 1 & 0 \\ 0 & 0 & 0 & 0 & 0 \end{bmatrix}$$

Hence $A = LU$ where $L = \begin{bmatrix} 5 & 0 & 0 & 0 \\ -3 & 8 & 0 & 0 \\ -2 & 4 & -2 & 0 \\ 1 & 8 & 0 & 1 \end{bmatrix}$ and $U = \begin{bmatrix} 1 & -1 & 2 & 0 & 1 \\ 0 & 0 & 1 & \frac{1}{4} & \frac{1}{2} \\ 0 & 0 & 0 & 1 & 0 \\ 0 & 0 & 0 & 0 & 0 \end{bmatrix}$

as the reader can verify. □

Example 4. *Find an LU-factorization for the invertible matrix*

$$A = \begin{bmatrix} 1 & 2 & -1 \\ 2 & -1 & 3 \\ 1 & 1 & 2 \end{bmatrix}.$$

Solution. The lower reduction to row-echelon form is

$$A = \begin{bmatrix} 1 & 2 & -1 \\ 2 & -1 & 3 \\ 1 & 1 & 2 \end{bmatrix} \rightarrow \begin{bmatrix} 1 & 2 & -1 \\ 0 & -5 & 5 \\ 0 & -1 & 3 \end{bmatrix}$$

$$\rightarrow \begin{bmatrix} 1 & 2 & -1 \\ 0 & 1 & -1 \\ 0 & 0 & 2 \end{bmatrix} \rightarrow \begin{bmatrix} 1 & 2 & -1 \\ 0 & 1 & -1 \\ 0 & 0 & 1 \end{bmatrix} = U.$$

Hence $L = \begin{bmatrix} 1 & 0 & 0 \\ 2 & -5 & 0 \\ 1 & -1 & 2 \end{bmatrix}.$ □

1.7.3 Dealing with Row Interchanges

Every matrix A can be carried to a row-echelon matrix U by row operations. If we do not insist that U is reduced, this can be done without adding any multiples of a row of A to a row above it. If no row interchanges are required,

1.7. LU-FACTORIZATION

we have seen that A admits an LU-factorization. But some matrices have no LU-factorization (for example $A = \begin{bmatrix} 0 & 1 \\ 1 & 0 \end{bmatrix}$) and so require at least one row interchange. However, it turns out that if all the necessary row interchanges are carried out first, the resulting matrix requires no interchanges and so has an LU-factorization.

Theorem 2. *Suppose an $m \times n$ matrix A is carried to a row-echelon matrix U using no row operations that add a multiple of a row to a row above it. Let P_1, $P_2, \cdots, P_{s-1}, P_s$ denote the elementary matrices corresponding (in order) to the row interchanges used. Write $P = P_s P_{s-1} \cdots P_2 P_1$ (if no interchanges are used take $P = I_m$). Then:*

(1) *PA is the matrix obtained by doing these interchanges (in order) to A.*

(2) *PA has an LU-factorization.*

The proof of Theorem 2 is omitted.

If $P_1, P_2, \cdots, P_{s-1}, P_s$ are elementary matrices corresponding to row interchanges, the matrix $P = P_s P_{s-1} \cdots P_2 P_1$ is called a **permutation matrix**. Such a matrix is obtained from the identity matrix by doing the same interchanges in order to I (starting with the one corresponding to P_1). Thus P is obtained from I by placing the rows in a different order (hence the name), and so has exactly one 1 in each row and column. We regard the identity matrix as a permutation matrix.

Example 5. If $A = \begin{bmatrix} 0 & 0 & -2 \\ 2 & 4 & 2 \\ 1 & -1 & 4 \end{bmatrix}$, find a permutation matrix P such that PA has an LU-factorization, and then find the factorization.

Solution. First carry A to row-echelon form:

$$A = \begin{bmatrix} 0 & 0 & -2 \\ 2 & 4 & 2 \\ 1 & -1 & 4 \end{bmatrix} \rightarrow \begin{bmatrix} 1 & 2 & 1 \\ 0 & 0 & 1 \\ 1 & -1 & 4 \end{bmatrix}$$

$$\rightarrow \begin{bmatrix} 1 & 2 & 1 \\ 0 & 0 & 1 \\ 0 & -3 & 3 \end{bmatrix} \rightarrow \begin{bmatrix} 1 & 2 & 1 \\ 0 & 1 & -1 \\ 0 & 0 & 1 \end{bmatrix}$$

Only two row interchanges are needed, first rows 1 and 2, and then rows 2 and 3. Hence, as in Theorem 2, the required permutation matrix is

$$P = P_2 P_1 = \begin{bmatrix} 1 & 0 & 0 \\ 0 & 0 & 1 \\ 0 & 1 & 0 \end{bmatrix} \begin{bmatrix} 0 & 1 & 0 \\ 1 & 0 & 0 \\ 0 & 0 & 1 \end{bmatrix} = \begin{bmatrix} 0 & 1 & 0 \\ 0 & 0 & 1 \\ 1 & 0 & 0 \end{bmatrix}.$$

If we do these interchanges in order to A, the result is PA. Applying the LU-algorithm to PA gives

$$PA = \begin{bmatrix} 2 & 4 & 2 \\ 1 & -1 & 4 \\ 0 & 0 & -2 \end{bmatrix} \to \begin{bmatrix} 1 & 2 & 1 \\ 0 & -3 & 3 \\ 0 & 0 & -2 \end{bmatrix}$$

$$\to \begin{bmatrix} 1 & 2 & 1 \\ 0 & 1 & -1 \\ 0 & 0 & -2 \end{bmatrix} \to \begin{bmatrix} 1 & 2 & 1 \\ 0 & 1 & -1 \\ 0 & 0 & 1 \end{bmatrix} = U.$$

Hence $PA = LU$ where $L = \begin{bmatrix} 2 & 0 & 0 \\ 1 & -3 & 0 \\ 0 & 0 & -2 \end{bmatrix}$ and $U = \begin{bmatrix} 1 & 2 & 1 \\ 0 & 1 & -1 \\ 0 & 0 & 1 \end{bmatrix}$. $\square$

Consider the system $AX = B$ of linear equations where A is $m \times n$. If P is any permutation matrix, the matrix PA is obtained from A by writing the rows in a different order. Hence the system of equations

$$(PA)X = PB$$

is identical to the system $AX = B$ except for the order in which the equations are written, and so the two systems have the same solutions. Since the permutation matrix P can be chosen so that PA has an LU-factorization, the system $(PA)X = PB$ can be solved by forward and backward substitution.

Exercises

1. In each case use the given LU-factorization of A to solve the system $AX = B$ by solving $LY = B$ and $UX = Y$ by backward and forward substitution respectively.

   (a)
   $$A = \begin{bmatrix} 2 & 0 & 0 & 0 \\ 3 & -1 & 0 & 0 \\ 7 & 0 & 1 & 0 \\ 2 & -3 & 5 & 2 \end{bmatrix} \begin{bmatrix} 2 & -1 & 0 & 0 & 3 \\ 0 & 5 & 7 & -3 & 2 \\ 0 & 0 & -1 & 0 & 3 \\ 0 & 0 & 0 & 0 & 0 \end{bmatrix}, B = \begin{bmatrix} 1 \\ 0 \\ -2 \\ 0 \end{bmatrix}$$

   (b)
   $$A = \begin{bmatrix} 1 & 0 & 0 & 0 \\ 5 & -3 & 0 & 0 \\ 2 & 0 & -1 & 0 \\ -1 & 7 & 0 & 1 \end{bmatrix} \begin{bmatrix} 1 & -3 & 7 & 11 & -4 \\ 0 & 2 & -5 & 6 & 1 \\ 0 & 0 & 0 & 0 & 0 \\ 0 & 0 & 0 & 0 & 0 \end{bmatrix}, B = \begin{bmatrix} 2 \\ -1 \\ 7 \\ 2 \end{bmatrix}$$

2. In each case find a LU-factorization of the given matrix.

1.7. LU-FACTORIZATION

(a) $\begin{bmatrix} 2 & 4 & 2 \\ 1 & -1 & 3 \\ -1 & 7 & -7 \end{bmatrix}$
(b) $\begin{bmatrix} -3 & 6 & 9 \\ 2 & -4 & 1 \\ -1 & 2 & -7 \end{bmatrix}$

(c) $\begin{bmatrix} 2 & 6 & -2 & 0 & 2 \\ 3 & 9 & -3 & 3 & 1 \\ -1 & -3 & 1 & -3 & 1 \end{bmatrix}$
(d) $\begin{bmatrix} 3 & -9 & 6 & 0 & -9 \\ -2 & 6 & 1 & 1 & 1 \\ -1 & 3 & 0 & 4 & 3 \end{bmatrix}$

(e) $\begin{bmatrix} -5 & 10 & 0 & -15 \\ 2 & -3 & 1 & 5 \\ 1 & -2 & 4 & 0 \\ 1 & 1 & 1 & 1 \end{bmatrix}$
(f) $\begin{bmatrix} 1 & -3 & 2 & 0 \\ 2 & -5 & -3 & 1 \\ 0 & -1 & 3 & 7 \\ -1 & 3 & 2 & 5 \end{bmatrix}$

(g) $\begin{bmatrix} 2 & 2 & -2 & 4 & 2 \\ 1 & -1 & 0 & 2 & 1 \\ 3 & 1 & -2 & 6 & 3 \\ 1 & 3 & -2 & 2 & 1 \end{bmatrix}$
(h) $\begin{bmatrix} -1 & -3 & 1 & 0 & -1 \\ 1 & 4 & 1 & 1 & 1 \\ 1 & 2 & -3 & -1 & 1 \\ 0 & -2 & -4 & -2 & 0 \end{bmatrix}$

3. In your own words, write the procedure to find an LU-factorization of a matrix A, and describe how it is used to solve a system $AX = B$.

4. In each case find a permutation matrix P such that PA has an LU-factorization, and find that factorization.

(a) $A = \begin{bmatrix} -1 & 4 & 5 & 2 \\ 0 & 0 & 0 & -1 \\ 1 & -2 & -2 & 0 \\ 0 & -1 & -1 & 0 \end{bmatrix}$
(b) $A = \begin{bmatrix} 0 & 0 & 0 & 3 \\ 0 & 0 & 2 & -4 \\ 0 & -1 & 0 & 5 \\ 1 & 3 & -2 & 3 \end{bmatrix}$

5. In each case either show that the statement is true or give an example showing that it is false.

    (a) The sum of two upper triangular matrices is again upper triangular.

    (b) Every invertible matrix has an LU-factorization.

    (c) Every permutation matrix is invertible.

6. If $A = \begin{bmatrix} 0 & 1 \\ 1 & 0 \end{bmatrix}$, show that $A = LU$ is impossible for *any* lower triangular 2×2 matrix L and upper triangular 2×2 matrix U. In particular, show that A has no LU-factorization.

7. If A is lower triangular and invertible, give an LU-factorization of A.

8. Show that any row interchange can be accomplished by row operations of other types.

9. Show that any multiple of a row can be added to a row above it by row operations of other types.

10. (a) Show that every permutation matrix is invertible.

    (b) Show the the inverse of a permutation matrix is again a permutation matrix.

11. If A is invertible and has an LU-factorization $A = LU$, show that L and U are uniquely determined by A.

12. A triangular matrix is called *unit triangular* if it is square and every main diagonal element is 1.

    (a) If A has an LU-factorization, show that $A = LU$ where L is unit lower triangular and U is upper triangular.

    (b) Show that the factorization in (a) is unique if A is invertible.

1.8 APPLICATION TO MARKOV CHAINS[19]

A Markov chain is a mathematical model that provides insight and information about a wide variety of applications, but is simple enough to be described using only linear equations and matrix multiplication. The model requires an acquaintance with probability theory, but this is no hindrance as we can proceed informally.

The probability of getting a head when flipping a fair coin is $\frac{1}{2}$ because, in a long series of flips, the coin will fall heads half the time. In general, the **probability** of an event is the long-term proportion of the time that the event will occur. For example, the probability of obtaining a 6 on the roll of a die is $\frac{1}{6}$ because each of the 6 faces of the die is equally likely to appear. Probabilities of events are thus numbers between 0 and 1, with impossible events having probability 0, and certain events having probability 1. Our use of probabilities arises in applications like the following.

1.8.1 An Example

Consider the weather in a certain city. We assume for simplicity that it is always in one of two *states*: sunny or cloudy. The weather evolves in daily *stages*, changing from one state to another. We are interested in finding the answers to questions of the following type:

> Question 1. *If it is sunny on Monday, what is the probability that it is sunny on Thursday?*
>
> Question 2. *What is the long-term proportion of the time that it is sunny?*

To answer such questions, we need to know the probability that the weather will change or remain the same from one day to the next. Suppose that long-term weather records show that:

---
[19]The material in this section will not be used elsewhere in the text.

1.8. APPLICATION TO MARKOV CHAINS

(1) If it is sunny one day, it is equally likely to be sunny or cloudy next day.

(2) If it is cloudy one day, the probabilities it is sunny and cloudy the next day are $\frac{3}{4}$ and $\frac{1}{4}$ respectively.

With the help of matrix algebra, this information provides answers to the above questions.

The given weather data are summarized in the following table.

		Present day	
		S	C
Next Day	S	1/2	3/4
	C	1/2	1/4

Note that, if it is sunny on a given day, the first column lists the probabilities for the weather on the next day. Similarly, the second column gives the probabilities for the weather the day after a cloudy day. In particular, each column sums to 1 because, in this model, it *must* be either sunny or cloudy next day. The corresponding matrix of probabilities

$$P = \begin{bmatrix} 1/2 & 3/4 \\ 1/2 & 1/4 \end{bmatrix}$$

is called the *probability transition matrix* for this model of the weather.

Of course the weather each day is not determined; all we can hope to know is the *probability* that it is sunny or cloudy on any given day. Hence we introduce the *state vector* for day m:

$$S_m = \begin{bmatrix} s_1 \\ s_2 \end{bmatrix}$$

Here s_1 and s_2 are the probabilities that the weather is sunny and cloudy, respectively, on day m. In particular S_0 represents the initial situation. Note that $s_1 + s_2 = 1$ because the weather must be in some state on day m. The remarkable thing is that we can use matrix multiplication to compute these state vectors. More precisely, the following matrix equations hold:

$$S_{m+1} = PS_m \quad \text{for each } m = 0, 1, \cdots. \tag{*}$$

In other words, the state vector S_{m+1} for a given day can be computed by multiplying S_m by the transition matrix P. To apply this to our situation, choose the initial day to be Monday. Then $S_0 = \begin{bmatrix} 1 \\ 0 \end{bmatrix}$ because we are assuming that the weather is in state S on Monday. With this, we can apply (*) repeatedly to compute

$$S_1 = PS_0 = \begin{bmatrix} 1/2 & 3/4 \\ 1/2 & 1/4 \end{bmatrix} \begin{bmatrix} 1 \\ 0 \end{bmatrix} = \begin{bmatrix} 1/2 \\ 1/2 \end{bmatrix},$$

$$S_2 = PS_1 = \begin{bmatrix} 1/2 & 3/4 \\ 1/2 & 1/4 \end{bmatrix} \begin{bmatrix} 1/2 \\ 1/2 \end{bmatrix} = \begin{bmatrix} 5/8 \\ 3/8 \end{bmatrix},$$

$$S_3 = PS_2 = \begin{bmatrix} 1/2 & 3/4 \\ 1/2 & 1/4 \end{bmatrix} \begin{bmatrix} 5/8 \\ 3/8 \end{bmatrix} = \begin{bmatrix} 19/32 \\ 13/32 \end{bmatrix}.$$

Since we are taking Monday to be day 0, Thursday is day 3. Hence the probability that it is sunny Thursday is 19/32, answering Question 1 above. Of course we have also determined that the probability it is cloudy on Thursday is 13/32.

Turning to Question 2, we compute successively

$$S_4 = \begin{bmatrix} .6015625 \\ .3984375 \end{bmatrix}, S_5 = \begin{bmatrix} .599609375 \\ .400390625 \end{bmatrix}, S_6 = \begin{bmatrix} .600097656 \\ .399902344 \end{bmatrix}, \cdots.$$

Clearly S_m is getting closer and closer to $\begin{bmatrix} .6 \\ .4 \end{bmatrix}$ as m increases. It follows that, in the long run, the weather is sunny with probability .6, and cloudy with probability .4. This answers Question 2.

The vector $S = \begin{bmatrix} .6 \\ .4 \end{bmatrix}$ is called the *steady-state* vector for the weather. There is another say to compute it that is much better (and simpler) than computing several state vectors S_1, S_2, $\cdots$ and observing their behavior. Assume that S_m is indeed very close to some fixed vector S for all sufficiently large m. Then S_{m+1} is also very close to S, and so the equation $S_{m+1} = PS_m$ is closely approximated by $S = PS$. This can be written

$$(I - P)S = 0$$

which is a system of linear homogeneous equations for S. In the above model of the weather, reduction of the augmented matrix to reduced form is

$$\begin{bmatrix} 1/2 & -3/4 & 0 \\ -1/2 & 3/4 & 0 \end{bmatrix} \rightarrow \begin{bmatrix} 1 & -3/2 & 0 \\ 0 & 0 & 0 \end{bmatrix}$$

so the general solution to the system $(I - P)S = 0$ is $\begin{bmatrix} \frac{3}{2}t \\ t \end{bmatrix}$ where t is a parameter. Since the entries of S must sum to 1, we must have $\frac{3}{2}t + t = 1$, and so $t = \frac{2}{5}$. This gives $S = \begin{bmatrix} 3/5 \\ 2/5 \end{bmatrix} = \begin{bmatrix} .6 \\ .4 \end{bmatrix}$, which is what we had before.

1.8.2 Markov Chains

This model of the weather is an example of a Markov chain. In general, a **Markov**[20] **chain** is a system that evolves through a series of **stages** (the consecutive days in the above discussion) and, at any stage, must be in one of a finite number of **states** (sunny or cloudy). To analyze the chain, we must know each **transition probability** p_{ij}, that is the probability that, if the chain is in state j at some stage, it will be in state i after one more transition. If the chain has n states, the $n \times n$ matrix $P = [p_{ij}]$ is called the **transition matrix** for the chain. The 3×3 case is illustrated as follows:

	PRESENT STAGE		
	State 1	State 2	State 3
NEXT STAGE State 1	p_{11}	p_{12}	p_{13}
State 2	p_{21}	p_{22}	p_{23}
State 3	p_{31}	p_{32}	p_{33}

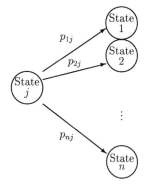

Observe that each column of P sums to 1 because, starting in state j at present, the chain *must* go to some state at the next stage. In general a square matrix is called **stochastic** if the entries are all non-negative and each column sum is 1.

In general, columns whose entries are probabilities that sum to 1 are called **probability vectors**. At the mth stage, the chain is described by the **state vector**

$$S_m = \begin{bmatrix} s_1^{(m)} \\ s_2^{(m)} \\ \vdots \\ s_n^{(m)} \end{bmatrix}$$

where $s_k^{(m)}$ is the probability that the chain is in state k after m transitions. Note that each S_m is a probability vector because the chain must be in *some* state after m transitions. A probability vector S is called a **steady state vector** for the Markov chain if the state vectors S_m get closer and closer to S as m increases. Thus the entries of S are the long term probabilities that the the chain will be in the various states.

It can happen that a Markov chain does not have a steady state vector. One condition that guarantees that a steady state vector exists is that the chain

---

[20]Named after the Russian mathematician Andreĭ Andreevič Markov (1856-1922).

is **regular**, that is some power P^k of the transition matrix has all its entries positive. For small chains this can be checked graphically as follows: Represent the states as the vertices $v_1, v_2, \cdots, v_n$ of a directed graph (called the **graph** of the chain) and draw an arrow from state v_j to v_i when it is possible to go from state v_j to state v_i (that is when the probability $p_{ij} > 0$). Then the chain is regular if there exists an integer k such that it is possible to go from any vertex to any other vertex in exactly k transitions.

Thus, for example, the weather chain discussed above is regular because it is possible to go from any state to any other state in *one* transition. The graph is given in the diagram.

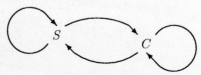

The behavior of regular Markov chains is summarized in the following theorem.

Theorem 1. *Consider a Markov chain with transition matrix P.*

(1) $S_{m+1} = PS_m$ *for all* $m = 0, 1, 2, \cdots$

(2) *Suppose the chain is regular. Then the steady state vector S is the unique solution to the homogeneous system*

$$(I - P)S = 0$$

whose entries sum to 1.

We give a proof of (1) of Theorem 1 at the end of this section which illustrates how it is that matrix multiplication arises here. Part (2) of Theorem 1 will not be proved in these notes.[21]

Example 1. *Consider a mouse in the maze shown. If it is in any cell, assume that it is equally likely to*

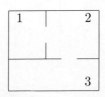

stay in the cell, leave by the first exit, leave by the second exit, etc.

(a) *If it starts in cell 1, what is the probability it is in each of the cells after three moves.*

(b) *What proportion of the time does it spend in each of the cells?*

---

[21] A proof can be found in Kemeny, Mirkil, Snell and Thompson, *Finite Mathematical Structures*, Prentice-Hall, 1958.

1.8. APPLICATION TO MARKOV CHAINS

Solution. The states here are the cells 1, 2, and 3, and the stages are the successive moves. If the mouse is in cell 1, it has two options: stay or go to cell 2. By assumption these are equally likely, so both have probability $\frac{1}{2}$. Since it has probability 0 of going to cell 3, this gives the first column of the transition matrix $P = \begin{bmatrix} 1/2 & 1/3 & 0 \\ 1/2 & 1/3 & 1/2 \\ 0 & 1/3 & 1/2 \end{bmatrix}$. If the mouse is in cell 2 it can stay, go to cell 1, or go to cell 3, and these all have probability $\frac{1}{3}$ (being equally likely). This gives column 2 of P, and column 3 is found in a similar way.

If the mouse starts in cell 1, we have $S_0 = \begin{bmatrix} 1 & 0 & 0 \end{bmatrix}^T$. Hence we compute

$$S_1 = PS_0 = \begin{bmatrix} 1/2 \\ 1/2 \\ 0 \end{bmatrix}, \quad S_2 = PS_1 = \begin{bmatrix} 5/12 \\ 5/12 \\ 2/12 \end{bmatrix} \quad S_3 = PS_2 = \begin{bmatrix} 25/72 \\ 31/72 \\ 16/72 \end{bmatrix}$$

Thus the probabilities the mouse is in cells 1, 2 and 3 after three moves are $\frac{25}{72}$, $\frac{31}{72}$ and $\frac{16}{72}$ respectively.

The graph of this chain is

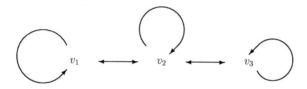

so any cell is reachable from any other cell in exactly 2 transitions. Hence the chain is regular and we can find the steady-state vector S by solving the system of linear equations

$$(I - P)S = 0$$

The reduction of the augmented matrix to reduced form is as follows:

$$\begin{bmatrix} 1/2 & -1/3 & 0 & 0 \\ -1/2 & 2/3 & -1/2 & 0 \\ 0 & -1/3 & 1/2 & 0 \end{bmatrix} \rightarrow \begin{bmatrix} 1/2 & -1/3 & 0 & 0 \\ 0 & 1/3 & -1/2 & 0 \\ 0 & 0 & 0 & 0 \end{bmatrix}$$

$$\rightarrow \begin{bmatrix} 1 & 0 & -1 & 0 \\ 0 & 1 & -3/2 & 0 \\ 0 & 0 & 0 & 0 \end{bmatrix}$$

Thus the general solution to $(I - P)S = 0$ is $\begin{bmatrix} t & \frac{3}{2}t & t \end{bmatrix}^T$. This is a probability vector if the entries sum to 1, that is if $t + \frac{3}{2}t + t = 1$. This gives $t = \frac{2}{7}$, so the steady-state vector is $S = \begin{bmatrix} 2/7 & 3/7 & 2/7 \end{bmatrix}^T$. Thus the mouse spends $\frac{3}{7}$ of his time in cell 2, and $\frac{2}{7}$ of his time in each of cells 1 and 3. □

For any regular Markov chain, it can be shown that the solution S of the homogeneous system $(I - P)S = 0$ always involves exactly one parameter t (as in Example 1). Then t is determined by the requirement that the entries of the steady-state vector S must sum to 1.

We conclude with one final observation. Let P be the transition matrix of a Markov chain, and let $S_1, S_2, \cdots S_m, \cdots$ be the state vectors. Then $S_1 = PS_0$ by Theorem 1, so $S_2 = PS_1 = P(PS_0) = P^2 S_0$, again by Theorem 1. Next $S_3 = PS_2 = P(P^2 S_0) = P^3 S_0$. The pattern is clear:

$$S_m = P^m S_0 \quad \text{for each } m = 1, 2, 3, \cdots.$$

Hence the behavior of the chain is entirely determined by the initial state vector S_0 and the powers P^m of the transition matrix. In Section 2.3 we develop methods to calculate the powers of any square matrix.

1.8.3 Proof of (1) of Theorem 1.

Think of the chain as having been run N times where N is large. Recall that the mth state vector is

$$S_m = [s_1^{(m)} \quad s_2^{(m)} \quad \cdots \quad s_n^{(m)}]^T$$

where $s_i^{(m)}$ is the probability that it is in state i at stage m. Looking at stage $m+1$, we have that, approximately,

$$s_i^{(m+1)} N \text{ is the number of times it is in state } i \text{ at stage } m+1.$$

We count this number a different way. To get to state i at stage $m+1$ the chain had to be in *some* state j at stage m. Since the approximate number of times it is in state j at stage m is $s_j^{(m)} N$, and since p_{ij} is the probability that it goes from state j to state i in one transition, the number of times it got to state i via state j is $p_{ij}(s_j^{(m)} N)$. Summing over j gives the total number of times it got to state i. This is the number we got before, so

$$s_i^{(m+1)} N = p_{i1} s_1^{(m)} N + p_{i2} s_2^{(m)} N + \cdots + p_{in} s_n^{(m)} N$$

Cancelling N gives $s_i^{(m+1)} = p_{i1} s_1^{(m)} + p_{i2} s_2^{(m)} + \cdots + p_{in} s_n^{(m)}$. This asserts that entry i of the column matrix S_{m+1} is equal to entry i of the matrix product PS_m. Hence $S_{m+1} = PS_m$, which is (1) of Theorem 1.

Exercises

1. A city has only rainy or sunny days. Assume that there is an 80% probability of having a sunny day today, an 80% probability of having sun the day following a sunny day, and a 60% probability of having rain the day

1.8. APPLICATION TO MARKOV CHAINS

following a rainy day. If there is an 80% probability of having a sunny day today, calculate the probabilities of rain and shine for each of the following 10 days, calculate the long term probabilities of rain and shine, and find out how many days it will take to reach this steady state vector with 2 decimals accuracy.

2. In each case find the steady-state vector and, assuming the chain starts in state 1, find the probability that it is in state 2 after 3 transitions.

    (a) $P = \begin{bmatrix} .5 & .3 \\ .5 & .7 \end{bmatrix}$
    (b) $P = \begin{bmatrix} \frac{1}{2} & 1 \\ \frac{1}{2} & 0 \end{bmatrix}$

    (c) $P = \begin{bmatrix} 0 & \frac{1}{2} & \frac{1}{4} \\ 1 & 0 & \frac{1}{4} \\ 0 & \frac{1}{2} & \frac{1}{2} \end{bmatrix}$
    (d) $P = \begin{bmatrix} .4 & .1 & .5 \\ .2 & .6 & .2 \\ .4 & .3 & .3 \end{bmatrix}$

    (e) $P = \begin{bmatrix} .8 & .0 & .2 \\ .1 & .6 & .1 \\ .1 & .4 & .7 \end{bmatrix}$
    (f) $P = \begin{bmatrix} .1 & .3 & .3 \\ .3 & .1 & .6 \\ .6 & .6 & .1 \end{bmatrix}$

3. In each case determine whether the given probability transition matrix is regular by drawing the graph.

    (a) $P = \begin{bmatrix} 0 & 0 & \frac{1}{2} \\ 1 & 0 & \frac{1}{2} \\ 0 & 1 & 0 \end{bmatrix}$
    (b) $P = \begin{bmatrix} \frac{1}{2} & 0 & \frac{1}{3} \\ \frac{1}{4} & 1 & \frac{1}{3} \\ \frac{1}{4} & 0 & \frac{1}{3} \end{bmatrix}$

    (c) $P = \begin{bmatrix} \frac{1}{2} & 0 & 0 \\ \frac{1}{2} & \frac{1}{4} & \frac{2}{3} \\ 0 & \frac{3}{4} & \frac{1}{3} \end{bmatrix}$
    (d) $P = \begin{bmatrix} 0 & 0 & \frac{1}{3} \\ \frac{1}{2} & 0 & \frac{1}{2} \\ \frac{1}{2} & 1 & \frac{1}{6} \end{bmatrix}$

4. A man eats one of three soups for lunch each day—beef, chicken or vegetable. He never eats the same soup two days in a row. If he eats beef soup one day he is equally likely to eat each of the others next day; if he does not eat beef soup one day he is twice as likely to eat it next day as the alternative.

    (a) If he has beef soup one day, what is the probability he has it again two days later.

    (b) Find the long-run probabilities he eats each of the soups.

5. A wolf pack always hunts in one of four regions, R_1, R_2, R_3 or R_4. Its hunting habits are as follows:

    (1) If it hunts in some region, it is as likely as not to hunt there the next day.

    (2) If it hunts in R_1, it never hunts in R_2 or in R_3 the next day.

    (3) If it hunts in R_2 or R_3, it is equally likely to hunt in each of the other regions the next day.

(4) If it hunts in R_4, it is equally likely to hunt in regions R_2 and R_3 on the next day, but twice as likely to hunt in R_1 than in either R_2 or R_3.

(a) Find the proportion of the time the pack hunts in R_1, R_2, R_3, and R_4.

(b) If the pack hunts in R_1 on Monday, find the probability it hunts there on Thursday.

6. A gambler always plays one of three slot machines, A, B or C. He never plays the same machine on two successive days. If he plays A, he plays C next day. If he plays B or C, he is twice as likely to play A next day as the alternative.

    (a) What proportion of his time does he spend playing A, B and C?

    (b) If he plays C one day, find the probability he plays it again three days later.

7. A lab rat has a choice of three foods P, Q and R each day. On any given day it has a 60% chance of choosing the same food as it chose the previous day, and is equally likely to choose either of the other foods.

    (a) If it chooses P one day, find the probability it chooses Q three days later.

    (b) What percentage of its meals are food P, Q and R?

8. Assume that there are three classes—upper U, middle M and lower L—and that social mobility is modeled as follows:

    (1) Of children of U parents, 70% remain U while 20% become L and 10% become M.

    (2) Of children of M parents, 80% remain M while 15% become L and 15% become U.

    (3) Of children of L parents, 60% remain L while 10% become U and 30% become M.

    (a) Find the probability that the grandchild of lower-class parents becomes upper-class

    (b) Find the long-term breakdown of society into classes.

9. John makes it to work on time one Monday out of four. On other work days his behavior is as follows: If he is late one day he is twice as likely to be on time next day as to be late. If he is on time one day he is equally likely to be on time or late next day. Find the probability of his being on time Thursdays.

10. The governor says she will call an election. This gossip is passed from person to person with probability $p \neq 0$ that the information is passed incorrectly at any stage. Assume that when a person hears the news he

1.8. APPLICATION TO MARKOV CHAINS

or she passes it to one person who does not know. Find the long-term probability that a person will hear that there is going to be an election.

11. Suppose you play a game with a friend in which you either win or lose $1 at each round with equal probabilities. You start with $1 and you decide in advance to quit if you ever go broke or get $4. Assume that your friend never quits.

   (a) If the states are 0, 1, 2, 3 and 4 (representing your wealth), show that the transition matrix is not regular.

   (b) Find the probability that you go broke after exactly 3 matches.

   (c) Find a steady state vector by calculating S_n for large values of n, and make an argument whether as a casino owner you would profit from playing such a game in the long run against your customers.

12. In each case either show that the statement is true or give an example showing that it is false. Throughout, P denotes a probability transition matrix.

   (a) If all rows but the last in P are known, then the last row is also known.

   (b) If P has a 1 on the main diagonal, it is regular.

   (c) If P is regular, then it is invertible.

   (d) If P is invertible, then it is regular.

13. Let $E = \begin{bmatrix} 1 & 1 & \cdots & 1 \end{bmatrix}$ denote the $1 \times n$ matrix with each entry 1.

   (a) If an $n \times n$ matrix P has non-negative entries, show that P is stochastic if and only if $EP = E$.

   (b) Show that the product of two stochastic matrices is stochastic.

   (c) If P is stochastic, show that there exists a column $S \neq 0$ such that $PS = S$. [In fact S can be chosen to be a probability vector.]

14. Let $P = \begin{bmatrix} 0 & 0 & 1 \\ 1 & 0 & 0 \\ 0 & 1 & 0 \end{bmatrix}$. Is the corresponding chain regular? Draw the graph of the chain. Can you get from any vector to any other vector? Why does the graphical test fail? Show that $PS = S$ when $S = \begin{bmatrix} \frac{1}{3} & \frac{1}{3} & \frac{1}{3} \end{bmatrix}^T$. Do the state vectors get closer and closer to S?

Chapter 2

DETERMINANTS AND DIAGONALIZATION

2.1 THE LAPLACE EXPANSION

In Section 1.6 we defined the determinant of a 2×2 matrix $A = \begin{bmatrix} a & b \\ c & d \end{bmatrix}$ as follows:

$$det A = det \begin{bmatrix} a & b \\ c & d \end{bmatrix} = ad - bc$$

We then showed (Example 5 §1.6) that A has an inverse if $det A \neq 0$, and gave a formula for the inverse in that case. One goal of this chapter is to do this for *any* square matrix A.

There is no problem if A is 1×1, say $A = [a]$. In that case we define

$$det[a] = a$$

and note that $[a]$ is invertible if (and only if) $a \neq 0$, and that then the formula for the inverse is $[a]^{-1} = [\frac{1}{a}]$.

So what do we do if A is $n \times n$ where $n \geq 3$? The formula for the determinant of a 3×3 matrix turns out to be:

$$det \begin{bmatrix} a & b & c \\ d & e & f \\ g & h & i \end{bmatrix} = aei + bfg + cdh - ceg - afh - bdi \qquad (*)$$

and the expression for the inverse is even more involved. The situation is much more complicated for $n \times n$ matrices where $n > 3$.

2.1.1 Laplace Expansion

There is, however, an inductive way to define the determinant $det A$ of any square matrix A. The idea is to define the determinant of a 3×3 matrix in terms of determinants of 2×2 matrices, then give the determinant of any 4×4 matrix in terms of determinants of 3×3 matrices, and so on.

To motivate the procedure[1], we collect the terms in (*) involving the first row entries a, b and c, and observe that the multipliers are 2×2 determinants:

$$\begin{aligned} det \begin{bmatrix} a & b & c \\ d & e & f \\ g & h & i \end{bmatrix} &= aei + bfg + cdh - ceg - afh - bdi \\ &= a(ei - fh) - b(di - fg) + c(dh - eg) \\ &= a\, det \begin{bmatrix} e & f \\ h & i \end{bmatrix} - b\, det \begin{bmatrix} d & f \\ g & i \end{bmatrix} + c\, det \begin{bmatrix} d & e \\ g & h \end{bmatrix} \end{aligned}$$

This last expression can be described as follows: To compute the determinant of a 3×3 matrix A, multiply each entry in the first row by ± 1 times the determinant of a certain 2×2 matrix, namely the one obtained from A by deleting the row and column of the entry in question. Here the signs alternate down the row (starting with $+1$).[2]

Example 1. *Find the determinant of*

$$A = \begin{bmatrix} 2 & -3 & 7 \\ 5 & -1 & 6 \\ 4 & 0 & 8 \end{bmatrix}.$$

Solution. Using the above procedure, we get:

$$\begin{aligned} det\, A &= 2 det \begin{bmatrix} -1 & 6 \\ 0 & 8 \end{bmatrix} - (-3) det \begin{bmatrix} 5 & 6 \\ 4 & 8 \end{bmatrix} + 7 det \begin{bmatrix} 5 & -1 \\ 4 & 0 \end{bmatrix} \\ &= 2(-8 - 0) + 3(40 - 24) + 7(0 - (-4)) \\ &= -16 + 48 + 28 = 60. \end{aligned}$$

□

In this form the procedure can be generalized. To do so, it is convenient to introduce some terminology. Suppose that we have already defined how to

---

[1] Another way to generalize (*) is to notice that the right side is the sum of all products of entries of the matrix which have exactly one factor from each row and column, and which have a sign $+$ or $-$. When it is clarified which products get which sign, this gives a very satisfactory definition of the determinant.

[2] One way to remember the formula (*) is to adjoin columns 1 and 2 of A on the right to obtain $\begin{bmatrix} a & b & c & a & b \\ d & e & f & d & e \\ g & h & i & g & h \end{bmatrix}$. Then the terms aei, bfg and cdh in (*) are the products down to the right, starting at a, b and c, while the terms $-ceg$, $-afh$ and $-bdi$ come from products down to the left, starting at c, a and b. I hesitate to reveal this rule because it does **not**, repeat **not**, extend to $n \times n$ matrices for $n > 3$.

2.1. THE LAPLACE EXPANSION

compute the determinant of any $(n-1) \times (n-1)$ matrix. Given an $n \times n$ matrix A, let A_{ij} denote the $(n-1) \times (n-1)$ matrix obtained from A by deleting row i and column j. Then we define the (i,j)-**cofactor** $C_{ij}(A)$ of A by

$$C_{ij}(A) = (-1)^{i+j} det(A_{ij}) \quad \text{for each } i \text{ and } j.$$

Here $(-1)^{i+j}$ is called the **sign** of the (i,j)-position in the matrix. The following diagram is useful for remembering the sign of a position.

$$\begin{bmatrix} +1 & -1 & +1 & -1 & \cdots \\ -1 & +1 & -1 & +1 & \cdots \\ +1 & -1 & +1 & -1 & \cdots \\ -1 & +1 & -1 & +1 & \cdots \\ \vdots & \vdots & \vdots & \vdots & \end{bmatrix}$$

Note that the signs alternate along each row or column, starting with $+1$ in the upper left corner.

Since we already know how to compute 2×2 determinants, we can find the cofactors of any 3×3 matrix.

Example 2. Find $C_{12}(A)$, $C_{32}(A)$, and $C_{33}(A)$ for the matrix

$$A = \begin{bmatrix} 2 & -3 & 7 \\ 5 & -1 & 6 \\ 4 & 0 & 8 \end{bmatrix}.$$

Solution. The sign of the $(1,2)$-position is -1, and A_{12} is obtained from A by deleting row 1 and column 2. Hence

$$C_{12}(A) = -det(A_{12}) = -det\begin{bmatrix} 5 & 6 \\ 4 & 8 \end{bmatrix} = -(40 - 24) = -16.$$

Similarly,

$$C_{32}(A) = -det(A_{32}) = -det\begin{bmatrix} 2 & 7 \\ 5 & 6 \end{bmatrix} = -(12 - 35) = 23,$$

$$C_{33}(A) = +det(A_{33}) = det\begin{bmatrix} 2 & -3 \\ 5 & -1 \end{bmatrix} = -2 + 15 = 13.$$

Clearly there are $9 = 3^2$ different cofactors for A, and they all can be computed in a similar way. □

Again, suppose that we have already defined how to compute the determinant of any $(n-1) \times (n-1)$ matrix. Then if $A = [a_{ij}]$ is any $n \times n$ matrix, we define the **determinant** $det A$ of A as follows:

$$det A = a_{11}C_{11}(A) + a_{12}C_{12}(A) + \cdots + a_{1n}C_{1n}(A).$$

In other words, $det A$ is obtained (as in (*)) by multiplying each entry in row 1 of A by the corresponding cofactor, and adding the results. This is called the Laplace expansion of A along row 1. More generally:

If A is any square matrix, the **Laplace expansion** along any row (or column) is a number defined as follows: *Multiply each entry in the row (or column) by the corresponding cofactor, and add the results.*

Thus a square matrix A has many Laplace expansions, one along each row or column. The astonishing thing about these expansions is that they are all equal. This common value is the determinant $det A$.

Theorem 1. Laplace Expansion Theorem. *If A is any square matrix, the determinant $det A$ is equal to the Laplace expansion along any row or column of A.*

The proof of Theorem 1 is beyond the scope of this book.

Note that if a row or column of a square matrix A consists *entirely* of zeros, then $det A = 0$ by the Laplace expansion along that row or column. This observation will be used frequently, and we record it for reference.

Corollary. *If a square matrix A has a row or column of zeros, then $det A = 0$.*

Example 3. *Compute the determinant of* $A = \begin{bmatrix} 5 & -1 & 7 \\ 9 & -3 & 6 \\ 4 & 8 & 0 \end{bmatrix}$ *using the Laplace expansion along column 2 and then by using the expansion along row 3.*

Solution. The expansion along column 2 is as follows:

$$\begin{aligned} det A &= (-1)C_{12}(A) + (-3)C_{22}(A) + 8\, C_{32}(A) \\ &= (-1)\left\{-det \begin{bmatrix} 9 & 6 \\ 4 & 0 \end{bmatrix}\right\} + (-3)\left\{det \begin{bmatrix} 5 & 7 \\ 4 & 0 \end{bmatrix}\right\} + 8\left\{-det \begin{bmatrix} 5 & 7 \\ 9 & 6 \end{bmatrix}\right\} \\ &= (-1)(24) + (-3)(-28) + 8(33) \\ &= 324. \end{aligned}$$

Similarly, the expansion along row 3 is

$$det A = 4\, det \begin{bmatrix} -1 & 7 \\ -3 & 6 \end{bmatrix} + 8\left\{-det \begin{bmatrix} 5 & 7 \\ 9 & 6 \end{bmatrix}\right\} + 0\, det \begin{bmatrix} 5 & -1 \\ 9 & -3 \end{bmatrix}$$

$$= 4 \cdot 15 + 8 \cdot 33 + 0 = 324.$$

Of course the two values are equal as Theorem 1 guarantees. □

When computing the determinant of a 4×4 matrix, the Laplace expansion involves computing determinants of 3×3 matrices. These in turn are evaluated by calculating 2×2 determinants. Here is an example.

2.1. THE LAPLACE EXPANSION

Example 4. *Compute* $det A$ *where* $A = \begin{bmatrix} 2 & -1 & 0 & 3 \\ 1 & 0 & 5 & 7 \\ 7 & 9 & 0 & 2 \\ 4 & 0 & 0 & 8 \end{bmatrix}$.

Solution. Since (by Theorem 1) we may expand $det A$ along any row or column, we choose one that simplifies the calculation. One way to do this is to choose a row or column with as many zeros as possible. Hence we use column 3 in A because it contains three zeros.

$$det A = 0\, C_{13}(A) + 5\, C_{23}(A) + 0\, C_{33}(A) + 0\, C_{43}(A)$$

$$= 5\, C_{23}(A) = -5\, det \begin{bmatrix} 2 & -1 & 3 \\ 7 & 9 & 2 \\ 4 & 0 & 8 \end{bmatrix}$$

where we used the fact that the sign of the $(2,3)$-position is -1. This reduces the calculation to finding a 3×3 determinant, and again we choose a row or column with as many zeros as possible. Using row 3 gives

$$det A = -5 \left\{ 4\, det \begin{bmatrix} -1 & 3 \\ 9 & 2 \end{bmatrix} + 0 + 8\, det \begin{bmatrix} 2 & -1 \\ 7 & 9 \end{bmatrix} \right\}$$

$$= -5\, \{4(-29) + 8(25)\} = -420$$

as required. □

2.1.2 Elementary Operations and Determinants

One way to create zeros in a matrix (and so simplify the computation of the determinant) is by using elementary row operations. The next theorem shows that row operations have a simple effect on the determinant, and that *column* operations can also be used. This can be combined with the Laplace expansion to greatly reduce the computing time involved in finding determinants.

Theorem 2. *Let A denote any $n \times n$ matrix.*

(1) *If B is obtained from A by interchanging two different rows (columns), then $det B = -det A$.*

(2) *If B is obtained from A by multiplying a row (column) by a number k, then $det B = k\, det A$.*

(3) *If B is obtained from A by adding a multiple of some row (column) of A to a different row (column), then $det B = det A$.*

Proof. We prove the theorem for row operations; the proof for column operations is analogous.

(1). We proceed by induction on n. The cases $n = 1$ and $n = 2$ are left to the reader. If $n > 2$ expand $det B$ along a row *other* than the two that were interchanged. The entries of this row are the same for both A and B, but the cofactors in B are the negatives of those in A by induction, because the corresponding $(n-1) \times (n-1)$ matrices have two rows interchanged. It follows that $det B = -det A$.

(2). Expand $det B$ along the row that is multiplied by k. The entries in this row of B are k times the entries of the same row of A, and the corresponding cofactors are the same as those for A (because B and A differ only in this row). It follows that $det B = k \, det A$.

(3). Let B be obtained from $A = [a_{ij}]$ by adding u times row p to row q. Thus row q of B is

$$[a_{q1} + u\,a_{p1},\ a_{q2} + u\,a_{p2},\ \cdots,\ a_{qn} + u\,a_{pn}]$$

The cofactors of the positions in this row are the same as those for A (they do not involve row q), that is $C_{qj}(B) = C_{qj}(A)$ for each j. Hence expanding $det B$ along row q gives

$$\begin{aligned}
det B &= (a_{q1} + u\,a_{p1})C_{q1}(B) + \cdots + (a_{qn} + u\,a_{pn})C_{qn}(B) \\
&= [a_{q1}C_{q1}(B) + \cdots + a_{qn}C_{qn}(B)] + u[a_{p1}C_{q1}(B) + \cdots + a_{pn}C_{qn}(B)] \\
&= [a_{q1}C_{q1}(A) + \cdots + a_{qn}C_{qn}(A)] + u[a_{p1}C_{q1}(A) + \cdots + a_{pn}C_{qn}(A)] \\
&= det A + u\, det A_1
\end{aligned}$$

where A_1 is the matrix obtained from A by replacing row q by row p, and both expansions are along row q. It remains to show that $det A_1 = 0$. But A_1 has two identical rows, so A_1 is unchanged if these rows are interchanged. Hence $det A_1 = -det A_1$ by (1), whence $det A_1 = 0$. Thus $det B = det A$ as required. □

The proof of (3) in Theorem 2 establishes the following useful fact.

Corollary. *If a square matrix A has two identical rows (or columns), then $det A = 0$.*

The properties in Theorem 2 can be stated in other words as follows:

(1) If two distinct rows (columns) are interchanged, the determinant changes sign.

(2) A common factor in any row (column) can be "taken out" of the determinant.

(3) Adding a multiple of any row (column) to another row (column) has no effect on the determinant.

These facts are very useful in computing determinants because they reduce the amount of computation required (and for other reasons as well). This is illustrated by the following three examples.

2.1. THE LAPLACE EXPANSION

Example 5. *Compute* $\det A$ *where*

$$A = \begin{bmatrix} 5 & -3 & 8 \\ -2 & 0 & 4 \\ 7 & 11 & 1 \end{bmatrix}.$$

Solution. We could expand along row 2 (or column 2) and utilize the zero entry, but we can easily obtain *another* zero in row 2 by adding twice column 1 to column 3. Since this does not change the determinant (by Theorem 2), we obtain

$$\det A = \det \begin{bmatrix} 5 & -3 & 18 \\ -2 & 0 & 0 \\ 7 & 11 & 15 \end{bmatrix}$$

$$= -2 \left\{ -\det \begin{bmatrix} -3 & 18 \\ 11 & 15 \end{bmatrix} \right\} + 0 + 0$$

$$= 2 \det \begin{bmatrix} -3 & 18 \\ 11 & 15 \end{bmatrix}$$

using the Laplace expansion along row 2. We could simply evaluate at this point to get $\det A = 2(-45 - 198) = -486$. However we can simplify the computation by using the fact that there is a common factor 3 in column 2 which we can "take out" by Theorem 2 (2). The result is

$$\det A = 2 \det \begin{bmatrix} -3 & 18 \\ 11 & 15 \end{bmatrix} = 2 \cdot 3 \det \begin{bmatrix} -3 & 6 \\ 11 & 5 \end{bmatrix} = 6(-15 - 66) = -486$$

as required. □

Example 6. *Compute* $\det A$ *if*

$$\det \begin{bmatrix} a & b & c \\ p & q & r \\ x & y & z \end{bmatrix} = 5 \quad \text{and} \quad A = \begin{bmatrix} a+2x & b+2y & c+2z \\ 3x+4p & 3y+4q & 3z+4r \\ -2p & -2q & -2r \end{bmatrix}.$$

Solution. We reduce $\det A$ to the known determinant using Theorem 2 several times:

$$\det A = -2 \det \begin{bmatrix} a+2x & b+2y & c+2z \\ 3x+4p & 3y+4q & 3z+4r \\ p & q & r \end{bmatrix}$$

$$= -2 \det \begin{bmatrix} a+2x & b+2y & c+2z \\ 3x & 3y & 3z \\ p & q & r \end{bmatrix}$$

where the last equality results from subtracting 4 times row 3 from row 2. Continuing:

$$\det A = -2 \cdot 3 \det \begin{bmatrix} a+2x & b+2y & c+2z \\ x & y & z \\ p & q & r \end{bmatrix} = -6 \det \begin{bmatrix} a & b & c \\ x & y & z \\ p & q & r \end{bmatrix}.$$

Finally we interchange rows 2 and 3 (and use Theorem 2) to get

$$det A = -6(-1) det \begin{bmatrix} a & b & c \\ p & q & r \\ x & y & z \end{bmatrix} = 6 \cdot 5 = 30$$

because the given determinant is 5. □

We shall see in Section 2.2 that a square matrix A is *not* invertible if and only if $det A = 0$. Hence it is frequently important to be able to determine when $det A = 0$. If the entries of A are functions of x, it is thus helpful to have $det A$ in factored form. Theorem 2 is often useful.

Example 7. *Find the values of x such that $det A = 0$ where $A = \begin{bmatrix} 1 & x & x \\ x & 1 & x \\ x & x & 1 \end{bmatrix}$.*

Solution. As in Gaussian elimination we begin by using the 1 in column 1 to "clean up" column 1:

$$det A = det \begin{bmatrix} 1 & x & x \\ 0 & 1-x^2 & x-x^2 \\ 0 & x-x^2 & 1-x^2 \end{bmatrix} = det \begin{bmatrix} 1-x^2 & x-x^2 \\ x-x^2 & 1-x^2 \end{bmatrix}.$$

If we evaluate this we obtain $det A = 1 - 3x^2 + 2x^3$, so finding x such that $det A = 0$ means factoring the cubic $1 - 3x^2 + 2x^3$, a nontrivial task. However, it is often easier to factor each of the entries before evaluating the determinant, and then use Theorem 2. In the present example we use the fact that $1 - x^2 = (1+x)(1-x)$ and $x - x^2 = x(1-x)$ to get

$$\begin{aligned} det A &= det \begin{bmatrix} (1+x)(1-x) & x(1-x) \\ x(1-x) & (1+x)(1-x) \end{bmatrix} \\ &= (1-x)^2 det \begin{bmatrix} 1+x & x \\ x & 1+x \end{bmatrix} \\ &= (1-x)^2(1+2x) \end{aligned}$$

where we took the common factor $(1-x)$ out of *each* column. Hence the condition $det A = 0$ becomes $(1-x)^2(1+2x) = 0$, that is $x = 1$ or $x = -\frac{1}{2}$. □

The fact that a common factor can be taken out of any row or column of a matrix (as in Example 7) leads to the following useful result.

Theorem 3. *If A is an $n \times n$ matrix, then $det(kA) = k^n det A$ for any scalar k.*

Proof. The matrix kA has k as a common factor in each of its n rows. Hence (by (2) of Theorem 2) we can take out a common factor k from each of the n rows. The result is $det(kA) = k^n det A$. □

2.1. THE LAPLACE EXPANSION

Recall that a matrix A is called upper (lower) triangular if every entry below (above) the main diagonal is zero, and A is called triangular if it is either upper or lower triangular. The next example reveals that the determinant of an upper triangular matrix is very easy to compute.

Example 8. If $A = \begin{bmatrix} a & x & y & z \\ 0 & b & u & v \\ 0 & 0 & c & w \\ 0 & 0 & 0 & d \end{bmatrix}$, show that $\det A = abcd$.

Solution. Use column 1 expansions to get

$$\det A = a \det \begin{bmatrix} b & u & v \\ 0 & c & w \\ 0 & 0 & d \end{bmatrix} = ab \det \begin{bmatrix} c & w \\ 0 & d \end{bmatrix} = abcd.$$

Hence $\det A$ is the product of the main diagonal elements. □

The procedure in Example 8 works on any triangular matrix (use row operations in the lower triangular case), and so proves

Theorem 4. *If a square matrix A is triangular then $\det A$ is the product of the entries on the main diagonal.*

Note that Theorem 4 implies that

$$\det I = 1 \quad \text{for any identity matrix } I$$

because I is upper (and lower) triangular with 1's on the main diagonal.

Theorem 4 shows that finding the determinant of a triangular matrix is easy. This is important in computer calculations because the Gaussian Algorithm gives a routine method of carrying a matrix to (upper) triangular form using row operations.

The last result of this section is not surprising in view of the fact that row and column operations play identical roles in the Laplace Expansion Theorem.

Theorem 5. *If A is any $n \times n$ matrix then $\det A^T = \det A$.*

Proof. We proceed by induction on n, the cases $n = 1$ and $n = 2$ being easily verified. If $n \geq 3$ it is enough to show that the Laplace expansion of $\det A^T$ along row 1 is identical to the Laplace expansion of $\det A$ along column 1. But the entries of column 1 of A^T are the same as the entries of row 1 of A, and the corresponding cofactors are also the same by induction because they are determinants of $(n-1) \times (n-1)$ matrices which are transposes of each other. It follows that $\det A^T = \det A$. □

Exercises

1. If $A = \begin{bmatrix} 5 & -1 & 4 \\ 2 & 3 & -7 \\ -3 & 9 & 4 \end{bmatrix}$, compute:

   (a) $C_{23}(A)$ (b) $C_{31}(A)$ (c) $C_{22}(A)$ (d) $C_{12}(A)$.

2. Explain why the sign of the lower right entry of a square matrix A is always $+1$.

3. Compute the determinants of the following matrices along various rows and columns, and compare your results.

   (a) $\begin{bmatrix} 3 & -1 \\ 2 & 5 \end{bmatrix}$ (b) $\begin{bmatrix} 6 & -9 \\ -14 & 21 \end{bmatrix}$

   (c) $\begin{bmatrix} a^2 & ab \\ ab & b^2 \end{bmatrix}$ (d) $\begin{bmatrix} a+1 & a \\ a & a-1 \end{bmatrix}$

   (e) $\begin{bmatrix} 3 & -1 & 0 \\ 2 & 3 & 5 \\ -2 & 1 & 4 \end{bmatrix}$ (f) $\begin{bmatrix} 7 & -1 & 2 \\ 5 & 2 & 0 \\ 3 & -4 & 9 \end{bmatrix}$

   (g) $\begin{bmatrix} 1 & b & c \\ b & c & 1 \\ c & 1 & b \end{bmatrix}$ (h) $\begin{bmatrix} 0 & a & b \\ a & 0 & c \\ b & c & 0 \end{bmatrix}$

   (i) $\begin{bmatrix} -3 & -1 & 5 & 1 \\ 1 & 0 & -3 & 2 \\ 3 & 4 & 0 & 1 \\ -2 & 3 & 4 & 0 \end{bmatrix}$ (j) $\begin{bmatrix} 4 & 3 & 0 & 1 \\ -1 & 1 & 1 & 2 \\ 3 & 0 & 2 & -1 \\ -1 & 2 & 2 & 1 \end{bmatrix}$

   (k) $\begin{bmatrix} 1 & 3 & -1 & 1 \\ -1 & 1 & -3 & 1 \\ 5 & 2 & 8 & 2 \\ 2 & 4 & 0 & -1 \end{bmatrix}$ (l) $\begin{bmatrix} 0 & 0 & 0 & a \\ 0 & 0 & b & x \\ 0 & c & y & z \\ d & r & s & t \end{bmatrix}$

4. Explain in your own words how to find the determinant of a matrix using the Laplace expansion.

5. Explain why a matrix with a row (or column) of zeros has determinant zero.

6. In each case, compute the determinant of the given matrix by first adding a multiple of some row (or column) to another row (or column). Explain why this makes the computation by hand a bit simpler.

   (a) $A = \begin{bmatrix} 90 & 71 & 38 \\ -7 & 3 & 5 \\ 29 & -12 & -20 \end{bmatrix}$

2.1. THE LAPLACE EXPANSION

(b) $B = \begin{bmatrix} -35 & 5 & 7 \\ -10 & -11 & 2 \\ 66 & 13 & -13 \end{bmatrix}$

7. For various $n \times n$ matrices A and B, compute the following pairs of numbers:
   (a) $det(A + B)$ and $det A + det B$
   (b) $det(AB)$ and $det A \, det B$
   (c) $det A$ and $det A^{-1}$

   Is it possible to make a conjecture about either $det(A + B)$, $det(AB)$, or $det A^{-1}$?

8. In each case either show that the statement is true or give an example showing that it is false. Throughout, A and B denote square matrices.
   (a) If $det A = 0$ then A has two identical rows.
   (b) $det(-A) = -det A$.
   (c) $det(A + B) = det A + det B$.
   (d) If A is 2×2, $det(5A) = 25 \, det A$.
   (e) If $det A = det B$, then $A = B$.
   (f) If the main diagonal of A consists of zeros then $det A = 0$.

9. Prove Theorem 4 by induction on the size of the matrix.

10. Consider the matrix $A = \begin{bmatrix} 2 & -4 & 0 \\ 3 & 6 & 5 \\ -2 & 1 & 4 \end{bmatrix}$, and calculate its determinant $det A$. In each of the following cases, first try to guess what $det B$ will be, and verify your claim by a direct calculation.
    (a) B is the matrix obtained from A by interchanging rows 1 and 3.
    (b) B is the matrix obtained from A by multiplying row 2 by $1/3$.
    (c) B is the matrix obtained from A by multiplying row 2 by 3.
    (d) B is the matrix obtained from A by adding row 1 to row 3.
    (e) B is the matrix obtained from A by adding -2 times row 1 to row 2.
    (f) B is the matrix obtained by transposing A.

11. If B is obtained from A by multiplying row 1 by 5, and $det B = 10$, what is $det A$?

12. In each case, calculate the determinant by transforming the matrices to upper triangular form by elementary row operations (a row-echelon form for example), and using Theorem 2. Then verify your computation by using the Laplace expansion, and make an estimate about which method is more efficient.

(a) $A = \begin{bmatrix} -1 & -9 & 6 \\ 2 & -2 & 1 \\ 7 & 6 & -9 \end{bmatrix}$ (b) $A = \begin{bmatrix} 3 & -6 & 2 \\ 7 & -1 & 0 \\ 7 & 5 & 11 \end{bmatrix}$

(c) $A = \begin{bmatrix} 2 & -3 & 1 & 4 \\ 12 & -5 & 11 & 3 \\ 7 & 6 & -9 & 4 \\ 2 & -7 & 6 & 9 \end{bmatrix}$ (d) $A = \begin{bmatrix} 4 & 7 & 2 & -2 \\ 8 & 9 & 2 & -2 \\ 3 & -3 & 1 & -8 \\ 0 & -2 & 3 & 1 \end{bmatrix}$

13. Find the real numbers x and y such that $det A = 0$ if:

(a) $A = \begin{bmatrix} 0 & x & y \\ y & 0 & x \\ x & y & 0 \end{bmatrix}$ (b) $A = \begin{bmatrix} 1 & -x & -x \\ x & -2 & -x \\ x & x & -3 \end{bmatrix}$

(c) $A = \begin{bmatrix} 1 & x & x^2 & x^3 \\ x & x^2 & x^3 & 1 \\ x^2 & x^3 & 1 & x \\ x^3 & 1 & x & x^2 \end{bmatrix}$ (d) $A = \begin{bmatrix} x & 0 & 0 & y \\ y & x & 0 & 0 \\ 0 & y & x & 0 \\ 0 & 0 & y & x \end{bmatrix}$

14. Compute $det \begin{bmatrix} cos\theta & -sin\theta \\ sin\theta & cos\theta \end{bmatrix}$ for any angle θ.

15. Find $det A$ if A is 3×3 and $det(2A) = 6$. What if A is 4×4?

16. Under what conditions is $det(-A) = det A$? Support your answer.

17. By direct calculation, show that $det A = det A^T$ for any 2×2 matrix $A = \begin{bmatrix} a & b \\ c & d \end{bmatrix}$.

18. (a) Show that $det(A + B^T) = det(A^T + B)$ for any $n \times n$ matrices A and B.

    (b) What is wrong with the following argument:

    $$det(A + B^T) = det A + det B^T = det A^T + det B = det(A^T + B).$$

19. Show that $det \begin{bmatrix} 0 & 1 & 1 & 1 \\ 1 & 0 & x & x \\ 1 & x & 0 & x \\ 1 & x & x & 0 \end{bmatrix} = -3x^2$.

20. Show that $det \begin{bmatrix} 1 & x & x^2 \\ 1 & y & y^2 \\ 1 & z & z^2 \end{bmatrix} = (y-x)(z-x)(z-y)$. [This is the 3×3 Vandermonde determinant. There is an $n \times n$ version given in Section 2.7.]

21. (a) Show that $det \begin{bmatrix} x & -1 & 0 & 0 \\ 0 & x & -1 & 0 \\ 0 & 0 & x & -1 \\ a & b & c & x+d \end{bmatrix} = a + bx + cx^2 + dx^3 + x^4$. [This matrix is called the **companion matrix** of the polynomial $a + bx + cx^2 + dx^3$.]

(b) Now write the 5×5 companion matrix for $a + bx + cx^2 + dx^3 + ex^4$ (last row is $[a \ b \ c \ d \ x+e]$) and verify your claim. Can you generalize this to the $n \times n$ case?

2.2 DETERMINANTS AND INVERSES

We can now prove the promised theorem that a square matrix A has an inverse if and only if $det A \neq 0$. Furthermore, we derive a formula for A^{-1} using determinants, and give a formula (called Cramer's Rule) for the solution of any set of linear equations with A as coefficient matrix.

2.2.1 The Product Theorem

We begin with a remarkable theorem about the determinant of the product of two matrices. The proof is given at the end of this section.

Theorem 1. Product Theorem. *If A and B are $n \times n$ matrices then $det(AB) = det A \, det B$.*

Example 1. For the matrix product $\begin{bmatrix} 2 & -1 \\ 3 & 5 \end{bmatrix} \begin{bmatrix} 4 & 0 \\ -7 & 2 \end{bmatrix} = \begin{bmatrix} 15 & -2 \\ -23 & 10 \end{bmatrix}$, the corresponding determinant product is $13 \cdot 8 = 104$, which is correct as Theorem 1 asserts. □

The product theorem extends to products of more than two matrices. For example

$$det(ABC) = det(A(BC)) = det A \, det(BC) = det A \, det B \, det C.$$

In general, if $A_1, A_2, \cdots, A_k$ are square matrices of the same size, then

$$det(A_1 A_2 \cdots A_k) = det A_1 \, det A_2 \cdots det A_k.$$

In particular, taking $A_i = A$ for each i gives

$$det(A^k) = (det A)^k \quad \text{for each } k \geq 1.$$

We will use these formulas frequently below.

Our first application of the product theorem is to give the determinant test for invertibility.

Theorem 2. *Let A be an $n \times n$ matrix. Then:*

(1) *A is invertible if and only if* $\det A \neq 0$.

(2) *If A is invertible then* $\det(A^{-1}) = \frac{1}{\det A}$.

Proof. If A is invertible then $AA^{-1} = I$ so the product theorem gives
$$1 = \det I = \det(AA^{-1}) = \det A \det(A^{-1})$$
This shows that $\det A \neq 0$, proving half of (1), and it also proves (2). For the rest of (1), assume that $\det A \neq 0$; we must show that A is invertible. By Theorem 5 §1.5 it suffices to show that A can be carried to the identity matrix by row operations. Certainly A can be carried to a reduced row-echelon matrix R by row operations. Each row operation multiplies the determinant by a nonzero constant by Theorem 2 §2.1. Hence $\det A = c \det R$ for some constant c, so $\det R \neq 0$. In particular R has no row of zeros, so $R = I$ because it is square and reduced row-echelon. This is what we wanted. □

Example 2. *If* $\det A = 2$ *and* $\det B = -3$, *compute* $\det(A^3 B^{-1} A^T B^2)$

Solution. We use Theorems 1 and 2 together with Theorem 5 §2.1:
$$\begin{aligned} \det(A^3 B^{-1} A^T B^2) &= \det(A^3) \det(B^{-1}) \det(A^T) \det(B^2) \\ &= (\det A)^3 \frac{1}{\det B} \det A (\det B)^2 \\ &= (\det A)^4 \det B \\ &= -48. \end{aligned}$$
□

Example 3. *Find the values of c for which* $A = \begin{bmatrix} c & -1 & 2 \\ c & c & 1 \\ 1 & c & 1 \end{bmatrix}$ *has an inverse.*

Solution. We compute $\det A$ by first subtracting c times column 3 from column 2:
$$\det A = \det \begin{bmatrix} c & -1 & 2 \\ c & c & 1 \\ 1 & c & 1 \end{bmatrix} = \det \begin{bmatrix} c & -1-2c & 2 \\ c & 0 & 1 \\ 1 & 0 & 1 \end{bmatrix}$$
$$= (1 + 2c) \det \begin{bmatrix} c & 1 \\ 1 & 1 \end{bmatrix} = (1 + 2c)(c - 1).$$

Hence $\det A = 0$ if and only if $c = -\frac{1}{2}$ or $c = 1$, so A^{-1} exists for any value of c except $-\frac{1}{2}$ and 1. □

Theorem 4 §2.1 asserts that the determinant of any triangular matrix is the product of the main diagonal entries. This extends to **block upper (lower) triangular** matrices, that is matrices which have square blocks on the main diagonal and zero blocks below (respectively above) the main diagonal. Then the extended result asserts that the determinant of a block triangular matrix

2.2. DETERMINANTS AND INVERSES

is the product of the determinants of the main diagonal blocks. We state it here but only for matrices with two diagonal blocks. The proof depends on the product theorem.

Theorem 3. *Let A and B be square matrices (possibly of different sizes). Then*

$$\det \begin{bmatrix} A & X \\ 0 & B \end{bmatrix} = \det A \det B \quad \text{and} \quad \det \begin{bmatrix} A & 0 \\ X & B \end{bmatrix} = \det A \det B.$$

Proof. We factor the first matrix as $\begin{bmatrix} A & X \\ 0 & B \end{bmatrix} = \begin{bmatrix} I & 0 \\ 0 & B \end{bmatrix} \begin{bmatrix} A & X \\ 0 & I \end{bmatrix}$. Then the product theorem gives

$$\det \begin{bmatrix} A & X \\ 0 & B \end{bmatrix} = \det \begin{bmatrix} I & 0 \\ 0 & B \end{bmatrix} \det \begin{bmatrix} A & X \\ 0 & I \end{bmatrix}.$$

The first result follows because $\det \begin{bmatrix} I & 0 \\ 0 & B \end{bmatrix} = \det B$ by repeated expansions along row 1, and also $\det \begin{bmatrix} A & X \\ 0 & I \end{bmatrix} = \det A$ by repeated expansions along the last row. The second formula follows from the first and the fact that $\det C = \det(C^T)$ for any matrix C. $\square$

Example 4. *Find $\det A$ if* $A = \begin{bmatrix} a & x & b & y \\ c & z & d & w \\ 0 & p & 0 & q \\ 0 & r & 0 & s \end{bmatrix}.$

Solution. First interchange columns 2 and 3:

$$\det A = -\det \begin{bmatrix} a & b & x & y \\ c & d & z & w \\ 0 & 0 & p & q \\ 0 & 0 & r & s \end{bmatrix}$$

$$= -\det \begin{bmatrix} a & b \\ c & d \end{bmatrix}, \det \begin{bmatrix} p & q \\ r & s \end{bmatrix} = -(ad-bc)(ps-qr)$$

using the block upper triangular partitioning with two 2×2 diagonal blocks. $\square$

2.2.2 Adjoint of a Matrix

The Laplace expansion theorem has an important matrix form which we describe next. Let A denote an arbitrary $n \times n$ matrix. The **adjoint**[3] of A is defined to be the transpose of the matrix of cofactors:

$$\operatorname{adj} A = [C_{ij}(A)]^T.$$

---
[3]The adjoint is also called the **adjugate** or the **classical adjoint** of the matrix.

Note that $adj\,A$ is also an $n \times n$ matrix.

Example 5. *Find the adjoint of* $A = \begin{bmatrix} 3 & 0 & -1 \\ 4 & 7 & 3 \\ -2 & 8 & 5 \end{bmatrix}$.

Solution. Writing $C_{ij}(A) = C_{ij}$ for simplicity, the adjoint is

$$adj\,A = \begin{bmatrix} C_{11} & C_{12} & C_{13} \\ C_{21} & C_{22} & C_{23} \\ C_{31} & C_{32} & C_{33} \end{bmatrix}^T = \begin{bmatrix} 11 & -26 & 46 \\ -8 & 13 & -24 \\ 7 & -13 & 21 \end{bmatrix}^T$$

$$= \begin{bmatrix} 11 & -8 & 7 \\ -26 & 13 & -13 \\ 46 & -24 & 21 \end{bmatrix}.$$

$\square$

Example 6. If $A = \begin{bmatrix} a & b \\ c & d \end{bmatrix}$ and we write $C_{ij} = C_{ij}(A)$, we have

$$adj\,A = \begin{bmatrix} C_{11} & C_{12} \\ C_{21} & C_{22} \end{bmatrix}^T = \begin{bmatrix} det[d] & -det[c] \\ -det[b] & det[a] \end{bmatrix}^T$$

$$= \begin{bmatrix} d & -c \\ -b & a \end{bmatrix}^T = \begin{bmatrix} d & -b \\ -c & a \end{bmatrix}.$$

This agrees with the adjoint used in Example 5 §1.5 to give a formula for the inverse of a 2×2 matrix. $\square$

The adjoint of a square matrix A is "almost" the inverse of A in that it is only off by a numerical factor (the determinant). This is part (1) of the next theorem.

Theorem 4. *Let A be any square matrix.*

(1) (Adjoint Formula) $A(adj\,A) = (det\,A)\,I = (adj\,A)A$.

(2) *If* $det\,A \neq 0$ *then* $A^{-1} = \frac{1}{det\,A}adj\,A$.

Proof. (1). The proof is given at the end of this section.
(2) If we multiply (1) through by the number $\frac{1}{det\,A}$, the result is

$$A \cdot [\tfrac{1}{det\,A}adj\,A] = I = [\tfrac{1}{det\,A}adj\,A] \cdot A.$$

This shows that $\frac{1}{det\,A}adj\,A$ is indeed the inverse of A. $\square$

2.2. DETERMINANTS AND INVERSES

Example 7. *Find the inverse of the matrix* $A = \begin{bmatrix} 3 & 0 & -1 \\ 4 & 7 & 3 \\ -2 & 8 & 5 \end{bmatrix}$ *in Example 5.*

Solution. Since $\det A = -13$ as the reader can check, Example 5 gives

$$A^{-1} = \frac{1}{-13} \operatorname{adj} A = -\frac{1}{13} \begin{bmatrix} 11 & -8 & 7 \\ -26 & 13 & -13 \\ 46 & -24 & 21 \end{bmatrix}.$$

$\square$

Example 8. *Find the (2,3)-entry of A^{-1} if $A = \begin{bmatrix} 5 & -1 & 0 \\ 6 & 9 & -4 \\ 2 & 7 & 3 \end{bmatrix}$.*

Solution. We have $\det A = 301$ as the reader can verify. The (2,3)-entry of $A^{-1} = \frac{1}{301}[C_{ij}(A)]^T$ is the (3,2)-entry of the transpose matrix $\frac{1}{301}[C_{ij}(A)]$; that is it equals $\frac{1}{301}C_{32}(A) = \frac{1}{301}\left\{-\det\begin{bmatrix} 5 & 0 \\ 6 & -4 \end{bmatrix}\right\} = \frac{20}{301}.$ $\square$

Example 9. *Find a formula for the inverse of the matrix* $A = \begin{bmatrix} c & -1 & 2 \\ c & c & 1 \\ 1 & c & 1 \end{bmatrix}$ *in Example 3 for those values of c for which it exists.*

Solution. We have $\det A = (1 + 2c)(c - 1)$ by Example 3, and we compute

$$\operatorname{adj} A = \begin{bmatrix} C_{11} & C_{12} & C_{13} \\ C_{21} & C_{22} & C_{23} \\ C_{31} & C_{32} & C_{33} \end{bmatrix}^T = \begin{bmatrix} 0 & -c+1 & c^2-c \\ 1+2c & c-2 & -c^2-1 \\ -1-2c & c & c^2+c \end{bmatrix}^T$$

$$= \begin{bmatrix} 0 & 1+2c & -1-2c \\ -c+1 & c-2 & c \\ c^2-c & -c^2-1 & c^2+c \end{bmatrix}.$$

Hence if $c \neq 1$ and $c \neq -\frac{1}{2}$, we have

$$A^{-1} = \frac{1}{(1+2c)(c-1)} \begin{bmatrix} 0 & 1+2c & -1-2c \\ -c+1 & c-2 & c \\ c^2-c & -c^2-1 & c^2+c \end{bmatrix}.$$

Of course this can be checked by multiplying by A. $\square$

114 CHAPTER 2. DETERMINANTS AND DIAGONALIZATION

2.2.3 Cramer's Rule

We now use the adjoint formula to derive an old theorem about determinants called Cramer's rule[4]. If A is an invertible matrix, we know from (Theorem 2 §1.5) that the system of equations $AX = B$ has the unique solution $X = A^{-1}B$. Cramer's rule uses determinants to give a formula for the solution. In order to state the rule, we need some notation: If A is an $n \times n$ matrix and B is an $n \times 1$ column, let

$A_i(B)$ denote the $n \times n$ matrix obtained from A by replacing column i by B.

With this we can state Cramer's rule.

Theorem 5. Cramer's Rule. *Consider the system $AX = B$ of linear equations where A is invertible. If $X = [x_1, x_2, \cdots, x_n]^T$ where $x_1, x_2, \cdots, x_n$ are the variables, then*

$$x_i = \frac{det[A_i(B)]}{det A} \quad \text{for each } i = 1, 2, \cdots, n.$$

Proof. We write $C_{ij} = C_{ij}(A)$ for simplicity, and write $B = [b_1 \; b_2 \; \cdots \; b_n]^T$. The solution to $AX = B$ is given by $X = A^{-1}B$, so the adjoint formula for A^{-1} gives

$$X = A^{-1}B = \left(\frac{1}{det A} adj A\right) B = \frac{1}{det A} \begin{bmatrix} C_{11} & C_{21} & \cdots & C_{n1} \\ C_{12} & C_{22} & \cdots & C_{n2} \\ \vdots & \vdots & & \vdots \\ C_{1n} & C_{2n} & \cdots & C_{nn} \end{bmatrix} \begin{bmatrix} b_1 \\ b_2 \\ \vdots \\ b_n \end{bmatrix}$$

Since entry i of X is x_i, this gives

$$x_i = \frac{1}{det A} [b_1 C_{1i} + b_2 C_{2i} + \cdots + b_n C_{ni}] = \frac{det[A_i(B)]}{det A}$$

where $det[A_i(B)]$ is expanded along column i. $\square$

Example 10. *Use Cramer's rule to find x_2 if*

$$\begin{cases} 5x_1 & -7x_2 & +8x_3 & = & 23 \\ 2x_1 & +6x_2 & -9x_3 & = & 61 \\ -x_1 & -4x_2 & +3x_3 & = & -19 \end{cases}$$

---

[4]Cramer's Rule was known and used long before the Swiss mathematician Gabriel Cramer (1704-1752) popularized it in his book *Introduction à l'analyse des lignes courbes algébriques*, published in 1750. The rule was widely used for many years but is less important today.

2.2. DETERMINANTS AND INVERSES

Solution. We have $A = \begin{bmatrix} 5 & -7 & 8 \\ 2 & 6 & -9 \\ -1 & -4 & 3 \end{bmatrix}$ and $A_2(B) = \begin{bmatrix} 5 & 23 & 8 \\ 2 & 61 & -9 \\ -1 & -19 & 3 \end{bmatrix}$ where we are thinking of the system in matrix form $AX = B$. One verifies that $det A = -127$ and $det[A_2(B)] = 313$, so

$$x_2 = \frac{det[A_2(B)]}{det A} = \frac{313}{-127} = -\frac{313}{127}.$$

Of course x_1 and x_3 can be obtained in the same way. □

Suppose a system of linear equations is given (as in Example 10) with invertible coefficient matrix. At first appearance it seems to be an efficient procedure to solve for one variable using Cramer's rule without having to find the other variables. However, finding that one variable by Cramer's rule requires computing two determinants, and each of them takes approximately the same amount of computer time as solving the whole system using Gaussian elimination. Except for small systems, Cramer's rule is *not* a practical method for solving systems of linear equations; its virtue is theoretical.

2.2.4 Proof of the Product Theorem

If A and B are $n \times n$ matrices, we must show that

$$det(AB) = det A \, det B. \qquad (*)$$

Recall (Section 1.6) that an elementary matrix E is one obtained by doing a single elementary row operation to an identity matrix. Recall further (Lemma 1 §1.6) that doing this operation to a matrix B results in EB. By looking at the three types of elementary matrices separately, Theorem 2 §2.1 gives

$$det(EB) = det E \, det B, \text{ whenever } E \text{ is elementary}. \qquad (**)$$

Now suppose that $A \to C$ by row operations with elementary matrices $E_1, E_2, \cdots, E_k$ in order. Using Lemma 1 §1.6 several times, the reduction is $A \to E_1 A \to E_2 E_1 A \to \cdots \to E_k \cdots E_2 E_1 A = C$. This combines with (**) to give

Lemma 1. *If $A \to C$ by elementary row operations, then $C = E_k \cdots E_2 E_1 A$ where the E_i are elementary matrices, and $det C = det E_k \cdots det E_2 \, det E_1 \, det A$.*

Lemma 2. *If A has no inverse, then $det A = 0$.*

Proof. Let $A \to R$ by row operations where R is in reduced row-echelon form. Then $R = E_k \cdots E_2 E_1 A$ where the E_i are elementary by Lemma 1. But A is not invertible, so $R \neq I$ by Theorem 5 §1.5. Thus R has a row of zeros, and so $0 = det R = det E_k \cdots det E_2 \, det E_1 \, det A$, again by Lemma 1. Since $det E_i \neq 0$ for each i, this implies that $det A = 0$. □

Now we can prove (*) by distinguishing two cases.

Case 1. A has no inverse. In this case it suffices to show that AB also has no inverse, because then $det A = 0 = det(AB)$ by Lemma 2, so (*) holds because each side is zero. But if $(AB)^{-1}$ exists then $A[B(AB)^{-1}] = I$, so A is invertible by Theorem 5 §1.5, contrary to our assumption.

Case 2. A is invertible. Then A is a product of elementary matrices by Theorem 2 §1.6, say $A = E_1 E_2 \cdots E_k$. Hence Lemma 1 gives

$$\begin{aligned} det A &= det(E_1 E_2 \cdots E_k I) \\ &= det E_1 det E_2 \cdots det E_k det I \\ &= det E_1 det E_2 \cdots det E_k. \end{aligned}$$

Finally we obtain

$$det(AB) = det(E_1 E_2 \cdots E_k B) = det E_1 det E_2 \cdots det E_k \, det B = det A \, det B$$

again by Lemma 1. Thus (*) holds in this case too.

2.2.5 Proof of the Adjoint Formula

If A is any $n \times n$ matrix, we must show that

$$A \, (adj \, A) = (det A) \, I = (adj \, A) \, A.$$

We prove the left equality; the other is similar.

Write $A \, (adj \, A) = [u_{ij}]$, so u_{ij} denotes the (i,j)-entry of the matrix product $A \, (adj \, A)$. We must show that

$$u_{ij} = \begin{cases} det A & \text{if } i = j \\ 0 & \text{if } i \neq j \end{cases}$$

Now write $A = [a_{ij}]$, and let $C_{ij} = C_{ij}(A)$ denote the (i,j)-cofactor of A. By the definition of matrix multiplication, the number u_{ij} is the dot product of row i of A and column j of $adj \, A$. Now row i of A is $[a_{i1} \; a_{i2} \; \cdots \; a_{in}]$. Since column j of $adj \, A = [C_{ij}]^T$ is the transpose of row j of the cofactor matrix $[C_{ij}]$, we obtain

$$u_{ij} = [a_{i1} \; a_{i2} \; \cdots \; a_{in}] \begin{bmatrix} C_{j1} \\ C_{j2} \\ \vdots \\ C_{jn} \end{bmatrix} = a_{i1} C_{j1} + a_{i2} C_{j2} + \cdots + a_{in} C_{jn}. \quad (***)$$

If $i = j$, the sum in (***) is the Laplace expansion of $det A$ along row i, so we have $u_{ii} = det A$ for each i. If $i \neq j$, let A' denote the matrix obtained from A by replacing row j by row i. Then row j of A' is $[a_{i1} \; a_{i2} \; \cdots \; a_{in}]$, and the cofactors are $C_{jk}(A') = C_{jk}(A) = C_{jk}$ for $k = 1, 2, \cdots, n$. Hence, (***) is the expansion of $det A'$ along row j, that is $u_{ij} = det A'$. But $det \, A' = 0$ because A' has two identical rows, so $u_{ij} = 0$ if $i \neq j$. This proves that $A(adj \, A) = (det A) \, I$. A similar argument shows that $(adj \, A) \, A = (det A) \, I$. This is what we wanted.

2.2. DETERMINANTS AND INVERSES

Exercises

1. Test the Product Theorem for determinants; that is create some matrices A and B of your choice; then calculate the matrix product AB; calculate the determinants $det(A)$, $det(B)$, $det(AB)$; and verify that $det(AB) = det(A)det(B)$. [ILAW: Explorations 2.2.1 and/or 2.1.1.]

2. Explain in your own words why the Product Theorem implies that $det(A^{-1}) = 1/det(A)$ whenever A is invertible.

3. Test the connection between the determinant of a matrix and that of its inverse; that is create some square matrix A of your choice, then calculate its inverse A^{-1} and the determinants $det(A)$ and $det(A^{-1})$, and verify that $det(A^{-1}) = 1/det(A)$. [ILAW: Explorations 2.2.1 and/or 2.1.1.]

4. What can be said about the value of $det A$ where A is an $n \times n$ such that:
   (a) $A^2 = I$ (b) $A^3 = I$ (c) $A^2 = 3A$ (d) $A = -A^T$
   (e) $A^2 + I = 0$ (f) $A^3 = A$ (g) $A^{-1} = A^T$.

5. If $A^k = 0$ for some $k \geq 1$, show that A has no inverse.

6. Assume $det A = -2$, $det B = 3$ and $det C = -1$, where A, B and C are $n \times n$ matrices. Compute:
   (a) $det[A^3 B^{-1} C^T B^2 A^{-1}]$ (b) $det[B^T A^{-1} B^{-1} C A^2 (C^{-1})^T]$

7. If A is 3×3 and $det(2A^{-1}) = 5 = det[A^2(B^T)^{-1}]$, find $det A$ and $det B$.

8. Find the adjoint and determinant of the following matrices, and verify your result by verifying that $A\,adj(A) = det(A)I$ in each case. Whenever possible, use this information to calculate the inverse of the matrix. [ILAW: Use Exploration 2.2.2.]

   (a) $A = \begin{bmatrix} 2 & -1 & 0 \\ 1 & 2 & -2 \\ -1 & -1 & 3 \end{bmatrix}$ (b) $A = \begin{bmatrix} -2 & 3 & 1 \\ 2 & -3 & 1 \\ 0 & 5 & 1 \end{bmatrix}$

   (c) $A = \begin{bmatrix} 8 & -11 & 0 & 4 \\ 11 & 2 & -2 & 9 \\ -1 & 12 & 3 & 3 \\ 12 & 3 & 8 & 0 \end{bmatrix}$ (d) $A = \begin{bmatrix} -2 & 3 & 6 & 2 \\ 7 & -3 & 3 & 9 \\ 3 & 5 & 7 & 12 \\ -5 & 10 & 21 & 20 \end{bmatrix}$

9. Explain what is the difference between the adjoint of a matrix, and the matrix of cofactors.

10. In each case either show that the statement is true or give an example showing that it is false. Throughout, A denotes a square matrix.
    (a) If $det A \neq 0$ and $AB = AC$, then $B = C$.
    (b) If $adj A$ exists then A is invertible.

(c) If $A^T = -A$ then $det A = -1$.

(d) $det(3A) = 3 det A$.

(e) $det(AB) = det(BA)$.

(f) If A is invertible and $adj(A) = A^{-1}$, then $det(A) = 1$.

(g) $det(I^n) = 1^n$.

(h) If $adj(A) = 0$, then $A = 0$.

11. In each case use Theorem 4(2) to find the (2,3)-entry of A^{-1}.

(a) $A = \begin{bmatrix} 3 & -1 & 2 \\ 5 & 5 & -2 \\ 1 & 2 & 3 \end{bmatrix}$ (b) $A = \begin{bmatrix} -2 & 3 & 6 \\ 7 & -3 & 3 \\ 3 & 5 & 7 \end{bmatrix}$

12. If A and B are invertible $n \times n$ matrices, evaluate:

(a) $det(A^{-1}BA)$ (b) $det(B^{-1}A^{-1}BA)$.

13. By computing the determinant, show that $A = \begin{bmatrix} 1 & a & b \\ -a & 1 & c \\ -b & -c & 1 \end{bmatrix}$ has an inverse for any a, b and c.

14. In each case: (1) Find the values of the number c such that A has an inverse, and (2) Find A^{-1} for those values of c.

(a) $A = \begin{bmatrix} 1 & c & 0 \\ 2 & 0 & c \\ c & -1 & 1 \end{bmatrix}$ (b) $A = \begin{bmatrix} 1 & -c & c \\ 1 & 1 & -1 \\ c & -c & 1 \end{bmatrix}$

15. In each case solve for y by Cramer's rule.

(a) $\begin{aligned} 2x - 5y + 7z &= 9 \\ -x + 4y + 2z &= -2 \\ 3x + 3y - 6z &= 5 \end{aligned}$

(b) $\begin{aligned} 3x - 2y + 4z &= 5 \\ 5x + 3y + z &= 8 \\ -2x + 6y + 7z &= -3 \end{aligned}$

16. If A, B and C are matrices such that $det A = 3$, $det B = -1$ and $det C = 2$, then compute the determinant of:

(a) $\begin{bmatrix} A & X & Y \\ 0 & B & Z \\ 0 & 0 & C \end{bmatrix}$ (b) $\begin{bmatrix} A & 0 & 0 \\ X & B & 0 \\ Z & Y & C \end{bmatrix}$

17. As in Example 4, use Theorem 3 to find $det A$.

(a) $A = \begin{bmatrix} 2 & 5 & 1 & -1 \\ -3 & -4 & 2 & 1 \\ 0 & 0 & -1 & -2 \\ 0 & 0 & -2 & 5 \end{bmatrix}$ (b) $A = \begin{bmatrix} 0 & -1 & 2 & 0 \\ 0 & 3 & 1 & 0 \\ -1 & 2 & -7 & 5 \\ 0 & -3 & 6 & 0 \end{bmatrix}$

2.3. DIAGONALIZATION AND EIGENVALUES

18. If A is $n \times n$, use Theorem 1 to show that $det(kA) = k^n \, detA$ for all scalars k (This is Theorem 3 §2.1). [*Hint*: First show that $kA = (kI)\, A$.]

19. (a) If $A = \begin{bmatrix} 0 & 1 \\ -1 & 0 \end{bmatrix}$, show that $A^2 = -I$.

    (b) Show that there is no 3×3 matrix A with the property that $A^2 = -I$.

20. If A is $n \times n$ and invertible, show that $det(adj\, A) = (detA)^{n-1}$.

21. If A is 3×3 and $detA = 2$, compute $det[-A^2(adj\, A)^{-1}]$.

22. If $detA = 2$, evaluate $det[A^{-1} + adj\, A]$.

23. (a) If $A = UB$ where $detU = 1$, show that $detA = detB$.

    (b) If A and B are invertible and $detA = detB$, show that $A = UB$ for some invertible matrix U such that $detU = 1$

2.3 DIAGONALIZATION AND EIGENVALUES

A central problem in the applications of linear algebra is to describe systems that are changing with time. As Example 1 below shows, this often comes down to finding a way to efficiently calculate powers $A, A^2, A^3, \cdots$ of a square matrix A. In this section we outline one method of doing this. The technique used is called diagonalization, and is one of the most important ideas in linear algebra.

2.3.1 Diagonalization

We begin with an example that motivates the method.

Example 1. We are interested in how the population of an endangered species of birds changes over time. Since the number of males is approximately equal to the number of females, we count only females. Each female remains a juvenile for one year and then becomes an adult. Only adults have offspring. After k years the female population can be expressed as a sum $a_k + j_k$ where

a_k denotes the number of adults after k years

j_k denotes the number of juveniles after k years

We assume that we know the initial values a_0 and j_0, and we are interested in computing a_k and j_k for $k = 1, 2, \cdots$. To do this we model the population growth as follows.

We assume that $j_{k+1} = 2a_k$, that is that the number of juveniles hatched in any year is on average twice the number of adult females alive the year before. We also assume that $\frac{1}{2}$ of the adult birds in any year survive to the next year, but only $\frac{1}{4}$ of the juveniles survive into adulthood. Hence the number a_{k+1} of adults in any year is given by $a_{k+1} = \frac{1}{2}a_k + \frac{1}{4}j_k$. Then the numbers a_k and j_k in successive years are related by the following equations:

$$\begin{aligned} a_{k+1} &= \tfrac{1}{2}a_k + \tfrac{1}{4}j_k \\ j_{k+1} &= 2a_k \end{aligned}$$

If we write $V_k = \begin{bmatrix} a_k \\ j_k \end{bmatrix}$ and $A = \begin{bmatrix} \frac{1}{2} & \frac{1}{4} \\ 2 & 0 \end{bmatrix}$, these equations take the simple matrix form

$$V_{k+1} = AV_k \quad \text{for each } k = 1, 2, \cdots. \tag{$*$}$$

The column V_k is called the *population profile* for the species. Since the initial profile $V_0 = \begin{bmatrix} a_0 \\ j_0 \end{bmatrix}$ is assumed to be known, (*) gives successively: $V_1 = AV_0$, $V_2 = AV_1 = A^2V_0$, $V_3 = AV_2 = A^3V_0$, etc. In general we get

$$V_k = A^k V_0 \quad \text{for each } k = 1, 2, \cdots.$$

Thus to find V_k (and hence a_k and j_k) it is sufficient to compute the powers A^k of the matrix A. In this section we look at an important method of doing this. This example is an illustration of a linear dynamical system, and we return to it in the following section. □

Direct computation of the powers A^k of a square matrix A is time-consuming for the large matrices that occur in practice. Furthermore, we are frequently only interested in approximating the long-term behavior of $V_k = A^k V_0$ as k increases, where V_0 is some fixed column. Hence we adopt an indirect method, currently used in industry, which lets us easily compute the powers A^k explicitly, and which also gives information about the long-term behavior. The idea is to first **diagonalize** the matrix A, that is to find an invertible matrix P such that

$$P^{-1}AP = D \quad \text{is a diagonal matrix.} \tag{$**$}$$

The reason this works is two-fold: First, the powers D^k of the diagonal matrix D are easy to compute (more about this later). Second, (**) enables us to compute powers A^k of the matrix A in terms of powers D^k of D. Indeed, we can solve (**) for A to get $A = PDP^{-1}$. Squaring this gives

$$A^2 = (PDP^{-1})(PDP^{-1}) = PD^2P^{-1}.$$

Using this we can compute A^3 as follows:

$$A^3 = AA^2 = (PDP^{-1})(PD^2P^{-1}) = PD^3P.$$

Continuing in this way (and noting that this works even if D is not diagonal), we obtain

Theorem 1. *If $A = PDP^{-1}$ then $A^k = PD^kP^{-1}$ for each $k = 1, 2, \cdots$.*

Hence computing A^k comes down to finding the invertible matrix P in equation (**). To do this it is necessary to first compute certain numbers (called eigenvalues) associated with the matrix A.

2.3. DIAGONALIZATION AND EIGENVALUES

2.3.2 Eigenvalues and Eigenvectors

If A is an $n \times n$ matrix, a number λ is called an **eigenvalue** of A if

$$AX = \lambda X \quad \text{for some column } X \neq 0.$$

Such a nonzero column X is called an **eigenvector** of A corresponding to the eigenvalue λ. Note that the condition $AX = \lambda X$ is automatically satisfied if $X = 0$, so the requirement that $X \neq 0$ is critical.

Example 2. Consider the 2×2 matrix $A = \begin{bmatrix} 5 & -2 \\ 4 & -1 \end{bmatrix}$. Then $\lambda = 3$ is an eigenvalue of A with eigenvector $X = \begin{bmatrix} 1 \\ 1 \end{bmatrix}$ because $X \neq 0$ and $AX = \begin{bmatrix} 5 & -2 \\ 4 & -1 \end{bmatrix} \begin{bmatrix} 1 \\ 1 \end{bmatrix} = \begin{bmatrix} 3 \\ 3 \end{bmatrix} = 3 \begin{bmatrix} 1 \\ 1 \end{bmatrix} = 3X$. $\square$

The matrix in Example 2 has another eigenvalue in addition to $\lambda = 3$. To find it, we develop the following general procedure which works for any $n \times n$ matrix A. By definition a number λ is an eigenvalue of A if and only if

$$AX = \lambda X \quad \text{for some } X \neq 0.$$

If I is the identity matrix of the same size as A, this is equivalent to asking that the homogeneous linear system

$$(\lambda I - A)X = 0$$

has a nontrivial solution $X \neq 0$. By Theorem 5 §1.5 this happens if and only if the matrix $\lambda I - A$ is not invertible, and this in turn holds if and only if the determinant of the coefficient matrix is zero:

$$det(\lambda I - A) = 0.$$

This last condition prompts the following definition: If A is an $n \times n$ matrix, the **characteristic polynomial** $c_A(x)$ of A is defined by

$$c_A(x) = det(xI - A).$$

Note that $c_A(x)$ is indeed a polynomial[5] in the variable x, and it has degree n if A is an $n \times n$ matrix (this is illustrated in the examples below).

The above discussion shows that a number λ is an eigenvalue of A if and only if $c_A(\lambda) = 0$, that is if and only if λ is a **root** of the characteristic polynomial $c_A(x)$. Moreover, the calculation reveals a method for finding the eigenvectors

---

[5] A **polynomial** is an expression of the form $a_0 + a_1 x + a_2 x^2 + \cdots + a_n x^n$ where the a_i are numbers and x is a variable. If $a_n \neq 0$, the number n is called the **degree** of the polynomial, and a_n is called the **leading coefficient**. Thus $2 - 5x + 2x^2 - x^4$ is a polynomial of degree 4 with leading coefficient -1. Polynomials are discussed in Appendix A.3.

X of A corresponding to λ. Indeed, X is determined by the condition that $(\lambda I - A)X = 0$, and so the eigenvectors X are the (nonzero) solutions to this system of homogeneous linear equations. We record these observations in

Theorem 2. *Let A be an $n \times n$ matrix.*

(1) *The eigenvalues λ of A are the roots of the characteristic polynomial $c_A(x)$ of A.*

(2) *The eigenvectors X corresponding to λ are the nonzero solutions to the homogeneous system*
$$(\lambda I - A)X = 0$$
of linear equations with $\lambda I - A$ as coefficient matrix. □

Observe that once a number λ is known to be an eigenvalue of A, finding the solutions to the equations $(\lambda I - A)X = 0$ (and hence finding the eigenvectors) is a routine application of Gaussian elimination. Here are two examples.

Example 3. *Find the characteristic polynomial of the matrix $A = \begin{bmatrix} 5 & -2 \\ 4 & -1 \end{bmatrix}$ discussed in Example 2, and then find all the eigenvalues and their eigenvectors.*

Solution. Since $xI - A = \begin{bmatrix} x & 0 \\ 0 & x \end{bmatrix} - \begin{bmatrix} 5 & -2 \\ 4 & -1 \end{bmatrix} = \begin{bmatrix} x-5 & 2 \\ -4 & x+1 \end{bmatrix}$, we get

$$\begin{aligned} c_A(x) &= \det \begin{bmatrix} x-5 & 2 \\ -4 & x+1 \end{bmatrix} \\ &= (x-5)(x+1) + 8 = x^2 - 4x + 3 = (x-3)(x-1). \end{aligned}$$

Hence the roots of $c_A(x)$ are $\lambda_1 = 3$ and $\lambda_2 = 1$, so these are the eigenvalues of A. Note that $\lambda_1 = 3$ was the eigenvalue mentioned in Example 2, but we have found a new one: $\lambda_2 = 1$.

To find the eigenvectors corresponding to $\lambda_2 = 1$, observe that in this case

$$\lambda_2 I - A = \begin{bmatrix} \lambda_2 - 5 & 2 \\ -4 & \lambda_2 + 1 \end{bmatrix} = \begin{bmatrix} -4 & 2 \\ -4 & 2 \end{bmatrix}$$

so the solutions to $(\lambda_2 I - A)X = 0$ are $X = t \begin{bmatrix} \frac{1}{2} \\ 1 \end{bmatrix}$ where t is any real number.

Hence the eigenvectors X corresponding to λ_2 are $X = t \begin{bmatrix} \frac{1}{2} \\ 1 \end{bmatrix}$ where $t \neq 0$ is arbitrary. A convenient choice is $X_2 = \begin{bmatrix} 1 \\ 2 \end{bmatrix}$, when $t = 2$. Similarly, $\lambda_1 = 3$ gives rise to the eigenvectors $X = t \begin{bmatrix} 1 \\ 1 \end{bmatrix}, t \neq 0$, which includes the observation in Example 2. □

2.3. DIAGONALIZATION AND EIGENVALUES

Note that there are *many* eigenvectors of a square matrix A associated with a given eigenvalue λ, in fact *every* nonzero solution X of $(\lambda I - A) X = 0$ is an eigenvector. Of course the eigenvalue λ is chosen so that there *must* be nonzero solutions.

Example 4. *Find the characteristic polynomial, eigenvalues and eigenvectors for* $A = \begin{bmatrix} 1 & 1 & 1 \\ 0 & 2 & -1 \\ 0 & -3 & 0 \end{bmatrix}$.

Solution. Here the characteristic polynomial is given by

$$\begin{aligned} c_A(x) &= \det \begin{bmatrix} x-1 & -1 & -1 \\ 0 & x-2 & 1 \\ 0 & 3 & x \end{bmatrix} \\ &= (x-1) \det \begin{bmatrix} x-2 & 1 \\ 3 & x \end{bmatrix} = (x-1)(x+1)(x-3) \end{aligned}$$

so the eigenvalues are $\lambda_1 = 1$, $\lambda_2 = -1$, and $\lambda_3 = 3$. To find the eigenvectors for $\lambda_1 = 1$, compute

$$\lambda_1 I - A = \begin{bmatrix} \lambda_1 - 1 & -1 & -1 \\ 0 & \lambda_1 - 2 & 1 \\ 0 & 3 & \lambda_1 \end{bmatrix} = \begin{bmatrix} 0 & -1 & -1 \\ 0 & -1 & 1 \\ 0 & 3 & 1 \end{bmatrix}.$$

We want the (nonzero) solutions to $(\lambda_1 I - A)X = 0$. The augmented matrix becomes

$$\begin{bmatrix} 0 & -1 & -1 & 0 \\ 0 & -1 & 1 & 0 \\ 0 & 3 & 1 & 0 \end{bmatrix} \to \begin{bmatrix} 0 & 1 & 1 & 0 \\ 0 & 0 & 2 & 0 \\ 0 & 0 & -2 & 0 \end{bmatrix} \to \begin{bmatrix} 0 & 1 & 0 & 0 \\ 0 & 0 & 1 & 0 \\ 0 & 0 & 0 & 0 \end{bmatrix}$$

using row operations. Hence the solutions X to $(\lambda_1 I - A)X = 0$ are $X = t \begin{bmatrix} 1 \\ 0 \\ 0 \end{bmatrix}$, where $t \neq 0$ is arbitrary, so we can use the basic solution $X_1 = \begin{bmatrix} 1 \\ 0 \\ 0 \end{bmatrix}$ as an eigenvector corresponding to $\lambda_1 = 1$. Similarly eigenvectors corresponding to $\lambda_2 = -1$, and $\lambda_3 = 3$ are $X_2 = \begin{bmatrix} -2 \\ 1 \\ 3 \end{bmatrix}$ and $X_3 = \begin{bmatrix} 0 \\ 1 \\ -1 \end{bmatrix}$, respectively, as the reader can verify. □

There are several comments that must be made at this point:

Remark 1. The eigenvalues of a real matrix need not be real numbers (as in the above examples). For example, the characteristic polynomial of the matrix $A = \begin{bmatrix} 0 & -1 \\ 1 & 0 \end{bmatrix}$ is $x^2 + 1$, so the eigenvalues of A are the nonreal complex roots $\lambda = i$ and $\lambda = -i$. We return to this in Section 2.5.

Remark 2. An $n \times n$ matrix always has n (possibly complex) eigenvalues, but they may not be distinct (as they are in the above examples). For example, the matrix $A = \begin{bmatrix} 1 & 1 \\ 0 & 1 \end{bmatrix}$ has characteristic polynomial $c_A(x) = (x-1)^2$, so there is only one eigenvalue $\lambda = 1$. However λ is a *double root* of $c_A(x)$, and we say that λ has *multiplicity* 2. We shall have more to say about such repeated eigenvalues later (Theorem 5).

Remark 3. In practice, the eigenvalues of a square matrix A are usually not computed as the roots of the characteristic polynomial. There are iterative numerical methods (for example the power method, or the QR-algorithm) that are more efficient for large matrices. These algorithms (and others) are described in books on computational linear algebra.

2.3.3 Diagonalization

An $n \times n$ matrix D is called a **diagonal matrix** if all its entries off the main diagonal are zero, that is if D has the form

$$D = \begin{bmatrix} \lambda_1 & 0 & \cdots & 0 \\ 0 & \lambda_2 & \cdots & 0 \\ \vdots & \vdots & \ddots & \vdots \\ 0 & 0 & \cdots & \lambda_n \end{bmatrix} = diag(\lambda_1, \lambda_2, \cdots, \lambda_n)$$

where $\lambda_1, \lambda_2, \cdots, \lambda_n$ are numbers. We use the notation $D = diag(\lambda_1, \lambda_2, \cdots, \lambda_n)$ for convenience. Calculations with diagonal matrices are very easy. For example, if $n = 2$ we have

$$\begin{bmatrix} \lambda_1 & 0 \\ 0 & \lambda_2 \end{bmatrix} \begin{bmatrix} \mu_1 & 0 \\ 0 & \mu_2 \end{bmatrix} = \begin{bmatrix} \lambda_1\mu_1 & 0 \\ 0 & \lambda_2\mu_2 \end{bmatrix}$$

and

$$\begin{bmatrix} \lambda_1 & 0 \\ 0 & \lambda_2 \end{bmatrix} + \begin{bmatrix} \mu_1 & 0 \\ 0 & \mu_2 \end{bmatrix} = \begin{bmatrix} \lambda_1 + \mu_1 & 0 \\ 0 & \lambda_2 + \mu_2 \end{bmatrix}.$$

Thus each operation is carried out by doing the same operation on corresponding diagonal entries.

In general, if $D = diag(\lambda_1, \lambda_2, \cdots, \lambda_n)$ and $E = diag(\mu_1, \mu_2, \cdots, \mu_n)$ are two diagonal matrices, their product DE and sum $D+E$ are again diagonal, and are obtained as in the 2×2 case by doing the same operations to the corresponding diagonal elements:

$$DE = diag(\lambda_1\mu_1, \lambda_2\mu_2, \cdots, \lambda_n\mu_n)$$

$$D + E = diag(\lambda_1 + \mu_1, \lambda_2 + \mu_2, \cdots, \lambda_n + \mu_n)$$

The simplicity of these formulas is one reason the following notion is so important.

2.3. DIAGONALIZATION AND EIGENVALUES

A square $n \times n$ matrix A is called **diagonalizable** if

$$P^{-1}AP \quad \text{is diagonal for some invertible } n \times n \text{ matrix } P.$$

In this case the invertible matrix P is called a **diagonalizing matrix** for A. To discover when such a matrix P exists, we let $X_1, X_2, \cdots, X_n$ denote the columns of P and look for ways to determine when such X_i exist and how to compute them. To this end write P in terms of its columns as follows:

$$P = [\ X_1 \quad X_2 \quad \cdots \quad X_n\ ].$$

The condition $P^{-1}AP = D$ holds for some diagonal matrix D if and only if

$$AP = PD$$

and we proceed by writing each side of this equation in terms of its columns. If we write $D = diag(\lambda_1, \lambda_2, \cdots, \lambda_n)$, where the λ_i are numbers to be determined, the equation $AP = PD$ becomes

$$A[\ X_1 \quad X_2 \quad \cdots \quad X_n\ ] = [\ X_1 \quad X_2 \quad \cdots \quad X_n\ ] \begin{bmatrix} \lambda_1 & 0 & \cdots & 0 \\ 0 & \lambda_2 & \cdots & 0 \\ \vdots & \vdots & \ddots & \vdots \\ 0 & 0 & \cdots & \lambda_n \end{bmatrix}.$$

By the definition of matrix multiplication each side simplifies as follows

$$[\ AX_1 \quad AX_2 \quad \cdots \quad AX_n\ ] = [\ \lambda_1 X_1 \quad \lambda_2 X_2 \quad \cdots \quad \lambda_n X_n\ ].$$

Comparing columns shows that $AX_1 = \lambda_1 X_1$, $AX_2 = \lambda_2 X_2$, $\cdots$ etc., and so

$$P^{-1}AP = D \quad \text{if and only if} \quad AX_i = \lambda_i X_i \quad \text{for each } i.$$

In other words, $P^{-1}AP = D$ holds if and only if the diagonal entries of D are eigenvalues of A, and the columns of P are corresponding eigenvectors. This proves the following fundamental result.

Theorem 3. *Let A be an $n \times n$ matrix.*

(1) *A is diagonalizable if and only if it has eigenvectors $X_1, X_2, \cdots, X_n$ such that the matrix $P = [\ X_1 \quad X_2 \quad \cdots \quad X_n\ ]$ is invertible.*

(2) *When this is the case, $P^{-1}AP = diag(\lambda_1, \lambda_2, \cdots, \lambda_n)$ where, for each i, λ_i is the eigenvalue of A corresponding to X_i.*

Example 5. *Diagonalize the matrix $A = \begin{bmatrix} 1 & 1 & 1 \\ 0 & 2 & -1 \\ 0 & -3 & 0 \end{bmatrix}$ discussed in Example 4.*

Solution. By Example 4, the eigenvalues of A are $\lambda_1 = 1$, $\lambda_2 = -1$ and $\lambda_3 = 3$, with corresponding eigenvectors $X_1 = \begin{bmatrix} 1 \\ 0 \\ 0 \end{bmatrix}$, $X_2 = \begin{bmatrix} -2 \\ 1 \\ 3 \end{bmatrix}$ and $X_3 = \begin{bmatrix} 0 \\ 1 \\ -1 \end{bmatrix}$. Since the matrix $P = [\, X_1 \; X_2 \; X_3 \,] = \begin{bmatrix} 1 & -2 & 0 \\ 0 & 1 & 1 \\ 0 & 3 & -1 \end{bmatrix}$ is invertible, Theorem 3 guarantees that $P^{-1}AP = \begin{bmatrix} \lambda_1 & 0 & 0 \\ 0 & \lambda_2 & 0 \\ 0 & 0 & \lambda_3 \end{bmatrix} = \begin{bmatrix} 1 & 0 & 0 \\ 0 & -1 & 0 \\ 0 & 0 & 3 \end{bmatrix}$.
The reader can verify this directly. $\square$

Remark 4. In Example 5, suppose we let $Q = [\, X_2 \; X_1 \; X_3 \,]$ be the matrix formed from the eigenvectors X_1, X_2, and X_3 of A, but in a *different order* than that used to form P. Then $Q^{-1}AQ = D = diag(\lambda_2, \lambda_1, \lambda_3)$ is diagonal by Theorem 3, but the eigenvalues are in the *new* order. Hence we can choose the diagonalizing matrix P so that the eigenvalues λ_i appear in any order we want along the main diagonal of D.

Remark 5. On the other hand, we can multiply any column of P by a nonzero constant without changing D (because the resulting column is still an eigenvector). This is sometimes convenient for simplifying the form of P.

Unfortunately, not every square matrix A is diagonalizable. However, if the eigenvalues of A are all distinct (as in Example 5), the matrix P of eigenvectors *must* be invertible (and so A is diagonalizable). We state this result in the next theorem; the proof is deferred to Section 4.7.

Theorem 4. *If A is an $n \times n$ matrix with n distinct eigenvalues, then A is diagonalizable.*

On the other hand, many matrices with repeated eigenvalues *are* diagonalizable. The following example exhibits such a matrix.

Example 6. *Diagonalize the matrix* $A = \begin{bmatrix} 0 & 1 & 1 \\ 1 & 0 & 1 \\ 1 & 1 & 0 \end{bmatrix}$.

Solution. To compute the characteristic polynomial of A first add rows 2 and 3 to row 1:

$$\begin{aligned} c_A(x) &= det \begin{bmatrix} x & -1 & -1 \\ -1 & x & -1 \\ -1 & -1 & x \end{bmatrix} = det \begin{bmatrix} x-2 & x-2 & x-2 \\ -1 & x & -1 \\ -1 & -1 & x \end{bmatrix} \\ &= det \begin{bmatrix} x-2 & 0 & 0 \\ -1 & x+1 & 0 \\ -1 & 0 & x+1 \end{bmatrix} = (x-2)(x+1)^2. \end{aligned}$$

2.3. DIAGONALIZATION AND EIGENVALUES

Hence the eigenvalues are $\lambda_1 = 2$ and $\lambda_2 = -1$, with λ_2 repeated twice (we say that λ_2 has *multiplicity* two). Thus Theorem 4 does not apply. However A is diagonalizable. For $\lambda_1 = 2$ the system of equations $(\lambda_1 I - A)X = 0$ has general solution $X = t \begin{bmatrix} 1 \\ 1 \\ 1 \end{bmatrix}$, so the basic solution $X_1 = \begin{bmatrix} 1 \\ 1 \\ 1 \end{bmatrix}$ is an eigenvector corresponding to $\lambda_1 = 2$. Turning to the repeated eigenvalue $\lambda_2 = -1$, we must solve $(\lambda_2 I - A)X = 0$. By Gaussian Elimination, the general solution is $X = s \begin{bmatrix} -1 \\ 1 \\ 0 \end{bmatrix} + t \begin{bmatrix} -1 \\ 0 \\ 1 \end{bmatrix}$ where s and t are arbitrary. Hence the basic solutions

$X_2 = \begin{bmatrix} -1 \\ 1 \\ 0 \end{bmatrix}$ and $Y_2 = \begin{bmatrix} -1 \\ 0 \\ 1 \end{bmatrix}$ are *both* eigenvectors of A corresponding to $\lambda_2 = -1$, and they arise naturally from the Gaussian Algorithm. If we take

$P = [\, X_1 \; X_2 \; Y_2 \,] = \begin{bmatrix} 1 & -1 & -1 \\ 1 & 1 & 0 \\ 1 & 0 & 1 \end{bmatrix}$, we find that P is invertible. Hence $P^{-1}AP = diag(2, -1, -1)$ by Theorem 3. □

Example 6 is typical of the situation for any diagonalizable matrix. To describe the general case, we need some terminology. An eigenvalue λ of a square matrix A is said to have **multiplicity** m if it occurs m times as a root of the characteristic polynomial $c_A(x)$. When the homogeneous system $(\lambda I - A)X = 0$ is solved, any set of basic solutions is called a set of **basic eigenvectors** corresponding to λ. Recall that the number of basic eigenvectors equals the number of parameters involved in the solution of the system $(\lambda I - A)X = 0$.

Thus the eigenvalue $\lambda_2 = -1$ in Example 6 has multiplicity 2. In that case, the Gaussian Algorithm yields *two* basic eigenvectors corresponding to $\lambda_2 = -1$, arising from the fact that the general solution to the system $(\lambda_2 I - A)X = 0$ involves *two* parameters. This works in general: A matrix is diagonalizable if and only if every eigenvalue of multiplicity m yields m basic eigenvectors. This is the content of the following theorem; the proof is given in Section 4.7.

Theorem 5. *A square matrix A is diagonalizable if and only if it satisfies the following condition:*

> *The multiplicity of every eigenvalue λ of A equals the number of basic eigenvectors corresponding to λ, which is the number of parameters in the solution of $(\lambda I - A)X = 0$.*

In this case the basic solutions of the system $(\lambda I - A)X = 0$ become columns in the invertible diagonalizing matrix P such that $P^{-1}AP$ is diagonal.

In fact, it turns out that the number of basic eigenvectors corresponding to an eigenvalue λ can *never exceed* the multiplicity of λ. Thus a square matrix is *not* diagonalizable if and only if it has an eigenvalue λ of multiplicity m for which the system $(\lambda I - A)X = 0$ has *fewer* than m basic solutions. For this reason, non-diagonalizable matrices are sometimes called **defective**.

The following procedure summarizes Theorem 5.

Diagonalization Algorithm. *To diagonalize an $n \times n$ matrix A:*

Step 1. *Find the distinct eigenvalues λ of A.*

Step 2. *Compute the basic eigenvectors corresponding to each of these eigenvalues λ as the basic solutions of the homogeneous system $(\lambda I - A)X = 0$.*

Step 3. *The matrix A is diagonalizable if and only if there are n basic eigenvectors in all.*

Step 4. *If A is diagonalizable, the $n \times n$ matrix P with these basic eigenvectors as columns is a diagonalizing matrix for A, that is P is invertible and $P^{-1}AP$ is diagonal.*

Remark 6. The diagonalization algorithm is valid even if the eigenvalues are complex numbers that are not real. In this case the eigenvectors will be also be complex, but we will not pursue this here.

Example 7. *Show that the matrix $A = \begin{bmatrix} 1 & 1 \\ 0 & 1 \end{bmatrix}$ is not diagonalizable.*

Solution 1. The characteristic polynomial of A is $c_A(x) = (x-1)^2$, so A has only one eigenvalue $\lambda_1 = 1$, which is of multiplicity 2. Here the system of equations $(\lambda_1 I - A)X = 0$ has general solution $X = t \begin{bmatrix} 1 \\ 0 \end{bmatrix}$, so there is only one basic solution $\begin{bmatrix} 1 \\ 0 \end{bmatrix}$. Hence A is not diagonalizable by Theorem 5.

Solution 2. If A were diagonalizable, there would be an invertible matrix P such that $P^{-1}AP = D$ is diagonal. By Theorem 3, the diagonal entries of D must be eigenvalues of A, so $D = I$ because 1 is the only eigenvalue of A. Hence $P^{-1}AP = I$, so $A = PIP^{-1} = I$, a contradiction. Hence A is not diagonalizable. □

2.3.4 Similar Matrices

If A and B are $n \times n$ matrices, we say that A and B are **similar**, and write $A \sim B$, if $B = P^{-1}AP$ for some invertible matrix P. In particular, a matrix A is called diagonalizable if it is similar to a diagonal matrix. The language of similarity is used throughout linear algebra.

If $A \sim B$, then necessarily $B \sim A$. To see why, suppose that $B = P^{-1}AP$. Then $A = PBP^{-1} = Q^{-1}BQ$ where $Q = P^{-1}$. Since Q is invertible, this proves the second of the following properties of similarity (the others are Exercise 18):

2.3. DIAGONALIZATION AND EIGENVALUES

1. $A \sim A$ for all square matrices A.

2. If $A \sim B$ then $B \sim A$. (*)

3. If $A \sim B$ and $B \sim C$ then $A \sim C$.

These properties are often expressed by saying that the similarity relation $\sim$ is an *equivalence relation* on the set of $n \times n$ matrices. Moreover, similarity is compatible with inverses, transposes and powers in the following sense:

$$\text{If } A \sim B \text{ then } \begin{cases} A^{-1} \sim B^{-1} \\ A^T \sim B^T \\ A^k \sim B^k \text{ for all } k \geq 0 \end{cases} \quad (**)$$

The proofs are routine matrix computations (Exercise 18). In addition, similar matrices share many properties, some of which are collected in the next theorem for reference.

Theorem 6. *If A and B are similar $n \times n$ matrices, then:*

(1) $det A = det B$.

(2) $c_A(x) = c_B(x)$.

(3) *A and B have the same eigenvalues.*

Proof. Let $B = P^{-1}AP$ for some invertible matrix P.
(1). $det B = det(P^{-1}) det A det P = det A$ because $det(P^{-1}) = 1/det P$.
(2). One verifies that $xI - B = P^{-1}(xI - A)P$. Hence (1) gives

$$c_B(x) = det(xI - B) = det\{P^{-1}(xI - A)P\} = det(xI - A) = c_A(x).$$

(3). The eigenvalues of a matrix are the roots of its characteristic polynomial, so (2) applies. □

The class of diagonalizable matrices is well behaved because these matrices are so closely related to diagonal matrices. Here are two illustrations.

Example 8. *If A is similar to B and either A or B is diagonalizable, show that the other is also diagonalizable.*

Solution. We have $A \sim B$. Suppose that A is diagonalizable, say $A \sim D$ where D is diagonal. Since $B \sim A$ by (2) of (*), we have $B \sim A$ and $A \sim D$. Hence $B \sim D$ by (3) of (*), so B is diagonalizable too. A similar argument works if we assume instead that B is diagonalizable. □

Example 9. *If A is diagonalizable, show that A^T, A^{-1} (if it exists), and A^k (for each $k \geq 1$) are all diagonalizable.*

Solution. Since A is diagonalizable, we have $A \sim D$ where D is a diagonal matrix. Hence (**) gives $A^T \sim D^T$, $A^{-1} \sim D^{-1}$, and $A^k \sim D^k$, and the result follows because $D^T = D$, D^{-1} and D^k are all diagonal. $\square$

Suppose that A is diagonalizable, say $P^{-1}AP = D$ is diagonal. Theorem 6 shows that A and D have the same eigenvalues. But the eigenvalues of the diagonal matrix D are just the main diagonal entries (verify) so the diagonal entries of D must be the eigenvalues of A in some order. This means that A inherits many properties of D, and hence of its eigenvalues. The following example illustrates how this happens.

Example 10. *Let A be a diagonalizable 2×2 matrix. If $\lambda^4 = 5\lambda$ for each eigenvalue λ of A, show that $A^4 = 5A$.*

Solution. Let $P^{-1}AP = D = \begin{bmatrix} \lambda_1 & 0 \\ 0 & \lambda_2 \end{bmatrix}$ where λ_1 and λ_2 are the eigenvalues of A and P is invertible. Then

$$D^4 = \begin{bmatrix} \lambda_1^4 & 0 \\ 0 & \lambda_2^4 \end{bmatrix} = \begin{bmatrix} 5\lambda_1 & 0 \\ 0 & 5\lambda_2 \end{bmatrix} = 5 \begin{bmatrix} \lambda_1 & 0 \\ 0 & \lambda_2 \end{bmatrix} = 5D.$$

Since $A = PDP^{-1}$ (obtained by solving $P^{-1}AP = D$ for A) this gives

$$A^4 = (PDP^{-1})^4 = PD^4P^{-1} = P(5D)P^{-1} = 5(PDP^{-1}) = 5A$$

using Theorem 1. $\square$

The result in Example 10 can be reformulated as follows: Write $p(x) = x^4 - 5x$. Then the example asserts that, if $p(\lambda) = 0$ for every eigenvalue λ of A then $p(A) = 0$. This holds much more generally.

If $p(x) = a_0 + a_1x + a_2x^2 + \cdots + a_kx^k$ is any polynomial, the **evaluation** of $p(x)$ at the (square) matrix A is defined to be

$$p(A) = a_0I + a_1A + a_2A^2 + \cdots + a_kA^k$$

where I is the identity matrix of the same size as A. Then the method used in Example 10 shows that A is diagonalizable, and if if $p(x)$ is any polynomial such that $p(\lambda) = 0$ for all eigenvalues λ of A, then necessarily $p(A) = 0$. In particular, the characteristic polynomial $c_A(x)$ certainly has the property that $c_A(\lambda) = 0$ for each eigenvalue of A, so we must have that $c_A(A) = 0$. In other words, this shows that $c_A(A) = 0$ for every diagonalizable matrix A. In fact, this holds for *every* square matrix A, and in full generality is called the

Cayley-Hamilton Theorem. *If A is a square matrix then $c_A(A) = 0$.*

The proof is given in Section 5.6.

2.3. DIAGONALIZATION AND EIGENVALUES

Exercises

1. Consider the matrix $A = \begin{bmatrix} 1 & 1 & 1 \\ 1 & 1 & -1 \\ 1 & 1 & 1 \end{bmatrix}$.

   (a) Directly calculate the matrix power A^{20}.

   (b) Use the fact that A is diagonalizable to compute A^{20}; that is there are matrices P and D such that $P^{-1}AP = D$ where

   $$P^{-1}AP = \begin{bmatrix} -1 & 0 & 1 \\ 1 & 1 & 0 \\ -1 & -1 & 1 \end{bmatrix} \begin{bmatrix} 1 & 1 & 1 \\ 1 & 1 & -1 \\ 1 & 1 & 1 \end{bmatrix} \begin{bmatrix} -1 & 1 & 1 \\ 1 & 0 & -1 \\ 0 & 1 & 1 \end{bmatrix}$$

   $$= \begin{bmatrix} 0 & 0 & 0 \\ 0 & 2 & 0 \\ 0 & 0 & 1 \end{bmatrix} = D.$$

   (c) Explain the benefits of diagonalizing a matrix to compute its powers.

2. In each case find the characteristic polynomial, eigenvalues and eigenvectors, and if possible find an invertible matrix P such that $P^{-1}AP = D$.

   (a) $A = \begin{bmatrix} 1 & 3 \\ 2 & 2 \end{bmatrix}$
   (b) $A = \begin{bmatrix} 2 & -1 \\ -4 & -1 \end{bmatrix}$

   (c) $A = \begin{bmatrix} 7 & 0 & 5 \\ 0 & 5 & 0 \\ -4 & 0 & -2 \end{bmatrix}$
   (d) $A = \begin{bmatrix} 3 & -4 & 2 \\ 1 & -2 & 2 \\ 1 & -5 & 5 \end{bmatrix}$

   (e) $A = \begin{bmatrix} 1 & 1 & 0 \\ 1 & 1 & -1 \\ 0 & 1 & 1 \end{bmatrix}$
   (f) $A = \begin{bmatrix} 7 & 4 & -16 \\ 2 & 5 & -8 \\ 2 & 2 & -5 \end{bmatrix}$

3. Write your own short description of how to find eigenvalues of a matrix A, the corresponding eigenvectors, and an invertible matrix P such that $P^{-1}AP$ is a diagonal matrix.

4. If $P^{-1}AP = D$ is a diagonal matrix, what are the diagonal entries of D and the columns of P?

5. (a) Show that $A = \begin{bmatrix} 1 & 3 \\ -3 & -5 \end{bmatrix}$ is not diagonalizable.

   (b) Show that $A = \begin{bmatrix} 1 & c \\ 0 & 1 \end{bmatrix}$ is not diagonalizable unless $c = 0$.

6. In each case determine whether A is diagonalizable. Give reasons for your answer.

(a) $A = \begin{bmatrix} 5 & -2 & 2 \\ 4 & 1 & 4 \\ 4 & -4 & 9 \end{bmatrix}$ (b) $A = \begin{bmatrix} 2 & 2 & -1 \\ 1 & 1 & 0 \\ 1 & -2 & -2 \end{bmatrix}$

7. In each case either show that the statement is true or give an example showing that it is false. Throughout A represents a square matrix.

   (a) If A has real eigenvalues then it is diagonalizable.

   (b) If A is diagonalizable then it has distinct eigenvalues.

   (c) If all of the eigenvalues of A are real and distinct then A is diagonalizable.

   (d) If A is diagonalizable then its transpose A^T is also diagonalizable.

   (e) Every invertible matrix is diagonalizable.

   (f) Every diagonalizable matrix is symmetric.

8. If λ is an eigenvalue of A, and if α is a number, show that $\lambda - \alpha$ is an eigenvalue of $A_1 = A - \alpha I$. How do the eigenvectors compare?

9. Find the eigenvalues of $A = \begin{bmatrix} \cos\theta & -\sin\theta \\ \sin\theta & \cos\theta \end{bmatrix}$.

10. If X and Y are eigenvectors of A corresponding to λ, show that the same is true of $X + Y$ and aX for any number a (provided that $X + Y$ and aX are not zero).

11. If λ is an eigenvalue of A, show that:

    (a) $k\lambda$ is an eigenvalue of kA for each real number k.

    (b) λ^2 is an eigenvalue of A^2.

    (c) $3 - 2\lambda + 5\lambda^3$ is an eigenvalue of $3I - 2A + 5A^3$.

12. Let A denote an invertible matrix.

    (a) If λ is an eigenvalue of A, show that $\lambda \neq 0$, and that $\frac{1}{\lambda}$ is an eigenvalue of A^{-1}.

    (b) Show that every eigenvalue μ of A^{-1} has the form $\mu = \frac{1}{\lambda}$ where λ is some eigenvalue of A.

13. If $D = diag(\lambda_1, \lambda_2, \cdots, \lambda_n)$ and $E = diag(\mu_1, \mu_2, \cdots, \mu_n)$, show that

    (a) $D + E = diag(\lambda_1 + \mu_1, \lambda_2 + \mu_2, \cdots, \lambda_n + \mu_n)$

    (b) $DE = diag(\lambda_1\mu_1, \lambda_2\mu_2, \cdots, \lambda_n\mu_n)$

    (c) $D^k = diag(\lambda_1^k, \lambda_2^k, \cdots, \lambda_n^k)$ for each integer $k \geq 1$.

14. (a) Show that A and its transpose A^T have the same characteristic polynomial. [Hint: $xI = (xI)^T$]

    (b) If A is diagonalizable, show that A^T is diagonalizable.

15. If A is diagonalizable with eigenvalues $\lambda_1, \lambda_2, \cdots, \lambda_n$, show that $\det A = \lambda_1 \lambda_2 \cdots \lambda_n$.

16. (a) If $A^{-1} = A$ and λ is an eigenvalue of A, show that $\lambda = \pm 1$.

    (b) If A is diagonalizable and every eigenvalue is ± 1, show that $A^{-1} = A$.

17. (a) If $A^2 = A$ and λ is an eigenvalue of A, show that $\lambda = 0, 1$.

    (b) If A is diagonalizable and every eigenvalue is 0 or 1, show that $A^2 = A$.

18. Square matrices A and B are called **similar** (denoted $A \sim B$) if $B = P^{-1}AP$ for some invertible matrix P. Show that:

    (a) If $B = QAQ^{-1}$ for some invertible matrix Q, then $A \sim B$.

    (b) $A \sim A$ for all square matrices A.

    (c) If $A \sim B$ then $B \sim A$.

    (d) If $A \sim B$ and $B \sim C$ then $A \sim C$.

    (e) If $A \sim B$ then $\det A = \det B$.

    (f) If $A \sim B$ then $A^{-1} \sim B^{-1}$.

    (g) If $A \sim B$ then $A^k \sim B^k$ for all integers $k \geq 1$.

    (h) If $A \sim B$ then $A^T \sim B^T$.

19. If $A = \begin{bmatrix} a & b \\ c & d \end{bmatrix}$, show that:

    (a) $c_A(x) = x^2 - (\operatorname{tr} A)x + \det A$ where $\operatorname{tr} A = a + d$ is called the **trace** of A.

    (b) The eigenvalues of A are $\lambda = \frac{1}{2}[(a+d) \pm \sqrt{(a-d)^2 + 4bc}]$.

20. (a) If D is diagonal with distinct eigenvalues, and if $CD = DC$, show that C is also diagonal.

    (b) If A is diagonalizable with distinct eigenvalues, and if $BA = AB$, show that B is also diagonalizable.

2.4 LINEAR DYNAMICAL SYSTEMS

We began Section 2.3 with an example from ecology which models the evolution of the population of a species of birds as time goes on. As promised, we now complete the example and extend it to other situations.

2.4.1 Linear Dynamical Systems

The bird population was described by computing the female population profile $V_k = \begin{bmatrix} a_k \\ j_k \end{bmatrix}$ of the species, where a_k and j_k represent the number of adult and juvenile females present k years after the initial values a_0 and j_0 were observed. The model assumes that these numbers are related by the following equations:

$$\begin{aligned} a_{k+1} &= \tfrac{1}{2}a_k + \tfrac{1}{4}j_k \\ j_{k+1} &= 2a_k \end{aligned}$$

If we write $A = \begin{bmatrix} \frac{1}{2} & \frac{1}{4} \\ 2 & 0 \end{bmatrix}$ this system takes the following matrix form

$$V_{k+1} = AV_k \text{ for each } k = 1, 2, \cdots.$$

Hence $V_1 = AV_0$, whence $V_2 = AV_1 = A^2V_0$, and in general

$$V_k = A^k V_0 \text{ for } k = 1, 2, \cdots.$$

We can now use our diagonalization techniques to determine the population profile V_k for all values of k in terms of the initial values.

Example 1. *Assuming that the initial values were $a_0 = 100$ adult females and $j_0 = 40$ juvenile females, compute a_k and j_k for $k = 1, 2, \cdots$.*

Solution. The characteristic polynomial of the matrix $A = \begin{bmatrix} \frac{1}{2} & \frac{1}{4} \\ 2 & 0 \end{bmatrix}$ is $c_A(x) = x^2 - \tfrac{1}{2}x - \tfrac{1}{2} = (x-1)(x+\tfrac{1}{2})$, so the eigenvalues are $\lambda_1 = 1$ and $\lambda_2 = -\tfrac{1}{2}$ with corresponding eigenvectors $X_1 = \begin{bmatrix} 1 \\ 2 \end{bmatrix}$ and $X_2 = \begin{bmatrix} -1 \\ 4 \end{bmatrix}$ respectively (as the reader can verify). Hence a diagonalizing matrix is $P = \begin{bmatrix} 1 & -1 \\ 2 & 4 \end{bmatrix}$ and we obtain

$$P^{-1}AP = D \quad \text{where} \quad D = \begin{bmatrix} 1 & 0 \\ 0 & -\frac{1}{2} \end{bmatrix}.$$

This gives $A = PDP^{-1}$ so, for each $k \geq 0$, we can compute A^k explicitly:

$$\begin{aligned} A^k &= PD^kP^{-1} = \begin{bmatrix} 1 & -1 \\ 2 & 4 \end{bmatrix} \begin{bmatrix} 1 & 0 \\ 0 & (-\frac{1}{2})^k \end{bmatrix} \tfrac{1}{6} \begin{bmatrix} 4 & 1 \\ -2 & 1 \end{bmatrix} \\ &= \tfrac{1}{6} \begin{bmatrix} 4 + 2(-\frac{1}{2})^k & 1 - (-\frac{1}{2})^k \\ 8 - 8(-\frac{1}{2})^k & 2 + 4(-\frac{1}{2})^k \end{bmatrix}. \end{aligned}$$

Hence we obtain

$$\begin{aligned} \begin{bmatrix} a_k \\ j_k \end{bmatrix} &= V_k = A^k V_0 = \tfrac{1}{6} \begin{bmatrix} 4 + 2(-\frac{1}{2})^k & 1 - (-\frac{1}{2})^k \\ 8 - 8(-\frac{1}{2})^k & 2 + 4(-\frac{1}{2})^k \end{bmatrix} \begin{bmatrix} 100 \\ 40 \end{bmatrix} \\ &= \tfrac{1}{6} \begin{bmatrix} 440 + 160(-\frac{1}{2})^k \\ 880 - 640(-\frac{1}{2})^k \end{bmatrix}. \end{aligned}$$

Equating top and bottom entries we obtain exact formulas for a_k and j_k:

2.4. LINEAR DYNAMICAL SYSTEMS

$$a_k = \tfrac{220}{3} + \tfrac{80}{3}(-\tfrac{1}{2})^k \quad \text{and} \quad j_k = \tfrac{440}{3} - \tfrac{320}{3}(-\tfrac{1}{2})^k \quad \text{for } k = 1, 2, \cdots.$$

In practice, the exact values of a_k and j_k are usually not required. What is needed is a measure of how these numbers behave for large values of k. This is easy to obtain here. Since $(-\tfrac{1}{2})^k$ is nearly zero for large k, we have the following approximate values

$$a_k \approx \frac{220}{3} \quad \text{and} \quad j_k \approx \frac{440}{3} \quad \text{if } k \text{ is large.}$$

Hence, in the long term, the female population stabilizes with approximately twice as many juveniles as adults. □

The population model in Example 1 is an example of a **dynamical system**, that is a sequence of columns $V_0, V_1, V_2, \cdots$ where the initial column V_0 is known, and the other columns are determined by the condition

$$V_{k+1} = AV_k \quad \text{for each } k \geq 0$$

where A is a square matrix.[6] The condition $V_{k+1} = AV_k$ is called a **matrix recurrence** for the columns V_k. While this example is from ecology, dynamical systems arise in many parts of science and engineering, and also in other areas such as economics.

Since V_0 is known, the matrix recurrence $V_{k+1} = AV_k$ gives successively

$$V_1 = AV_0,$$
$$V_2 = AV_1 = A(AV_0) = A^2 V_0,$$
$$V_3 = AV_2 = A(A^2 V_0) = A^3 V_0,$$
$$\vdots$$

Continuing in this way, we get

$$V_k = A^k V_0 \text{ for each } k = 1, 2, \cdots. \quad (*)$$

Hence we can find the columns V_k if we can compute the powers A^k of the matrix A. As we have seen, this can be done if A is diagonalizable; indeed (*) can be used (as in Example 1) to give a nice "formula" for the columns V_k in this case.

Assume that A is diagonalizable with eigenvalues $\lambda_1, \lambda_2, \cdots, \lambda_n$ and corresponding eigenvectors $X_1, X_2, \cdots, X_n$. If $P = \begin{bmatrix} X_1 & X_2 & \cdots & X_n \end{bmatrix}$ is the diagonalizing matrix with the X_i as columns, then P is invertible and

$$P^{-1}AP = D = diag(\lambda_1, \lambda_2, \cdots, \lambda_n)$$

by Theorem 3 §2.3. Hence $A = PDP^{-1}$ so (*) and Theorem 1 §2.3 give

$$V_k = A^k V_0 = (PDP^{-1})^k V_0 = (PD^k P^{-1})V_0 = PD^k(P^{-1}V_0)$$

---

[6]A Markov Chain (Section 1.8) is a dynamical system where the matrix recursion $S_{k+1} = PS_k$ relates the state vectors S_k, and P is the transition matrix.

for each $k = 1, 2, \cdots$. For convenience, we denote the column $P^{-1}V_0$ arising here as follows

$$P^{-1}V_0 = \begin{bmatrix} b_1 \\ b_2 \\ \vdots \\ b_n \end{bmatrix} = \begin{bmatrix} b_1 & b_2 & \cdots & b_n \end{bmatrix}^T.$$

Then matrix multiplication gives (see Theorem 3 §1.4)

$$\begin{aligned} V_k &= PD^k(P^{-1}V_0) \\ &= \begin{bmatrix} X_1 & X_2 & \cdots & X_n \end{bmatrix} \begin{bmatrix} \lambda_1^k & 0 & \cdots & 0 \\ 0 & \lambda_2^k & \cdots & 0 \\ \vdots & \vdots & \ddots & \vdots \\ 0 & 0 & \cdots & \lambda_n^k \end{bmatrix} \begin{bmatrix} b_1 \\ b_2 \\ \vdots \\ b_n \end{bmatrix} \\ &= \begin{bmatrix} \lambda_1^k X_1 & \lambda_2^k X_2 & \cdots & \lambda_n^k X_n \end{bmatrix} \begin{bmatrix} b_1 \\ b_2 \\ \vdots \\ b_n \end{bmatrix} \\ &= b_1 \lambda_1^k X_1 + b_2 \lambda_2^k X_2 + \cdots + b_n \lambda_n^k X_n \end{aligned}$$

for each $k \geq 0$. This is a useful explicit formula for the columns V_k. Note that, in particular, $V_0 = b_1 X_1 + b_2 X_2 + \cdots + b_n X_n$.

However such an exact formula for V_k is often not required in practice; all that is needed is to *estimate* V_k for large values of k (as was done in Example 1). This can be easily done if the diagonalizable $n \times n$ matrix A has a largest eigenvalue.

2.4.2 Dominant Eigenvalues

An eigenvalue λ of an $n \times n$ matrix A is called a **dominant eigenvalue** of A if

$$|\lambda| > |\mu| \quad \text{for all eigenvalues } \mu \neq \lambda$$

where $|\lambda|$ denotes the absolute value[7] of the number λ. For example, $\lambda_1 = 1$ is dominant in Example 1 because $\lambda_2 = -\frac{1}{2}$ is smaller in absolute value.

Returning to the above discussion, suppose A has a dominant eigenvalue which, for convenience, we assume to be of multiplicity 1. By choosing the order in which the columns X_i are placed in P, we may assume that λ_1 is dominant among the eigenvalues $\lambda_1, \lambda_2, \cdots, \lambda_n$ of A. In this case, take λ_1^k out as a common factor in the expression $V_k = b_1 \lambda_1^k X_1 + b_2 \lambda_2^k X_2 + \cdots + b_n \lambda_n^k X_n$ for V_k. The result is

$$V_k = \lambda_1^k \left[b_1 X_1 + b_2 \left(\frac{\lambda_2}{\lambda_1}\right)^k X_2 + \cdots + b_n \left(\frac{\lambda_n}{\lambda_1}\right)^k X_n \right]$$

---

[7]The **absolute value** $|a|$ of a number a is defined as follows: $|a| = \begin{cases} a & \text{if } a \geq 0 \\ -a & \text{if } a < 0 \end{cases}$
Thus, for example, $|3| = 3$ and $|-2| = 2$. It is useful to note that the absolute value is also given by the formula $|a| = \sqrt{a^2}$, where $\sqrt{a^2}$ denotes the positive square root of a^2.

2.4. LINEAR DYNAMICAL SYSTEMS

for each $k \geq 0$. Since λ_1 is dominant, we have $|\lambda_i| < |\lambda_1|$ for each $i \geq 2$, so each of the numbers $\left(\frac{\lambda_i}{\lambda_1}\right)^k$ become small in absolute value as k increases. Hence V_k is approximately equal to the first term $\lambda_1^k b_1 X_1$, and we write this as $V_k \approx \lambda_1^k b_1 X_1$. These observations are summarized in the following theorem (together with the above exact formula for V_k).

Theorem 1. *Consider the dynamical system with matrix recurrence*

$$V_{k+1} = AV_k \quad \text{for } k \geq 0$$

where A and V_0 are given. Assume that A is a diagonalizable $n \times n$ matrix with eigenvalues $\lambda_1, \lambda_2, \cdots, \lambda_n$ and corresponding eigenvectors $X_1, X_2, \cdots, X_n$, and let $P = \begin{bmatrix} X_1 & X_2 & \cdots & X_n \end{bmatrix}$ be the diagonalizing matrix. Then an exact formula for the V_k is

$$V_k = b_1 \lambda_1^k X_1 + b_2 \lambda_2^k X_2 + \cdots + b_n \lambda_n^k X_n \quad \text{for each } k \geq 0$$

where the coefficients b_i come from $P^{-1} V_0 = \begin{bmatrix} b_1 & b_2 & \cdots & b_n \end{bmatrix}^T$. Moreover, if A has dominant eigenvalue λ_1, of multiplicity 1, then[8] *V_k is approximated by*

$$V_k \approx \lambda_1^k b_1 X_1 \quad \text{for sufficiently large } k.$$

Remark. Suppose the dynamical system in Theorem 1 has a dominant eigenvalue λ_1 of multipicity 1. If $\lambda_1 = 1$, Theorem 1 shows that the V_k become constant in the long run. If $|\lambda_1| < 1$ then the powers λ_1^k will become close to zero for large k, so the columns V_k will converge to 0. On the other hand, if $|\lambda_1| > 1$, the V_k will become large as k increases. We give some examples of this behavior below.

In Example 1 the number a_k and j_k of adult and juvenile female birds alive were related by the equations

$$\begin{aligned} a_{k+1} &= \tfrac{1}{2} a_k + \tfrac{1}{4} j_k \\ j_{k+1} &= 2 a_k. \end{aligned}$$

This means that $\tfrac{1}{2}$ of the adults and $\tfrac{1}{4}$ of the juveniles alive in year k survive to year $k+1$, and that each female produces 2 juveniles per year on average. In Example 1 we showed that the species population becomes stable under these conditions. We now examine what happens if we lower the juvenile survival rate to $\tfrac{1}{8}$. It turns out that this drives the species to extinction.

Example 2. *Now assume that the numbers a_k and j_k of adult and juvenile birds satisfy the equations*

$$\begin{aligned} a_{k+1} &= \tfrac{1}{2} a_k + \tfrac{1}{8} j_k \\ j_{k+1} &= 2 a_k. \end{aligned}$$

---

[8] Suppose λ_1 and λ_2 are both dominant (possibly equal), that is $|\lambda_1| = |\lambda_2| > \lambda_i$ for each $i \geq 3$. Then $V_k \approx \lambda_1^k b_1 X_1 + \lambda_2^k b_2 X_2$ for sufficiently large k. A similar analysis applies if λ_1, λ_2 and λ_3 are all dominant.

If $a_0 = 100$ and $j_0 = 40$ as in Example 1, estimate the population profile $V_k = \begin{bmatrix} a_k \\ j_k \end{bmatrix}$ for large k.

Solution. Now $V_{k+1} = AV_k$ where $A = \begin{bmatrix} \frac{1}{2} & \frac{1}{8} \\ 2 & 0 \end{bmatrix}$. The characteristic polynomial is $c_A(x) = x^2 - \frac{1}{2}x - \frac{1}{4}$, so the eigenvalues are $\lambda_1 = \frac{1}{4}\left[1 + \sqrt{5}\right] = .809$ and $\lambda_2 = \frac{1}{4}\left[1 - \sqrt{5}\right] = -.309$. Corresponding eigenvectors are $X_1 = \begin{bmatrix} .809 \\ 2 \end{bmatrix}$ and $X_2 = \begin{bmatrix} -.309 \\ 2 \end{bmatrix}$, so the diagonalizing matrix is $P = \begin{bmatrix} .809 & -.309 \\ 2 & 2 \end{bmatrix}$. Since we are still assuming that $V_0 = \begin{bmatrix} 100 \\ 40 \end{bmatrix}$, the coefficients b_1 and b_2 are given by

$$\begin{bmatrix} b_1 \\ b_2 \end{bmatrix} = P^{-1}V_0 = \frac{1}{2.236}\begin{bmatrix} 2 & .309 \\ -2 & .809 \end{bmatrix}\begin{bmatrix} 100 \\ 40 \end{bmatrix} = \begin{bmatrix} 94.97 \\ -74.97 \end{bmatrix}$$

Since λ_1 is dominant and of multiplicity 1, an estimate for V_k is given by

$$\begin{bmatrix} a_k \\ j_k \end{bmatrix} = V_k \cong b_1 \lambda_1^k X_1 = 94.97(.809)^k \begin{bmatrix} .809 \\ 2 \end{bmatrix}$$

Comparing entries gives

$$a_k \approx 76.83(.809)^k \quad \text{and} \quad j_k \approx 189.94(.809)^k$$

for sufficiently large k. We thus obtain the total female population $a_k + j_k \cong 266.77\,(.809)^k$ so, since $.809 < 1$, the population becomes smaller and smaller as k increases. In other words, the species becomes extinct. Note that this also shows that the breakdown of the (decreasing) female population into adults and juveniles stabilizes with approximately $\frac{76.83}{266.77} = .288 = 28.8\%$ adults and $\frac{189.94}{266.77} = .712 = 71.2\%$ juveniles. □

Much of the computation in Examples 1 and 2 can in fact be carried out for a much more general model. As before, let a_k and j_k denote the number of adult and juvenile female birds alive in year k, and assume they satisfy the equations

$$\begin{aligned} a_{k+1} &= \alpha a_k + \beta j_k \\ j_{k+1} &= m a_k \end{aligned}$$

for $k \geq 0$. Here α and β are the *adult and juvenile survival rates*, that is the proportions of adult and juvenile females respectively to survive from one year to the next. Similarly, m is the *reproduction rate*, that is the number of juveniles hatched (per adult female) in any year. We assume that α, β and m are all positive.

Writing $V_k = \begin{bmatrix} a_k \\ j_k \end{bmatrix}$ and $A = \begin{bmatrix} \alpha & \beta \\ m & 0 \end{bmatrix}$ as before, the above equations become the matrix recurrence $V_{k+1} = AV_k$. The characteristic polynomial of A

2.4. LINEAR DYNAMICAL SYSTEMS

is $c_A(x) = \det \begin{bmatrix} x-\alpha & -\beta \\ -m & x \end{bmatrix} = x^2 - \alpha x - \beta m$. Hence the quadratic formula gives the eigenvalues

$$\lambda_1 = \tfrac{1}{2}(\alpha + r) \quad \text{and} \quad \lambda_2 = \tfrac{1}{2}(\alpha - r)$$

where, for convenience, we write $r = \sqrt{\alpha^2 + 4\beta m}$. Since the eigenvalues λ_i satisfy the characteristic polynomial, we have $\lambda_i^2 = \alpha\lambda_i + \beta m$ for each i. This shows that corresponding eigenvectors are

$$X_1 = \begin{bmatrix} \lambda_1 \\ m \end{bmatrix} \quad \text{and} \quad X_2 = \begin{bmatrix} \lambda_2 \\ m \end{bmatrix}$$

(because $AX_i = \lambda_i X_i$ for each i as the reader can verify). Furthermore, $r = \sqrt{\alpha^2 + 4\beta m} > \alpha$, so $\lambda_2 < 0$. Thus

$$|\lambda_2| = -\lambda_2 = \tfrac{1}{2}(r - \alpha) < \tfrac{1}{2}(\alpha + r) = \lambda_1 = |\lambda_1|$$

so λ_1 is the dominant eigenvalue and has multiplicity 1 no matter what choice of α, β and m is made. Hence Theorem 1 gives

$$V_k \approx b_1 \lambda_1^k X_1 = b_1 \lambda_1^k \begin{bmatrix} \lambda_1 \\ m \end{bmatrix}$$

where b_1 is computed as in Theorem 1 ($b_1 > 0$ as the reader can verify). Thus, in this model, the magnitude of $\lambda_1 > 0$ determines the ultimate population size as follows:

If $\lambda_1 > 1$ the population becomes very large as k increases
If $\lambda_1 = 1$ the population stabilizes as k increases
If $\lambda_1 < 1$ the population becomes very small as k increases

In the last case, the population will become extinct unless something happens to change the values of α, β and m.

2.4.3 A Predator-Prey Model.

Examples 1 and 2 describe a single population and its evolution over time. The following example uses a matrix recurrence to study the dynamics between two interrelated populations.

Example 3. *Let h_k and m_k denote the number of hawks and mice respectively in a certain region in year k. Assume that they are related as follows:*

$$\begin{aligned} h_{k+1} &= \tfrac{1}{2} h_k + \tfrac{1}{100} m_k \\ m_{k+1} &= -\tfrac{50}{4} h_k + \tfrac{5}{4} m_k \end{aligned} \quad \text{for } k \geq 0.$$

Find the limiting values of the populations of hawks and mice if the populations begin with 50 hawks and 1600 mice.

Solution. If we write $V_k = \begin{bmatrix} h_k \\ m_k \end{bmatrix}$ and $A = \begin{bmatrix} \frac{1}{2} & \frac{1}{100} \\ -\frac{50}{4} & \frac{5}{4} \end{bmatrix}$, then the equations become $V_{k+1} = AV_k$ and we have a matrix recurrence. The characteristic polynomial is $x^2 - \frac{7}{4}x + \frac{3}{4} = (x-1)(x-\frac{3}{4})$, so the eigenvalues are $\lambda_1 = 1$ and $\lambda_2 = \frac{3}{4}$ with corresponding eigenvectors $X_1 = \begin{bmatrix} 1 \\ 50 \end{bmatrix}$ and $X_2 = \begin{bmatrix} 1 \\ 25 \end{bmatrix}$. Since λ_1 is dominant here we can estimate the limiting value of V_k. Because $P = \begin{bmatrix} 1 & 1 \\ 50 & 25 \end{bmatrix}$ is a diagonalizing matrix, we obtain $\begin{bmatrix} b_1 \\ b_2 \end{bmatrix} = P^{-1}V_0 = \begin{bmatrix} -1 & \frac{1}{25} \\ 2 & \frac{-1}{25} \end{bmatrix} \begin{bmatrix} 50 \\ 1600 \end{bmatrix} = \begin{bmatrix} 14 \\ 36 \end{bmatrix}$ as in Theorem 1. Hence, for large k,

$$V_k \approx \lambda_1^k b_1 X_1 = 14 \begin{bmatrix} 1 \\ 50 \end{bmatrix}$$

Thus the populations stabilize at approximately 14 hawks and 700 mice. □

Of course, if the matrix A varies in Example 3 the populations may not stabilize. One way to guarantee stability is to insist that $\lambda_1 = 1$ is a dominant eigenvalue. It turns out that this can be done in a whole class of models.

To see what happens in general, assume that the numbers h_k and m_k of hawks and mice are related as follows:

$$\begin{array}{rcl} h_{k+1} & = & ph_k + \gamma m_k \\ m_{k+1} & = & -\delta h_k + qm_k \end{array} \quad \text{for } k \geq 0 \quad (**)$$

where p and q are the survival rate parameters, and γ and δ are the correlation parameters. We are looking for a "stable range" for these parameters, that is for values of the parameters that will ensure that the two populations will reach equilibrium levels. The following conditions on the parameters are inherent in the model:

$p > 0$ because more hawks one year means more the next year.

$p < 1$ because few mice one year means the number of hawks decreases.

$q > 1$ because few hawks one year means the number of mice increases.

$\gamma > 0$ because more mice one year means more hawks the next year.

$\delta > 0$ because more hawks one year means fewer mice the next year.

If we write $A = \begin{bmatrix} p & \gamma \\ -\delta & q \end{bmatrix}$ and $V_k = \begin{bmatrix} h_k \\ m_k \end{bmatrix}$, the equations $(**)$ become the matrix recursion

$$V_{k+1} = AV_k \quad \text{for } k = 0, 1, 2, \cdots$$

2.4. LINEAR DYNAMICAL SYSTEMS

where the initial situation V_0 is assumed to be known. We compute the characteristic polynomial

$$c_A(x) = \det \begin{bmatrix} x - p & -\gamma \\ \delta & x - q \end{bmatrix} = x^2 - (p+q)x + (pq + \gamma\delta)$$

This has roots

$$\tfrac{1}{2}\left\{(p+q) \pm \sqrt{(p+q)^2 - 4(pq + \gamma\delta)}\right\} = \tfrac{1}{2}\left\{(p+q) \pm \sqrt{(q-p)^2 - 4\gamma\delta}\right\}$$

by the quadratic formula so, if we write

$$r = \sqrt{(q-p)^2 - 4\gamma\delta} = \sqrt{(p+q)^2 - 4(pq + \gamma\delta)},$$

the eigenvalues are

$$\lambda_1 = \tfrac{1}{2}[(p+q) + r] \quad \text{and} \quad \lambda_2 = \tfrac{1}{2}[(p+q) - r].$$

We want r to be a real number, so we require that $4\gamma\delta \leq (q-p)^2$. In addition we want $r > 0$ because we want the eigenvalues to be distinct (so A is diagonalizable). Hence we insist that

$$4\gamma\delta < (q-p)^2. \tag{***}$$

Hence $\lambda_1 > \lambda_2 > 0$ (because $r > p + q > 0$), so λ_1 is the dominant eigenvalue. By Theorem 1 the columns V_k become constant for large k if $\lambda_1 = 1$, that is if $p + q + r = 2$. This gives $(p + q - 2)^2 = r^2 = (q-p)^2 - 4\gamma\delta$, which simplifies to the equation

$$\gamma\delta = (1 - p)(q - 1)$$

as the reader can verify. This condition leads to equilibrium.

Equilibrium Condition. *Choose numbers p and q such that $0 < p < 1$, $q > 1$ and $p + q < 2$, and then choose $\gamma > 0$ and $\delta > 0$ satisfying*

$$\gamma\delta = (1 - p)(q - 1)$$

If we use the model given by (**), the populations of hawks and mice become constant for large k.

Proof. Using the condition that $\gamma\delta = (1 - p)(q - 1)$, we have

$$(q - p)^2 - 4\gamma\delta = [2 - (p + q)]^2.$$

Since $p + q < 2$, it follows that (***) holds and $r = 2 - (p + q)$. This means that $\lambda_1 = \tfrac{1}{2}[(p + q) + r] = 1$, so $V_k = \begin{bmatrix} h_k \\ m_k \end{bmatrix}$ becomes constant for large k by Theorem 1, as required. $\square$

The equilibrium constraints on p and q are that
$$0 < p < 1$$
$$1 < q$$
$$p + q < 2$$

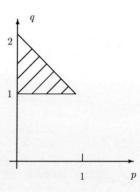

The region of acceptable pairs (p, q) is shown in the diagram. Having chosen such p and q, any positive numbers γ and δ can be chosen so that $\gamma\delta = (1-p)(q-1)$ and the populations in the resulting model will be in equilibrium.

In fact we can find formulas for the limiting values for h_k and m_k. First

$$X_1 = \begin{bmatrix} \gamma \\ \lambda_1 - p \end{bmatrix} \quad \text{and} \quad X_2 = \begin{bmatrix} \gamma \\ \lambda_2 - p \end{bmatrix}$$

are eigenvectors corresponding to λ_1 and λ_2 respectively because $AX_i = \lambda_i X_i$ for each i. (This is because λ_i is a root of the characteristic equation so $\lambda_i^2 - (p+q)\lambda_i + (pq + \gamma\delta) = 0$.) Then Theorem 1 gives

$$\begin{bmatrix} h_k \\ m_k \end{bmatrix} = V_k \approx \lambda_1^k b_1 X_1 = b_1 X_1$$

where b_1 is obtained from $[b_1 \; b_2]^T = P^{-1} V_0$, and $P = [X_1 \; X_2]$ is the diagonalizing matrix for A. Since $\lambda_1 = 1$ for equilibrium, this gives

$$h_k = b_1 \gamma \quad \text{and} \quad m_k = b_1(1-p).$$

Thus, in this model the limiting ratio of hawks to mice is $\frac{\gamma}{1-p}$.

2.4.4 Graphical Description of Solutions.

If a dynamical system $V_{k+1} = AV_k$ is given, the sequence $V_0, V_1, V_2, \cdots$ is called the **trajectory** of the system starting at V_0. It is instructive to obtain a graphical plot of the system by writing $V_k = \begin{bmatrix} x_k \\ y_k \end{bmatrix}$ and plotting the successive values as points in the plane, identifying V_k with the point $P(x_k, y_k)$. We give several examples which illustrate properties of dynamical systems. For convenience we assume that the matrix A is simple, usually diagonal, for ease of calculation.

2.4. LINEAR DYNAMICAL SYSTEMS

Example 4. Let $A = \begin{bmatrix} \frac{1}{2} & 0 \\ 0 & \frac{1}{3} \end{bmatrix}$. Then the eigenvalues are $\frac{1}{2}$ and $\frac{1}{3}$, with corresponding eigenvectors $X_1 = \begin{bmatrix} 1 \\ 0 \end{bmatrix}$ and $X_2 = \begin{bmatrix} 0 \\ 1 \end{bmatrix}$. The general solution is

$$V_k = b_1(\tfrac{1}{2})^k \begin{bmatrix} 1 \\ 0 \end{bmatrix} + b_2(\tfrac{1}{3})^k \begin{bmatrix} 0 \\ 1 \end{bmatrix}$$

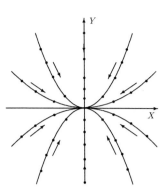

for $k = 0, 1, 2, \cdots$ by Theorem 1, where the coefficients b_1 and b_2 depend[9] on the initial point V_0. Several trajectories are plotted in the diagram and, for each choice of V_0, the trajectories converge toward the origin because both eigenvalues are less than 1 in absolute value. For this reason, the origin is called an **attractor** for the system.

Example 5. Let $A = \begin{bmatrix} \frac{3}{2} & 0 \\ 0 & \frac{4}{3} \end{bmatrix}$. Here the eigenvalues are $\frac{3}{2}$ and $\frac{4}{3}$, with corresponding eigenvectors $X_1 = \begin{bmatrix} 1 \\ 0 \end{bmatrix}$ and $X_2 = \begin{bmatrix} 0 \\ 1 \end{bmatrix}$ as before. The general solution is

$$V_k = b_1(\tfrac{3}{2})^k \begin{bmatrix} 1 \\ 0 \end{bmatrix} + b_2(\tfrac{4}{3})^k \begin{bmatrix} 0 \\ 1 \end{bmatrix}$$

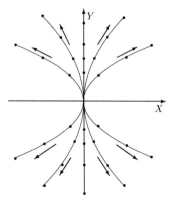

for $k = 0, 1, 2, \cdots$ Since both eigenvalues are greater than 1 in absolute value, the trajectories diverge away from the origin for every choice of initial point V_0. For this reason, the origin is called a **repellor** for the system.

---

[9]In fact $P = I$ here so $V_0 = \begin{bmatrix} b_1 \\ b_2 \end{bmatrix}$.

Example 6. Let $A = \begin{bmatrix} 1 & -\frac{1}{2} \\ -\frac{1}{2} & 1 \end{bmatrix}$. Now the eigenvalues are $\frac{3}{2}$ and $\frac{1}{2}$, with corresponding eigenvectors $X_1 = \begin{bmatrix} -1 \\ 1 \end{bmatrix}$ and $X_2 = \begin{bmatrix} 1 \\ 1 \end{bmatrix}$. The general solution is

$$V_k = b_1(\tfrac{3}{2})^k \begin{bmatrix} -1 \\ 1 \end{bmatrix} + b_2(\tfrac{1}{3})^k \begin{bmatrix} 1 \\ 1 \end{bmatrix}$$

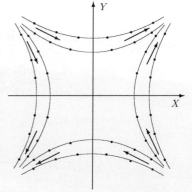

for $k = 0, 1, 2, \cdots$. In this case $\frac{3}{2}$ is the dominant eigenvalue so, if $b_1 \neq 0$, we have $V_k \approx b_1(\tfrac{3}{2})^k \begin{bmatrix} -1 \\ 1 \end{bmatrix}$ for large k and V_k is approaching the line $y = x$. However, if $b_1 = 0$ then $V_k = b_2(\tfrac{1}{3})^k \begin{bmatrix} -1 \\ 1 \end{bmatrix}$ and so approaches the origin along the line $y = -x$. In general the trajectories appear as in the diagram, and the origin is called a **saddle point** for the dynamical system in this case.

Example 7. Let $A = \begin{bmatrix} 0 & \frac{1}{2} \\ -\frac{1}{2} & 0 \end{bmatrix}$. Now the characteristic polynomial is $c_A(x) = x^2 + \frac{1}{4}$, so the eigenvalues are the complex numbers $\frac{i}{2}$ and $-\frac{i}{2}$ where $i^2 = -1$. Hence A is not diagonalizable as a real matrix[10]. However the trajectories are not difficult to describe. If we start with $V_0 = \begin{bmatrix} 1 \\ 1 \end{bmatrix}$ then the trajectory begins as

$$V_1 = \begin{bmatrix} \frac{1}{2} \\ -\frac{1}{2} \end{bmatrix}, \quad V_2 = \begin{bmatrix} -\frac{1}{4} \\ -\frac{1}{4} \end{bmatrix}, \quad V_3 = \begin{bmatrix} -\frac{1}{8} \\ \frac{1}{8} \end{bmatrix},$$
$$V_4 = \begin{bmatrix} \frac{1}{16} \\ \frac{1}{16} \end{bmatrix}, \quad V_5 = \begin{bmatrix} \frac{1}{32} \\ -\frac{1}{32} \end{bmatrix}, \quad V_6 = \begin{bmatrix} -\frac{1}{64} \\ -\frac{1}{64} \end{bmatrix}, \cdots$$

These points are plotted in the diagram. Here each trajectory spirals in to the origin, so the origin is an attractor. Note that the two (complex) eigenvalues have absolute value less than 1 here. If they had absolute value greater than 1, the trajectories would spiral out from the origin.

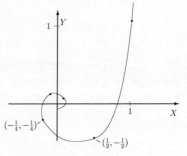

---

[10]Corresponding eigenvectors are $X_1 = \begin{bmatrix} 1 \\ i \end{bmatrix}$ and $X_2 = \begin{bmatrix} 1 \\ -i \end{bmatrix}$ so the theory works for the complex numbers. We return to this in Section 2.5.

2.4. LINEAR DYNAMICAL SYSTEMS

Exercises

1. Consider the female bird population model where the adult survival rate is $1/2$, the juvenile survival rate is $1/3$, and the reproduction rate is 2; that is the growth matrix A is $\begin{bmatrix} \frac{1}{2} & \frac{1}{3} \\ 2 & 0 \end{bmatrix}$.

    (a) Using an initial population of 100 adults and 60 juveniles, calculate the population for the next few years.

    (b) Does the population become extinct? Stabilize? Diverge?

    (c) Does it depend on the initial population?

    (d) In the long term, what percentage of the total female population is made up of juveniles?

2. Consider once again the female bird population model where the adult survival rate is $3/10$, the juvenile survival rate is $1/3$, and the reproduction rate is 2, and the initial population is 100 adults and 50 juveniles.

    (a) Write a matrix recurrence for the bird population in the form $V_{k+1} = AV_k = A^k V_0$, where A is the growth matrix.

    (b) Find the eigenvalues of A, corresponding eigenvectors, and an invertible matrix P which diagonalizes the matrix A.

    (c) Find an exact formula for V_k in term of the eigenvalues and eigenvectors.

    (d) Use the formula found in (c) to calculate the population 5 years from now.

    (e) Find the dominant eigenvalue.

    (f) Find an approximating formula for V_k using the dominant eigenvalue.

    (g) Use the formula found in (f) to approximate the population 10 years from now.

    (h) Compare the results found in (d) and (g).

    (i) Does the population become extinct? Stabilize? Diverge?

    (j) Does it depend on the initial population?

    (k) In the long term, what percentage of the total female population is made up of juveniles?

3. Consider the female bird population model where the adult survival rate is $2/5$, and the reproduction rate is 4. Find the required value of the juvenile survival rate β to ensure that the population will stabilize in the long term.

4. In the general bird population model, assume that the adult survival rate α and the reproduction rate m are fixed. If β denotes the juvenile survival rate, show that the population:

(a) Stabilizes if $\beta = \frac{1-\alpha}{m}$;

(b) Becomes extinct if $\beta < \frac{1-\alpha}{m}$;

(c) Becomes very large if $\beta > \frac{1-\alpha}{m}$.

5. The population of an endangered hawk species and its natural prey the field mouse, originally with 30 hawks and 2000 mice, is represented by the following model:

$$\begin{aligned} h_{k+1} &= .55h_k + .005m_k \\ m_{k+1} &= -18h_k + 1.2m_k \end{aligned}$$

where k_k and m_k represent the hawk and mice population respectively in year k.

(a) Make sure you understand what each parameter means.

(b) Express the model in the matrix form $V_{k+1} = AV_k$, and write $V_0 = \begin{bmatrix} 30 \\ 2000 \end{bmatrix}$.

(c) Find the eigenvalues and corresponding eigenvectors of A, and use them to diagonalize the matrix A, and write a general formula to calculate V_k.

(d) Calculate the predicted population every few years for the next 30 years or so, and reason using eigenvalues whether an equilibrium will be reached.

(e) In 5 years, biologists try to improve the hawk population by introducing 1000 mice in the hawk hunting area. Predict and compare the new hawk population 10 years from now.

(f) From that point on, the hawk survival rate drops to 0.50. The biologists are alarmed that this may wipe out the hawk population. Test to see what the model is predicting, and explain what is happening.

6. Describe the trajectories of the dynamical system with matrix

$$A = \begin{bmatrix} 0 & 1 \\ -1 & 0 \end{bmatrix}.$$

2.5 COMPLEX EIGENVALUES

For the most part the matrices we have been dealing with have had real eigenvalues. However there are important matrices for which this is not the case. For example the matrix $A = \begin{bmatrix} 0 & -1 \\ 1 & 0 \end{bmatrix}$ has characteristic polynomial $c_A(x) = x^2 + 1$, and this has no real root because $x^2 \geq 0$ for every real number x. Nonetheless, we can find eigenvalues for A and diagonalize it, but we have to use complex numbers to do it. This section develops the basic properties of these numbers.

2.5. COMPLEX EIGENVALUES

It turns out the whole thing depends on the existence of a number i such that $i^2 = -1$. Because such a number i cannot be real, it is usually called an "imaginary" number. But it is no more imaginary than the number -1, or the number 2 for that matter. Numbers exist only in our minds and so are *all* imaginary. The integers $1, 2, 3, \cdots$ were undoubtedly invented to count things (enemy warriors, or animals to hunt), the positive real numbers were used to measure distances and areas, and the negative numbers enabled people to systematically denote phenomena like debt. In each case the numbers were invented (some would say "discovered") to describe some aspect of the world. This proved to be very useful because the algebraic properties of numbers could then be used to describe "laws" of nature. For example, $A = lw$ is the area of a rectangle of length l and width w; and $f = ma$ relates the force f, mass m and acceleration a of a particle moving in a straight line (Newton's second law of motion). Thus it is important that the new numbers being created satisfy the rules of arithmetic.

2.5.1 Complex Numbers

In any event, the eighteenth century mathematicians introduced a new (non-real) number i with the property that

$$i^2 = -1$$

and they called expressions of the form

$$a + ib \quad \text{where } a \text{ and } b \text{ are real}$$

complex numbers. The set of all complex numbers is denoted $\mathbb{C}$. The rules of real arithmetic hold for these complex numbers with the single exception that $i^2 = -1$.

Given a complex number $z = a + ib$, the real number a is called the **real part** of z, and b is the **imaginary part** of z. Every real number a is a complex number, namely $a = a + 0i$, and complex numbers of the form $bi = 0 + bi$ are sometimes called **pure imaginary** numbers. The number i itself is called the **imaginary unit**.

If two complex numbers are equal, they must have the same real and imaginary parts. That is

$$\text{If } a + bi = a' + b'i \quad \text{then} \quad a = a' \text{ and } b = b'.$$

Indeed, if b were not equal to b' then we could solve for $i = \frac{a-a'}{b'-b}$, so i would be a real number, a contradiction. Hence $b = b'$, and so $a = a'$ (because $a + bi = a' + bi$).

Because of this, it is possible to display the complex numbers geometrically as the points in the plane. This is accomplished by identifying the complex number $z = a + bi$ with the point $P(a, b)$.

148   CHAPTER 2. DETERMINANTS AND DIAGONALIZATION

Some examples are given in the diagram. When this is done the plane is called the **complex plane**. Note that the real numbers are identified with the points on the X-axis in the usual way. For this reason, the X-axis is called the **real axis**. Similarly, the Y-axis is called the **imaginary axis**.

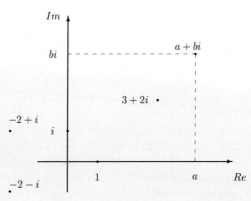

Complex numbers are added and multiplied in much the same way as linear polynomials $a + bx$, except that $i^2 = -1$. Thus

If $z = a + ib$ and $w = a' + ib'$ then $\begin{cases} z + w &= (a + a') + (b + b'), \\ zw &= (aa' - bb') + (ab' + ba')i. \end{cases}$

Example 1. If $z = 3 - 2i$ and $w = -5 + 7i$, compute $z + w$, $z - w$, $5z$, zw and z^2.

Solution.

$$\begin{aligned} z + w &= (3 - 5) + (-2 + 7)i = -2 + 5i. \\ z - w &= (3 + 5) + (-2 - 7)i = 8 - 9i. \\ 5z &= 15 - 10i. \\ zw &= (-15 - 14i^2) + (21 + 10)i = -1 + 31i. \\ z^2 &= (9 + 4i^2) + (-6 - 6)i = 5 - 12i. \end{aligned}$$

$\square$

Before describing the process of dividing one complex number by another we must introduce two important notions. If $z = a + ib$ is any complex number, the **conjugate** of z is another complex number, denoted $\overline{z}$, given by

$$\overline{z} = a - ib,$$

that is $\overline{z}$ is formed by negating the imaginary part of z. Thus, in the complex plane, z and $\overline{z}$ are reflections of each other in the real axis (see the diagram).

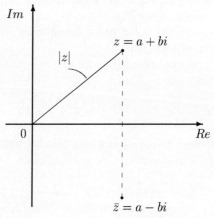

Here are four useful properties of conjugation:

2.5. COMPLEX EIGENVALUES

C1 $\overline{(z \pm w)} = \overline{z} \pm \overline{w}$.
C2 $\overline{(zw)} = \overline{z}\,\overline{w}$.
C3 $\overline{(\overline{z})} = z$.
C4 z is real if and only if $\overline{z} = z$.

The routine verifications of C1, C2 and C3 are left as Exercise 6. As to C4, if $z = a + 0i$ is real then $\overline{z} = a - 0i = z$. Conversely, if $z = a + bi$ and $\overline{z} = z$, then $a - bi = a + bi$. Hence equating imaginary parts gives $-b = b$. This means $b = 0$, whence $z = a + 0i = a$ is real.

Given any complex number $z = a + bi$, the distance from z to the origin 0 is

$$\sqrt{a^2 + b^2}$$

by Pythagoras' Theorem[11]. This number is called the **absolute value** or **modulus** of the complex number $z = a + bi$, and is denoted (see the diagram)

$$|z| = \sqrt{a^2 + b^2}.$$

It is routine to verify that

$$z\overline{z} = a^2 + b^2$$

which gives the following useful formula:

$$|z|^2 = z\overline{z} \text{ for every complex number } z.$$

Note further that, if $z = a + 0i$ is real then $|z| = \sqrt{a^2}$ is the absolute value of z viewed as a real number.

We list some basic properties of the absolute value.

A1 $|z| \geq 0$ for all complex numbers z.
A2 $|z| = 0$ if and only if $z = 0$.
A3 $|zw| = |z|\,|w|$.

A1 and A2 follow immediately from the fact that $|z|$ is the distance from z to the origin 0. To see why A3 holds, we use the fact (above) that $|z|^2 = z\overline{z}$ holds for every complex number z. Combining this with C2, we compute

$$|zw|^2 = (zw)\overline{(zw)} = zw\overline{z}\,\overline{w} = (z\overline{z})(w\overline{w}) = |z|^2\,|w|^2 = (|z|\,|w|)^2.$$

Now A3 follows by taking positive square roots.

We can now describe division by a complex number.

Example 2. *Compute* $\frac{3-5i}{2+7i}$.

Solution. The idea is to multiply top and bottom by the conjugate of the denominator (and so turn the denominator into a real number):

$$\frac{3-5i}{2+7i} = \frac{(3-5i)(2-7i)}{(2+7i)(2-7i)} = \frac{-29-31i}{2^2+7^2} = -\frac{29}{53} - \frac{31}{53}i.$$

□

---

[11] $\sqrt{p}$ denotes the *positive* square root of the real number $p \geq 0$.

2.5.2 Quadratics

While the introduction of the imaginary unit i has been motivated by the problem of finding a root of the equation $x^2 + 1 = 0$, the complex numbers easily give the roots of any real quadratic. In fact, if a, b and c are real numbers with $a \neq 0$, the quadratic equation $ax^2 + bx + c = 0$ has roots given by the famous **quadratic formula**:

$$x = \frac{-b \pm \sqrt{b^2 - 4ac}}{2a}.$$

The quantity $d = b^2 - 4ac$ is called the **discriminant** of the quadratic $ax^2 + bx + c$. It determines the nature of the roots:
 If $d > 0$ there are two real roots.
 If $d = 0$ there is one (repeated) real root.
 If $d < 0$ there is no real root.
When $d < 0$ the quadratic $ax^2 + bx + c$ is said to be **irreducible,** and the quadratic formula gives two complex roots which are conjugates of each other:

$$\lambda = \frac{1}{2a}(-b + i\sqrt{|d|}) \text{ and } \overline{\lambda} = \frac{1}{2a}(-b - i\sqrt{|d|})$$

The converse of this is also true: Given any nonreal complex number λ, then λ and $\overline{\lambda}$ are the roots of some real irreducible quadratic. Indeed, the quadratic

$$x^2 - (\lambda + \overline{\lambda})x + (\lambda\overline{\lambda}) = (x - \lambda)(x - \overline{\lambda})$$

is irreducible (the roots λ and $\overline{\lambda}$ are not real) and has real coefficients ($\lambda\overline{\lambda} = |\overline{\lambda}|^2$ and $\lambda + \overline{\lambda}$ is twice the real part of λ).

Example 3. *Find a real irreducible quadratic with $\lambda = 3 - 4i$ as a root.*

Solution. We have $\lambda + \overline{\lambda} = 6$ and $|\overline{\lambda}|^2 = 25$, so $x^2 - 6\lambda + 25$ is irreducible with roots λ and $\overline{\lambda} = 3 + 4i$. □

2.5.3 Complex Linear Algebra

Everything we have done for real matrices can be done for complex matrices. The methods are the same; the only difference is that the arithmetic is carried out with complex numbers rather than real ones. We give three examples.

Example 4. *Solve the system*
$$\begin{aligned} x + (1+i)y &= 1 - 2i \\ ix - y + iz &= 2 \\ (i-1)x - y - z &= 2i + 3 \end{aligned}$$

2.5. COMPLEX EIGENVALUES

Solution. The augmented matrix is reduced as follows:

$$\begin{bmatrix} 1 & 1+i & 0 & 1-2i \\ i & -1 & i & 2 \\ i-1 & -1 & -1 & 2+3i \end{bmatrix} \to \begin{bmatrix} 1 & 1+i & 0 & 1-2i \\ 0 & -i & i & -i \\ 0 & 1 & -1 & 1 \end{bmatrix}$$

$$\to \begin{bmatrix} 1 & 1+i & 0 & 1-2i \\ 0 & 1 & -1 & 1 \\ 0 & 0 & 0 & 0 \end{bmatrix}$$

$$\to \begin{bmatrix} 1 & 0 & 1+i & -3i \\ 0 & 1 & -1 & 1 \\ 0 & 0 & 0 & 0 \end{bmatrix}.$$

Hence we take $z = t$ where t is an arbitrary (complex) parameter, and then solve for $x = -3i - (1+i)t$, and $y = 1 + t$. □

Example 5. *Find the inverse of* $A = \begin{bmatrix} i & -1 \\ -i & 2 \end{bmatrix}$.

Solution. We have $\det A = 2i - (-1)(-i) = i \neq 0$. So A has an inverse and (the complex version of) Example 5 §1.5 gives

$$A^{-1} = \frac{1}{i} \begin{bmatrix} 2 & 1 \\ i & i \end{bmatrix} = (-i) \begin{bmatrix} 2 & 1 \\ i & i \end{bmatrix} = \begin{bmatrix} -2i & -i \\ 1 & 1 \end{bmatrix}.$$

The reader can check this directly. Of course the inverse can also be found using the matrix inversion algorithm. □

Example 6. *Diagonalize the matrix* $A = \begin{bmatrix} 0 & -1 \\ 1 & 0 \end{bmatrix}$.

Solution. The characteristic polynomial of A is $c_A(x) = \det(xI - A) = \det \begin{bmatrix} x & 1 \\ -1 & x \end{bmatrix} = x^2 + 1$, so the eigenvalues are $\lambda_1 = i$ and $\lambda_2 = -i$. The augmented matrix of the system $(\lambda_1 I - A)X = 0$ is reduced as follows:

$$\begin{bmatrix} i & 1 & 0 \\ -1 & i & 0 \end{bmatrix} \to \begin{bmatrix} -1 & i & 0 \\ 0 & 0 & 0 \end{bmatrix} \to \begin{bmatrix} 1 & -i & 0 \\ 0 & 0 & 0 \end{bmatrix}.$$

Hence an eigenvector corresponding to $\lambda_1 = i$ is $X_1 = \begin{bmatrix} i \\ 1 \end{bmatrix}$. Similarly $X_2 = \begin{bmatrix} -i \\ 1 \end{bmatrix}$ is an eigenvector for $\lambda_2 = -i$. Hence the complex version of Theorem 3 §2.3 shows that $P = \begin{bmatrix} X_1 & X_2 \end{bmatrix} = \begin{bmatrix} i & -i \\ 1 & 1 \end{bmatrix}$ is invertible and

$$P^{-1}AP = \begin{bmatrix} \lambda_1 & 0 \\ 0 & \lambda_2 \end{bmatrix} = \begin{bmatrix} i & 0 \\ 0 & -i \end{bmatrix}.$$

As these examples illustrate, calculations in complex linear algebra are really not much different from their real counterparts, except for the fact that the arithmetic uses complex numbers.

2.5.4 Roots of Polynomials

However there is one major advantage of the complex case. As we have seen, real matrices can have no real eigenvalues. Perhaps the most important reason for the study of complex linear algebra is that this situation does not arise: Every complex matrix has an eigenvalue (possibly complex). This is a consequence of the following remarkable theorem. (The proof is beyond the scope of this book.)

Fundamental Theorem of Algebra[12]. *Every non-constant polynomial with complex coefficients has a complex root.*

If $f(x)$ is a polynomial with complex coefficients of degree $n \geq 1$, and if λ_1 is a root, then the factor theorem[13] asserts that

$$f(x) = (x - \lambda_1)\, g(x) \text{ where } g(x) \text{ is a polynomial of degree } n - 1.$$

Suppose that λ_2 is a root of $g(x)$, again by the fundamental theorem. Then $g(x) = (x - \lambda_2)\, h(x)$, so

$$f(x) = (x - \lambda_1)(x - \lambda_2)\, h(x).$$

This process continues until the last polynomial to appear is linear. This last factor can be written in the form $u \cdot (x - \lambda_n)$, so the fundamental theorem takes the following form:

Theorem 1. *Every complex polynomial $f(x)$ of degree $n \geq 1$ has the form*

$$f(x) = u \cdot (x - \lambda_1)(x - \lambda_2) \cdots (x - \lambda_n)$$

where the complex numbers $\lambda_1, \lambda_2, \cdots \lambda_n$ are the roots of $f(x)$ and need not be all distinct, and $u \neq 0$ is the coefficient of x^n in $f(x)$.

In the case of a quadratic polynomial $f(x)$, the degree is 2 and the quadratic formula provides the two roots λ_1 and λ_2. For polynomials of higher degree it is not so simple to find the roots, but computer programs exist to do the calculations.

---

[12] This Theorem was first proved by Gauss at age 20, and was published in his doctoral dissertation in 1799.
[13] See Appendix A.3.

2.5. COMPLEX EIGENVALUES

Of course, the characteristic polynomial $c_A(x)$ of any complex matrix A factors completely as in Theorem 1. Since the roots of $c_A(x)$ are just the eigenvalues of A, we obtain

Theorem 2. *Every $n \times n$ complex matrix A has n complex eigenvalues (possibly with some repeated).*

We have seen several illustrations of Theorem 2 (with real eigenvalues) in Section 2.3.

2.5.5 Symmetric Matrices

Most of the applications of linear algebra involve a real matrix A and, while A will have complex eigenvalues by Theorem 2, it is always of interest to know when the eigenvalues are in fact real. This turns out to be the case whenever A is symmetric:

Theorem 3. *Let A be a symmetric real matrix. If λ is any eigenvalue of A, then λ is real.*

This important theorem will be used extensively in Chapter 4. Surprisingly enough, the theory of *complex* eigenvalues can be used to prove this useful result about *real* eigenvalues. The proof is given at the end of this section; we verify the 2×2 case.

Example 7. *Verify Theorem 3 when A is 2×2, real and symmetric.*

Solution. If $A = \begin{bmatrix} a & b \\ b & c \end{bmatrix}$ is real then $c_A(x) = (x-a)(x-c) - b^2 = x^2 - (a+c)x + (ac - b^2)$. The discriminant of $c_A(x)$ is

$$d = (a+c)^2 - 4(ac - b^2) = (a-c)^2 + 4b^2.$$

Hence $d \geq 0$ because a, b and c are real, so $c_A(x)$ has real roots. □

2.5.6 Polar Form[14]

Up to this point we have been describing complex numbers z by writing them in the form $z = a + ib$ where (a, b) are the coordinates of z as a point in the complex plane.

---

[14]This material is not required elsewhere in this book.

However, to describe multiplication of two complex numbers, it is convenient to represent them using polar coordinates in the plane. Write the absolute value of $z = a + ib$ as

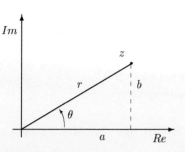

$$r = |z| = \sqrt{a^2 + b^2}.$$

If $z \neq 0$, the angle θ shown in the diagram is called an **argument** of z and is denoted

$$\theta = arg(z).$$

The angle θ is not uniquely determined by z and the angle $\theta \pm 2\pi k$ would do equally well for any integer k.[15] The numbers r and θ are related to the real and imaginary parts a and b of z by

$$a = r\,cos\theta \quad \text{and} \quad b = r\,sin\theta.$$

Hence the complex number $z = a + ib$ has the form

$$z = r(cos\theta + i\,sin\theta).$$

There is a famous expression (called **Euler's formula**[16]) for the complex numbers $cos\theta + i\,sin\theta$:

$$e^{i\theta} = cos\theta + i\,sin\theta \quad \text{for any angle } \theta.$$

Here $e = 2.71828\cdots$ is the constant familiar from calculus. Then the **polar form** of z is

$$z = re^{i\theta} \quad \text{where } r = |z| \quad \text{and} \quad \theta = arg(z).$$

Of course it is not unique because the argument θ can be changed by adding or subtracting multiples of 2π.

Example 8. *Find the polar form of* $z = -1 + i$ *and of* $w = \sqrt{3} - i$.

Solution. These complex numbers are plotted in the diagram. We have $|z| = \sqrt{(-1)^2 + 1^2} = \sqrt{2}$, and the triangle determined by z and the real axis has sides of length 1 and hypotenuse $\sqrt{2}$. Hence the acute angles in the triangle are both $\frac{\pi}{4} = 45°$, so one argument for z is $arg(z) = \frac{3\pi}{4}$ as can be seen in the diagram. Thus the polar form of z is $z = \sqrt{2}e^{3\pi i/4}$.

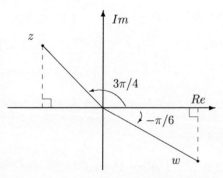

---

[15]Recall that 2π radians represents one full circle, so the angle $\theta \pm 2\pi k$ is obtained from θ by adding (or subtracting) k full revolutions. For a review of general angles, see Appendix A.1.

[16]Named after the great Swiss mathematician Leonhard Euler (1707-1783). We have the Maclaurin expansion $e^x = 1 + x + \frac{1}{2!}x^2 + \frac{1}{3!}x^3 + \cdots$, and substituting $x = i\theta$ gives Euler's formula when we remember that $sinx = x - \frac{1}{3!}x^3 + \frac{1}{5!}x^5 - \cdots$ and $cosx = 1 - \frac{1}{2!}x^2 + \frac{1}{4!}x^4 - \cdots$.

2.5. COMPLEX EIGENVALUES

Turning to w, we see that $|w| = \sqrt{(\sqrt{3})^2 + (-1)^2} = 2$. The triangle determined by w and the real axis has acute angles $\frac{\pi}{6} = 30°$ and $\frac{\pi}{3} = 60°$, so $arg(w) = -\frac{\pi}{6}$. Hence the polar form of w is $w = 2\,e^{-\pi i/6}$. □

Because $cos^2\theta + sin^2\theta = 1$ for any angle θ, the numbers $e^{i\theta} = cos\theta + i\,sin\theta$ all have absolute value 1:

$$\left|e^{i\theta}\right| = 1 \text{ for any angle } \theta.$$

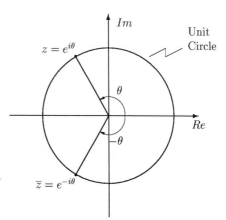

In other words, each $e^{i\theta}$ lies on the **unit circle** (center at the origin and radius 1). On the other hand, every complex number z on the unit circle has the form $z = e^{i\theta}$ for some angle θ (see the diagram):

$$|z| = 1 \text{ if and only if}$$
$$z = e^{i\theta} \text{ for some angle } \theta.$$

Furthermore, if $|z| = 1$ then $z\bar{z} = |z|^2 = 1$, so $\bar{z} = 1/z$. The diagram also shows that:

If $z = e^{i\theta}$, then $\bar{z} = e^{-i\theta} = 1/z$.

Our main purpose for introducing the polar form is to clearly describe the product of two complex numbers. This is achieved in the following result.

Theorem 4. Multiplication Rule. *If $z = re^{i\theta}$ and $w = se^{i\varphi}$ in polar form, then*

$$re^{i\theta}\,se^{i\varphi} = rse^{i(\theta+\varphi)}.$$

Thus to multiply two complex numbers simply multiply the absolute values and add the angles.

Proof. It clearly suffices to show that $e^{i\theta}e^{i\varphi} = e^{i(\theta+\varphi)}$. This is a consequence of the trigonometric identities for $cos(\theta + \varphi)$ and $sin(\theta + \varphi)$:

$$\begin{aligned}
e^{i\theta}e^{i\varphi} &= (cos\theta + i\,sin\theta)(cos\varphi + i\,sin\varphi) \\
&= (cos\theta cos\varphi - sin\theta sin\varphi) + i(cos\theta sin\varphi + sin\theta cos\varphi) \\
&= cos(\theta + \varphi) + i\,sin(\theta + \varphi) \\
&= e^{i(\theta+\varphi)}.
\end{aligned}$$
□

Example 9. *Multiply $(-1 + i)(\sqrt{3} - i)$ using the multiplication rule.*

Solution. In Example 8 we found that the polar forms for these numbers are $-1 + i = \sqrt{2}\, e^{3\pi i/4}$ and $\sqrt{3} - i = 2\, e^{-\pi i/6}$. Hence the multiplication rule gives

$$\begin{aligned}(-1+i)(\sqrt{3}-i) &= \sqrt{2}e^{3\pi i/4}\, 2\, e^{-\pi i/6} = 2\sqrt{2}\, e^{(3\pi/4 - \pi/6)i} \\ &= 2\sqrt{2}\, e^{7\pi i/12} = 2\sqrt{2}(\cos \tfrac{7\pi}{12} + \sin \tfrac{7\pi}{12})\end{aligned}$$

in polar form. Of course we can also multiply the left side in Cartesian form to obtain $(1 - \sqrt{3}) + (1 + \sqrt{3})i = 2\sqrt{2}(\cos\tfrac{7\pi}{12} + \sin\tfrac{7\pi}{12})$. Equating real and imaginary parts gives the (somewhat unexpected) formulas $\cos \tfrac{7\pi}{12} = \tfrac{1-\sqrt{3}}{2\sqrt{2}}$ and $\sin \tfrac{7\pi}{12} = \tfrac{1+\sqrt{3}}{2\sqrt{2}}$. □

The multiplication rule takes a particularly nice form when we use it to compute powers of a complex number. If $z = re^{i\theta}$ in polar form then

$$z^2 = (re^{i\theta})(re^{i\theta}) = r^2 e^{i2\theta}.$$

Next compute z^3 as $z^3 = zz^2$ to get

$$z^3 = zz^2 = (re^{i\theta})(r^2 e^{i2\theta}) = r^3 e^{i3\theta},$$

again by the multiplication rule. The pattern is clear and the general result is

Theorem 5. DeMoivre's Theorem[17]**.** *If $z = re^{i\theta}$ in polar form then*

$$z^n = r^n e^{in\theta} \text{ for } n = 0, \pm 1, \pm 2, \cdots.$$

Proof. The above argument proves it for $n = 1, 2, \cdots$, and it is clear if $n = 0$. If n is negative, write $n = -m$ where $m > 0$. Then $z^n = \tfrac{1}{z^m} = \tfrac{1}{r^m e^{im\theta}} = r^{-m}(e^{im\theta})^{-1} = r^{-m} e^{-(im\theta)} = r^{-m} e^{i(-m)\theta} = r^n e^{in\theta}$. □

Example 10. *Compute $(1 - i)^{27}$.*

Solution. We have $1 - i = \sqrt{2} e^{-\pi i/4}$, so DeMoivre's theorem gives

$$(1 - i)^{27} = (\sqrt{2})^{27} e^{-27\pi i/4}.$$

Since $(\sqrt{2})^2 = 2$, we have $(\sqrt{2})^{27} = 2^{13}\sqrt{2}$. We can simplify the expression $e^{-\frac{27\pi}{4}i}$ by using the fact that $e^{2\pi ki} = 1$ for all integers k (even negative values). Observe that $-\tfrac{27\pi}{4} = -\tfrac{3\pi}{4} - 6\pi$. Hence

$$e^{-27\pi i/4} = e^{-3\pi i/4} e^{-6\pi i} = e^{-3\pi i/4} \cdot 1 = e^{-3\pi i/4} = -\tfrac{1}{\sqrt{2}}(1 + i).$$

Finally $(1 - i)^{27} = (\sqrt{2})^{27} e^{-27\pi i/4i} = 2^{13}\sqrt{2}\{-\tfrac{1}{\sqrt{2}}(1 + i)\} = -2^{13}(1 + i)$. □

DeMoivre's theorem also allows us to take roots of complex numbers. Here is an example.

---

[17]Abraham DeMoivre (1667-1754) was a French mathematician and statistician.

2.5. COMPLEX EIGENVALUES

Example 11. Find all fourth roots of $8(1 + \sqrt{3}i)$; that is find all complex numbers z such that $z^4 = 8(1 + \sqrt{3}i)$.

Solution. The idea is to write $z = re^{i\theta}$, and find r and θ. We have $8(1+\sqrt{3}i) = 8(2e^{\pi i/3})$, so the condition that $z^4 = 8(1 + \sqrt{3}i)$ becomes

$$r^4 e^{4\theta i} = 16 e^{\pi i/3}. \tag{*}$$

Taking absolute values gives $r^4 = 16$, so $r = \sqrt[4]{16} = 2$ because $r > 0$. Even though the complex numbers in (*) are equal, their angles may differ by a multiple of 2π:

$$4\theta = \frac{\pi}{3} + 2\pi k \text{ for some } k = 0, \pm 1, \pm 2, \pm 3, \cdots.$$

Hence $\theta = \frac{\pi}{12} + \frac{\pi}{2}k$ for any integer k, and it appears that there are infinitely many possibilities for θ, one for each choice of k. However, the various possibilities for θ are obtained from $\frac{\pi}{12}$ by adding multiples of $\frac{\pi}{2}$. Hence the distinct ones correspond to $k = 0, 1, 2,$ and 3; all the others differ from one of these by a multiple of 2π and so yield nothing new (see the diagram).

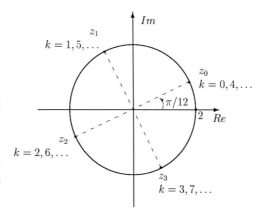

So the four possibilities for z are

$$z_0 = 2e^{\pi i/12}, \ z_1 = 2e^{7\pi i/12}, \ z_2 = 2e^{13\pi i/12}, \ z_3 = 2e^{19\pi i/12}.$$

These are plotted in the diagram. They are equally spaced on the circle $|z| = 2$. □

The technique in Example 11 works to find all nth roots of any complex number w; that is all complex numbers z such that $z^n = w$. It turns out that there are n of them, equally spaced on a circle of radius $\sqrt[n]{|w|}$, so the angle between consecutive values is $\frac{2\pi}{n}$.

Possibly the most important case is that of the complex numbers z such that $z^n = 1$, the so called nth **roots of unity**. Here, if we write $z = re^{i\theta}$ we obtain $r^n e^{n\theta i} = 1 e^{0i}$, so equating absolute values gives $r = 1$, and equating angles gives $n\theta = 0 + 2\pi k$ for $k = 0, \pm 1, \pm 2, \pm 3, \cdots$. Thus $\theta = \frac{2\pi}{n}k$, so the roots are all given by taking $k = 0, 1, 2, \cdots, n-1$.

Theorem 6. *The nth roots of unity (that is the solutions to $z^n = 1$) are the complex numbers*

$e^{2\pi ki/n}$ where $k = 0, 1, 2, 3, \cdots, n-1$.

These nth roots of unity are equally spaced on the unit circle. Every value of k yields a value of $\theta = \frac{2\pi}{n}k$ that differs from one of these by a multiple of 2π. Of course $k = 0$ gives the root $z = 1$, so all the roots are obtained by dividing the unit circle into n equal sectors, starting at $z = 1$. The 5th roots of unity are plotted in the diagram.

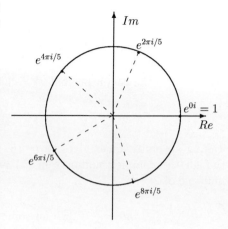

2.5.7 Proof of Theorem 3

If Z is a complex matrix, the **conjugate matrix** $\overline{Z}$ is defined to be the matrix obtained from Z by conjugating every entry. Thus if $Z = [z_{ij}]$ then $\overline{Z} = [\overline{z}_{ij}]$. For example, if

$$Z = \begin{bmatrix} -i+2 & 5 \\ i & 3+4i \end{bmatrix} \text{ then } \overline{Z} = \begin{bmatrix} i+2 & 5 \\ -i & 3-4i \end{bmatrix}.$$

It follows from C1 and C2 above that, if Z and W are two complex matrices and λ is a complex number then

$$\overline{Z + W} = \overline{Z} + \overline{W}$$
$$\overline{ZW} = \overline{Z}\,\overline{W}$$
$$\overline{(\lambda Z)} = \overline{\lambda}\,\overline{Z}$$

Now suppose that A is a real symmetric matrix. Observe that $\overline{A} = A$ because A is real (using C4 above). If λ is an eigenvalue of A, we show that λ is real by showing that $\overline{\lambda} = \lambda$ (again by C4). Let X be a (possibly complex) eigenvector corresponding to λ, so that $X \neq 0$ and $AX = \lambda X$. Define $c = X^T\overline{X}$. If we write $X = [z_1 \; z_2 \; \cdots \; z_n]^T$ where the z_i are complex numbers, we have

$$\begin{aligned} c &= X^T\overline{X} = [z_1 \; z_2 \; \cdots \; z_n]\begin{bmatrix} \overline{z_1} \\ \overline{z_2} \\ \vdots \\ \overline{z_n} \end{bmatrix} \\ &= z_1\overline{z_1} + z_2\overline{z_2} + \cdots + z_n\overline{z_n} \\ &= |z_1|^2 + |z_2|^2 + \cdots + |z_n|^2. \end{aligned}$$

Thus c is a real number, and $c > 0$ because at least one of the $z_i \neq 0$ (as $X \neq 0$). We show that $\overline{\lambda} = \lambda$ by verifying that $\lambda c = \overline{\lambda} c$. We have

$$\lambda c = \lambda(X^T\overline{X}) = (\lambda X)^T \overline{X} = (AX)^T \overline{X} = X^T A^T \overline{X}.$$

2.5. COMPLEX EIGENVALUES

At this point we use the hypothesis that A is symmetric and real. This means $A^T = A = \overline{A}$ so we continue

$$\lambda c = X^T A^T \overline{X} = X^T(\overline{A}\,\overline{X}) = X^T\overline{(AX)} = X^T\overline{(\lambda X)} = X^T(\overline{\lambda}\,\overline{X}) = \overline{\lambda}X^T\overline{X} = \overline{\lambda}c$$

as required. The technique in this proof will be used again in Chapter 4.

Exercises

Throughout these exercises, z and w denote complex numbers.

1. Write each of the following in the form $a + bi$.
   (a) $(2 - 3i) - 3(5 + 7i) + 4$
   (b) $(4 - i)(2i - 5) + |3 - 2i|$
   (c) i^{11}
   (d) $(2 - 3i)^3$
   (e) $\frac{3-i}{2i+5}$
   (f) $\frac{3-5i}{7-i}$
   (g) $\frac{1-i}{2-3i} - \frac{1+2i}{5+i}$
   (h) $\frac{3-i}{2-i} + 3\frac{5i}{2i+1}$
   (i) $\frac{|2-3i|}{|2+3i|}$
   (j) $\left|\frac{3+4i}{|3+4i|}\right|$

2. In each case solve for the real number x.
   (a) $(2 + xi)(3 - 2i) = 12 + 5i$
   (b) $(2 + xi)^2 = 4$

3. In each case solve for the complex number z.
   (a) $iz + (1 - i)^2 = 5 - 2i$
   (b) $(i + z) - 3i(2 - z) = iz + 1$
   (c) $z(1 + i) = \bar{z} - (3 + 2i)$
   (d) $z(2 - i) = (\bar{z} + 1)(1 + i)$

4. Find all complex numbers z such that $|z| = 1$ and $z/\bar{z} = -1$.

5. (a) Show that $z + \bar{z}$ is a real number for each complex number z.
   (b) What can you say about $z - \bar{z}$?

6. Let $z = a + bi$ and $w = a' + b'i$ be arbitrary complex numbers. In each case verify the identity by computing each side separately and comparing.
   (a) C1. $\overline{(z \pm w)} = \bar{z} \pm \bar{w}$
   (b) C2. $\overline{(zw)} = \bar{z}\bar{w}$
   (c) C3. $\overline{(\bar{z})} = z$

7. In each case either show that the statement is true or give an example showing that it is false.
   (a) If $a^2 + b^2 = 0$ where a and b are real numbers, then $a = 0$ and $b = 0$.
   (b) If $z^2 + w^2 = 0$ where z and w are complex numbers, then $z = 0$ and $w = 0$.
   (c) If $z = 1/z$ then $z = \pm 1$.
   (d) If $|z| = 1$ then $z = 1, -1, i$ or $-i$.
   (e) If $|z| = |w|$ then $z = \pm w$.

8. Find the roots of the following quadratic polynomials:
   (a) $x^2 - 1$
   (b) $x^2 + 1$
   (c) $3x^2 - 5x + 10$
   (d) $3x^2 + 5x + 10$
   (e) $-3x^2 - 5x - 10$
   (f) $-3x^2 + 5x + 10$

9. In each case find a real irreducible quadratic with λ as a root:
   (a) $\lambda = 1 - 2i$
   (b) $\lambda = 3i$
   (c) $\lambda = -2i$
   (d) $\lambda = 2 - 5i$

10. Solve each of the following systems of equations.

    (a) $\begin{aligned} x + iy - iy &= 3+i \\ -ix + 2y + iz &= 2 \\ (i-1)x - (1+2i)y + 2z &= i-1 \end{aligned}$

    (b) $\begin{aligned} x + (1+i)y - iy &= 2 \\ (1+i)x + (1-i)z &= 3+i \\ -x + (1-i)y + iz &= i-1 \end{aligned}$

11. Find the inverse of the following matrices if possible.

    (a) $\begin{bmatrix} 1+i & -2 \\ 2i & -1-i \end{bmatrix}$
    (b) $\begin{bmatrix} 1 & 1-i \\ 2+i & 3+i \end{bmatrix}$

    (c) $\begin{bmatrix} 1+2i & -i & 3+5i \\ 2 & 1+4i & 3-i \\ -i & 5+i & 4-2i \end{bmatrix}$
    (d) $\begin{bmatrix} 1-i & 1-i & 1+i \\ 2+i & 3+i & 4+i \\ -3i & 2-i & 5+5i \end{bmatrix}$

12. In each case find if possible an invertible matrix P such that $P^{-1}AP = D$ is diagonal.

    (a) $\begin{bmatrix} -2 & -1 \\ 5 & 2 \end{bmatrix}$
    (b) $\begin{bmatrix} 1 & i \\ -i & 1 \end{bmatrix}$

    (c) $\begin{bmatrix} -3 & 5 & 4 \\ -2 & 3 & 2 \\ 0 & 0 & 1 \end{bmatrix}$
    (d) $\begin{bmatrix} -5 & 6 & 2 \\ -3 & 4 & 1 \\ -5 & 5 & 2 \end{bmatrix}$

    (e) $\begin{bmatrix} 2 & 51 & -53 \\ 10 & -32 & -24 \\ -11 & 19 & -78 \end{bmatrix}$
    (f) $\begin{bmatrix} -91 & 70 & -23 \\ 32 & 14 & 16 \\ 25 & 19 & -19 \end{bmatrix}$

    (g) $\begin{bmatrix} -1-i & -3+i & -2+2i \\ 1+i & -3+2i & -i \\ -3+5i & 2+3i & 3+5i \end{bmatrix}$
    (h) $\begin{bmatrix} 2+i & 2-i & -4+i \\ 3-i & 1+2i & 6+5i \\ 5-i & 6+i & 2+7i \end{bmatrix}$

13. Let $A = \begin{bmatrix} \cos\theta & -\sin\theta \\ \sin\theta & \cos\theta \end{bmatrix}$ where θ is any angle.

    (a) Show that the characteristic polynomial of A is $c_A(x) = x^2 - 2\cos\theta\, x + 1$.

2.5. COMPLEX EIGENVALUES

(b) Show that the eigenvalues of A are $\lambda_1 = \cos\theta + i\sin\theta$ and $\lambda_2 = \cos\theta - i\sin\theta$.

(c) Find an invertible matrix P such that $P^{-1}AP = \begin{bmatrix} \lambda_1 & 0 \\ 0 & \lambda_2 \end{bmatrix}$.

14. If Z and W are complex matrices, show that $\overline{ZW} = \bar{Z}\bar{W}$.

15. If λ is an eigenvalue (possibly not real) of a real matrix A, show that $\bar{\lambda}$ is also an eigenvalue.

16. If λ is a root of a real polynomial $f(x)$, show that $\bar{\lambda}$ is also a root.

17. Show that $|\bar{z}| = |z|$ for any complex number z.

18. Describe the set of all complex numbers z such that $|z| = 1$.

19. Show that $\left|\frac{z}{w}\right| = \frac{|z|}{|w|}$ whenever $w \neq 0$.

20. Express each of the following in polar form.

    (a) $2 - 2i$ (b) $-2i$
    (c) $-\sqrt{3} + i$ (d) $-3 - \sqrt{3}i$
    (e) $5i$ (f) $5(1 + i)$

21. Express each of the following in the form $a + bi$.

    (a) $3e^{\pi i}$ (b) $e^{5\pi i/3}$
    (c) $2e^{\pi i/4}$ (d) $\sqrt{2}e^{-\pi i/4}$
    (e) $e^{15\pi/4}$ (f) $2\sqrt{3}e^{-\pi i/3}$

22. Express each of the following in the form $a + bi$.

    (a) $(1 - \sqrt{3}i)^5$ (b) $(\sqrt{3} + i)^{-4}$
    (b) $(1 + i)^{12}$ (d) $(1 - i)^{10}$
    (e) $(1 - i)^6(\sqrt{3} + i)^3$ (f) $(\sqrt{3} - i)^9(2 - 2i)^5$

23. Find all: (a) Fourth roots of unity. (b) Cube roots of unity.

24. Find all complex numbers z such that:

    (a) $z^4 = -1$. (b) $z^4 = 2(\sqrt{3}i - 1)$
    (c) $z^3 = -27i$ (d) $z^6 = -64$

25. Use DeMoivre's theorem to show that:

    (a) $cos(2\theta) = cos^2\theta - sin^2\theta$ and $sin(2\theta) = 2\cos\theta \sin\theta$.
    (b) $cos(3\theta) = cos^3\theta - 3\cos\theta \sin^2\theta$ and $sin(2\theta) = 3\cos^2\theta \sin\theta - sin^3\theta$.

26. Show that $cos\theta = \frac{1}{2}(e^{i\theta} + e^{-i\theta})$ and $sin\theta = \frac{1}{2i}(e^{i\theta} - e^{-i\theta})$ for any angle θ.

2.6 LINEAR RECURRENCES

A sequence $x_0, x_1, x_2, \cdots$ of numbers is said to be given **recursively** if each number in the sequence is completely determined by those that come before it. Such sequences arise frequently in mathematics and computer science, and also occur in other parts of science. The simplest recursive sequence occurs when x_{k+1} is a fixed multiple of x_k for each k, say $x_{k+1} = ax_k$. If x_0 is specified, the remaining x_k can be computed successively:

$$\begin{aligned} x_1 &= ax_0 \\ x_2 &= ax_1 = a^2 x_0 \\ x_3 &= ax_2 = a^3 x_0 \\ &\vdots \end{aligned}$$

Clearly $x_k = a^k x_0$ holds for each $k \geq 0$, giving an explicit formula for x_k as a function of k (when x_0 is given).

2.6.1 Linear Recurrences

Such formulas are not always so easy to find. Suppose the numbers $x_0, x_1, x_2, \cdots$ are given by a **linear recurrence relation**

$$x_{k+2} = x_{k+1} + 2x_k \quad \text{for} \quad k \geq 0$$

where x_0 and x_1 are specified. If $x_0 = 1$ and $x_1 = 2$ the sequence is easily determined:

$$\begin{aligned} x_0 &= 1 \\ x_1 &= 2 \\ x_2 &= x_1 + 2x_0 = 4 \\ x_3 &= x_2 + 2x_1 = 8 \\ x_4 &= x_3 + 2x_2 = 16 \\ &\vdots \end{aligned}$$

It is clear that $x_k = 2^k$ holds for $k = 0, 1, 2, 3$ and 4, and this formula satisfies the recurrence $x_{k+2} = x_{k+1} + 2x_k$ as is readily checked.

However if we take $x_0 = 1$ and $x_1 = 1$ instead, the sequence continues $x_2 = 3, x_3 = 5, x_4 = 11, x_5 = 21, \cdots$. In this case, the sequence is uniquely determined but no formula is apparent. Nontheless, a simple device transforms the recurrence into a matrix recurrence, which we can solve easily.

Example 1. *Determine the sequence* $x_0, x_1, x_2, \cdots$ *if* $x_0 = 1$, $x_1 = 1$ *and* $x_{k+2} = x_{k+1} + 2x_k$ *for each* $k \geq 0$.

Solution. The idea is to consider the sequence $V_0, V_1, V_2, \cdots$ instead of $x_0, x_1, x_2, \cdots$ where

$$V_k = \begin{bmatrix} x_k \\ x_{k+1} \end{bmatrix}.$$

2.6. LINEAR RECURRENCES

Then $V_0 = \begin{bmatrix} x_0 \\ x_1 \end{bmatrix} = \begin{bmatrix} 1 \\ 1 \end{bmatrix}$ is specified, and the numerical recurrence $x_{k+2} = x_{k+1} + 2x_k$ transforms into a matrix recurrence as follows:

$$V_{k+1} = \begin{bmatrix} x_{k+1} \\ x_{k+2} \end{bmatrix} = \begin{bmatrix} x_{k+1} \\ x_{k+1} + 2x_k \end{bmatrix} = \begin{bmatrix} 0 & 1 \\ 2 & 1 \end{bmatrix} \begin{bmatrix} x_k \\ x_{k+1} \end{bmatrix} = AV_k$$

where $A = \begin{bmatrix} 0 & 1 \\ 2 & 1 \end{bmatrix}$. Thus Theorem 1§2.4 applies if the matrix A is diagonalizable. We have

$$c_A(x) = \det \begin{bmatrix} x & -1 \\ -2 & x-1 \end{bmatrix} = x(x-1) - 2 = (x-2)(x+1).$$

Hence the eigenvalues $\lambda_1 = 2$ and $\lambda_2 = -1$ are distinct so A is indeed diagonalizable. Corresponding eigenvectors are $X_1 = \begin{bmatrix} 1 \\ 2 \end{bmatrix}$ and $X_2 = \begin{bmatrix} 1 \\ -1 \end{bmatrix}$ as the reader can check, so $P = \begin{bmatrix} 1 & 1 \\ 2 & -1 \end{bmatrix}$ is a diagonalizing matrix for A. We compute the coefficients b_i in Theorem 1 §2.4:

$$\begin{bmatrix} b_1 \\ b_2 \end{bmatrix} = P^{-1}V_0 = \frac{1}{-3}\begin{bmatrix} -1 & -1 \\ -2 & 1 \end{bmatrix}\begin{bmatrix} 1 \\ 1 \end{bmatrix} = \begin{bmatrix} \frac{2}{3} \\ \frac{1}{3} \end{bmatrix}.$$

Thus Theorem 1 §2.4 gives

$$\begin{bmatrix} x_k \\ x_{k+1} \end{bmatrix} = V_k = b_1 \lambda_1^k X_1 + b_2 \lambda_2^k X_2 = \tfrac{2}{3} 2^k \begin{bmatrix} 1 \\ 2 \end{bmatrix} + \tfrac{1}{3}(-1)^k \begin{bmatrix} 1 \\ -1 \end{bmatrix}.$$

Equating top entries yields

$$x_k = \tfrac{1}{3}\left(2^{k+1} + (-1)^k\right) \qquad \text{for } k \geq 0$$

which is the desired formula for the x_k.

Note that we can write $x_k = \tfrac{1}{3}2^{k+1}\left(1 + \tfrac{1}{2}(\tfrac{-1}{2})^k\right) \approx \tfrac{1}{3}2^{k+1}$ for large k. (This amounts to the observation that $\lambda_1 = 2$ is the dominant eigenvalue here.) This formula is quite useful in this case because the approximation is good even for small k. For example, we get $x_6 \approx \tfrac{1}{3}2^{6+1} = \tfrac{128}{3} = 42.66$, which compares favorably with the true value $x_5 = 43$. In fact, for large k, x_k is the integer nearest to $\tfrac{1}{3}2^{k+1}$. □

Linear recurrences arise frequently in combinatorial problems where the object is to count the number of ways to do something. Here is an example.

Example 2. *An urban planner wants to determine the number x_k of ways that a row of k parking spaces can be filled with cars and trucks if trucks take up two spaces each. Find a formula for x_k and estimate it for large k.*

Solution. Clearly $x_0 = 1$ and $x_1 = 1$, while $x_2 = 2$ since there can only be two cars or one truck. We have $x_3 = 3$ (the 3 configurations are ccc, cT and Tc) and $x_4 = 5$ ($cccc$, ccT, cTc, Tcc, and TT). We claim that

$$x_{k+2} = x_k + x_{k+1} \quad \text{for every } k \geq 0. \tag{*}$$

Indeed, every way to fill $k + 2$ spaces falls into one of two categories: Either a car is parked in the first space (and the remaining $k+1$ spaces are filled in x_{k+1} ways), or a truck is parked in the first two spaces (with the other k spaces filled in x_k ways). Hence there are $x_{k+1} + x_k$ ways to fill the $k + 2$ spaces. This is (*).

The recurrence (*) determines x_k for every $k \geq 2$ since x_0 and x_1 are given. In fact the first few values are

$$\begin{aligned}
x_0 &= 1 \\
x_1 &= 1 \\
x_2 &= x_0 + x_1 = 2 \\
x_3 &= x_1 + x_2 = 3 \\
x_4 &= x_2 + x_3 = 5 \\
x_5 &= x_3 + x_4 = 8 \\
x_6 &= x_4 + x_5 = 13 \\
&\vdots
\end{aligned}$$

As no formula for x_k is apparent here, we turn to diagonalization. If we write $V_k = \begin{bmatrix} x_k \\ x_{k+1} \end{bmatrix}$ the recurrence (*) becomes a matrix recursion for the V_k:

$$\begin{aligned}
V_{k+1} = \begin{bmatrix} x_{k+1} \\ x_{k+2} \end{bmatrix} &= \begin{bmatrix} x_{k+1} \\ x_k + x_{k+1} \end{bmatrix} \\
&= \begin{bmatrix} 0 & 1 \\ 1 & 1 \end{bmatrix} \begin{bmatrix} x_k \\ x_{k+1} \end{bmatrix} \\
&= A V_k \quad \text{for all } k \geq 0
\end{aligned}$$

where $A = \begin{bmatrix} 0 & 1 \\ 1 & 1 \end{bmatrix}$. Moreover A is diagonalizable here. The characteristic polynomial is $c_A(x) = \det \begin{bmatrix} x & -1 \\ -1 & x-1 \end{bmatrix} = x^2 - x - 1$. This has roots $\frac{1}{2}[1 \pm \sqrt{5}]$ by the quadratic formula, so A has distinct eigenvalues

$$\lambda_1 = \tfrac{1}{2}[1 + \sqrt{5}] \quad \text{and} \quad \lambda_2 = \tfrac{1}{2}[1 - \sqrt{5}].$$

Since $\lambda_i^2 = \lambda_i + 1$ for each i, corresponding eigenvectors are $X_1 = \begin{bmatrix} 1 \\ \lambda_1 \end{bmatrix}$ and $X_2 = \begin{bmatrix} 1 \\ \lambda_2 \end{bmatrix}$ respectively as the reader can verify. Hence the diagonalizing

2.6. LINEAR RECURRENCES

matrix is $P = \begin{bmatrix} 1 & 1 \\ \lambda_1 & \lambda_2 \end{bmatrix}$, so we compute the coefficients b_1 and b_2 (in Theorem 1 §2.4) as follows

$$\begin{bmatrix} b_1 \\ b_2 \end{bmatrix} = P^{-1}V_0 = \frac{1}{-\sqrt{5}}\begin{bmatrix} \lambda_2 & -1 \\ -\lambda_1 & 1 \end{bmatrix}\begin{bmatrix} 1 \\ 1 \end{bmatrix} = \frac{1}{\sqrt{5}}\begin{bmatrix} \lambda_1 \\ -\lambda_2 \end{bmatrix}$$

where we used the fact that $\lambda_1 + \lambda_2 = 1$. Thus Theorem 1 §2.4 gives

$$\begin{bmatrix} x_k \\ x_{k+1} \end{bmatrix} = V_k = b_1\lambda_1^k X_1 + b_2\lambda_2^k X_2 = \frac{\lambda_1}{\sqrt{5}}\lambda_1^k\begin{bmatrix} 1 \\ \lambda_1 \end{bmatrix} - \frac{\lambda_2}{\sqrt{5}}\lambda_2^k\begin{bmatrix} 1 \\ \lambda_2 \end{bmatrix}.$$

Comparing first entries gives an exact formula for the numbers x_k:

$$x_k = \tfrac{1}{\sqrt{5}}[\lambda_1^{k+1} - \lambda_2^{k+1}] \qquad \text{for each } k \geq 0.$$

Finally, observe that λ_1 is dominant here (in fact $\lambda_1 = 1.618$ and $\lambda_2 = -0.618$ to three figures) so λ_2^{k+1} is negligible compared with λ_1^{k+1} if k is large. Thus

$$x_k \approx \tfrac{1}{\sqrt{5}}\lambda_1^{k+1} \qquad \text{for each } k \geq 0.$$

(This also follows from Theorem 1 §2.4.) This is a good approximation, even for as small a value as $k = 12$. Indeed, repeated use of the recurrence $x_{k+2} = x_k + x_{k+1}$ gives the exact value $x_{12} = 233$, while the approximation is $x_{12} \approx \frac{(1.618)^{13}}{\sqrt{5}} = 232.94$. $\square$

The sequence $x_0, x_1, x_2, \cdots$ in Example 2 was first discussed in 1202 by Leonardo Pisano of Pisa, also known as Fibonacci, and is now called the **Fibonacci sequence**. It is completely determined by the conditions $x_0 = 1$, $x_1 = 1$ and the recurrence $x_{k+2} = x_k + x_{k+1}$ for each $k \geq 0$. These numbers have been studied for centuries, and have many interesting properties (there is even a journal devoted exclusively to them). For example, biologists have discovered that the arrangement of leaves around the stems of some plants follow a Fibonacci pattern.

We conclude with an example showing that nonlinear recurrences can be complicated.

Example 3. Suppose a sequence $x_0, x_1, x_2, \cdots$ satisfies the following recurrence:

$$x_{k+1} = \begin{cases} \tfrac{1}{2}x_k & \text{if } x_k \text{ is even} \\ 3x_k + 1 & \text{if } x_k \text{ is odd} \end{cases}$$

If $x_0 = 1$, the sequence is $1, 4, 2, 1, 4, 2, 1, \cdots$ and so continues to cycle indefinitely. The same thing happens if $x_0 = 7$. Then the sequence is

$$7, 22, 11, 34, 17, 52, 26, 13, 40, 20, 10, 5, 16, 8, 4, 2, 1, \cdots$$

and it again cycles. However, it is not known whether every choice of x_0 will lead eventually to 1. It is quite possible that, for some x_0, the sequence will continue to produce different values indefinitely, or will repeat a value and cycle without reaching 1. No one knows for sure. $\square$

Exercises

1. Consider a sequence $x_0, x_1, x_2, \cdots$ satisfying the recurrence $x_{k+2} = 6x_k + x_{k+1}$.

    (a) If $x_0 = 1$ and $x_1 = 3$, compute the first five terms of the sequence and guess a formula for x_k.

    (b) If $x_0 = 1$ and $x_1 = 1$, find a formula for x_k.

2. Solve the following linear recurrences.

    (a) $x_{k+2} = 3x_k + 2x_{k+1}$, where $x_0 = 1$ and $x_1 = 1$.

    (b) $x_{k+2} = 2x_k - x_{k+1}$, where $x_0 = 1$ and $x_1 = 2$.

    (c) $x_{k+2} = 2x_k + x_{k+1}$, where $x_0 = 0$ and $x_1 = 1$.

    (d) $x_{k+2} = 6x_k - x_{k+1}$, where $x_0 = 1$ and $x_1 = 1$.

3. Solve the following linear recurrences.

    (a) $x_{k+3} = 6x_{k+2} - 11x_{k+1} + 6x_k$, where $x_0 = 1$, $x_1 = 0$ and $x_2 = 1$.

    (b) $x_{k+3} = -2x_{k+2} + x_{k+1} + 2x_k$, where $x_0 = 1$, $x_1 = 0$ and $x_2 = 1$.

4. Solve the following linear recurrences.

    (a) $x_{k+3} = x_{k+2} + x_{k+1} + x_k$, where $x_0 = 1$, $x_1 = 1$ and $x_2 = 1$.

    (b) $x_{k+3} = 2x_{k+2} - x_{k+1} - 2x_k$, where $x_0 = 1$, $x_1 = 1$ and $x_2 = 1$.

    (c) $x_{k+5} = 2x_{k+4} - x_{k+3} - 2x_{k+2} - x_{k+1} - x_k$, where $x_0 = 1$, $x_1 = 0$, $x_2 = 1$, $x_3 = 1$, and $x_4 = 1$.

5. (a) Consider the recurrence $x_{k+2} = ax_k + bx_{k+1}$. Show that the corresponding matrix A has characteristic polynomial $x^2 - bx - a$, eigenvalues $\lambda_1 = \frac{1}{2}\left[b + \sqrt{b^2 + 4a}\right]$ and $\lambda_2 = \frac{1}{2}\left[b - \sqrt{b^2 + 4a}\right]$, and corresponding eigenvectors $X_i = \begin{bmatrix} 1 \\ \lambda_i \end{bmatrix}$ for each i.

    (b) If $x_0 = 0$, $x_1 = 1$, use the diagonalizing matrix $P = [X_1 \; X_2]$ for A to get $V_k = \frac{1}{\sqrt{b^2+4a}}\left[\lambda_1^k X_1 - \lambda_2^k X_2\right]$, and hence $x_k = \frac{1}{\sqrt{b^2+4a}}\left[\lambda_1^k - \lambda_2^k\right]$ for all $k \geq 0$.

6. Consider $x_{k+2} = -5x_k + 2x_{k+1}$; $x_0 = 0$, $x_1 = 1$. Show that the eigenvalues are $\lambda_1 = 1 + 2i$, $\lambda_2 = 1 - 2i$, with eigenvectors $X_1 = \begin{bmatrix} 1 \\ 1+2i \end{bmatrix}$, $X_2 = \begin{bmatrix} 1 \\ 1-2i \end{bmatrix}$. Use Exercise 5 to get the formula

$$x_k = \frac{i}{4}\left[(1-2i)^k - (1+2i)^k\right] \quad \text{for } k \geq 0.$$

7. A man must climb a flight of k steps. He always takes one or two steps at a time. Thus he can climb 3 steps in the following ways: 111, 12, or 21. Find the number s_k of ways he can climb the flight of k steps.

2.6. LINEAR RECURRENCES

8. How many "words" of k letters can be made from the letters $\{a, b\}$ if there are no adjacent a's.

9. How many sequences of k flips of a coin are there with no HH?

10. Find the number x_k of ways to make a stack of k poker chips if only red, blue and gold chips are used and no two gold chips are adjacent.

11. A nuclear reactor contains α- and β-particles. In every second each α-particle splits into three β-particles, and each β-particle splits into an α-particle and two β-particles. If there is a single α-particle in the reactor at time $t = 0$, how many α-particles are there at $t = 20$ seconds?

12. The annual yield of wheat in a certain country has been found to equal the average of the yield in the previous two years. If the yields in 1990 and 1991 were 10 and 12 million tons respectively, find a formula for the yield k years after 1990. What is the long-term average yield?

13. Find the general solution to the recurrence $x_{k+1} = rx_k + c$ where r and c are constants.

14. Consider a linear recurrence $x_{k+2} = ax_{k+1} + bx_k + c$ where c may not be zero.

    (a) If $a + b \neq 1$ show that p can be found such that, if we set $y_k = x_k + p$, then $y_{k+2} = ay_{k+1} + by_k$. [Hence the sequence x_k can be found provided y_k can be found by the methods of this section (or otherwise).]

    (b) Use (a) to solve the recurrence $x_{k+2} = x_{k+1} + 6x_k + 5$ where $x_0 = 1$ and $x_1 = 1$.

15. Consider the linear recurrence

    $$x_{k+2} = ax_{k+1} + bx_k + c(k) \qquad (*)$$

    where $c(k)$ is a function of k. Also consider the related recurrence

    $$x_{k+2} = ax_{k+1} + bx_k. \qquad (**)$$

    Suppose that $x_k = p_k$ is a particular solution of $(*)$.

    (a) If q_k is any solution of $(**)$, show that $q_k + p_k$ is a solution of $(*)$.

    (b) Show that every solution of $(*)$ arises as in (a) as the sum of a solution of $(**)$ plus the particular solution p_k of $(**)$.

2.7 POLYNOMIAL INTERPOLATION

In experimental science it is frequently necessary to use known data to estimate some quantity and there are many methods to do this. In this section, we present the method of polynomial interpolation.

2.7.1 Polynomial Interpolation

Example 1. An engineer wants to estimate the deflection of a wooden beam under a load of 2200 kg.

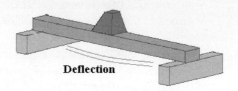
Deflection

He knows the data in the following table

LOAD (KG)	1000	2000	3000
DEFLECTION (MM)	1.5	2.1	3.4

and wants to "interpolate" to find the estimate he needs. A common way to do this is to find a quadratic $p(x)$ which fits the data exactly in the sense that

$$p(1000) = 1.5$$
$$p(2000) = 2.1$$
$$p(3000) = 3.4$$

Then the estimate would be $p(2200)$. If $p(x) = r_0 + r_1 x + r_2 x^2$, these conditions become three equations in the unknown coefficients r_0, r_1 and r_2:

$$r_0 + 1000\, r_1 + 1000^2 r_2 = 1.5$$
$$r_0 + 2000\, r_1 + 2000^2 r_2 = 2.1$$
$$r_0 + 3000\, r_1 + 3000^2 r_2 = 3.4$$

The solution is $r_0 = 1.6$, $r_1 = -.00045$ and $r_2 = .00000035$, so the polynomial is

$$p(x) = 1.6 - .00045x + .00000035x^2.$$

The engineer uses $p(2200) = 2.304$ as the estimate of the deflection. □

It often happens that two variables x and y are known to be related but the actual function $y = f(x)$ relating them is not known. Suppose that for certain values $x_1, x_2, \cdots, x_n$ of x, the corresponding values $y_1, y_2, \cdots, y_n$ are known (say from tables or experimental measurements), and it is desired to estimate the value of y corresponding to some other value of x. As in Example 1, one way to do this is to find a polynomial $p(x)$ that "fits" the data in the sense

2.7. POLYNOMIAL INTERPOLATION

that $p(x_i) = y_i$ for each $i = 1, 2, \cdots, n$. Then the estimate for y is $p(x)$. Such a polynomial always exists if the x_i are distinct.

Theorem 1. *Suppose n data pairs $(x_1, y_1), (x_2, y_2), \cdots, (x_n, y_n)$ are given where the x_i are distinct. Then there exists a unique polynomial*

$$p(x) = r_0 + r_1 x + r_2 x^2 + \cdots + r_{n-1} x^{n-1}$$

such that $p(x_i) = y_i$ for each $i = 1, 2, \cdots, n$.

Proof. We consider the case $n = 4$; the general case is entirely similar. The given conditions that $p(x_i) = y_i$ for each $i = 1, 2, 3$ and 4 give equations

$$\begin{aligned} r_0 + r_1 x_1 + r_2 x_1^2 + r_3 x_1^3 &= y_1 \\ r_0 + r_1 x_2 + r_2 x_2^2 + r_3 x_2^3 &= y_2 \\ r_0 + r_1 x_3 + r_2 x_3^2 + r_3 x_3^3 &= y_3 \\ r_0 + r_1 x_4 + r_2 x_4^2 + r_3 x_4^3 &= y_4 \end{aligned}$$

In matrix form this system can be written as $AR = Y$ where

$$A = \begin{bmatrix} 1 & x_1 & x_1^2 & x_1^3 \\ 1 & x_2 & x_2^2 & x_2^3 \\ 1 & x_3 & x_3^2 & x_3^3 \\ 1 & x_4 & x_4^2 & x_4^3 \end{bmatrix}, \quad R = \begin{bmatrix} r_0 \\ r_1 \\ r_2 \\ r_3 \end{bmatrix} \text{ and } Y = \begin{bmatrix} y_1 \\ y_2 \\ y_3 \\ y_4 \end{bmatrix}.$$

Hence it suffices to show that the matrix A is invertible when the x_i are distinct (then $R = A^{-1} Y$ is the column of coefficients of the polynomial we want).

We show that A is invertible by verifying that the only solution to the homogeneous system $AX = 0$ is the trivial solution $X = 0$ (Theorem 5 §1.5). If we write $X = \begin{bmatrix} t_0 & t_1 & t_2 & t_3 \end{bmatrix}^T$, consider the polynomial $f(x) = t_0 + t_1 x + t_2 x^2 + t_3 x^3$ with the entries of X as coefficients. Then we have

$$AX = \begin{bmatrix} 1 & x_1 & x_1^2 & x_1^3 \\ 1 & x_2 & x_2^2 & x_2^3 \\ 1 & x_3 & x_3^2 & x_3^3 \\ 1 & x_4 & x_4^2 & x_4^3 \end{bmatrix} \begin{bmatrix} t_0 \\ t_1 \\ t_2 \\ t_3 \end{bmatrix} = \begin{bmatrix} f(x_1) \\ f(x_2) \\ f(x_3) \\ f(x_4) \end{bmatrix}.$$

Hence the requirement that $AX = 0$ means $f(x_i) = 0$ for each $i = 1, 2, 3$ and 4. In other words, $f(x)$ is a polynomial of degree at most 3 which has 4 distinct roots x_1, x_2, x_3 and x_4. This can only happen if $f(x)$ is identically zero[18], that is if $t_0 = t_1 = t_2 = t_3 = 0$. This means that $X = 0$ as required. □

The polynomial $p(x)$ in Theorem 1 is called the **interpolating polynomial** for the data $(x_1, y_1), (x_2, y_2), \cdots, (x_n, y_n)$. It is an exact fit in the sense that

---

[18] Since $f(x_1) = 0$ the factor theorem asserts that $f(x) = (x - x_1) g(x)$ where $g(x)$ is another polynomial. Since $f(x)$ has *four* distinct roots, continuing this process means that all four terms $(x - x_1)$, $(x - x_2)$, $(x - x_3)$, and $(x - x_4)$ are factors of $f(x)$, contrary to the fact that $f(x)$ has degree at most 3. See Appendix A.3.

$p(x_i) = y_i$ holds for each i. Finding $p(x)$ may not be feasible if n is large, and polynomials of lower degree which "best fit" the data can be found, for example by the method of least squares (see Section 4.6.5).

Matrices like A in the proof of Theorem 1, and their determinants, arise in other contexts and so have independent interest. Given numbers $x_1, x_2, \cdots, x_n$ the matrix

$$\begin{bmatrix} 1 & x_1 & x_1^2 & \cdots & x_1^{n-1} \\ 1 & x_2 & x_2^2 & \cdots & x_2^{n-1} \\ 1 & x_3 & x_3^2 & \cdots & x_3^{n-1} \\ \vdots & \vdots & \vdots & & \vdots \\ 1 & x_n & x_n^2 & \cdots & x_n^{n-1} \end{bmatrix}$$

(or its transpose) is called the **Vandermonde matrix**[19] corresponding to the numbers $x_1, x_2, \cdots, x_n$, and its determinant is called the **Vandermonde determinant**. There is a simple formula for this determinant. The following example treats the case $n = 3$.

Example 2. *Show that* $\det \begin{bmatrix} 1 & x_1 & x_1^2 \\ 1 & x_2 & x_2^2 \\ 1 & x_3 & x_3^2 \end{bmatrix} = (x_2 - x_1)(x_3 - x_1)(x_3 - x_2)$.

Solution. We begin by using row operations to introduce zeros into column 1:

$$\begin{aligned} \det \begin{bmatrix} 1 & x_1 & x_1^2 \\ 1 & x_2 & x_2^2 \\ 1 & x_3 & x_3^2 \end{bmatrix} &= \det \begin{bmatrix} 1 & x_1 & x_1^2 \\ 0 & x_2 - x_1 & x_2^2 - x_1^2 \\ 0 & x_3 - x_1 & x_3^2 - x_1^2 \end{bmatrix} \\ &= \det \begin{bmatrix} 1 & x_1 & x_1^2 \\ 0 & x_2 - x_1 & (x_2 - x_1)(x_2 + x_1) \\ 0 & x_3 - x_1 & (x_3 - x_1)(x_3 + x_1) \end{bmatrix} \end{aligned}$$

Now take the common factor out of each of rows 2 and 3:

$$\begin{aligned} \det \begin{bmatrix} 1 & x_1 & x_1^2 \\ 1 & x_2 & x_2^2 \\ 1 & x_3 & x_3^2 \end{bmatrix} &= (x_2 - x_1)(x_3 - x_1) \det \begin{bmatrix} 1 & x_1 & x_1^2 \\ 0 & 1 & x_2 + x_1 \\ 0 & 1 & x_3 + x_1 \end{bmatrix} \\ &= (x_2 - x_1)(x_3 - x_1) \det \begin{bmatrix} 1 & x_2 + x_1 \\ 1 & x_3 + x_1 \end{bmatrix} \\ &= (x_2 - x_1)(x_3 - x_1)(x_3 - x_2). \end{aligned}$$

The result in Example 2 is typical of what happens for any Vandermonde determinant. The result is stated in full generality in the following theorem.

---

[19] After A.T. Vandermonde (1735-1796), a French mathematician.

2.7. POLYNOMIAL INTERPOLATION

Theorem 2. Vandermonde Determinant. *Let $x_1, x_2, \cdots, x_n$ be real numbers where $n \geq 2$. Then the corresponding Vandermonde determinant is given by*

$$\det \begin{bmatrix} 1 & x_1 & x_1^2 & \cdots & x_1^{n-1} \\ 1 & x_2 & x_2^2 & \cdots & x_2^{n-1} \\ 1 & x_3 & x_3^2 & \cdots & x_3^{n-1} \\ \vdots & \vdots & \vdots & & \vdots \\ 1 & x_n & x_n^2 & \cdots & x_n^{n-1} \end{bmatrix} = \prod_{1 \leq j < i \leq n} (x_i - x_j)$$

where $\prod_{1 \leq j < i \leq n}(x_i - x_j)$ means the product of all factors $(x_i - x_j)$ where $j < i$ and both i and j are between 1 and n.

We omit the proof. To illustrate the theorem we state the expansion of the 4×4 Vandermonde determinant:

$$\det \begin{bmatrix} 1 & x_1 & x_1^2 & x_1^3 \\ 1 & x_2 & x_2^2 & x_2^3 \\ 1 & x_3 & x_3^2 & x_3^3 \\ 1 & x_4 & x_4^2 & x_4^3 \end{bmatrix} =$$

$$(x_2 - x_1)(x_3 - x_1)(x_3 - x_2)(x_4 - x_1)(x_4 - x_2)(x_4 - x_3).$$

Exercises

1. In each case find the interpolating polynomial for the given data, and use the polynomial to estimate the value of y corresponding to the given value of x.

    (a) $(0, 2), (1, 3), (3, 8)$; $x = 2$.
    (b) $(0, 5), (1, 3), (2, 5)$; $x = 1.5$.
    (c) $(0, 1), (1, 1), (-1, 4), (2, 5)$; $x = \frac{1}{2}$.
    (d) $(0, 1), (1, 1), (-1, 2), (-2, -3)$; $x = -\frac{1}{2}$.

2. Some observations of the growth of a stock are given in the following table:

Week	1	2	3
Value ($)	1.22	1.49	2.45

    (a) Use polynomial interpolation to estimate the value of the stock in weeks 4 and 10.

    (b) The value turned out to be $3.75 in week 4, now use polynomial interpolation again to estimate the value of the stock in weeks 5 and 10.

    (c) Argue whether this method provides valuable information or not.

3. Consider a ski jump starting at a height of 40 meters, finishing at a height of 5 meters, covering from start to finish a horizontal distance of 35 meters.

   (a) Using polynomial interpolation, design the jump so that, 30 horizontal meters from takeoff, the skier is 7.5 meters high.

   (b) If you know a bit of Calculus, design the jump so that further the angle at takeoff is $0°$ with the horizontal.

   (c) Specify now that the jump forms a straight straight line descent for the first 25 horizontal and 20 vertical meters. Design the jump so that the line segment and the remaining curve have the same first derivative (tangent) at the point of intersection.

   (d) Do as in (c), but require also that the second derivatives also match at the point of intersection.

4. *Cubic Spline* In some applications, a better choice of a curve through several points is obtained by joining cubic (i.e., degree 3) polynomials through consecutive points with matching first and second derivatives at the given points.

   So a *cubic spline* through data points $(0, 1)$, $(1, 3)$, and $(3, 2)$ for example, is formed by 2 cubic polynomials of the form

   $$s_1 = a_1 x^3 + b_1 x^2 + c_1 x + d_1 \text{ and } s_2 = a_2 x^3 + b_2 x^2 + c_2 x + d_2$$

   satisfying the 6 equations (in the 8 unknowns $a_1, a_2, b_1, b_2, c_1, c_2, d_1, d_2$):

   $$s_1(0) = 1, \; s_1(1) = 3, \; s_2(1) = 3, \; s_2(3) = 2, \; s_1'(1) = s_2'(1), \; s_1''(1) = s_2''(1)$$

   To uniquely determine the two cubics, we also require some conditions at the endpoints, for example a vanishing second derivative:

   $$s_1''(0) = 0 \text{ and } s_2''(2) = 0$$

   (a) Solve the above system of 8 equations in 8 unknowns to find the two required cubic polynomials.

   (b) Find the cubic spline through the data points of Example 1, and use it to estimate the deflection of the beam under a load of 2200 kg. Compare with the results of polynomial interpolation and argue why there is a difference.

5. Let A denote the Vandermonde matrix corresponding to the numbers $x_1, x_2, \cdots, x_n$. In the proof of Theorem 1 it was shown that A is invertible if the x_i are distinct. Show, conversely, that if A is invertible then the x_i must be distinct.

2.8 SYSTEMS OF DIFFERENTIAL EQUATIONS

Many problems in Science and Engineering come down to solving differential equations. In this brief section we sketch how diagonalization can be used to solve certain systems of first order, ordinary differential equations. Of course some familiarity with calculus is required.

Recall that a real valued function f is a rule assigning a uniquely determined real number $f(x)$ to every real number x. Sometimes $f(x)$ is specified by a formula, for example $f(x) = 2x^2 + 3$ for all x. But more elaborate functions also arise such as the exponential function $f(x) = e^x$, or the cosine function $f(x) = cosx$. Note that two functions f and g are *equal* if they have the same effect on every real number x, that is if $f(x) = g(x)$ holds for all real x. The *sum* $f + g$ and the *scalar product* af (where a is a number) are new functions defined by

$$(f+g)(x) = f(x) + g(x) \quad \text{for all } x,$$
$$af(x) = a\,f(x) \quad \text{for all } x.$$

These are called **pointwise** addition and scalar multiplication, respectively; they are the usual operations that occur in calculus and algebra.

A function f is called **differentiable** if its derivative f' exists; we refer the reader to calculus texts for the precise definition of the derivative. All we need is the fact that, if f and g are both differentiable so also are $f + g$ and af (for any constant a), and the derivatives are given by

$$(f+g)' = f' + g' \quad \text{and} \quad (af)' = a\,f'.$$

These facts are proved in every calculus text.

2.8.1 The Diagonalization Method

Let f denote a differentiable function of a real variable x. The simplest first order[20] differential equation is

$$f' = af \quad \text{where } a \text{ is a constant.}$$

The chain rule of differentiation shows that $f(x) = e^{ax}$ satisfies this equation, and some elementary calculus shows that every solution is a multiple of e^{ax}.

Lemma 1. *All solutions of the differential equation $f' = af$ are given by*

$$f(x) = ce^{ax} \quad \text{for some constant } c.$$

Proof. If f is any solution, define a function g by $g(x) = f(x)e^{-ax}$. The product rule of differentiation shows that $g'(x) = 0$, so $g(x) = c$ is a constant function. Hence $f(x) = ce^{ax}$. □

---

[20] A *first order* differential equation is one involving only the function and its derivative, and no second or higher derivatives occur.

The problem we want to address here how to use Lemma 1 to find all solutions of a system of first order differential equations.

More precisely, if $f_1, f_2, \cdots, f_n$, are differentiable functions, we want to solve the system

$$\begin{array}{rcl} f_1' & = & a_{11}f_1 + a_{12}f_2 + \cdots + a_{1n}f_n \\ f_2' & = & a_{21}f_1 + a_{22}f_2 + \cdots + a_{2n}f_n \\ & \vdots & \\ f_n' & = & a_{n1}f_1 + a_{n2}f_2 + \cdots + a_{nn}f_n \end{array} \qquad (*)$$

where the a_{ij} are constants. That is we want to describe all differentiable functions $f_1, f_2, \cdots, f_n$ that satisfy thess equations. This system can easily be written in matrix form. Let

$$\mathbf{f} = \begin{bmatrix} f_1 \\ f_2 \\ \vdots \\ f_n \end{bmatrix} \quad \mathbf{f}' = \begin{bmatrix} f_1' \\ f_2' \\ \vdots \\ f_n' \end{bmatrix} \quad \text{and} \quad A = \begin{bmatrix} a_{11} & a_{12} & \cdots & a_{1n} \\ a_{21} & a_{22} & \cdots & a_{2n} \\ \vdots & \vdots & & \vdots \\ a_{n1} & a_{n2} & \cdots & a_{nn} \end{bmatrix}.$$

Then the system (*) of differential equations becomes the single matrix equation

$$\mathbf{f}' = A\mathbf{f}. \qquad (**)$$

Given the matrix A we ask for all columns $\mathbf{f}$ of functions that satisfy (**). A variation on the technique used on dynamical systems will solve the problem if A is a diagonalizable matrix. We illustrate the method with an example where $n = 2$.

Example 1. *Find all solutions to the system*

$$\begin{array}{rcl} f_1' & = & 2f_1 - f_2 \\ f_2' & = & 6f_1 - 5f_2 \end{array}.$$

Then find a solution that satisfies $f_1(0) = 2$ and $f_2(0) = 3$.

Solution. This system has the form $\mathbf{f}' = A\mathbf{f}$ where $A = \begin{bmatrix} 2 & -1 \\ 6 & -5 \end{bmatrix}$ and $f = \begin{bmatrix} f_1 \\ f_2 \end{bmatrix}$. The characteristic polynomial of A is $c_A(x) = (x-1)(x+4)$, so the eigenvalues are $\lambda_1 = 1$ and $\lambda_2 = -4$, with corresponding eigenvectors $X_1 = \begin{bmatrix} 1 \\ 1 \end{bmatrix}$ and $X_2 = \begin{bmatrix} 1 \\ 6 \end{bmatrix}$ as is easily verified. Hence the matrix $P = [X_1 \ X_2] = \begin{bmatrix} 1 & 1 \\ 1 & 6 \end{bmatrix}$ diagonalizes A, that is

$$P^{-1}AP = D = diag(\lambda_1, \lambda_2) = \begin{bmatrix} \lambda_1 & 0 \\ 0 & \lambda_2 \end{bmatrix}.$$

2.8. SYSTEMS OF DIFFERENTIAL EQUATIONS

The key to our method is to consider new functions $\mathbf{g} = \begin{bmatrix} g_1 \\ g_2 \end{bmatrix}$ given by

$$\mathbf{f} = P\mathbf{g}, \quad \text{equivalently} \quad \mathbf{g} = P^{-1}\mathbf{f}.$$

In other words,

$$\begin{bmatrix} f_1 \\ f_2 \end{bmatrix} = \begin{bmatrix} 1 & 1 \\ 1 & 6 \end{bmatrix} \begin{bmatrix} g_1 \\ g_2 \end{bmatrix}, \quad \text{that is} \quad \begin{matrix} f_1 = g_1 + g_2 \\ f_2 = g_1 + 6g_2 \end{matrix}.$$

It follows that $f_1' = g_1' + g_2'$ and $f_2' = g_1' + 6g_2'$, and in matrix form this is

$$\mathbf{f}' = \begin{bmatrix} f_1' \\ f_2' \end{bmatrix} = \begin{bmatrix} g_1' + g_2' \\ g_1' + 6g_2' \end{bmatrix} = \begin{bmatrix} 1 & 1 \\ 1 & 6 \end{bmatrix} \begin{bmatrix} g_1' \\ g_2' \end{bmatrix} = P\mathbf{g}'.$$

Thus the system $\mathbf{f}' = A\mathbf{f}$ becomes $P\mathbf{g}' = A(P\mathbf{g})$, that is $\mathbf{g}' = P^{-1}AP\mathbf{g} = D\mathbf{g}$. In matrix form this is

$$\begin{bmatrix} g_1' \\ g_2' \end{bmatrix} = \begin{bmatrix} 1 & 0 \\ 0 & -4 \end{bmatrix} \begin{bmatrix} g_1 \\ g_2 \end{bmatrix}, \quad \text{or} \quad \begin{matrix} g_1' = g_1 \\ g_2' = -4g_2 \end{matrix}.$$

By Lemma 1, the only solutions to these equations are $g_1 = c_1 e^x$ and $g_2 = c_2 e^{-4x}$ where c_1 and c_2 are constants. Since $f_1 = g_1 + g_2$ and $f_2 = g_1 + 6g_2$, this gives all solutions to the system $\mathbf{f}' = A\mathbf{f}$:

$$\begin{matrix} f_1(x) = c_1 e^x + c_2 e^{-4x} \\ f_2(x) = c_1 e^x + 6c_2 e^{-4x} \end{matrix} \quad \text{where } c_1 \text{ and } c_2 \text{ are arbitrary constants.}$$

The matrix form of these equations reveals the role of the eigenvectors:

$$\begin{bmatrix} f_1(x) \\ f_2(x) \end{bmatrix} = \begin{bmatrix} c_1 e^x + c_2 e^{-4x} \\ c_1 e^x + 6c_2 e^{-4x} \end{bmatrix}$$
$$= c_1 \begin{bmatrix} 1 \\ 1 \end{bmatrix} e^x + c_2 \begin{bmatrix} 1 \\ 6 \end{bmatrix} e^{-4x}$$
$$= c_1 X_1 e^x + c_2 X_2 e^{-4x}.$$

If we write $\mathbf{f}(x) = \begin{bmatrix} f_1(x) \\ f_2(x) \end{bmatrix}$, this takes the compact form $\mathbf{f}(x) = c_1 X_1 e^x + c_2 X_2 e^{-4x}$.

Finally, if we insist that $f_1(0) = 2$ and $f_2(0) = -3$, we see that $c_1 = 3$ and $c_2 = -1$. Hence $f_1(x) = 3e^x - e^{-4x}$ and $f_2(x) = 3e^x - 6e^{-4x}$ is a solution satisfying these additional requirements. □

The technique in Example 1 works in general provided the matrix of coefficients is diagonalizable. For convenience, given a column $\mathbf{f} = \begin{bmatrix} f_1 \\ f_2 \\ \vdots \\ f_n \end{bmatrix}$ of functions we write $\mathbf{f}(x) = \begin{bmatrix} f_1(x) \\ f_2(x) \\ \vdots \\ f_n(x) \end{bmatrix}$ for all real x.

Theorem 1. *Consider a system $\mathbf{f}' = A\mathbf{f}$ of first order differential equations where $\mathbf{f}$ is a column of differentiable functions and A is a diagonalizable matrix. If $\lambda_1, \lambda_2, \cdots \lambda_n$ are the eigenvalues of A, with corresponding eigenvectors $X_1, X_2, \cdots X_n$, then every solution to $\mathbf{f}' = A\mathbf{f}$ is given in the form*

$$\mathbf{f}(x) = c_1 X_1 e^{\lambda_1 x} + c_2 X_2 e^{\lambda_2 x} + \cdots + c_n X_n e^{\lambda_n x}$$

where the c_i are arbitrary constants.

Proof. If $P = [X_1 \ X_2 \ \cdots \ X_n]$ then $P^{-1}AP = D = diag(\lambda_1, \lambda_2, \cdots, \lambda_n)$ is diagonal. Define a new column of functions $\mathbf{g}$ by $\mathbf{g} = P^{-1}\mathbf{f}$, equivalently $\mathbf{f} = P\mathbf{g}$. Write $\mathbf{f} = [f_1 \ f_2 \ \cdots \ f_n]^T$ and $P = [p_{ij}]$. For each $i = 1, 2, \cdots, n$ the equation $\mathbf{f} = P\mathbf{g}$ gives

$$f_i = p_{i1}g_1 + p_{i2}g_2 + \cdots + p_{in}g_n, \quad \text{whence} \quad f_i' = p_{i1}g_1' + p_{i2}g_2' + \cdots + p_{in}g_n'.$$

It follows that $\mathbf{f}' = P\mathbf{g}'$. When this is substituted in the matrix equation $\mathbf{f}' = A\mathbf{f}$, the result is $P\mathbf{g}' = AP\mathbf{g}$, whence $\mathbf{g}' = P^{-1}AP\mathbf{g} = D\mathbf{g}$. If we write $\mathbf{g} = [g_1 \ g_2 \ \cdots \ g_n]^T$, comparing entries gives

$$g_i' = \lambda_i g_i \text{ for each } i, \text{ so that } g_i(x) = c_i e^{\lambda_i x} \text{ for all } x.$$

Using Theorem 3 §1.4, it follows that

$$\begin{aligned}
\mathbf{f}(x) = P\mathbf{g}(x) &= [X_1 \ X_2 \ \cdots \ X_n] \begin{bmatrix} g_1(x) \\ g_2(x) \\ \vdots \\ g_n(x) \end{bmatrix} \\
&= g_1(x)X_1 + g_2(x)X_2 + \cdots + g_n(x)X_n \\
&= c_1 X_1 e^{\lambda_1 x} + c_2 X_2 e^{\lambda_2 x} + \cdots + c_n X_n e^{\lambda_n x}.
\end{aligned}$$

This completes the proof. □

For each i, the individual function $f_i(x)$ can be read off from the ith component of $\mathbf{f}(x)$. Here is an example.

Example 2. *Find all solutions to the system*

$$\begin{aligned}
f_1' &= f_2 + f_3 \\
f_2' &= f_1 + f_3 \\
f_3' &= f_1 + f_2
\end{aligned}$$

Then find the solution such that $f_1(0) = -1$ and $f_2(0) = f_3(0) = 1$.

Solution. This system has the form $\mathbf{f}' = A\mathbf{f}$ where $A = \begin{bmatrix} 0 & 1 & 1 \\ 1 & 0 & 1 \\ 1 & 1 & 0 \end{bmatrix}$. This matrix is diagonalizable; it was analyzed in Example 6 §2.3, where the eigenvalues were found to be $\lambda_1 = 2$ and $\lambda_2 = \lambda_3 = -1$, with corresponding eigenvectors

2.8. SYSTEMS OF DIFFERENTIAL EQUATIONS

$X_1 = [1\ 1\ 1]^T$, $X_2 = [-1\ 1\ 0]^T$, and $X_3 = [-1\ 0\ 1]^T$. Hence Theorem 1 applies and every solution has the form

$$\mathbf{f}(x) = c_1 \begin{bmatrix} 1 \\ 1 \\ 1 \end{bmatrix} e^{2x} + c_2 \begin{bmatrix} -1 \\ 1 \\ 0 \end{bmatrix} e^{-x} + c_3 \begin{bmatrix} -1 \\ 0 \\ 1 \end{bmatrix} e^{-x}$$

for some choice of c_1, c_2 and c_3. Hence

$$\begin{aligned} f_1(x) &= c_1 e^{2x} - c_2 e^{-x} - c_3 e^{-x} \\ f_2(x) &= c_1 e^{2x} + c_2 e^{-x} \\ f_3(x) &= c_1 e^{2x} + c_3 e^{-x} \end{aligned}$$

If we require that $f_1(0) = -1$ and $f_2(0) = f_3(0) = 1$, then $c_1 = \frac{1}{3}$ and $c_2 = c_3 = \frac{2}{3}$. □

We conclude with a comment about differential equations of higher order. The second derivative of a function f is defined by $f'' = (f')'$. Consider the second order differential equation

$$f'' - af' - bf = 0.$$

If we put $f_1 = f$, and then put $f_2 = f' - af = f_1' - af_1$, then the given equation reads $f_2' - bf_1 = 0$. Hence these functions satisfy the following system of first order equations:

$$\begin{aligned} f_1' &= af_1 + f_2 \\ f_2' &= bf_1 \end{aligned}$$

Hence solving this system is equivalent to solving $f'' - af' - bf = 0$. Similarly, an nth order differential equation (one involving differentiating f as many as n times) leads to a system of n first order equations.

Exercises

1. In each case find all solutions to the system of differential equations, and find a particular solution that satisfies the given conditions.

   (a) $\begin{aligned} f_1' &= 2f_1 + 4f_2 \\ f_2' &= 3f_1 + 3f_2 \end{aligned}$ $\begin{aligned} f_1(0) &= 0 \\ f_2(0) &= 1 \end{aligned}$

   (b) $\begin{aligned} f_1' &= -f_1 + 5f_2 \\ f_2' &= f_1 + 3f_2 \end{aligned}$ $\begin{aligned} f_1(0) &= 1 \\ f_2(0) &= -1 \end{aligned}$

   (c) $\begin{aligned} f_1' &= 4f_2 + 4f_3 \\ f_2' &= f_1 + f_2 - 2f_3 \\ f_3' &= -f_1 + f_2 + 4f_3 \end{aligned}$ $f_1(0) = f_2(0) = f_3(0) = 1$

   (d) $\begin{aligned} f_1' &= 2f_1 + f_2 + 2f_3 \\ f_2' &= 2f_1 + 2f_2 - 2f_3 \\ f_3' &= 3f_1 + f_2 + f_3 \end{aligned}$ $f_1(0) = f_2(0) = f_3(0) = 1$

2. Set up a linear system of three first order differential equations that solves the single third order equation $f''' - af'' - bf' - cf = 0$.

Chapter 3

VECTOR GEOMETRY

3.1 GEOMETRIC VECTORS

The word geometry in Greek means *measurement of the earth*, and the practical uses of geometry go back to antiquity. In ancient Greece all of mathematics was regarded as geometry, but in this chapter we will be primarily concerned with lines and planes in space. Our approach will be to regard points as column matrices (called vectors in this context) and then use matrix algebra to simplify calculations.

3.1.1 Coordinate Systems

The Greeks practised **synthetic geometry**, which means dealing with geometric figures without using a coordinate system. The use of coordinates was initiated by René Descartes[1] in 1637, and enabled him to use algebraic equations to describe geometric figures. This method, called **analytic geometry**, allows the computational power of arithmetic to be used to simplify many geometric calculations, and was fundamental to the development of analysis (calculus) in the 18th century. On the other hand, the method clarifies the study of equations by applying the intuitive, conceptual power of geometry to their graphs.

In the plane, coordinates are introduced as follows: Choose a point O called the **origin**, choose two perpendicular lines through O called the X-**axis**, and the Y-**axis**, and introduce a number scale on each axis with 0 at the origin. Then each point P determines a unique pair (x, y) of numbers called the **coordinates** of P as shown in Figure 1. We write $P = P(x, y)$ when the coordinates of P are to be emphasized.

---

[1] René Descartes (1596-1650) was a French philosopher and mathematician. His geometrical methods were first published in his *La géométrié*, Paris, 1637.

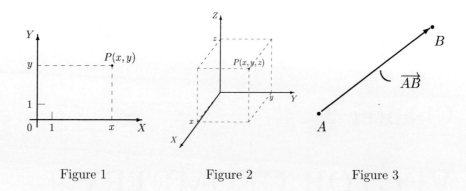

Figure 1   Figure 2   Figure 3

Coordinates in space are introduced in the same way except that *three* mutually perpendicular axes through the origin O are chosen as in Figure 2 (called the X-**axis**, the Y-**axis** and the Z-**axis**). Then a point P determines a unique *triple* (x, y, z) of numbers (the **coordinates** of P), and we denote P as $P(x, y, z)$.

3.1.2 Vectors and Scalars

We are all familiar with quantities that are completely specified by a single number. Examples are time, mass, temperature and pressure. These are called **scalar** quantities. However some quantities are not scalars. For example, the velocity of an airplane is not completely described by saying that the airplane is going 400 kilometers per hour; its *direction* must be given too, say *east* at 400 km/hr. A quantity which requires both a **magnitude** and a **direction** to specify it is called a **vector** quantity. Thus the vector velocity of our airplane has magnitude 400 km/hr (usually called the *speed*) and it has easterly direction. In order to say that two airplanes have the same velocity, both the speeds and the directions must be the same. In general:

> *Two vector quantities are the same if and only if they have the same magnitude and the same direction.*

In this chapter we will be concerned only with vectors arising in geometry. If A and B are two geometrical points, the directed line segment from A to B is called the **geometric vector** (or simply the **vector**) from A to B, and is denoted $\overrightarrow{AB}$. As the name implies, this is a vector quantity with direction from A to B and magnitude the distance between A and B. The vector $\overrightarrow{AB}$ is represented by an arrow from A to B as shown in Figure 3. The point A is called the **tail** of the vector $\overrightarrow{AB}$, and B is called the **tip** of $\overrightarrow{AB}$. The magnitude of a geometric vector $\overrightarrow{AB}$ is called its **length** and is denoted $\left\lVert \overrightarrow{AB} \right\rVert$.

Throughout this chapter, vectors will arise in both two and three dimensions. We will usually discuss the three-dimensional case, but the two situations are analogous.

3.1. GEOMETRIC VECTORS

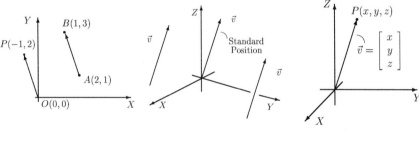

Figure 4   Figure 5   Figure 6

Two vectors can be the same even though the tips and tails are different. For example $\overrightarrow{AB} = \overrightarrow{OP}$ in Figure 4 because they have the same length and the same direction (they both proceed one unit to the left and two units up).[2] Thus the same vector can be translated from one position to another; what is important is that the length and direction remain the same, and not where the tips and tails are located. For this reason, we shall often denote vectors as $\vec{u}$, $\vec{v}$, $\vec{w}$, etc. which make no reference to the tips or tails.

3.1.3 Matrix Form

Thus the same vector $\vec{v}$ can have *any* point as its tail, as illustrated in Figure 5. In particular, $\vec{v}$ is said to be in **standard position** if its tail is at the origin O. Then $\vec{v} = \overrightarrow{OP}$ as in Figure 6 where $P(x, y, z)$ is the uniquely determined point at its tip, and we write $\vec{v}$ as the 3×1 matrix[3]

$$\vec{v} = \begin{bmatrix} x \\ y \\ z \end{bmatrix} = [x \ y \ z]^T.$$

In this case we say that $\vec{v}$ is written[4] in **matrix form**, that x, y, and z are the **components** of $\vec{v}$, and that $\vec{v}$ is the **position vector** of the point $P = P(x, y, z)$. Thus every vector $\vec{v}$ is the position vector of a (uniquely determined) point P, and the coordinates of P are the same as the components of $\vec{v}$. Furthermore, every point P *has* a position vector $\vec{v} = \overrightarrow{OP}$.

In the same way, a two-dimensional vector $\vec{v}$ has matrix form $\vec{v} = \begin{bmatrix} x \\ y \end{bmatrix}$ where $\vec{v}$ is the position vector of $P = P(x, y)$. Thus, for example, the vector $\vec{v} = \overrightarrow{AB} = \overrightarrow{OP}$ in Figure 4 has matrix form $\vec{v} = \begin{bmatrix} -1 \\ 2 \end{bmatrix}$.

---

[2] Fractions provide another example where quantities can be the same but *look* different: $\frac{6}{9}$ and $\frac{14}{21}$ certainly look different, but they are equal fractions (in fact both equal $\frac{2}{3}$).

[3] Since a vector $\vec{v}$ is a matrix, it should be denoted (as in Section 1.1) by an upper case letter, say $\vec{v} = V$. However we use the notation $\vec{v}$ for vectors because it is consistent with the notation $\overrightarrow{AB}$.

[4] The more compact notation $[x \ y \ z]^T$ is used for convenience of printing.

Because we can write vectors as matrices, we can speak of sums and scalar multiples of vectors. This yields a very powerful tool in geometry because these matrix operations have natural geometric meanings. In particular, geometric calculations can be performed easily using matrix algebra. This method is called **vector geometry**, and its study is the main theme of this chapter.

Let $\vec{v} = \begin{bmatrix} x \\ y \\ z \end{bmatrix} = [x \ y \ z]^T$ and $\vec{w} = \begin{bmatrix} x_1 \\ y_1 \\ z_1 \end{bmatrix} = [x_1 \ y_1 \ z_1]^T$ be vectors

and let a be a scalar. We define the **zero vector** $\vec{0}$, the **negative** $-\vec{v}$, the **sum** $\vec{v} + \vec{w}$ and the **scalar product** $a\vec{v}$ as the corresponding matrices:

$$\vec{0} = \begin{bmatrix} 0 \\ 0 \\ 0 \end{bmatrix}, \ -\vec{v} = \begin{bmatrix} -x \\ -y \\ -z \end{bmatrix}, \ \vec{v} + \vec{w} = \begin{bmatrix} x + x_1 \\ y + y_1 \\ z + z_1 \end{bmatrix}, \ a\vec{v} = \begin{bmatrix} ax \\ ay \\ az \end{bmatrix}$$

Because they are matrix operations, vector addition and scalar multiplication obey the basic rules of matrix arithmetic (Theorems 1 and 2, Section 1.1). We repeat them here for convenience: If $\vec{u}$, $\vec{v}$, and $\vec{w}$ are (two- or) three-dimensional vectors, and if a and b are scalars, we have:

$$\begin{array}{ll} \vec{u} + \vec{v} = \vec{v} + \vec{u} & (a + b)\vec{v} = a\vec{v} + b\vec{v} \\ \vec{u} + (\vec{v} + \vec{w}) = (\vec{u} + \vec{v}) + \vec{w} & a(\vec{u} + \vec{v}) = a\vec{u} + a\vec{v} \\ \vec{0} + \vec{v} = \vec{v} & a(b\vec{v}) = (ab)\vec{v} \\ \vec{v} + (-\vec{v}) = \vec{0} & 1\vec{v} = \vec{v} \end{array}$$

We will use these rules without comment below. They are useful for doing geometric calculations because, as mentioned above, these matrix operations have simple geometric interpretations. In fact the point of view in vector geometry is this: If it is desired to find a point $P(x, y, z)$, find it's position vector $\vec{p} = [x \ y \ z]^T$ instead. This allows us to bring the full power of matrix algebra to bear on the geometric problem. Hence, the rest of this section is devoted to clarifying what these matrix operations mean geometrically.

3.1.4 Geometric Descriptions

We are going to describe sums and scalar products of vectors solely in terms of the lengths and directions of the vectors involved. Such descriptions are called **intrinsic** because they makes no reference to the coordinate system, a fact which is very important in applications (they were, in fact, the original definitions for vector operations in physics). In addition, we provide other useful translations between geometric and matrix language.

We begin with intrinsic descriptions of vector equality and of the zero vector. These are given in parts (1) and (3) of the following theorem. The theorem also contains two important results about the magnitude of a vector $\vec{v}$, not the least of which is a formula in part (2) for calculating the magnitude $\|\vec{v}\|$ from the components of $\vec{v}$. Recall that the absolute value of a number a is denoted $|a|$.

3.1. GEOMETRIC VECTORS

Theorem 1. Let $\vec{v} = \begin{bmatrix} x \\ y \\ z \end{bmatrix}$ and $\vec{w} = \begin{bmatrix} x_1 \\ y_1 \\ z_1 \end{bmatrix}$ be vectors.

(1) $\vec{v} = \vec{w}$ as vectors if and only if $x = x_1$, $y = y_1$ and $z = z_1$.

(2) $\|\vec{v}\| = \sqrt{x^2 + y^2 + z^2}$.[5]

(3) $\vec{v} = \vec{0}$ if and only if $\|\vec{v}\| = 0$.

(4) $\|a\vec{v}\| = |a|\,\|\vec{v}\|$ for any scalar a.

Proof. (1). If $\vec{v}$ and $\vec{w}$ are positioned with their tails at the origin as in Figure 7, the tips of $\vec{v}$ and $\vec{w}$ are the points $P(x, y, z)$ and $P_1(x_1, y_1, z_1)$ respectively. Then it is clear from Figure 7 that $\vec{v} = \vec{w}$ (same length and direction) if and only if P and P_1 are equal points, that is if and only if $x = x_1$, $y = y_1$, and $z = z_1$.

(2). We have $\vec{v} = \overrightarrow{OP}$ as in Figure 8, where $P = P(x, y, z)$. Since OQP is a right triangle, Pythagoras' Theorem[6] gives $\|\vec{v}\|^2 = h^2 + z^2$, and it gives $x^2 + y^2 = h^2$ when applied to the right triangle ORQ. Then (2) follows by eliminating h^2.

(3). Since $\vec{0} = \overrightarrow{OO}$ is the position vector of the origin O, it is clear that $\|\vec{0}\| = 0$. Conversely, if $\|\vec{v}\| = 0$ for some vector $\vec{v} = [x\ y\ z]^T$, then $\sqrt{x^2 + y^2 + z^2} = 0$ by (2). Hence $x = y = z = 0$, and so $\vec{v} = \vec{0}$.

(4). If $\vec{v} = [x\ y\ z]^T$ then $a\vec{v} = [ax\ by\ cz]^T$ by definition, so (2) gives

$$\|a\vec{v}\|^2 = (ax)^2 + (ay)^2 + (az)^2 = a^2(x^2 + y^2 + z^2) = a^2 \|\vec{v}\|^2.$$

Now (4) follows by taking positive square roots because $\sqrt{a^2} = |a|$. □

Example 1. If $\vec{v} = \begin{bmatrix} 3 \\ -1 \\ 2 \end{bmatrix}$ and $\vec{w} = \begin{bmatrix} 4 \\ -3 \end{bmatrix}$, then $\|\vec{v}\| = \sqrt{3^2 + (-1)^2 + 2^2} = \sqrt{14}$ and $\|\vec{w}\| = \sqrt{4^2 + (-3)^2} = \sqrt{25} = 5$. □

Theorem 1 gives intrinsic descriptions of vector equality, the zero vector and the negative of a vector. The descriptions of vector equality and of the zero vector come from parts (1) and (3) of Theorem 1.

Vector Equality. *Two vectors are equal as 3×1 matrices if and only if they are equal as vectors (same length and direction).*

Zero Vector. $\vec{0}$ *is the* only *vector of length zero. Note that, since $\vec{0}$ has length zero, it makes no sense to talk about the* direction *of the zero vector.*

---

[5] Throughout this book $\sqrt{a}$ means the *positive* square root of a.

[6] Pythagoras' Theorem states that if a and b are the sides of a right angled triangle with hypotenuse h, then $a^2 + b^2 = h^2$. A proof is given at the end of this section.

Negative of a Vector. *If $\vec{v}$ is a vector, then $-\vec{v}$ is the vector with the same length as $\vec{v}$ but opposite direction.*

As to the negative, we have $-\vec{v} = (-1)\vec{v}$, so that $\|-\vec{v}\| = |-1|\,\|-\vec{v}\| = \|-\vec{v}\|$ by (4) of Theorem 1. Hence $\vec{v}$ and $-\vec{v}$ have the same length. Moreover, if $\vec{v} = \overrightarrow{OP}$, then $-\vec{v} = \overrightarrow{PO}$, and $\overrightarrow{PO}$ has the *opposite* direction of $\vec{v}$. This is illustrated (for a two-dimensional vector) in Figure 9.

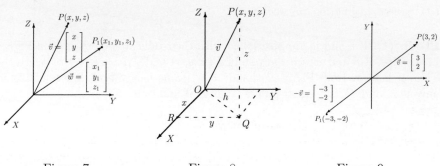

Figure 7 Figure 8 Figure 9

3.1.5 Vector Addition

If two vectors $\vec{v}$ and $\vec{w}$ are positioned with a common tail, they create a parallelogram,[7] called the parallelogram **determined** by $\vec{v}$ and $\vec{w}$. Two illustrations of this appear in Figure 10. Since this parallelogram is completely determined by the lengths and directions of $\vec{v}$ and $\vec{w}$, the following "law" provides an intrinsic description of the sum $\vec{v} + \vec{w}$.

Parallelogram Law of Vector Addition. *In the parallelogram determined by two vectors $\vec{v}$ and $\vec{w}$, the vector $\vec{v} + \vec{w}$ is the diagonal with the same tail as $\vec{v}$ and $\vec{w}$ (see Figure 11).*

Proof. We may assume that $\vec{v}$ and $\vec{w}$ are in standard position because the parallelogram they determine is the same for any choice of common tail. We verify the parallelogram law in the two-dimensional case; the three dimensional case is analogous. Hence write $\vec{v} = \begin{bmatrix} x_1 \\ y_1 \end{bmatrix}$ and $\vec{w} = \begin{bmatrix} x_2 \\ y_2 \end{bmatrix}$, so that $\vec{v} + \vec{w} = \begin{bmatrix} x_1 + x_2 \\ y_1 + y_2 \end{bmatrix}$. Then the tips of $\vec{v}$, $\vec{w}$ and $\vec{v} + \vec{w}$ are $P_1(x_1, y_1)$, $P_2(x_2, y_2)$ and $P(x_1 + x_2, y_1 + y_2)$ respectively in Figure 12, and we must verify that the figure OP_1PP_2 is a parallelogram. But this follows from the fact that the right triangles OAP_1 and P_2BP in Figure 12 are identical. In fact, both have vertical

---

[7]A **parallelogram** is a four-sided figure with straight sides such that opposite sides are parallel and of equal length. In particular, a rectangle is a parallelogram in which all the interior angles are right angles.

3.1. GEOMETRIC VECTORS

and horizontal sides of length $|y_1|$ and $|x_1|$ respectively. (Figure 12 depicts the situation where x_1 and y_1 are positive.) □

We note in passing that the parallelogram law for vector addition can be shown by experiment to be valid for such non-geometrical vectors as velocity and force. Hence it is a powerful tool in physics.

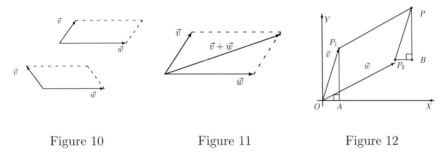

Figure 10   Figure 11   Figure 12

The parallelogram law provides an intrinsic description of the sum $\vec{v} + \vec{w}$ by describing the length and direction of $\vec{v} + \vec{w}$ in terms of the parallelogram determined by $\vec{v}$ and $\vec{w}$. However, there is another way to look at vector addition which is more convenient than the parallelogram law in most situations, and is very useful in applications. It is described as follows:

Tip-to-Tail Method of Adding Vectors. *Given two vectors $\vec{v}$ and $\vec{w}$, place the tail of $\vec{w}$ at the tip of $\vec{v}$ as in Figure 13. Then $\vec{v} + \vec{w}$ is the vector from the tail of $\vec{v}$ to the tip of $\vec{w}$.*

This can be described as the "first $\vec{v}$ then $\vec{w}$" method of computing $\vec{v} + \vec{w}$, and it really does give $\vec{v} + \vec{w}$ because the diagram in Figure 13 is the "top half" of the parallelogram determined by $\vec{v}$ and $\vec{w}$. Clearly these vectors could also be added as "first $\vec{w}$ then $\vec{v}$" to get $\vec{w} + \vec{v} = \vec{v} + \vec{w}$ as in Figure 14.

However the most important fact is that two (or more) vectors can be added graphically by placing them tip-to-tail in a sequence. This is illustrated for three vectors in Figure 15, and gives a very useful "picture" of the sum of several vectors as a "tip-to-tail" sequence.

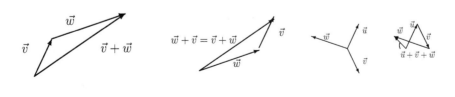

Figure 13   Figure 14   Figure 15

Example 2. *Find the net effect of a 1-km walk northeast and a 1-km walk east.*

Solution. Let $\vec{w}_1$ and $\vec{w}_2$ denote the northeast and east walks, respectively. If walk $\vec{w}_1$ is taken first, followed by $\vec{w}_2$, the result is $\vec{w}_1 + \vec{w}_2$ as in the first diagram in Figure 16. On the other hand, if $\vec{w}_2$ is taken first, the result is $\vec{w}_2 + \vec{w}_1$ as in the second diagram. The net result is the walk $\vec{w}_1 + \vec{w}_2 = \vec{w}_2 + \vec{w}_1$. □

The tip-to-tail method of adding vectors takes an easily remembered form for geometric vectors when the $\overrightarrow{AB}$ notation is used. Indeed, if A, B and C are points, it asserts that

$$\overrightarrow{AB} + \overrightarrow{BC} = \overrightarrow{AC}$$

as is illustrated in Figure 17. Thus adding vectors in this notation becomes a matter of "following the letters". Here is an illustration of how this can be used in geometry.

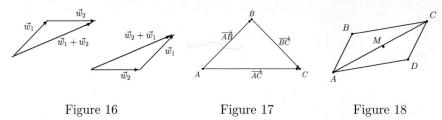

Figure 16　　　　　Figure 17　　　　　Figure 18

Example 3. *Show that the diagonals of any parallelogram bisect each other.*

Solution. Let A, B, C and D be the four vertices of the parallelogram in order, as in Figure 18, and let M be the midpoint of the diagonal AC. We must show that M is also the midpoint of the diagonal BD. To do this it suffices to show that $\overrightarrow{BM} = \overrightarrow{MD}$. Indeed, the fact that these vectors have the same direction shows that M is *on* the diagonal BD; the fact that they have the same length shows that M is the *midpoint* of BD.

Now observe that $\overrightarrow{BA} = \overrightarrow{CD}$ because $ABCD$ is a parallelogram, and $\overrightarrow{AM} = \overrightarrow{MC}$ because M is the midpoint of AC. Hence tip-to-tail addition (twice) gives

$$\overrightarrow{BM} = \overrightarrow{BA} + \overrightarrow{AM} = \overrightarrow{CD} + \overrightarrow{MC} = \overrightarrow{MC} + \overrightarrow{CD} = \overrightarrow{MD}.$$

This is what we wanted. □

There is a simple geometric way to visualize the **difference** $\vec{v} - \vec{w}$ of two vectors $\vec{v}$ and $\vec{w}$. Position $\vec{v}$ and $\vec{w}$ with a common tail A as in Figure 19, and let B and C be the respective tips. Then adding tip-to-tail gives $\vec{w} + \overrightarrow{CB} = \vec{v}$ so $\overrightarrow{CB} = \vec{v} - \vec{w}$. Hence we obtain an intrinsic description of vector subtraction.

Vector Subtraction. *$\vec{v} - \vec{w}$ is the vector from the tip of $\vec{w}$ to the tip of $\vec{v}$.*

Thus in the parallelogram determined by $\vec{v}$ and $\vec{w}$, both $\vec{v} - \vec{w}$ and $\vec{v} + \vec{w}$ appear as diagonals (Figure 20).

3.1. GEOMETRIC VECTORS

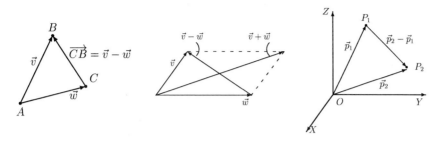

Figure 19   Figure 20   Figure 21

One of the reasons for the importance of vector subtraction is that it gives a simple way to find the vector from one point to another in matrix form. This appears in the following theorem along with a formula for the distance between two points.

Theorem 2. Let $P_1(x_1, y_1, z_1)$ and $P_2(x_2, y_2, z_2)$ be two points. Then:

(1) $\overrightarrow{P_1 P_2} = \begin{bmatrix} x_2 - x_1 \\ y_2 - y_1 \\ z_2 - z_1 \end{bmatrix}$.

(2) The distance between P_1 and P_2 is $\sqrt{(x_2 - x_1)^2 + (y_2 - y_1)^2 + (z_2 - z_1)^2}$.

Proof. (1). Write $\vec{p_1} = \overrightarrow{OP_1}$ and $\vec{p_2} = \overrightarrow{OP_2}$ as in Figure 21, so that $\vec{p_1} = [x_1 \ y_1 \ z_1]^T$ and $\vec{p_2} = [x_2 \ y_2 \ z_2]^T$. Then the above discussion gives $\overrightarrow{P_1 P_2} = \vec{p_2} - \vec{p_1} = [x_2 - x_1 \ y_2 - y_1 \ z_2 - z_1]^T$, as required.

(2). The distance is $\left\| \overrightarrow{P_1 P_2} \right\| = \sqrt{(x_2 - x_1)^2 + (y_2 - y_1)^2 + (z_2 - z_1)^2}$ using (1) and Theorem 1. □

Of course the two-dimensional version of Theorem 2 is also valid: If $P_1(x_1, y_1)$ and $P_2(x_2, y_2)$ are points, then $\overrightarrow{P_1 P_2} = \begin{bmatrix} x_2 - x_1 \\ y_2 - y_1 \end{bmatrix}$, and the distance between P_1 and P_2 is $\left\| \overrightarrow{P_1 P_2} \right\| = \sqrt{(x_2 - x_1)^2 + (y_2 - y_1)^2}$.

Example 4. Given points $P_1(3, -1, 2)$ and $P_2(1, 2, 0)$, the vector from P_1 to P_2 is

$$\overrightarrow{P_1 P_2} = [1 - 3 \ \ 2 - (-1) \ \ 0 - 2]^T = [-2 \ 3 \ -2]^T$$

and the distance between P_1 and P_2 is $\left\| \overrightarrow{P_1 P_2} \right\| = \sqrt{(-2)^2 + 3^2 + (-2)^2} = \sqrt{4 + 9 + 4} = \sqrt{17}$. □

188 CHAPTER 3. VECTOR GEOMETRY

3.1.6 Scalar Multiplication

As for vector addition, we want to give an intrinsic description of the scalar product $a\vec{v}$ by describing its length and direction in terms of the scalar a and the length and direction of the vector $\vec{v}$. The length $\|a\vec{v}\| = |a|\, \|\vec{v}\|$ of the vector $a\vec{v}$ has already been found in Theorem 1. To describe the direction of $a\vec{v}$, we need only deal with the case $a\vec{v} \neq \vec{0}$ because the zero vector $\vec{0}$ has no direction. Hence we restrict our attention to the case $a\vec{v} \neq \vec{0}$ in part (2) of the following result.

Geometric Description of Scalar Multiplication. *If a is a scalar and $\vec{v}$ is a vector, then:*

(1) *The length of $a\vec{v}$ is $\|a\vec{v}\| = |a|\, \|\vec{v}\|$.*

(2) *If $a\vec{v} \neq \vec{0}$, the direction of $a\vec{v}$ is* $\begin{cases} \text{the same as } \vec{v} \text{ if } a > 0 \\ \text{opposite to } \vec{v} \text{ if } a < 0 \end{cases}$.

Proof. (1). This is part (4) of Theorem 1.

(2). Each component of $a\vec{v}$ equals a times the corresponding component of $\vec{v}$. If $a > 0$ this means that $a\vec{v}$ has the same direction as $\vec{v}$, as is illustrated in Figure 22 for the vector $\vec{v} = [2\ 1]^T$. If $a < 0$ write $b = -a$. Then $b > 0$ so $b\vec{v}$ and $\vec{v}$ have the same direction (by what has just been proved). But $a\vec{v} = -(b\vec{v})$, so $a\vec{v}$ has direction opposite to $b\vec{v}$, and so opposite to $\vec{v}$. □

This description of scalar multiplication is illustrated in Figure 23 where several scalar multiples of a vector $\vec{v}$ are displayed.

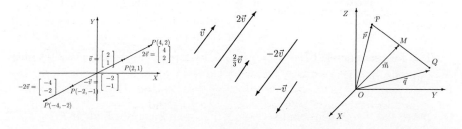

Figure 22 Figure 23 Figure 24

We give two examples illustrating how useful the above geometric description of scalar multiplication is in geometry. Recall that the position vector of a point $P(x, y, z)$, is the vector $\vec{p} = \overrightarrow{OP} = [x\ y\ z]^T$ from the origin to P.

Example 5. *Find the coordinates of the midpoint $M(x, y, z)$ between the points $P(3, -1, 5)$ and $Q(-1, 2, 3)$.*

Solution. Let $\vec{p} = [3\ -1\ 5]^T$, $\vec{q} = [-1\ 2\ 3]^T$ and $\vec{m} = [x\ y\ z]^T$ be the position vectors of P, Q and M respectively, as in Figure 24. We find the point

3.1. GEOMETRIC VECTORS

M by computing its position vector $\vec{m}$. We have $\overrightarrow{PQ} = \vec{q} - \vec{p}$. Hence, as M is the midpoint of PQ, we obtain $\overrightarrow{PM} = \frac{1}{2}\overrightarrow{PQ} = \frac{1}{2}(\vec{q} - \vec{p})$. Thus

$$\vec{m} = \overrightarrow{OM} = \vec{p} + \overrightarrow{PM} = \vec{p} + \tfrac{1}{2}(\vec{q} - \vec{p}) = \tfrac{1}{2}(\vec{p} + \vec{q}) = \frac{1}{2}\begin{bmatrix} 3-1 \\ -1+2 \\ 5+3 \end{bmatrix} = \begin{bmatrix} 1 \\ \tfrac{1}{2} \\ 4 \end{bmatrix}.$$

It follows that $M = M(1, \tfrac{1}{2}, 4)$ is the required midpoint. □

The method in Example 5 works in general: The position vector $\vec{m}$ of the midpoint between *any* two points P and Q is the *average* $\vec{m} = \tfrac{1}{2}(\vec{p} + \vec{q})$ of the position vectors $\vec{p}$ and $\vec{q}$ of the two points.

Example 6. *A quadrilateral is a figure with straight sides and four vertices. Show that the midpoints of the sides of any quadrilateral are the vertices of a parallelogram.*

Solution. Let the vertices of the quadrilateral be A, B, C and D in order, and let E, F, G and H be the midpoints as shown in Figure 25. It suffices to show that $\overrightarrow{EF} = \overrightarrow{HG}$ since then the corresponding sides are parallel (because $\overrightarrow{EF}$ and $\overrightarrow{HG}$ have the same direction) and of equal length (because $\left\|\overrightarrow{EF}\right\| = \left\|\overrightarrow{HG}\right\|$).

Now observe that $\overrightarrow{EB} = \tfrac{1}{2}\overrightarrow{AB}$ because E is the midpoint of AB, and similarly $\overrightarrow{BF} = \tfrac{1}{2}\overrightarrow{BC}$. Hence tip-to-tail vector addition gives

$$\overrightarrow{EF} = \overrightarrow{EB} + \overrightarrow{BF} = \tfrac{1}{2}\overrightarrow{AB} + \tfrac{1}{2}\overrightarrow{BC} = \tfrac{1}{2}\left(\overrightarrow{AB} + \overrightarrow{BC}\right) = \tfrac{1}{2}\overrightarrow{AC}.$$

A similar argument shows that $\overrightarrow{HG} = \tfrac{1}{2}\overrightarrow{AC}$ too, so $\overrightarrow{EF} = \tfrac{1}{2}\overrightarrow{AC} = \overrightarrow{HG}$ as required. □

The notion of parallel lines has been a central topic in geometry for centuries. We define it in vector terms as follows: Two nonzero vectors are called **parallel** if they have the same or opposite direction. We conclude this section by giving an important connection between parallel vectors and scalar multiplication.

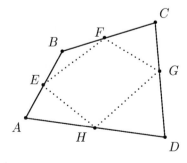

Figure 25

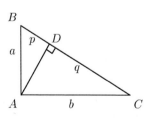

Figure 26

Theorem 3. *The following conditions are equivalent for nonzero vectors $\vec{v}$ and $\vec{w}$:*

(1) *$\vec{v}$ and $\vec{w}$ are parallel.*

(2) *Each of $\vec{v}$ and $\vec{w}$ is a scalar multiple of the other.*

(3) *One of $\vec{v}$ and $\vec{w}$ is a scalar multiple of the other.*

Proof. (2)$\Rightarrow$(3) is obvious, and (3)$\Rightarrow$(1) by Part (2) of the above geometric definition of scalar multiplication. So we show that (1)$\Rightarrow$(2). Hence assume that $\vec{v}$ and $\vec{w}$ are parallel; there are two cases:
Case 1. $\vec{v}$ and $\vec{w}$ have the same direction. In this case we show that $\vec{v} = t\vec{w}$ where $t = \frac{\|\vec{v}\|}{\|\vec{w}\|}$. Since t is positive $\vec{v}$ and $t\vec{w}$ have the same direction (the common direction of $\vec{v}$ and $\vec{w}$). Moreover, Theorem 1 gives $\|t\vec{w}\| = |t|\,\|\vec{w}\| = t\,\|\vec{w}\| = \|\vec{v}\|$ so $\vec{v}$ and $t\vec{w}$ also have the same length. Hence $\vec{v} = t\vec{w}$ as required.

Case 2. $\vec{v}$ and $\vec{w}$ have opposite directions. In this case $\vec{v} = -t\vec{w}$ where $t = \frac{\|\vec{v}\|}{\|\vec{w}\|}$. The details (similar to Case 1) are left to the reader. □

Example 7. *Show that $\vec{v} = [7 \ -28 \ 35]^T$ and $\vec{w} = [-5 \ 20 \ -25]^T$ are parallel, while $\vec{u} = [2 \ -3 \ 6]^T$ and $\vec{z} = [1 \ -2 \ 3]^T$ are not parallel.*

Solution. We have $\vec{v} = -\frac{7}{5}\vec{w}$, so $\vec{v}$ and $\vec{w}$ are parallel (and have opposite direction). If $\vec{u}$ and $\vec{z}$ are parallel then one is a scalar multiple of the other, say $\vec{u} = t\vec{z}$ for some scalar t. Comparing components gives $2 = t$, $-3 = -2t$ and $6 = 3t$. Clearly, no such number t exists. Hence $\vec{u}$ and $\vec{z}$ are not parallel. □

3.1.7 A Proof of Pythagoras' Theorem

Pythagoras (c 550 BC) founded a semireligious school whose motto was "Everything is number". Their studies were called *mathema* ("what is learned") and consisted of music, astronomy, geometry and arithmetic. The Pythagorean theorem was known and used earlier, but Pythagoras is credited with the first logical, deductive proof of the result. Many proofs of the the theorem are known; the one we give depends on the basic properties of similar triangles.

Theorem 4. Pythagoras' Theorem. *Given a right angled triangle with hypotenuse h and sides a and b, then*

$$a^2 + b^2 = h^2.$$

Proof. Suppose that the triangle has vertices A, B and C with the right angle at A, as in Figure 26. Let D be the foot of a perpendicular from A to the hypotenuse and let p and q be the lengths of BD and DC respectively. Then ABC and DBA are similar triangles, so $\frac{p}{a} = \frac{a}{h}$. This means $a^2 = ph$. In the

3.1. GEOMETRIC VECTORS

same way, the similarity of ABC and DAC gives $\frac{q}{b} = \frac{b}{h}$, whence $b^2 = qh$. But then
$$a^2 + b^2 = ph + qh = (p+q)h = h^2$$
because $p + q = h$. This proves Pythagoras' theorem. □

Exercises

1. Consider the points $P(6, 7)$ and $Q(-12, 3)$ in the 2-dimensional XY-plane.

   (a) Locate these two points. Explain the difference with the points $P'(7, 6)$ and $Q'(3, -12)$.

   (b) Form the vector $\overrightarrow{PQ}$ from the point P to the point Q. Explain the difference with the vector $\overrightarrow{QP}$ from the point Q to the point P.

   (c) Translate the vector $\overrightarrow{PQ}$ to standard position, then give the coordinates of its terminal point, its components and matrix form.

   (d) Find the length of the vector $\overrightarrow{PQ}$. What is the connection with the distance between the two points P and Q?

   (e) Repeat the preceding questions with the new points $P(2, -1, 4)$ and $Q(-8, -7, 3)$ in 3-dimensional XYZ-space.

2. Consider the points $P(3, -2)$ and $Q(-1, 6)$ in the 2-dimensional XY-plane.

   (a) Represent the position vectors $\vec{p}$ and $\vec{q}$ of P and Q in matrix form, and calculate their lengths.

   (b) Draw and compute the matrix forms and lengths of the vectors $\vec{p} + \vec{q}$, $\vec{p} - \vec{q}$, $3\vec{p}$, $-2\vec{q}$, and $3\vec{p} - 2\vec{q}$.

   (c) Repeat the preceding questions with the new points $P(4, 1, -3)$ and $Q(-2, 4, -8)$ in 3-dimensional XYZ-space.

3. In each case compute $\overrightarrow{PQ}$ and $\left\|\overrightarrow{PQ}\right\|$.

   (a) $P(2, -1, 3)$ and $Q(0, 1, 1)$ (b) $P(0, 2, -3)$ and $Q(4, -1, 2)$
   (c) $P(2, 0, -5)$ and $Q(0, 5, 7)$ (d) $P(1, 2, 3)$ and $Q(3, 2, 1)$

4. (a) Given the point $P(3, -2, 1)$, find the point Q such that $\overrightarrow{PQ} = [-1\ 2\ 5]^T$.

   (b) Given the point $Q(0, -1, 3)$, find the point P such that $\overrightarrow{PQ} = [3\ 0\ -2]^T$.

5. Let $\vec{u} = [2\ -1\ 3]^T$, $\vec{v} = [0\ 2\ -3]^T$, and $\vec{w} = [-5\ 1\ 2]^T$. Compute each of the following.

   (a) $3\vec{u} - 5\vec{v} + 7\vec{w}$ (b) $7\vec{u} + 2\vec{v} - 5\vec{w}$
   (c) $\left\|3\vec{u} - \frac{1}{2}\vec{v} + \vec{w}\right\|$ (d) $\left\|\frac{1}{2}(\vec{u} + \vec{v} + 2\vec{w})\right\|$

6. (a) Find a vector $\vec{u}$ of length 1 in the same direction as $\vec{v} = [3 \ -1 \ 2]^T$.

    (b) If $\vec{v} \neq \vec{0}$, show that $\vec{u} = \frac{1}{\|\vec{v}\|}\vec{v}$ is a vector of length 1 in the same direction as $\vec{v}$.

7. In each case simplify the expression.

    (a) $3(2\vec{w} - 2\vec{u} - \vec{v}) - 2(\vec{u} + \vec{w} - 5\vec{v}) + 2(4\vec{u} + 4\vec{v} - 2\vec{w})$

    (b) $5(\vec{u} - \vec{v} - 2\vec{w}) + 3(2\vec{v} + 3\vec{w} - 3\vec{u}) - 2(3\vec{u} + \vec{v} + 8\vec{w})$

8. Let $\vec{u} = [-2 \ 3 \ -5]^T$ and $\vec{v} = [1 \ -2 \ 2]^T$. In each case find the vector $\vec{x}$.

    (a) $3\vec{u} - 5\vec{x} = \frac{3}{2}(4\vec{x} - \vec{v})$

    (b) $3\vec{x} - \|\vec{v}\|\vec{x} = 5(\vec{u} - 3\vec{x})$

    (c) $2\vec{x} - 3\vec{v} = \|\vec{u}\|^2(\vec{u} + 4\vec{x})$

    (d) $4(5\vec{x} - 2\vec{u}) = 3(5\vec{v} + 6\vec{x})$

9. Let $\vec{u} = [2 \ -1 \ 3]^T$, $\vec{v} = [0 \ 2 \ -3]^T$, and $\vec{w} = [6 \ 1 \ 3]^T$. In each case determine whether or not numbers a, b and c exist such that $\vec{x} = a\vec{u} + b\vec{v} + c\vec{w}$. Support your answer.

    (a) $\vec{x} = [1 \ 0 \ 0]^T$ (b) $\vec{x} = [2 \ -1 \ 3]^T$
    (c) $\vec{x} = [9 \ 3 \ 0]^T$ (d) $\vec{x} = [1 \ 0 \ 1]^T$

10. In each case determine if $\vec{v}$ and $\vec{w}$ are parallel. Support your answer.

    (a) $\vec{u} = [10 \ -15 \ 25]^T$ and $\vec{v} = [14 \ -21 \ 35]^T$

    (b) $\vec{u} = [3 \ -2 \ 1]^T$ and $\vec{v} = [-6 \ 4 \ -2]^T$

    (c) $\vec{u} = [2 \ 3 \ -5]^T$ and $\vec{v} = [-8 \ 12 \ 20]^T$

    (d) $\vec{u} = [-3 \ 5 \ 7]^T$ and $\vec{v} = [6 \ -10 \ 0]^T$

11. Find all vectors $\vec{u}$ that are parallel to $\vec{v} = [3 \ 5 \ -1]^T$ and satisfy $\|\vec{u}\| = 3\|\vec{v}\|$.

12. In each case find all points Q at a distance 5 from P and such that $\overrightarrow{PQ}$ is parallel to $\vec{d}$.

    (a) $P = P(5, 0, -3)$, $\vec{d} = [1 \ 2 \ -2]^T$

    (b) $P = P(1, 2, -5)$, $\vec{d} = [0 \ -3 \ 2]^T$

13. In each case P, Q and R are three vertices of a parallelogram $PQRS$. Find the other vertex S.

    (a) $P(3, -1, -1)$, $Q(1, -2, 0)$, $R(1, -1, 2)$

    (b) $P(2, 0, -1)$, $Q(-2, 4, 1)$, $R(3, -1, 0)$

14. In each case find the midpoint between the given points.

    (a) $P(2, -3, 5)$, $Q(-1, 7, -3)$

    (b) $P(1, 2, 3)$, $Q(-3, 5, -3)$

15. Given $P(1, 1, 3)$ and $Q(0, -3, 5)$ find:
    (a) The point A that is $\frac{1}{3}$ the way from P to Q.
    (b) The point B that is $\frac{1}{4}$ the way from Q to P.

16. Let P and Q be two points with position vectors $\vec{p}$ and $\vec{q}$ respectively. If $0 \le r \le 1$, show that the point A that is fraction r of the way from P to Q has position vector $\vec{a} = (1-r)\vec{p} + r\vec{q}$.

17. In each case either show that the statement is true or give an example showing that it is false. Throughout, $\vec{u}$, $\vec{v}$ and $\vec{w}$ denote vectors.
    (a) The zero vector $\vec{0}$ is the only vector of length 0.
    (b) If $\|\vec{v} - \vec{w}\| = \vec{0}$ then $\vec{v} = \vec{w}$.
    (c) If $\vec{v} = -\vec{v}$ then $\vec{v} = \vec{0}$.
    (d) If $\|\vec{v}\| = \|\vec{w}\|$ then $\vec{v} = \vec{w}$.
    (e) If $\|\vec{v}\| = \|\vec{w}\|$ then $\vec{v} = \pm\vec{w}$.
    (f) If $\vec{w} = t\vec{v}$ for some scalar t then $\vec{v}$ and $\vec{w}$ have the same direction.
    (g) If $\vec{v}$ and $\vec{v} + \vec{w}$ are parallel, then $\vec{v}$ and $\vec{w}$ are parallel.
    (h) $\|(-5)\vec{v}\| = -5\|\vec{v}\|$ for all $\vec{v}$.
    (i) If $\|\vec{v}\| = \|2\vec{v}\|$ then $\vec{v} = \vec{0}$.
    (j) $\|\vec{v} + \vec{w}\| = \|\vec{v}\| + \|\vec{w}\|$ for all $\vec{v}$ and $\vec{w}$.

18. Consider a triangle with vertices A, B and C, and let E and F be the midpoints of sides AB and BC.
    (a) Explain why $\overrightarrow{EB} = \frac{1}{2}\overrightarrow{AB}$ and $\overrightarrow{BF} = \frac{1}{2}\overrightarrow{BC}$.
    (b) Show that the line joining E and F is parallel to AC and half as long.

19. Let A, B, C, D, E, and F be the six vertices in order of a regular hexagon. Show that
$$\overrightarrow{AB} + \overrightarrow{AC} + \overrightarrow{AD} + \overrightarrow{AE} + \overrightarrow{AF} = 3\overrightarrow{AD}.$$

20. Let $P = P(x, y)$ be an arbitrary point in the plane with position vector $\vec{p} = [x \ y]^T$. Let $\mathcal{C}$ denote the circle with center at the origin and radius $r > 0$.
    (a) If P is on the circle $\mathcal{C}$, explain geometrically why $\|\vec{p}\| = r$.
    (b) If $\|\vec{p}\| = r$, explain geometrically why P is on the circle $\mathcal{C}$.
    (c) Use (a) and (b) to show that the equation of the circle $\mathcal{C}$ is $x^2 + y^2 = r^2$.

21. Let $P = P(x, y, z)$ be an arbitrary point in space with position vector $\vec{p} = [x \ y \ z]^T$. Let $\mathcal{S}$ denote the sphere with center at the origin and radius $r > 0$.
    (a) If P is on the sphere $\mathcal{S}$, explain geometrically why $\|\vec{p}\| = r$.

(b) If $\|\vec{p}\| = r$, explain geometrically why P is on the sphere $\mathcal{S}$.

(c) Use (a) and (b) to show that the equation of the sphere $\mathcal{S}$ is $x^2 + y^2 + z^2 = r^2$.

3.2 DOT PRODUCT AND PROJECTIONS

The concept of perpendicular lines is fundamental in geometry. It is required for the statement of Pythagoras' theorem and it is basic to the study of trigonometry. In order to use vectors to study perpendicularity, we develop a simple test for when two nonzero vectors have perpendicular directions. The test depends on whether a certain number (called the dot product of the two vectors) is zero or not. As is often the case with a good idea, the dot product has other applications: It can be used to find the angle between *any* two nonzero vectors, and it leads to a method of solving certain optimization problems in geometry. For example, we will be able (in Section 3.3) to find the point on a plane that is closest to an arbitrary point.

3.2.1 The Dot Product

Given two vectors $\vec{v}$ and $\vec{w}$, the fact that they are column matrices means that the matrix product $\vec{v}^T \vec{w}$ is a 1×1 matrix, which we regard as a number. This number is called the **dot product** $\vec{v} \bullet \vec{w}$ of $\vec{v}$ and $\vec{w}$. In terms of components, the definition is as follows:

$$\text{If } \vec{v} = \begin{bmatrix} x_1 \\ y_1 \\ z_1 \end{bmatrix} \text{ and } \vec{w} = \begin{bmatrix} x_2 \\ y_2 \\ z_2 \end{bmatrix} \text{ then } \vec{v} \bullet \vec{w} = \vec{v}^T \vec{w} = x_1 x_2 + y_1 y_2 + z_1 z_2.$$

$$\text{If } \vec{v} = \begin{bmatrix} x_1 \\ y_1 \end{bmatrix} \text{ and } \vec{w} = \begin{bmatrix} x_2 \\ y_2 \end{bmatrix} \text{ then } \vec{v} \bullet \vec{w} = \vec{v}^T \vec{w} = x_1 x_2 + y_1 y_2.$$

In words, the dot product of two vectors is formed by multiplying corresponding components and adding the results. Because $\vec{v} \bullet \vec{w}$ is a *number* (even though $\vec{v}$ and $\vec{w}$ are vectors), it is sometimes called the **scalar product** of $\vec{v}$ and $\vec{w}$.

Example 1. If $\vec{v} = \begin{bmatrix} 6 \\ -2 \\ 5 \end{bmatrix}$ and $\vec{w} = \begin{bmatrix} 2 \\ -1 \\ -3 \end{bmatrix}$, then $\vec{v} \bullet \vec{w} = 6 \cdot 2 + (-2)(-1) + 5(-3) = -1$.

The dot product $\vec{v} \bullet \vec{w}$ is the central concept in this section, and the following theorem collects several of its basic properties.

Theorem 1. *Let $\vec{u}$, $\vec{v}$ and $\vec{w}$ denote vectors, and let a be a number. Then:*

(1) *$\vec{v} \bullet \vec{w}$ is a number.*

3.2. DOT PRODUCT AND PROJECTIONS

(2) $\vec{v} \bullet \vec{w} = \vec{w} \bullet \vec{v}$.

(3) $\vec{v} \bullet \vec{0} = 0 = \vec{0} \bullet \vec{w}$.

(4) $\vec{v} \bullet \vec{v} = \|\vec{v}\|^2$.

(5) $(a\vec{v}) \bullet \vec{w} = a(\vec{v} \bullet \vec{w}) = \vec{v} \bullet (a\vec{w})$.

(6) $\vec{v} \bullet (\vec{u} + \vec{w}) = \vec{v} \bullet \vec{u} + \vec{v} \bullet \vec{w}$.

(7) $\vec{v} \bullet (\vec{u} - \vec{w}) = \vec{v} \bullet \vec{u} - \vec{v} \bullet \vec{w}$.

Proof. (1) holds by the definition of $\vec{v} \bullet \vec{w}$, and (4) comes from (2) of Theorem 1, § 3.1. To prove (2) write $\vec{v} = [x_1 \ y_1 \ z_1]^T$ and $\vec{w} = [x_2 \ y_2 \ z_2]^T$. Then:

$$\vec{v} \bullet \vec{w} = x_1 x_2 + y_1 y_2 + z_1 z_2 = x_2 x_1 + y_2 y_1 + z_2 z_1 = \vec{w} \bullet \vec{v}$$

proving (2), The rest of the assertions in the theorem are properties of matrix arithmetic. For example (6) is verified as follows:

$$\vec{v} \bullet (\vec{u} + \vec{w}) = \vec{v}^T (\vec{u} + \vec{w}) = \vec{v}^T \vec{u} + \vec{v}^T \vec{w} = \vec{v} \bullet \vec{u} + \vec{v} \bullet \vec{w}.$$

The proofs of (3), (5) and (7) are similar and are left as Exercise 17. □

The properties in Theorem 1 can be combined to enable calculations like the following:

$$3\vec{u} \bullet (\vec{v} - 2\vec{w} + 5\vec{z}) = 3(\vec{u} \bullet \vec{v}) - 6(\vec{u} \bullet \vec{w}) + 15(\vec{u} \bullet \vec{z}).$$

Such computations will be used without comment below. Here is an illustration.

Example 2. *Verify that* $\|\vec{v} + \vec{w}\|^2 = \|\vec{v}\|^2 + 2(\vec{v} \bullet \vec{w}) + \|\vec{w}\|^2$ *for any vectors $\vec{v}$ and $\vec{w}$.*

Solution. We apply Theorem 1 several times:

$$\begin{aligned} \|\vec{v} + \vec{w}\|^2 &= (\vec{v} + \vec{w}) \bullet (\vec{v} + \vec{w}) \\ &= \vec{v} \bullet \vec{v} + \vec{v} \bullet \vec{w} + \vec{w} \bullet \vec{v} + \vec{w} \bullet \vec{w} \\ &= \|\vec{v}\|^2 + 2(\vec{v} \bullet \vec{w}) + \|\vec{w}\|^2. \end{aligned}$$

This is what we wanted. □

3.2.2 Angles

Suppose two nonzero vectors $\vec{v}$ and $\vec{w}$ are positioned with a common tail. Then they determine a unique angle θ in the range [8]

$$0 \leq \theta \leq \pi$$

called the **angle between** the vectors $\vec{v}$ and $\vec{w}$. Figure 1 illustrates the cases where θ is acute (less than $\frac{\pi}{2}$) and obtuse (greater than $\frac{\pi}{2}$). If either $\vec{v}$ or $\vec{w}$ is the zero vector, the angle between them is not defined.

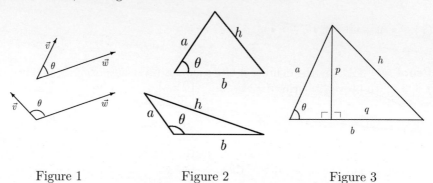

Figure 1   Figure 2   Figure 3

There is a close connection between the dot product $\vec{v} \bullet \vec{w}$ and the angle between $\vec{v}$ and $\vec{w}$. The relationship depends upon an extension of Pythagoras' theorem known as the

Law of Cosines. *If a, b and h are the sides of a triangle and θ is the angle opposite h (as in Figure 2), then*

$$h^2 = a^2 + b^2 - 2ab\cos\theta.$$

Proof: We consider the case when θ is acute; the obtuse case is similar. Referring to Figure 3, right triangle trigonometry gives $p = a\sin\theta$ and $q = b - a\cos\theta$, so Pythagoras' theorem gives

$$h^2 = p^2 + q^2 = a^2\sin^2\theta + [b^2 - 2ab\cos\theta + a^2\cos^2\theta] = a^2 + b^2 - 2ab\cos\theta$$

where we used the fact that $\sin^2\theta + \cos^2\theta = 1$ for any angle θ. $\square$

Note that the law of cosines reduces to Pythagoras' theorem if θ is a right angle because then $\theta = \frac{\pi}{2}$ and so $\cos\theta = \cos\frac{\pi}{2} = 0$.

With the law of cosines in hand we can derive an important connection between angles and dot products.

---

[8] We use radian measure for angles. Thus π equals $180°$, $\frac{\pi}{2}$ equals $90°$, etc. See Appendix A.1 for a discussion of angles.

3.2. DOT PRODUCT AND PROJECTIONS

Theorem 2. *If θ is the angle between the nonzero vectors $\vec{v}$ and $\vec{w}$, then*

$$\vec{v} \bullet \vec{w} = \|\vec{v}\| \, \|\vec{w}\| \cos\theta.$$

Proof. We compute $\|\vec{v} - \vec{w}\|^2$ two ways and compare the results. If we apply the law of cosines to the triangle in Figure 4, the result is

$$\|\vec{v} - \vec{w}\|^2 = \|\vec{v}\|^2 + \|\vec{w}\|^2 - 2\|\vec{v}\| \, \|\vec{w}\| \cos\theta.$$

On the other hand, Theorem 1 gives

$$\begin{aligned}\|\vec{v} - \vec{w}\|^2 &= (\vec{v} - \vec{w}) \bullet (\vec{v} - \vec{w}) \\ &= \vec{v} \bullet \vec{v} - \vec{v} \bullet \vec{w} - \vec{w} \bullet \vec{v} + \vec{w} \bullet \vec{w} \\ &= \|\vec{v}\|^2 - 2\vec{v} \bullet \vec{w} + \|\vec{w}\|^2.\end{aligned}$$

Comparing these expressions proves Theorem 2. □

Theorem 2 gives an intrinsic description of the dot product because the expression $\|\vec{v}\| \, \|\vec{w}\| \cos\theta$ depends only on the lengths of the vectors $\vec{v}$ and $\vec{w}$ and the angle θ between them, and does not depend on the coordinate system used to describe the vectors. Moreover, if $\vec{v}$ and $\vec{w}$ are nonzero it provides a formula for the cosine of the angle θ between them:

$$\cos\theta = \frac{\vec{v} \bullet \vec{w}}{\|\vec{v}\| \, \|\vec{w}\|}.$$

Here is how the formula is used.

Example 3. *Find the angle θ between $\vec{v} = \begin{bmatrix} -1 \\ -2 \\ 1 \end{bmatrix}$ and $\vec{w} = \begin{bmatrix} 2 \\ 1 \\ 1 \end{bmatrix}$.*

Solution. Using Theorem 2, we compute

$$\cos\theta = \frac{\vec{v} \bullet \vec{w}}{\|\vec{v}\| \, \|\vec{w}\|} = \frac{-2 - 2 + 1}{\sqrt{1+4+1}\sqrt{4+1+1}} = \frac{-3}{\sqrt{6}\sqrt{6}} = -\frac{1}{2}.$$

Now recall that $\cos\theta$ and $\sin\theta$ are defined [9] so that $(\cos\theta, \sin\theta)$ is the point on the unit circle determined by the angle θ (drawn counterclockwise from the positive X-axis as in Figure 5). In the present example we have $\cos\theta = -\frac{1}{2}$ and $0 \leq \theta \leq \pi$, so it follows that $\theta = \frac{2\pi}{3}$ or $120°$. □

Let θ denote the angle between the nonzero vectors $\vec{v}$ and $\vec{w}$, so that $0 \leq \theta \leq \pi$. It follows from the formula $\cos\theta = \frac{\vec{v} \bullet \vec{w}}{\|\vec{v}\| \, \|\vec{w}\|}$ that $\vec{v} \bullet \vec{w}$ and $\cos\theta$ have the same sign. But $\cos\theta$ is the X-coordinate of the point on the unit circle determined by θ (see Figure 5), so $\cos\theta$ is positive, zero or negative if and only if $\theta < \frac{\pi}{2}, \theta = \frac{\pi}{2}$

---

[9] A review of basic trigonometry is given in Appendix A.1.

or $\theta > \frac{\pi}{2}$ respectively. Hence we obtain the following relationship between the sign of $\vec{v} \bullet \vec{w}$ and the size of θ:

$$\begin{array}{lll}\vec{v} \bullet \vec{w} > 0 & \text{if and only if} & \theta \text{ is } \textbf{acute } (0 \leq \theta < \frac{\pi}{2}) \\ \vec{v} \bullet \vec{w} = 0 & \text{if and only if} & \theta \text{ is a } \textbf{right angle } (\theta = \frac{\pi}{2}) \\ \vec{v} \bullet \vec{w} < 0 & \text{if and only if} & \theta \text{ is } \textbf{obtuse } (\frac{\pi}{2} < \theta \leq \pi)\end{array}$$

The second of these three cases is the most important. The vectors $\vec{v}$ and $\vec{w}$ are called **orthogonal** if the angle θ between them is a right angle, that is if $\theta = \frac{\pi}{2}$. The above discussion proves the following result.

Theorem 3. *Let $\vec{v}$ and $\vec{w}$ denote nonzero vectors. Then:*

$$\vec{v} \text{ and } \vec{w} \text{ are orthogonal if and only if } \vec{v} \bullet \vec{w} = 0.$$

Theorem 3 will be referred to frequently because perpendicular lines play such a basic role in geometry. The reason, of course, is that one can test whether two lines are perpendicular by checking if two nonzero vectors pointing along the lines are orthogonal. The following three examples provide illustrations.

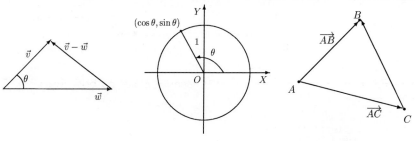

Figure 4   Figure 5   Figure 6

Example 4. *Show that the points $A(3, -1, 2)$, $B(5, -2, 5)$ and $C(2, 0, 3)$, are the vertices of a right angled triangle.*

Solution. We have $\overrightarrow{AB} = [2 \ -1 \ 3]^T$, $\overrightarrow{AC} = [-1 \ 1 \ 1]^T$ and $\overrightarrow{BC} = [-3 \ 2 \ -2]^T$. Now observe that $\overrightarrow{AB} \bullet \overrightarrow{AC} = -2 - 1 + 3 = 0$, so the vectors $\overrightarrow{AB}$ and $\overrightarrow{AC}$ are orthogonal by Theorem 3. Since these vectors point along the sides AB and AC of the triangle (Figure 6), this means that the angle at vertex A is a right angle. $\square$

Example 5. *A **rhombus** is a parallelogram in which all sides are of equal length. Show that the diagonals of any rhombus are perpendicular.*

Solution. Let $\vec{v}$ and $\vec{w}$ denote the vectors along adjacent sides of the rhombus as in Figure 7. Then $\vec{v} + \vec{w}$ and $\vec{v} - \vec{w}$ point along the diagonals, and we compute

$$\begin{aligned}(\vec{v} - \vec{w}) \bullet (\vec{v} + \vec{w}) &= \vec{v} \bullet \vec{v} + \vec{v} \bullet \vec{w} - \vec{w} \bullet \vec{v} - \vec{w} \bullet \vec{w} \\ &= \|\vec{v}\|^2 - \|\vec{w}\|^2 \\ &= 0\end{aligned}$$

3.2. DOT PRODUCT AND PROJECTIONS

because $\|\vec{v}\| = \|\vec{w}\|$ (it is a rhombus). Hence $\vec{v} + \vec{w}$ and $\vec{v} - \vec{w}$ are orthogonal by Theorem 3, and so the diagonals are perpendicular. □

Example 6. *The line through a vertex of a triangle which is perpendicular to the opposite side is called an* **altitude** *of the triangle. Show that the three altitudes of any triangle are concurrent (that is they meet at a common point).*

Solution. Let the vertices be A, B and C, and let P be the point of intersection of the altitudes through A and B, as in Figure 8. It suffices to show that $\overrightarrow{PC}$ is orthogonal to $\overrightarrow{AB}$; that is, $\overrightarrow{PC} \bullet \overrightarrow{AB} = 0$. We have $\overrightarrow{AB} = \overrightarrow{AC} - \overrightarrow{BC}$, so

$$\overrightarrow{PC} \bullet \overrightarrow{AB} = \overrightarrow{PC} \bullet (\overrightarrow{AC} - \overrightarrow{BC}) = \overrightarrow{PC} \bullet \overrightarrow{AC} - \overrightarrow{PC} \bullet \overrightarrow{BC}$$

Writing $\overrightarrow{PC}$ two different ways on the right side, we get

$$\begin{aligned}
\overrightarrow{PC} \bullet \overrightarrow{AB} &= (\overrightarrow{PB} + \overrightarrow{BC}) \bullet \overrightarrow{AC} - (\overrightarrow{PA} + \overrightarrow{AC}) \bullet \overrightarrow{BC} \\
&= \overrightarrow{PB} \bullet \overrightarrow{AC} + \overrightarrow{BC} \bullet \overrightarrow{AC} - \overrightarrow{PA} \bullet \overrightarrow{BC} - \overrightarrow{AC} \bullet \overrightarrow{BC} \\
&= 0 + \overrightarrow{BC} \bullet \overrightarrow{AC} - 0 - \overrightarrow{AC} \bullet \overrightarrow{BC} \\
&= 0
\end{aligned}$$

where $\overrightarrow{PB} \bullet \overrightarrow{AC} = 0$ and $\overrightarrow{PA} \bullet \overrightarrow{BC} = 0$ because P lies on both altitudes. This is what we wanted. □

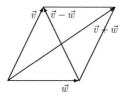
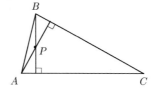

Figure 7   Figure 8

3.2.3 Projections

In applications of vectors, it is frequently useful to write a vector as the sum of two orthogonal vectors. Here is an example.

Example 7. *Suppose a ten kilogram block is placed on a flat surface inclined $30°$ to the horizontal as in Figure 9. Neglecting friction, how much force is required to keep the block from sliding down the surface?*

Solution. Let $\vec{w}$ denote the weight (force due to gravity) exerted on the block. Then $\|\vec{w}\| = 10$ and the direction of $\vec{w}$ is vertically down as in Figure 9.

The idea is to write $\vec{w}$ as a sum $\vec{w} = \vec{w}_1 + \vec{w}_2$ where $\vec{w}_1$ is parallel to the inclined surface, and $\vec{w}_2$ is perpendicular to the surface. Since there is no friction, the force required is $-\vec{w}_1$ because the force $\vec{w}_2$ has no effect parallel to the surface. As the angle between $\vec{w}$ and $\vec{w}_2$ is $30°$ in Figure 9, we have $\frac{\|\vec{w}_1\|}{\|\vec{w}\|} = \sin 30° = \frac{1}{2}$. Hence $\|\vec{w}_1\| = \frac{1}{2}\|\vec{w}\| = \frac{1}{2}10 = 5$. Thus the required force has a magnitude of 5 kilograms, directed up the surface. □

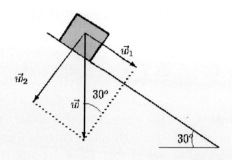

Figure 9

The key idea in the solution of Example 7 was to write $\vec{w}$ as a sum of two vectors, one parallel to the surface and one perpendicular to the surface. This type of decomposition arises frequently in applications, and there is a simple way to find it. The general formulation of the problem is as follows.

Suppose that vectors $\vec{v}$ and $\vec{d}$ are given with $\vec{d} \neq \vec{0}$. We want to be able to write $\vec{v}$ as a sum

$$\vec{v} = \vec{v}_1 + \vec{v}_2$$

where $\vec{v}_1$ is parallel to $\vec{d}$ and $\vec{v}_2$ is orthogonal to $\vec{d}$ (see Figure 10). Since $\vec{v}_1$ is parallel to $\vec{d}$ it must have the form

$$\vec{v}_1 = t\vec{d}$$

where t is some scalar. This scalar t is determined by the fact that $\vec{v}_2$ is orthogonal to $\vec{d}$, that is $\vec{v}_2 \bullet \vec{d} = 0$. Since $\vec{v}_2 = \vec{v} - \vec{v}_1$, this condition takes the form

$$0 = \vec{v}_2 \bullet \vec{d} = (\vec{v} - \vec{v}_1) \bullet \vec{d} = (\vec{v} - t\vec{d}) \bullet \vec{d} = \vec{v} \bullet \vec{d} - t(\vec{d} \bullet \vec{d}) = \vec{v} \bullet \vec{d} - t\|\vec{d}\|^2.$$

In other words, the requirement is that $t\|\vec{d}\|^2 = \vec{v} \bullet \vec{d}$. Since $\|\vec{d}\|^2 \neq \vec{0}$ (because $\vec{d} \neq \vec{0}$) this gives

$$t = \frac{\vec{v} \bullet \vec{d}}{\|\vec{d}\|^2}$$

Hence $\vec{v}_1 = \left(\frac{\vec{v} \bullet \vec{d}}{\|\vec{d}\|^2}\right)\vec{d}$, so $\vec{v}_1$ and $\vec{v}_2 = \vec{v} - \vec{v}_1$ are uniquely determined by $\vec{v}$ and $\vec{d}$. The vector $\vec{v}_1$ is called the **projection of $\vec{v}$ on $\vec{d}$**, written

$$\vec{v}_1 = proj_{\vec{d}}(\vec{v}).$$

3.2. DOT PRODUCT AND PROJECTIONS

This proves the following useful theorem.

Theorem 4. Projection Theorem. Let $\vec{v}$ and $\vec{d} \neq \vec{0}$ be vectors.

(1) The projection of $\vec{v}$ on $\vec{d}$ is given by

$$proj_{\vec{d}}(\vec{v}) = \left(\frac{\vec{v} \cdot \vec{d}}{\|\vec{d}\|^2}\right)\vec{d}.$$

(2) The vector $\vec{v} - proj_{\vec{d}}(\vec{v})$ is orthogonal to $\vec{d}$.

(3) The vector $\vec{v}$ can be written uniquely in the form $\vec{v} = \vec{v}_1 + \vec{v}_2$ where $\vec{v}_1$ is parallel to $\vec{d}$ and $\vec{v}_2$ is orthogonal to $\vec{d}$. In fact $\vec{v}_1 = proj_{\vec{d}}(\vec{v})$ and $\vec{v}_2 = \vec{v} - \vec{v}_2$.

Observe that the projection $\vec{v}_1 = proj_{\vec{d}}(\vec{v})$ has the same direction as $\vec{d}$ if the angle θ between $\vec{v}$ and $\vec{d}$ is less than $\frac{\pi}{2}$, and $\vec{v}_1$ has the opposite direction to $\vec{d}$ if θ is greater than $\frac{\pi}{2}$. These two situations are illustrated in Figure 11. Of course, if $\vec{v}$ and $\vec{d}$ are orthogonal (that is if $\theta = \frac{\pi}{2}$) then $\vec{v}_1 = \vec{0}$.

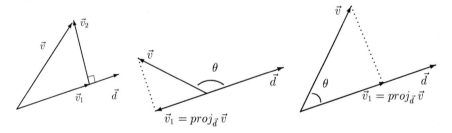

Figure 10   Figure 11

Example 8. If $\vec{v} = \begin{bmatrix} 1 \\ 2 \\ -3 \end{bmatrix}$ and $\vec{d} = \begin{bmatrix} 4 \\ -1 \\ 1 \end{bmatrix}$, express $\vec{v}$ in the form $\vec{v} = \vec{v}_1 + \vec{v}_2$ where $\vec{v}_1$ is parallel to $\vec{d}$ and $\vec{v}_2$ is orthogonal to $\vec{d}$.

Solution. By the Projection Theorem, $\vec{v}_1$ is the projection of $\vec{v}$ on $\vec{d}$; that is

$$\vec{v}_1 = proj_{\vec{d}}(\vec{v}) = \left(\frac{\vec{v} \cdot \vec{d}}{\|\vec{d}\|^2}\right)\vec{d} = \left(\frac{4 - 2 - 3}{4^2 + (-1)^2 + 1^2}\right)\vec{d} = \left(\frac{-1}{18}\right)\vec{d} = \frac{1}{18}\begin{bmatrix} -4 \\ 1 \\ -1 \end{bmatrix}$$

Since $\vec{v} = \vec{v}_1 + \vec{v}_2$, we can compute $\vec{v}_2$ as follows:

$$\vec{v}_2 = \vec{v} - \vec{v}_1 = \begin{bmatrix} 1 \\ 2 \\ -3 \end{bmatrix} - \tfrac{1}{18} \begin{bmatrix} -4 \\ 1 \\ -1 \end{bmatrix}$$

$$= \tfrac{1}{18} \left(\begin{bmatrix} 18 \\ 36 \\ -54 \end{bmatrix} - \begin{bmatrix} -4 \\ 1 \\ -1 \end{bmatrix} \right)$$

$$= \tfrac{1}{18} \begin{bmatrix} 22 \\ 35 \\ -53 \end{bmatrix}$$

A good check on the arithmetic is to verify that $\vec{v}_2$ is actually orthogonal to $\vec{d}$. In this case, we have $\vec{v}_2 \bullet \vec{d} = \tfrac{1}{18} (22 \cdot 4 + 35 \, (-1) + (-53) \cdot 1) = 0$ so this vector $\vec{v}_2$ is indeed orthogonal to $\vec{d}$. □

Exercises

1. In each case compute $\vec{v} \bullet \vec{w}$.
    (a) $\vec{v} = [5 \; -3]^T$, $\vec{w} = [2 \; 6]^T$
    (b) $\vec{v} = [4 \; 0]^T$, $\vec{w} = [1 \; -1]^T$
    (c) $\vec{v} = [2 \; -1 \; 3]^T$, $\vec{w} = [1 \; 2 \; -5]^T$
    (d) $\vec{v} = [3 \; -1 \; 4]^T$, $\vec{w} = \vec{v}$
    (e) $\vec{v} = [2 \; 2 \; -1]^T$, $\vec{w} = -3[1 \; -1 \; 5]^T$
    (f) $\vec{v} = [3 \; -1 \; 5]^T$, $\vec{w} = [4 \; 2 \; -2]^T$
    (g) $\vec{v} = [-5 \; 7 \; 2]^T$, $\vec{w} = [9 \; 7 \; -2]^T$
    (h) $\vec{v} = [5 \; -6 \; 2]^T$, $\vec{w} = [-3 \; 4 \; -2]^T$

2. In each case compute $\vec{v} \bullet \vec{v}$ and $\|\vec{v}\|$.
    (a) $\vec{v} = [2 \; -1 \; 3]^T$ (b) $\vec{v} = [1 \; -2 \; 2]^T$
    (c) $\vec{v} = \tfrac{\sqrt{2}}{10}[3 \; -5 \; 4]^T$ (d) $\vec{v} = [1 \; -1 \; 1]^T$

3. For each of the following pairs of vectors, determine whether the angle between $\vec{v}$ and $\vec{w}$ is acute, obtuse or a right angle.
    (a) $\vec{v} = [5 \; -3]^T$, $\vec{w} = [2 \; 6]^T$
    (b) $\vec{v} = [4 \; 0]^T$, $\vec{w} = [1 \; -1]^T$
    (c) $\vec{v} = [2 \; -1 \; 3]^T$, $\vec{w} = [1 \; 2 \; -5]^T$
    (d) $\vec{v} = [3 \; -1 \; 4]^T$, $\vec{w} = \vec{v}$
    (e) $\vec{v} = [2 \; 2 \; -1]^T$, $\vec{w} = -3[1 \; -1 \; 5]^T$
    (f) $\vec{v} = [3 \; -1 \; 5]^T$, $\vec{w} = [4 \; 2 \; -2]^T$
    (g) $\vec{v} = [-5 \; 7 \; 2]^T$, $\vec{w} = [9 \; 7 \; -2]^T$
    (h) $\vec{v} = [5 \; -6 \; 2]^T$, $\vec{w} = [-3 \; 4 \; -2]^T$

4. In each case find the angle between $\vec{v}$ and $\vec{w}$.

3.2. DOT PRODUCT AND PROJECTIONS

(a) $\vec{v} = [3\ 1]^T$, $\vec{w} = [1\ 2]^T$ (b) $\vec{v} = [1\ 1]^T$, $\vec{w} = [-1\ 1]^T$
(c) $\vec{v} = [2\ 1\ 1]^T$, $\vec{w} = [1\ 1\ 0]^T$ (d) $\vec{v} = [2\ 3\ 4]^T$, $\vec{w} = [3\ -4\ 7]^T$
(e) $\vec{v} = [7\ 3\ -1]^T$, $\vec{w} = [1\ -1\ 4]^T$ (f) $\vec{v} = [0\ -1\ 1]^T$, $\vec{w} = [1\ 1\ 0]^T$

5. Find all a, b and c such that $\vec{u} = [a\ b\ c]^T$ is orthogonal to both vectors $\vec{v} = [2\ -1\ 3]^T$ and $\vec{w} = [1\ 1\ -2]^T$.

6. Consider the triangle with vertices $A(2,-3,0)$, $B(5,1,-2)$, and $C(7,3,5)$.

    (a) Show that ABC is a right-angled triangle.

    (b) Find the lengths of the three sides and verify Pythagoras' theorem.

7. Show that the line through $A(1,-2,1)$ and $B(0,2,5)$ is perpendicular to the line through $C(2,1,-1)$ and $D(2,3,-3)$.

8. In each case consider the line through A and B, and the line through C and D, and determine if these lines are parallel, perpendicular, or neither. Support your answer.

    (a) $A(2,-1,3)$, $B(3,1,-1)$, $C(1,1,0)$, $D(0,-1,4)$
    (b) $A(5,0,-2)$, $B(1,1,2)$, $C(3,-2,5)$, $D(4,-2,3)$
    (c) $A(3,2,1)$, $B(1,1,1)$, $C(-5,-1,0)$, $D(4,-3,2)$
    (d) $A(-1,2,5)$, $B(9,-3,2)$, $C(0,-3,2)$, $D(1,-4,7)$
    (e) $A(5,4,-1)$, $B(8,3,3)$, $C(2,4,6)$, $D(1,5,7)$
    (f) $A(3,0,-2)$, $B(6,-5,2)$, $C(4,4,-3)$, $D(-2,14,-11)$

9. In each case compute the projection of $\vec{v}$ along $\vec{w}$.

    (a) $\vec{v} = [-2\ 3]^T$, $\vec{w} = [1\ 2]^T$
    (b) $\vec{v} = [-1\ 3]^T$, $\vec{w} = [5\ 0]^T$
    (c) $\vec{v} = [3\ -1\ 1]^T$, $\vec{w} = [2\ 0\ -3]^T$
    (d) $\vec{v} = [-2\ 0\ 12]^T$, $\vec{w} = [5\ -4\ 4]^T$
    (e) $\vec{v} = [6\ -1\ 2]^T$, $\vec{w} = [-1\ 2\ 4]^T$
    (f) $\vec{v} = [5\ -1\ 1]^T$, $\vec{w} = [1\ -3\ -2]^T$

10. In each case write $\vec{v}$ as a sum $\vec{v} = \vec{v}_1 + \vec{v}_2$ where $\vec{v}_1$ is parallel to $\vec{w}$ and $\vec{v}_2$ is orthogonal to $\vec{w}$.

    (a) $\vec{v} = [1\ -1\ 3]^T$, $\vec{w} = [2\ -1\ 1]^T$
    (b) $\vec{v} = [3\ 1\ 2]^T$, $\vec{w} = [-2\ 1\ 4]^T$
    (c) $\vec{v} = [2\ -1\ 0]^T$, $\vec{w} = [3\ 2\ -1]^T$
    (d) $\vec{v} = [3\ -2\ 1]^T$, $\vec{w} = [1\ 4\ -1]^T$

11. Let A, B, C and D be the vertices in order of a parallelogram $ABCD$. Given $A(1,-1,2)$, $C(2,1,0)$ and the midpoint $M(1,0,-3)$ of side AB, find $\overrightarrow{BD}$.

12. Let A, B and C be the vertices of a triangle, and let M be the midpoint of BC. Given $A(2,-1,3)$, $\overrightarrow{BC} = [4\ 1\ -2]^T$ and $\overrightarrow{AM} = [2\ 0\ -1]^T$, find B and C.

13. Show that if the diagonals of a parallelogram are perpendicular, it is necessarily a rhombus. (See Example 5).

14. Let A and B be the endpoints of a diameter of a circle as in the diagram. If C is any point on the circle, show that the line segments AC and BC are perpendicular.

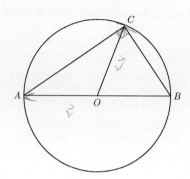

15. Consider the rectangular solid with sides of length 1, 1, and $\sqrt{2}$. Find the angle between a diagonal and one of the longest sides.

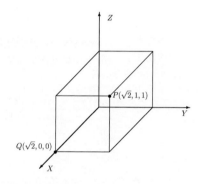

16. In each case either show that the statement is true or give an example showing that it is false. Throughout $\vec{u}$, $\vec{v}$ and $\vec{w}$ represent vectors.

    (a) If $\vec{v} \bullet \vec{w} = 0$ then either $\vec{v} = \vec{0}$ or $\vec{w} = \vec{0}$.

    (b) If $\vec{v}$ and $\vec{w}$ are orthogonal then $2\vec{v}$ and $-3\vec{w}$ are also orthogonal.

    (c) If $-\vec{v}$ is orthogonal to $\vec{w}$ then $\vec{v}$ is parallel to $\vec{w}$.

    (d) If the angle between $\vec{v}$ and $\vec{w}$ is acute then the angle between $-\vec{v}$ and $\vec{w}$ is obtuse.

    (e) If the projection $proj_{\vec{d}}(\vec{v}) = \vec{0}$ then $\vec{v} = \vec{0}$.

17. Prove the following parts of Theorem 1:

    (a) Part (3) (b) Part (5) (c) Part (7).

18. If $\vec{v}$ and $\vec{w}$ are orthogonal, show that $\|\vec{v} + \vec{w}\|^2 = \|\vec{v}\|^2 + \|\vec{w}\|^2$.

19. Show that $(\vec{v} + \vec{w}) \bullet (\vec{v} - \vec{w}) = \|\vec{v}\|^2 - \|\vec{w}\|^2$.

20. Show that $\|\vec{v} + \vec{w}\|^2 + \|\vec{v} - \vec{w}\|^2 = 2(\|\vec{v}\|^2 + \|\vec{w}\|^2)$.

21. If each of $\vec{u}$, $\vec{v}$, and $\vec{w}$ is orthogonal to the other two, show that $\|\vec{u} + \vec{v} + \vec{w}\|^2 = \|\vec{u}\|^2 + \|\vec{v}\|^2 + \|\vec{w}\|^2$.

22. Verify the following two properties of projection.
    (a) $proj_{\vec{d}}(\vec{v} + \vec{w}) = proj_{\vec{d}}(\vec{v}) + proj_{\vec{d}}(\vec{w})$ for all vectors $\vec{v}$ and $\vec{w}$.
    (b) $proj_{\vec{d}}(a\vec{v}) = a\, proj_{\vec{d}}(\vec{v})$ for all scalars a and vectors $\vec{v}$.
    [Because of these properties, the function $proj_{\vec{d}}(\ )$ is said to be a **linear transformation**; this will be treated further in Section 3.5.]

23. If $\vec{v}$ is any nonzero vector, let α, β and γ be the angles $\vec{v}$ makes with the positive X-, Y- and Z-axes respectively. Then $cos\alpha$, $cos\beta$ and $cos\gamma$ are called the **direction cosines** of the vector $\vec{v}$.
    (a) If $\vec{v} = [a\ b\ c]^T$, show that $cos\alpha = \frac{a}{\|\vec{v}\|}$, $cos\beta = \frac{b}{\|\vec{v}\|}$ and $cos\gamma = \frac{c}{\|\vec{v}\|}$.
    (b) Show that $cos^2\alpha + cos^2\beta + cos^2\gamma = 1$.

24. Consider the line $y = mx$ through the origin with slope m.
    (a) Show that $\vec{d} = [1\ m]^T$ is a vector pointing along the line.
    (b) Show that the lines $y = m_1 x$ and $y = m_2 x$ are perpendicular if and only if $m_1 m_2 = -1$.

25. Assume that $\vec{d}$ and $\vec{d'}$ are parallel, nonzero vectors. Use part (1) of Theorem 4 to show that $proj_{\vec{d}}(\vec{v}) = proj_{\vec{d'}}(\vec{v})$ for every vector $\vec{v}$.

26. (a) Show that $|\vec{v} \bullet \vec{w}| \leq \|\vec{v}\|\,\|\vec{w}\|$ for all vectors $\vec{v}$ and $\vec{w}$. This is called the *Cauchy-Schwarz inequality*.
    (b) Show that $|\vec{v} \bullet \vec{w}| = \|\vec{v}\|\,\|\vec{w}\|$ if and only if $\vec{v}$ and $\vec{w}$ are parallel.

3.3 LINES AND PLANES

We now turn our attention to describing two of the most commonly encountered objects in geometry: lines and planes. We take a vector approach, and the main tool is the concept of the position vector of a point.

3.3.1 Position Vectors

Let $P = P(x, y, z)$ be the point in space with coordinates (x, y, z). The **position vector** $\vec{p}$ of P was defined in Section 3.1 to be the vector from the origin to P:

$$\vec{p} = \overrightarrow{OP} = \begin{bmatrix} x \\ y \\ z \end{bmatrix} = [x\ y\ x]^T.$$

Similarly, the position vector of $P(x, y)$ in the plane is $\vec{p} = \begin{bmatrix} x \\ y \end{bmatrix} = [x\ y]^T$. Thus every point has a unique position vector. Conversely every vector is the position vector of a unique point (place the vector with its tail at the origin and use the point at its tip). Hence position vectors are just a different way to describe points.

The introduction of position vectors provides a new perspective in geometry: To compute the *coordinates* x, y, and z of a point $P(x, y, z)$, we find instead the position vector $\vec{p} = [x \ y \ x]^T$ of P, and read off its *components* x, y, and z. They are the same three numbers, but this change of emphasis allows vector methods to be used in geometry, and so gives a new approach to the subject. This vector perspective is particularly effective for studying lines and planes, and so is central to this section.

3.3.2 Lines

Given a point P, it is evident geometrically that there is exactly one line[10] through P which is parallel to a given nonzero vector. We are going to use this observation to give a precise description of lines in space. With this in mind, a nonzero vector $\vec{d}$ is called a **direction vector** for the line if it is parallel to the line, that is if $\vec{d}$ is parallel to $\overrightarrow{AB}$ for some distinct points A and B on the line. Thus $\overrightarrow{AB}$ itself is a direction vector for the line, and there are many others. Indeed any nonzero scalar multiple of a direction vector is again a direction vector.

A direction vector for a line in space describes the orientation of the line. This role is played by the slope for lines in the plane, and the following example relates the two concepts for such lines.

Example 1. *Show that* $\vec{d} = \begin{bmatrix} 1 \\ m \end{bmatrix}$ *is a direction vector for any line in the plane with slope* m.

Solution. Such a line has equation $y = mx + b$ for some number b. Taking $x = 0$ and $x = 1$ gives two points $P(0, b)$ and $Q(1, m + b)$ on the line. Thus the vector $\vec{d} = \overrightarrow{PQ} = \begin{bmatrix} 1 - 0 \\ (m + b) - b \end{bmatrix} = \begin{bmatrix} 1 \\ m \end{bmatrix}$ is parallel to the line (see Figure 1) and so will serve as a position vector for the line. □

---

[10]The term "line" means "straight line" in this book.

3.3. LINES AND PLANES

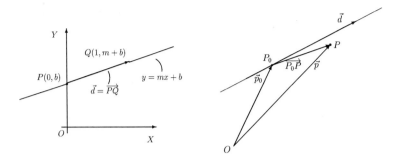

Figure 1   Figure 2

Now suppose a point $P_0(x_0, y_0, z_0)$ and a nonzero vector $\vec{d} = [a \ b \ c]^T$ are given. We want to determine the unique line through P_0 which has $\vec{d}$ as direction vector; that is, we want to give a condition that an arbitrary point $P = P(x, y, z)$ lies on this line. Let $\vec{p_0} = [x_0 \ y_0 \ z_0]^T$ and $\vec{p} = [x \ y \ z]^T$ be the position vectors of P_0 and P respectively, so that $\overrightarrow{P_0P} = \vec{p} - \vec{p_0}$ (see Figure 2). The key observation is that P lies on the line if and only if $\vec{p} - \vec{p_0} = \overrightarrow{P_0P}$ is parallel to $\vec{d}$, that is if and only if

$$\vec{p} - \vec{p_0} = t\vec{d} \quad \text{for some scalar } t.$$

In other words, the point P lies on the line if and only if its position vector $\vec{p}$ has the form $\vec{p} = \vec{p_0} + t\vec{d}$ for some scalar t. This equation has a name.

Vector Equation of a Line. *Consider the line with direction vector $\vec{d} \neq \vec{0}$ which passes through the point with position vector $\vec{p_0}$. The point with position vector $\vec{p}$ lies on this line if and only if*

$$\vec{p} = \vec{p_0} + t\vec{d} \text{ for some scalar } t.$$

In component form, the vector equation of the line becomes

$$\begin{bmatrix} x \\ y \\ z \end{bmatrix} = \begin{bmatrix} x_0 \\ y_0 \\ z_0 \end{bmatrix} + t \begin{bmatrix} a \\ b \\ c \end{bmatrix}$$

Equating components gives another description of the line.

Scalar Equations of a Line. *The line through the point $P_0(x_0, y_0, z_0)$ with direction vector $\vec{d} = [a \ b \ c]^T \neq \vec{0}$ is given by*

$$\begin{aligned} x &= x_0 + ta \\ y &= y_0 + tb \quad t \text{ any scalar} \\ z &= z_0 + tc \end{aligned}$$

In other words, a point $P(x, y, z)$ lies on the line if and only if a real number t exists such that $x = x_0 + ta$, $y = y_0 + tb$ and $z = z_0 + tc$.

Note that a direction vector $\vec{d}$ can be "read off" from the scalar equations of a line because the components of $\vec{d}$ are the coefficients of the parameter t. This is analogous to "reading off" the slope of the line $y = 3 - 4x$ as the coefficient -4 of x.

The following examples illustrate how these equations are used to describe lines and solve problems about lines.

Example 2. *Determine whether $A(1, -5, 15)$ and $B(-1, 1, 4)$ lie on the line with scalar equations* $\begin{cases} x = -2 + t \\ y = 4 - 3t \\ z = -3 + 6t \end{cases}$

Solution. Every point P on the line has the form $P(-2+t, 4-3t, -3+6t)$ for some t. Hence A is on the line if a number t can be found such that $-2 + t = 1$, $4 - 3t = -5$ and $-3 + 6t = 15$. Since $t = 3$ satisfies these equations, the point A is indeed on the line. As to B, it lies on the line if $-2 + t = -1$, $4 - 3t = 1$ and $-3 + 6t = 4$ for some t. The first two equations are satisfied for $t = 1$, but this does not satisfy $-3 + 6t = 4$. Hence no such number t exists, and so B is not on the line. □

Example 3. If $P(3, -1, 4)$ and $\vec{d} = \begin{bmatrix} 5 & -2 & 8 \end{bmatrix}^T$, the line through P with direction vector $\vec{d}$ has vector equation $\vec{p} = \begin{bmatrix} 3 \\ -1 \\ 4 \end{bmatrix} + t \begin{bmatrix} 5 \\ -2 \\ 8 \end{bmatrix}$ and scalar equations $\begin{cases} x = 3 + 5t \\ y = -1 - 2t \\ z = 4 + 8t \end{cases}$. □

Example 4. *Find scalar equations of the line through the points $A(2, 0, -3)$ and $B(1, 1, -5)$.*

Solution. Because both A and B lie on the line, the vector

$$\overrightarrow{AB} = \begin{bmatrix} 1 - 2 \\ 1 - 0 \\ -5 - (-3) \end{bmatrix} = \begin{bmatrix} -1 \\ 1 \\ -2 \end{bmatrix}$$

is a vector along the line. Thus we may use $\vec{d} = \overrightarrow{AB}$ as a direction vector for the line. Using $A(2, 0, -3)$ as the point on the line, the scalar equations are $\begin{cases} x = 2 - t \\ y = t \\ z = -3 - 2t \end{cases}$. □

3.3. LINES AND PLANES

Example 5. *Determine whether the lines*

$$\begin{array}{lll} x = 2-5t & & x = t \\ y = -1+t & \text{and} & y = -3+2t \\ z = 3t & & z = 5+t \end{array}$$

are perpendicular.

Solution. $\vec{d_1} = [-5 \ \ 1 \ \ 3]^T$ and $\vec{d_2} = [1 \ \ 2 \ \ 1]^T$ are direction vectors, and they are orthogonal because $\vec{d_1} \bullet \vec{d_2} = -5 + 2 + 3 = 0$. Hence the lines are perpendicular. □

Example 6. *Find scalar equations of the line through $A(4, -1, 5)$ which is parallel to the line* $\begin{cases} x = 5 \\ y = 2-t \\ z = -1+3t \end{cases}$

Solution. A direction vector of the given line is $\vec{d} = [0 \ \ -1 \ \ 3]^T$, and this will do as a direction vector of the line we want because the lines are parallel. Since the point $A(4, -1, 5)$ lies on the line we want, the scalar equations are
$\begin{cases} x = 4 \\ y = -1-t \\ z = 5+3t \end{cases}$. □

Example 7. *Determine whether the lines*

$$\begin{array}{lll} x = 2-t & & x = 3+2s \\ y = 5+2t & \text{and} & y = 2-3s \\ z = -1+3t & & z = -6-4s \end{array}$$

intersect and, if they do, find the point of intersection.

Solution. Suppose $P(x, y, z)$ is a point of intersection. Then it lies on *both* lines so there must exist real numbers t and s such that

$$\begin{bmatrix} 2-t \\ 5+2t \\ -1+3t \end{bmatrix} = \begin{bmatrix} x \\ y \\ z \end{bmatrix} = \begin{bmatrix} 3+2s \\ 2-3s \\ -6-4s \end{bmatrix}$$

Equating components gives three equations in the two variables s and t:

$$\begin{array}{rcl} 2-t & = & 3+2s \\ 5+2t & = & 2-3s \\ -1+3t & = & -6-4s \end{array}$$

These equations have a unique solution $s = 1$ and $t = -3$, so the lines meet at the corresponding point $P(5, -1, -10)$. (Of course, if these equations for s and t did not have a solution, the lines would not intersect.) □

Two lines in the plane are parallel if they have equal slopes. The next example uses these vector methods to derive a well known test for when two lines in the plane are perpendicular.

Example 8. *Show that two lines in the plane with slopes m_1 and m_2 are perpendicular if and only if $m_1 m_2 = -1$.*

Solution. The two lines have direction vectors $\vec{d_1} = \begin{bmatrix} 1 \\ m_1 \end{bmatrix}$ and $\vec{d_2} = \begin{bmatrix} 1 \\ m_2 \end{bmatrix}$ by Example 1. Hence the lines are perpendicular if and only if $\vec{d_1}$ and $\vec{d_2}$ are orthogonal, that is (by Theorem 3 §3.2) if and only if $\vec{d_1} \bullet \vec{d_2} = 0$. But this last condition reads $1 + m_1 m_2 = 0$. □

In applications of geometry, it is frequently required to solve an optimization problem, that is to maximize or minimize some quantity. Such problems can sometimes be solved using projections; the following example provides an illustration.

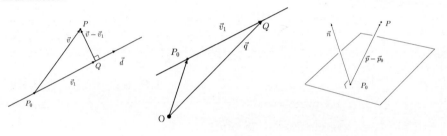

Figure 3   Figure 4   Figure 5

Example 9. *Find the shortest distance from the point $P(3,5,6)$ to the line through $P_0(6,1,-3,)$ with direction vector $\vec{d} = [-1 \ 3 \ 0]^T$. Then find the point Q on the line that is closest to P.*

Solution. Write $\vec{v} = \overrightarrow{P_0 P} = [-3 \ 4 \ 9]^T$. Referring to Figure 3, the required distance is $\|\vec{v} - \vec{v_1}\|$ where $\vec{v_1}$ is the projection of $\vec{v}$ on $\vec{d}$. We can compute $\vec{v_1}$ as follows:

$$\vec{v_1} = proj_{\vec{d}}(\vec{v}) = \left(\frac{\vec{v} \bullet \vec{d}}{\|\vec{d}\|^2} \right) \vec{d} = \left(\frac{3 + 12 + 0}{1 + 9 + 0} \right) \vec{d} = \tfrac{3}{2} \begin{bmatrix} -1 \\ 3 \\ 0 \end{bmatrix}.$$

Then the shortest distance from P to the line is

$$\left\| \overrightarrow{QP} \right\| = \|\vec{v} - \vec{v_1}\| = \left\| \begin{bmatrix} -3 \\ 4 \\ 9 \end{bmatrix} - \tfrac{3}{2} \begin{bmatrix} -1 \\ 3 \\ 0 \end{bmatrix} \right\| = \left\| \tfrac{1}{2} \begin{bmatrix} -3 \\ -1 \\ 18 \end{bmatrix} \right\| = \tfrac{1}{2}\sqrt{334}.$$

Observe that we did not have to compute the point Q to find the distance. However, it is easy to find Q. Let $\vec{p_0} = [6 \ 1 \ -3]^T$ and $\vec{q} = [x \ y \ z]^T$ be the

3.3. LINES AND PLANES

position vectors of P_0 and Q respectively. Then Figure 4 gives

$$\vec{q} = \vec{p}_0 + \vec{v}_1 = \begin{bmatrix} 6 \\ 1 \\ -3 \end{bmatrix} + \begin{bmatrix} -1 \\ 3 \\ 0 \end{bmatrix} \frac{3}{2} = \begin{bmatrix} \frac{9}{2} \\ \frac{11}{2} \\ -3 \end{bmatrix}$$

It follows that $Q = Q(\frac{9}{2}, \frac{11}{2}, -3)$ is the point on the line closest to P. $\square$

3.3.3 Planes

The orientation of a line in space is described by specifying a direction vector for the line. Similarly, the orientation of a plane in space can be described by a vector as follows. A nonzero vector $\vec{n}$ is called a **normal** to a plane if it is orthogonal to every vector in the plane. As for direction vectors, there are many normals to a given plane; in particular any nonzero scalar multiple of a normal is again a normal.

Given a point $P_0(x_0, y_0, z_0)$ and a nonzero vector $\vec{n} = [a \ b \ c]^T$, it is clear geometrically that there is exactly one plane with normal $\vec{n}$ which contains the point P_0. Our goal now is to describe this unique plane. In other words, given an arbitrary point $P(x, y, z)$, we must give a condition (which turns out to be an equation) that P lies in the plane.

Let $\vec{p}_0$ and $\vec{p}$ be the position vectors of P_0 and P respectively. Then (see Figure 5) P lies in the plane if and only if $\vec{n}$ is orthogonal to $\overrightarrow{P_0P} = \vec{p} - \vec{p}_0$; that is if and only if $\vec{n} \bullet (\vec{p} - \vec{p}_0) = 0$. This is the

Vector Equation of a Plane. *Consider the plane with normal $\vec{n} \neq \vec{0}$ which contains the point $P_0(x_0, y_0, z_0)$. The point with position vector $\vec{p}$ lies in this plane if and only if*

$$\vec{n} \bullet (\vec{p} - \vec{p}_0) = 0.$$

Now let $\vec{n}$, $\vec{p}_0$ and $\vec{p}$ be given in matrix form, say $\vec{n} = [a \ b \ c]^T$, $\vec{p}_0 = [x_0 \ y_0 \ z_0]^T$ and $\vec{p} = [x \ y \ z]^T$. Then $\vec{p} - \vec{p}_0 = [x - x_0 \ y - y_0 \ z - z_0]^T$, so the vector equation $\vec{n} \bullet (\vec{p} - \vec{p}_0) = 0$ becomes a scalar description of the plane.

Scalar Equation of a Plane. *The plane through $P_0(x_0, y_0, z_0)$ with normal $\vec{n} = [a \ b \ c]^T \neq \vec{0}$ has equation*

$$a(x - x_0) + b(y - y_0) + c(z - z_0) = 0.$$

In other words, a point $P(x, y, z)$ lies in the plane if and only if x, y and z satisfy this equation.

Example 10. The plane through $P_0(-2, 3, 5)$ with normal $\vec{n} = [7 \ -4 \ 8]^T$ has equation

$$7(x - (-2)) - 4(y - 3) + 8(z - 5) = 0.$$

This simplifies to $7x - 4y + 8x = 14$. $\square$

The scalar equation $a(x - x_0) + b(y - y_0) + c(z - z_0) = 0$ of a plane can be written as follows:
$$ax + by + cz = ax_0 + by_0 + cz_0.$$
The quantity $ax_0 + by_0 + cz_0$ on the right side is a constant depending on P_0 and $\vec{n}$. If we write it as k, we have proved part (1) of the following theorem.

Theorem 1. Let $\vec{n} = [a\ b\ c]^T \neq \vec{0}$ be a fixed nonzero vector.

(1) *Every plane with normal $\vec{n}$ has equation $ax + by + cz = k$ for some constant k.*

(2) *For every constant k, the graph of the equation $ax + by + cz = k$ is a plane with normal $\vec{n}$.*

Proof. It remains to prove (2). Since $\vec{n} \neq \vec{0}$, one of the components a, b and c is nonzero. If $a \neq 0$, the equation $ax + by + cz = k$ can be written
$$a\left(x - \frac{k}{a}\right) + b(y - 0) + c(z - 0) = 0$$
and so is the equation of the plane through $P_0(\frac{k}{a}, 0, 0)$ with normal $\vec{n}$. A similar argument works if $b \neq 0$ or $c \neq 0$. □

In particular, if the equation $ax + by + cz = k$ of a plane is given, the components of a normal $\vec{n} = [a\ b\ c]^T$ of the plane can be "read off" as the coefficients of x, y and z in the equation.

Example 11. *Find the equation of the plane through $P(4, 0, -1)$ which is parallel to the plane with equation $3x - 2y + z = 9$.*

Solution. The given plane has normal $\vec{n} = [3\ -2\ 1]^T$ by Theorem 1, so $\vec{n}$ will do as a normal for the plane we want (because the two planes are parallel). Hence our plane has equation $3x - 2y + z = k$ for some constant k by Theorem 1. The constant k is determined by the fact that $P(4, 0, -1)$ lies in the plane. Indeed, this means that $3 \cdot 4 - 2 \cdot 0 + (-1) = k$, whence $k = 11$. So the required equation is $3x - 2y + z = 11$. □

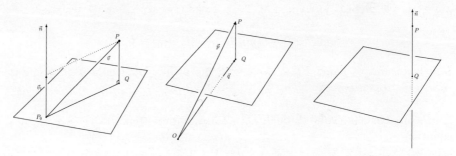

Figure 6　　　　　　Figure 7　　　　　　Figure 8

3.3. LINES AND PLANES

Example 12. *Find the shortest distance from the point $P(1,5,5)$ to the plane with equation $2x - y + 3z = 5$, and find the point Q in the plane closest to P.*

There are two approaches to this problem, and we present both to illustrate the techniques.

Solution 1. Choose any point P_0 in the plane, say $P_0(1,0,1)$, and compute $\vec{v} = \overrightarrow{P_0P} = [0\ 5\ 4]^T$. Position the normal $\vec{n} = [2\ -1\ 3]^T$ with its tail at P_0 as in Figure 6, and compute the projection $\vec{v}_1$ of $\vec{v}$ on $\vec{n}$:

$$\vec{v}_1 = proj_{\vec{n}}(\vec{v}) = \frac{\vec{v} \bullet \vec{n}}{\|\vec{n}\|^2}\vec{n} = \frac{0 - 5 + 12}{4 + 1 + 9}\vec{n} = \tfrac{1}{2}\vec{n}.$$

Now observe that $\overrightarrow{QP} = \vec{v}_1$ by Figure 6. Hence the required distance is

$$\left\|\overrightarrow{QP}\right\| = \|\vec{v}_1\| = \left\|\tfrac{1}{2}\vec{n}\right\| = \tfrac{1}{2}\|\vec{n}\| = \tfrac{1}{2}\sqrt{14}.$$

Note that we did not have to find Q to be able to compute the shortest distance. However, it is not difficult to find the point Q. Let $\vec{p}$ and $\vec{q}$ be the position vectors of P and Q respectively. Then $\vec{p} = \vec{q} + \overrightarrow{QP} = \vec{q} + \vec{v}_1$ by Figure 7, so

$$\vec{q} = \vec{p} - \vec{v}_1 = \begin{bmatrix} 1 \\ 5 \\ 5 \end{bmatrix} - \tfrac{1}{2}\begin{bmatrix} 2 \\ -1 \\ 3 \end{bmatrix} = \begin{bmatrix} 0 \\ \tfrac{11}{2} \\ \tfrac{7}{2} \end{bmatrix}.$$

Hence $Q = Q(0, \tfrac{11}{2}, \tfrac{7}{2})$.

Solution 2. The point Q is the point of intersection of the given plane and the line through P which is perpendicular to the plane (see Figure 8). But $\vec{n} = [2\ -1\ 3]^T$ is a direction vector for this line, so the line is given by the equation

$$[x\ y\ z]^T = [1\ 5\ 5]^T + t[2\ -1\ 3]^T, \quad t \text{ a scalar.}$$

Since Q lies on this line, we have $Q = Q(1+2t, 5-t, 5+3t)$ for some t. But Q is also on the plane, its coordinates must satisfy the equation of the plane:

$$2(1 + 2t) - (5 - t) + 3(5 + 3t) = 5$$

Solving this for t gives $t = -\tfrac{1}{2}$, so $Q = Q(1-1, 5+\tfrac{1}{2}, 5-\tfrac{3}{2}) = Q(0, \tfrac{11}{2}, \tfrac{7}{2})$ as before. Moreover, having found Q, the required distance is

$$\left\|\overrightarrow{QP}\right\| = \|\vec{p} - \vec{q}\| = \left\|[1\ \tfrac{-1}{2}\ \tfrac{3}{2}]^T\right\| = \left\|\tfrac{1}{2}[2\ -1\ 3]\right\| = \tfrac{1}{2}\sqrt{14}.$$

Again this agrees with Solution 1. □

Exercises

1. Explain the difference between a point and its position vector, the coordinates of a point and the components of its position vector.

2. In each case find the scalar equations of the line:

    (a) Parallel to $[0\ -1\ 3]^T$ and passing through $P(4, -3, 5)$.

    (b) Parallel to $[3\ -2\ 0]^T$ and passing through $P(-3, 0, 7)$.

    (c) Passing through $P(5, -7, 2)$ and $Q(0, 0, 4)$.

    (d) Passing through $P(9, -3, 1)$ and $Q(-2, 6, -4)$.

    (e) Passing through $P(8, 4, -1)$ and parallel to the line $[x\ y\ z]^T = [2\ 9\ -3]^T + t[5\ 1\ 6]^T$.

    (f) Passing through $P(1, 2, 0)$ and parallel to the line $[x\ y\ z]^T = [0\ 0\ 1]^T + t[-7\ 3\ -5]^T$.

    (g) Passing through $P(6, 0, -2)$ parallel to the line through $Q(1, 1, 1)$ and $R(2, 2, 2)$.

    (h) Passing through $P(2, -5, 7)$ parallel to the line through $Q(0, 1, 2)$ and $R(-5, 7, 3)$.

    (i) Passing through $P(2, -5, 7)$ perpendicular to the plane $3x - 2y + 5z = 7$.

    (j) Passing through $P(1, 2, -9)$ perpendicular to the plane $4y - 3z = 2$.

    (k) Passing through $P(6, 0, -3)$, intersecting the line $[x\ y\ z]^T = [1\ 2\ -3]^T + t[1\ -2\ 0]^T$ and perpendicular to it.

    (l) Passing through $P(-7, 2, 7)$, intersecting the line $[x\ y\ z]^T = [0\ -1\ 2]^T + t[5\ 1\ 1]^T$ and perpendicular to it.

3. Does the line through the points $A(2, -1, 4)$ and $B(-1, 3, 5)$ contain the point $P(1, 2, 4)$? The point $Q(-7, 11, 7)$? Support your answer.

4. In each case determine if the given points all lie on the same line. Support your answer.

    (a) $A(2, 1, 3)$, $B(-1, 2, 7)$, $C(5, 0, 4)$

    (b) $A(1, 2, -1)$, $B(0, 5, 1)$, $C(2, -1, -3)$.

5. Describe the equations of all lines parallel to the X-axis.

6. Find all points C on the line through $A(1, -1, 2)$ and $B(2, 0, 1)$ such that $\left\|\overrightarrow{AC}\right\| = 2\left\|\overrightarrow{BC}\right\|$.

7. In each case either find the point of intersection of the two lines or show why no such point exists.

3.3. LINES AND PLANES

(a) $\quad x = 2-t \qquad x = 3-2s$
$\quad\quad y = 3-t \qquad y = s$
$\quad\quad z = 1 \qquad\quad z = 5-3s$

(b) $\quad x = 1-3t \qquad x = 1+s$
$\quad\quad y = 7-2t \qquad y = 2-s$
$\quad\quad z = 3+t \qquad\quad z = 1-s$

(c) $[x\ y\ z]^T = [3\ 1\ 2]^T + t[-1\ 1\ -1]^T$
$\quad\ [x\ y\ z]^T = [3\ 1\ 5]^T + t[1\ -1\ 0]^T$

(d) $[x\ y\ z]^T = [0\ 1\ 2]^T + t[3\ -1\ -1]^T$
$\quad\ [x\ y\ z]^T = [1\ 3\ 5]^T + t[-2\ 1\ 1]^T$

8. In each case find the shortest distance from the point P to the line, and find the point Q on the line closest to P.

   (a) $P(1, 2, -3);\ [x\ y\ z]^T = [1\ 2\ 0]^T + t[-3\ 5\ 4]^T$
   (b) $P(-1, 0, 1);\ [x\ y\ z]^T = [3\ -1\ 4]^T + t[3\ -2\ 0]^T$
   (c) $P(-3, 0, 5);\ [x\ y\ z]^T = [3\ -1\ 2]^T + t[2\ -1\ 2]^T$
   (d) $P(1, 1, -1);\ [x\ y\ z]^T = [4\ 2\ -3]^T + t[1\ -5\ 0]^T$

9. Does the plane $2x - 3y + z = 6$ contain the point $P(1, -2, -2)$? The point $Q(5, -6, 0)$? Support your answer.

10. In each case find the equation of the plane:

    (a) Passing through $P(3, -7, 5)$ and parallel to the plane with equation $3x - y + 2z = 5$.

    (b) Passing through $P(1, -2, 4)$ and parallel to the plane with equation $x - 2y + 3z = -6$.

    (c) Containing $P(2, -3, 0)$ and perpendicular to the line
    $[x\ y\ z]^T = [2\ -5\ 3]^T + t[6\ -6\ 5]^T$.

    (d) Containing $P(1, -1, 4)$ and perpendicular to the line
    $[x\ y\ z]^T = [7\ 0\ 5]^T + t[2\ 1\ -7]^T$.

    (e) Each point of which is the same distance from the points $A(1, -3, 7)$ and $B(3, -5, 9)$.

    (f) Each point of which is the same distance from the points $A(6, 0, 2)$ and $B(1, 1, -3)$.

11. Give a rule to determine if the plane with equation $ax + by + cz = k$ passes through the origin.

12. Describe the equations of all planes perpendicular to the Z-axis.

13. In each case determine if the given line lies in the given plane. Support your answer.

    (a) Line through $P(1, 2, -3)$ with direction vector $\vec{d} = [1\ 1\ -1]^T$, plane $2x - y + z = -3$.

    (b) Line through $P(-1, 3, 1)$ with direction vector $\vec{d} = [0\ 2\ 5]^T$, plane $x - y + z = 1$.

14. In each case determine if the given plane contains the given line. Support your answer.

    (a) Plane $2x - 3y + z = 1$, line $[x \ y \ z]^T = [3 \ 2 \ 1]^T + t[2 \ 0 \ -4]^T$.

    (b) Plane $x + 2y - 3z = 5$, line $[x \ y \ z]^T = [3 \ 7 \ 4]^T + t[1 \ 1 \ 1]^T$.

15. In each case find the point of intersection of the plane and the line.

    (a) $x - 3y + 2z = 7$ and $[x \ y \ z]^T = [2 \ 0 \ -1]^T + t[4 \ -1 \ 2]^T$.

    (b) $2x + 4y - 3z = 1$ and $[x \ y \ z]^T = [5 \ -3 \ 1]^T + t[0 \ 2 \ -3]^T$.

16. Show that every plane containing both $A(1, -3, 2)$ and $B(4, 5, 0)$ must also contain $C(7, 13, -2)$.

17. In each case find the shortest distance from the point P to the plane, and find the point Q on the plane closest to P.

    (a) $P(1, 0, -2); \ 3x - y + 4z = 5$ (b) $P(3, -1, 1); \ 3x - y + 4z = 5$

    (c) $P(5, 1, -3); \ 2x - 3y + z = 1$ (d) $P(-7, -10, 9); \ x - 3y + 2z = 6$

18. In each case, find the solution of the system of linear equations that is closest to the given point.

    (a) $\begin{aligned} 2x - 3y + 5z &= 2 \\ x + 2y - z &= 1 \\ 3x - y + z &= 4 \end{aligned}$ and the point $P(2, 3, 0)$.

    (b) $\begin{aligned} x - y + 3z &= 1 \\ -x + 3y - z &= 3 \\ y + z &= 2 \end{aligned}$ and the point $P(-1, 2, 0)$.

    (c) $\begin{aligned} x - y + 3z &= -1 \\ -x + 3y - z &= 3 \\ x + y + 5z &= 1 \end{aligned}$ and the point $P(-4, 0, 1)$.

    (d) $\begin{aligned} x - 3y + 5z &= 1 \\ -x + 3y - 5z &= -1 \end{aligned}$ and the point $P(-1, 4, 0)$.

19. In each case either show that the statement is true or give an example showing that it is false.

    (a) If a line is parallel to a plane it never intersects the plane.

    (b) Any three planes no two of which are parallel meet in a single point.

    (c) If the plane $ax + by + cz = k$ passes through the origin then $k = 0$.

    (d) Every plane has exactly one equation of the form $ax + by + cz = k$.

    (e) If two lines do not intersect, they do not both lie in the same plane.

    (f) If a line is parallel to the normal of a plane, then it is parallel to the plane.

    (g) The line of intersection of two (non-parallel) planes is perpendicular to both normals.

(h) A line perpendicular to the normal of a plane must be parallel to the plane.

20. If a plane contains two distinct points A and B, show that it contains the line through A and B.

21. Let $\vec{d} = [a \ b \ c]^T$ be a vector for which a, b and c are *all* nonzero. Show that the scalar equations of the line through an arbitrary point $P_0(x_0, y_0, z_0)$ with $\vec{d}$ as direction vector $\vec{d}$ can be written in the form $\frac{x-x_0}{a} = \frac{y-y_0}{b} = \frac{z-z_0}{c}$. This is called the *symmetric form* of the equations.

22. (a) Show that every plane with normal $\vec{n} \neq \vec{0}$ has vector equation $\vec{p} \bullet \vec{n} = k$ for some scalar k, where $\vec{p} = [x \ y \ z]^T$.

    (b) If $\vec{a}$ is the position vector of a point A, show that the shortest distance from A to the plane with equation $\vec{p} \bullet \vec{n} = k$ is $\dfrac{|k - (\vec{a} \bullet \vec{n})|}{\|\vec{n}\|}$.

23. Consider the two parallel planes $ax + by + cz = k$ and $ax + by + cz = k_1$. Show that the distance between them is $\dfrac{|k_1 - k|}{\|\vec{n}\|}$ where $\vec{n} = [a \ b \ c]^T$ is the common normal.

3.4 THE CROSS PRODUCT

It is clear geometrically that there is a unique plane containing any three points A, B and C that are not all on a line. In determining the equation of this plane, the problem is finding a normal vector for the plane. Since any nonzero vector which is orthogonal to both $\overrightarrow{AB}$ and $\overrightarrow{AC}$ will do as a normal, what is needed is a systematic method for finding a nonzero vector orthogonal to any two nonparallel vectors. The cross product solves this problem, and has a number of other useful properties as well.

3.4.1 The Cross Product

Given vectors $\vec{v} = \begin{bmatrix} x_1 \\ y_1 \\ z_1 \end{bmatrix}$ and $\vec{w} = \begin{bmatrix} x_2 \\ y_2 \\ z_2 \end{bmatrix}$, the vector

$$\vec{v} \times \vec{w} = \begin{bmatrix} y_1 z_2 - y_2 z_1 \\ -(x_1 z_2 - x_2 z_1) \\ x_1 y_2 - x_2 y_1 \end{bmatrix} \quad (*)$$

is called the **cross product** of $\vec{v}$ and $\vec{w}$. Observe that $\vec{v} \times \vec{w}$ is a *vector* (as opposed to the *scalar* $\vec{v} \bullet \vec{w}$) and, for this reason, $\vec{v} \times \vec{w}$ is sometimes called the **vector product** of $\vec{v}$ and $\vec{w}$.

The reader can verify directly that $\vec{v} \bullet (\vec{v} \times \vec{w}) = 0 = \vec{w} \bullet (\vec{v} \times \vec{w})$, and hence that $\vec{v} \times \vec{w}$ is indeed orthogonal to both $\vec{v}$ and $\vec{w}$. This is verified in another way

218 CHAPTER 3. VECTOR GEOMETRY

below (Theorem 2) using a description of $\vec{v} \times \vec{w}$ which employs determinants. This description has other uses as well, not the least of which being a simple way to *remember* the above formula for $\vec{v} \times \vec{w}$.

The determinant description of the cross product utilizes the **coordinate vectors**

$$\vec{i} = \begin{bmatrix} 1 \\ 0 \\ 0 \end{bmatrix}, \vec{j} = \begin{bmatrix} 0 \\ 1 \\ 0 \end{bmatrix} \text{ and } \vec{k} = \begin{bmatrix} 0 \\ 0 \\ 1 \end{bmatrix}.$$

These are vectors of length 1[11] pointing along the positive X-, Y- and Z-axes respectively, and the name comes from that fact that any vector $\begin{bmatrix} x \\ y \\ z \end{bmatrix}$ can be written in the form

$$\begin{bmatrix} x \\ y \\ z \end{bmatrix} = x\vec{i} + y\vec{j} + z\vec{k}.$$

This enables us to give a useful description of the cross product using determinants. It comes from the observation that the components of $\vec{v} \times \vec{w}$ in (*) can be written as 2×2 determinants:

$$\text{If } \vec{v} = \begin{bmatrix} x_1 \\ y_1 \\ z_1 \end{bmatrix} \text{ and } \vec{w} = \begin{bmatrix} x_2 \\ y_2 \\ z_2 \end{bmatrix} \text{ then } \vec{v} \times \vec{w} = \begin{bmatrix} \det \begin{bmatrix} y_1 & y_2 \\ z_1 & z_2 \end{bmatrix} \\ -\det \begin{bmatrix} x_1 & x_2 \\ z_1 & z_2 \end{bmatrix} \\ \det \begin{bmatrix} x_1 & x_2 \\ y_1 & y_2 \end{bmatrix} \end{bmatrix} \quad (**)$$

This prompts the following way of writing the cross product.

Determinant Description of the Cross Product.

If $\vec{v} = \begin{bmatrix} x_1 \\ y_1 \\ z_1 \end{bmatrix}$ and $\vec{w} = \begin{bmatrix} x_2 \\ y_2 \\ z_2 \end{bmatrix}$ are vectors, the cross product is given by

$$\vec{v} \times \vec{w} = \det \begin{bmatrix} \vec{i} & x_1 & x_2 \\ \vec{j} & y_1 & y_2 \\ \vec{k} & z_1 & z_2 \end{bmatrix}$$
$$= \det \begin{bmatrix} y_1 & y_2 \\ z_1 & z_2 \end{bmatrix} \vec{i} - \det \begin{bmatrix} x_1 & x_2 \\ z_1 & z_2 \end{bmatrix} \vec{j} + \det \begin{bmatrix} x_1 & x_2 \\ y_1 & y_2 \end{bmatrix} \vec{k}.$$

Here the determinant is expanded along the first column by cofactors, so the components of $\vec{v} \times \vec{w}$ are the coefficients of $\vec{i}$, $\vec{j}$ and $\vec{k}$ respectively.

---
[11]Vectors of length 1 are called **unit vectors**. They will reappear in Chapter 4.

3.4. THE CROSS PRODUCT

Example 1. If $\vec{v} = \begin{bmatrix} 4 \\ -2 \\ 1 \end{bmatrix}$ and $\vec{w} = \begin{bmatrix} 5 \\ 0 \\ 3 \end{bmatrix}$, then

$$\vec{v} \times \vec{w} = \det \begin{bmatrix} \vec{i} & 4 & 5 \\ \vec{j} & -2 & 0 \\ \vec{k} & 1 & 3 \end{bmatrix} = \begin{bmatrix} -6 \\ -7 \\ 10 \end{bmatrix}.$$

Note that $\vec{v} \times \vec{w}$ is orthogonal to both $\vec{v}$ and $\vec{w}$ in Example 1 because, as is easily verified, $\vec{v} \bullet (\vec{v} \times \vec{w}) = 0$ and $\vec{w} \bullet (\vec{v} \times \vec{w}) = 0$. This happens in every case as we shall see. In fact there is an easy way to compute the number $\vec{u} \bullet (\vec{v} \times \vec{w})$ for *any* vector $\vec{u}$: It is the determinant of the matrix $[\vec{u} \ \vec{v} \ \vec{w}]$ which has $\vec{u}$, $\vec{v}$ and $\vec{w}$ as its columns.

Theorem 1. If $\vec{u}$, $\vec{v}$ and $\vec{w}$ are vectors then $\vec{u} \bullet (\vec{v} \times \vec{w}) = \det[\vec{u} \ \vec{v} \ \vec{w}]$.

Proof. Write $\vec{u} = \begin{bmatrix} x_0 \\ y_0 \\ z_0 \end{bmatrix}$, $\vec{v} = \begin{bmatrix} x_1 \\ y_1 \\ z_1 \end{bmatrix}$ and $\vec{w} = \begin{bmatrix} x_2 \\ y_2 \\ z_2 \end{bmatrix}$ in matrix form. Then (**) gives

$$\begin{aligned} \vec{u} \bullet (\vec{v} \times \vec{w}) &= x_0 \det \begin{bmatrix} y_1 & y_2 \\ z_1 & z_2 \end{bmatrix} - y_0 \det \begin{bmatrix} x_1 & x_2 \\ z_1 & z_2 \end{bmatrix} + z_0 \det \begin{bmatrix} x_1 & x_2 \\ y_1 & y_2 \end{bmatrix} \\ &= \det \begin{bmatrix} x_0 & x_1 & x_2 \\ y_0 & y_1 & y_2 \\ z_0 & z_1 & z_2 \end{bmatrix} \end{aligned}$$

where this last determinant is expanded along the first column. □

The number $\vec{u} \bullet (\vec{v} \times \vec{w})$ is called the **scalar triple product** of the vectors $\vec{u}$, $\vec{v}$ and $\vec{w}$, and has a geometric meaning that we will return to later. For now we use it to confirm that the cross product solves the problem of finding a vector orthogonal to two given vectors.

Theorem 2. For any non-zero vectors $\vec{v}$ and $\vec{w}$, the cross product $\vec{v} \times \vec{w}$ is orthogonal to both $\vec{v}$ and $\vec{w}$.

Proof. Using Theorem 1 with $\vec{u} = \vec{v}$ we obtain $\vec{v} \bullet (\vec{v} \times \vec{w}) = \det[\vec{v} \ \vec{v} \ \vec{w}] = 0$ because the matrix $[\vec{v} \ \vec{v} \ \vec{w}]$ has two identical columns.[12] Hence $\vec{v}$ and $\vec{v} \times \vec{w}$ are orthogonal. Similarly, $\vec{w} \bullet (\vec{v} \times \vec{w}) = 0$, so $\vec{w}$ and $\vec{v} \times \vec{w}$ are also orthogonal vectors. □

The following three examples illustrate how Theorem 2 is applied to the study of planes and lines. The next example returns to the problem which motivated the cross product.

---

[12] As noted earlier, the fact that $\vec{v} \bullet (\vec{v} \times \vec{w}) = 0$ can be verified directly without any reference to determinants.

Example 2. *Find the equation of the plane containing the points $A(3, 0, -1)$, $B(1, 1, 1)$ and $C(2, -3, 1)$.*

Solution. The vectors $\overrightarrow{AB} = [-2 \ 1 \ 2]^T$ and $\overrightarrow{AC} = [-1 \ -3 \ 2]^T$ lie in the plane (see Figure 1) so, if their cross product is nonzero, it will do as a normal. Compute:

$$\overrightarrow{AB} \times \overrightarrow{AC} = \det \begin{bmatrix} \vec{i} & -2 & -1 \\ \vec{j} & 1 & -3 \\ \vec{k} & 2 & 2 \end{bmatrix} = \begin{bmatrix} 8 \\ 2 \\ 7 \end{bmatrix}.$$

(As a check on the arithmetic, we verify that $\overrightarrow{AB} \bullet (\overrightarrow{AB} \times \overrightarrow{AC}) = 0$ and $\overrightarrow{AC} \bullet (\overrightarrow{AB} \times \overrightarrow{AC}) = 0$.) Using $\overrightarrow{AB} \times \overrightarrow{AC}$ as a normal, the plane has equation $8x + 2y + 7z = k$ for some number k. Since the point $A(3, 0, -1)$ lies on the plane, we obtain $k = 8 \cdot 3 + 2 \cdot 0 + 7(-1) = 17$. Thus the required equation is $8x + 2y + 7z = 17$. □

Example 3. *Find the equation of the plane through $P(1, 0, -2)$ which contains the line* $\begin{cases} x = 2 - t \\ y = -3 \\ z = t \end{cases}$.

Solution. The points $P(1, 0, -2)$ and $Q(2, -3, 0)$ are both in the plane (Q is on the line) so the vector $\vec{v} = \overrightarrow{QP} = [-1 \ 3 \ -2]^T$ lies in the plane (see Figure 2). Since the direction vector $\vec{d} = [-1 \ 0 \ 1]^T$ of the line also lies in the plane, $\vec{v} \times \vec{d}$ is perpendicular to the plane. We have $\vec{v} \times \vec{d} = \det \begin{bmatrix} \vec{i} & -1 & -1 \\ \vec{j} & 3 & 0 \\ \vec{k} & -2 & 1 \end{bmatrix} = \begin{bmatrix} 3 \\ 3 \\ 3 \end{bmatrix}.$

Thus we may take $\vec{n} = [1 \ 1 \ 1]^T$ as the normal to the plane (any nonzero scalar multiple of $\vec{v} \times \vec{d}$ will do), so the equation of the plane is $x + y + z = k$ for some number k. Since $P(1, 0, -2)$ is in the plane, we get $k = 1 + 0 - 2 = -1$. Thus the required equation is $x + y + z = -1$.

As a check, note that every point R on the given line has coordinates of the form $R(2 - t, -3, t)$ for some value of t, so R lies in the plane $x + y + z = -1$ because $(2 - t) - 3 + t = -1$ for all t. Thus the *entire line* lies in the plane, as required. □

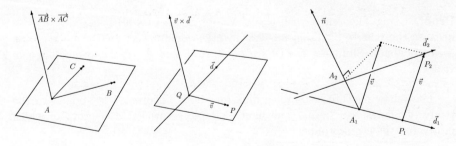

Figure 1   Figure 2   Figure 3

3.4. THE CROSS PRODUCT

Example 4. *Find the shortest distance between the lines*

$$\begin{cases} x = 3+t \\ y = -1+t \\ z = 2 \end{cases} \quad \text{and} \quad \begin{cases} x = t \\ y = 5-2t \\ z = 2+t \end{cases}$$

and find the points A_1 and A_2 on the lines which are closest together.

Solution. The points $P_1(3,-1,2)$ and $P_2(0,5,2)$ lie on the respective lines, and their direction vectors are $\vec{d_1} = [1 \ 1 \ 0]^T$ and $\vec{d_2} = [1 \ -2 \ 1]^T$. Hence $\vec{n} = \vec{d_1} \times \vec{d_2} = [1 \ -1 \ -3]^T$ is a vector perpendicular to both lines. We must find the distance between the points A_1 and A_2 in Figure 3. But A_1 is in the plane through P_1 with normal $\vec{n}$, while A_2 is in the plane through P_2 with normal $\vec{n}$, so we must compute the distance between these parallel planes. Observe that vector $\vec{v} = \overrightarrow{P_1P_2} = [-3 \ 6 \ 0]^T$ goes from one plane to the other. Hence if the tail of $\vec{v}$ is moved to A_1 as in Figure 3, the tip of $\vec{v}$ remains in the other plane. Thus the required distance is the length $\|\vec{v_1}\|$ where $\vec{v_1}$ is the projection of $\vec{v}$ along $\vec{n}$. We have $\vec{v_1} = proj_{\vec{n}}(\vec{v}) = \left(\frac{\vec{v} \bullet \vec{n}}{\|\vec{n}\|^2}\right) \vec{n} = \frac{-9}{11}\vec{n}$, so the distance is $\|\vec{v_1}\| = \left|\frac{-9}{11}\right| \|\vec{n}\| = \frac{9}{11}\sqrt{11}$.

This computation gives the distance without finding the points A_1 and A_2. However, these points can be located as follows: Observe that the coordinates of A_1 are $A_1 = A_1(3+t, -1+t, 2)$ for some t because it lies on the first line, and similarly $A_2 = A_2(s, 5-2s, 2+s)$ for some s. Hence $\overrightarrow{A_1A_2} = \begin{bmatrix} s-3-t \\ 6-2s-t \\ s \end{bmatrix}$, and the parameters s and t are determined by the fact that $\overrightarrow{A_1A_2}$ is orthogonal to both $\vec{d_1}$ and $\vec{d_2}$. In fact this means

$$\begin{cases} \vec{d_1} \bullet (\overrightarrow{A_1A_2}) = 0 \\ \vec{d_2} \bullet (\overrightarrow{A_1A_2}) = 0 \end{cases} \quad \text{which leads to the equations} \quad \begin{cases} s+2t = 3 \\ 6s+t = 15 \end{cases}.$$

The solution is $s = \frac{27}{11}$ and $t = \frac{3}{11}$, whence $A_1 = A_1(\frac{36}{11}, \frac{-8}{11}, 2)$ and $A_2 = A_2(\frac{27}{11}, \frac{1}{11}, \frac{49}{11})$. The reader can verify that $\|\overrightarrow{A_1A_2}\| = \frac{9}{11}\sqrt{11}$ as before. □

3.4.2 Other Properties of the Cross Product

We begin with a theorem which collects several algebraic properties of the cross product.

Theorem 3. *Let $\vec{u}, \vec{v}$ and $\vec{w}$ be vectors. Then:*

(1) *$\vec{v} \times \vec{w}$ is a vector.*

(2) *$\vec{v} \times \vec{0} = \vec{0} = \vec{0} \times \vec{v}$.*

(3) *$\vec{v} \times \vec{v} = \vec{0}$.*

(4) $\vec{w} \times \vec{v} = -(\vec{v} \times \vec{w})$.

(5) $(a\vec{v}) \times \vec{w} = a(\vec{v} \times \vec{w}) = \vec{v} \times (a\vec{w})$ *for any scalar a.*

(6) $\vec{v} \times (\vec{u} + \vec{w}) = (\vec{v} \times \vec{u}) + (\vec{v} \times \vec{w})$.

(7) $(\vec{u} + \vec{w}) \times \vec{v} = (\vec{u} \times \vec{v}) + (\vec{w} \times \vec{v})$.

Proof. We have already noted (1). Now recall that if $\vec{v} = [x_1 \; y_1 \; z_1]^T$ and $\vec{w} = [x_2 \; y_2 \; z_2]^T$, then

$$\vec{v} \times \vec{w} = \det \begin{bmatrix} \vec{i} & x_1 & x_2 \\ \vec{j} & y_1 & y_2 \\ \vec{k} & z_1 & z_2 \end{bmatrix}.$$

Hence (2) and (3) follow because a determinant is zero if a column is zero or if two columns are equal, (4) follows because a determinant changes sign if two columns are interchanged, and (5) follows because multiplying a column of a determinant by a scalar a multiplies the value of the determinant by a. Finally (6) and (7) follow from the definition of the cross product and the verifications are left as Exercise 17. $\square$

We have not yet given an intrinsic description of the cross product that does not depend on the choice of a coordinate system. The following theorem leads to such a description, and provides a fundamental connection between the dot and cross products.

Theorem 4. The Lagrange[13] Identity. *If $\vec{v}$ and $\vec{w}$ are any two vectors then*

$$\|\vec{v} \times \vec{w}\|^2 = \|\vec{v}\|^2 \|\vec{w}\|^2 - (\vec{v} \bullet \vec{w})^2 \, .$$

Proof. Write $\vec{v} = [x_1 \; y_1 \; z_1]^T$ and $\vec{w} = [x_2 \; y_2 \; z_2]^T$ in component form. Then the identity becomes

$$(y_1 z_2 - y_2 z_1)^2 + (x_1 z_2 - x_2 z_1)^2 + (x_1 y_2 - x_2 y_1)^2$$
$$= (x_1^2 + y_1^2 + z_1^2)(x_2^2 + y_2^2 + z_2^2) - (x_1 x_2 + y_1 y_2 + z_1 z_2)^2$$

We leave the routine verification as an exercise. $\square$

If $\vec{v}$ and $\vec{w}$ are nonzero vectors, the Lagrange identity can be used to give an intrinsic description of the length of the vector $\vec{v} \times \vec{w}$. Recall (Theorem 2 §3.2) that $\vec{v} \bullet \vec{w} = \|\vec{v}\| \|\vec{w}\| \cos\theta$ where θ is the angle between $\vec{v}$ and $\vec{w}$. If this is substituted into the right side of the Lagrange identity, the result is

$$\begin{aligned} \|\vec{v} \times \vec{w}\|^2 &= \|\vec{v}\|^2 \|\vec{w}\|^2 - (\|\vec{v}\| \|\vec{w}\| \cos\theta)^2 \\ &= \|\vec{v}\|^2 \|\vec{w}\|^2 (1 - \cos^2\theta) \\ &= (\|\vec{v}\| \|\vec{w}\| \sin\theta)^2 \end{aligned}$$

---

[13] Joseph Louis Lagrange (1736-1813) was one of the great mathematicians. He contributed to many parts of mathematics, and is best remembered for his work in mechanics.

3.4. THE CROSS PRODUCT

where we have used the fact that $sin^2\theta = 1 - cos^2\theta$. Because $sin\,\theta$ is positive on the range $0 \leq \theta \leq \pi$, taking positive square roots gives

$$\|\vec{v} \times \vec{w}\| = \|\vec{v}\|\,\|\vec{w}\|\,sin\,\theta$$

This is an important (intrinsic) formula for the magnitude of $\vec{v}\times\vec{w}$. In addition, it has a nice geometric interpretation in terms of the parallelogram determined by $\vec{v}$ and $\vec{w}$ (see Figure 4). In fact, the base of the parallelogram has length $\|\vec{v}\|$ while the height is $\|\vec{w}\|\,sin\,\theta$. Hence the area of the parallelogram is $\|\vec{w}\|\,(\|\vec{v}\|\,sin\,\theta) = \|\vec{v} \times \vec{w}\|$, and we have proved part (1) of

Theorem 5. *If θ is the angle between the nonzero vectors $\vec{v}$ and $\vec{w}$, then:*

(1) $\|\vec{v} \times \vec{w}\| = \|\vec{v}\|\,\|\vec{w}\|\,sin\,\theta$ = *The area of the parallelogram determined by $\vec{v}$ and $\vec{w}$.*

(2) *$\vec{v}$ and $\vec{w}$ are parallel if and only if $\vec{v} \times \vec{w} = \vec{0}$.*

Proof. Part (1) is proved above. By Theorem 1 Section 3.1, we have $\vec{v} \times \vec{w} = \vec{0}$ if and only if $\|\vec{v} \times \vec{w}\| = 0$. By (1), this holds if and only if the parallelogram determined by $\vec{v}$ and $\vec{w}$ has zero area. But this happens if and only if $\vec{v}$ and $\vec{w}$ have the same or opposite direction; that is if and only if $\vec{v}$ and $\vec{w}$ are parallel vectors. □

Note the analogue between (2) of Theorem 5 and the fact that vectors $\vec{v}$ and $\vec{w}$ are orthogonal if and only if $\vec{v} \bullet \vec{w} = 0$ (Theorem 3, Section 3.2).

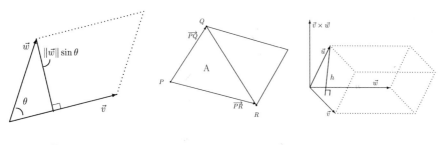

Figure 4   Figure 5   Figure 6

Example 5. *Determine the area of the triangle with vertices $P(1,-1,1)$, $Q(2,0,3)$ and $R(1,1,-3)$.*

Solution. We have $\overrightarrow{PQ} = [1\ \ 1\ \ 2]^T$ and $\overrightarrow{PR} = [0\ \ 2\ \ -4]^T$, so $\overrightarrow{PQ} \times \overrightarrow{PR} = [-8\ \ 4\ \ 2]^T = 2[-4\ \ 2\ \ 1]^T$. The area A of the triangle PQR is half the area of the parallelogram determined by $\overrightarrow{PQ}$ and $\overrightarrow{PR}$ (see Figure 5), so we have $A = \frac{1}{2}\left\|\overrightarrow{PQ} \times \overrightarrow{PR}\right\| = \frac{1}{2}\left[2\sqrt{16+4+1}\right] = \sqrt{21}$. □

If three vectors $\vec{u}$, $\vec{v}$ and $\vec{w}$ are given they determine a geometric solid called a **parallelepiped** (see Figure 6), whose six faces are parallelograms. It is often of interest to find the volume of such a solid. The base of the parallelepiped is the parallelogram determined by $\vec{v}$ and $\vec{w}$ and so the base has area $A = \|\vec{v} \times \vec{w}\|$ by Theorem 5. By Figure 6 the volume of the parallelepiped is hA where h is the length of the projection of $\vec{u}$ on $\vec{v} \times \vec{w}$. Hence Theorem 4 §3.2 gives

$$h = \left| \frac{\vec{u} \bullet (\vec{v} \times \vec{w})}{\|\vec{v} \times \vec{w}\|^2} \right| \|\vec{v} \times \vec{w}\| = \frac{|\vec{u} \bullet (\vec{v} \times \vec{w})|}{\|\vec{v} \times \vec{w}\|} = \frac{|\vec{u} \bullet (\vec{v} \times \vec{w})|}{A}$$

Thus the volume is $hA = |\vec{u} \bullet (\vec{v} \times \vec{w})|$. This, combined with Theorem 1 proves an important formula.

Theorem 6. *Consider the parallelepiped determined by three vectors $\vec{u}$, $\vec{v}$ and $\vec{w}$.*

The volume of the parallelepiped is $|\vec{u} \bullet (\vec{v} \times \vec{w})| = |det[\vec{u}\ \vec{v}\ \vec{w}]|$

where $[\vec{u}\ \vec{v}\ \vec{w}]$ is the matrix with $\vec{u}$, $\vec{v}$ and $\vec{w}$ as its columns.

3.4.3 Geometrical Meaning of the Cross Product

As for the dot product, our definition of $\vec{v} \times \vec{w}$ is given in terms of the components of $\vec{v}$ and $\vec{w}$ and so appears to depend on the choice of coordinate system. In fact this is not so. To begin with, Theorem 5 gives an intrinsic description of the magnitude of $\vec{v} \times \vec{w}$:

$\|\vec{v} \times \vec{w}\|$ is the area of the parallelogram determined by $\vec{v}$ and $\vec{w}$.

Note that this holds even if $\vec{v}$ or $\vec{w}$ is zero.

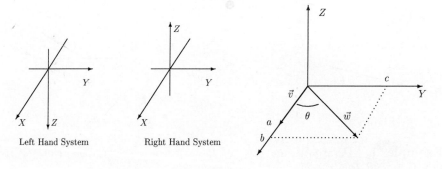

Figure 7   Figure 8

To give a coordinate-free description of the *direction* of $\vec{v} \times \vec{w}$ we must first clarify how coordinate axes are chosen in space. The procedure is as follows: An origin is selected and two perpendicular lines (the X- and Y-axes) are chosen through the origin. Then a positive direction is chosen arbitrarily on each of these axes. Now the Z-axis is taken to be the (unique) line through the origin perpendicular to both the X- and Y-axes. All that remains is to select the

3.4. THE CROSS PRODUCT

positive direction on this Z-axis. There are two possibilities (shown in Figure 7) and it is a convention to choose a **right-hand system**. This means that the positive Z-direction is so chosen that, if the Z-axis is grasped in the right hand with the thumb pointing in the positive direction, the fingers will curl around from the positive X-axis to the positive Y-axis (through a right angle).

With the convention about right-hand systems established, the direction of $\vec{v} \times \vec{w}$ can be given by the following rule:

Right-Hand Rule. Let $\vec{v}$ and $\vec{w}$ be nonzero vectors which are not parallel (so $\vec{v} \times \vec{w} \neq \vec{0}$), and let θ be the angle between $\vec{v}$ and $\vec{w}$. If the vector $\vec{v} \times \vec{w}$ is grasped in the right hand and the fingers curl around from $\vec{v}$ to $\vec{w}$ through the angle θ, the thumb points in the direction of $\vec{v} \times \vec{w}$.

To indicate why this is so, choose coordinate axes in such a way that the initial points of $\vec{v}$ and $\vec{w}$ are at the origin, $\vec{v}$ points along the positive X-axis, and $\vec{w}$ is in the X-Y-plane on the same side of the X-axis as the positive Y-axis (see Figure 8). In this coordinate system we have $\vec{v} = [a \ 0 \ 0]^T$ and $\vec{w} = [b \ c \ 0]^T$ where $a > 0$ and $c > 0$. Hence

$$\vec{v} \times \vec{w} = \det \begin{bmatrix} \vec{i} & a & b \\ \vec{j} & 0 & c \\ \vec{k} & 0 & 0 \end{bmatrix} = \begin{bmatrix} 0 \\ 0 \\ ac \end{bmatrix}$$

and so $\vec{v} \times \vec{w}$ has the positive Z-direction because $ac > 0$. This is the direction given by the right hand rule.

Exercises

1. In each case compute $\vec{v} \times \vec{w}$.
    (a) $\vec{v} = [2 \ 1 \ -3]^T$ and $\vec{w} = [1 \ 2 \ 1]^T$
    (b) $\vec{v} = [3 \ -5 \ 1]^T$ and $\vec{w} = [1 \ 1 \ 2]^T$
    (c) $\vec{v} = [3 \ 2 \ -1]^T$ and $\vec{w} = [-6 \ -4 \ 2]^T$
    (d) $\vec{v} = [1 \ 1 \ 1]^T$ and $\vec{w} = [4 \ 2 \ -3]^T$

2. In each case find the area of the triangle with vertices A, B and C.
    (a) $A(1, -1, 2)$, $B(3, 3, 2)$, $C(5, 0, -4)$
    (b) $A(1, 1, 1)$, $B(2, 3, -5)$, $C(0, 2, 2)$

3. If $\vec{i}$, $\vec{j}$ and $\vec{k}$ are the coordinate vectors, show that $\vec{i} \times \vec{j} = \vec{k}$, $\vec{j} \times \vec{k} = \vec{i}$ and $\vec{k} \times \vec{i} = \vec{j}$.

4. In each case, find the scalar equations of the line:
    (a) Containing $P(7, 1, 0)$ and perpendicular to the lines $[x \ y \ z]^T = [9 \ -1 \ 4]^T + t[-2 \ 1 \ 3]^T$ and $[x \ y \ z]^T = [1 \ 1 \ 1]^T + t[6 \ -2 \ 3]^T$.

(b) Containing $P(2,-1,3)$ and perpendicular to the lines
$[x\ y\ z]^T = [4\ -1\ 2]^T + t[7\ 0\ 1]^T$ and $[x\ y\ z]^T = [-2\ 0\ 1]^T + t[2\ 3\ 0]^T$.

(c) Of intersection of the planes $2x - 3y + z = 0$ and $3x - 2y + 5z = 4$.

(d) Of intersection of the planes $2x - y - 3z = 1$ and $-x - 3y + 5z = 0$.

5. In each case find the equation of the plane:

   (a) Passing through $A(3,1,2)$, $B(5,-1,3)$, and $C(-4,2,0)$.

   (b) Passing through $A(6,-1,1)$, $B(1,0,0)$, and $C(21,3,2)$.

   (c) Passing through $A(6,1,0)$ and perpendicular to the line $[x\ y\ z]^T = [4\ 0\ -3]^T + t[1\ -1\ 4]^T$.

   (d) Passing through $A(1,-1,3)$ and perpendicular to the line $[x\ y\ z]^T = [5\ 5\ 5]^T + t[8\ -2\ 0]^T$.

   (e) Passing through $P(2,0,3)$ and parallel to the plane through the points $A(1,1,-5)$, $B(0,1,-2)$, and $C(1,0,6)$.

   (f) Passing through $P(2,-1,5)$ and parallel to the plane through the points $A(3,-7,1)$, $B(2,0,-1)$, and $C(1,3,0)$.

   (g) Containing $A(2,1,-1)$ and the line
   $[x\ y\ z]^T = [3\ -1\ 0]^T + t[0\ -1\ 2]^T$.

   (h) Containing $A(0,-2,3)$ and the line
   $[x\ y\ z]^T = [1\ -5\ 2]^T + t[-1\ 3\ 5]^T$.

   (i) Containing the lines $[x\ y\ z]^T = [2\ 1\ 0]^T + t[1\ 1\ -2]^T$ and $[x\ y\ z]^T = [1\ 6\ 2]^T + t[2\ -1\ -4]^T$.

   (j) Containing the lines $[x\ y\ z]^T = [1\ 1\ -2]^T + t[3\ 0\ 4]^T$ and $[x\ y\ z]^T = [7\ -1\ 7]^T + t[3\ -2\ 5]^T$.

6. In each case determine if the given points all lie on the same plane? Support your answer.

   (a) $A(1,0,1)$, $B(2,3,0)$, $C(-1,1,2)$, $D(5,1,1)$
   (b) $A(0,2,1)$, $B(1,4,1)$, $C(2,0,-1)$, $D(-1,3,2)$
   (c) $A(2,1,0)$, $B(3,0,3)$, $C(3,3,3)$, $D(0,1,-6)$
   (d) $A(1,-3,2)$, $B(2,-4,4)$, $C(3,-2,2)$, $D(5,1,1)$

7. In each case find the shortest distance between the two parallel lines.

   (a) $[x\ y\ z]^T = [2\ 1\ 0]^T + t[1\ 1\ -2]^T$ and
   $[x\ y\ z]^T = [3\ 3\ 0]^T + t[1\ 1\ -2]^T$.

   (b) $[x\ y\ z]^T = [1\ 1\ 1]^T + t[2\ 0\ -3]^T$ and
   $[x\ y\ z]^T = [3\ 1\ 2]^T + t[2\ 0\ -3]^T$.

8. In each case find the shortest distance between the two nonparallel lines.

   (a) $[x\ y\ z]^T = [3\ 0\ -1]^T + t[1\ 1\ 0]^T$ and
   $[x\ y\ z]^T = [2\ -3\ 5]^T + t[2\ 1\ -1]^T$.

3.4. THE CROSS PRODUCT

(b) $[x \ y \ z]^T = [3 \ 1 \ -2]^T + t[2 \ 1 \ -3]^T$ and
$[x \ y \ z]^T = [2 \ 3 \ -1]^T + t[1 \ 1 \ 1]^T$.

(c) $[x \ y \ z]^T = [7 \ 8 \ 1]^T + t[3 \ -2 \ 5]^T$ and
$[x \ y \ z]^T = [6 \ 1 \ 5]^T + t[-1 \ -3 \ -2]^T$.

(d) $[x \ y \ z]^T = [5 \ 6 \ 7]^T + t[-1 \ -2 \ -3]^T$ and
$[x \ y \ z]^T = [3 \ -1 \ 2]^T + t[0 \ 0 \ 1]^T$.

9. In each case find the volume of the parallelepiped determined by $\vec{u}$, $\vec{v}$ and $\vec{w}$.

   (a) $\vec{u} = [1 \ 1 \ 2]^T$, $\vec{v} = [2 \ 0 \ 1]^T$, $\vec{w} = [-1 \ 1 \ 2]^T$.

   (b) $\vec{u} = [3 \ 0 \ 1]^T$, $\vec{v} = [-3 \ 1 \ 2]^T$, $\vec{w} = [1 \ 1 \ 1]^T$.

10. In each case either show that the statement is true or give an example showing that it is false.

    (a) If $\vec{v} \times \vec{w} = \vec{0}$ then either $\vec{v} = \vec{0}$ or $\vec{w} = \vec{0}$.

    (b) If $\vec{v} \times \vec{w} = \vec{w} \times \vec{v}$ where $\vec{v} \neq \vec{0}$ and $\vec{w} \neq \vec{0}$, then $\vec{v}$ and $\vec{w}$ are parallel.

11. Simplify $(a\vec{v} + b\vec{w}) \times (c\vec{v} + d\vec{w})$.

12. If $\vec{u} + \vec{v} + \vec{w} = \vec{0}$, show that $\vec{u} \times \vec{v} = \vec{v} \times \vec{w} = \vec{w} \times \vec{u}$.

13. Show that the volume of the parallelepiped determined by $\vec{v}$, $\vec{w}$ and $\vec{v} \times \vec{w}$ is $\|\vec{v} \times \vec{w}\|^2$.

14. Show that three points A, B and C all lie on some line if and only if $\vec{AB} \times \vec{AC} = \vec{0}$.

15. Show that four points A, B, C and D all lie in some plane if and only if $\vec{AB} \bullet (\vec{AC} \times \vec{AD}) = 0$.

16. If $\vec{u}$, $\vec{v}$ and $\vec{w}$ are vectors, show that $|\vec{u} \bullet (\vec{v} \times \vec{w})| = |\vec{v} \bullet (\vec{w} \times \vec{u})| = |\vec{w} \bullet (\vec{u} \times \vec{v})|$.

17. Use the definition of the cross product to: (a) Prove (6) of Theorem 3. (b) Prove (7) of Theorem 3.

18. Show that the shortest distance from a point P to the line through P_0 with direction vector $\vec{d}$ is
$$\frac{\|\vec{P_0P} \times \vec{d}\|}{\|\vec{d}\|}.$$

19. Consider the plane $ax + by + cz = k$ with normal $\vec{n} = [a \ b \ c]^T \neq \vec{0}$. Show that the shortest distance from $P_0(x_0, y_0, z_0)$ to this plane is
$$\frac{|ax_0 + by_0 + cz_0 - k|}{\|\vec{n}\|}.$$

3.5 MATRIX TRANSFORMATIONS OF $\mathbb{R}^2$

Apart from their use in solving systems of linear equations, matrices have so far been treated rather abstractly with the emphasis on their algebraic properties. In this section, the vector approach to geometry reveals a concrete, geometrical way to view 2×2 matrices. This discovery not only provides a new interpretation of matrix algebra but also applies matrices to give new geometrical insights.

Throughout this section we denote the set of real numbers by $\mathbb{R}$, and we write $\mathbb{R}^2$ for the Euclidean plane. It is convenient to identify points in $\mathbb{R}^2$ with their position vectors; more precisely, we make no distinction between a point $P(x,y)$ and its position vector $\vec{v} = \begin{bmatrix} x \\ y \end{bmatrix}$, and refer to it simply as a "vector" in $\mathbb{R}^2$.

3.5.1 Transformations

To illustrate how 2×2 matrices arise as geometrical transformations of $\mathbb{R}^2$, consider *reflection* in the X-axis. This operation carries the vector $\begin{bmatrix} x \\ y \end{bmatrix}$ to its reflection $\begin{bmatrix} x \\ -y \end{bmatrix}$, and so can be described geometrically as in Figure 1. Now observe that

$$\begin{bmatrix} x \\ -y \end{bmatrix} = \begin{bmatrix} 1 & 0 \\ 0 & -1 \end{bmatrix} \begin{bmatrix} x \\ y \end{bmatrix}$$

so reflecting $\begin{bmatrix} x \\ y \end{bmatrix}$ in the X-axis can be achieved by multiplying by the matrix $\begin{bmatrix} 1 & 0 \\ 0 & -1 \end{bmatrix}$.

Reflection in the X-axis is an example of a **transformation** T of $\mathbb{R}^2$, that is a function[14]

$$T : \mathbb{R}^2 \to \mathbb{R}^2$$

which carries every vector $\vec{v} = \begin{bmatrix} x \\ y \end{bmatrix}$ in $\mathbb{R}^2$ to another vector in $\mathbb{R}^2$, denoted $T(\vec{v})$. Thus to describe a transformation T we must specify the vector $T(\vec{v})$ for every vector $\vec{v}$ in $\mathbb{R}^2$. This is referred to as **defining** T, or as specifying the **action** of T. In particular, if T is reflection in the X-axis as above then T is defined by $T\begin{bmatrix} x \\ y \end{bmatrix} = \begin{bmatrix} x \\ -y \end{bmatrix}$ for all $\begin{bmatrix} x \\ y \end{bmatrix}$ in $\mathbb{R}^2$.

Matrices provide an important way of defining transformations of $\mathbb{R}^2$. If A is any 2×2 matrix, multiplication by A gives a transformation

$$T : \mathbb{R}^2 \to \mathbb{R}^2 \quad \text{defined by} \quad T(\vec{v}) = A\vec{v} \text{ for every } \vec{v} \text{ in } \mathbb{R}^2.$$

---

[14] Functions $f : \mathbb{R} \to \mathbb{R}$ are basic in High School algebra or calculus. They are often defined by a formula, say $f(x) = x^2 + 1$ for all x in $\mathbb{R}$, but other functions are used (for example $f(x) = cosx$ or $f(x) = e^x$). It is conventional to use the term "transformation" for functions $f : \mathbb{R}^2 \to \mathbb{R}^2$.

3.5. MATRIX TRANSFORMATIONS OF $\mathbb{R}^2$

This is called the **matrix transformation induced** by A. Thus reflection in the X-axis is the matrix transformation induced by $A = \begin{bmatrix} 1 & 0 \\ 0 & -1 \end{bmatrix}$.

If $A = 0$ is the zero matrix, the corresponding matrix transformation T is given by $T(\vec{v}) = A\vec{v} = \vec{0}$ for every $\vec{v}$ in $\mathbb{R}^2$. This is called the **zero transformation** and is denoted $T = 0$. Another example is the **identity transformation** $1_{\mathbb{R}^2} : \mathbb{R}^2 \to \mathbb{R}^2$ induced by the identity matrix. It is given by

$$1_{\mathbb{R}^2}(\vec{v}) = \vec{v} \quad \text{for all } \vec{v} \text{ in } \mathbb{R}^2.$$

It turns out that many important geometric transformations are in fact matrix transformations. Here is another example.

Example 1. *Show that the counterclockwise rotation about the origin through a right angle is a matrix transformation of $\mathbb{R}^2$, and find the corresponding matrix.*

Solution. The transformation is shown in Figure 2. Since the two right triangles are congruent, the rotation carries $\begin{bmatrix} x \\ y \end{bmatrix}$ to $\begin{bmatrix} -y \\ x \end{bmatrix} = \begin{bmatrix} 0 & -1 \\ 1 & 0 \end{bmatrix} \cdot \begin{bmatrix} x \\ y \end{bmatrix}$, so the matrix is $A = \begin{bmatrix} 0 & -1 \\ 1 & 0 \end{bmatrix}$. □

We will see later that *any* rotation about the origin is a matrix transformation of $\mathbb{R}^2$.

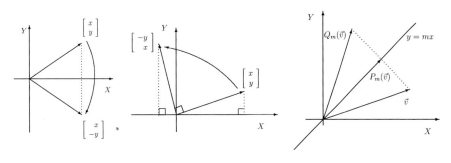

Figure 1   Figure 2   Figure 3

The line through the origin with slope m has equation $y = mx$. Given m, we define two transformations P_m and Q_m of $\mathbb{R}^2$ as follows: Given a vector $\vec{v}$ in $\mathbb{R}^2$, let

$$P_m(\vec{v}) \text{ denote the \textbf{projection} of } \vec{v} \text{ on the line } y = mx$$

and let

$$Q_m(\vec{v}) \text{ denote the \textbf{reflection} of } \vec{v} \text{ in the line } y = mx$$

as shown in Figure 3. Geometrically, $Q_m(\vec{v})$ is the "mirror image" of $\vec{v}$ in the line $y = mx$ while, if $\vec{d}$ is any direction vector for the line, then $P_m(\vec{v}) = proj_{\vec{d}}(\vec{v})$ is the usual projection.

Theorem 1. *Consider the line $y = mx$ through the origin with slope m. Then P_m and Q_m are both matrix transformations of $\mathbb{R}^2$, indeed*

$$P_m \quad \text{is induced by the matrix} \quad \frac{1}{1+m^2}\begin{bmatrix} 1 & m \\ m & m^2 \end{bmatrix},$$

$$Q_m \quad \text{is induced by the matrix} \quad \frac{1}{1+m^2}\begin{bmatrix} 1-m^2 & 2m \\ 2m & m^2-1 \end{bmatrix}.$$

Proof. The line $y = mx$ contains the origin and the point $\begin{bmatrix} 1 \\ m \end{bmatrix}$. Thus $\vec{d} = \begin{bmatrix} 1 \\ m \end{bmatrix}$ is a direction vector for the line. If we write $\vec{v} = \begin{bmatrix} x \\ y \end{bmatrix}$, we have the formula $P_m(\vec{v}) = proj_{\vec{d}}(\vec{v}) = \left(\frac{\vec{v} \cdot \vec{d}}{\|\vec{d}\|^2}\right)\vec{d}$. Hence

$$\begin{aligned} P_m(\vec{v}) &= \frac{x+ym}{1+m^2}\begin{bmatrix} 1 \\ m \end{bmatrix} = \frac{1}{1+m^2}\begin{bmatrix} x+my \\ mx+m^2y \end{bmatrix} \\ &= \frac{1}{1+m^2}\begin{bmatrix} 1 & m \\ m & m^2 \end{bmatrix}\begin{bmatrix} x \\ y \end{bmatrix} \end{aligned}$$

as required. Turning to Q_m, Figure 3 gives $Q_m(\vec{v}) = \vec{v} + 2[P_m(\vec{v}) - \vec{v}] = 2P_m(\vec{v}) - \vec{v}$. Hence some matrix arithmetic gives

$$\begin{aligned} Q_m(\vec{v}) &= \frac{2}{1+m^2}\begin{bmatrix} 1 & m \\ m & m^2 \end{bmatrix}\begin{bmatrix} x \\ y \end{bmatrix} - \begin{bmatrix} x \\ y \end{bmatrix} \\ &= \frac{1}{1+m^2}\begin{bmatrix} 1-m^2 & 2m \\ 2m & m^2-1 \end{bmatrix}\begin{bmatrix} x \\ y \end{bmatrix}. \end{aligned}$$

Thus P_m and Q_m are matrix transformations. $\square$

Example 2: *Find the projection and reflection of the vector $\vec{v} = \begin{bmatrix} 3 \\ -1 \end{bmatrix}$ on the line of slope $\frac{1}{2}$ through the origin.*

Solution: Using our derived formulas for P_m and Q_m, we have:

$$P_{\frac{1}{2}}\begin{bmatrix} 3 \\ -1 \end{bmatrix} = \frac{1}{1+(\frac{1}{2})^2}\begin{bmatrix} 1 & \frac{1}{2} \\ \frac{1}{2} & (\frac{1}{2})^2 \end{bmatrix}\begin{bmatrix} 3 \\ -1 \end{bmatrix} = \begin{bmatrix} 2 \\ 1 \end{bmatrix}.$$

and

$$Q_{\frac{1}{2}}\begin{bmatrix} 3 \\ -1 \end{bmatrix} = \frac{1}{1+(\frac{1}{2})^2}\begin{bmatrix} 1-(\frac{1}{2})^2 & 2(\frac{1}{2}) \\ 2(\frac{1}{2}) & (\frac{1}{2})^2-1 \end{bmatrix}\begin{bmatrix} 3 \\ -1 \end{bmatrix} = \begin{bmatrix} 1 \\ 3 \end{bmatrix}.$$

3.5. MATRIX TRANSFORMATIONS OF $\mathbb{R}^2$

$\square$

We hasten to note that not all transformations of $\mathbb{R}^2$ are matrix transformations. If $\vec{w}$ is a fixed vector, define the transformation $T_{\vec{w}} : \mathbb{R}^2 \to \mathbb{R}^2$ by $T_{\vec{w}}(\vec{v}) = \vec{v} + \vec{w}$ for all $\vec{v}$ in $\mathbb{R}^2$. Then $T_{\vec{w}}$ is called **translation** by $\vec{w}$, and it is not a matrix transformation unless $\vec{w} = \vec{0}$ as the next example shows.

Example 3. *If $\vec{w} \neq \vec{0}$, show that translation by $\vec{w}$ is not a matrix transformation.*

Solution. If $T_{\vec{w}}$ were induced by a matrix A then $A\vec{v} = T_{\vec{w}}(\vec{v}) = \vec{v} + \vec{w}$ would hold for each $\vec{v}$ in $\mathbb{R}^2$. In particular, taking $\vec{v} = \vec{0}$ gives $\vec{w} = \vec{0} + \vec{w} = T_{\vec{w}}(\vec{0}) = A\vec{0} = \vec{0}$, contrary to the assumption that $\vec{w} \neq \vec{0}$. $\square$

3.5.2 Linear Transformations

Let $T : \mathbb{R}^2 \to \mathbb{R}^2$ be the matrix transformation induced by the 2×2 matrix A; that is T is given by $T(\vec{v}) = A\vec{v}$ for all vectors $\vec{v}$ in $\mathbb{R}^2$. Then T has the following properties:

$$(T1.) \quad T(\vec{v} + \vec{w}) = T(\vec{v}) + T(\vec{w}) \quad \text{for all } \vec{v} \text{ and } \vec{w} \text{ in } \mathbb{R}^2 .$$

$$(T2.) \quad T(a\vec{v}) = aT(\vec{v}) \quad \text{for all } \vec{v} \text{ in } \mathbb{R}^2 \text{ and all numbers } a.$$

In fact, (T1) is just the distributive property of matrix multiplication:

$$T(\vec{v} + \vec{w}) = A(\vec{v} + \vec{w}) = A\vec{v} + A\vec{w} = T(\vec{v}) + T(\vec{w}).$$

Similarly $T(a\vec{v}) = A(a\vec{v}) = a(A\vec{v}) = aT(\vec{v})$, giving (T2).

Transformations $T : \mathbb{R}^2 \to \mathbb{R}^2$ which enjoy properties (T1) and (T2) are called **linear transformations**. Thus the above discussion shows that every matrix transformation is linear. The remarkable thing is that the converse is true:

Every linear transformation is actually a matrix transformation.

Moreover, a simple method exists for finding the corresponding matrix. To derive it, consider the **coordinate vectors**

$$\vec{i} = \begin{bmatrix} 1 \\ 0 \end{bmatrix} \quad \text{and} \quad \vec{j} = \begin{bmatrix} 0 \\ 1 \end{bmatrix}$$

and call the pair of vectors $\{\vec{i}, \vec{j}\}$ the **standard basis** of $\mathbb{R}^2$. The importance of the coordinate vectors $\vec{i}$ and $\vec{j}$ (and the reason for the name) lies in the fact that each vector in $\mathbb{R}^2$ can be uniquely expressed in terms of them:

$$\begin{bmatrix} x \\ y \end{bmatrix} = \begin{bmatrix} x \\ 0 \end{bmatrix} + \begin{bmatrix} 0 \\ y \end{bmatrix} = x\vec{i} + y\vec{j} \quad \text{for all} \begin{bmatrix} x \\ y \end{bmatrix} \text{ in } \mathbb{R}^2.$$

In other words, the coordinates of the vector arise as the coefficients of $\vec{i}$ and $\vec{j}$. We will use this observation below.

Now consider an arbitrary linear transformation $T : \mathbb{R}^2 \to \mathbb{R}^2$. Then $T(\vec{i})$ and $T(\vec{j})$ are columns in $\mathbb{R}^2$, so write them as follows:

$$T(\vec{i}) = \begin{bmatrix} a \\ c \end{bmatrix} \quad \text{and} \quad T(\vec{j}) = \begin{bmatrix} b \\ d \end{bmatrix}.$$

Since T is linear, the properties (T1) and (T2) apply to T and we obtain

$$\begin{aligned} T\begin{bmatrix} x \\ y \end{bmatrix} &= T(x\vec{i} + y\vec{j}) = T(x\vec{i}) + T(y\vec{j}) \\ &= xT(\vec{i}) + yT(\vec{j}) \\ &= x\begin{bmatrix} a \\ c \end{bmatrix} + y\begin{bmatrix} b \\ d \end{bmatrix} \\ &= \begin{bmatrix} xa + yb \\ xc + yd \end{bmatrix} \\ &= \begin{bmatrix} a & b \\ c & d \end{bmatrix} \begin{bmatrix} x \\ y \end{bmatrix}. \end{aligned}$$

Hence T is indeed a matrix transformation, the one induced by $A = \begin{bmatrix} a & b \\ c & d \end{bmatrix}$. Moreover the first and second columns of A are $T(\vec{i}) = \begin{bmatrix} a \\ c \end{bmatrix}$ and $T(\vec{j}) = \begin{bmatrix} b \\ d \end{bmatrix}$ respectively, so it is natural to denote A in terms of its columns as follows:

$$A = \begin{bmatrix} T(\vec{i}) & T(\vec{j}) \end{bmatrix}.$$

This proves Theorem 2.

Theorem 2. *A transformation $T : \mathbb{R}^2 \to \mathbb{R}^2$ is linear if and only if it is a matrix transformation. In this case, if $\{\vec{i}, \vec{j}\}$ is the standard basis of $\mathbb{R}^2$, then:*

$$T \text{ is induced by the matrix } A = \begin{bmatrix} T(\vec{i}) & T(\vec{j}) \end{bmatrix}.$$

Here $\begin{bmatrix} T(\vec{i}) & T(\vec{j}) \end{bmatrix}$ *denotes the matrix with columns $T(\vec{i})$ and $T(\vec{j})$.*

Because of Theorem 2, we will use the phrases "linear transformation" and "matrix transformation" interchangeably.

Example 4. *Let $T : \mathbb{R}^2 \to \mathbb{R}^2$ be a linear transformation. If $T\begin{bmatrix} 1 \\ 2 \end{bmatrix} = \begin{bmatrix} 3 \\ -1 \end{bmatrix}$ and $T\begin{bmatrix} 1 \\ 0 \end{bmatrix} = \begin{bmatrix} 5 \\ 4 \end{bmatrix}$, find $T\begin{bmatrix} x \\ y \end{bmatrix}$ for any $\begin{bmatrix} x \\ y \end{bmatrix}$ in $\mathbb{R}^2$.*

Solution 1. We first find the matrix $A = \begin{bmatrix} T(\vec{i}) & T(\vec{j}) \end{bmatrix}$ of T. We are given the first column $T(\vec{i}) = \begin{bmatrix} 5 \\ 4 \end{bmatrix}$. For convenience, write $\vec{w} = \begin{bmatrix} 1 \\ 2 \end{bmatrix}$, so we are also given that $T(\vec{w}) = \begin{bmatrix} 3 \\ -1 \end{bmatrix}$. Now simple matrix arithmetic gives

3.5. MATRIX TRANSFORMATIONS OF $\mathbb{R}^2$

$$\vec{j} = \tfrac{1}{2}(\vec{w} - \vec{i}).$$

If we apply T to both sides, the linearity of T gives

$$T(\vec{j}) = \tfrac{1}{2}T(\vec{w} - \vec{i}) = \tfrac{1}{2}[T(\vec{w}) - T(\vec{i})] = \tfrac{1}{2}\left[\begin{bmatrix} 3 \\ -1 \end{bmatrix} - \begin{bmatrix} 5 \\ 4 \end{bmatrix}\right] = \begin{bmatrix} -1 \\ -\tfrac{5}{2} \end{bmatrix}.$$

Hence $A = \begin{bmatrix} T(\vec{i}) & T(\vec{j}) \end{bmatrix} = \begin{bmatrix} 5 & -1 \\ 4 & -\tfrac{5}{2} \end{bmatrix}$ by Theorem 2, so

$$T\begin{bmatrix} x \\ y \end{bmatrix} = A\begin{bmatrix} x \\ y \end{bmatrix} = \begin{bmatrix} 5 & -1 \\ 4 & -\tfrac{5}{2} \end{bmatrix}\begin{bmatrix} x \\ y \end{bmatrix} = \begin{bmatrix} 5x - y \\ 4x - \tfrac{5}{2}y \end{bmatrix}.$$

This is what we wanted.

Solution 2. We claim that $\begin{bmatrix} x \\ y \end{bmatrix} = s\begin{bmatrix} 1 \\ 2 \end{bmatrix} + t\begin{bmatrix} 1 \\ 0 \end{bmatrix}$ for numbers s and t (depending on x and y). In fact, equating entries and solving for s and t gives

$$\begin{bmatrix} x \\ y \end{bmatrix} = \tfrac{y}{2}\begin{bmatrix} 1 \\ 2 \end{bmatrix} + \tfrac{2x-y}{2}\begin{bmatrix} 1 \\ 0 \end{bmatrix} \quad \text{for all } x, y.$$

Since T has properties (T1) and (T2), we get

$$T\begin{bmatrix} x \\ y \end{bmatrix} = \tfrac{y}{2}T\begin{bmatrix} 1 \\ 2 \end{bmatrix} + \tfrac{2x-y}{2}T\begin{bmatrix} 1 \\ 0 \end{bmatrix} = \tfrac{y}{2}\begin{bmatrix} 3 \\ -1 \end{bmatrix} + \tfrac{2x-y}{2}\begin{bmatrix} 5 \\ 4 \end{bmatrix} = \begin{bmatrix} 5x - y \\ 4x - \tfrac{5}{2}y \end{bmatrix}$$

as before. Thus, $T\begin{bmatrix} x \\ y \end{bmatrix} = \begin{bmatrix} 5 & -1 \\ 4 & -\tfrac{5}{2} \end{bmatrix}\begin{bmatrix} x \\ y \end{bmatrix}$, and we obtain the matrix $A = \begin{bmatrix} 5 & -1 \\ 4 & -\tfrac{5}{2} \end{bmatrix}$ of T. □

The rotations about the origin are another important class of transformations of $\mathbb{R}^2$. Given a vector $\vec{v}$ in $\mathbb{R}^2$, let

$R_\theta(\vec{v})$ denote the **counterclockwise rotation** of $\vec{v}$ through the angle θ.

The action of R_θ is shown in Figure 4. It is somewhat surprising that R_θ is actually a matrix transformation. To see why, the idea is to first show that R_θ is linear and then apply Theorem 2. Of course this also produces the matrix of the transformation R_θ. (In fact this is often the most convenient way to obtain the matrix of a geometric transformation.)

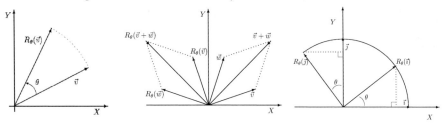

Figure 4   Figure 5   Figure 6

Theorem 3. *The rotation $R_\theta : \mathbb{R}^2 \to \mathbb{R}^2$ is a linear transformation. In fact R_θ is the matrix transformation induced by the matrix*

$$\begin{bmatrix} \cos\theta & -\sin\theta \\ \sin\theta & \cos\theta \end{bmatrix}.$$

Proof. To see that R_θ is linear, consider two vectors $\vec{v}$ and $\vec{w}$ in $\mathbb{R}^2$. Then $\vec{v}+\vec{w}$ is the diagonal of the parallelogram determined by $\vec{v}$ and $\vec{w}$ (see Figure 5), The effect of R_θ is to rotate the *entire parallelogram* to obtain the parallelogram determined by $R_\theta(\vec{v})$ and $R_\theta(\vec{w})$, with diagonal $R_\theta(\vec{v}+\vec{w})$. But this diagonal is $R_\theta(\vec{v})+R_\theta(\vec{w})$ by the parallelogram law, and it follows that

$$R_\theta(\vec{v}+\vec{w}) = R_\theta(\vec{v})+R_\theta(\vec{w}).$$

A similar argument shows that $R_\theta(a\vec{v}) = aR_\theta(\vec{v})$ for any scalar a, so R_θ is a linear transformation.

With this established, we can use Theorem 2. Indeed, from Figure 6 we see that

$$R_\theta(\vec{i}) = \begin{bmatrix} \cos\theta \\ \sin\theta \end{bmatrix} \quad \text{and} \quad R_\theta(\vec{j}) = \begin{bmatrix} -\sin\theta \\ \cos\theta \end{bmatrix}$$

so Theorem 2 shows that R_θ is induced by the matrix

$$[R_\theta(\vec{i})\ R_\theta(\vec{j})] = \begin{bmatrix} \cos\theta & -\sin\theta \\ \sin\theta & \cos\theta \end{bmatrix}.$$

□

In particular, the rotations through $\frac{\pi}{2}$ and π respectively are given by[15]

$$R_{\frac{\pi}{2}}\begin{bmatrix} x \\ y \end{bmatrix} = \begin{bmatrix} \cos\frac{\pi}{2} & -\sin\frac{\pi}{2} \\ \sin\frac{\pi}{2} & \cos\frac{\pi}{2} \end{bmatrix}\begin{bmatrix} x \\ y \end{bmatrix} = \begin{bmatrix} 0 & -1 \\ 1 & 0 \end{bmatrix}\begin{bmatrix} x \\ y \end{bmatrix} = \begin{bmatrix} -y \\ x \end{bmatrix}$$

and

$$R_\pi\begin{bmatrix} x \\ y \end{bmatrix} = \begin{bmatrix} \cos\pi & -\sin\pi \\ \sin\pi & \cos\pi \end{bmatrix}\begin{bmatrix} x \\ y \end{bmatrix} = \begin{bmatrix} -1 & 0 \\ 0 & -1 \end{bmatrix}\begin{bmatrix} x \\ y \end{bmatrix} = \begin{bmatrix} -x \\ -y \end{bmatrix}$$

using Theorem 3. The first of these again confirms the result in Example 1, and the second shows that rotating a vector through an angle π is the same as negating the vector, a fact that is clear without Theorem 3.

3.5.3 Effect on the Unit Square

If $\vec{v}$ is a vector in $\mathbb{R}^2$ then every vector $\vec{u}$ between $\vec{0}$ and $\vec{v}$ has the form $\vec{u} = s\vec{v}$ for some scalar s with $0 \leq s \leq 1$ (see Figure 7). Thus if $\vec{v}$ and $\vec{w}$ are two vectors, Figure 8 shows that each vector in the parallelogram determined by $\vec{v}$ and $\vec{w}$ has the form

$$s\vec{v}+t\vec{w} \quad \text{where } 0 \leq s \leq 1 \quad \text{and} \quad 0 \leq t \leq 1.$$

---

[15] Angles and the trigonometric functions are discussed in Appendix A.1.

3.5. MATRIX TRANSFORMATIONS OF $\mathbb{R}^2$

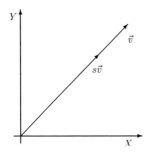

Figure 7   Figure 8

This observation has useful consequences.

Indeed, let $T : \mathbb{R}^2 \to \mathbb{R}^2$ be the matrix transformation induced by A. If $s\vec{v} + t\vec{w}$ is any vector in the parallelogram determined by $\vec{v}$ and $\vec{w}$, then (since T is linear)

$$T(s\vec{v} + t\vec{w}) = T(s\vec{v}) + T(t\vec{w}) = sT(\vec{v}) + tT(\vec{w}).$$

Geometrically this means that

$T(s\vec{v} + t\vec{w})$ is a point in the parallelogram determined by $T(\vec{v})$ and $T(\vec{w})$.

The parallelogram determined by $T(\vec{v})$ and $T(\vec{w})$ is called the **image** under T of the parallelogram determined by $\vec{v}$ and $\vec{w}$. The situation is described in Figure 9.

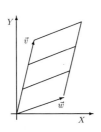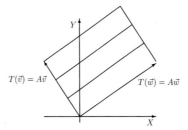

Figure 9

The parallelogram determined by the coordinate vectors $\vec{i}$ and $\vec{j}$ is actually a square, called the **unit square**. Its image under the matrix transformation T is the parallelogram determined by $T(\vec{i})$ and $T(\vec{j})$. This image is important because (as Theorem 2 implies) T is completely determined by $T(\vec{i})$ and $T(\vec{j})$. Thus the image of the unit square gives a graphic sense of how the transformation acts. The next two examples illustrate this.

Example 5. If $a > 0$, the matrix transformation $T\begin{bmatrix} x \\ y \end{bmatrix} = \begin{bmatrix} ax \\ y \end{bmatrix}$ induced by the matrix $A = \begin{bmatrix} a & 0 \\ 0 & 1 \end{bmatrix}$ is called an X-**expansion** if $a > 1$ and an X-

compression if $0 < a < 1$. Here $T(\vec{i}) = A\vec{i} = a\vec{i}$ and $T(\vec{j}) = A\vec{j} = \vec{j}$, so the effect of T on the unit square is shown in Figure 10. Similarly, if $b > 0$ the matrix $\begin{bmatrix} 1 & 0 \\ 0 & b \end{bmatrix}$ gives rise to Y-**expansions** and Y-**compressions** . □

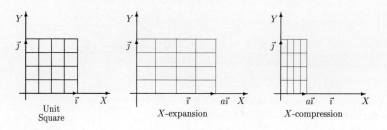

Figure 10

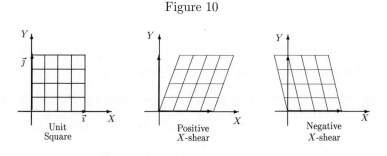

Figure 11

Example 6. If a is a number, the matrix transformation $T\begin{bmatrix} x \\ y \end{bmatrix} = \begin{bmatrix} x + ay \\ y \end{bmatrix}$ induced by the matrix $A = \begin{bmatrix} 1 & a \\ 0 & 1 \end{bmatrix}$ is called an X-**shear**, (**positive** if $a > 0$ and **negative** if $a < 0$). In this case, $T(\vec{i}) = A\vec{i} = \vec{i}$ and $T(\vec{j}) = A\vec{j} = \begin{bmatrix} a \\ 1 \end{bmatrix}$. The effect of T on the unit square is as illustrated in Figure 11. Similarly, the matrix $\begin{bmatrix} 1 & 0 \\ a & 1 \end{bmatrix}$ induces a Y-**shear** for any number a. □

Theorem 2 shows that a matrix transformation is completely determined by its effect on the unit square (equivalently by its effect on the standard basis). The next theorem reveals the role of the determinant.

Theorem 4. Let $T : \mathbb{R}^2 \to \mathbb{R}^2$ be the matrix transformation induced by the 2×2 matrix A. Then the image under T of the unit square has area $|det(A)|$.

Proof. Write $A = \begin{bmatrix} a & b \\ c & d \end{bmatrix}$. For the moment, identify the vector $\begin{bmatrix} x \\ y \end{bmatrix}$ in $\mathbb{R}^2$

3.5. MATRIX TRANSFORMATIONS OF $\mathbb{R}^2$

with the vector $\begin{bmatrix} x \\ y \\ 0 \end{bmatrix}$ in $\mathbb{R}^3$. Thus the image of the unit square under T is the parallelogram determined by $T(\vec{i}) = A\vec{i} = \begin{bmatrix} a \\ c \end{bmatrix} = \begin{bmatrix} a \\ c \\ 0 \end{bmatrix}$ and $T(\vec{j}) = A\vec{j} = \begin{bmatrix} b \\ d \end{bmatrix} = \begin{bmatrix} b \\ d \\ 0 \end{bmatrix}$. By Theorem 5 in Section 3.4, this parallelogram has area

$$\|A\vec{i} \times A\vec{j}\| = \left\| \begin{bmatrix} 0 \\ 0 \\ ad - bc \end{bmatrix} \right\| = \sqrt{(ad-bc)^2} = |(ad-bc)| = |det(A)|.$$

This is what we wanted. □

Example 7. The image of the unit square under a rotation R_θ or a reflection Q_m is again a square of side 1, and so has area 1. Hence the determinant of the matrix of each of these transformations must have absolute value 1 by Theorem 4. In fact the matrices are $\begin{bmatrix} cos\,\theta & -sin\,\theta \\ sin\,\theta & cos\,\theta \end{bmatrix}$ and $\frac{1}{1+m^2}\begin{bmatrix} 1-m^2 & 2m \\ 2m & m^2-1 \end{bmatrix}$ respectively, so the determinants are 1 and -1 respectively, as is easily verified. (We return to this in Theorem 7 below.)

On the other hand, the image of the unit square under the projection P_m on the line $y = mx$ is the parallelogram determined by $P_m(\vec{i})$ and $P_m(\vec{j})$. But these vectors are parallel (both are parallel to the line $y = mx$), and so the parallelogram they determine has area 0. Thus the matrix $\frac{1}{1+m^2}\begin{bmatrix} 1 & m \\ m & m^2 \end{bmatrix}$ of P_m must have determinant 0, a fact that is easily verified directly. □

3.5.4 Composition of Matrix Transformations

It often happens that the effect of a transformation can be achieved by doing a sequence of other (often simpler) transformations one after the other. Here is an example.

Example 8. *Show that the effect of first reflecting in the X-axis and then reflecting in the line $y = x$ is the same as rotation through the angle $\frac{\pi}{2}$.*

Solution. The effect of reflecting in the X-axis is $\begin{bmatrix} x \\ y \end{bmatrix} \to \begin{bmatrix} x \\ -y \end{bmatrix}$, and the effect of reflection in the line $y = x$ is $\begin{bmatrix} x \\ y \end{bmatrix} \to \begin{bmatrix} y \\ x \end{bmatrix}$ by Theorem 1 (with $m = 1$). Hence the combined effect is

$$\begin{bmatrix} x \\ y \end{bmatrix} \to \begin{bmatrix} x \\ -y \end{bmatrix} \to \begin{bmatrix} -y \\ x \end{bmatrix}.$$

But this net effect $\begin{bmatrix} x \\ y \end{bmatrix} \to \begin{bmatrix} -y \\ x \end{bmatrix}$ can be achieved by rotating through the angle $\frac{\pi}{2}$ as was observed in Example 1. □

The result of applying two transformations one after the other is described as follows. If $S : \mathbb{R}^2 \to \mathbb{R}^2$ and $T : \mathbb{R}^2 \to \mathbb{R}^2$ are two transformations of $\mathbb{R}^2$, their **composite** $S \circ T$ is the transformation

$$S \circ T : \mathbb{R}^2 \to \mathbb{R}^2 \quad \text{defined by} \quad [S \circ T](\vec{v}) = S[T(\vec{v})] \quad \text{for all } \vec{v} \text{ in } \mathbb{R}^2.$$

Thus the effect of $S \circ T$ can be described as "first T, then S" (note the order!). This is illustrated in Figure 12.

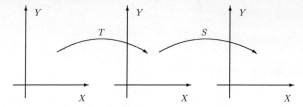

Figure 12

In this notation, the observation in Example 8 can be described as follows: If T is reflection in the X-axis and S is reflection in the line $y = x$, then $S \circ T$ is rotation through $\frac{\pi}{2}$. Now observe that (by Example 8)

$$S \text{ has matrix } A = \begin{bmatrix} 0 & 1 \\ 1 & 0 \end{bmatrix}$$

$$T \text{ has matrix } B = \begin{bmatrix} 1 & 0 \\ 0 & -1 \end{bmatrix}$$

and that

$$S \circ T \text{ has matrix } \begin{bmatrix} 0 & -1 \\ 1 & 0 \end{bmatrix} = \begin{bmatrix} 0 & 1 \\ 1 & 0 \end{bmatrix} \begin{bmatrix} 1 & 0 \\ 0 & -1 \end{bmatrix} = AB$$

In other words, the matrix of $S \circ T$ is the product AB of the matrices of A of S and B of T. This is no coincidence as the next theorem shows.

Theorem 5. *Let $S : \mathbb{R}^2 \to \mathbb{R}^2$ and $T : \mathbb{R}^2 \to \mathbb{R}^2$ be matrix transformations with matrices A and B respectively. Then*

$S \circ T$ is the matrix transformation with matrix AB.

Proof. Given any vector $\vec{v}$ in $\mathbb{R}^2$, we have $S(\vec{v}) = A\vec{v}$ and $T(\vec{v}) = B\vec{v}$. Hence

$$[S \circ T](\vec{v}) = S[T(\vec{v})] = S[B(\vec{v})] = A[B(\vec{v})] = (AB)\vec{v}$$

3.5. MATRIX TRANSFORMATIONS OF $\mathbb{R}^2$

for all $\vec{v}$ in $\mathbb{R}^2$. Hence $S \circ T$ is the matrix transformation induced by the matrix AB. □

Theorem 5 is useful in two ways. First, it enables the effect of the composite of two geometric transformations to be computed using matrix multiplication. But it also provides a very useful geometric interpretation of matrix multiplication: The matrix product AB corresponds to the transformation resulting from first applying B, and then applying A (again note the order). Thus the study of matrices can cast light on geometrical transformations, and vice versa.

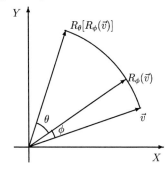

Figure 13

Example 9. Consider the rotations R_θ and R_ϕ through the angles θ and ϕ. The composite $R_\theta \circ R_\phi$ is the transformation resulting from first rotating through ϕ and then rotating through θ, and so is the rotation through the angle $\theta + \phi$ as is illustrated in Figure 13. Thus

$$R_\theta \circ R_\phi = R_{\theta+\phi}$$

Hence Theorem 5 shows that the corresponding relationship holds for the matrices of these transformations:

$$\begin{bmatrix} \cos\theta & -\sin\theta \\ \sin\theta & \cos\theta \end{bmatrix} \begin{bmatrix} \cos\phi & -\sin\phi \\ \sin\phi & \cos\phi \end{bmatrix} = \begin{bmatrix} \cos(\theta+\phi) & -\sin(\theta+\phi) \\ \sin(\theta+\phi) & \cos(\theta+\phi) \end{bmatrix}.$$

If we perform the matrix multiplication on the left, and then compare first column entries, we obtain

$$\begin{aligned} \cos\theta\cos\phi - \sin\theta\sin\phi &= \cos(\theta+\phi) \\ \sin\theta\cos\phi + \cos\theta\sin\phi &= \sin(\theta+\phi) \end{aligned}$$

These are the two basic identities from which most of trigonometry can be derived. □

3.5.5 Inverse of a Matrix Transformation

Another application of Theorem 5 provides a geometrical interpretation of the inverse of a matrix. The matrix transformation T induced by a matrix A is called an **invertible** linear transformation if A is an invertible matrix, and in this case the matrix transformation induced by the inverse matrix A^{-1} is called the **inverse** of the transformation T, and is denoted T^{-1}. Hence T^{-1} is defined by

$$T^{-1}(\vec{v}) = A^{-1}\vec{v} \text{ for all } \vec{v} \text{ in } \mathbb{R}^2.$$

Moreover Theorem 5 shows that $T \circ T^{-1}$ has matrix $AA^{-1} = I_2$, and it follows that $T[T^{-1}(\vec{v})] = (T \circ T^{-1})(\vec{v}) = I_2 \vec{v} = \vec{v}$ for all $\vec{v}$ in $\mathbb{R}^2$. Similarly $T^{-1} \circ T$ has matrix $A^{-1}A = I_2$, and we obtain

$$T[T^{-1}(\vec{v})] = \vec{v} \quad \text{and} \quad T^{-1}[T(\vec{v})] = \vec{v} \quad \text{for all } \vec{v} \text{ in } \mathbb{R}^2.$$

These are called the **fundamental identities** relating T and T^{-1}. They show that T^{-1} "reverses" the action of the transformation T in the following sense: If T carries a vector $\vec{v}$ to $\vec{w} = T(\vec{v})$, then T^{-1} carries $\vec{w}$ right back to $\vec{v}$. Similarly T reverses the action of T^{-1}. Thus the basic relationship between T and T^{-1} is that each reverses the action of the other.

Recall that the matrix transformation induced by the identity matrix I is called the identity transformation $1_{\mathbb{R}^2} : \mathbb{R}^2 \to \mathbb{R}^2$. It is given by $1_{\mathbb{R}^2}(\vec{v}) = I\vec{v} = \vec{v}$ for each $\vec{v}$ in $\mathbb{R}^2$. Using this, the fundamental identities can by reformulated as follows:

$$T \circ T^{-1} = 1_{\mathbb{R}^2} \quad \text{and} \quad T^{-1} \circ T = 1_{\mathbb{R}^2}$$

This leads to a useful characterization of when a linear transformation is invertible.

Theorem 6. Let $T : \mathbb{R}^2 \to \mathbb{R}^2$ be the linear transformation induced by a 2×2 matrix A. The following conditions are equivalent:

(1) T is invertible (that is A is invertible).

(2) There exists a linear transformation $S : \mathbb{R}^2 \to \mathbb{R}^2$ such that $T \circ S = 1_{\mathbb{R}^2}$ and $S \circ T = 1_{\mathbb{R}^2}$.

In this case, the matrix of S is A^{-1}, so S is the inverse T^{-1} of T.

Proof. If (1) holds then (2) follows with $S = T^{-1}$. Conversely, suppose that (2) is true. Let S be induced by the matrix B. By Theorem 5, the equations $T \circ S = 1_{\mathbb{R}^2}$ and $S \circ T = 1_{\mathbb{R}^2}$ translate to $AB = I$ and $BA = I$. Thus A is invertible and $A^{-1} = B$. This proves (2) and the last statement. $\square$

This geometric view of the inverse of a linear transformation provides a way to find the inverse of a matrix. More precisely, if A is an invertible matrix:

(1) Let T be the linear transformation induced by A.

(2) Obtain the linear transformation T^{-1} which "reverses" the action of T.

(3) Then A^{-1} is the matrix of T^{-1}.

Here is an example.

3.5. MATRIX TRANSFORMATIONS OF $\mathbb{R}^2$

Example 10. *If θ is any angle, find the inverse of the matrix*

$$A = \begin{bmatrix} \cos\theta & -\sin\theta \\ \sin\theta & \cos\theta \end{bmatrix}$$

by examining the corresponding matrix transformation.

Solution. By Theorem 3, A is the matrix of the rotation R_θ through the angle θ. Hence A^{-1} will be the matrix of the inverse transformation $[R_\theta]^{-1}$. But $[R_\theta]^{-1}$ must be the rotation $R_{-\theta}$ through the angle $-\theta$ because $R_{-\theta}$ "reverses" the action of R_θ. Hence A^{-1} is the matrix of $R_{-\theta}$, so (by Theorem 3 again with θ replaced by $-\theta$)

$$A^{-1} = \begin{bmatrix} \cos(-\theta) & -\sin(-\theta) \\ \sin(-\theta) & \cos(-\theta) \end{bmatrix} = \begin{bmatrix} \cos\theta & \sin\theta \\ -\sin\theta & \cos\theta \end{bmatrix}.$$

Of course this can be verified directly. $\square$

3.5.6 Computer Graphics[16]

Computer generated graphics arise in a variety of applications, ranging from word processors, to Star Wars animations, to wire-frame images of a car. The images are described by collections of straight line segments, so manipulating an image means manipulating all the line segments. Matrix transformations are important here because matrix images of line segments are again[17] line segments.

It is clearly important to be able to translate images on the screen by some fixed vector $\vec{w}$, that is to apply the transformation $T_{\vec{w}} : \mathbb{R}^2 \to \mathbb{R}^2$ given by $T_{\vec{w}}(\vec{v}) = \vec{v} + \vec{w}$ for all $\vec{v}$ in $\mathbb{R}^2$. The problem is that translations are not matrix transformations (see Example 3). However, there is a clever way around this. The idea is to represent a point $\vec{v} = \begin{bmatrix} x \\ y \end{bmatrix}$ as a 3×1 column $\begin{bmatrix} x \\ y \\ 1 \end{bmatrix}$, called the **homogeneous coordinates** of $\vec{v}$. Then translation by $\vec{w} = \begin{bmatrix} p \\ q \end{bmatrix}$ can by achieved by multiplying by a 3×3 matrix:

$$\begin{bmatrix} 1 & 0 & p \\ 0 & 1 & q \\ 0 & 0 & 1 \end{bmatrix} \begin{bmatrix} x \\ y \\ 1 \end{bmatrix} = \begin{bmatrix} x+p \\ y+q \\ 1 \end{bmatrix} = \begin{bmatrix} \vec{v}+\vec{w} \\ 1 \end{bmatrix} = \begin{bmatrix} T_{\vec{w}}(\vec{v}) \\ 1 \end{bmatrix}.$$

---

[16] This section is not referred to elsewhere in this book.

[17] If a vector $\vec{v} = \begin{bmatrix} x \\ y \end{bmatrix}$ lies on the line through $\vec{v}_0$ with direction vector $\vec{d}$, then $\vec{v} = \vec{v}_0 + t\vec{d}$ for some t. Hence, the matrix image $A\vec{v} = A\vec{v}_0 + tA\vec{d}$ lies on the line through $A\vec{v}_0$ with direction vector $A\vec{d}$. (Note that thye image segment is a single point $A\vec{v}_0$ if $A\vec{d} = \vec{v}_0$.)

On the other hand, the matrix transformation induced by $A = \begin{bmatrix} a & b \\ c & d \end{bmatrix}$ is also given by a 3×3 matrix:

$$\begin{bmatrix} a & b & 0 \\ c & d & 0 \\ 0 & 0 & 1 \end{bmatrix} \begin{bmatrix} x \\ y \\ 1 \end{bmatrix} = \begin{bmatrix} ax + by \\ cx + dy \\ 1 \end{bmatrix} = \begin{bmatrix} A\vec{v} \\ 1 \end{bmatrix}$$

Thus two-dimensional graphics comes down to millions of 3×3 matrix multiplications per second. □

3.5.7 Isometries[18]

A linear transformation $T : \mathbb{R}^2 \to \mathbb{R}^2$ is called an **isometry** if it preserves the distance between any two vectors; that is if the distance between $T(\vec{v})$ and $T(\vec{w})$ equals the distance between $\vec{v}$ and $\vec{w}$ for all vectors $\vec{v}$ and $\vec{w}$. Since the distance between vectors $\vec{v}$ and $\vec{w}$ is $\|\vec{v} - \vec{w}\|$, this means that

$$\|T(\vec{v}) - T(\vec{w})\| = \|\vec{v} - \vec{w}\| \quad \text{for all } \vec{v} \text{ and } \vec{w}.$$

If T is induced by the matrix A, this condition becomes

$$\|A\vec{v} - A\vec{w}\| = \|\vec{v} - \vec{w}\| \quad \text{for all } \vec{v} \text{ and } \vec{w}. \tag{*}$$

It is clear geometrically that every rotation about the origin is an isometry, as is every reflection about a line through the origin. The surprising fact is that these are the only possibilities.

To see why, let $T : \mathbb{R}^2 \to \mathbb{R}^2$ be an isometry and let $A = \begin{bmatrix} a & b \\ c & d \end{bmatrix}$ be the matrix of T. If we set $\vec{w} = \vec{0}$ in equation (*), the result is $\|A\vec{v}\| = \|\vec{v}\|$ for all $\vec{v}$. Taking $\vec{v} = \begin{bmatrix} 1 \\ 0 \end{bmatrix}$, $\vec{v} = \begin{bmatrix} 0 \\ 1 \end{bmatrix}$ and $\vec{v} = \begin{bmatrix} 1 \\ 1 \end{bmatrix}$ results in the equations

$$a^2 + c^2 = 1, \quad b^2 + d^2 = 1 \quad \text{and} \quad ab + cd = 0$$

respectively. These in turn show that $A^T A = I$ as the reader can verify, whence

$$1 = det(I) = det(A^T A) = det(A^T) det(A) = (det\, A)^2.$$

It follows that $det\, A = 1$ or $det\, A = -1$, and it turns out that T is a rotation if $det\, A = 1$ and T is a reflection if $det\, A = -1$. We examine each case separately.

Case 1. $det\, A = 1$. Then $1 = ad - bc$ and multiplying this by c gives

$$c = acd - bc^2 = a(-ab) - bc^2 = -b(a^2 + c^2) = -b.$$

In the same way, multiplying $1 = ad - bc$ by d yields $d = a$. Thus A has the form

$$A = \begin{bmatrix} a & -c \\ c & a \end{bmatrix} \quad \text{where } a^2 + c^2 = 1.$$

---
[18]While this section is mentioned later, it can be omitted with no loss of continuity.

3.5. MATRIX TRANSFORMATIONS OF $\mathbb{R}^2$

The condition $a^2 + c^2 = 1$ means that the point $P(a, c)$ lies on the unit circle. Hence $a = cos\,\theta$ and $c = sin\,\theta$ where θ is the angle determined by $P(a, c)$, so we have

$$A = \begin{bmatrix} cos\,\theta & -sin\,\theta \\ sin\,\theta & cos\,\theta \end{bmatrix} \text{ for some angle } \theta.$$

Hence T is the rotation R_θ through the angle θ by Theorem 3.

Case 2. $det\,A = -1$. Now $1 = bc - ad$ and an argument analogous to that in Case 1 yields that $A = \begin{bmatrix} a & c \\ c & -a \end{bmatrix}$ where $a^2 + c^2 = 1$. Again, $a = cos\,\theta$ and $c = sin\,\theta$ where $0 \leq \theta < 2\pi$, so

$$A = \begin{bmatrix} cos\,\theta & sin\,\theta \\ sin\,\theta & -cos\,\theta \end{bmatrix} \text{ for some angle } \theta.$$

If $\theta = \pi$ then $A = \begin{bmatrix} -1 & 0 \\ 0 & 1 \end{bmatrix}$ so T is reflection in the Y-axis. If $\theta \neq \pi$ then $cos(\theta/2) \neq 0$ because $\theta/2 \neq \pi/2$ and $0 \leq \theta/2 < \pi$. Hence $tan(\theta/2) = \frac{sin(\theta/2)}{cos(\theta/2)}$ is defined, and we claim that T is the reflection in the line through the origin with slope $m = tan(\theta/2)$. In fact, Theorem 1 (and some standard trigonometry) shows that this reflection has matrix

$$\frac{1}{1+m^2} \begin{bmatrix} 1-m^2 & 2m \\ 2m & m^2-1 \end{bmatrix} = \begin{bmatrix} cos\,\theta & sin\,\theta \\ sin\,\theta & -cos\,\theta \end{bmatrix} = A.$$

It is worth noting that, in this case, $y = mx$ is the line through the origin which makes an angle $\theta/2$ with the positive X-axis.

This discussion is summarized as part of the following theorem.

Theorem 7. *Every isometry $T : \mathbb{R}^2 \to \mathbb{R}^2$ is either a rotation or a reflection. Moreover, if T is induced by the matrix A, then either $det\,A = 1$ or $det\,A = -1$, and*

$$T \text{ is a rotation if and only } det\,A = 1,$$

$$T \text{ is a reflection if and only } det\,A = -1.$$

Proof. All that remains is to show that $det\,A = 1$ if T is a rotation, and $det\,A = -1$ if T is a reflection. But these follow by Theorem 3 and Theorem 1, respectively. □

Exercises

1. In each case show that $T : \mathbb{R}^2 \to \mathbb{R}^2$ is a matrix transformation, and find the matrix of T:

(a) $T\begin{bmatrix} x \\ y \end{bmatrix} = \begin{bmatrix} 2x+y \\ 3x-2y \end{bmatrix}$ (b) $T\begin{bmatrix} x \\ y \end{bmatrix} = \begin{bmatrix} 3x-y \\ 2y+x \end{bmatrix}$

(c) $T\begin{bmatrix} x \\ y \end{bmatrix} = \begin{bmatrix} 0 \\ 5y+6x \end{bmatrix}$ (d) $T\begin{bmatrix} x \\ y \end{bmatrix} = \begin{bmatrix} y-x \\ x+y \end{bmatrix}$

2. In each case, assume that $T: \mathbb{R}^2 \to \mathbb{R}^2$ is a linear transformation, and use the information to determine $T\begin{bmatrix} x \\ y \end{bmatrix}$ for all $\begin{bmatrix} x \\ y \end{bmatrix}$ in $\mathbb{R}^2$, and also find the matrix of T

   (a) $T\begin{bmatrix} 1 \\ 0 \end{bmatrix} = \begin{bmatrix} 2 \\ -3 \end{bmatrix}$ and $T\begin{bmatrix} 0 \\ 1 \end{bmatrix} = \begin{bmatrix} 4 \\ -2 \end{bmatrix}$

   (b) $T\begin{bmatrix} 1 \\ 2 \end{bmatrix} = \begin{bmatrix} 1 \\ -3 \end{bmatrix}$ and $T\begin{bmatrix} 3 \\ -1 \end{bmatrix} = \begin{bmatrix} 2 \\ 1 \end{bmatrix}$

   (c) $T\begin{bmatrix} 1 \\ 1 \end{bmatrix} = \begin{bmatrix} 2 \\ -1 \end{bmatrix}$ and $T\begin{bmatrix} 1 \\ 0 \end{bmatrix} = \begin{bmatrix} -1 \\ 1 \end{bmatrix}$

   (d) $T\begin{bmatrix} 1 \\ -1 \end{bmatrix} = \begin{bmatrix} 1 \\ 0 \end{bmatrix}$ and $T\begin{bmatrix} 0 \\ 1 \end{bmatrix} = \begin{bmatrix} 2 \\ -3 \end{bmatrix}$

3. Give the matrix of each of the following transformations and sketch the image of the unit square.

   (a) Reflection in the line $y = -x$.
   (b) Reflection in the line $y = 2x$.
   (c) Rotation through $\pi/4$.
   (d) Rotation through $-\pi/2$.
   (e) Projection on the Y-axis.
   (f) Projection on the line $y = -x$.

4. In each case show that T is not a matrix transformation.

   (a) $T\begin{bmatrix} x \\ y \end{bmatrix} = \begin{bmatrix} xy \\ 0 \end{bmatrix}$ (b) $T\begin{bmatrix} x \\ y \end{bmatrix} = \begin{bmatrix} 0 \\ y^2 \end{bmatrix}$

5. In each case show that $T: \mathbb{R}^2 \to \mathbb{R}^2$ is a matrix transformation, and find the matrix of T.

   (a) $T(\vec{v}) = a\vec{v}$ for all $\vec{v}$ in $\mathbb{R}^2$. (Here a is a fixed number.)

   (b) $T\begin{bmatrix} x \\ y \end{bmatrix} = x\vec{u} + y\vec{w}$ for all $\begin{bmatrix} x \\ y \end{bmatrix}$ in $\mathbb{R}^2$. (Here $\vec{u} = \begin{bmatrix} p \\ q \end{bmatrix}$ and $\vec{w} = \begin{bmatrix} r \\ s \end{bmatrix}$ are fixed in $\mathbb{R}^2$.)

6. In each case, find the matrix of T, $T\begin{bmatrix} 1 \\ 1 \end{bmatrix}$, $T\begin{bmatrix} 2 \\ -1 \end{bmatrix}$, and sketch the image of the unit square.

(a) T is reflection in the line $y = -2x$.
(b) T is reflection in the line $y = 3x$.
(c) T is rotation through $\pi/3$.
(d) T is rotation through $-\pi/4$.
(e) T is projection on the line $y = 3x$.
(f) T is projection on the line $x + 2y = 0$.

7. In each case show that that T is either projection on a line, reflection in a line, or rotation through an angle, and find the line or angle.

(a) $T \begin{bmatrix} x \\ y \end{bmatrix} = \frac{1}{5} \begin{bmatrix} x + 2y \\ 2x + 4y \end{bmatrix}$ (b) $T \begin{bmatrix} x \\ y \end{bmatrix} = \frac{1}{2} \begin{bmatrix} x - y \\ y - x \end{bmatrix}$

(c) $T \begin{bmatrix} x \\ y \end{bmatrix} = \frac{1}{\sqrt{2}} \begin{bmatrix} -x - y \\ x - y \end{bmatrix}$ (d) $T \begin{bmatrix} x \\ y \end{bmatrix} = \frac{1}{5} \begin{bmatrix} -3x + 4y \\ 4x + 3y \end{bmatrix}$

(e) $T \begin{bmatrix} x \\ y \end{bmatrix} = \begin{bmatrix} -y \\ -x \end{bmatrix}$ (f) $T \begin{bmatrix} x \\ y \end{bmatrix} = \frac{1}{2} \begin{bmatrix} x - \sqrt{3}y \\ \sqrt{3}x + y \end{bmatrix}$

8. Let R and S be the matrix transformations induced by matrices A and B respectively. In each case show that T is a matrix transformation and find its matrix in terms of A and B.

(a) $T(\vec{v}) = R(\vec{v}) + S(\vec{v})$ for all $\vec{v}$ in $\mathbb{R}^2$.

(b) $T(\vec{v}) = k R(\vec{v})$ for all $\vec{v}$ in $\mathbb{R}^2$. (Here k is a fixed number.)

9. Given $a > 0$, let $T : \mathbb{R}^2 \to \mathbb{R}^2$ be the transformation which carries each point at distance r from the origin radially out (or in if $a < 1$) to the point at distance ar from the origin. Show that T is a matrix transformation, find its matrix, and sketch the image of the unit square. [T is called a **dilation** if $a > 1$ and a **contraction** if $a < 1$.]

10. Let L be the line through the origin with direction vector $\vec{d} \neq \vec{0}$, and let P_L denote the projection on L. Consider the formula $P_L(\vec{v}) = \frac{\vec{v} \cdot \vec{d}}{\|\vec{d}\|^2} \vec{d}$ for all $\vec{v}$ in $\mathbb{R}^2$.

(a) Use the formula to show that $P_L(\vec{v})$ is the same no matter which (nonzero) direction vector $\vec{d}$ is used for L.

(b) Use the formula to show directly that P_L is linear.

11. Show that the following hold for all matrix transformations $T : \mathbb{R}^2 \to \mathbb{R}^2$.

(a) $T(\vec{0}) = \vec{0}$.

(b) $T(-\vec{v}) = -T(\vec{v})$ for all **v** in $\mathbb{R}^2$.

(c) $T(a\vec{v} + b\vec{w}) = aT(\vec{v}) + bT(\vec{w})$ for all $\vec{v}$ and $\vec{w}$ in $\mathbb{R}^2$.

12. Let L denote the line through the origin with (nonzero) direction vector $\vec{d} = \begin{bmatrix} a \\ b \end{bmatrix}$.

(a) Show that the matrix of the projection on L is $\frac{1}{a^2+b^2}\begin{bmatrix} a^2 & ab \\ ab & b^2 \end{bmatrix}$.

(b) Show that the matrix of the reflection in L is

$$\frac{1}{a^2+b^2}\begin{bmatrix} a^2 - b^2 & 2ab \\ 2ab & b^2 - a^2 \end{bmatrix}.$$

13. Let T be the matrix transformation induced by an invertible 2×2 matrix A. In each case, interpret T^{-1} geometrically and so (as in Example 9) find A^{-1}.

    (a) $A = \begin{bmatrix} 2 & 0 \\ 0 & 1 \end{bmatrix}$ (b) $A = \begin{bmatrix} 1 & 5 \\ 0 & 1 \end{bmatrix}$

    (c) $A = \frac{1}{5}\begin{bmatrix} -3 & 4 \\ 4 & 3 \end{bmatrix}$ (d) $A = \frac{1}{\sqrt{2}}\begin{bmatrix} 1 & -1 \\ 1 & 1 \end{bmatrix}$

    (e) $A = \begin{bmatrix} 0 & -1 \\ -1 & 0 \end{bmatrix}$ (f) $A = \frac{1}{1+m^2}\begin{bmatrix} 1-m^2 & 2m \\ 2m & m^2-1 \end{bmatrix}$

14. In each case find a rotation or reflection which equals the given transformation.

    (a) Reflection in the Y-axis followed by rotation through $\pi/2$.

    (b) Rotation through π followed by reflection in the X-axis.

    (c) Rotation through $\pi/2$ followed by reflection in the line $y = x$.

    (d) Reflection in the X-axis followed by rotation through $\pi/2$.

    (e) Reflection in the line $y = x$ followed by reflection in the X-axis.

15. Express reflection in the line $y = -x$ as the composite of a rotation followed by reflection in the line $y = x$.

16. Determine the effect of the following transformations.

    (a) Rotation through $\pi/2$, followed by projection on the Y-axis, followed by reflection in the line $y = x$.

    (b) Projection on the line $y = x$ followed by projection on the line $y = -x$.

    (c) Projection on the X-axis followed by reflection in the line $y = x$.

    (d) Y-Expansion by 2, followed by the rotation through $\pi/2$.

17. Given any line $y = mx$ through the origin, show that $P_m \circ Q_m = P_m = Q_m \circ P_m$ in two ways: (a) Geometrically; (b) Using Theorem 1.

18. Given two perpendicular lines $y = mx$ and $y = m_1 x$ through the origin, show that $P_m \circ P_{m_1} = 0$ in two ways: (a) Geometrically; (b) Using Theorem 1.

19. If T is any projection, show that $T \circ T = T$ in two ways: (a) Geometrically; (b) Using Theorem 1.

3.5. MATRIX TRANSFORMATIONS OF $\mathbb{R}^2$

20. If T is any reflection, show that $T^{-1} = T$ in two ways: (a) Geometrically; (b) Using Theorem 1.

21. (a) Show that the composite of two isometries is an isometry.

    (b) Show that the composite of two reflections is a rotation.

22. Show that every reflection is reflection in the X-axis followed by a rotation.

23. Let $area(\vec{v}, \vec{w})$ denote the area of the parallelogram determined by the vectors $\vec{v}$ and $\vec{w}$.

    (a) Show that $area(A\vec{v}, A\vec{w}) = |det(A)| \, |det[\vec{v}, \vec{w}]|$ where $[\vec{v}, \vec{w}]$ denotes the matrix with columns $\vec{v}$ and $\vec{w}$.

    (b) Use the result in (a) to deduce Theorem 4.

24. A 2×2 matrix A is called **orthogonal** if A is invertible and $A^{-1} = A^T$.

    (a) Show that the matrix of any isometry is orthogonal.

    (b) If A is orthogonal, show that the corresponding matrix transformation is an isometry.

Chapter 4

THE VECTOR SPACE $\mathbb{R}^n$

4.1 SUBSPACES AND SPANNING

In Chapter 3 we studied 3-dimensional geometry by regarding each point $P = P(x, y, z)$ in space as a 3×1 column matrix $\vec{p} = [x\ y\ z]^T$. This enabled us to give simple descriptions of planes and lines in space, and to study orthogonality and projections. In the same way, we regarded the plane as the set of all 2×1 column matrices, and described geometric transformations such as rotations and reflections using matrix multiplication. In this chapter we extend these geometrical ideas to n dimensional space where $n > 3$. This leads to some of the most important concepts in linear algebra, and has a number of important applications as the next example illustrates.

4.1.1 An Example

Example 1. An engineer would like to be able to predict the response of an amplifier for any input voltage x between 2 and 6 millivolts. She inputs trial voltages x_i of $2, 3, 4, 5,$ and 6 mv and measures the response y_i in the table. When the data pairs are plotted it is apparent (see the diagram) that the points (x_i, y_i) are nearly on a straight line. So the question is: What is the straight line that best fits the data?

Trial i	1	2	3	4	5
Input x_i	2	3	4	5	6
Response y_i	4.8	7.5	8.6	11.1	12.3

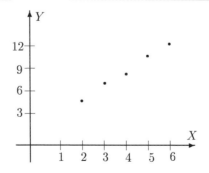

We know that every (non-vertical) line has the form $y = mx + b$ for some constants m and b. Hence, given any choice of m and b, we get predicted responses $y'_1, \cdots, y'_5$, where we write $y'_i = mx_i + b$ for convenience. Thus the task is to find the values of m and b such that the observed responses $y_1, \cdots, y_5$ are as close as possible to the predicted responses $y'_1, \cdots, y'_5$.

But what does "as close as possible" mean?

At this point recall that the distance between two vectors $\vec{v} = [v_1 \ v_2 \ v_3]^T$ and $\vec{w} = [w_1 \ w_2 \ w_3]^T$ in $\mathbb{R}^3$ is given by

$$\|\vec{v} - \vec{w}\| = \sqrt{(v_1 - w_1)^2 + (v_2 - w_2)^2 + (v_3 - w_3)^2}.$$

Now think of the observed responses y_i and the predicted values y'_i as columns $Y = [y_1 \ y_2 \ y_3 \ y_4 \ y_5]^T$ and $Y' = [y'_1 \ y'_2 \ y'_3 \ y'_4 \ y'_5]^T$. If we think of the "distance" between Y and Y' as the number

$$\sqrt{(y_1 - y'_1)^2 + (y_2 - y'_2)^2 + (y_3 - y'_3)^2 + (y_4 - y'_4)^2 + (y_5 - y'_5)^2},$$

then we are asking for the values of m and b that will make this distance as small as possible. In this form the problem can be solved using matrix theory. For obvious reasons, this technique is called the "method of least squares". The idea can be traced back to Gauss and we will return to it in Section 4.6 below. □

Example 1 suggests that geometric notions like distance should be extended to the set of all $n \times 1$ columns. This is the purpose of this chapter.

4.1.2 Euclidean n-space

As in Section 3.5, let $\mathbb{R}$ denote the set of real numbers. Recall that points in the plane and in space are represented by ordered pairs (x, y) and ordered triples (x, y, z) respectively. If $n \geq 1$, an ordered n-tuple

$$(a_1, a_2, \cdots, a_n)$$

of n real numbers is called an n-**vector**, or simply a **vector**. The set of *all* n-vectors is called **Euclidean** n-**space**, and a special notation is used for it:

$$\mathbb{R}^n \text{ denotes the set of all } n\text{-vectors.}$$

As for ordered pairs and triplets, two n-vectors are equal if and only if corresponding entries are equal, that is if they are equal as matrices. Hence we write these n-vectors as column matrices (and frequently as row matrices), and denote them as matrices X, Y etc. However, in $\mathbb{R}^2$ and $\mathbb{R}^3$ we will continue to use the notation $\vec{v}$ from Chapter 3.

Thus $\mathbb{R}^1$ is just the set $\mathbb{R}$ of real numbers, while $\mathbb{R}^2$ and $\mathbb{R}^3$ represent the plane and space respectively. The fact that we are writing points in $\mathbb{R}^3$ as column matrices is consistent with our convention in Section 3.5 to identify a point $P = P(x, y, z)$ in $\mathbb{R}^3$ with its position vector $\vec{p} = [x \ y \ z]^T$. Writing n-vectors as column matrices amounts to adopting this convention in $\mathbb{R}^n$ for every n. Hence, for example, we will speak of the *vector* $\vec{p} = [x \ y \ z]^T$ in $\mathbb{R}^3$, rather than the *point* $P(x, y, z)$.

4.1. SUBSPACES AND SPANNING

4.1.3 Subspaces of $\mathbb{R}^n$

Consider the line L through the origin in $\mathbb{R}^3$ with direction vector $\vec{d} \neq \vec{0}$. Then every vector $\vec{p}$ on the line L has the form $\vec{p} = \vec{0} + t\vec{d}$ for some scalar t, so we have[1]
$$L = \{t\vec{d} \mid t \text{ any real number}\}.$$
Of course the zero vector $\vec{0}$ is in L (because $\vec{0} = 0\vec{d}$). However L has two other properties that are worthy of note: Sums and scalar multiples of vectors in L are again in L. Indeed, if $\vec{v} = t\vec{d}$ and $\vec{v}_1 = t_1\vec{d}$ are in L, then
$$\vec{v} + \vec{v}_1 = t\vec{d} + t_1\vec{d} = (t + t_1)\vec{d} \quad \text{and} \quad r\vec{v} = r(t\vec{d}) = (rt)\vec{d}$$
are both in L for all scalars r. For this reason, the line L is an example of a *subspace* of $\mathbb{R}^3$. This is an important concept in linear algebra, and the general definition is as follows:

A set U of vectors in $\mathbb{R}^n$ is called a **subspace** of $\mathbb{R}^n$ if it has the following three properties:

S1. *The zero vector 0 is in U.*
S2. *If X and Y are in U, then $X + Y$ is also in U.*
S3. *If X is in U, then rX is also in U for any scalar r.*

When property S2 holds for U we say that U is **closed under addition,** and we say that U is **closed under scalar multiplication** if S3 holds.

Example 2. Clearly $\mathbb{R}^n$ is a subspace of itself. The set $\{0\}$ consisting of only the zero vector in $\mathbb{R}^n$ is also a subspace of $\mathbb{R}^n$ because $0 + 0 = 0$ and $r0 = 0$ for any scalar r. This is called the **zero subspace** of $\mathbb{R}^n$. Any subspace of $\mathbb{R}^n$ other than $\{0\}$ or $\mathbb{R}^n$ is called a **proper subspace** of $\mathbb{R}^n$. □

Example 3. Show that $U = \left\{ \begin{bmatrix} 2s - t \\ s - 3t \\ -s + t \\ t \end{bmatrix} \mid s \text{ and } t \text{ real} \right\}$ is a subspace of $\mathbb{R}^4$.

Solution. If $X_1 = \begin{bmatrix} 2s_1 - t_1 \\ s_1 - 3t_1 \\ -s_1 + t_1 \\ t_1 \end{bmatrix}$ and $X_2 = \begin{bmatrix} 2s_2 - t_2 \\ s_2 - 3t_2 \\ -s_2 + t_2 \\ t_2 \end{bmatrix}$ are in U then

$X_1 + X_2 = \begin{bmatrix} 2s_3 - t_3 \\ s_3 - 3t_3 \\ -s_3 + t_3 \\ t_3 \end{bmatrix}$ where $s_3 = s_1 + s_2$ and $t_3 = t_1 + t_2$. Hence $X_1 + X_2$

is in U. The verification that rX_1 is in U for any scalar r is similar and is left to the reader. Clearly 0 is in U (take $s = t = 0$), so U is a subspace of $\mathbb{R}^4$. □

---

[1] We are using set notation here. In general the notation $\{p \mid q\}$ means the set of all objects p which satisfy condition q.

The reader should not get the impression that every subset of $\mathbb{R}^n$ is a subspace.

Example 4. *Consider the subsets $U_1 = \{[x\ y\ z]^T \mid x \geq 0\}$ and $U_2 = \{[x\ y\ z]^T \mid x^2 = y^2\}$ of $\mathbb{R}^3$. Show that neither U_1 nor U_2 is a subspace of $\mathbb{R}^3$.*

Solution. Both U_1 and U_2 contain $0 = [0\ 0\ 0]^T$. However U_1 is not closed under scalar multiplication: For example $[1\ 0\ 0]^T$ is in U_1 but $(-1)[1\ 0\ 0]^T = [-1\ 0\ 0]^T$ is not in U_1. Hence U_1 is not a subspace, even though it is closed under addition (as the reader can verify). As to U_2, it is not closed under addition because, for example, $[1\ 1\ 0]^T$ and $[-1\ 1\ 0]^T$ are both in U_2 but their sum $[1\ 1\ 0]^T + [-1\ 1\ 0]^T = [0\ 2\ 0]^T$ is not in U_2, so U_2 is not a subspace. (Note that U_2 is actually closed under scalar multiplication.) □

Example 5. *Show that lines through the origin and planes through the origin are all subspaces of $\mathbb{R}^3$.*

Solution. The line $L = \{t\vec{d} \mid t \text{ real}\}$ through the origin with direction vector $\vec{d}$ was verified above to be a subspace of $\mathbb{R}^3$.

Let P denote the plane through the origin with normal $\vec{n} = [a\ b\ c]^T$ (using the notation of Chapter 3). Then P has equation $ax + by + cz = 0$, where the right side is zero because P contains the origin. Now let $\vec{v} = [x\ y\ z]^T$ and $\vec{v}_1 = [x_1\ y_1\ z_1]^T$ be two vectors in the plane P. Then $\vec{v} + \vec{v}_1 = [x+x_1\ y+y_1\ z+z_1]^T$ is also in the plane because

$$\begin{aligned}a(x+x_1) + b(y+y_1) + c(z+z_1) &= (ax+by+cz) + (ax_1+by_1+cz_1)\\&= 0 + 0 = 0.\end{aligned}$$

Hence S2 holds for P. Similarly $r\vec{v} = [rx\ ry\ rz]^T$ is in the plane for any scalar r, as the reader can verify, so S3 also holds for P. Since S1 clearly holds, P is a subspace of $\mathbb{R}^3$. □

It turns out (Example 8 §4.3) that lines and planes through the origin are the *only* proper subspaces of $\mathbb{R}^3$. Thus the geometry of planes and lines through the origin in space is captured by the subspace concept.

However subspaces have other uses, one of which is describing various features of an $m \times n$ matrix A. In fact there are two important subspaces associated with A, the **null space** of A, denoted $null A$, and the **image** of A, denoted $im A$. They are defined as follows:

$$\begin{aligned}null A &= \{X \text{ in } \mathbb{R}^n \mid AX = 0\}.\\im A &= \{Y \text{ in } \mathbb{R}^m \mid Y = AX \text{ for some } X \text{ in } \mathbb{R}^n\}.\end{aligned}$$

Thus $null A$ is the set of all solutions to the homogeneous system $AX = 0$, and $im A$ is the set of all columns Y in $\mathbb{R}^m$ such that the system $AX = Y$ has a solution. Observe that $null A$ is a subset of $\mathbb{R}^n$ while $im A$ is a subset of $\mathbb{R}^m$. In fact they are both subspaces.

4.1. SUBSPACES AND SPANNING

Example 6. *For any $m \times n$ matrix A, show that $null\, A$ is a subspace of $\mathbb{R}^n$, and that $im\, A$ is a subspace of $\mathbb{R}^m$.*

Solution. A column X is in $null\, A$ if it satisfies the *condition* that $AX = 0$. Hence $0 \in null\, A$ because it satisfies the condition: $A0 = 0$. Thus S1 holds for $null\, A$. Suppose that X and X_1 are in $null\, A$, that is $AX = 0$ and $AX_1 = 0$. Then

$$A(X + X_1) = AX + AX_1 = 0 + 0 = 0 \quad \text{and} \quad A(rX) = r(AX) = r0 = 0$$

show, respectively, that $X + X_1$ and rX are in $null\, A$ for all scalars r. Hence S2 and S3 are also satisfied for $null\, A$.

Turning to $im\, A$, note that it consists of the vectors Y of the *form* $Y = AX$ for some $X \in \mathbb{R}^n$. Thus $0 \in im\, A$ because it has the required form: $0 = A0$, so S1 is satisfied for $im\, A$. Let Y and Y_1 be vectors in $im\, A$, so that $Y = AX$ and $Y_1 = AX_1$ for some X and X_1 in $\mathbb{R}^n$. Then

$$Y + Y_1 = AX + AX_1 = A(X + X_1) \quad \text{and} \quad rY = r(AX) = A(rX)$$

show that $Y + Y_1$ and rY are both in $im\, A$ for all r. Thus S2 and S3 are both satisfied for $im\, A$. □

The image and null space of a matrix A are essential features of A and will arise frequently in this chapter.

Example 7. Let A denote an $n \times n$ matrix. If λ is a number, let

$$E_\lambda(A) = \{X \text{ in } \mathbb{R}^n \mid AX = \lambda X\}.$$

This is a subspace of $\mathbb{R}^n$ as is easily verified (in fact it is the null space of the matrix $(\lambda I - A)$. The subspace $E_\lambda(A)$ is called the **eigenspace** of A corresponding to λ, and several definitions in Section 2.3 can be stated in terms of $E_\lambda(A)$. The number λ is called an **eigenvalue** of A if $E_\lambda(A) \neq \{0\}$, and in that case the nonzero vectors in $E_\lambda(A)$ are called the **eigenvectors** of A corresponding to λ. Recall that these eigenvalues and eigenvectors are closely associated with the process of diagonalizing A; we return to this in Section 4.7. □

4.1.4 Spanning Sets

The Gaussian algorithm can be regarded as a procedure for finding the null space of a matrix A. To illustrate, consider the matrix

$$A = \begin{bmatrix} 1 & -2 & 1 & 1 \\ -1 & 2 & 0 & 1 \\ 2 & -4 & 1 & 0 \end{bmatrix}.$$

The null space of A consists of all columns X in $\mathbb{R}^4$ such that $AX = 0$, that is all solutions $X = [x_1\ x_2\ x_3\ x_4]^T$ to the system $AX = 0$ of (homogeneous) equations with A as coefficient matrix:

$$\begin{array}{rcrcrcrcl} x_1 & - & 2x_2 & + & x_3 & + & x_4 & = & 0 \\ -x_1 & + & 2x_2 & & & + & x_4 & = & 0 \\ 2x_1 & - & 4x_2 & + & x_3 & & & = & 0 \end{array} \quad (*)$$

This system is solved in Example 2 §1.3, and the general solution X is presented there as a sum of scalar multiples of the basic solutions $X_1 = [2\ 1\ 0\ 0]^T$ and $X_2 = [1\ 0\ -2\ 1]^T$:

$$X = sX_1 + tX_2 \text{ where } s \text{ and } t \text{ are parameters.}$$

This way of combining the vectors X_1 and X_2 is important, and some terminology is commonly used to describe it.

Given vectors $X_1, X_2, \cdots, X_k$ in $\mathbb{R}^n$, a vector of the form

$$X = t_1 X_1 + t_2 X_2 + \cdots + t_k X_k$$

where the t_i are scalars is called a **linear combination** of the X_i. The scalar t_i is called the **coefficient** of X_i in the linear combination. The set of *all* such linear combinations is called the **span** of the X_i, and is denoted $span\{X_1, X_2, \cdots, X_k\}$. More formally,

$$span\{X_1, X_2, \cdots, X_k\} = \{t_1 X_1 + t_2 X_2 + \cdots + t_k X_k \mid t_i \text{ in } \mathbb{R}\}.$$

In particular, if X and Y are two vectors in $\mathbb{R}^n$ then

$$span\{X, Y\} = \{sX + tY \mid s \text{ and } t \text{ real numbers}\}$$

consists of all sums of scalar multiples of X and Y. Similarly

$$span\{X\} = \{sX \mid s \text{ a real number}\}.$$

Example 8. *Determine whether $X = [3\ -1\ 2]^T$ and $Y = [-1\ 13\ -3]^T$ are in $span\{X_1, X_2\}$ where $X_1 = [1\ 5\ 0]^T$ and $X_2 = [1\ -1\ 1]^T$*

Solution. To test if X is in $span\{X_1, X_2\}$ we must discover whether X is a linear combination of X_1 and X_2, that is whether scalars s and t exist such that $X = sX_1 + tX_2$. This condition is

$$X = [3\ -1\ 2]^T = s[1\ 5\ 0]^T + t[1\ -1\ 1]^T$$

and comparing entries gives equations $3 = s + t$, $-1 = 5s - t$ and $2 = t$. These equations clearly have no solution, so X is not in $span\{X_1, X_2\}$.

On the other hand, a similar procedure using Y in place of X yields equations $-1 = s + t$, $13 = 5s - t$ and $-3 = t$. These have solution $s = 2$ and $t = -3$,

4.1. SUBSPACES AND SPANNING

whence $Y = 2X_1 - 3X_2$ as the reader can verify. Hence Y is in $span\{X_1, X_2\}$. □

If $\vec{d}$ is a nonzero vector in $\mathbb{R}^3$, then (as discussed prior to Example 2 above)

$$span\{\vec{d}\} = \{t\vec{d} \mid t \text{ real}\}$$

is the line through the origin with direction vector $\vec{d}$. In particular $span\{\vec{d}\}$ is a **subspace** of $\mathbb{R}^3$. In fact the span of *any* set of vectors is a subspace as the next theorem shows.

We need the following notation. If U and V are sets we say that U is **contained** in V, and write $U \subseteq V$, if every element of U is also an element of V (that is, if U is a subset of V). This gives a useful test for set equality: $U = V$ if and only if both $U \subseteq V$ and $V \subseteq U$ (that is U and V have the same elements).

Theorem 1. *Let $X_1, X_2, \cdots, X_k$ be any vectors in $\mathbb{R}^n$.*

(1) $span\{X_1, X_2, \cdots, X_k\}$ is a subspace of $\mathbb{R}^n$ which contains each of the vectors $X_1, X_2, \cdots, X_k$.

(2) If the vectors $X_1, X_2, \cdots, X_k$ all lie in some subspace V, then $span\{X_1, X_2, \cdots, X_k\} \subseteq V$.

Proof. (1). For convenience, write $U = span\{X_1, X_2, \cdots, X_k\}$. Condition S1 clearly holds for U since $0 = 0X_1 + \cdots + 0X_k$ is in U. Given two vectors $X = t_1X_1 + t_2X_2 + \cdots + t_kX_k$ and $Y = s_1X_1 + s_2X_2 + \cdots + s_kX_k$ in U, their sum is

$$X + Y = (s_1 + t_1)X_1 + (s_2 + t_2)X_2 + \cdots + (s_k + t_k)X_k$$

which is again in U. Hence S2 holds for U, and the proof for S3 is similar. Thus U is a subspace. It contains each vector X_i because X_i is the linear combination with the coefficient of X_i equal to 1, and all the other coefficients zero.

(2). If V is a subspace containing all the X_i, then V contains every linear combination of the X_i by S2 and S3. In other words, V contains $span\{X_1, X_2, \cdots, X_k\}$. □

Note that (2) of Theorem 1 asserts that $U = span\{X_1, X_2, \cdots, X_k\}$ is the *smallest* subspace containing all the vectors X_i because U is contained in every subspace that contains each of the X_i.

The next example illustrates how Theorem 1 is used.

Example 9. *If X and Y are vectors in $\mathbb{R}^n$, show that $span\{X, Y\} = span\{X+Y, X-Y\}$.*

Solution. Clearly both $X + Y$ and $X - Y$ are in $span\{X, Y\}$, so $span\{X+Y, X-Y\} \subseteq span\{X, Y\}$ by Theorem 1. On the other hand, we have

$$X = \tfrac{1}{2}(X+Y) + \tfrac{1}{2}(X-Y) \text{ and } Y = \tfrac{1}{2}(X+Y) - \tfrac{1}{2}(X-Y)$$

which show that both X and Y are in $span\{X+Y, X-Y\}$. Thus $span\{X,Y\}$ $\subseteq span\{X+Y, X-Y\}$, again by Theorem 1. Since we have verified both containments, $span\{X,Y\} = span\{X+Y, X-Y\}$. □

If a subspace U has the form $U = span\{X_1, X_2, \cdots, X_k\}$, we say that the vectors $X_1, X_2, \cdots, X_k$ are a **spanning set** for U, or that U is **spanned** by the X_i. Many important subspaces have naturally occurring spanning sets. Here are three examples.

Example 10. *Show that $\mathbb{R}^n = span\{E_1, E_2, \cdots, E_n\}$ where $E_1, E_2, \cdots, E_n$ are the columns of the $n \times n$ identity matrix.*

Solution. Given $X = [x_1 \ x_2 \ \cdots \ x_n]^T$ in $\mathbb{R}^n$, we can write X as follows:

$$X = \begin{bmatrix} x_1 \\ x_2 \\ \vdots \\ x_n \end{bmatrix} = \begin{bmatrix} x_1 \\ 0 \\ \vdots \\ 0 \end{bmatrix} + \begin{bmatrix} 0 \\ x_2 \\ \vdots \\ 0 \end{bmatrix} + \cdots + \begin{bmatrix} 0 \\ 0 \\ \vdots \\ x_n \end{bmatrix}$$

$$= x_1 \begin{bmatrix} 1 \\ 0 \\ \vdots \\ 0 \end{bmatrix} + x_2 \begin{bmatrix} 0 \\ 1 \\ \vdots \\ 0 \end{bmatrix} + \cdots + x_n \begin{bmatrix} 0 \\ 0 \\ \vdots \\ 1 \end{bmatrix}.$$

In other words, $X = x_1 E_1 + x_2 E_2 + \cdots + x_n E_n$ is in $span\{E_1, E_2, \cdots, E_n\}$. Since X could be any vector in $\mathbb{R}^n$, this shows that $\mathbb{R}^n \subseteq span\{E_1, E_2, \cdots, E_n\}$. On the other hand, each $E_i \in \mathbb{R}^n$ so Theorem 1 gives $span\{E_1, E_2, \cdots, E_n\} \subseteq \mathbb{R}^n$. Hence we have equality: $span\{E_1, E_2, \cdots, E_n\} = \mathbb{R}^n$. □

If A is an $m \times n$ matrix, recall that we defined $null(A) = \{X \mid AX = 0\}$ to be the set of all solutions to the homogeneous system $AX = 0$ of linear equations with A as coefficient matrix. The next example gives a systematic way to find a spanning set for $null(A)$.

Example 11. *Let A be an $m \times n$ matrix, and let $\{X_1, X_2, \cdots, X_k\}$ denote the basic solutions of the homogeneous system $AX = 0$ given by Gaussian elimination (Section 1.3). Show that the set $nullA$ of all solutions of the system is*

$$nullA = span\{X_1, X_2, \cdots, X_k\}.$$

Solution. Theorem 2 §1.3 shows that every solution X of $AX = 0$ is a linear combination of the basic solutions $X_1, X_2, \cdots, X_k$. In our present terminology, it shows that $nullA \subseteq span\{X_1, X_2, \cdots, X_k\}$. On the other hand, each X_i is in the subspace $nullA$ (being a solution), so $span\{X_1, X_2, \cdots, X_k\} \subseteq nullA$ by Theorem 1. Hence we have equality. □

As our third example of a spanning set, we show that imA is spanned by the columns of A.

4.1. SUBSPACES AND SPANNING

Example 12. If A is any $m \times n$ matrix, show that $imA = \text{span}\{C_1, C_2, \cdots, C_n\}$ where $C_1, C_2, \cdots, C_n$ are the columns of A.

Solution. Write $A = [C_1\ C_2\ \cdots\ C_n]$ in block form as a row of columns. If $X = [x_1\ x_2\ \cdots\ x_n]^T$ is any vector in $\mathbb{R}^n$ then

$$AX = [C_1\ C_2\ \cdots\ C_n]\begin{bmatrix} x_1 \\ x_2 \\ \vdots \\ x_n \end{bmatrix} = x_1C_1 + x_2C_2 + \cdots + x_nC_n$$

by block multiplication (Theorem 3 §1.4). Since $imA = \{AX \mid X \in \mathbb{R}^n\}$, this shows that each vector in imA is in $\text{span}\{C_1, C_2, \cdots, C_n\}$, and vice-versa. In other words, $imA = \text{span}\{C_1, C_2, \cdots, C_n\}$. $\square$

Exercises

1. Give four examples of vectors belonging to each Euclidean n-space.
   (a) $\mathbb{R}^2$.   (b) $\mathbb{R}^3$.   (c) $\mathbb{R}^5$.   (d) $\mathbb{R}^1$.

2. Which Euclidean n-space do the following vectors belong to? That is, what is the value of n of $\mathbb{R}^n$ for the following vectors?
   (a) $(-2, -3)$
   (b) $(-2, -3, 0, 0, 0, 0)$
   (c) $(1, 5, 4, 2)$
   (d) 100
   (e) $(8, -9, 3)$
   (f) $(0, 0, 8, -9, 3)$

3. Identify which pairs of vectors are equal. If the two vectors are not equal, describe why.
   (a) $(4, 5, -9)$ and $(4, -5, -9)$
   (b) $(10, 2, 7)$ and $(10, 2, 7, 1)$
   (c) $(-1, 4)$ and $(4, -1)$
   (d) $(3, -9, -1, 0)$ and $(3, -9, -1, 0)$
   (e) $(-4, -7, 4)$ and $(-4, -7, 4, 0, 0)$

4. Find $nullA$ and imA for the following matrices.
   (a) $A = \begin{bmatrix} 1 & 0 & 0 & 0 \\ 0 & 1 & 0 & 0 \\ 0 & 0 & 1 & 0 \\ 0 & 0 & 0 & 1 \end{bmatrix}$

(b) $A = \begin{bmatrix} 0 & 0 & 0 & 0 \\ 0 & 0 & 0 & 0 \\ 0 & 0 & 0 & 0 \\ 0 & 0 & 0 & 0 \end{bmatrix}$

(c) $A = \begin{bmatrix} 0 & 2 & 0 & -5 \\ 0 & 1 & 4 & 0 \\ 0 & 0 & 1 & 0 \\ 0 & 0 & 0 & 1 \end{bmatrix}$

(d) $A = \begin{bmatrix} 5 & 0 & -7 & 1 \\ 0 & 1 & 4 & 0 \\ 0 & 0 & 1 & 0 \\ 5 & 0 & 0 & 1 \end{bmatrix}$

(e) $A = \begin{bmatrix} -1 & -4 & 1 \\ 7 & -9 & 0 \\ 10 & 3 & -3 \\ -9 & 1 & 2 \end{bmatrix}$

5. In each case determine whether U is a subspace of $\mathbb{R}^3$. Support your answer.

   (a) $U = \{[s\ t\ 1]^T \mid s \text{ and } t \text{ in } \mathbb{R}\}$.
   (b) $U = \{[0\ t\ 0]^T \mid t \text{ in } \mathbb{R}\}$.
   (c) $U = \{[r\ s\ t]^T \mid r,\ s, \text{ and } t \text{ in } \mathbb{R},\ 3r - 2s + t = 0\}$.
   (d) $U = \{[r\ 0\ s]^T \mid r^2 + s^2 = 0,\ r \text{ and } s \text{ in } \mathbb{R}\}$.
   (e) $U = \{[r-1\ 3s\ t]^T \mid r,\ s, \text{ and } t \text{ in } \mathbb{R}\}$.
   (f) $U = \{[r\ r\ t]^T \mid r \text{ and } t \text{ in } \mathbb{R}\}$.

6. In each case determine if X lies in $U = span\{Y, Z\}$. If X is in U write it as a linear combination of Y and Z; if X is not in U, show why not.

   (a) $X = [2\ -1\ 0\ 1]^T$, $Y = [1\ 0\ 0\ 1]^T$ and $Z = [0\ 1\ 0\ 1]^T$.
   (b) $X = [1\ 2\ 15\ 11]^T$, $Y = [2\ -1\ 0\ 2]^T$ and $Z = [1\ -1\ -3\ 1]^T$.
   (c) $X = [8\ 3\ -13\ 20]^T$, $Y = [2\ 1\ -3\ 5]^T$ and $Z = [-1\ 0\ 2\ -3]^T$.
   (d) $X = [2\ 5\ 8\ 3]^T$, $Y = [2\ -1\ 0\ 5]^T$ and $Z = [-1\ 2\ 2\ -3]^T$.

7. In each case determine if the given vectors span $\mathbb{R}^4$. Support your answer.

   (a) $\{[1\ 1\ 1\ 1]^T,\ [0\ 1\ 1\ 1]^T,\ [0\ 0\ 1\ 1]^T,\ [0\ 0\ 0\ 1]^T\}$.
   (b) $\{[1\ 3\ -5\ 0]^T,\ [-2\ 1\ 0\ 0]^T,\ [0\ 2\ 1\ -1]^T,\ [3\ 6\ -3\ -2]^T\}$.

8. In each case either show that the statement is true or give an example showing that it is false.

   (a) If U is a subspace of $\mathbb{R}^n$ and $X + Y$ is in U, then X and Y are both in U.

   (b) If U is a subspace of $\mathbb{R}^n$ and rX is in U for all r in $\mathbb{R}$, then X is in U.

4.1. SUBSPACES AND SPANNING

(c) If U is a nonempty set and $sX + tY$ is in U whenever X and Y are in U and for all s and t in $\mathbb{R}$, then U is a subspace.

(d) If U is a subspace of $\mathbb{R}^n$ and X is in U, then $-X$ is in U.

9. Is it possible that $\{[1 \ -1 \ 0]^T, \ [2 \ 0 \ 3]^T\}$ can span the subspace $U = \{[r \ 0 \ s]^T \mid r \text{ and } s \text{ in } \mathbb{R}\}$? Defend your answer.

10. If X is any vector in $\mathbb{R}^n$, show that $span\{X\} = span\{aX\}$ for any $a \neq 0$ in $\mathbb{R}$.

11. Describe $span\{0\}$.

12. Is $\mathbb{R}^2$ a subspace of $\mathbb{R}^3$? Defend your answer.

13. Suppose $X_1, X_2, \cdots, X_k$ are vectors in $\mathbb{R}^n$. If $Y = a_1X_1 + a_2X_2 + \cdots + a_kX_k$ where $a_1 \neq 0$, show that $span\{X_1, X_2, \cdots, X_k\} = span\{Y, X_2, \cdots, X_k\}$.

14. Suppose that $U = span\{X_1, X_2, \cdots, X_k\}$ where each X_i is in $\mathbb{R}^n$. If $AX_i = 0$ for each i for some $m \times n$ matrix A, show that $AY = 0$ for each Y in U.

15. Let A be an $m \times n$ matrix.

    (a) If U is an invertible $m \times m$ matrix, show that $null(UA) = null\,A$.

    (b) If V is an invertible $n \times n$ matrix, show that $im(AV) = im\,A$.

16. Let U be a subspace of $\mathbb{R}^n$.

    (a) If aX is in U where $a \neq 0$ is a number and X is in $\mathbb{R}^n$, show that X is in U.

    (b) If X and $X + Y$ are in U where X and Y are in $\mathbb{R}^n$, show that Y is in U.

17. Let $\mathcal{X}$ and $\mathcal{Y}$ denote finite sets of vectors in $\mathbb{R}^n$. If $\mathcal{X} \subseteq \mathcal{Y}$, show that $span\{\mathcal{X}\} \subseteq span\{\mathcal{Y}\}$.

18. Let A and B denote two $m \times n$ matrices. If $U = \{X \text{ in } \mathbb{R}^n \mid AX = BX\}$, show that U is a subspace of $\mathbb{R}^n$.

19. If A is an $n \times n$ matrix and λ is a number, write $E_\lambda(A) = \{X \text{ in } \mathbb{R}^n \mid AX = \lambda X\}$ as in Example 7.

    (a) Show that $E_\lambda(A)$ is a subspace of $\mathbb{R}^n$ for any choice of A and λ.

    (b) Explain how these subspaces $E_\lambda(A)$ are related to the eigenvalues and eigenvectors of A.

20. Let A be an $m \times n$ matrix. For which columns B is $U = \{X \text{ in } \mathbb{R}^n \mid AX = B\}$ a subspace of $\mathbb{R}^n$? Support your answer.

21. If $V \neq 0$ is a vector in $\mathbb{R}^n$, determine all subspaces of $span\{V\}$.

22. Let U be a *nonempty* subset of $\mathbb{R}^n$. Show that U is a subspace of $\mathbb{R}^n$ if and only if $S2$ and $S3$ hold.

23. If U and W are subspaces of $\mathbb{R}^n$, define their **intersection** $U \cap W$ and **sum** $U + W$ as follows:

$U \cap W = \{X \text{ in } \mathbb{R}^n \mid X \text{ is in both } U \text{ and } W\}$.
$U + W = \{X \text{ in } \mathbb{R}^n \mid X \text{ is a sum of a vector in } U \text{ and a vector in } W\}$.

Show that:

(a) $U \cap W$ is a subspace of $\mathbb{R}^n$.

(b) $U + W$ is a subspace of $\mathbb{R}^n$.

(c) If $U \cap W = \{0\}$, show that each vector X in $U + W$ has a unique representation as a sum $X = Y + Z$ where Y is in U and Z is in W.

4.2 LINEAR INDEPENDENCE

Some spanning sets for a subspace U of $\mathbb{R}^n$ are better than others. If $U = \text{span}\{X_1, X_2, \cdots, X_k\}$ then every vector in U can be written as a linear combination of the X_i in at least one way. Our interest here is in spanning sets for which each vector in U has *exactly one* representation as a linear combination of these vectors.

4.2.1 Independent Sets of Vectors

Suppose that two linear combinations are equal:

$$r_1 X_1 + r_2 X_2 + \cdots + r_k X_k = s_1 X_1 + s_2 X_2 + \cdots + s_k X_k.$$

We are looking for a condition on the vectors X_i such that this equation will imply that $r_i = s_i$ for each i. Taking all terms to the left side gives

$$(r_1 - s_1)X_1 + (r_2 - s_2)X_2 + \cdots + (r_k - s_k)X_k = 0$$

so the required condition is that this equation forces all the coefficients $r_i - s_i$ to be zero. With this in mind we make the following definition. A set $\{X_1, X_2, \cdots, X_k\}$ of vectors is called **linearly independent** (or simply **independent**) if it satisfies the following condition:

$$\text{If } t_1 X_1 + t_2 X_2 + \cdots + t_k X_k = 0 \text{ then } t_1 = t_2 = \cdots = t_k = 0.$$

We record the result of the above discussion for reference.

Theorem 1. *If $\{X_1, X_2, \cdots, X_k\}$ is an independent set of vectors, then every vector X in $\text{span}\{X_1, X_2, \cdots, X_k\}$ has a unique representation as a linear combination of the X_i.*

It is useful to state the definition of independence in a slightly different form. Call a linear combination of vectors **trivial** if every coefficient is zero. Then:

4.2. LINEAR INDEPENDENCE

> *A set of vectors is independent if and only if the only linear combination which vanishes is the trivial one.*

(Here we say that a linear combination **vanishes** if it equals the zero vector.) Hence the procedure for checking that a set of vectors is independent is as follows:

Independence Test. *To verify that a set $\{X_1, X_2, \cdots, X_k\}$ of vectors in $\mathbb{R}^n$ is linearly independent, proceed as follows*:

Step 1. *Set a linear combination of the vectors equal to zero*: $t_1 X_1 + t_2 X_2 + \cdots + t_k X_k = 0$.

Step 2. *Show that the only way this can happen is the trivial one with all variables $t_i = 0$.*

Of course, if some nontrivial linear combination exists in Step 2, the vectors X_i are <u>not</u> linearly independent.

The following example illustrates how this condition is used to verify independence in a concrete situation.

Example 1. Show that the set of vectors $\{[2\ 0\ -2\ 3]^T, [2\ 1\ 0\ 7]^T, [1\ 1\ 3\ 0]^T\}$ in $\mathbb{R}^4$ is linearly independent.

Solution. Suppose a linear combination vanishes:

$$r[2\ 0\ -2\ 3]^T + s[2\ 1\ 0\ 7]^T + t[1\ 1\ 3\ 0]^T = 0 = [0\ 0\ 0\ 0]^T.$$

Equating corresponding entries gives four equations:

$$\begin{aligned} 2r + 2s + t &= 0 \\ s + t &= 0 \\ -2r + 3t &= 0 \\ 3r + 7s &= 0 \end{aligned}$$

Using methods from Chapter 1, it is easily verified that the only solution is the trivial one $r = s = t = 0$. Hence the vectors are linearly independent by the Independence Test. □

The solution of Example 1 is typical of the way the Independence Test shows that a set of vectors (columns or rows) in $\mathbb{R}^n$ is linearly independent. However, in many situations matrix algebra can be used to verify that a set of vectors is independent. In every case the idea is to set a linear combination of the vectors equal to zero, and show somehow that this forces all the coefficients to be zero. Here are six examples.

Example 2. If $E_1, E_2, \cdots, E_n$ are the columns of the $n \times n$ identity matrix, show that $\{E_1, E_2, \cdots, E_n\}$ is an independent set of vectors in $\mathbb{R}^n$.

Solution. Suppose a linear combination vanishes: $t_1E_1 + t_2E_2 + \cdots + t_nE_n = 0$ where $t_1, t_2, \cdots, t_n$ are scalars. We have $t_1E_1+t_2E_2+\cdots+t_nE_n = [t_1\ t_2\ \cdots\ t_n]^T$ as the reader can verify, and it follows that $[t_1\ t_2\ \cdots\ t_n]^T = 0$. This means that $t_1 = t_2 = \cdots = t_n = 0$, so $\{E_1, E_2, \cdots, E_n\}$ is independent by the Independence Test. $\square$

Example 3. *Show that $\{X\}$ is an independent set for any nonzero vector $X \neq 0$ in $\mathbb{R}^n$.*

Solution. Suppose $tX = 0$ where t is a scalar. Then $t = 0$ because $X \neq 0$ (Theorem 3 §1.1). In other words, the only linear combination from $\{X\}$ that vanishes is the trivial one. $\square$

Example 4. *If $\{X, Y\}$ is linearly independent, show that $\{2X + Y, 3X - 5Y\}$ is also independent.*

Solution. Suppose a linear combination of the vectors $2X + Y$ and $3X - 5Y$ vanishes, say
$$s(2X + Y) + t(3X - 5Y) = 0.$$
We must show that this is the trivial linear combination, that is $s = t = 0$. If we collect terms in X and Y on the left side, the result is a linear combination of X and Y that vanishes:
$$(2s + 3t)X + (s - 5t)Y = 0.$$
Hence the independence of $\{X, Y\}$ requires that this is the trivial linear combination, that is both coefficients must be zero. Thus $2s + 3t = 0$ and $s - 5t = 0$, and the reader can verify that these equations have only the trivial solution $s = t = 0$. This is what we wanted. $\square$

Example 5. *Show that no set of vectors containing the zero vector can be linearly independent.*

Solution. Given a set $\{0, X_1, X_2, \cdots, X_k\}$ of vectors containing the zero vector 0, we have a linear combination
$$1 \cdot 0 + 0X_1 + 0X_2 + \cdots + 0X_k = 0$$
which vanishes but is nontrivial. Hence $\{0, X_1, X_2, \cdots, X_k\}$ is not linearly independent. $\square$

Example 6. *Let $\{X_1, X_2, \cdots, X_k\}$ be a linearly independent set in $\mathbb{R}^n$. If U is any invertible $n \times n$ matrix, show that the set $\{UX_1, UX_2, \cdots, UX_k\}$ is also linearly independent.*

Solution. Suppose $t_1(UX_1) + t_2(UX_2) + \cdots + t_k(UX_k) = 0$. Then $U(t_1X_1 + t_2X_2 + \cdots + t_kX_k) = 0$, so left multiplication by U^{-1} gives $t_1X_1 + t_2X_2 +$

4.2. LINEAR INDEPENDENCE

$\cdots + t_k X_k = 0$. Hence $t_1 = t_2 = \cdots = t_k = 0$ by the independence of $\{X_1, X_2, \cdots, X_k\}$. □

The next example reveals another reason for the importance of row-echelon matrices, and will be referred to later.

Example 7. *Show that the nonzero rows of any row-echelon matrix R are linearly independent.*

Solution. Let $Y_1, Y_2, \cdots, Y_r$ denote the nonzero rows of R, and suppose a linear combination vanishes:

$$t_1 Y_1 + t_2 Y_2 + \cdots + t_r Y_r = 0.$$

For convenience write $C = t_1 Y_1 + t_2 Y_2 + \cdots + t_r Y_r$. We must show that each coefficient t_i is zero. Suppose the first leading 1 in R lies in column j. Then the jth entry of C is t_1 because all the other Y's have j^{th} entry zero (the first leading 1 has zeros below it in R). Hence the condition $C = 0$ forces $t_1 = 0$.

Thus $C = t_2 Y_2 + \cdots + t_r Y_r = 0$. Now regard $Y_2, \cdots, Y_r$ as the rows of the (smaller) row echelon matrix obtained from R by deleting row 1. Then the same argument shows that $t_2 = 0$. Continue in this way to conclude that $t_i = 0$ for each $i = 1, 2, \cdots, r$. This is what we wanted. □

4.2.2 Invertibility of Matrices

The invertibility of an $n \times n$ matrix A is closely related to the independence (spanning) of the columns or rows of A. The whole thing depends on the following observation. Write $A = [C_1 \; C_2 \; \cdots \; C_n]$ where $C_1, C_2, \cdots, C_n$ are the columns of A and, given scalars $x_1, x_2, \cdots x_n$, write $X = [x_1 \; x_2 \; \cdots \; x_n]^T$. Then block multiplication (Theorem 3 §1.4) gives

$$AX = [C_1 \; C_2 \; \cdots \; C_n] \begin{bmatrix} x_1 \\ x_2 \\ \vdots \\ x_n \end{bmatrix} = x_1 C_1 + x_2 C_2 + \cdots + x_n C_n. \quad (*)$$

This equation gives the equivalence of (1), (2) and (3) in the following useful theorem.

Theorem 2. *The following are equivalent for an $n \times n$ matrix A:*

(1) *A is invertible.*

(2) *The columns of A are linearly independent in $\mathbb{R}^n$.*

(3) *The columns of A span $\mathbb{R}^n$.*

(4) *The rows of A are linearly independent in $\mathbb{R}^n$.*

(5) *The rows of A span $\mathbb{R}^n$.*

Proof. As above, let $C_1, C_2, \cdots, C_n$ denote the columns of A.

(1)$\Leftrightarrow$(2). By Theorem 5 §1.5, A is invertible if and only if the system $AX = 0$ has only the trivial solution $X = 0$. By (*) this holds if and only if $\{C_1, C_2, \cdots, C_n\}$ is linearly independent.

(1)$\Leftrightarrow$(3). By Theorem 5 §1.5, A is invertible if and only if $AX = B$ has a solution for every B in $\mathbb{R}^n$. By (*) this holds if and only if $span\{C_1, C_2, \cdots, C_n\} = \mathbb{R}^n$.

(1)$\Leftrightarrow$(4). The independence of the rows of A in condition (4) is equivalent to the independence of the *columns* of A^T, and hence to the invertibility of A^T (by (1)$\Leftrightarrow$(2) applied to A^T). But A^T is invertible if and only if A is invertible (by the Corollary to Theorem 3 §1.5).

(1)$\Leftrightarrow$(5). This is analogous to the proof of (1)$\Leftrightarrow$(4). $\square$

Example 8. *Show that the set $\{[1\ 0\ -2\ 4]^T, [5\ -3\ 7\ 0]^T, [2\ 8\ -1\ 6]^T, [1\ 3\ 0\ 2]^T\}$ is linearly independent.*

Solution. The matrix $A = \begin{bmatrix} 1 & 5 & 2 & 1 \\ 0 & -3 & 8 & 3 \\ -2 & 7 & -1 & 0 \\ 4 & 0 & 6 & 2 \end{bmatrix}$ with these vectors as its columns is invertible because $det A = 36 \neq 0$. Hence the result follows from Theorem 2. $\square$

4.2.3 Linear Dependence

A set $\{X_1, X_2, \cdots, X_k\}$ of vectors in $\mathbb{R}^n$ is called **linearly dependent** (or simply **dependent**) if it is *not* linearly independent. Hence our condition for independence becomes:

> *A set of vectors is linearly dependent if and only if some nontrivial linear combination vanishes.*

There is a useful test for linear dependence which we record for reference.

Theorem 3. *A set $\{X_1, X_2, \cdots, X_k\}$ of vectors in $\mathbb{R}^n$ is linearly dependent if and only if one of the vectors X_i is a linear combination of the others.*

Proof. Suppose that $\{X_1, X_2, \cdots, X_k\}$ is dependent. Then some nontrivial linear combination vanishes, say $t_1 X_1 + t_2 X_2 + \cdots + t_k X_k = 0$ where some $t_i \neq 0$. If $t_1 \neq 0$ then $X_1 = -\frac{t_2}{t_1} X_2 - \cdots - \frac{t_k}{t_1} X_k$ so X_1 is a linear combination of the rest. Similarly, if $t_i \neq 0$ then X_i is a linear combination of the others.

Conversely, assume that one of the vectors is a linear combination of the others, say $X_1 = r_2 X_2 + \cdots + r_k X_k$. Then $1 X_1 - r_2 X_2 - \cdots - r_k X_k = 0$ is a

4.2. LINEAR INDEPENDENCE

nontrivial linear combination that vanishes, so $\{X_1, X_2, \cdots, X_k\}$ is dependent. A similar argument works if any X_i is a linear combination of the rest. □

Theorem 3 has a geometric interpretation in $\mathbb{R}^3$ which clarifies the notion of linear independence. Let $\vec{v}$ and $\vec{w}$ be two nonzero vectors in $\mathbb{R}^3$. Then Theorem 3 asserts that the set $\{\vec{v}, \vec{w}\}$ is linearly dependent if and only if one of $\vec{v}$ and $\vec{w}$ is a scalar multiple of the other, that is if and only if they are parallel. Hence we have

Corollary. Let $\vec{v}$ and $\vec{w}$ be nonzero vectors in $\mathbb{R}^3$.

(1) $\{\vec{v}, \vec{w}\}$ is linearly dependent if and only if $\vec{v}$ and $\vec{w}$ are parallel.

(2) $\{\vec{v}, \vec{w}\}$ is linearly independent if and only if $\vec{v}$ and $\vec{w}$ are not parallel.

Now we can give a complete geometrical description of the span of two nonzero vectors in $\mathbb{R}^3$.

Theorem 4. *Assume that $\vec{v}$ and $\vec{w}$ are two nonzero vectors in $\mathbb{R}^3$. There are two cases*:

Case 1. *If $\vec{v}$ and $\vec{w}$ are parallel (that is $\{\vec{v}, \vec{w}\}$ is linearly dependent in $\mathbb{R}^3$) then $span\{\vec{v}, \vec{w}\}$ is the line through the origin with direction vector $\vec{v}$ (or $\vec{w}$). This is illustrated in Figure 1.*

Case 2. *If $\vec{v}$ and $\vec{w}$ are not parallel (that is $\{\vec{v}, \vec{w}\}$ is linearly independent in $\mathbb{R}^3$) then $span\{\vec{v}, \vec{w}\}$ is the plane through the origin with normal $\vec{n} = \vec{v} \times \vec{w}$. In this case $span\{\vec{v}, \vec{w}\}$ is the unique plane through the origin containing $\vec{v}$ and $\vec{w}$. This is illustrated in Figure 2.*

Proof of Case 1. In this case each of $\vec{v}$ and $\vec{w}$ is a scalar multiple of the other by Theorem 3, and so $span\{\vec{v}, \vec{w}\} = span\{\vec{v}\} = span\{\vec{w}\}$ is the line through the origin with direction vector $\vec{v}$ (or $\vec{w}$).

Proof of Case 2. We have $\vec{n} = \vec{v} \times \vec{w} \neq \vec{0}$ because $\vec{v}$ and $\vec{w}$ are not parallel (Theorem 5 § 3.4). For convenience, write $U = span\{\vec{v}, \vec{w}\}$ and let P denote the plane through the origin with normal $\vec{n}$. Then $P = \{\vec{p} \mid \vec{n} \bullet \vec{p} = 0\}$ so P is a subspace of $\mathbb{R}^3$ (verify). Moreover, $\vec{v}$ and $\vec{w}$ are in P by Theorem 2 §3.4 so $U \subseteq P$ by Theorem 1 §4.1. So it remains to show that $P \subseteq U$. Let $\vec{p}$ be a vector in P; we must show that $\vec{p}$ is in U. If $[\vec{p}\ \vec{v}\ \vec{w}]$ denotes the matrix with $\vec{p}$, $\vec{v}$ and $\vec{w}$ as its columns, then Theorem 1 §3.4 gives

$$det\,[\vec{p}\ \vec{v}\ \vec{w}] = \vec{p} \bullet (\vec{v} \times \vec{w}) = \vec{p} \bullet \vec{n} = 0.$$

Hence $[\vec{p}\ \vec{v}\ \vec{w}]$ is not an invertible matrix, so $\{\vec{p}, \vec{v}, \vec{w}\}$ is linearly dependent by Theorem 2. This means that there is a linear combination $a\vec{p}+b\vec{v}+c\vec{w} = \vec{0}$ where a, b and c are not all zero. But $a \neq 0$ because $\{\vec{v}, \vec{w}\}$ is linearly independent, so $\vec{p} = \frac{-b}{a}\vec{v} + \frac{-c}{a}\vec{w}$ is in U. This shows that $P \subseteq U$, as required. □

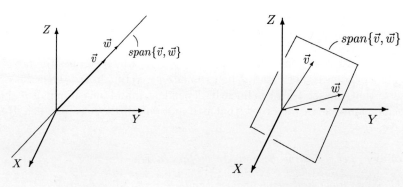

Figure 1 Figure 2

Exercises

1. Show that each of the following sets of vectors is independent.

    (a) $\{[-1\ 2]^T, [3\ 4]^T\}$ in $\mathbb{R}^2$.

    (b) $\{[1\ 1\ 1]^T, [2\ 1\ 0]^T\}$ in $\mathbb{R}^3$.

    (c) $\{[3\ 1\ -2]^T, [-2\ -3\ -1]^T, [1\ 3\ 3]^T\}$ in $\mathbb{R}^3$.

    (d) $\{[1\ 0\ 1\ 0]^T, [0\ 1\ 0\ 1]^T, [3\ -1\ 2\ -1]^T\}$ in $\mathbb{R}^4$.

2. Which of the following subsets are independent? If the set is dependent, give an example of a nontrivial linear combination that equals zero.

    (a) $\{[1\ 5]^T, [0\ -2]^T\}$ in $\mathbb{R}^2$.

    (b) $\{[2\ -6]^T, [-1\ 3]^T\}$ in $\mathbb{R}^2$.

    (c) $\{[1\ -1\ 0]^T, [3\ 2\ -1]^T, [5\ 0\ -1]^T\}$ in $\mathbb{R}^3$.

    (d) $\{[3\ 0\ -1]^T, [2\ 1\ -1]^T\}$ in $\mathbb{R}^3$.

    (e) $\{[1\ -1\ 1\ -1]^T, [2\ 0\ 1\ 0]^T, [0\ -2\ 1\ -2]^T\}$ in $\mathbb{R}^4$.

    (f) $\{[1\ 2\ 3\ 4]^T, [-1\ 0\ 2\ 2]^T, [-4\ 2\ 0\ -1]^T\}$ in $\mathbb{R}^4$.

    (g) $\{[1\ 1\ 0\ 0]^T, [1\ 0\ 1\ 0]^T, [0\ 0\ 1\ 1]^T, [0\ 1\ 0\ 1]^T\}$ in $\mathbb{R}^4$.

3. The goal of the next 4 questions is to become familiar with Theorem 2. In this question, use Theorem 2 to determine if A is invertible.

    (a) $A = \begin{bmatrix} 5 & 10 \\ 1 & 2 \end{bmatrix}$

    (b) $A = \begin{bmatrix} 1 & 1 & 0 \\ 0 & -2 & 5 \\ 3 & 0 & 4 \end{bmatrix}$

4.2. LINEAR INDEPENDENCE

(c) $A = \begin{bmatrix} 3 & 4 & -1 & 1 \\ -3 & -3 & 0 & 0 \\ 2 & -2 & 1 & -2 \\ 1 & -8 & -6 & 1 \end{bmatrix}$

(d) $A = \begin{bmatrix} 3 & -1 & 4 & -1 & -4 \\ 5 & 6 & 6 & 13 & 1 \\ 6 & 6 & -6 & 0 & 0 \\ 3 & 0 & 0 & -3 & -3 \\ 2 & 0 & 9 & 7 & -2 \end{bmatrix}$

4. Use Theorem 2 to determine whether or not these sets of vectors are independent.

   (a) $\{[3\ 1]^T, [-2\ 2]^T\}$ in $\mathbb{R}^2$.
   (b) $\{[1\ 1\ -1]^T, [1\ -1\ 1]^T, [0\ 0\ 1]^T\}$ in $\mathbb{R}^3$.
   (c) $\{[5\ -2\ 1]^T, [-5\ 0\ -5]^T, [4\ -2\ 0]^T\}$ in $\mathbb{R}^3$.
   (d) $\{[1\ 0\ -2\ 5]^T, [4\ 4\ -3\ 2]^T, [0\ 1\ 0\ -3]^T, [1\ 3\ 3\ 1]^T\}$ in $\mathbb{R}^4$.

5. Using Theorem 2, determine if the following sets of vectors span the indicated Euclidean space ($\mathbb{R}^n$).

   (a) $\{[4\ -7]^T, [2\ -5]^T\}$, $\mathbb{R}^2$.
   (b) $\{[0\ -6\ -6]^T, [8\ -3\ 5]^T, [-9\ 7\ -2]^T\}$, $\mathbb{R}^3$.
   (c) $\{[-8\ -1\ 3\ 6]^T, [-5\ -5\ 4\ -9]^T, [7\ 5\ 9\ 0]^T, [6\ 1\ -16\ 3]^T\}$, $\mathbb{R}^4$.
   (d) $\{[2\ 1\ 7\ -2]^T, [3\ 5\ 4\ 5]^T, [4\ -4\ -3\ -3]^T, [-5\ 0\ 6\ -4]^T\}$, $\mathbb{R}^4$.

6. For which sets of vectors in Exercises 1 and 2 could you use Theorem 2 to determine their independence?

7. In each case either show that the statement is true or give an example showing that it is false. Throughout, $X, Y, Z, X_1, X_2, \cdots, X_n$ denote vectors in $\mathbb{R}^n$.

   (a) If $\{X, Y\}$ is independent then $\{X, Y, X + Y\}$ is independent.
   (b) If $\{X, Y, Z\}$ is independent, then $\{X, Y\}$ is independent.
   (c) If $\{X, Y\}$ is dependent, then $\{X, Y, Z\}$ is dependent.
   (d) If all of $X_1, X_2, \cdots, X_n$ are nonzero, then $\{X_1, X_2, \cdots, X_n\}$ is independent.
   (e) If one of $X_1, X_2, \cdots, X_n$ is zero, then $\{X_1, X_2, \cdots, X_n\}$ is dependent.
   (f) If $aX + bY + cZ = 0$ where a, b and c are in $\mathbb{R}$, then $\{X, Y, Z\}$ is independent.
   (g) If $\{X, Y, Z\}$ is independent then $aX + bY + cZ = 0$ for some a, b and c in $\mathbb{R}$.
   (h) If $\{X_1, X_2, \cdots, X_n\}$ is dependent then $t_1 X_1 + t_2 X_2 + \cdots + t_n X_n = 0$ for t_i in $\mathbb{R}$ not all zero.

(i) If $\{X_1, X_2, \cdots, X_n\}$ is independent then $t_1 X_1 + t_2 X_2 + \cdots + t_n X_n = 0$ for some t_i in $\mathbb{R}$.

8. Suppose that $\vec{u}$, $\vec{v}$, and $\vec{w}$ are nonzero vectors in $\mathbb{R}^3$ with the property that $\vec{u} \bullet \vec{v} = \vec{u} \bullet \vec{w} = \vec{v} \bullet \vec{w} = 0$. Show that $\{\vec{u}, \vec{v}, \vec{w}\}$ is linearly independent.

9. Let $\{X, Y, Z, W\}$ be an independent set in $\mathbb{R}^n$. Which of the following sets is independent? Support your answer.

   (a) $\{X - Y, Y - Z, Z - X\}$

   (b) $\{X + Y, Y + Z, Z + X\}$

   (c) $\{X - Y, Y - Z, Z - W, W - X\}$

   (d) $\{X + Y, Y + Z, Z + W, W + X\}$

10. If A is an $n \times n$ matrix, show that $det A = 0$ if and only if some column of A is a linear combination of the other columns.

11. Let A be any $m \times n$ matrix, and let $B_1, B_2, B_3, \cdots, B_k$ be columns in $\mathbb{R}^m$ such that the system $AX_i = B_i$ has a solution X_i for each i. If $\{B_1, B_2, B_3, \cdots, B_k\}$ is independent in $\mathbb{R}^m$, show that $\{X_1, X_2, X_3, \cdots, X_k\}$ is independent in $\mathbb{R}^n$.

12. If $\{X_1, X_2, X_3, \cdots, X_k\}$ is independent, show that $\{X_1, X_1 + X_2, X_1 + X_2 + X_3, \cdots, X_1 + X_2 + \cdots + X_k\}$ is also independent.

13. If $\{Y, X_1, X_2, X_3, \cdots, X_k\}$ is independent, show that $\{Y + X_1, Y + X_2, Y + X_3, \cdots, Y + X_k\}$ is also independent.

14. Suppose $\{X_1, X_2, X_3, \cdots, X_k\}$ is independent in $\mathbb{R}^n$. If Y is not in $span\{X_1, X_2, X_3, \cdots, X_k\}$, show that $\{Y, X_1, X_2, X_3, \cdots, X_k\}$ is independent.

15. (a) If $\{X_1, X_2, X_3, X_4, X_5, X_6\}$ is an independent set of vectors, show that the subset $\{X_2, X_3, X_5\}$ is also independent.

    (b) If $\mathcal{X}$ is any independent set in $\mathbb{R}^n$, show that any nonempty subset $\mathcal{Y} \subseteq \mathcal{X}$ is also independent.

16. (a) If $\{X_1, X_2, X_3, X_4\}$ is a dependent set of vectors, show that any superset $\{X_1, X_2, X_3, X_4, Y, Z\}$ is also dependent.

    (b) If $\mathcal{X}$ is any dependent set in $\mathbb{R}^n$, show that any superset $\mathcal{Y} \supseteq \mathcal{X}$ is also dependent.

4.3 DIMENSION

Consider the following four classes of subspaces of $\mathbb{R}^3$:

4.3. DIMENSION

The zero subspace $\{0\}$.
The lines through the origin.
The planes through the origin.
The space $\mathbb{R}^3$ itself.

It is common geometrical language to say that $\mathbb{R}^3$ is 3-dimensional, that planes are all 2-dimensional, and that lines are all 1-dimensional. Thus the idea of "dimension" gives a measure of the "size" of a subspace of $\mathbb{R}^3$. In this section we show how to define the dimension of any subspace of $\mathbb{R}^n$ and, as a byproduct, show that the above list contains *all* the subspaces of $\mathbb{R}^3$. In doing so, we introduce one of the most important concepts in linear algebra, the idea of a basis of a subspace.

4.3.1 Fundamental Theorem

The notion of a basis rests on the following remarkable relationship between the sizes of spanning and independent sets: *The number of independent vectors in a subspace U cannot exceed the number of vectors in a spanning set for U.* The importance of this theorem is difficult to exaggerate.

Theorem 1. Fundamental Theorem. *Let U be a subspace of $\mathbb{R}^n$.*

If U is spanned by m vectors and U contains k linearly independent vectors, then $k \leq m$.

Because of its importance, we give two proofs of this theorem at the end of this section.

As its name suggests, the Fundamental Theorem has a wide variety of applications. Since $\mathbb{R}^n$ can be spanned by the n columns of the identity matrix (Example 10 §4.1), taking $U = \mathbb{R}^n$ in Theorem 1 gives:

Corollary. *No linearly independent set in $\mathbb{R}^n$ can contain more than n vectors.*

Our main use of Theorem 1 depends on the following concept. If U is a subspace of $\mathbb{R}^n$, a set $\{X_1, X_2, \cdots, X_k\}$ of vectors in U is called a **basis** of U if:

1. $\{X_1, X_2, \cdots, X_k\}$ is linearly independent.
2. $U = span\{X_1, X_2, \cdots, X_k\}$.

Possibly the most important consequence of the Fundamental Theorem is the following important result: Any two bases[2] of a subspace must have the same number of elements.

---
[2] The plural of "basis" is "bases".

Theorem 2. Invariance Theorem. *If* $\{X_1, X_2, \cdots, X_k\}$ *and* $\{Y_1, Y_2, \cdots, Y_m\}$ *are two bases of a subspace* U *of* $\mathbb{R}^n$, *then* $k = m$.

Proof. We have $k \leq m$ by the Fundamental Theorem because $\{X_1, X_2, \cdots, X_k\}$ is independent in U and $\{Y_1, Y_2, \cdots, Y_m\}$ spans U. Similarly, $m \leq k$, whence $m = k$. □

The whole point of Theorem 2 is that it allows us to make the following definition: If U is a subspace of $\mathbb{R}^n$, the number of vectors in a basis of U is called the **dimension** of U, and is denoted $dim U$. The remarkable thing about the dimension is that (by Theorem 2) it can be determined by counting the number of vectors in *any* basis of U. Because of this, the dimension is called an **invariant** of the subspace U.

If $\{E_1, E_2, \cdots, E_n\}$ is the set of columns of the $n \times n$ identity matrix, then $\mathbb{R}^n = span\{E_1, E_2, \cdots, E_n\}$ by Example 10 §4.1, and $\{E_1, E_2, \cdots, E_n\}$ is linearly independent by Example 2 §4.2. Hence $\{E_1, E_2, \cdots, E_n\}$ is a basis of $\mathbb{R}^n$, called the **standard basis** of $\mathbb{R}^n$. In particular, this shows:

Example 1. $dim(\mathbb{R}^n) = n$ for each $n \geq 1$, and $\{E_1, E_2, \cdots, E_n\}$ is a basis .

For $n = 2$ and $n = 3$ this is compatible with our sense that the Euclidean plane $\mathbb{R}^2$ is two-dimensional and Euclidean space $\mathbb{R}^3$ is three-dimensional. Note that if $n = 1$ it says that $dim \mathbb{R} = 1$. In other words, the Euclidean line $\mathbb{R}$ has dimension 1, and the standard basis is $\{1\}$.

Returning to subspaces of $\mathbb{R}^n$, we define the dimension of the zero subspace $\{0\}$ to be zero:
$$dim\{0\} = 0.$$
This amounts to saying that $\{0\}$ has an empty basis, that is a basis containing no vectors. (This makes sense because the zero vector 0 cannot belong to *any* linearly independent set by Example 5 §4.2.)

Example 2. *Show that the lines through the origin in* $\mathbb{R}^3$ *are precisely the subspaces of dimension* 1.

Solution. If L is the line through the origin with direction vector $\vec{d}$, then $L = \{t\vec{d} \mid t \text{ in } \mathbb{R}\} = span\{\vec{d}\}$. Since $\vec{d} \neq \vec{0}$, the set $\{\vec{d}\}$ is linearly independent, and so is a basis of L. Thus $dim L = 1$. Conversely, if $U \subseteq \mathbb{R}^3$ is a subspace of dimension 1, let $\{\vec{v}\}$ be a basis. Then $U = span\{\vec{v}\} = \{t\vec{v} \mid t \text{ in } \mathbb{R}\}$, and so U is the line through the origin with direction vector $\vec{v}$. □

We will see below (Example 8) that the planes through the origin are precisely the subspaces of $\mathbb{R}^3$ of dimension 2.

Note that the argument in Example 2 also shows more generally that the 1-dimensional subspaces U of $\mathbb{R}^n$ are precisely the subspaces of the form $U = span\{X\} = \{tX \mid t \text{ in } \mathbb{R}\}$ for some $X \neq 0$ in $\mathbb{R}^n$.

4.3. DIMENSION

When working with a subspace U, it is often convenient to simplify a basis by multiplying some of the basis vectors by nonzero scalars. The next example shows that this will always produce another basis.

Example 3. *Assume that $\{X_1, X_2, \cdots, X_k\}$ is a basis of a subspace U of $\mathbb{R}^n$. If $a_1, a_2, \cdots, a_k$ are nonzero scalars, show that $\{a_1X_1, a_2X_2, \cdots, a_kX_k\}$ is also a basis of U.*

Solution. Suppose that a linear combination of the new set of vectors vanishes:
$$t_1(a_1X_1) + t_2(a_2X_2) + \cdots + t_k(a_kX_k) = 0.$$
Then $t_1a_1 = t_2a_2 = \cdots = t_ka_k = 0$ because $\{X_1, X_2, \cdots, X_k\}$ is independent. Since each $a_i \neq 0$, this forces $t_1 = t_2 = \cdots = t_k = 0$. Hence $\{a_1X_1, a_2X_2, \cdots, a_kX_k\}$ is linearly independent.

It remains to show that $U = span\{a_1X_1, a_2X_2, \cdots, a_kX_k\}$. Write $V = span\{a_1X_1, a_2X_2, \cdots, a_kX_k\}$ for convenience. Then $V \subseteq U$ by Theorem 1 §4.1 because each a_iX_i is in U. Moreover, the same theorem shows that $U \subseteq V$ because $X_i = \frac{1}{a_i}(a_iX_i)$ is in V for each i. Hence $U = V$ and we are done. $\square$

4.3.2 Existence of Bases

In order to calculate the dimension of a nonzero subspace U of $\mathbb{R}^n$ it is necessary to find a basis of U. However, at this point we do not know whether U *has* a basis. In fact it does, and one way to verify this is to find a way to create larger and larger linearly independent sets in U until a basis is reached. The next result contains the key idea.

Lemma 1. Independent Lemma. *Suppose $\{X_1, X_2, \cdots, X_k\}$ is a linearly independent set of vectors in $\mathbb{R}^n$. If Y is any vector of $\mathbb{R}^n$ which is* not *in $span\{X_1, X_2, \cdots, X_k\}$, then the larger set $\{Y, X_1, X_2, \cdots, X_k\}$ is also linearly independent.*

Proof. Suppose that
$$tY + t_1X_1 + t_2X_2 + \cdots + t_kX_k = 0 \qquad (*)$$
is a linear combination from $\{Y, X_1, X_2, \cdots, X_k\}$ that vanishes; we must show that $t = t_1 = t_2 = \cdots = t_k = 0$. First we have $t = 0$ because, otherwise, $Y = \frac{-1}{t}(t_1X_1 + t_2X_2 + \cdots + t_kX_k)$ is in $span\{X_1, X_2, \cdots, X_k\}$, contrary to our assumption. Hence $t = 0$ and (*) becomes $t_1X_1 + t_2X_2 + \cdots + t_kX_k = 0$. But then $t_1 = t_2 = \cdots = t_k = 0$ because $\{X_1, X_2, \cdots, X_k\}$ is independent. $\square$

The Independent Lemma gives a geometrical interpretation of what it means for three vectors $\vec{u}$, $\vec{v}$ and $\vec{w}$ in $\mathbb{R}^3$ to be linearly independent.

Example 4. Observe first that if $\{\vec{u}, \vec{v}, \vec{w}\}$ is any linearly independent set of vectors, then $\{\vec{v}, \vec{w}\}$ must also be linearly independent (Exercise 15 §4.2). Hence $\vec{v}$ and $\vec{w}$ are nonparallel vectors in $\mathbb{R}^3$ as in the diagram. Then $span\{\vec{v}, \vec{w}\}$ is the plane through the origin containing $\vec{v}$ and $\vec{w}$ (by Theorem 4 §4.2). Hence the Independent Lemma asserts that if $\vec{u}$ is any vector *not* in the plane containing $\vec{v}$ and $\vec{w}$, then $\{\vec{u}, \vec{v}, \vec{w}\}$ is a linearly independent set of vectors.

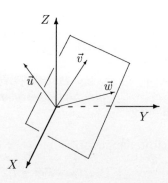

With the Independent Lemma in hand, we can show that every nonzero subspace of $\mathbb{R}^n$ does indeed have a basis.

Theorem 3. *Let $U \neq \{0\}$ be any nonzero subspace of $\mathbb{R}^n$.*

(1) *U has a basis, and $dim U \leq n$.*

(2) *Any linearly independent subset of U can be enlarged to a basis of U.*

(3) *Any spanning set of U contains a basis of U.*

Proof. We first prove (2), then (1) and (3).

(2). Suppose that $\{X_1, X_2, \cdots, X_k\}$ is any independent subset of U; we must show that it is part of a basis of U. If $U = span\{X_1, X_2, \cdots, X_k\}$ we are done because $\{X_1, X_2, \cdots, X_k\}$ is itself a basis. Otherwise choose a vector X_{k+1} in U which is not in $span\{X_1, X_2, \cdots, X_k\}$. Then $\{X_1, X_2, \cdots, X_k, X_{k+1}\}$ is a larger independent subset of U by the Independent Lemma. Now repeat the process. If $U = span\{X_1, X_2, \cdots, X_k, X_{k+1}\}$ we are done. Otherwise enlarge it again. Since the Corollary of the Fundamental Theorem shows that U contains no independent set with more than n vectors, this enlarging process cannot continue indefinitely. So at some stage we reach an independent set which spans U, that is we reach a basis of U. This proves (2).

(1). As $U \neq \{0\}$, it contains at least one linearly independent subset (for example $\{X\}$ where X is any nonzero vector in U). Hence U has a basis by (2), and this basis cannot contain more than n vectors by the Corollary of the Fundamental Theorem. In other words, $dim U \leq n$.

(3). Let $U = span\{X_1, X_2, \cdots, X_k\}$. If $\{X_1, X_2, \cdots, X_k\}$ is independent, it is a basis of U and we are done. Otherwise one of the X_i is a linear combination of the others by Theorem 3 §4.2. After possible relabeling, assume that X_1 is in $span\{X_2, X_3, \cdots, X_k\}$. Hence $U = span\{X_2, X_3, \cdots, X_k\}$ by Theorem 1 §4.1. Again, if $\{X_2, X_3, \cdots, X_k\}$ is independent, we are done. If not, continue the process. Either we are done at some stage or, after possible relabeling, we reach $U = span\{X_k\}$. But then $\{X_k\}$ is a basis because $X_k \neq 0$ (since $U \neq \{0\}$). □

4.3. DIMENSION

Example 5. *If $\{C_1, C_2, \cdots, C_k\}$ is a linearly independent set of columns in $\mathbb{R}^n$, show that the C_i are the first k columns of some invertible $n \times n$ matrix.*

Solution. Since $\{C_1, C_2, \cdots, C_k\}$ is independent in $\mathbb{R}^n$, (2) of Theorem 3 shows that it can be enlarged to a basis $\{C_1, C_2, \cdots, C_k, C_{k+1}, \cdots, C_n\}$ of $\mathbb{R}^n$. Hence the matrix $A = [C_1 \; C_2 \; \cdots \; C_k \; C_{k+1} \; \cdots \; C_n]$ is invertible by Theorem 2 §4.2, as required. □

Note that, apart from the Independent Lemma, Theorem 3 does not specify how to construct the new columns $C_{k+1}, \cdots, C_n$ in Example 5. In fact there are many choices. Theorem 3 asserts only the *existence* of such columns.

In the following sections we will make frequent use of the next result, each part of which is an easy consequence of Theorem 3.

Theorem 4. *Let U and V denote subspaces of $\mathbb{R}^n$.*

(1) *If $U \subseteq V$ then $dim U \leq dim V$.*

(2) *If $U \subseteq V$ and $dim U = dim V$, then $U = V$.*

(3) *If $dim U = d$, then any set of d linearly independent vectors in U is automatically a basis of U.*

(4) *If $dim U = d$, then any spanning set for U containing d vectors is automatically a basis of U.*

Proof. (1). If $U = \{0\}$ then $dim U = 0 \leq dim V$. Otherwise, any basis of U is an independent set in V, and so (by the Fundamental Theorem) contains no more vectors than a basis of V. In other words, $dim U \leq dim V$.

(2). Write $d = dim U = dim V$, and let $\{X_1, X_2, \cdots, X_d\}$ be a basis of U. If $U \neq V$ choose Y in V such that Y is not in $U = span\{X_1, X_2, \cdots, X_d\}$. By the Independent Lemma the set $\{X_1, X_2, \cdots, X_d, Y\}$ is a linearly independent set in V containing $d+1$ vectors. But this contradicts the Fundamental Theorem because V has a spanning set (in fact a basis) of d vectors. Hence $U \neq V$ is impossible, so $U = V$ after all.

(3). Let $\{X_1, X_2, \cdots, X_d\}$ be an independent set in U; we must show that it spans U. Write $W = span\{X_1, X_2, \cdots, X_d\}$. Then $W \subseteq U$ by Theorem 1 §4.1 because each X_i is in U, and $dim W = d$ because $\{X_1, X_2, \cdots, X_d\}$ is a basis of W. Hence $W = U$ by (2). Thus $U = W = span\{X_1, X_2, \cdots, X_d\}$, so $\{X_1, X_2, \cdots, X_d\}$ is a basis of U.

(4). If $U = span\{X_1, X_2, \cdots, X_d\}$, then $\{X_1, X_2, \cdots, X_d\}$ contains a basis of U by Theorem 2. This basis contains $d = dim U$ vectors, and so must be all of $\{X_1, X_2, \cdots, X_d\}$. In other words, $\{X_1, X_2, \cdots, X_d\}$ is a basis of U. □

Parts (3) and (4) of Theorem 4 are "labor-saving" results since they show that, in a subspace U of dimension d, to verify that a set of d vectors is a basis it suffices to verify either independence or spanning; the other is then automatic. Of course it is necessary that the number of vectors in the set equals $dim U$.

Example 6. *Show that* $\{[2\ 0\ -1]^T, [1\ 4\ 9]^T, [-1\ 5\ 6]^T\}$ *is a basis of* $\mathbb{R}^3$.

Solution. By Theorem 4 it suffices to show that $\{[2\ 0\ -1]^T, [1\ 4\ 9]^T, [-1\ 5\ 6]^T\}$ is linearly independent (spanning would do as well). This follows from Theorem 2 §4.2 if we can show that the matrix A with these vectors as columns is invertible. But, using row operations as in Chapter 2,

$$det A = det \begin{bmatrix} 2 & 1 & -1 \\ 0 & 4 & 5 \\ -1 & 9 & 6 \end{bmatrix} = det \begin{bmatrix} 0 & 19 & 11 \\ 0 & 4 & 5 \\ -1 & 9 & 6 \end{bmatrix} = -51 \neq 0.$$

Hence A is indeed invertible. □

Part (2) of Theorem 4 is also a "labor-saving" result, and is frequently the easiest way to show that two subspaces of the same dimension are equal: Simply show that one is contained in the other. This is illustrated in the following example (where we give a different proof of part of Theorem 4 §4.2).

Example 7. *Let $\vec{v}$ and $\vec{w}$ be linearly independent (that is non-parallel) vectors in $\mathbb{R}^3$, and let $P = \text{span}\{\vec{v}, \vec{w}\}$ denote the plane spanned by $\vec{v}$ and $\vec{w}$. Show that P is the plane (through the origin) with normal $\vec{n} = \vec{v} \times \vec{w}$.*

Solution. Note first that $\vec{n} \neq \vec{0}$ by Theorem 5 §3.4 because $\vec{v}$ and $\vec{w}$ are not parallel. Let Q denote the plane through the origin with $\vec{n}$ as normal. Then Q has equation $ax + by + cz = 0$, that is

$$Q = \left\{[x\ y\ z]^T \mid ax + by + cz = 0\right\} = \{\vec{x} \mid \vec{n} \bullet \vec{x} = 0\}$$

where we denote $\vec{x} = [x\ y\ z]^T$. Since $\vec{n}$ is orthogonal to both $\vec{v}$ and $\vec{w}$ (Theorem 2 §3.4), we have $\vec{n} \bullet \vec{v} = 0$ and $\vec{n} \bullet \vec{w} = 0$. This means that both $\vec{v}$ and $\vec{w}$ are in Q, and so that $P \subseteq Q \subseteq \mathbb{R}^3$. Since $dim P = 2$, Theorem 4 gives

$$dim P \leq dim Q \leq dim \mathbb{R}^3, \text{ that is } 2 \leq dim Q \leq 3.$$

Now observe that $dim Q$ is an integer, so this means that $dim Q = 2$ or $dim Q = 3$, whence $Q = P$ or $Q = \mathbb{R}^3$ by (2) of Theorem 4. But $Q \neq \mathbb{R}^3$ because $\vec{n}$ is not in Q (as $\vec{n} \neq \vec{0}$), so $Q = P$ as required. □

We conclude with a characterization (mentioned earlier) of all subspaces of $\mathbb{R}^3$.

Example 8. *Show that the only subspaces of $\mathbb{R}^3$ are the zero space $\{0\}$, lines through the origin, planes through the origin, and $\mathbb{R}^3$ itself.*

Solution. Let U denote a subspace of $\mathbb{R}^3$. Then $dim U \leq dim(\mathbb{R}^3) = 3$ so, since $dim U$ is a non-negative integer, $dim U$ is one of 0, 1, 2 or 3. We examine each case separately:

4.3. DIMENSION

$dimU = 0$.   Then $U = \{0\}$.

$dimU = 1$.   If $\{\vec{d}\}$ is a basis of U, then $U = span\{\vec{d}\} = \{t\vec{d} \mid t$ a real number$\}$ is the line through the origin with direction vector $\vec{d}$.

$dimU = 2$.   If $\{\vec{v}, \vec{w}\}$ is a basis of U, then U is the plane through the origin with normal $\vec{v} \times \vec{w}$ containing $\vec{v}$ and $\vec{w}$ by Example 7 (or Theorem 4 §4).

$dimU = 3$.   Since $U \subseteq \mathbb{R}^3$ and $dimU = 3 = dim(\mathbb{R}^3)$, we have $U = \mathbb{R}^3$ by (2) of Theorem 4.   □

4.3.3 Proof of the Fundamental Theorem

If U is a subspace of $\mathbb{R}^n$ which is spanned by m vectors, and if U contains k linearly independent vectors, we must show that $k \leq m$. Let $U = span\{X_1, X_2, \cdots, X_m\}$, and let $\{Y_1, Y_2, \cdots, Y_k\}$ be an independent set in U.

Proof 1. Since each Y_j is in $span\{X_1, X_2, \cdots X_m\}$, it is a linear combination of $X_1, X_2, \cdots, X_m$, say[3]

$$Y_j = a_{1j}X_1 + a_{2j}X_2 + \cdots + a_{mj}X_m = \Sigma_{i=1}^m a_{ij}X_i \text{ for each } j = 1, 2, \cdots, k.$$

The coefficients here form a $m \times k$ matrix $A = [a_{ij}]$. We must show that $k \leq m$; we do so by showing that $k > m$ leads to a contradiction. Indeed, if $k > m$ the system $AX = 0$ has a nontrivial solution $X = [x_1\ x_2\ \cdots\ x_k]^T \neq 0$ by Theorem 1 §1.3. Using the entries x_j of X as coefficients, we compute the following linear combination of the Y_j :

$$\begin{aligned}\Sigma_{j=1}^k x_j Y_j &= \Sigma_{j=1}^k x_j (\Sigma_{i=1}^m a_{ij} X_i) \\ &= \Sigma_{j=1}^k \Sigma_{i=1}^m a_{ij} x_j X_i \\ &= \Sigma_{i=1}^m (\Sigma_{j=1}^k a_{ij} x_j) X_i \\ &= \Sigma_{i=1}^m (0) X_i = 0 \end{aligned}$$

where the sum $\Sigma_{j=1}^k a_{ij} x_j = 0$ for each i because it is entry i of $AX = 0$. Thus $\Sigma_{j=1}^k x_j Y_j = 0$, a contradiction because the Y_j are independent and not all the x_j are zero (since $X \neq 0$).   □

Proof 2. We assume that $k > m$ and show that this leads to a contradiction. As $U = span\{X_1, X_2, \cdots, X_m\}$, let $Y_1 = a_1X_1 + a_2X_2 + \cdots + a_mX_m$. As $Y_1 \neq 0$ not all of the a_i are zero, say $a_1 \neq 0$ (after relabeling the X_i). Then $U = span\{Y_1, X_2, X_3, \cdots, X_m\}$ as the reader can verify. Hence write $Y_2 = b_1Y_1 + c_2X_2 + c_3X_3 + \cdots + c_mX_m$. Then some $c_i \neq 0$ because $\{Y_1, Y_2\}$ is independent so, as before, $U = span\{Y_1, Y_2, X_3, \cdots, X_m\}$ after possible relabeling of the X_i. As $k > m$ this procedure continues until all the vectors $X_1, X_2, \cdots, X_m$ are replaced by $Y_1, Y_2, \cdots, Y_m$. In particular $U = span\{Y_1, Y_2, \cdots, Y_m\}$. But then

---

[3] We again use summation notation for convenience. For example $a_1 + a_2 + a_3 + a_4 = \Sigma_{i=1}^4 a_i$ and $a_3X_3 + a_4X_4 + a_5X_5 = \Sigma_{i=3}^5 a_iX_i$.

Y_{m+1} is a linear combination of $Y_1, Y_2, \cdots, Y_m$ contrary to the independence of the Y_j. □

If $U = span\{X_1, X_2, \cdots, X_m\}$, and if $\{Y_1, Y_2, \cdots, Y_k\}$ is independent in U, Proof 2 shows not only that $k \leq m$ but also that k of the (spanning) vectors $X_1, X_2, \cdots, X_m$ can be replaced by the (independent) vectors $Y_1, Y_2, \cdots, Y_k$ and the resulting set will still span U. In this form the result is called the **Steinitz Exchange Lemma**.

Exercises

1. The following sets of vectors are all bases of the subspace they span. Give the dimension of each subspace.
   (a) $\{[8 \ 4 \ 0]^T, [-7 \ 7 \ 3]^T\}$.
   (b) $\{[3 \ 0 \ -5]^T, [2 \ -1 \ 0]^T, [-2 \ 4 \ -9]^T\}$.
   (c) $\{[-9 \ 0 \ -1 \ 6]^T, [5 \ 7 \ -3 \ -2]^T\}$.
   (d) $\{[-1 \ 6 \ 2 \ 6]^T, [-2 \ 4 \ 0 \ -4]^T, [-6 \ 5 \ 0 \ 1]^T\}$.
   (e) $\{[1 \ 0 \ 0 \ 0]^T, [0 \ 0 \ 1 \ 0]^T, [0 \ 1 \ 0 \ 0]^T, [0 \ 0 \ 0 \ 1]^T\}$.

2. Show that the following sets of vectors are bases of the indicated space.
   (a) $\{[1 \ 1 \ 0]^T, [1 \ 0 \ 1]^T, [0 \ 1 \ 1]^T\}$ in $\mathbb{R}^3$.
   (b) $\{[-1 \ 1 \ 1]^T, [1 \ -1 \ 1]^T, [1 \ 1 \ -1]^T\}$ in $\mathbb{R}^3$.
   (c) $\{[1 \ 2 \ 1 \ 2]^T, [-1 \ 0 \ 2 \ 1]^T, [0 \ 0 \ 1 \ 1]^T, [0 \ 1 \ -1 \ 1]^T\}$ in $\mathbb{R}^4$.
   (d) $\{[1 \ 1 \ 0 \ 0]^T, [1 \ 0 \ 0 \ 1]^T, [0 \ 1 \ 1 \ 0]^T, [0 \ 1 \ 0 \ 1]^T\}$ in $\mathbb{R}^4$.

3. Verify that the following sets of vectors span the given space. Could they form a basis for the space? Support your answer.
   (a) $\{[-5 \ -2 \ -2]^T, [7 \ -9 \ 3]^T, [4 \ -8 \ 9]^T, [8 \ 4 \ 7]^T\}$ in $\mathbb{R}^3$.
   (b) $\{[2 \ 3 \ 5 \ -4]^T, [6 \ 1 \ -4 \ -7]^T, [3 \ 6 \ 3 \ 2]^T, [1 \ 0 \ 0 \ 0]^T,$
   $[6 \ -3 \ -2 \ 0]^T, [7 \ -4 \ 2 \ 9]^T\}$ in $\mathbb{R}^4$.
   (c) $\{[-4 \ 3 \ 2]^T, [-1 \ 1 \ 11]^T, [-3 \ 2 \ -9]^T\}$ in $\mathbb{R}^3$.
   (d) $\{[1 \ -6 \ 3 \ -7]^T, [5 \ 1 \ 5 \ 9]^T, [3 \ 9 \ 0 \ 2]^T, [2 \ 0 \ 3 \ 0]^T\}$ in $\mathbb{R}^4$.

4. The following pairs of vector sets claim to be two bases of the same subspace. Can you determine if this is false without any calculations?
   (a) $\{[3 \ -1]\}$ and $\{[2 \ 0], [-5 \ 3]\}$ of a subspace in $\mathbb{R}^2$.
   (b) $\{[6 \ 0 \ -9], [-4 \ 2 \ 3], [-3 \ 8 \ 7]\}$ and $\{[6 \ 4 \ 3], [2 \ 9 \ -1]\}$ of a subspace in $\mathbb{R}^3$.
   (c) $\{[2 \ 1 \ 0], [-3 \ -7 \ 3]\}$ and $\{[1 \ 6 \ -3], [5 \ 8 \ -3]\}$ of a subspace in $\mathbb{R}^3$.

(d) $\{[6\ -2\ -8\ 0], [7\ 0\ 9\ 0]\}$ and $\{[-2\ -5\ -4\ 6],$
$[6\ 3\ 7\ -8], [9\ 9\ 0\ 2], [-7\ -1\ 0\ 9]\}$ of a subspace in $\mathbb{R}^4$.

5. Generate the standard basis for the given space.
   (a) $\mathbb{R}^2$. (b) $\mathbb{R}^3$. (c) $\mathbb{R}^5$. (d) $\mathbb{R}$.

6. Find a basis and calculate the dimension of the following subspaces of $\mathbb{R}^4$. Can you find more than one basis for each subspace?
   (a) $span\{[1\ -1\ 2\ 0]^T, [2\ 3\ 0\ 3]^T, [1\ 9\ -6\ 6]^T\}$.
   (b) $span\{[2\ 1\ 0\ -1]^T, [-1\ 1\ 1\ 2]^T, [2\ 7\ 4\ 5]^T\}$.
   (c) $span\{[-1\ 2\ 1\ 0]^T, [2\ 0\ 3\ -1]^T, [4\ 4\ 11\ -3]^T, [3\ -2\ 2\ -1]^T\}$.
   (d) $span\{[-2\ 0\ 3\ 1]^T, [1\ 2\ -1\ 0]^T, [-2\ 8\ 5\ 3]^T, [1\ 2\ 2\ 1]^T\}$.

7. Find a basis and calculate the dimension of the following subspaces of $\mathbb{R}^4$.
   (a) $U = \{[a\ a+b\ a-b\ b]^T \mid a$ and b in $\mathbb{R}\}$.
   (b) $U = \{[a+b\ a-b\ b\ a]^T \mid a$ and b in $\mathbb{R}\}$.
   (c) $U = \{[a\ b\ c+a\ c]^T \mid a, b$ and c in $\mathbb{R}\}$.
   (d) $U = \{[a-b\ b+c\ a\ b+c]^T \mid a, b$ and c in $\mathbb{R}\}$.
   (e) $U = \{[a\ b\ c\ d]^T \mid a+b-c+d = 0$ in $\mathbb{R}\}$.

8. In each case find a basis for and calculate the dimension of $null\, A$:
   (a) $A = \begin{bmatrix} 1 & 2 & -3 & 4 & 0 \\ -1 & 0 & 2 & -4 & 1 \\ -1 & 4 & 0 & -4 & 3 \end{bmatrix}$
   (b) $A = \begin{bmatrix} 1 & 1 & -2 & 0 & 4 \\ 2 & 1 & 0 & 1 & -5 \\ 7 & 2 & 6 & -10 & -7 \end{bmatrix}$

9. In each case show that U is a subspace of $\mathbb{R}^4$, find a basis of U and calculate the dimension of U.
   (a) $U = \{X$ in $\mathbb{R}^4 \mid X^T A = 0\}$ where $A = \begin{bmatrix} 1 & 1 & 0 & 3 \\ -1 & 2 & 2 & -1 \end{bmatrix}^T$.
   (b) $U = \{X$ in $\mathbb{R}^4 \mid X^T A = 0\}$ where $A = \begin{bmatrix} 0 & 1 & 2 & -1 \\ 1 & -1 & 0 & 3 \end{bmatrix}^T$.
   (c) $U = \{X$ in $\mathbb{R}^4 \mid X^T A = (BX)^T\}$ where $A = \begin{bmatrix} 1 & 0 & 1 & 0 \\ 0 & 1 & 0 & 1 \end{bmatrix}^T$ and $B = \begin{bmatrix} 1 & 2 & 3 & 4 \\ 0 & 1 & 1 & -1 \end{bmatrix}$.
   (d) $U = \{X$ in $\mathbb{R}^4 \mid X^T A = (BX)^T\}$ where $A = \begin{bmatrix} 2 & 0 & 1 & -3 \\ -1 & 1 & 1 & 0 \end{bmatrix}^T$ and $B = \begin{bmatrix} 2 & 1 & 0 & -1 \\ 1 & 1 & -1 & -1 \end{bmatrix}$.

10. In each case either show that the statement is true or give an example showing that it is false.

    (a) Every set of four nonzero vectors in $\mathbb{R}^4$ is a basis.

    (b) No basis of $\mathbb{R}^3$ can contain a vector with a component 0.

    (c) $\mathbb{R}^3$ has a basis of the form $\{X, X+Y, Y\}$ where X and Y are vectors.

    (d) Every basis of $\mathbb{R}^5$ contains one column of I_5.

    (e) Every nonempty subset of a basis of $\mathbb{R}^3$ is again a basis of $\mathbb{R}^3$.

    (f) If $\{X_1, X_2, X_3, X_4\}$ and $\{Y_1, Y_2, Y_3, Y_4\}$ are bases of $\mathbb{R}^4$, then $\{X_1+Y_1, X_2+Y_2, X_3+Y_3, X_4+Y_4\}$ is also a basis of $\mathbb{R}^4$.

11. Given that $\{[3\ -1\ 2\ 1], [0\ 2\ -1\ 0], [4\ -1\ 3\ -3]\}$ is a basis for some subspace U of $\mathbb{R}^4$, is $\{[-6\ 2\ -4\ -2], [0\ -4\ 2\ 0], [-8\ 2\ -6\ 6]\}$ also a basis of U?

12. Show that every nonzero vector in $\mathbb{R}^n$ is in some basis of $\mathbb{R}^n$.

13. (a) Find a basis of $\mathbb{R}^4$ containing the vector $[11\ -19\ 203\ -131]^T$.

    (b) If X is *any* vector in $\mathbb{R}^4$, can you find a basis containing X? Defend your answer.

    (c) Find a condition on vectors X and Y in $\mathbb{R}^4$ such that there is a basis of $\mathbb{R}^4$ containing both X and Y if and only if the condition is satisfied. Defend your answer.

14. Let $\{X, Y, Z, W\}$ denote vectors in $\mathbb{R}^3$. Find a condition on this set of vectors such that it contains a basis of $\mathbb{R}^3$ if and only if the condition is satisfied. Defend your answer.

15. (a) Find a basis of vectors in $\mathbb{R}^3$ consisting of vectors whose components sum to 1.

    (b) Can you find a basis as in (a) if we ask that the components sum to 0? Defend your answer.

16. Suppose that $\{X, Y, Z, W\}$ is a basis of $\mathbb{R}^4$. Show that:

    (a) $\{X + aW, Y, Z, W\}$ is also a basis of $\mathbb{R}^4$ for any choice of the scalar a.

    (b) $\{X + W, Y + W, Z + W, W\}$ is also a basis of $\mathbb{R}^4$.

    (c) $\{X, X + Y, X + Y + Z, X + Y + Z + W\}$ is also a basis of $\mathbb{R}^4$.

17. Suppose that the columns $\{X, Y\}$ constitute a basis of $\mathbb{R}^2$, and let $A = \begin{bmatrix} a & c \\ b & d \end{bmatrix}$ denote a matrix.

    (a) If A is invertible, show that $\{aX + bY, cX + dY\}$ is a basis of $\mathbb{R}^2$.

    (b) If $\{aX + bY, cX + dY\}$ is a basis of $\mathbb{R}^2$, show that A is invertible.

4.4. RANK

18. Suppose that $\{X_1, X_2, X_3, \cdots, X_n\}$ is a basis of $\mathbb{R}^n$. If A is an invertible matrix, show that $\{AX_1, AX_2, AX_3, \cdots, AX_n\}$ is also a basis of $\mathbb{R}^n$.

19. Let U and W denote subspaces of $\mathbb{R}^n$, and assume that $U \subseteq W$. If $dimW = 1$, show that either $U = \{0\}$ or $U = W$.

20. Let U and W denote subspaces of $\mathbb{R}^n$, and assume that $U \subseteq W$. If $dimU = n - 1$, show that either $W = U$ or $W = \mathbb{R}^n$.

21. Let U and W denote subspaces of $\mathbb{R}^n$, and suppose that $dimU = 2$. Show that either $U \subseteq W$ or $dim(U \cap W) \leq 1$. (See Exercise 23 §4.1.) [Hint: $U \cap W$ is a subspace of U. Use Theorem 4.]

22. Let A denote an $m \times n$ matrix. Show that:

    (a) $dim(nullA) = dim(null(AV))$ for every invertible $n \times n$ matrix V.

    (b) $dim(imA) = dim(im(UA))$ for every invertible $m \times m$ matrix U.

23. Let U be a subspace of $\mathbb{R}^n$, and let $\{X_1, X_2, \cdots, X_k\}$ be a maximal independent subset of U (that is no independent subset of U contains more than k vectors). Show that $\{X_1, X_2, \cdots, X_k\}$ is a basis of U. [Hint: If $span\{X_1, X_2, \cdots, X_k\} \neq U$ choose Y in U which is not in $span\{X_1, X_2, \cdots, X_k\}$, and apply Lemma 1.]

24. Let $U \neq \{0\}$ be a subspace of $\mathbb{R}^n$, and let $\{X_1, X_2, \cdots, X_k\}$ be a minimal spanning subset of U (that is $U = span\{X_1, X_2, \cdots, X_k\}$ and no spanning set of U contains fewer than k vectors). Show that $\{X_1, X_2, \cdots, X_k\}$ is a basis of U. [Hint: If $\{X_1, X_2, \cdots, X_k\}$ is not independent, apply Theorem 3 §4.2.]

25. Let U and W be any two subspaces of $\mathbb{R}^n$, and consider the subspaces
    $U \cap W = \{X \text{ in } \mathbb{R}^n \mid X \text{ is in both } U \text{ and } W\}$
    $U + W = \{X \text{ in } \mathbb{R}^n \mid X \text{ is a sum of a vector in } U \text{ and a vector in } W\}$

    as in Exercise 23 §4.1. Show that $dim(U+W) = dimU + dimW - dim(U \cap W)$. [Hint: Let $\{X_1, \cdots, X_d\}$ be a basis of $U \cap W$, and by Theorem 3, extend it to bases $\{X_1, \cdots, X_d, Y_1, \cdots, Y_k\}$ and $\{X_1, \cdots, X_d, Z_1, \cdots, Z_m\}$ of U and W respectively. Show that $\{X_1, \cdots, X_d, Y_1, \cdots, Y_k, Z_1, \cdots, Z_m\}$ is a basis of $U + W$.]

4.4 RANK

There are a number of subspaces associated with an $m \times n$ matrix A which are important for the analysis of A. In this section we will exhibit bases of these spaces, calculate their dimensions, and relate these dimensions to the rank of the matrix A.

4.4.1 Row and Column Spaces

If A is an $m \times n$ matrix, we have already discussed the null space $null A$ and the image $im A$. We are going to find natural bases for both these spaces, and so calculate their dimension. To do this, and for other reasons as well, it is essential to consider two other subspaces associated with A. They are defined as follows:

> The **column space** $colA$ of an $m \times n$ matrix A is the subspace of $\mathbb{R}^m$ spanned by the columns of A.
>
> The **row space** $rowA$ of an $m \times n$ matrix A is the subspace of $\mathbb{R}^n$ spanned by the rows of A.

Observe that in discussing $rowA$ we are regarding the elements of $\mathbb{R}^n$ as rows.

One reason for the importance of the spaces $rowA$ and $colA$ is that they are unchanged when row (respectively column) operations are performed on the matrix A.

Lemma 1. *Let A denote an $m \times n$ matrix.*

(1) *If $A \to B$ using a sequence of row operations, then $rowB = rowA$.*

(2) *If $A \to B$ using a sequence of column operations, then $colB = colA$.*

Proof. (1). It is enough to prove it for a single row operation (since B is obtained from A by a sequence of row operations). We do this for a row operation of type III; a similar argument works for the other types of row operations.

Hence, suppose that B is obtained by adding u times row p of A to row q. Then each row of B is a linear combination of the rows of A, so $rowB \subseteq rowA$ by Theorem 1 §4.1. On the other hand we can carry $B \to A$ by *subtracting* u times row p of B from row q. Hence $rowA \subseteq rowB$ follows in the same way. Thus $rowA = rowB$ as required.

(2). This is proved in the same way using column operations instead of row operations. □

In particular, suppose $A \to R$ by row operations where R is a row-echelon matrix, and suppose further that R has k leading 1's. Then R has k nonzero rows, say $Y_1, Y_2, \cdots, Y_k$, so $rowR = span\{Y_1, Y_2, \cdots, Y_k\}$. Moreover $\{Y_1, Y_2, \cdots, Y_k\}$ is linearly independent by Example 7 §4.2 so it is a basis of $rowR$. Since $rowR = rowA$ by Lemma 1, we have

$$k = dim(rowR) = dim(rowA).$$

Hence the number k of leading 1's is the same *no matter how A is carried to row-echelon form*[4]. We express this by saying that the number of leading 1's is

---
[4] This was asserted without proof in Section 1.2.

4.4. RANK

an *invariant* of A, called the **rank** of A, and denoted $rank A$. This is consistent with the usage in Section 1.2, and the above discussion is summarized in the following theorem.

Theorem 1. *Let A be an $m \times n$ matrix. Then*
$$rank A = dim(row A).$$
Moreover, if $A \to R$ by row operations where R is a row-echelon matrix, the nonzero rows of R are a basis of $row A$.

Theorem 1 has an additional virtue: It provides a routine method for finding a basis for a subspace U of $\mathbb{R}^n$ when a spanning set is given. Indeed, if $U = span\{X_1, X_2, \cdots, X_k\}$ where the X_i are written as rows, simply carry the matrix with the X_i as its rows to row-echelon form R. Then the nonzero rows of R are a basis of U.

Example 1. *Find a basis of $U = span\{[1 \ 1 \ -2 \ 4], [2 \ 5 \ 4 \ -2], [1 \ 7 \ 14 \ -16]\}$.*

Solution. Observe that $U = row A$ where $A = \begin{bmatrix} 1 & 1 & -2 & 4 \\ 2 & 5 & 4 & -2 \\ 1 & 7 & 14 & -16 \end{bmatrix}$. The reduction of A to row-echelon form is

$$\begin{bmatrix} 1 & 1 & -2 & 4 \\ 2 & 5 & 4 & -2 \\ 1 & 7 & 14 & -16 \end{bmatrix} \to \begin{bmatrix} 1 & 1 & -2 & 4 \\ 0 & 3 & 8 & -10 \\ 0 & 6 & 16 & -20 \end{bmatrix}$$

$$\to \begin{bmatrix} 1 & 1 & -2 & 4 \\ 0 & 3 & 8 & -10 \\ 0 & 0 & 0 & 0 \end{bmatrix}$$

$$\to \begin{bmatrix} 1 & 1 & -2 & 4 \\ 0 & 1 & \frac{8}{3} & -\frac{10}{3} \\ 0 & 0 & 0 & 0 \end{bmatrix}.$$

Hence $\{[1 \ 1 \ -2 \ 4], [0 \ 1 \ \frac{8}{3} \ -\frac{10}{3}]\}$ is a basis of $row A = U$. We note that $\{[1 \ 1 \ -2 \ 4], [0 \ 3 \ 8 \ -10]\}$ is also a basis (see Example 3 §4.3), possibly more convenient. □

4.4.2 The Rank Theorem

Theorem 1 can also be used to find a basis of $col A$: Carry A^T to row-echelon form R and use the transposes of the nonzero rows of R. (Equivalently, carry A to "column-echelon form" C and use the nonzero columns of C.) However there is an easy way to select a basis of $col A$ directly from the columns of A. This is part of the following very important theorem: If A is any $m \times n$ matrix, the subspace $row\,A$ of $\mathbb{R}^n$ has the *same dimension* as the subspace $col\,A$ of $\mathbb{R}^m$.

Theorem 2. Rank Theorem. *Let A be an $m \times n$ matrix. Then*

$$dim(row\,A) = dim(col\,A) = rank\,A.$$

Moreover, if $A \to R$ by row operations where R is a row-echelon matrix, and if the leading 1's are in columns $j_1, j_2, \cdots, j_r$ of R, then the corresponding columns $j_1, j_2, \cdots, j_r$ of A are a basis of $col\,A$.

Proof. Since $r = rank\,A$, the displayed equations follow from the last sentence and Theorem 1. To prove the last sentence, write $A = [C_1\ C_2\ \cdots\ C_n]$ where C_j denotes column j of A. We use the fact that $R = UA$ for some invertible matrix U (Theorem 1 §1.6). Hence

$$\begin{aligned}R\ &= UA\ = U[C_1\ C_2\ \cdots\ C_n]\\ &= [UC_1\ UC_2\ \cdots\ UC_n]\end{aligned}$$

so UC_j is column j of R. Since R contains r leading 1's, let UC_{j_1}, UC_{j_2}, $\cdots$, UC_{j_r} be the columns of R that contain a leading 1. Then $\{UC_{j_1}, UC_{j_2}, \cdots, UC_{j_r}\}$ is a basis of $col\,R$ as the reader can verify. We use this to show that $\{C_{j_1}, C_{j_2}, \cdots, C_{j_r}\}$ is a basis of $col\,A$, as required.

Independence: Suppose $t_1 C_{j_1} + t_2 C_{j_2} + \cdots + t_r C_{j_r} = 0$ where each t_i is a real number. Then

$$0 = U0 = U(t_1 C_{j_1} + t_2 C_{j_2} + \cdots + t_r C_{j_r}) = t_1 UC_{j_1} + t_2 UC_{j_2} + \cdots + t_r UC_{j_r}.$$

Hence each $t_i = 0$ by the independence of $\{UC_{j_1}, UC_{j_2}, \cdots, UC_{j_r}\}$, and we have proved the independence of $\{C_{j_1}, C_{j_2}, \cdots, C_{j_r}\}$.

Spanning: Given X in $col\,A$, write $X = r_1 C_1 + r_2 C_2 + \cdots + r_n C_n$ where each r_j is a real number. Then $UX = r_1 UC_1 + r_2 UC_2 + \cdots + r_n UC_n$ is in $col\,R$ because each UC_j is in $col\,R$. But $\{UC_{j_1}, UC_{j_2}, \cdots, UC_{j_r}\}$ spans $col\,R$, so there exist $t_1, t_2, \cdots, t_r$ such that

$$UX = t_1 UC_{j_1} + t_2 UC_{j_2} + \cdots + t_r UC_{j_r}.$$

Hence $UX = U(t_1 C_{j_1} + t_2 C_{j_2} + \cdots + t_r C_{j_r})$, so left multiplication by U^{-1} gives $X = t_1 C_{j_1} + t_2 C_{j_2} + \cdots + t_r C_{j_r}$. Thus X is in $span\{C_{j_1}, C_{j_2}, \cdots, C_{j_r}\}$, as required. $\square$

Example 2. *If $A = \begin{bmatrix} 1 & -1 & 3 & -2 \\ 2 & -2 & 2 & -1 \\ -1 & 1 & 5 & -4 \end{bmatrix}$, find a basis for $row\,A$ and $col\,A$.*

4.4. RANK

Solution. One reduction of A to row-echelon form is as follows:

$$A = \begin{bmatrix} 1 & -1 & 3 & -2 \\ 2 & -2 & 2 & -1 \\ -1 & 1 & 5 & -4 \end{bmatrix}$$

$$\rightarrow \begin{bmatrix} 1 & -1 & 3 & -2 \\ 0 & 0 & -4 & 3 \\ 0 & 0 & 8 & -6 \end{bmatrix}$$

$$\rightarrow \begin{bmatrix} 1 & -1 & 3 & -2 \\ 0 & 0 & 1 & -\frac{3}{4} \\ 0 & 0 & 0 & 0 \end{bmatrix} = R.$$

Hence $\{[1\ -1\ 3\ -2], [0\ 0\ 1\ -\frac{3}{4}]\}$ is a basis of $rowA$ by Theorem 1. Since the leading ones are in columns 1 and 3 of the row-echelon matrix R, Theorem 2 shows that columns 1 and 3 of A constitute a basis $\left\{\begin{bmatrix} 1 \\ 2 \\ -1 \end{bmatrix}, \begin{bmatrix} 3 \\ 2 \\ 5 \end{bmatrix}\right\}$ of $colA$. Of course both bases contain $2 = rankA$ vectors as the Rank Theorem asserts. □

The Rank Theorem has many consequences. First, if A is an $m \times n$ matrix, then $rowA$ is spanned by the m rows of A, whence $rankA = dim(rowA) \leq m$ because the dimension of a subspace cannot exceed the number of vectors in any spanning set. Similarly $rankA \leq n$ because $rankA = dim(colA)$. We record this as

Corollary 1. *If A is an $m \times n$ matrix then $rankA \leq m$ and $rankA \leq n$.*

If A is an $n \times n$ matrix, then $rowA$ is spanned by n rows, so $dim(rowA) = n$ if and only if the rows are independent (using Theorem 4 §4.3). But the rows are independent if and only if A is invertible (by Theorem 2 §4.2), so we have

Corollary 2. *An $n \times n$ matrix A is invertible if and only if $rankA = n$.*

The fact that the columns of A are just the transposes of the rows of A^T means that $dim(colA) = dim(row(A^T))$, that is

Corollary 3. $rankA^T = rankA$ *for any matrix A.*

Corollary 4. *If A is an $m \times n$ matrix then $rankA = rank(UAV)$ for any invertible matrices U and V.*

Proof. Recall (Lemma 1 §1.6) that doing an elementary row operation to A is the same as left multiplying A by an elementary matrix. Since the invertible matrix U is a product of elementary matrices (Theorem 1 §1.6), it follows that

UA is obtained from A by a series of row operations. Hence $rank(UA) = rankA$ by Lemma 1. Now, write $UA = B$ for convenience, and apply Corollary 3, to get
$$rank(UAV) = rank(BV) = rank(BV)^T = rank(V^T B^T).$$
But V^T is invertible so the first part of this proof gives $rank(V^T B^T) = rankB^T = rankB = rankA$. Thus $rank(UAV) = rankA$ as required. □

The result in Corollary 4 should be compared to Theorem 3 §1.6.

If A is an $n \times n$ matrix then A is invertible if and only if $rankA = n$ by Corollary 2. In view of Corollary 1, this asserts that A is invertible if and only if $rankA$ is as large as possible. This raises a natural question about an arbitrary $m \times n$ matrix A: When do the two extreme cases $rankA = n$ and $rankA = m$ arise? The next two theorems show that these cases are naturally related to the independence and spanning properties of the columns of A (and hence to the systems of linear equations with A as coefficient matrix), and also to the invertibility of the (square and symmetric) matrices $A^T A$ and AA^T.

Theorem 3. *Let A be an $m \times n$ matrix. The following are equivalent:*

(1) *$AX = 0$ has only the trivial solution $X = 0$.*

(2) *The columns of A are linearly independent.*

(3) *$rankA = n$.*

(4) *$A^T A$ is an invertible matrix.*

Proof. Write $A = [C_1 \; C_2 \; \cdots \; C_n]$ where C_j is column j of A. Given scalars $x_1, x_2, \cdots x_n$, write $X = [x_1 \; x_2 \; \cdots \; x_n]^T$. Then block multiplication (Theorem 3 §1.4) gives

$$AX = [C_1 \; C_2 \; \cdots \; C_n] \begin{bmatrix} x_1 \\ x_2 \\ \vdots \\ x_n \end{bmatrix} = x_1 C_1 + x_2 C_2 + \cdots + x_n C_n. \qquad (*)$$

(1)⇔(2). Equation (*) shows that $AX = 0$ has only the trivial solution $X = 0$ if and only if $x_1 C_1 + x_2 C_2 + \cdots + x_k C_k = 0$ implies that $x_1 = x_2 = \cdots = x_k = 0$, that is if and only if $\{C_1, C_2, \cdots, C_n\}$ is independent. This is (1)⇔(2).

(2)⇔(3). We always have $rankA = dim(colA)$. Since part (2) implies that $dim(colA) = n$, this shows that (2)⇒(3). Conversely, if (3) holds, then $dim(colA) = rankA = n$. Since $\{C_1, C_2, \cdots, C_n\}$ is a spanning set for $colA$, (2) follows from Theorem 4 §4.3. Hence (3)⇒(2).

(1)⇔(4). If (4) holds and $AX = 0$, then $(A^T A)X = 0$ so $X = 0$ by (4). Hence (4)⇒(1). Conversely, assume that (1) holds. To prove that $A^T A$ is invertible, it suffices (by Theorem 5 §1.5) to show that $(A^T A)X = 0$ implies

4.4. RANK

that $X = 0$. So assume that $(A^T A)X = 0$. Write $AX = [y_1\ y_2\ \cdots\ y_n]^T$ and compute

$$y_1^2 + y_2^2 + \cdots + y_n^2 = (AX)^T(AX) = (X^T A^T)AX = X^T(A^T AX) = X^T 0 = 0.$$

It follows that each $y_i = 0$, and hence that $AX = 0$. This means $X = 0$ by (1), as required. □

Theorem 3 describes what happens when an $m \times n$ matrix has rank n. The following companion theorem discusses the situation when the rank is m.

Theorem 4. *The following are equivalent for an $m \times n$ matrix A:*

(1) $AX = B$ *has a solution for every column B in $\mathbb{R}^m$.*

(2) *The columns of A span $\mathbb{R}^m$.*

(3) $rank A = m$.

(4) AA^T *is an invertible matrix.*

Proof. We refer again to equation (*) in the proof of Theorem 3.

(1)⇔(2). Let B denote a column in $\mathbb{R}^m$. Then equation (*) shows that $B = AX$ for some X in $\mathbb{R}^m$ if and only if $B = x_1 C_1 + x_2 C_2 + \cdots + x_n C_n$ for some x_i in $\mathbb{R}$, that is if and only if B is in $span\{C_1, C_2, \cdots, C_n\}$. This proves (1)⇔(2).

(2)⇔(3). If (2) holds then $col A = \mathbb{R}^m$, so $rank A = dim(col A) = dim(\mathbb{R}^m) = m$. This proves that (2)⇒(3). Conversely, if (3) holds then $dim(col A) = m$. Since $col A \subseteq \mathbb{R}^m$, it follows from Theorem 4 §4.3 that $col A = \mathbb{R}^m$. Thus (3)⇒(2).

(3)⇔(4). Consider the matrix $B = A^T$. If we apply (3)⇔(4) in Theorem 3 to the $n \times m$ matrix B, we obtain $B^T B$ is invertible if and only if $rank B = m$. Since $rank B = rank A$ and $B^T B = AA^T$, this is (3)⇔(4) of the present theorem. □

Let A be any $m \times n$ matrix. Theorems 3 and 4 relate several important properties of A to the invertibility of the square matrices $A^T A$ and AA^T. Even if the columns of A are not independent or do not span $\mathbb{R}^m$ (as in these theorems), the matrices $A^T A$ and AA^T are both symmetric and as such have real eigenvalues by Theorem 3 §2.5. This focuses our attention on the study of symmetric matrices. In fact the eigenvalues of $A^T A$ and AA^T are all non-negative, and this leads to the so-called singular value decomposition of A, which provides basic information about A and is important in numerical analysis. We investigate this in Section 4.11.

4.4.3 Null Space and Image

In Section 4.1 we discussed two other subspaces associated with an $m \times n$ matrix A, the null space $nullA = \{X \mid X \text{ in } \mathbb{R}^n \text{ and } AX = 0\}$ and the image $imA = \{AX \mid X \text{ in } \mathbb{R}^n\}$. Using the rank, there are simple ways to find bases for these spaces. The next theorem does this for imA because the Rank Theorem does it for $colA$.

Theorem 5. *If A is any $m \times n$ matrix of rank r, then*

$$colA = imA = \{AX \mid X \text{ in } \mathbb{R}^n\}.$$

Hence the Rank Theorem provides a basis for imA. In particular,

$$dim(imA) = rankA.$$

Proof. Equation (*) in the proof of Theorem 3 shows that the vectors AX in imA are just the linear combinations of the columns of A, that is $imA = colA$.[5] In particular, $dim(imA) = dim(colA) = rankA$. □

Turning to the null space, a natural basis can be found using the Gaussian Algorithm. We first discuss a specific example which exhibits most of the salient features of the situation.

Example 3. *If $A = \begin{bmatrix} 1 & -2 & 1 & 1 \\ -1 & 2 & 0 & 1 \\ 2 & -4 & 1 & 0 \end{bmatrix}$ find a basis of $nullA$ and so find its dimension.*

Solution. If X is in $nullA$, then $AX = 0$, so X is given by solving the system $AX = 0$. Using Gaussian elimination, the augmented matrix is carried to reduced row-echelon form as follows:

$$\begin{bmatrix} 1 & -2 & 1 & 1 & 0 \\ -1 & 2 & 0 & 1 & 0 \\ 2 & -4 & 1 & 0 & 0 \end{bmatrix} \rightarrow \begin{bmatrix} 1 & -2 & 1 & 1 & 0 \\ 0 & 0 & 1 & 2 & 0 \\ 0 & 0 & -1 & -2 & 0 \end{bmatrix}$$

$$\rightarrow \begin{bmatrix} 1 & -2 & 0 & -1 & 0 \\ 0 & 0 & 1 & 2 & 0 \\ 0 & 0 & 0 & 0 & 0 \end{bmatrix}.$$

Hence, writing $X = [x_1 \ x_2 \ x_3 \ x_4]^T$, the leading variables are x_1 and x_3, and the non-leading variables x_2 and x_4 become parameters: $x_2 = s$ and $x_4 = t$. Then the equations corresponding to the reduced matrix determine the leading variables in terms of the parameters:

$$x_1 = 2s + t \quad \text{and} \quad x_3 = -2t.$$

---
[5]This observation was made earlier in Example 12 §4.1.

4.4. RANK

This means that the general solution is

$$X = [2s+t \ \ s \ \ -2t \ \ t]^T = s[2 \ 1 \ 0 \ 0]^T + t[1 \ 0 \ -2 \ 1]^T. \quad (**)$$

Hence X is in $span\{X_1, X_2\}$ where $X_1 = [2 \ 1 \ 0 \ 0]^T$ and $X_2 = [1 \ 0 \ -2 \ 1]^T$ are the basic solutions, and we have shown that $null(A) \subseteq span\{X_1, X_2\}$. But each of X_1 and X_2 are in $null A$ (they are solutions of the system $AX = 0$), so we have

$$null A = span\{X_1, X_2\}$$

by Theorem 1 §4.1. We claim further that $\{X_1, X_2\}$ is linearly independent. To see this, let $sX_1 + tX_2 = 0$ be a linear combination that vanishes. Then $(**)$ shows that $[2s+t \ \ s \ \ -2t \ \ t]^T = 0$, whence $s = t = 0$. Thus $\{X_1, X_2\}$ is a basis of $null(A)$, and so $dim(null A) = 2$. □

The calculation in Example 3 is typical of what happens in general. If A is an $m \times n$ matrix, it was shown in Theorem 2 §1.3 that the system $AX = 0$ has exactly $n - r$ basic solutions $X_1, X_2, \cdots, X_{n-r}$ where $r = rank A$, and that

$$null A = span\{X_1, X_2, \cdots, X_{n-r}\}.$$

Moreover, the general solution is a linear combination $X = t_1 X_1 + t_2 X_2 + \cdots + t_{n-r} X_{n-r}$ where each coefficient t_i is a parameter equal to a nonleading variable. Thus if this linear combination vanishes, that is if $X = 0$, then each $t_i = 0$ (as for s and t in Example 3, each t_i is a coefficient when X is expressed as a linear combination of the standard basis of $\mathbb{R}^n$). This proves that $\{X_1, X_2, \cdots, X_{n-r}\}$ is linearly independent, and so is a basis of $null A$. Theorem 6 summarizes this discussion.

Theorem 6. Let A denote an $m \times n$ matrix of rank r, and let $X_1, X_2, \cdots, X_{n-r}$ be the basic solutions of the homogeneous system $AX = 0$ that are produced by the Gaussian Algorithm. Then $\{X_1, X_2, \cdots, X_{n-r}\}$ is a basis of $null A$. In particular

$$dim(null A) = n - r.$$

Combining Theorems 5 and 6, we obtain a useful equation relating the dimensions of the image and null space of a matrix.

Corollary. If A is any $m \times n$ matrix then

$$dim(im A) + dim(null A) = n.$$

Exercises

1. Give the row space and column space of the following matrices.

   (a) $A = \begin{bmatrix} 4 & -6 \\ 7 & 6 \end{bmatrix}$
   (b) $A = \begin{bmatrix} -5 & -4 \\ 8 & 0 \\ -1 & -3 \end{bmatrix}$

   (c) $A = \begin{bmatrix} 0 & -4 & -8 & -4 \\ 8 & 2 & -8 & 0 \\ -3 & 1 & -1 & -2 \end{bmatrix}$
   (d) $A = \begin{bmatrix} 9 & -5 & 1 & -1 \\ 8 & -5 & -3 & 1 \\ 0 & 4 & 7 & 6 \\ 0 & 2 & -7 & 7 \end{bmatrix}$

2. Match each matrix on the left with a matrix on the right sharing the same row or column spaces.

   (a) $A = \begin{bmatrix} 0 & 1 & -3 & 2 \\ 0 & 2 & 1 & -3 \\ 2 & -5 & 0 & 5 \end{bmatrix}$
   (b) $A = \begin{bmatrix} 1 & 0 & 0 & 0 \\ 0 & 1 & 0 & 0 \\ 2 & 1 & 0 & 0 \end{bmatrix}$

   (c) $A = \begin{bmatrix} 1 & 0 & 1 & 1 \\ 0 & 1 & 1 & -1 \\ 2 & 1 & 3 & 1 \end{bmatrix}$
   (d) $A = \begin{bmatrix} 0 & -2 & 6 & -4 \\ 2 & -3 & 1 & 2 \\ 2 & -5 & 0 & 5 \end{bmatrix}$

   (e) $A = \begin{bmatrix} 0 & 0 & 1 & 1 \\ -1 & 1 & 0 & 2 \\ -1 & 1 & 2 & 4 \end{bmatrix}$
   (f) $A = \begin{bmatrix} 1 & -1 & 0 & -2 \\ 0 & 0 & 1 & 1 \\ 0 & 0 & 0 & 0 \end{bmatrix}$

3. Given $row A$ for some matrix A, what is the rank of A?

   (a) $rowA = span\{[1 \ 0 \ 0 \ 1]^T, [0 \ 1 \ 1 \ 0]^T, [0 \ 0 \ 0 \ 1]^T\}$.

   (b) $rowA = span\{[1 \ 2 \ -1 \ 3]^T, [2 \ -3 \ 1 \ 4]^T, [4 \ -13 \ 5 \ 6]^T\}$.

   (c) $rowA = span\{[-1 \ 0 \ 3 \ 2]^T, [3 \ 2 \ 0 \ -4]^T, [9 \ 8 \ 9 \ -10]^T, [4 \ 2 \ -3 \ -6]^T\}$.

4. In each case find a basis of the subspace U.

   (a) $U = span\{[1 \ 2 \ -1 \ 3]^T, [2 \ -3 \ 1 \ 4]^T, [4 \ -13 \ 5 \ 6]^T\}$.

   (b) $U = span\{[1 \ 2 \ -1 \ 0 \ 4]^T, [-3 \ 1 \ 0 \ 2 \ 5]^T, [-7 \ 7 \ -2 \ 6 \ 13]^T, [-2 \ 3 \ -1 \ 2 \ 9]^T\}$.

   (c) $U = span\{[-1 \ 0 \ 3 \ 2]^T, [3 \ 2 \ 0 \ -4]^T, [9 \ 8 \ 9 \ -10]^T, [4 \ 2 \ -3 \ -6]^T\}$.

5. In each case find bases for the row and column spaces of A and determine the rank of A.

4.4. RANK

(a) $A = \begin{bmatrix} 1 & -1 & 3 & 2 & 1 \\ -2 & 3 & 1 & 1 & 1 \\ -4 & 7 & 9 & 7 & 5 \\ 3 & -4 & 2 & 1 & 0 \end{bmatrix}$

(b) $A = \begin{bmatrix} 1 & 2 & 0 & 5 & 7 & 1 \\ 2 & 1 & 1 & -1 & 0 & 3 \\ -5 & 2 & -4 & 19 & 21 & 9 \\ 1 & -1 & 1 & -6 & -7 & 2 \end{bmatrix}$

6. In each case find a basis of the nullspace of A, compute $rank A$, and verify Theorem 6.

(a) $A = \begin{bmatrix} 3 & 1 & 1 \\ 2 & 0 & 1 \\ 4 & 2 & 1 \\ 1 & -1 & 1 \end{bmatrix}$

(b) $A = \begin{bmatrix} 3 & 5 & 5 & 2 & 0 \\ 1 & 0 & 2 & 2 & 1 \\ 1 & 1 & 1 & -2 & -2 \\ 2 & 0 & 4 & 4 & 2 \end{bmatrix}$

7. In each case either show that the statement is true or give an example showing that it is false. Throughout let A denote an $m \times n$ matrix.

   (a) If A has a row of zeros then $row A = \{0\}$.

   (b) If the rows of A are linearly independent then $row A = \mathbb{R}^n$.

   (c) If the rows of A span $\mathbb{R}^n$ then $row A = \mathbb{R}^n$.

   (d) If A has a row of zeros then $rank A = 0$.

   (e) If A has a row of zeros then $rank A < m$.

   (f) If B is $m \times n$ and $rank B = r = rank A$, then $rank(A + B) = r$.

8. Can a 3×4 matrix have independent rows? Independent columns? Explain.

9. If A is 4×3 and $rank A = 2$, can A have independent columns? Independent rows? Explain.

10. Can a nonsquare matix A have its rows independent and its columns independent? Explain.

11. Can the null space of a 3×6 matrix have dimension 2? Explain.

12. If a matrix A is not square, show that either the rows or the columns of A are dependent.

13. If A is an $m \times n$ matrix and $rank A = m$, show that $m \leq n$.

14. Let A and B denote matrices of size $k \times m$ and $m \times n$ respectively.

    (a) Show that each column of AB is a linear combination of the columns of A. [*Hint*: Equation (*) in the proof of Theorem 3].

    (b) Show that $rank(AB) \leq rank A$.

    (c) Show that $rank(AB) \leq rank B$. [*Hint*: Corollary 3 of Theorem 2.]

15. Let A and B denote matrices of sizes $m \times n$ and $n \times k$ respectively.

    (a) If $AB = 0$ show that $col B \subseteq null A$. [*Hint*: Theorem 3 § 1.4 (2).]

    (b) If $col B \subseteq null A$ show that $AB = 0$. [*Hint*: Theorem 3 § 1.4 (1).]

16. Consider the system of equations $AX = B$ where A is $m \times n$ and B is $m \times 1$.

    (a) Show that the system $AX = B$ has a solution if and only if B is in $col A$. [*Hint*: Equation (*) in the proof of Theorem 3.]

    (b) Show that the system $AX = B$ has a solution if and only if $rank A = rank[A\ B]$, where $[A\ B]$ is the augmented matrix of the system.

17. If I is the $m \times m$ identity matrix, show that $I + ZZ^T$ is invertible for every $m \times k$ matrix Z. [*Hint*: Factor $I + ZZ^T = [I\ Z]\begin{bmatrix} I \\ Z^T \end{bmatrix}$ in block form and use Theorem 4.]

18. Let A denote an $m \times n$ matrix.

    (a) If $rank A = m$ show that $AB = I_m$ for some $n \times m$ matrix B. [*Hint*: Theorem 4 (4).]

    (b) If $AB = I_m$ for some $n \times m$ matrix B, show that $rank A = m$. [*Hint*: Exercise 14 (b).]

19. Let A denote an $m \times n$ matrix.

    (a) Show that $null(A^T A) = null(A)$. [*Hint*: Use the proof of (1)⇔(4) in Theorem 3.}

    (b) Conclude that $rank(A^T A) = rank A$.

    (c) Show that $im(A^T) = im(A^T A)$. [*Hint*: Show that $im(A^T A) \subseteq im(A^T)$, and apply Theorem 4 § 4.3.]

20. If A is an $m \times n$ matrix, it can be proved that there exists a unique $n \times m$ matrix $A^\#$ such that $AA^\# A = A$, $A^\# AA^\# = A^\#$, $AA^\#$ and $A^\# A$ are both symmetric. The matrix $A^\#$ is called the **Moore-Penrose** inverse of A.

    (a) If A is square and invertible, show that $A^\# = A^{-1}$.

    (b) If $rank A = m$, show that $A^\# = A^T(AA^T)^{-1}$.

    (c) If $rank A = n$, show that $A^\# = (A^T A)^{-1} A^T$.

4.5 ORTHOGONALITY

Length and orthogonality are two of the most important concepts in geometry. In $\mathbb{R}^2$ and $\mathbb{R}^3$ these notions can both be defined using the dot product, and this observation is exploited in this section to extend these concepts to $\mathbb{R}^n$. This in turn leads to a natural extension to $\mathbb{R}^n$ of the notion of distance, to an extension of Pythagoras' Theorem, and to the idea of an orthogonal basis—one of the most useful concepts in linear algebra.

4.5.1 Dot Product, Length and Distance

Recall that, given $\vec{v} = [x_1 \ x_2 \ x_3]^T$ and $\vec{w} = [y_1 \ y_2 \ y_3]^T$ in $\mathbb{R}^3$, their dot product $\vec{v} \bullet \vec{w}$ and the length $\|\vec{v}\|$ of $\vec{v}$ are defined by

$$\vec{v} \bullet \vec{w} = x_1 y_1 + x_2 y_2 + x_3 y_3 \quad \text{and} \quad \|\vec{v}\| = \sqrt{x_1^2 + x_2^2 + x_3^2}.$$

These formulas extend in the obvious way to columns in $\mathbb{R}^n$. Accordingly, if $X = [x_1 \ x_2 \ \cdots \ x_n]^T$ and $Y = [y_1 \ y_2 \ \cdots \ y_n]^T$ are columns in $\mathbb{R}^n$, we define their **dot product** $X \bullet Y$ as follows:

$$X \bullet Y = x_1 y_1 + x_2 y_2 + \cdots + x_n y_n = X^T Y.$$

Note that $X^T Y$ is technically a 1×1 matrix, which we take to be a number.[6] The **length** of the column vector $X = [x_1 \ x_2 \ \cdots \ x_n]^T$ is defined to be

$$\|X\| = \sqrt{X \bullet X} = \sqrt{x_1^2 + x_2^2 + \cdots + x_n^2}$$

where $\sqrt{(\ )}$ indicates the positive square root.[7]

Example 1. Given $X = [3 \ -2 \ 0 \ 4 \ -1]$ and $Y = [1 \ 1 \ -1 \ 0 \ 6]$ in $\mathbb{R}^5$, we have $X \bullet Y = 3 - 2 + 0 + 0 - 6 = -5$, and $\|X\| = \sqrt{9 + 4 + 0 + 16 + 1} = \sqrt{30}$.

The following facts will be used frequently below.

Theorem 1. *Let X, Y and Z denote columns in $\mathbb{R}^n$. Then:*

(1) $X \bullet Y = Y \bullet X.$

(2) $X \bullet (Y + Z) = X \bullet Y + X \bullet Z.$

(3) $(aX) \bullet Y = a(X \bullet Y) = X \bullet (aY)$ *for all scalars a.*

(4) $\|X\|^2 = X \bullet X.$

(5) $\|X\| \geq 0$, *and* $\|X\| = 0$ *if and only if* $X = 0$.

---

[6] If X and Y are written as rows rather than columns, this becomes $X \bullet Y = XY^T$.

[7] Since $X \bullet X = x_1^2 + x_2^2 + \cdots + x_n^2 \geq 0$ for all choices of the real numbers x_i, the length $\|X\|$ is a non-negative real number for every column X in $\mathbb{R}^n$.

(6) $\|aX\| = |a|\,\|X\|$ for all scalars a.

Proof. Let $X = [x_1\ x_2\ \cdots\ x_n]^T$ and $Y = [y_1\ y_2\ \cdots\ y_n]^T$ be columns in $\mathbb{R}^n$. Then (1) follows from $X \bullet Y = x_1 y_1 + x_2 y_2 + \cdots + x_n y_n$, (2) and (3) follow from matrix arithmetic because $X \bullet Y = X^T Y$, (4) is clear from the definition, and (6) follows because $|a| = \sqrt{a^2}$ for all real numbers a. As to (5), $\|X\| \geq 0$ is clear, as is $\|0\| = 0$; finally, if $\|X\| = 0$ then $x_1^2 + x_2^2 + \cdots + x_n^2 = 0$, so each $x_i = 0$ because the x_i are real numbers. Thus $X = 0$. $\square$

Because of Theorem 1, computations with dot products in $\mathbb{R}^n$ are similar to those in $\mathbb{R}^3$. In particular it follows from (1) and (2) that the dot product of two sums $X_1 + \cdots + X_k$ and $Y_1 + \cdots + Y_m$ in $\mathbb{R}^n$ is the sum of the dot products of every X_i with every Y_j:

$$
\begin{aligned}
(X_1 + \cdots + X_k) &\bullet (Y_1 + \cdots + Y_m) \\
&= X_1 \bullet Y_1 + X_1 \bullet Y_2 + X_2 \bullet Y_1 + \cdots + X_k \bullet Y_m \\
&= \Sigma_{i,j} X_i \bullet Y_j.
\end{aligned}
$$

If this is combined with (3) in Theorem 1, we can do calculations like

$$
\begin{aligned}
(2X_1 - 3X_2) &\bullet (5Y_1 + 4Y_2) \\
&= 10(X_1 \bullet Y_1) + 8(X_1 \bullet Y_2) - 15(X_2 \bullet Y_1) - 12(X_2 \bullet Y_2).
\end{aligned}
$$

Such computations will be carried out without comment below.

Example 2. *If X and Y are in $\mathbb{R}^n$, show that*

$$\|X + Y\|^2 = \|X\|^2 + 2\,X \bullet Y + \|Y\|^2.$$

Solution. We use Theorem 1 several times:

$$
\begin{aligned}
\|X + Y\|^2 &= (X + Y) \bullet (X + Y) \\
&= X \bullet X + X \bullet Y + Y \bullet X + Y \bullet Y \\
&= \|X\|^2 + 2\,X \bullet Y + \|Y\|^2.
\end{aligned}
$$

Example 3. *Let $\mathbb{R}^n = \mathrm{span}\{F_1, F_2, \cdots, F_m\}$. If X in $\mathbb{R}^n$ satisfies $X \bullet F_i = 0$ for each i, show that $X = 0$.*

Solution. We show $X = 0$ by showing that $\|X\| = 0$ and applying (5) of Theorem 1. By hypothesis, write $X = t_1 F_1 + t_2 F_2 + \cdots + t_m F_m$ for some t_i in $\mathbb{R}$. Then

$$
\begin{aligned}
\|X\|^2 = X \bullet X &= X \bullet (t_1 F_1 + t_2 F_2 + \cdots + t_m F_m) \\
&= t_1(X \bullet F_1) + t_2(X \bullet F_2) + \cdots + t_m(X \bullet F_m) \\
&= 0
\end{aligned}
$$

4.5. ORTHOGONALITY

by hypothesis. Hence $\|X\| = 0$, and so $X = 0$ by Theorem 1. □

A vector E in $\mathbb{R}^n$ is called a **unit vector** if $\|E\| = 1$. Thus if $\{E_1, E_2, \cdots, E_n\}$ is the standard basis of $\mathbb{R}^n$ then each E_i is a unit vector (E_i is column i of the $n \times n$ identity matrix). We will use the following observation several times below.

Example 4. *If $X \neq 0$ in $\mathbb{R}^n$, show that $E = \frac{1}{\|X\|}X$ is the unique positive scalar multiple of E that is a unit vector.*

Solution. If $E = tX$ where $t > 0$, then $\|E\| = |t| \|X\| = t\|X\|$ by Theorem 1. Hence if $\|E\| = 1$ we have $t\|X\| = 1$, so $t = \frac{1}{\|X\|}$. □

If $\vec{v}$ and $\vec{w}$ are nonzero vectors in $\mathbb{R}^3$, we showed (Theorem 2 §3.2) that the angle θ between them is given by $cos\theta = \frac{\vec{v} \bullet \vec{w}}{\|\vec{v}\|\|\vec{w}\|}$. Since $|cos\theta| \leq 1$ for any angle θ, it follows that $|\vec{v} \bullet \vec{w}| \leq \|\vec{v}\| \|\vec{w}\|$, and this holds even if $\vec{v} = \vec{0}$ or $\vec{w} = \vec{0}$. This inequality has an important analogue in $\mathbb{R}^n$.

Theorem 2. Cauchy Inequality[8]. *If X and Y are vectors in $\mathbb{R}^n$, then*

$$|X \bullet Y| \leq \|X\| \|Y\|.$$

Moreover, equality holds if and only if one of X and Y is a scalar multiple of the other.

Proof. If either $X = 0$ or $Y = 0$ the inequality holds (in fact it is equality). Otherwise, write $\|X\| = a$ and $\|Y\| = b$ for convenience. Then $a > 0$ and $b > 0$, and a computation like that in Example 2 gives

$$\|bX - aY\|^2 = 2ab(ab - X \bullet Y) \quad \text{and} \quad \|bX + aY\|^2 = 2ab(ab + X \bullet Y). \quad (*)$$

It follows that $ab - X \bullet Y \geq 0$ and $ab + X \bullet Y \geq 0$, whence $-ab \leq X \bullet Y \leq ab$. This gives the Cauchy Inequality. If equality holds, we have $|X \bullet Y| = ab$, so either $X \bullet Y = ab$ or $X \bullet Y = -ab$. Using (*), these imply that $bX - aY = 0$ or $bX + aY = 0$, respectively. Thus one of X and Y is a scalar multiple of the other (even if $a = 0$ or $b = 0$). □

The Cauchy Inequality is often written in the equivalent form $(X \bullet Y)^2 \leq \|X\|^2 \|Y\|^2$. Thus, for example, in $\mathbb{R}^4$ we have

$$(a_1b_1 + a_2b_2 + a_3b_3 + a_4b_4)^2 \leq (a_1^2 + a_2^2 + a_3^2 + a_4^2)(b_1^2 + b_2^2 + b_3^2 + b_4^2)$$

for all choices of the numbers a_i and b_i.

---

[8] Augustin Louis Cauchy (1789-1857) was born in Paris and became a professor at the École Polytechnique at the age of 26. He was one of the great mathematicians and is best remembered for his work in analysis where he established standards of rigor that carry down to today's calculus texts.

There is an important consequence of the Cauchy Inequality. Given X and Y in $\mathbb{R}^n$, we use the inequality and Example 2 to compute

$$\begin{aligned} \|X+Y\|^2 &= \|X\|^2 + 2X \bullet Y + \|Y\|^2 \\ &\leq \|X\|^2 + 2\|X\|\|Y\| + \|Y\|^2 \\ &= (\|X\| + \|Y\|)^2. \end{aligned}$$

If we take positive square roots, this gives the

Corollary. Triangle inequality. *If X and Y are vectors in $\mathbb{R}^n$ then* $\|X+Y\| \leq \|X\| + \|Y\|$.

The reason for the name will be clear when we have introduced the notion of distance into $\mathbb{R}^n$.

If $\vec{v}$ and $\vec{w}$ denote two vectors in $\mathbb{R}^3$, the distance between $\vec{v}$ and $\vec{w}$ (regarded as points) is $\|\vec{v} - \vec{w}\|$ as in the diagram. The analogue for $\mathbb{R}^n$ is as follows: If X and Y are vectors in $\mathbb{R}^n$, the **distance** between X and Y is defined to be

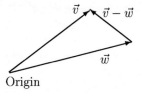

$$d(X,Y) = \|X - Y\|.$$

This distance function has all the intuitive properties of distance that are familiar from $\mathbb{R}^2$ and $\mathbb{R}^3$, including another version of the triangle inequality.

Theorem 3. *Let X, Y and Z denote vectors in $\mathbb{R}^n$. Then:*

(1) $d(X,Y) \geq 0$ *for all X and Y.*

(2) $d(X,Y) = 0$ *if and only if $X = Y$.*

(3) $d(X,Y) = d(Y,X)$.

(4) Triangle Inequality. $d(X,Y) \leq d(X,Z) + d(Z,Y)$.

Proof. (1) and (2) are translations of part (5) of Theorem 1, and (3) holds because $\|X\| = \|-X\|$ by part (6) of Theorem 1. To prove (4), we use the Corollary to Theorem 2:

$$\begin{aligned} d(X,Y) &= \|X - Y\| \\ &= \|(X - Z) + (Z - Y)\| \leq \|(X - Z)\| + \|(Z - Y)\| \\ &= d(Y,Z) + d(Z,Y). \end{aligned}$$

This proves (4). $\square$

Now the name "Triangle Inequality" in (4) makes sense: Consider the "triangle" in $\mathbb{R}^n$ with vertices X, Y and Z as depicted in the diagram. Then the length $d(X,Y)$ of one side of the triangle is smaller than the sum $d(X,Z) + d(Z,Y)$ of the lengths of the other two sides. Note that, as in $\mathbb{R}^3$, we regard a vector X in $\mathbb{R}^n$ as an arrow from the origin to the point X. While not totally accurate, this method of visualizing $\mathbb{R}^n$ does not lead one astray.

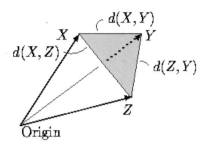

4.5.2 Orthogonal Sets and the Expansion Theorem

The concept of orthogonality is a central idea in geometry, occurring as it does in Pythagoras' Theorem, in the notion of the normal to a plane, and in the way we find the point on a line that is closest to a given point. Orthogonality turns out to play an even more important role in the geometry of $\mathbb{R}^n$ as we shall see. Moreover, as for length and distance, the definition is a natural extension of the situation in $\mathbb{R}^3$.

As in $\mathbb{R}^3$, two vectors X and Y in $\mathbb{R}^n$ are said to be **orthogonal** if

$$X \bullet Y = 0.$$

More generally, a set of vectors $\{X_1, X_2, \cdots, X_k\}$ is called an **orthogonal set** of vectors if

$$\begin{array}{ll} X_i \bullet X_j = 0 & \text{for all } i \neq j \\ X_i \neq 0 & \text{for all } i. \end{array}$$

(The reason for insisting that each X_i is nonzero is that we will be concerned primarily with orthogonal bases.) Note that $\{X\}$ is an orthogonal set for any $X \neq 0$ in $\mathbb{R}^n$. An orthogonal set of vectors $\{X_1, X_2, \cdots, X_k\}$ is called **orthonormal** if $\|X_i\| = 1$ for each i, that is if each X_i is a unit vector.

Example 5. The standard basis of $\mathbb{R}^n$ is an orthonormal set of vectors.

If $\{X_1, X_2, \cdots, X_k\}$ is an orthogonal set in $\mathbb{R}^n$, it is easy to verify that $\{a_1 X_1, a_2 X_2, \cdots, a_k X_k\}$ is also an orthogonal set for any choice of scalars $a_i \neq 0$. In particular, $\{\frac{1}{\|X_1\|} X_1, \frac{1}{\|X_2\|} X_2, \cdots, \frac{1}{\|X_k\|} X_k\}$ is an orthonormal set by Example 4, and we say it is the result of **normalizing** the orthogonal set $\{X_1, X_2, \cdots, X_k\}$.

Example 6. The set $\{[1\ -1\ 3\ 0]^T,\ [-2\ 1\ 1\ 1]^T,\ [1\ 1\ 0\ 1]^T,\ [1\ 10\ 3\ -11]^T\}$ is orthogonal in $\mathbb{R}^4$ as is easily verified. After normalizing, the corresponding orthonormal set is $\{\frac{1}{\sqrt{11}}[1\ -1\ 3\ 0]^T,\ \frac{1}{\sqrt{7}}[-2\ 1\ 1\ 1]^T,\ \frac{1}{\sqrt{3}}[1\ 1\ 0\ 1]^T,\ \frac{1}{\sqrt{231}}[1\ 10\ 3\ -11]^T\}$.

No discussion of geometry can omit mention of the most famous theorem about orthogonality: Pythagoras' Theorem. Suppose orthogonal vectors $\vec{v} \neq \vec{0}$ and $\vec{w} \neq \vec{0}$ are given in $\mathbb{R}^3$. Then Pythagoras' Theorem asserts that $\|\vec{v} + \vec{w}\|^2 = \|\vec{v}\|^2 + \|\vec{w}\|^2$ (see the diagram). More generally, if $\{X, Y\}$ is an orthogonal set in $\mathbb{R}^n$, then $X \bullet Y = 0$ so

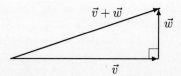

$$\|X + Y\|^2 = \|X\|^2 + 2(X \bullet Y) + \|Y\|^2$$
$$= \|X\|^2 + \|Y\|^2.$$

In fact, this works for orthogonal sets of more than two vectors n $\mathbb{R}^n$.

Theorem 4. Pythagoras' Theorem. *Let $\{X_1, X_2, \cdots, X_k\}$ be an orthogonal set[9] of vectors in $\mathbb{R}^n$. Then*

$$\|X_1 + X_2 + \cdots + X_k\|^2 = \|X_1\|^2 + \|X_2\|^2 + \cdots + \|X_k\|^2$$

Proof. The fact that $X_i \bullet X_j = 0$ whenever $i \neq j$ gives

$$\|X_1 + X_2 + \cdots + X_k\|^2$$
$$= (X_1 + X_2 + \cdots + X_k) \bullet (X_1 + X_2 + \cdots + X_k)$$
$$= X_1 \bullet X_1 + X_2 \bullet X_2 + \cdots + X_k \bullet X_k$$
$$+ (X_1 \bullet X_2 + X_1 \bullet X_3 + X_2 \bullet X_3 + \cdots)$$
$$= \|X_1\|^2 + \|X_2\|^2 + \cdots + \|X_k\|^2 + (0 + 0 + 0 + \cdots).$$

This proves the theorem. □

If $\{\vec{v}, \vec{w}\}$ is an orthogonal set in $\mathbb{R}^3$, then $\vec{v}$ and $\vec{w}$ are certainly not parallel (being orthogonal), so $\{\vec{v}, \vec{w}\}$ is linearly independent by the Corollary of Theorem 3 §4.2. The next theorem shows that this observation holds much more generally.

Theorem 5. *Every orthogonal set of vectors in $\mathbb{R}^n$ is linearly independent.*

Proof. If $\{X_1, X_2, \cdots, X_k\}$ is orthogonal, suppose a linear combination vanishes: $t_1 X_1 + t_2 X_2 + \cdots + t_k X_k = 0$. We must show that each $t_i = 0$. If we take the dot product with X_1, the result is

$$0 = X_1 \bullet 0 = X_1 \bullet (t_1 X_1 + t_2 X_2 + \cdots + t_k X_k)$$
$$= t_1 (X_1 \bullet X_1) + t_2 (X_1 \bullet X_2) + \cdots + t_k (X_1 \bullet X_k)$$
$$= t_1 \|X_1\|^2 + t_2 (0) + \cdots + t_k (0)$$
$$= t_1 \|X_1\|^2.$$

---

[9]This works even if some of the X_i are zero.

4.5. ORTHOGONALITY

Since $\|X_1\|^2 \neq 0$ (because $X_1 \neq 0$), it follows that $t_1 = 0$. Similarly, $t_i = 0$ for each i. □

Theorem 5 points immediately to the consideration of orthogonal bases. These turn out to be much more convenient than ordinary bases, for the following reason. If $\{X_1, X_2, \cdots, X_k\}$ is a basis of a subspace U, then $U = span\{X_1, X_2, \cdots, X_k\}$ so every vector X in U can be (uniquely) expressed as a linear combination $X = t_1 X_1 + t_2 X_2 + \cdots + t_k X_k$. The problem is that it is in general tedious to compute the coefficients t_i (this requires solving k equations in k variables). However, explicit formulas exist for the coefficients if the basis $\{X_1, X_2, \cdots, X_k\}$ is orthogonal.

Theorem 6. Expansion Theorem. Let $\{X_1, X_2, \cdots, X_k\}$ be an orthogonal basis of a subspace U of $\mathbb{R}^n$. If X is any vector in U, then

$$X = \frac{X \bullet X_1}{\|X_1\|^2} X_1 + \frac{X \bullet X_2}{\|X_2\|^2} X_2 + \cdots + \frac{X \bullet X_k}{\|X_k\|^2} X_k.$$

Proof. We have $X = t_1 X_1 + t_2 X_2 + \cdots + t_k X_k$ for some coefficients t_i because $U = span\{X_1, X_2, \cdots, X_k\}$. Take the dot product of both sides with X_1:

$$\begin{aligned} X \bullet X_1 &= (t_1 X_1 + t_2 X_2 + \cdots + t_k X_k) \bullet X_1 \\ &= t_1(X_1 \bullet X_1) + t_2(X_2 \bullet X_1) + \cdots + t_k(X_k \bullet X_1) \\ &= t_1 \|X_1\|^2 + t_2(0) + \cdots + t_k(0) \\ &= t_1 \|X_1\|^2. \end{aligned}$$

Since $\|X_1\|^2 \neq 0$, this gives $t_1 = \frac{X \bullet X_1}{\|X_1\|^2}$. Similarly, $t_i = \frac{X \bullet X_i}{\|X_i\|^2}$ for each i. □

Example 7. As in Example 6, let $X_1 = [1 \ -1 \ 3 \ 0]^T$, $X_2 = [-2 \ 1 \ 1 \ 1]^T$, $X_3 = [1 \ 1 \ 0 \ 1]^T$ and $X_4 = [1 \ 10 \ 3 \ -11]^T$. Then $\{X_1, X_2, X_3, X_4\}$ is orthogonal in $\mathbb{R}^4$, hence it is independent by Theorem 5, and so it is a basis because $dim(\mathbb{R}^4) = 4$. Given $X = [a \ b \ c \ d]^T$ in $\mathbb{R}^4$, we can express X as a linear combination of X_1, X_2, X_3, and X_4. In fact the Expansion Theorem gives $X = t_1 X_1 + t_2 X_2 + t_3 X_3 + t_4 X_4$ where

$$t_1 = \frac{X \bullet X_1}{\|X_1\|^2} = \frac{a-b+3c}{11} \quad t_2 = \frac{X \bullet X_2}{\|X_2\|^2} = \frac{-2a+b+c+d}{7}$$

$$t_3 = \frac{X \bullet X_3}{\|X_3\|^2} = \frac{a+b+d}{3} \quad t_4 = \frac{X \bullet X_4}{\|X_4\|^2} = \frac{a+10b+3c-11d}{231}.$$

The reader may wish to verify that in fact $X = t_1 X_1 + t_2 X_2 + t_3 X_3 + t_4 X_4$. □

4.5.3 The Gram-Schmidt Algorithm

The Expansion Theorem demonstrates the desirability of orthogonal bases, and more evidence will be given in the following section. However we must first show that every subspace of $\mathbb{R}^n$ *has* an orthogonal basis, and we accomplish this by giving a method for converting any basis into an orthogonal basis.

In showing that every subspace of $\mathbb{R}^n$ has a basis, it was essential in Section 4.3 to be able to enlarge any independent set. The key result was the Independent Lemma (Lemma 1 §4.3): If $\{X_1, X_2, \cdots, X_m\}$ is independent in $\mathbb{R}^n$ and Y is not in $span\{X_1, X_2, \cdots, X_m\}$, then $\{Y, X_1, X_2, \cdots, X_m\}$ is also an independent set. The analogue for orthogonal sets is

Lemma 1. Orthogonal Lemma. *Let $\{F_1, F_2, \cdots, F_k\}$ be an orthogonal set in $\mathbb{R}^n$. Given X in $\mathbb{R}^n$, write*

$$F_{k+1} = X - \left(\frac{X \bullet F_1}{\|F_1\|^2} F_1 + \frac{X \bullet F_2}{\|F_2\|^2} F_2 + \cdots + \frac{X \bullet F_k}{\|F_k\|^2} F_k \right).$$

Then:

(1) F_{k+1} *is orthogonal to each of* $F_1, F_2, \cdots, F_k$.

(2) *If X is not in $span\{F_1, F_2, \cdots, F_k\}$ then $F_{k+1} \neq 0$ and the larger set $\{F_1, F_2, \cdots, F_k, F_{k+1}\}$ is an orthogonal set.*

Proof. For convenience, write $c_i = \frac{X \bullet F_i}{\|F_i\|^2}$ for each i. Given $1 \leq i \leq k$, we have

$$\begin{aligned}
F_{k+1} \bullet F_i &= (X - c_1 F_1 - c_2 F_2 - \cdots - c_k F_k) \bullet F_i \\
&= X \bullet F_i - c_1(F_1 \bullet F_i) - c_2(F_2 \bullet F_i) - \cdots - c_k(F_k \bullet F_i) \\
&= X \bullet F_i - c_i \|F_i\|^2 \\
&= 0.
\end{aligned}$$

This proves (1), and (2) follows because $F_{k+1} \neq 0$ if X is not in $span\{F_1, F_2, \cdots, F_k\}$. $\square$

It is worth observing that if X is in $span\{F_1, F_2, \cdots, F_k\}$ in Lemma 1 then $F_{k+1} = 0$ by the Expansion Theorem.

The Orthogonal Lemma gives a method of extending an orthogonal set to a larger orthogonal set. In this way it provides a version for orthogonal sets of the fact that every independent subset of a subspace U is part of an basis of U.

Theorem 7. *If U is a nonzero subspace of $\mathbb{R}^n$, then every orthogonal subset $\{F_1, F_2, \cdots, F_k\}$ of U is part of an orthogonal basis of U. In particular U has an orthogonal basis.*

Proof. If $span\{F_1, F_2, \cdots, F_k\} = U$ then $\{F_1, F_2, \cdots, F_k\}$ is *already* an orthogonal basis. Otherwise, there exists X in U which is not in $span\{F_1, F_2, \cdots, F_k\}$. If F_{k+1} is as in the Orthogonal Lemma, then F_{k+1} is in U and $\{F_1, F_2, \cdots, F_k, F_{k+1}\}$ is orthogonal. If $span\{F_1, F_2, \cdots, F_k, F_{k+1}\} = U$ we are done. Otherwise, the process continues to create larger and larger orthogonal subsets of U, each containing $\{F_1, F_2, \cdots, F_k\}$. They are all independent by Theorem

4.5. ORTHOGONALITY

5, so we have an orthogonal basis when we reach such a set containing $dimU$ vectors. Finally, as $U \neq 0$ choose $F_1 \neq 0$ in U. Then $\{F_1\}$ is an orthogonal set, so the foregoing procedure produces an orthogonal basis of U (containing $\{F_1\}$). □

The method of proof of Theorem 7 actually gives more information. It leads to a procedure by which any basis $\{X_1, X_2, \cdots, X_m\}$ of a subspace U of $\mathbb{R}^n$ can be systematically modified to produce an orthogonal basis $\{F_1, F_2, \cdots, F_m\}$ of U. These new vectors F_i are constructed one at a time from the X_i. To start the process, take

$$F_1 = X_1.$$

Then X_2 is not in $span\{F_1\}$ because $\{X_1, X_2\}$ is independent so, with an eye on the Orthogonal Lemma, take

$$F_2 = X_2 - \frac{X_2 \bullet F_1}{\|F_1\|^2} F_1.$$

Thus $\{F_1, F_2\}$ is an orthogonal set by the Orthogonal Lemma. Moreover, $span\{F_1, F_2\} = span\{X_1, X_2\}$ as is easily verified. Hence X_3 is not in $span\{F_1, F_2\}$ because $\{X_1, X_2, X_3\}$ is independent. Now another application of the Orthogonal Lemma shows that $\{F_1, F_2, F_3\}$ is orthogonal where

$$F_3 = X_3 - \frac{X_3 \bullet F_1}{\|F_1\|^2} F_1 - \frac{X_3 \bullet F_2}{\|F_2\|^2} F_2.$$

Again one verifies that $span\{F_1, F_2, F_3\} = span\{X_1, X_2, X_3\}$, and the process continues. At the mth iteration we construct an orthogonal set $\{F_1, F_2, \cdots, F_m\}$ such that

$$span\{F_1, F_2, \cdots, F_m\} = span\{X_1, X_2, \cdots, X_m\} = U.$$

Hence $\{F_1, F_2, \cdots, F_m\}$ is the desired orthogonal basis of U. The procedure can be summarized as follows.

Theorem 8. Gram-Schmidt Algorithm.[10] *If $\{X_1, X_2, \cdots, X_m\}$ is any basis of a subspace U of $\mathbb{R}^n$, construct $F_1, F_2, \cdots, F_m$ in U successively as follows:*

$$\begin{aligned} F_1 &= X_1 \\ F_2 &= X_2 - \frac{X_2 \bullet F_1}{\|F_1\|^2} F_1 \\ F_3 &= X_3 - \frac{X_3 \bullet F_1}{\|F_1\|^2} F_1 - \frac{X_3 \bullet F_2}{\|F_2\|^2} F_2 \\ &\vdots \\ F_k &= X_k - \frac{X_k \bullet F_1}{\|F_1\|^2} F_1 - \frac{X_k \bullet F_2}{\|F_2\|^2} F_2 - \cdots - \frac{X_k \bullet F_{k-1}}{\|F_{k-1}\|^2} F_{k-1} \\ &\vdots \end{aligned}$$

---

[10] Erhardt Schmidt (1876-1959) was a German mathematician who first described the algorithm in 1907. Jörgen Pederson Gram (1850-1916) was a Danish actuary.

for each $k = 2, 3, \cdots, m$. Then:
(1) $\{F_1, F_2, \cdots, F_m\}$ is an orthogonal basis of U.
(2) $span\{F_1, F_2, \cdots, F_k\} = span\{X_1, X_2, \cdots, X_k\}$ for each $k = 1, 2, \cdots, m$.

Example 8. *Find an orthogonal basis for $span\{[1\ 1\ 0\ 1], [0\ 1\ -1\ 1], [1\ 0\ 1\ 1]\}$.*

Solution. Write $X_1 = [1\ 1\ 0\ 1]$, $X_2 = [0\ 1\ -1\ 1]$ and $X_3 = [1\ 0\ 1\ 1]$, and observe that $\{X_1, X_2, X_3\}$ is linearly independent. The Gram-Schmidt Algorithm gives:

$$\begin{aligned}
F_1 &= X_1 = [1\ 1\ 0\ 1]. \\
F_2 &= X_2 - \frac{X_2 \bullet F_1}{\|F_1\|^2} F_1 = [0\ 1\ -1\ 1] - \frac{2}{3}[1\ 1\ 0\ 1] \\
&= \frac{1}{3}[-2\ 1\ -3\ 1]. \\
F_3 &= X_3 - \frac{X_3 \bullet F_1}{\|F_1\|^2} F_1 - \frac{X_3 \bullet F_2}{\|F_2\|^2} F_2 \\
&= [1\ 0\ 1\ 1] - \frac{2}{3}[1\ 1\ 0\ 1] - \frac{(-4/3)}{(15/9)} \frac{1}{3}[-2\ 1\ -3\ 1] \\
&= \frac{1}{5}[-1\ -2\ 1\ 3].
\end{aligned}$$

Hence the algorithm gives the orthogonal basis $\{[1\ 1\ 0\ 1], \frac{1}{3}[-2\ 1\ -3\ 1], \frac{1}{5}[-1\ -2\ 1\ 3]\}$. In hand calculations, or for other reasons, it may be convenient to eliminate fractions and common factors, so $\{[1\ 1\ 0\ 1], [-2\ 1\ -3\ 1], [-1\ -2\ 1\ 3]\}$ is also an orthogonal basis. □

Note on Computation. Observe that the vector $\frac{X \bullet F_i}{\|F_i\|^2} F_i$ in the Gram-Schmidt Algorithm is unchanged if F_i is replaced by aF_i for any nonzero scalar a. Hence if a newly constructed F_i is multiplied by a nonzero scalar at some stage of the Gram-Schmidt Algorithm, the subsequent F's will be unchanged. This is useful in actual calculations. As an illustration, in Example 8 we could have used $F_2 = [-2\ 1\ -3\ 1]$ at the second stage of the algorithm. This would have made the hand calculation of F_3 easier, but there is another reason: Because computing F_2 involved a division (by 3 in Example 8), a computer would have carried some round-off error into the calculation of F_3. In a large calculation, this round-off error can accumulate and become a serious problem.

4.5.4 QR-Factorization

We conclude this section with the matrix version of the Gram-Schmidt process. The result is a factorization of any matrix with independent columns as the product of a matrix with *orthonormal* columns and an invertible upper triangular matrix with positive diagonal entries. This is important in applications, and is particularly useful in calculations because there are computer algorithms that accomplish the decomposition with good control over round-off error.

Suppose $A = [C_1\ C_2\ \cdots\ C_n]$ is an $m \times n$ matrix with linearly independent columns $C_1, C_2, \cdots, C_n$. The Gram-Schmidt algorithm can be applied to these

4.5. ORTHOGONALITY

columns to provide orthogonal columns $F_1, F_2, \cdots, F_n$ where

$$\begin{aligned}
F_1 &= C_1 \\
F_2 &= C_2 - \frac{C_2 \bullet F_1}{\|F_1\|^2} F_1 \\
F_3 &= C_3 - \frac{C_3 \bullet F_1}{\|F_1\|^2} F_1 - \frac{C_3 \bullet F_2}{\|F_2\|^2} F_2 \\
&\vdots \\
F_n &= C_n - \frac{C_n \bullet F_1}{\|F_1\|^2} F_1 - \frac{C_n \bullet F_2}{\|F_2\|^2} F_2 - \cdots - \frac{C_n \bullet F_{n-1}}{\|F_{n-1}\|^2} F_{n-1}.
\end{aligned}$$

Now write $Q_j = \frac{1}{\|F_j\|} F_j$ for each j. Then $Q_1, Q_2, \cdots, Q_n$ are orthonormal columns, and the above equations become

$$\begin{aligned}
\|F_1\| Q_1 &= C_1 \\
\|F_2\| Q_2 &= C_2 - (C_2 \bullet Q_1) Q_1 \\
\|F_3\| Q_3 &= C_3 - (C_3 \bullet Q_1) Q_1 - (C_3 \bullet Q_2) Q_2 \\
&\vdots \\
\|F_n\| Q_n &= C_n - (C_n \bullet Q_1) Q_1 - (C_n \bullet Q_2) Q_2 - \cdots - (C_n \bullet Q_{n-1}) Q_{n-1}.
\end{aligned}$$

Finally, use these equations to express each C_k as a linear combination of $Q_1, Q_2, \cdots, Q_n$:

$$\begin{aligned}
C_1 &= \|F_1\| Q_1 \\
C_2 &= (C_2 \bullet Q_1) Q_1 + \|F_2\| Q_2 \\
C_3 &= (C_3 \bullet Q_1) Q_1 + (C_3 \bullet Q_2) Q_2 + \|F_3\| Q_3 \\
&\vdots \\
C_n &= (C_n \bullet Q_1) Q_1 + (C_n \bullet Q_2) Q_2 + (C_n \bullet Q_3) Q_3 + \cdots + \|F_n\| Q_n.
\end{aligned}$$

Using block multiplication (Theorem 3 §1.4), these equations have a matrix form that gives the required factorization:

$$A = [C_1 \ C_2 \ C_3 \ \cdots \ C_n]$$
$$= [Q_1 \ Q_2 \ Q_3 \ \cdots \ Q_n] \begin{bmatrix} \|F_1\| & C_2 \bullet Q_1 & C_3 \bullet Q_1 & \cdots & C_n \bullet Q_1 \\ 0 & \|F_2\| & C_3 \bullet Q_2 & \cdots & C_n \bullet Q_2 \\ 0 & 0 & \|F_3\| & \cdots & C_n \bullet Q_3 \\ \vdots & \vdots & \vdots & \ddots & \vdots \\ 0 & 0 & 0 & \cdots & \|F_n\| \end{bmatrix} \quad (*)$$

Here the first factor $Q = [Q_1 \ Q_2 \ Q_3 \ \cdots \ Q_n]$ has orthonormal columns, and the second factor is an $n \times n$ upper triangular matrix R with positive diagonal entries (and so is invertible). We record this in the following theorem (the proof of uniqueness is deferred to the end of this section).

Theorem 9. **QR-Factorization.** *If A is an $m \times n$ matrix with linearly independent columns then A can be factored as*

$$A = QR$$

where Q is an $m \times n$ matrix with orthonormal columns, and R is an invertible, upper triangular matrix with positive entries on the main diagonal. Furthermore, this factorization is unique in the sense that if $A = Q_1 R_1$ is another such factorization then $Q_1 = Q$ and $R_1 = R$.

If a matrix A has independent rows and we apply QR-factorization to A^T, the result is:

Corollary. *If A has independent rows then $A = LP$ where P has orthonormal rows and L is an invertible lower triangular matrix with positive main diagonal entries.*

Example 9. *Find the QR-factorization of* $A = \begin{bmatrix} 1 & 1 & 0 \\ -1 & 0 & 1 \\ 0 & 1 & 1 \\ 0 & 0 & 1 \end{bmatrix}$.

Solution. The columns of A are $C_1 = [1\ -1\ 0\ 0]^T$, $C_2 = [1\ 0\ 1\ 1]^T$ and $C_3 = [0\ 1\ 1\ 1]^T$, and $\{C_1, C_2, C_3\}$ is independent. If we apply the Gram-Schmidt Algorithm to these columns C_i, the result is

$$\begin{array}{rcl rcl} F_1 & = & C_1 & = & [1\ -1\ 0\ 0]^T \\ F_2 & = & C_2 - \frac{1}{2}F_1 & = & [\frac{1}{2}\ \frac{1}{2}\ 1\ 0]^T \\ F_3 & = & C_3 + \frac{1}{2}F_1 - F_2 & = & [0\ 0\ 0\ 1]^T \end{array}$$

Hence let $Q_j = \frac{1}{\|F_j\|} F_j$ for each j. Then equation (*) preceding Theorem 9 gives $A = QR$ where

$$Q = [Q_1\ Q_2\ Q_2] = \begin{bmatrix} \frac{1}{\sqrt{2}} & \frac{1}{\sqrt{6}} & 0 \\ \frac{-1}{\sqrt{2}} & \frac{1}{\sqrt{6}} & 0 \\ 0 & \frac{2}{\sqrt{6}} & 0 \\ 0 & 0 & 1 \end{bmatrix} = \frac{1}{\sqrt{6}} \begin{bmatrix} \sqrt{3} & 1 & 0 \\ -\sqrt{3} & 1 & 0 \\ 0 & 2 & 0 \\ 0 & 0 & \sqrt{6} \end{bmatrix},$$

$$R = \begin{bmatrix} \|F_1\| & C_2 \bullet Q_1 & C_3 \bullet Q_1 \\ 0 & \|F_2\| & C_3 \bullet Q_2 \\ 0 & 0 & \|F_3\| \end{bmatrix} = \begin{bmatrix} \sqrt{2} & \frac{1}{\sqrt{2}} & \frac{-1}{\sqrt{2}} \\ 0 & \frac{\sqrt{3}}{\sqrt{2}} & \frac{\sqrt{3}}{\sqrt{2}} \\ 0 & 0 & 1 \end{bmatrix}$$

$$= \frac{1}{\sqrt{2}} \begin{bmatrix} 2 & 1 & -1 \\ 0 & \sqrt{3} & \sqrt{3} \\ 0 & 0 & \sqrt{2} \end{bmatrix}.$$

The reader can verify that indeed $A = QR$. □

Remark. If an $m \times n$ matrix A has independent columns then $A^T A$ is invertible (by Theorem 3 §4.4), and it is often desirable to compute the inverse of $A^T A$ (see Section 4.6 for example). This is simplified if we have a QR-factorization of A (and is one of the main reasons for the importance of Theorem 9). For if

4.5. ORTHOGONALITY

$A = QR$ is such a factorization, then $Q^T Q = I_n$ because Q has orthonormal columns (verify), so we obtain

$$A^T A = R^T Q^T Q R = R^T R.$$

Hence computing $(A^T A)^{-1}$ amounts to finding R^{-1}, and this is a routine matter because R is upper triangular. Thus the difficulty in computing $(A^T A)^{-1}$ lies in obtaining the QR-factorization of A.

4.5.5 Proof of Uniqueness in the QR-Factorization

Suppose $A = QR = Q_1 R_1$ are two QR-factorizations of the $m \times n$ matrix A. Write $Q = [C_1 \; C_2 \; \cdots \; C_n]$ and $Q_1 = [D_1 \; D_2 \; \cdots \; D_n]$ in terms of their columns, and observe first that $Q^T Q = I_n = Q_1^T Q_1$ because Q and Q_1 have orthonormal columns. Hence it suffices to show that $Q_1 = Q$ (then $R_1 = Q_1^T A = Q^T A = R$). Since $Q_1^T Q_1 = I_n$, the equation $QR = Q_1 R_1$ gives $Q_1^T Q = R_1 R^{-1}$; for convenience we write this matrix as

$$Q_1^T Q = R_1 R^{-1} = [t_{ij}].$$

This matrix is upper triangular with positive diagonal elements (since this is true for R and R_1), so $t_{ii} > 0$ for each i and $t_{ij} = 0$ if $i > j$. On the other hand, the (i,j)-entry of $Q_1^T Q$ is $D_i^T C_j = D_i \bullet C_j$, so we have $D_i \bullet C_j = t_{ij}$ for all i and j. But $\{C_1, C_2, \cdots, C_n\}$ is an orthonormal basis of $\mathbb{R}^n$, so the Expansion Theorem gives

$$\begin{aligned} D_i &= (D_i \bullet C_1)C_1 + (D_i \bullet C_2)C_2 + \cdots + (D_i \bullet C_i)C_i + \cdots + (D_i \bullet C_n)C_n \\ &= t_{i\,1}C_1 + t_{i\,2}C_2 + \cdots + t_{i\,i}C_i + \cdots + t_{i\,n}C_n \\ &= t_{i\,i}C_i + \cdots + t_{i\,n}C_n \end{aligned}$$

because $(D_i \bullet C_j)C_j = t_{i\,j} = 0$ if $i > j$. The first few equations here (in reverse order) are

$$\begin{aligned} D_n &= t_{n\,n}C_n \\ D_{n-1} &= t_{n-1\,n-1}C_{n-1} + t_{n-1\,n}C_n \\ D_{n-2} &= t_{n-2\,n-2}C_{n-2} + t_{n-2\,n-1}C_{n-1} + t_{n-2\,n}C_n \\ &\vdots \end{aligned}$$

The first of these equations gives $1 = \|D_n\| = \|t_{n\,n}C_n\| = |t_{n\,n}|\,\|C_n\| = t_{n\,n}$, whence $D_n = C_n$. But then

$$t_{n-1\,n} = D_{n-1} \bullet C_n = D_{n-1} \bullet D_n = 0,$$

so the second equation becomes $D_{n-1} = t_{n-1\,n-1}C_{n-1}$. As before it follows that $D_{n-1} = C_{n-1}$, and hence that $t_{n-2\,n} = 0$ and $t_{n-2\,n-1} = 0$. Hence the third equation reads $D_{n-2} = t_{n-2\,n-2}C_{n-2}$. Again this gives $D_{n-2} = C_{n-2}$, and the process continues in this way to get $D_i = C_i$ for all i. This means that $Q_1 = Q$, which is what we wanted.

Exercises

1. If $\|X\| = 2$, $\|Y\| = 1$ and $X \bullet Y = -1$, compute:
   - (a) $\|2X - 5Y\|$
   - (b) $\|3X - 4Y\|$
   - (c) $(2X - Y) \bullet (3X + 7Y)$
   - (d) $(3X + 4Y) \bullet (X - 5Y)$

2. In each case show that the set of vectors is orthogonal in $\mathbb{R}^4$.
   - (a) $\{[1\ 1\ -1\ 4]^T, [-1\ 2\ 5\ 1]^T, [28\ 5\ 5\ -7]^T\}$.
   - (b) $\{[4\ 5\ 2\ -1]^T, [1\ -1\ 0\ -1]^T, [2\ -1\ 0\ 3]^T\}$.

3. By normalizing the following sets, obtain an orthonormal basis of $\mathbb{R}^3$.
   - (a) $\{[1\ 0\ 2]^T, [-2\ 5\ 1]^T, [2\ 1\ -1]^T\}$.
   - (b) $\{[-3\ 1\ 2]^T, [1\ 1\ 1]^T, [1\ -5\ 4]^T\}$.

4. Let $U = span\{[1\ -2\ 1\ -1]^T, [2\ 1\ -1\ 1]^T\}$. Show that $Y = [1\ 3\ -2\ 2]^T$ is in U, and find all Z such that $\{Y, Z\}$ is an orthogonal basis of U.

5. In each case find all vectors $[a\ b\ c\ d]^T$ in $\mathbb{R}^4$ such that the given set is orthogonal.
   - (a) $\{[1\ -1\ 1\ 0]^T, [-1\ 3\ 4\ 1]^T, [2\ 1\ -1\ 3]^T, [a\ b\ c\ d]^T\}$.
   - (b) $\{[1\ 2\ 1\ -1]^T, [1\ 1\ -2\ 1]^T, [5\ -2\ 4\ 5]^T, [a\ b\ c\ d]^T\}$.

6. In each case show that the set of vectors is an orthogonal basis of $\mathbb{R}^3$, and use the Expansion Theorem to write $[a\ b\ c]^T$ as a linear combination of the basis vectors.
   - (a) $\{[3\ 1\ -1]^T, [1\ -2\ 1]^T, [1\ 4\ 7]^T\}$.
   - (b) $\{[-4\ 1\ 5]^T, [-1\ 1\ -1]^T, [2\ 3\ 1]^T\}$.

7. In each case use the Gram-Schmidt Algorithm to convert the given basis into an orthogonal basis.
   - (a) $\{[1\ -1]^T, [2\ 1]^T\}$ in $\mathbb{R}^2$
   - (b) $\{[2\ 1]^T, [1\ 2]^T\}$ in $\mathbb{R}^2$
   - (c) $\{[1\ -1\ 1]^T, [1\ 0\ 1]^T, [1\ 1\ 2]^T\}$ in $\mathbb{R}^3$
   - (d) $\{[0\ 1\ 1]^T, [1\ 1\ 1]^T, [1\ -2\ 2]^T\}$ in $\mathbb{R}^3$

8. Find the QR-factorization of each of the following matrices.
   - (a) $\begin{bmatrix} 1 & -1 \\ -1 & 0 \end{bmatrix}$
   - (b) $\begin{bmatrix} 2 & 1 \\ 1 & 1 \end{bmatrix}$
   - (c) $\begin{bmatrix} 1 & 1 & 1 \\ 1 & 1 & 0 \\ 1 & 0 & 0 \\ 0 & 0 & 0 \end{bmatrix}$
   - (d) $\begin{bmatrix} 1 & 1 & 0 \\ -1 & 0 & 1 \\ 0 & 1 & 1 \\ 1 & -1 & 0 \end{bmatrix}$

4.5. ORTHOGONALITY

9. In each case either show that the statement is true or give an example showing that it is false.

    (a) Every independent set in $\mathbb{R}^n$ is orthogonal.

    (b) If $\{X, Y\}$ is an orthogonal set in $\mathbb{R}^n$ then $\{X, X+Y\}$ is also orthogonal.

    (c) If $\{X, Y\}$ and $\{Z, W\}$ are both orthogonal in $\mathbb{R}^n$, then $\{X, Y, Z, W\}$ is also orthogonal.

    (d) If $\{X_1, X_2\}$ and $\{Y_1, Y_2, Y_3\}$ are orthogonal and $X_i \bullet Y_j = 0$ for all i and j, then $\{X_1, X_2, Y_1, Y_2, Y_3\}$ is orthogonal.

    (e) If $\{X_1, X_2, \cdots, X_n\}$ is orthogonal in $\mathbb{R}^n$ then $\mathbb{R}^n = span\{X_1, X_2, \cdots, X_n\}$.

10. If $\{X_1, \cdots, X_m\}$ is an orthogonal set in $\mathbb{R}^n$, show that $\{a_1 X_1, \cdots, a_m X_m\}$ is orthogonal for any numbers $a_i \neq 0$.

11. Let X and Y denote vectors in $\mathbb{R}^n$.

    (a) If X and Y are orthogonal show that $\|X - Y\| = \|X + Y\|$.

    (b) If $\|X - Y\| = \|X + Y\|$ show that X and Y are orthogonal.

12. Let X, Y and Z be vectors in $\mathbb{R}^n$.

    (a) If $\|X + Y\|^2 = \|X\|^2 + \|Y\|^2$ show that X and Y are orthogonal.

    (b) Show that $\|X + Y + Z\|^2 = \|X\|^2 + \|Y\|^2 + \|Z\|^2$ does not imply that $\{X, Y, Z\}$ is orthogonal by considering $X = [1\ 1\ 0]^T$, $Y = [0\ -1\ 1]^T$, and $Z = [1\ 1\ 0]^T$.

13. Let X and Y denote vectors in $\mathbb{R}^n$.

    (a) Show that $X \bullet Y = \frac{1}{4}\left\{\left[\|X+Y\|^2 - \|X-Y\|^2\right]\right\}$.

    (b) Show that $\|X\|^2 + \|Y\|^2 = \frac{1}{2}\left\{\left[\|X+Y\|^2 + \|X-Y\|^2\right]\right\}$.'

14. Let λ be an eigenvalue of $A^T A$ where A is an $n \times n$ matrix. Since $A^T A$ is symmetric, we know that λ is real from Theorem 3 §2.5. Show that $\lambda \geq 0$. [*Hint*: Compute $\|AX\|^2$ where X is an eigenvector.]

15. If Q is an $m \times n$ matrix with orthonormal columns, explain why $Q^T Q = I_n$.

16. If $\mathbb{R}^n = span\{F_1, F_2, \cdots, F_m\}$ and X in $\mathbb{R}^n$ satisfies $X \bullet F_i = 0$ for all i, show that $X = 0$. [*Hint*: Show that $\|X\| = 0$.]

17. If $\mathbb{R}^n = span\{F_1, F_2, \cdots, F_m\}$, and if X and Y in $\mathbb{R}^n$ satisfies $X \bullet F_i = Y \bullet F_i$ for all i, show that $X = Y$.

18. Let U be a subspace of $\mathbb{R}^n$ with $dim U = m \geq 2$, and assume that $\{F_1, F_2, \cdots, F_{m-1}\}$ is an orthonormal set in U.

    (a) Show that U has an orthonormal basis $\{F_1, F_2, \cdots, F_{m-1}, G\}$ for some G in U.

(b) If $\{F_1, F_2, \cdots, F_{m-1}, H\}$ is an orthonormal basis of U, determine H in terms of G.

19. Let $\{X_1, X_2, \cdots, X_m\}$ be an orthogonal basis of a subspace U of $\mathbb{R}^n$. If the Gram-Schmidt Algorithm is applied to this basis to produce an orthogonal basis $\{F_1, F_2, \cdots, F_m\}$ of U, describe the F_i in terms of the X_i.

20. Let $\{X_1, X_2, \cdots, X_n\}$ be a basis of $\mathbb{R}^n$. If $\{Y_1, Y_2, \cdots, Y_n\}$ are vectors in $\mathbb{R}^n$ which satisfy

$$X_i \bullet Y_j = \begin{cases} 1 & \text{if } i = j \\ 0 & \text{if } i \neq j \end{cases},$$

show that $\{Y_1, Y_2, \cdots, Y_n\}$ is also a basis of $\mathbb{R}^n$. What does this say about matrix inverses?

21. Let $\{G_1, G_2, \cdots, G_n\}$ be an orthogonal set in $\mathbb{R}^n$. Given X and Y in $\mathbb{R}^n$, show that

$$X \bullet Y = \frac{(X \bullet G_1)(Y \bullet G_1)}{\|G_1\|^2} + \frac{(X \bullet G_2)(Y \bullet G_2)}{\|G_2\|^2} + \cdots + \frac{(X \bullet G_n)(Y \bullet G_n)}{\|G_n\|^2}.$$

22. Let A be an $n \times n$ matrix of rank r. Show that there is an invertible $n \times n$ matrix U such that UA is a row-echelon matrix with orthogonal rows.

4.6 PROJECTIONS AND APPROXIMATION

Let U be a plane in $\mathbb{R}^3$ containing the origin, and let $\vec{v}$ be a vector in $\mathbb{R}^3$ (recall that we are identifying points with their position vectors). It is evident geometrically that there is a vector $\vec{p}$ in the plane U which is closest to $\vec{v}$ (see the diagram). Indeed the vector $\vec{p}$ is determined by the fact that the line through $\vec{v}$ and $\vec{p}$ must be perpendicular to the plane.

Now we make two observations. First, the plane is a *subspace* of $\mathbb{R}^3$; and second, the condition that $\vec{v} - \vec{p}$ is perpendicular to the plane means that $\vec{v} - \vec{p}$ is *orthogonal* to every vector in the plane. In this form the whole discussion makes sense in $\mathbb{R}^n$. In fact, we show in this section that, given a vector X and a subspace U of $\mathbb{R}^n$, there exists a vector P in U that is closest to X. This vector P is called the projection of X on the subspace U, and the result leads to a general method for solving certain minimization problems. This technique will be used throughout this chapter.

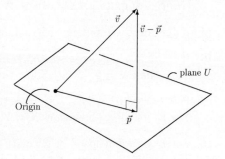

4.6.1 Orthogonal Complements

If U is a subspace of $\mathbb{R}^n$, the **orthogonal complement** $U^\perp$ is defined[11] to be the set of all vectors that are orthogonal to every vector in U:

$$U^\perp = \{X \text{ in } \mathbb{R}^n \mid X \bullet Y = 0 \text{ for all } Y \text{ in } U\}.$$

Example 1. If L is a line through the origin in $\mathbb{R}^3$ then $L^\perp$ is the plane through the origin perpendicular to L. Similarly, if P is a plane through the origin in $\mathbb{R}^3$ then $P^\perp$ is the line through the origin perpendicular to P.

In Example 1, note that $L^\perp$ and $P^\perp$ are both *subspaces* of $\mathbb{R}^3$. This holds much more generally. In fact, if U is any subspace of $\mathbb{R}^n$ then $U^\perp$ is again a subspace. For if X and X_1 are both in $U^\perp$ then $(X+X_1)\bullet Y = X\bullet Y + X_1 \bullet Y = 0+0 = 0$ for all Y in U, so $X+X_1$ is also in $U^\perp$. Similarly aX is in $U^\perp$ for all scalars a, as the reader can verify, so $U^\perp$ is indeed a subspace. Moreover there is a routine way to determine $U^\perp$ when a spanning set for U is known.

Theorem 1. *Let U denote a subspace of $\mathbb{R}^n$. Then*:

(1) $U^\perp$ *is also a subspace of $\mathbb{R}^n$.*

(2) *If $U = \text{span}\{X_1, \cdots, X_k\}$ then $U^\perp = \{X \mid X \bullet X_i = 0 \text{ for each } i = 1, 2, \cdots, k\}$.*

Proof. We verified (1) above. As to (2), suppose that $X \bullet X_i = 0$ for each i. If Y is in U, write it as $Y = t_1 X_1 + \cdots + t_k X_k$ where the t_i are scalars. Then

$$\begin{aligned} X \bullet Y &= X \bullet (t_1 X_1 + \cdots + t_k X_k) \\ &= t_1(X \bullet X_1) + \cdots + t_k(X \bullet X_k) \\ &= t_1(0) + \cdots + t_k(0) \\ &= 0. \end{aligned}$$

It follows that $\{X \mid X \bullet X_i = 0 \text{ for each } i\} \subseteq U^\perp$; the other inclusion follows because each X_i is in U. $\square$

When U has a spanning set of k vectors, the second part of Theorem 1 reduces the description of $U^\perp$ to solving a system of k linear equations. Here is an example.

Example 2. *Find $U^\perp$ if $U = \text{span}\{[1 \ -2 \ -2 \ 1 \ 0], [-2 \ 4 \ 3 \ -1 \ 2]\}$ in $\mathbb{R}^5$.*

Solution. By Theorem 1, a vector $X = [x_1 \ x_2 \ x_3 \ x_4 \ x_5]$ is in $U^\perp$ if and only if X is orthogonal to both $[1 \ -2 \ -2 \ 1 \ 0]$ and $[-2 \ 4 \ 3 \ -1 \ 2]$, that is if and only if

$$\begin{aligned} x_1 - 2x_2 - 2x_3 + x_4 &= 0 \\ -2x_1 + 4x_2 + 3x_3 - x_4 + 2x_5 &= 0 \end{aligned}$$

Gaussian elimination gives $X = r[2 \ 1 \ 0 \ 0 \ 0] + s[1 \ 0 \ 1 \ 1 \ 0] + t[4 \ 0 \ 2 \ 0 \ 1]$ where r, s and t are parameters. Hence $U^\perp = \text{span}\{[2 \ 1 \ 0 \ 0 \ 0], [1 \ 0 \ 1 \ 1 \ 0], [4 \ 0 \ 2 \ 0 \ 1]\}$. $\square$

---

[11] When speaking, $U^\perp$ is pronounced "U-perp".

4.6.2 Projections

Let L be a line through the origin in $\mathbb{R}^3$, and let $\vec{v}$ be a vector in $\mathbb{R}^3$. If $\vec{d}$ is any direction vector for L, let $\vec{p}$ denote the projection of $\vec{v}$ on $\vec{d}$ as discussed in Section 3.2 (see the diagram).

Clearly $\vec{p}$ can be characterized as the unique vector in L such that $\vec{v} - \vec{p}$ is perpendicular to L, that is such that

$\vec{v} - \vec{p}$ is in $L^\perp$.

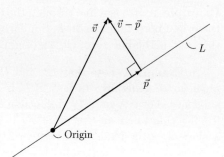

The fact that L is a subspace of $\mathbb{R}^3$ means that we can formulate this in $\mathbb{R}^n$.

Thus if U is a subspace of $\mathbb{R}^n$ and X is a vector in $\mathbb{R}^n$, we look for a vector P in U such that

$X - P$ is in $U^\perp$.

Remarkably, not only does such a vector P exist, there is an explicit formula for it. Indeed, given an orthogonal basis $\{F_1, F_2, \cdots, F_m\}$ of U, consider the vector

$$P = \frac{X \bullet F_1}{\|F_1\|^2} F_1 + \frac{X \bullet F_2}{\|F_2\|^2} F_2 + \cdots + \frac{X \bullet F_m}{\|F_m\|^2} F_m. \qquad (*)$$

Then P is in U because each F_i is in U, and $X - P$ is in $U^\perp$ because it is orthogonal to each F_i by the Orthogonal Lemma (Lemma 1 §4.5).

Even more is true: This vector P is the *same* for every choice of orthogonal basis of U. More precisely, if $\{G_1, G_2, \cdots, G_m\}$ is any other orthogonal basis of U, write

$$P' = \frac{X \bullet G_1}{\|G_1\|^2} G_1 + \frac{X \bullet G_2}{\|G_2\|^2} G_2 + \cdots + \frac{X \bullet G_m}{\|G_m\|^2} G_m$$

for the vector formed as in $(*)$ but using the G_i in place of the F_i. We claim that $P' = P$. To see this, note first that P' is in U and $X - P'$ is in $U^\perp$ by the Orthogonal Lemma, as before. Now write the vector $P - P'$ as follows:

$$P - P' = (X - P') - (X - P).$$

Then $P - P'$ is in $U^\perp$ (because $X - P$ and $X - P'$ are in $U^\perp$), and $P - P'$ is also in U (because P and P' are in U). But then $P - P'$ is orthogonal to *itself*! This means that

$$\|P - P'\|^2 = (P - P') \bullet (P - P') = 0.$$

Thus $P - P' = 0$, that is $P = P'$ as claimed. We record this observation as

Lemma 1. *Let U be a nonzero subspace of $\mathbb{R}^n$, and let X be any vector in $\mathbb{R}^n$. If $\{F_1, F_2, \cdots, F_m\}$ is an orthogonal basis of U, the vector*

$$P = \frac{X \bullet F_1}{\|F_1\|^2} F_1 + \frac{X \bullet F_2}{\|F_2\|^2} F_2 + \cdots + \frac{X \bullet F_m}{\|F_m\|^2} F_m$$

4.6. PROJECTIONS AND APPROXIMATION

depends only on U and X; that is, P is the same for every choice of orthogonal basis $\{F_1, F_2, \cdots, F_m\}$.

The vector P in Lemma 1 is called[12] the **projection** of X on U, and is denoted $proj_U(X)$. If $U = 0$ we define $proj_0(X) = 0$. The next example shows that this definition agrees with that in Section 3.2 when U is a line through the origin.

Example 3. Let L be the line through the origin in $\mathbb{R}^3$ with direction vector $\vec{d}$. Given $\vec{v}$ in $\mathbb{R}^3$, Theorem 4 §3.2 gives a formula for the projection of $\vec{v}$ on $\vec{d}$:

$$proj_{\vec{d}}(\vec{v}) = \frac{\vec{v} \bullet \vec{d}}{\|\vec{d}\|^2} \vec{d}.$$

This agrees with $proj_L(\vec{v})$ if we use $\{\vec{d}\}$ as the orthogonal basis of L.

The above discussion proves the most important theorem of this section.

Theorem 2. Projection Theorem. Let U be a subspace of $\mathbb{R}^n$, let X be a vector in $\mathbb{R}^n$. Then

$$proj_U(X) \text{ is in } U \quad \text{and} \quad X - proj_U(X) \text{ is in } U^\perp.$$

Let X be a vector in $\mathbb{R}^n$, let U be a subspace of $\mathbb{R}^n$, and write $P = proj_U(X)$. Then the situation in the Projection Theorem can be depicted as in the diagram (where the subspace U is shown as a plane). In particular, we see that

$$X = P + (X - P)$$

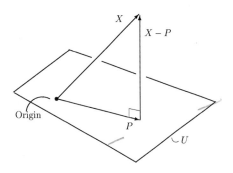

is one way to write a vector X in $\mathbb{R}^n$ as a sum of a vector P in U and a vector $X - P$ in $U^\perp$. In fact this is the *only* way to write X as such a sum. This observation is often useful and we record it in the following form.

Corollary. Let U be a subspace of $\mathbb{R}^n$, let X be a vector in $\mathbb{R}^n$. If

$$X = Y + Z$$

where Y is in U and Z is in $U^\perp$, then necessarily $Y = proj_U(X)$ and $Z = X - proj_U(X)$.

---

[12]This is sometimes called the **orthogonal projection** of X on U, but we prefer the simpler terminology.

Proof. Write $P = \text{proj}_U(X)$. We have $Y + Z = X = P + (X - P)$. Hence $Y - P = (X - P) - Z$, and this vector is in both U and $U^\perp$—in fact $Y - P$ is in U while $(X - P) - Z$ is in $U^\perp$. It follows that $Y = P$ and $(X - P) = Z$. This proves the Corollary. $\square$

Example 4. Let $U = \text{span}\{X_1, X_2\}$ where $X_1 = [1 \ -1 \ 0 \ 1]$ and $X_2 = [1 \ 0 \ 1 \ 0]$ in $\mathbb{R}^4$. If $X = [2 \ 1 \ -1 \ 1]$, express X as the sum of a vector in U and a vector in $U^\perp$.

Solution. Note first that, while $\{X_1, X_2\}$ is a basis of U, it is not orthogonal. The Gram-Schmidt Algorithm gives an orthogonal basis $\{F_1, F_2\}$ of U where

$$\begin{aligned} F_1 &= X_1 = [1 \ -1 \ 0 \ 1], \text{ and} \\ F_2 &= X_2 - \frac{X_2 \bullet F_1}{\|F_1\|^2} F_1 = X_2 - \tfrac{1}{3}F_1 = \tfrac{1}{3}[2 \ 1 \ 3 \ -1]. \end{aligned}$$

For convenience, we use the orthogonal basis $\{G_1, G_2\}$ where we take $G_1 = F_1 = [1 \ -1 \ 0 \ 1]$ and $G_2 = 3F_2 = [2 \ 1 \ 3 \ -1]$. We compute the projection P of X on U using $\{G_1, G_2\}$:

$$P = \text{proj}_U(X) = \frac{X \bullet G_1}{\|G_1\|^2} G_1 + \frac{X \bullet G_2}{\|G_2\|^2} G_2 = \tfrac{2}{3}G_1 + \tfrac{1}{15}G_2 = \tfrac{1}{5}[4 \ -3 \ 1 \ 3].$$

Then we compute $X - P = \tfrac{2}{5}[3 \ 4 \ -3 \ 1]$. Theorem 2 asserts that this vector $X - P$ is in $U^\perp$ (this can be verified by checking that it is orthogonal to the two generators X_1 and X_2 of U, and this is a good test of the arithmetic). Finally, since P is in U,

$$X = P + (X - P) = \tfrac{1}{5}[4 \ -3 \ 1 \ 3] + \tfrac{2}{5}[3 \ 4 \ -3 \ 1]$$

is the required decomposition of X. $\square$

The Projection Theorem has many consequences, among them the following two important facts about the orthogonal complement $U^\perp$ of a subspace U of $\mathbb{R}^n$. Since $U^\perp$ is also a subspace, it too has an orthogonal complement which, for convenience, we write as $U^{\perp\perp} = (U^\perp)^\perp$. Surprisingly, $U^{\perp\perp} = U$.

Theorem 3. *Let U be any subspace of $\mathbb{R}^n$. Then:*

(1) $\dim(U^\perp) = n - \dim U$.

(2) $U^{\perp\perp} = U$.

Proof. (1). If $U = 0$ then $U^\perp = \mathbb{R}^n$ and (1) holds. If $U^\perp = 0$ and X is in $\mathbb{R}^n$ then $X - \text{proj}_U(X)$ is in $U^\perp$ by the Projection Theorem, so $X = \text{proj}_U(X)$ is in U. Hence $U = \mathbb{R}^n$, and again (1) holds.

Thus we may assume that $U \neq 0$ and $U^\perp \neq 0$, so let $\{F_1, F_2, \cdots, F_m\}$ and $\{G_1, G_2, \cdots, G_k\}$ be orthogonal bases of U and $U^\perp$ respectively. Then

4.6. PROJECTIONS AND APPROXIMATION

$\{F_1, F_2, \cdots, F_m, G_1, G_2, \cdots, G_k\}$ is an orthogonal set in $\mathbb{R}^n$ (verify), and it spans $\mathbb{R}^n$ by the Projection Theorem. It is thus a basis of $\mathbb{R}^n$, so $dim(\mathbb{R}^n) = m + k$. Hence $n = dimU + dim(U^\perp)$, and (1) holds in this case too.

(2). We have $U \subseteq U^{\perp\perp}$ because every vector in U is orthogonal to every vector in $U^\perp$. Conversely, let Y be in $U^{\perp\perp}$, and write $P = proj_U(Y)$. Then P is in $U \subseteq U^{\perp\perp}$, so P is in $U^{\perp\perp}$. It follows that $Y - P$ is also in $U^{\perp\perp}$. But $Y - P$ is in $U^\perp$ by the Projection Theorem, and it follows that $Y - P$ is orthogonal to itself. Thus $Y - P = 0$, whence $Y = P$ is in U. This proves that $U^{\perp\perp} \subseteq U$, so $U^{\perp\perp} = U$. □

Let X be a vector in $\mathbb{R}^n$, and let $P = proj_U(X)$ be its projection on a subspace U. Then $X = (X - P) + P$ is a decomposition of X as the sum of a vector $X - P$ in $U^\perp$ and a vector P in $U = (U^\perp)^\perp$. Hence the Corollary to the Projection Theorem (for $U^\perp$ in place of U) shows that $X - P = proj_{U^\perp}(X)$. This proves the

Corollary. *Let U be a subspace of $\mathbb{R}^n$, and let X be a vector in $\mathbb{R}^n$. Then*

$$X = proj_U(X) + proj_{U^\perp}(X)$$

is the unique decomposition of X as the sum of a vector in U and a vector in $U^\perp$.

4.6.3 Approximation.

There is another facet of the Projection Theorem that is very important in applications as we shall see. If X is a vector in $\mathbb{R}^n$ and U is a subspace of $\mathbb{R}^n$, then the projection P of X on U is the vector in U that is *closest* to X.

This is intuitively clear from the diagram where P and Y are vectors in U and, as before, the subspace U is depicted as a plane. The following theorem makes it more precise. Recall that the distance between two vectors X and Y is $\|X - Y\|$.

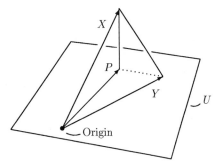

Theorem 4. Approximation Theorem. *Let U be a subspace of $\mathbb{R}^n$, let X be a vector in $\mathbb{R}^n$, and write $P = proj_U(X)$. Then P is the vector in U closest to X in the sense that*

$$\|X - P\| < \|X - Y\| \text{ for all } Y \neq P \text{ in } U.$$

Proof. Given Y in U where $Y \neq P$, write $X - Y$ in the form

$$X - Y = (X - P) + (P - Y).$$

Then $X - P$ is in $U^\perp$ by the Projection Theorem, and $P - Y$ is in U (because P is in U). Hence $X - P$ and $P - Y$ are orthogonal, so Pythagoras' Theorem gives

$$\|X - Y\|^2 = \|X - P\|^2 + \|P - Y\|^2 > \|X - P\|^2$$

because $P - Y \neq 0$. The theorem follows. □

Example 5. *Consider the plane U with equation $2x + 3y - z = 0$. Find the point in U which is closest to the point $X = [3 \ -1 \ 2]^T$.*

Solution. The plane U is the subspace of $\mathbb{R}^3$ whose points $[x \ y \ z]^T$ satisfy $z = 2x + 3y$. Hence

$$U = \{[s \ t \ 2s + 3t]^T \mid s, t \text{ in } \mathbb{R}\} = span\{[1 \ 0 \ 2]^T, [0 \ 1 \ 3]^T\}.$$

The Gram-Schmidt process gives an orthogonal basis $\{F_1, F_2\}$ for U where $F_1 = [1 \ 0 \ 2]^T$ and $F_2 = [-6 \ 5 \ 3]^T$. Hence the vector in U closest to X is

$$P = proj_U(X) = \frac{X \bullet F_1}{\|F_1\|^2} F_1 + \frac{X \bullet F_2}{\|F_2\|^2} F_2 = \tfrac{7}{5} F_1 + \tfrac{-17}{70} F_2 = \tfrac{1}{14}[40 \ -17 \ 29]^T.$$

As noted before, a good check on the arithmetic is to compute $X - P$ and check that it is in $U^\perp$. In the present case, $X - P = \tfrac{1}{14}[2 \ 3 \ -1]^T$, which is indeed in $U^\perp$ because it is orthogonal to both $[1 \ 0 \ 2]^T$ and $[0 \ 1 \ 3]^T$. □

4.6.4 Inconsistent Systems

A system of linear equations may be inconsistent, that is it may not have a solution. However, even when there is no solution, "best approximations" to a solution always exist, and finding them is one of the most useful applications of the Approximation Theorem. The theorem is applied by carefully choosing the subspace U so that the "best approximations" are the vectors in U closest to some particular vector.

More precisely, if A is an $m \times n$ matrix and B is a column in $\mathbb{R}^m$, consider the system

$$AX = B$$

of m linear equations in n variables. This system may have no solution, that is it may happen that $AZ \neq B$ for every column Z in $\mathbb{R}^n$. On the other hand, the distance $\|B - AZ\|$ between B and AZ is a measure of how close AZ is to B. Hence it is natural to ask if there are columns Z in $\mathbb{R}^n$ that are as close as possible to a solution in the sense that

$$\|B - AZ\|$$

4.6. PROJECTIONS AND APPROXIMATION

is the minimum value of $\|B - AX\|$ as X ranges over all columns in $\mathbb{R}^n$. The Approximation Theorem gives an affirmative answer. To see how, write

$$U = \{AX \mid X \text{ in } \mathbb{R}^n\} = im A.$$

This is a subspace of $\mathbb{R}^n$ as we have seen, and we want a vector in U as close as possible to B. But the Approximation Theorem shows that $proj_U(B)$ has exactly this property. Moreover, since $proj_U(B)$ is in U we have

$$proj_U(B) = AZ$$

for some Z in $\mathbb{R}^n$. Any such vector Z is called a **best approximation** to a solution of the system $AX = B$.

This appears to solve the problem, but there are two difficulties. First we need an orthogonal basis of U to compute $proj_U(B)$, and second we end up with AZ rather than Z. Fortunately, there is a routine way to find all best approximations Z directly. The key observation is that $B - AZ$ lies in $U^\perp$ (by the Projection Theorem), so it is orthogonal to every column AX in U. Thus, for all X in $\mathbb{R}^n$, we have

$$\begin{aligned} 0 &= (AX) \bullet (B - AZ) \\ &= (AX)^T (B - AZ) \\ &= X^T A^T (B - AZ) \\ &= X \bullet [A^T(B - AZ)]. \end{aligned}$$

In other words, $A^T(B - AZ)$ is orthogonal to every vector in $\mathbb{R}^n$, and so must be zero (being orthogonal to itself). Hence $(A^T A)Z = A^T B$, so every best approximation Z to a solution to the system $AX = B$ is a solution to the matrix equation

$$(A^T A)Z = A^T B.$$

This is a system of linear equations called the **normal equations** for Z, and it turns out that *every* solution is actually a best approximation.

Theorem 5. Let A be an $m \times n$ matrix, let B be any column in $\mathbb{R}^m$, and consider the system

$$AX = B$$

of m linear equations in n variables.

1. The solutions Z to the normal equations

   $$(A^T A)Z = A^T B$$

   are exactly the best approximations to a solution to $AX = B$ in the sense that $\|B - AZ\|$ is the minimum value of $\|B - AX\|$ as X ranges over all columns of $\mathbb{R}^n$.

2. The best approximation Z is unique if and only if $A^T A$ is invertible.

Proof. (1). We showed above that every best approximation Z is a solution to the normal equations. On the other hand, if Z satisfies these equations then, by reversing the above argument, we see that $B - AZ$ is in $U^\perp$, where $U = \{AX \mid X \text{ in } \mathbb{R}^n\}$. Thus we have $B = AZ + (B - AZ)$ where AZ is in U and $B - AZ$ is in $U^\perp$. This means that $AZ = proj_U(B)$ by the Corollary to the Projection Theorem, so Z is a best approximation to a solution of the system $AX = B$.

(2). If $A^T A$ is invertible then $Z = (A^T A)^{-1} A^T B$ is uniquely determined by A and B. Conversely, if Z is unique, we show that $A^T A$ is invertible by showing (Theorem 5 §1.5) that $(A^T A)X = 0$ implies $X = 0$. But if $(A^T A)X = 0$, then $(A^T A)(Z + X) = (A^T A)Z + 0 = A^T B$, so $Z + X = Z$ by the uniqueness, that is $X = 0$. This proves (2). $\square$

In connection with condition (2) of Theorem 5, recall Theorem 3 §4.4: If A is any $m \times n$ matrix, then $A^T A$ is invertible if and only if A has independent columns (equivalently, if $rank A = n$, or if $AX = 0$ has unique solution $X = 0$). Hence in this case there is a unique best approximation Z to a solution of the system $AX = B$.

Of course, having many best approximations is not necessarily a bad thing. It means that you have room to choose a best approximation that optimizes some other feature of the solution to the system.

Example 6. *Find the best approximations to a solution of the following inconsistent system of linear equations*:

$$\begin{array}{rcrcrcrcr} x_1 & - & x_2 & + & 3x_3 & & & = & 2 \\ -2x_1 & + & 2x_2 & & & + & 4x_4 & = & -1 \\ x_1 & + & x_2 & + & 9x_3 & + & 8x_4 & = & 3 \end{array}.$$

Solution. Here $A = \begin{bmatrix} 1 & -1 & 3 & 0 \\ -2 & 2 & 0 & 4 \\ -1 & 1 & 9 & 8 \end{bmatrix}$, $A^T A = \begin{bmatrix} 6 & -6 & -6 & -16 \\ -6 & 6 & 6 & 16 \\ -6 & 6 & 90 & 72 \\ -16 & 16 & 72 & 80 \end{bmatrix}$

and $B = \begin{bmatrix} 2 \\ -1 \\ 3 \end{bmatrix}$, so that $A^T B = [1 \ -1 \ 33 \ 20]^T$. Solving the normal equations $(A^T A)Z = A^T B$ by Gaussian elimination gives the general solution

$$Z = [\tfrac{4}{7} \ 0 \ \tfrac{17}{42} \ 0]^T + s[1 \ 1 \ 0 \ 0]^T + t[6 \ 0 \ -2 \ 3]^T$$

where s and t are parameters. Hence Z is a best approximation to a solution of $AX = B$ for any value of s and t, one possibility being $Z = [\tfrac{4}{7} \ 0 \ \tfrac{17}{42} \ 0]^T = \tfrac{1}{42}[24 \ 0 \ 17 \ 0]^T$. However, for all values of s and t we obtain

$$AZ = [\tfrac{25}{14} \ \tfrac{-8}{7} \ \tfrac{43}{14}]^T \approx [1.79 \ -1.14 \ 3.07]^T.$$

This is as close as possible to the given constants $[2 \ -1 \ 3]^T$. $\square$

4.6. PROJECTIONS AND APPROXIMATION 315

Example 7. *It is thought that the steel production s (in thousands of tons) in a country is linearly related to the number x_1 of mines and the number x_2 of steel mills in that country, that is that there exist constants a and b such that*

$$s = ax_1 + bx_2$$

holds for every country. The data in the table were collected for four countries A, B, C and D. Find the best approximation to the coefficients a and b.

Country	A	B	C	D
x_1	4	3	5	2
x_2	2	2	3	1
s	2.9	2.2	3.6	1.6

Solution. The given data yield four equations (one for each country) in the variables a and b:

$$\begin{aligned} 4a + 2b &= 2.9 \\ 3a + 2b &= 2.2 \\ 5a + 3b &= 3.6 \\ 2a + b &= 1.6 \end{aligned}$$

that is $AX = B$, where

$$A = \begin{bmatrix} 4 & 2 \\ 3 & 2 \\ 5 & 3 \\ 2 & 1 \end{bmatrix}, \quad X = \begin{bmatrix} a \\ b \end{bmatrix} \quad \text{and} \quad B = \begin{bmatrix} 2.9 \\ 2.2 \\ 3.6 \\ 1.6 \end{bmatrix}.$$

Here the system $AX = B$ has no solution. But the matrix $A^T A$ is invertible because A has independent columns, and the unique solution Z to the normal equations $(A^T A)Z = A^T B$ is $Z = \begin{bmatrix} .782 \\ -.091 \end{bmatrix}$, to three decimals. Hence $a = .782$, and $b = -.091$ are the best approximation, whence $s = .782 x_1 - .091 x_2$. □

4.6.5 Least Squares Approximation.

We began this chapter with an example where an engineer wants to be able to predict the response of an amplifier for any input voltage x between 2 and 6 millivolts. She inputs trial voltages x_i of $2, 3, 4, 5,$ and 6 mv and measures the response values y_i in the table.

Trial i	1	2	3	4	5
Input x_i	2	3	4	5	6
Response y_i	4.8	7.5	8.6	11.1	12.3

When the data pairs are plotted it is apparent (see the diagram) that the points (x_i, y_i) are nearly on a straight line. Theorem 5 enables us to find the straight line that best fits the data. Recall that every (non-vertical) line has the form $y = a_0 + a_1 x$ for some constants a_0 and a_1. Hence, given a_0 and a_1, we get predicted responses $y_1', \cdots, y_5'$, where we write $y_i' = a_0 + a_1 x_i$ for convenience. Thus the task is to find the values of a_0 and a_1 such that the observed responses $y_1, \cdots, y_5$ are as close as possible to the predicted responses $y_1', \cdots, y_5'$. This is where Theorem 5 comes in.

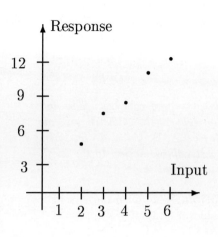

If we write $Y = [y_1 \ y_2 \ y_3 \ y_4 \ y_5]^T$ and $Y' = [y_1' \ y_2' \ y_3' \ y_4' \ y_5']^T$, the idea is to choose a_0 and a_1 such that

$$\|Y - Y'\| = \sqrt{(y_1 - y_1')^2 + (y_2 - y_2')^2 + (y_3 - y_3')^2 + (y_4 - y_4')^2 + (y_5 - y_5')^2}$$

is as small as possible (hence the name "least squares" approximation). Now observe that if we define

$$M = \begin{bmatrix} 1 & x_1 \\ 1 & x_2 \\ 1 & x_3 \\ 1 & x_4 \\ 1 & x_5 \end{bmatrix} \text{ and } Z = \begin{bmatrix} a_0 \\ a_1 \end{bmatrix},$$

then

$$MZ = \begin{bmatrix} y_1' \\ y_2' \\ y_3' \\ y_4' \\ y_5' \end{bmatrix} = Y'.$$

Hence we are to choose Z such that $\|Y - Y'\| = \|Y - MZ\|$ is as small as possible. In other words, Z is to be a best approximation to the system $MZ = Y$. Since Y and M are known matrices, this is solved by Theorem 5. Thus Z is given as any solution of the normal equations

$$(M^T M)Z = M^T Y.$$

In the present situation, solving these equations gives a unique best approximation $Z = [1.42 \ 1.86]^T$, so the best fitting line has equation $y = 1.42 + 1.86x$. Using this, the predicted values are

$$Y' = [5.14 \ 7.00 \ 8.86 \ 10.72 \ 12.58]^T$$

4.6. PROJECTIONS AND APPROXIMATION

compared to the observed data

$$Y = [4.8 \ 7.5 \ 8.6 \ 11.1 \ 12.3]^T.$$

These predicted values Y' are as close as possible (in the least squares sense) to the observed data Y using a straight line approximation.

The general situation is dealt with in the same way. Suppose that n data pairs have been collected giving corresponding values of two variables x and y:

$$(x_1, y_1), \ (x_2, y_2), \ (x_3, y_3), \ \cdots, (x_n, y_n).$$

The method used above will best fit a line $y = a_0 + a_1 x$ to these data. In fact, with very little extra effort, we can find a *polynomial*

$$y = a_0 + a_1 x + a_2 x^2 + \cdots + a_m x^m$$

of degree at most m that best fits the data (so the line is the case $m = 1$). For convenience write

$$f(x) = a_0 + a_1 x + a_2 x^2 + \cdots + a_m x^m.$$

Given any choice of the coefficients $a_0, a_1, \cdots, a_m$, we obtain predicted values $y'_i = f(x_i)$. If we write $Y = [y_1 \ y_2 \ y_3 \ \cdots \ y_n]^T$ and $Y' = [y'_1 \ y'_2 \ y'_3 \ \cdots \ y'_n]^T$ then, as before, the idea is to choose the coefficients $a_0, a_1, \cdots, a_m$ so that these predicted values y'_i are as close as possible to the observed values y_i. In other words, we choose the a_i so that the quantity

$$\|Y - Y'\| = \sqrt{(y_1 - y'_1)^2 + (y_2 - y'_2)^2 + (y_3 - y'_3)^2 + \cdots + (y_n - y'_n)^2}$$

is as small as possible. This "least squares" fit can always be achieved.

By analogy with the above computation, write

$$M = \begin{bmatrix} 1 & x_1 & x_1^2 & \cdots & x_1^m \\ 1 & x_2 & x_2^2 & \cdots & x_2^m \\ \vdots & \vdots & \vdots & \vdots & \vdots \\ 1 & x_n & x_n^2 & \cdots & x_n^m \end{bmatrix} \text{ and } Z = \begin{bmatrix} a_0 \\ a_1 \\ \vdots \\ a_m \end{bmatrix},$$

so that (as before) $MZ = Y'$. Hence $\|Y - Y'\| = \|Y - MZ\|$, so we want to find Z such that $\|Y - MZ\|$ is as small as possible. Thus the desired column Z is a best approximation to the linear system $MZ = Y$ where Y and M are known matrices. Hence Theorem 5 shows that any solution $Z = [a_0 \ a_1 \ \cdots \ a_m]^T$ of the normal equations $(M^T M)Z = M^T Y$ will provide a best approximation. The corresponding polynomial $a_0 + a_1 x + a_2 x^2 + \cdots + a_m x^m$ is called a **least squares approximating polynomial of degree** m for the data pairs. This discussion proves most of

318 CHAPTER 4. THE VECTOR SPACE $\mathbb{R}^n$

Theorem 6. Least Squares Theorem. *Given n data pairs (x_1, y_1), (x_2, y_2), (x_3, y_3), $\cdots$, (x_n, y_n), put*

$$Y = \begin{bmatrix} y_1 \\ y_2 \\ \vdots \\ y_n \end{bmatrix} \quad \text{and} \quad M = \begin{bmatrix} 1 & x_1 & x_1^2 & \cdots & x_1^m \\ 1 & x_2 & x_2^2 & \cdots & x_2^m \\ \vdots & \vdots & \vdots & \vdots & \vdots \\ 1 & x_n & x_n^2 & \cdots & x_n^m \end{bmatrix}.$$

Then a least squares approximating polynomial of degree m for these data is given by

$$y = a_0 + a_1 x + a_2 x^2 + \cdots + a_m x^m$$

where $Z = [a_0 \; a_1 \; \cdots \; a_m]^T$ is any solution of the normal equations

$$(M^T M) Z = M^T Y.$$

Furthermore, if at least $m+1$ of the x_i are distinct, then Z is unique and so the least squares polynomial is unique. (In fact $M^T M$ is invertible in this case, so $Z = (M^T M)^{-1} M^T Y$.)[13]

Proof. It remains to verify that Z is unique when at least $m+1$ of the x_i are distinct. It suffices to show that $M^T M$ is invertible (then $Z = (M^T M)^{-1} M^T Y$), and this follows if M has independent columns (Theorem 3 §4.4). Denote the columns of M by $C_0, C_1, C_2, \cdots, C_m$, and suppose a linear combination vanishes: $t_0 C_0 + t_1 C_1 + t_2 C_2 + \cdots + t_m C_m = 0$. If we write $g(x) = t_0 + t_1 x + t_2 x^2 + \cdots + t_m x^m$, this gives $g(x_i) = 0$ for each i. If at least $m+1$ of the x_i are distinct, this means that $g(x)$ is a polynomial of degree at most m which has more than m distinct roots. It is a theorem of algebra (see Appendix A.3), that this cannot happen unless all the coefficients t_i of $g(x)$ are zero. Hence $\{C_0, C_1, C_2, \cdots, C_m\}$ is independent. □

It is worth noting that the condition in Theorem 6 that $m+1$ of the x_i are distinct is not a severe restriction. It can be arranged, for example, in many experimental situations where the data pairs (x_i, y_i) are found by *choosing* the x_i and then observing the corresponding values y_i.

Example 8. *Newton's Laws of Motion imply that an object thrown vertically at a velocity of v meters per second will be at a height of $h = vt - \frac{1}{2} g t^2$ meters after t seconds, where g is the acceleration of gravity. The values in the table were observed. By fitting a quadratic $h = a_0 + a_1 t + a_2 t^2$ to these data, estimate the values of v and g.*

t	.5	1.0	1.5	2.0
h	23.7	45.1	64.0	80.4

---

[13]Even if $M^T M$ is invertible, it is rarely advisable to calculate Z from the formula $Z = (M^T M)^{-1} M^T Y$. It is nearly always more efficient numerically to solve the normal equations $(M^T M) Z = M^T Y$ by Gaussian elimination.

4.6. PROJECTIONS AND APPROXIMATION

Solution. Here $Y = \begin{bmatrix} 23.7 \\ 45.1 \\ 64.0 \\ 80.4 \end{bmatrix}$ and $M = \begin{bmatrix} 1 & .5 & .25 \\ 1 & 1.0 & 1.0 \\ 1 & 1.5 & 2.25 \\ 1 & 2.0 & 4.0 \end{bmatrix}$, where t and h are playing the role of x and y in Theorem 6. Solving the normal equations gives $h = -.175 + 50.25t - 4.98t^2$, whence the estimates are $v = 50.25$ and $-\frac{1}{2}g = -4.98$. The latter equation gives $g = 9.96$ (the true value is 9.81). □

Remark. There is an extension of Theorem 6 that should be mentioned. Given the data pairs $(x_1, y_1), (x_2, y_2), \cdots, (x_n, y_n)$, the theorem shows how to find a polynomial
$$p(x) = a_0 + a_1 x + a_2 x^2 + \cdots + a_m x^m$$
which approximates the data in the least squares sense. In fact the theorem describes how to find the coefficients a_i of the best fitting polynomial. Now view $p(x)$ as a linear combination of the functions $1, x, x^2, \cdots, x^m$. This suggests that perhaps this will work for linear combinations of other functions. If functions $f_0(x), f_1(x), f_2(x), \cdots, f_m(x)$ are given, we ask whether coefficients a_i can be found such that
$$f(x) = a_0 f_0(x) + a_1 f_1(x) + a_2 f_2(x) + \cdots + a_m f_m(x)$$
is a best approximation in the sense that $\|Y - Y'\|$ is as small as possible where $Y = [y_1 \ y_2 \ y_3 \ \cdots \ y_n]$ and $Y' = [f(x_1) \ f(x_2) \ f(x_3) \ \cdots \ f(x_n)]$. The answer is "yes", and the result is the analogue of Theorem 6 where we use a new matrix
$$M = \begin{bmatrix} f_0(x_1) & f_1(x_1) & \cdots & f_m(x_1) \\ f_0(x_2) & f_1(x_2) & \cdots & f_m(x_2) \\ \vdots & \vdots & & \vdots \\ f_0(x_n) & f_1(x_n) & \cdots & f_m(x_n) \end{bmatrix}.$$

As before, the coefficients $Z = [a_0 \ a_1 \ \cdots \ a_m]^T$ are found as the solution to the normal equations $(M^T M)Z = M^T Y$. However, the statement (at the end of Theorem 6) about the uniqueness of Z may fail for other choices of the functions $f_i(x)$.

Exercises

1. In each case write the vector X as the sum of a vector in U and a vector in $U^\perp$.
    (a) $X = [3 \ -1 \ 2]^T$, $U = span\{[1 \ 2 \ -1]^T, [2 \ 0 \ 1]^T\}$.
    (b) $X = [1 \ 1 \ 3]^T$, $U = span\{[-1 \ 0 \ 2]^T, [3 \ 1 \ 5]^T\}$.
    (c) $X = [3 \ 0 \ 2 \ 1]^T$,
    $U = span\{[1 \ 1 \ 1 \ 1]^T, [1 \ 1 \ -1 \ -1]^T, [1 \ -1 \ 1 \ -1]^T\}$.

(d) $X = [2\ 1\ 4\ -2]^T$,
$U = span\{[1\ 0\ 1\ 1]^T, [0\ 1\ -1\ 1]^T, [-2\ 0\ 1\ 1]^T\}$.

(e) $X = [a\ b\ c\ d]^T$,
$U = span\{[1\ 0\ 0\ 0]^T, [0\ 1\ 0\ 0]^T, [0\ 0\ 1\ 0]^T\}$.

(f) $X = [a\ b\ c\ d]^T$,
$U = span\{[1\ -1\ 2\ 0]^T, [-1\ 1\ 1\ 1]^T\}$.

2. Let $X = [1\ 1\ 0\ 0]^T$ and $U = span\{[1\ -1\ 0\ 2]^T, [2\ 2\ -1\ 0]^T\}$.

   (a) Compute $proj_U(X)$.

   (b) Show that $\{[1\ 3\ -1\ -2]^T, [7\ 1\ -2\ 6]^T\}$ is another orthogonal basis of U, and use it to compute $proj_U(X)$.

3. In each case find the vector in the subspace U that is closest to X.

   (a) $U = span\{[1\ 2\ -1]^T, [0\ 1\ 3]^T\}$, $X = [5\ -2\ 1]^T$.

   (b) $U = span\{[1\ -1\ 0]^T, [1\ 0\ -1]^T\}$, $X = [2\ 1\ 3]^T$.

   (c) $U = span\{[1\ 0\ 1\ 0]^T, [1\ 1\ 1\ 0]^T, [1\ 1\ 0\ 0]^T\}$, $X = [3\ -1\ 2\ 1]^T$.

   (d) $U = span\{[1\ -1\ 1\ 0]^T, [1\ 1\ 0\ 0]^T, [1\ 1\ 0\ 1]^T\}$, $X = [3\ 0\ 5\ -1]^T$.

4. Find the best approximation to a solution of each of the following systems of equations.

   (a) $\begin{aligned} x - y &= 3 \\ 2x + y &= -1 \\ x + 5y &= -4 \end{aligned}$ (b) $\begin{aligned} 2x + y &= 1 \\ x - y &= 2 \\ x + 5y &= 3 \end{aligned}$

   (c) $\begin{aligned} x + y + z &= 1 \\ 2x - y - z &= 0 \\ -x + 2y + 3z &= 2 \\ 2x + 2y + 3z &= 1 \end{aligned}$ (d) $\begin{aligned} x - y + 2z &= 0 \\ 3x + y + z &= -1 \\ 2x + y - 3z &= 3 \\ 3x + 4y - 4z &= 1 \end{aligned}$

5. Find the least squares approximating quadratic for the following sets of data pairs.

   (a) $(-1, 1), (0, 0), (2, 3), (3, 4)$

   (b) $(-1, 0), (0, 2), (1, 1), (3, -1)$

6. Find the least squares approximating cubic for the following sets of data pairs.

   (a) $(-1, 1), (0, -1), (1, 3), (3, -4)$

   (b) $(-2, 0), (0, 3), (1, 2), (3, -1), (4, 1)$

7. Find the least squares approximating function of the form $a_0 x + a_1 x^2 + a_2(-1)^x$ for the following set of data pairs: $(-1, 1), (0, 3), (1, 1), (2, 0)$. [*Hint:* See the Remark following Example 8.]

4.6. PROJECTIONS AND APPROXIMATION

8. Let $U = span\{Y_1, Y_2, \cdots, Y_m\}$ in $\mathbb{R}^n$ where the Y_i are written as rows, and let A denote the $m \times n$ matrix with the Y_i as its rows.
   (a) Show that $U^\perp = \{X \text{ in } \mathbb{R}^n \mid AX^T = 0\}$.
   (b) Use (a) to find $U^\perp$ if $U = span\{[1 \ -2 \ 1 \ 1]^T, [0 \ -1 \ 3 \ -2]^T\}$.

9. In each case either show that the statement is true or give an example showing that it is false.
   (a) If $proj_U(X)$ is in U then X is in U.
   (b) If $proj_U(X)$ is in $U^\perp$ then X is in $U^\perp$.
   (c) There is no vector X in both U and $U^\perp$.
   (d) If $proj_U(X) = 0$ then X is in $U^\perp$.
   (e) If X is in $U^\perp$ then $proj_U(X) = 0$.

10. Let U denote a subspace of $\mathbb{R}^n$. Show that a vector X in $\mathbb{R}^n$ is in U if and only if $proj_U(X) = X$.

11. Let U denote a subspace of $\mathbb{R}^n$.
    (a) Show that $U^\perp = \mathbb{R}^n$ if and only if $U = \{0\}$.
    (b) Show that $U^\perp = \{0\}$ if and only if $U = \mathbb{R}^n$.

12. If U is a subspace of $\mathbb{R}^n$, show that $U^\perp = \{X \text{ in } \mathbb{R}^n \mid proj_U(X) = 0\}$.

13. Let U be a subspace of $\mathbb{R}^n$. Show that:
    (a) $proj_U(X + Y) = proj_U(X) + proj_U(Y)$ for all X and Y in $\mathbb{R}^n$.
    (b) $proj_U(aX) = a\, proj_U(X)$ for all numbers a and all X in $\mathbb{R}^n$.
    (c) Show that $proj_U(proj_U(X)) = proj_U(X)$ for all X in $\mathbb{R}^n$.

14. If $\{F_1, F_2, \cdots, F_n\}$ is an orthogonal basis of $\mathbb{R}^n$ and $U = span\{F_1, F_2, \cdots, F_k\}$, show that $U^\perp = span\{F_{k+1}, \cdots, F_n\}$. [*Hint*: Apply the Expansion Theorem to a vector X in $\mathbb{R}^n$.]

15. If U is a subspace of $\mathbb{R}^n$, show how to find an $n \times n$ matrix A such that $U = \{X \text{ in } \mathbb{R}^n \mid AX = 0\}$. [*Hint*: Theorem 3 (2).]

16. Let U and W be supspaces of $\mathbb{R}^n$ and consider the subspaces
    $U \cap W = \{X \text{ in } \mathbb{R}^n \mid X \text{ is in both } U \text{ and } W\}$.
    $U + W = \{X \text{ in } \mathbb{R}^n \mid X \text{ is a sum of a vector in } U \text{ and a vector in } W\}$.
    (a) Show that $(U + W)^\perp = U^\perp \cap W^\perp$.
    (b) Show that for any two subsets A and B of $\mathbb{R}^n$ if $A \subseteq B$ then $B^\perp \subseteq A^\perp$.
    (c) Show that $(U \cap W)^\perp \supseteq U^\perp + W^\perp$.
    (d) Show that in fact $(U \cap W)^\perp = U^\perp + W^\perp$. [*Hint*: Use (a), Theorem 3 (1), and Exercise 25 §4.3 to show that $dim(U^\perp + W^\perp) = dim((U \cap W)^\perp)$. Then use (b) and Theorem 4 §4.3.]

17. A square matrix E is called a **projection matrix** if $E^2 = E = E^T$.

    (a) If A is $m \times n$ and AA^T is invertible, show that $E = A^T(AA^T)^{-1}A$ is a projection matrix.

    (b) If E is a projection matrix show that $I - E$ is also a projection matrix.

    (c) If E and F are projection matrices such that $EF = 0 = FE$, show that $E + F$ is a projection matrix.

4.7 ORTHOGONAL DIAGONALIZATION

In Section 2.3 we studied diagonalization of a square matrix A, and found important applications (for example to dynamical systems). We can now utilize the concepts of subspace, basis and dimension to clarify the diagonalization process and to prove some results that could not be demonstrated in Section 2.3 (Theorems 2 and 3). Then we integrate orthogonality into the discussion by introducing the idea of orthogonal diagonalization. This leads to the Principal Axis Theorem, one of the most useful results in linear algebra.

We begin with some results on diagonalization in general, and then move to the orthogonal case.

4.7.1 Diagonalization Revisited

Recall that a square matrix A is *diagonalizable* if there exists an invertible matrix P such that $P^{-1}AP = D$ is a diagonal matrix. Unfortunately, not all matrices are diagonalizable, for example $\begin{bmatrix} 1 & 1 \\ 0 & 1 \end{bmatrix}$. Determining whether A is diagonalizable is closely related to the eigenvalues and eigenvectors of A. Recall that a number λ is called an *eigenvalue* of A if $AX = \lambda X$ for some nonzero vector X in $\mathbb{R}^n$, and any such X is called an *eigenvector* of A corresponding to λ. Theorem 3 §2.3 asserts (in part) that A is diagonalizable if and only if it has n eigenvectors $X_1, \cdots, X_n$ such that the matrix $P = [X_1 \cdots X_n]$ with the X_i as columns is invertible. This is equivalent to asking that $\{X_1, \cdots, X_n\}$ is a basis of $\mathbb{R}^n$ consisting of eigenvectors of A. Hence we can restate Theorem 3 §2.3 as follows:

Theorem 1. *Let A be an $n \times n$ matrix.*

(1) *A is diagonalizable if and only if $\mathbb{R}^n$ has a basis $\{X_1, X_2, \cdots, X_n\}$ of eigenvectors of A.*

(2) *When this is the case, the matrix $P = [X_1 \cdots X_n]$ is invertible and $P^{-1}AP = diag(\lambda_1, \lambda_2, \cdots, \lambda_n)$ where, for each i, λ_i is the eigenvalue of A corresponding to X_i.*

Theorem 4 §2.3 asserts (without proof) that an $n \times n$ matrix A is diagonalizable if it has n distinct eigenvalues. The next result explains (and extends)

4.7. ORTHOGONAL DIAGONALIZATION

this by revealing an important connection between eigenvalues and linear independence: eigenvectors corresponding to distinct eigenvalues are necessarily linearly independent.

Theorem 2. Let $X_1, X_2, \cdots, X_k$ be eigenvectors corresponding to distinct eigenvalues $\lambda_1, \lambda_2, \cdots, \lambda_k$ of an $n \times n$ matrix A. Then $\{X_1, X_2, \cdots, X_k\}$ is a linearly independent set.

Proof. We use induction on k. If $k = 1$ then $\{X_1\}$ is independent because $X_1 \neq 0$. In general, suppose the theorem is true for some $k > 0$. Given eigenvectors $\{X_1, X_2, \cdots, X_{k+1}\}$, suppose a linear combination vanishes:

$$t_1 X_1 + t_2 X_2 + \cdots + t_{k+1} X_{k+1} = 0. \quad (*)$$

We must show that each $t_i = 0$. Left multiply (*) by A and use the fact that $AX_i = \lambda_i X_i$ to get

$$t_1 \lambda_1 X_1 + t_2 \lambda_2 X_2 + \cdots + t_{k+1} \lambda_{k+1} X_{k+1} = 0. \quad (**)$$

If we multiply (*) by λ_1 and subtract the result from (**), the first terms cancel and we obtain

$$t_2(\lambda_2 - \lambda_1) X_2 + t_3(\lambda_3 - \lambda_1) X_3 + \cdots + t_{k+1}(\lambda_{k+1} - \lambda_1) X_{k+1} = 0.$$

Since $\{X_2, X_3, \cdots, X_{k+1}\}$ is independent by the induction hypothesis, this gives
$$t_2(\lambda_2 - \lambda_1) = 0, \ t_3(\lambda_3 - \lambda_1) = 0, \ \cdots \ t_{k+1}(\lambda_{k+1} - \lambda_1) = 0,$$

and so $t_2 = t_3 = \cdots = t_{k+1} = 0$ because the λ_i are distinct. Hence (*) becomes $t_1 X_1 = 0$, which implies that $t_1 = 0$ because $X_1 \neq 0$. This is what we wanted. □

Theorem 2 completes some unfinished business from Section 2.3.

Corollary. (Theorem 4, Section 2.3.) If A is an $n \times n$ matrix with n distinct eigenvalues, then A is diagonalizable.

Proof. Choose one eigenvector for each of the n distinct eigenvalues. Then these eigenvectors are independent by Theorem 2, and so are a basis of $\mathbb{R}^n$ by Theorem 3 §4.3. Now use Theorem 1. □

Before proceeding, we prove an important lemma which formalizes a technique that is basic to diagonalization, and which will be used three times below.

Lemma 1. Let $\{X_1, X_2, \cdots, X_k\}$ be a linearly independent set of eigenvectors of an $n \times n$ matrix A, extend it to a basis $\{X_1, X_2, \cdots, X_k, \cdots, X_n\}$ of $\mathbb{R}^n$, and let

$$P = [X_1 \ X_2 \ \cdots \ X_n]$$

be the (invertible) matrix with the X_i as its columns. If $\lambda_1, \lambda_2, \cdots, \lambda_k$ are the (not necessarily distinct) eigenvalues of A corresponding to $X_1, X_2, \cdots, X_k$ respectively, then $P^{-1}AP$ has block form

$$P^{-1}AP = \begin{bmatrix} diag(\lambda_1, \lambda_2, \cdots, \lambda_k) & B \\ 0 & A_1 \end{bmatrix}$$

where B and A_1 are matrices of size $k \times (n-k)$ and $(n-k) \times (n-k)$ respectively.

Proof. If $\{E_1, E_2, \cdots, E_n\}$ is the standard basis of $\mathbb{R}^n$, note that PE_i is column i of P for each i, that is $PE_i = X_i$. Hence $P^{-1}X_i = E_i$ for each $1 \leq i \leq n$. On the other hand, observe that

$$P^{-1}AP = P^{-1}A[X_1 \ X_2 \ \cdots \ X_n] = [P^{-1}AX_1 \ P^{-1}AX_2 \ \cdots \ P^{-1}AX_n].$$

Hence, if $1 \leq i \leq k$, column i of $P^{-1}AP$ is

$$(P^{-1}A)X_i = P^{-1}(\lambda_i X_i) = \lambda_i(P^{-1}X_i) = \lambda_i E_i.$$

This describes the first k columns of $P^{-1}AP$, and Lemma 1 follows. $\square$

Note that Lemma 1 (with $k = n$) shows that an $n \times n$ matrix A is diagonalizable if $\mathbb{R}^n$ has a basis of eigenvectors of A, as in part (1) of Theorem 1.

If λ is an eigenvalue of an $n \times n$ matrix A, write

$$E_\lambda(A) = \{X \text{ in } \mathbb{R}^n \mid AX = \lambda X\}.$$

This is a subspace of $\mathbb{R}^n$ called the **eigenspace** of A corresponding to λ, and the eigenvectors corresponding to λ are just the nonzero members of $E_\lambda(A)$.[14] In fact $E_\lambda(A)$ is the null space of the matrix $(\lambda I - A)$:

$$E_\lambda(A) = \{X \mid (\lambda I - A)X = 0\} = null(\lambda I - A).$$

Hence, by Theorem 6 §4.4, the basic solutions of the homogeneous linear system $(\lambda I - A)X = 0$ given by the Gaussian Algorithm form a basis for $E_\lambda(A)$. In particular

$dim[E_\lambda(A)]$ is the number of basic solutions to the system $(\lambda I - A)X = 0$. (***)

Now recall that the *characteristic polynomial* $c_A(x)$ of an $n \times n$ matrix A is defined by $c_A(x) = det(xI - A)$, and that the *multiplicity* of an eigenvalue λ of A is the number of times λ occurs as a root of the characteristic polynomial. In other words,[15] the multiplicity of λ is the largest integer $m \geq 1$ such that

---

[14] Note that $E_\lambda(A)$ is a subspace for *any* number λ; the eigenvalues are the numbers λ for which $E_\lambda(A) \neq 0$.

[15] See Appendix A.3.

4.7. ORTHOGONAL DIAGONALIZATION

$c_A(x) = (x - \lambda)^m g(x)$ for some polynomial $g(x)$. Because of (***), the assertion (without proof) in Theorem 5 §2.3 is that a square matrix is diagonalizable if and only if the multiplicity of each eigenvalue λ equals $dim[E_\lambda(A)]$. We are going to prove this, and the proof requires the following result which is valid for *any* square matrix, diagonalizable or not.

Lemma 2. *Let λ be an eigenvalue of multiplicity m of a square matrix A. Then $dim[E_\lambda(A)] \leq m$.*

Proof. Write $dim[E_\lambda(A)] = d$. By the definition of multiplicity, it suffices to show that $c_A(x) = (x - \lambda)^d g(x)$ for some polynomial $g(x)$. To this end, let $\{X_1, X_2, \cdots, X_d\}$ be a basis of $E_\lambda(A)$. Then Lemma 1 shows that an invertible $n \times n$ matrix P exists such that

$$P^{-1}AP = \begin{bmatrix} \lambda I_d & B \\ 0 & A_1 \end{bmatrix}$$

in block form, where I_d denotes the $d \times d$ identity matrix. Now write $A' = P^{-1}AP$ and observe that $c_{A'}(x) = c_A(x)$ by Theorem 6 §2.3. But Theorem 3 §2.2 gives

$$\begin{aligned} c_{A'}(x) &= det(xI_n - A') \\ &= det \begin{bmatrix} (x - \lambda)I_d & -B \\ 0 & xI_{n-d} - A_1 \end{bmatrix} \\ &= det[(x - \lambda)I_d]det[xI_{n-d} - A_1] \\ &= (x - \lambda)^d c_{A_1}(x). \end{aligned}$$

Hence $c_A(x) = c_{A'}(x) = (x - \lambda)^d g(x)$ where $g(x) = c_{A_1}(x)$. This is what we wanted. □

It is impossible to ignore the question when equality holds in Lemma 2 for each eigenvalue λ. It turns out that this characterizes the diagonalizable matrices. This is the second piece of unfinished business from Section 2.3.

Theorem 3. (**Theorem 5, Section 2.3.**) *The following are equivalent for a square matrix A:*

(1) *A is diagonalizable.*

(2) *$dim[E_\lambda(A)]$ equals the multiplicity of λ for every eigenvalue λ of A.*

Proof. Let $\lambda_1, \lambda_2, \cdots, \lambda_k$ be the distinct eigenvalues of A and, for each i, let m_i denote the multiplicity of λ_i and write $d_i = dim[E_{\lambda_i}(A)]$. Then $c_A(x) = (x - \lambda_1)^{m_1}(x - \lambda_2)^{m_2} \cdots (x - \lambda_k)^{m_k}$ so $m_1 + \cdots + m_k = n$. Moreover, $d_i \leq m_i$ for each i by Lemma 2.

(1)⇒(2). By (1), $\mathbb{R}^n$ has a basis of n eigenvectors of A, so let t_i of them lie in $E_{\lambda_i}(A)$ for each i. Then $t_i \leq d_i$ for each i and so

$$n = t_1 + \cdots + t_k \leq d_1 + \cdots + d_k \leq m_1 + \cdots + m_k = n.$$

Hence $d_1 + \cdots + d_k = m_1 + \cdots + m_k$ so, since $d_i \leq m_i$ for each i, we must have $d_i = m_i$. This is (2).

(2)⇒(1). Let B_i denote a basis of $E_{\lambda_i}(A)$ for each i, and let $B = B_1 \cup \cdots \cup B_k$. Since each B_i contains m_i vectors by (2), and since the B_i are pairwise disjoint (the λ_i are distinct), it follows that B contains n vectors. So it suffices to show that B is linearly independent (then B is a basis of $\mathbb{R}^n$). Suppose a linear combination of the vectors in B vanishes, and let Y_i denote the sum of all terms that come from B_i. Then Y_i lies in $E_{\lambda_i}(A)$ for each i, so the nonzero Y_i are independent by Theorem 2 (as the λ_i are distinct). Since the sum of the Y_i is zero, it follows that $Y_i = 0$ for each i. Hence all coefficients of terms in Y_i are zero (because B_i is independent). This shows that B is independent. □

Examples of how Theorem 3 is used can be found in Section 2.3.

4.7.2 Orthogonal Matrices

Theorem 1 relates diagonalizability to the existence of a basis of eigenvectors. Since the most convenient bases of $\mathbb{R}^n$ are the orthogonal ones, this suggests two questions: Which $n \times n$ matrices admit an orthogonal basis $\{X_1, \cdots, X_n\}$ of eigenvectors? And what is special about $P = [X_1 \cdots X_n]$ in this case? Both these questions have interesting and important answers which we develop in this section.

We know (Theorem 2 §4.2) that an $n \times n$ matrix A is invertible if and only if its columns (respectively its rows) form a basis of $\mathbb{R}^n$. It is thus natural to investigate the situation where the columns (or the rows) of A are actually an orthogonal basis of $\mathbb{R}^n$. It turns out, however, that the appropriate condition is that the columns are *orthonormal* (that is, they are orthogonal and each is a unit vector). Surprisingly, this is equivalent to the condition that the rows are orthonormal.

Theorem 4. *The following conditions are equivalent for an $n \times n$ matrix P:*

(1) *P is invertible and $P^{-1} = P^T$.*

(2) *The columns of P are orthonormal.*

(3) *The rows of P are orthonormal.*

Proof. First note that (1) is equivalent to $P^T P = I$ by Corollary 1 to Theorem 5 §1.5. Now write $P = [C_1 \ C_2 \ \cdots \ C_n]$ where $C_1, C_2, \cdots, C_n$ denote the columns of P. Since row i of P^T is C_i^T, it follows that the (i,j)-entry of $P^T P$ is $C_i^T C_j = C_i \bullet C_j$. Thus $P^T P = I$ holds if and only if $C_i \bullet C_j = 0$ when $i \neq j$ and $C_i \bullet C_j = 1$ when $i = j$, that is if and only if the C_i are orthonormal. This proves (1)⇔(2), and (1)⇔(3) follows in a similar way. □

A square matrix P is called an **orthogonal matrix**[16] if it satisfies one (and hence all) of the conditions in Theorem 4. Clearly every identity matrix is

---
[16] In view of Theorem 4, *orthonormal* matrix might be a better name. However the term *orthogonal matrix* is standard.

4.7. ORTHOGONAL DIAGONALIZATION

orthogonal. We begin by listing some simple properties of these matrices that will be referred to below. The routine verifications are left to the reader.

Corollary. *Let P and Q denote $n \times n$ matrices. Then:*

(1) *If P is an orthogonal matrix, then either $det P = 1$ or $det P = -1$.*

(2) *If P is an orthogonal matrix, then $P^{-1} = P^T$ is an orthogonal matrix.*

(3) *If P and Q are orthogonal matrices, then PQ is an orthogonal matrix.*

The next example exhibits orthogonal 2×2 matrices with determinant 1 and -1, and also shows that orthogonal matrices have geometrical significance.

Example 1. The matrices $A = \begin{bmatrix} cos\theta & -sin\theta \\ sin\theta & cos\theta \end{bmatrix}$ and $B = \begin{bmatrix} cos\theta & sin\theta \\ sin\theta & -cos\theta \end{bmatrix}$ are orthogonal for any angle θ because $sin^2\theta + cos^2\theta = 1$. In fact these are the *only* 2×2 orthogonal matrices. This is verified in the discussion leading to Theorem 7 §3.5 where it is also shown that, if these matrices are viewed as transformations of $\mathbb{R}^2$, then A is rotation around the origin through the angle θ, and B is reflection in the line through the origin making an angle $\frac{\theta}{2}$ with the positive X-axis. □

A 2×2 matrix P, viewed as a transformation of $\mathbb{R}^2$, is called an *isometry* if it preserves distance, that is if $\|P\vec{v} - P\vec{w}\| = \|\vec{v} - \vec{w}\|$ for all vectors $\vec{v}$ and $\vec{w}$ in $\mathbb{R}^2$. Hence Theorem 7 §3.5 shows that P is an isometry if and only if P is a rotation or a reflection, and Example 1 shows that this happens if and only if P is orthogonal. Similarly, the isometries of $\mathbb{R}^n$ are precisely the orthogonal matrices (see Section 4.9). In the case of $\mathbb{R}^3$, the isometries are either rotations (about a line through the origin), reflections (in a plane through the origin), or a composition of two of these.

Example 2. It is not enough that the columns are merely orthogonal for a matrix to be an orthogonal matrix. For example, $A = \begin{bmatrix} 1 & 1 & 1 \\ 1 & -1 & 1 \\ -1 & 0 & 2 \end{bmatrix}$ has orthogonal columns, but the rows are not orthogonal. However, if the columns are normalized the result is the orthogonal matrix $P = \begin{bmatrix} \frac{1}{\sqrt{3}} & \frac{1}{\sqrt{2}} & \frac{1}{\sqrt{6}} \\ \frac{1}{\sqrt{3}} & -\frac{1}{\sqrt{2}} & \frac{1}{\sqrt{6}} \\ -\frac{1}{\sqrt{3}} & 0 & \frac{2}{\sqrt{6}} \end{bmatrix}$.

Note that now the rows are *also* orthonormal. □

The eigenvalues of an orthogonal matrix P need not be real. Indeed, the rotation $P = \begin{bmatrix} cos\theta & -sin\theta \\ sin\theta & cos\theta \end{bmatrix}$ in Example 1 has eigenvalues $cos\theta \pm i\, sin\theta$. These eigenvalues have absolute value $|cos\theta \pm i\, sin\theta| = 1$ as complex numbers, and the next theorem shows that this is a general property of orthogonal matrices.

Theorem 5. *If P is an orthogonal matrix then the (possibly complex) eigenvalues λ of P have absolute value $|\lambda| = 1$. In particular, the* only *real eigenvalues of P are 1 and -1.*

Proof. Recall (see §2.5) that if $Z = [z_{ij}]$ is a matrix with complex entries z_{ij}, the conjugate $\overline{Z} = [\overline{z}_{ij}]$ is defined to be the matrix obtained from Z by conjugating every entry. Recall further that if Z and W are two complex matrices then $\overline{Z + W} = \overline{Z} + \overline{W}$, $\overline{ZW} = \overline{Z}\,\overline{W}$ and $\overline{\lambda Z} = \overline{\lambda}\,\overline{Z}$ for each complex number λ.

Now suppose that λ is a complex eigenvalue of P, say $PX = \lambda X$ where $X \neq 0$ is a column in $\mathbb{C}^n$. Since P is real, we have $\overline{P} = P$ and so we compute

$$\begin{aligned}|\lambda|^2 (X^T \overline{X}) &= (\lambda \overline{\lambda}) X^T \overline{X} \\ &= (\lambda X)^T \overline{(\lambda X)} \\ &= (PX)^T \overline{(PX)} \\ &= X^T P^T \, \overline{P}\,\overline{X} \\ &= X^T P^T \, P\overline{X} = X^T I \overline{X} \\ &= X^T \overline{X}.\end{aligned}$$

Hence it suffices to show that $X^T \overline{X} \neq 0$. But if $X = [x_1 \, x_2 \, \cdots \, x_n]^T$ where the x_i are complex numbers, then $X^T \overline{X} = |x_1|^2 + |x_2|^2 + \cdots + |x_n|^2 \neq 0$ because some $x_i \neq 0$ (as $X \neq 0$). $\square$

A major virtue of orthogonal matrices is that they are easy to invert—simply take the transpose. This makes the $n \times n$ version of the QR factorization (Theorem 9 §4.5) even more useful. We restate it here for reference.

Theorem 6. QR Factorization. *Every invertible matrix A can be factored as*

$$A = QR$$

where Q is an orthogonal matrix and R is an invertible, upper triangular matrix with positive entries on the main diagonal. Furthermore, this factorization is unique in the sense that if $A = Q_1 R_1$ is another such factorization then $Q_1 = Q$ and $R_1 = R$.

It is important to note that a systematic procedure for finding QR-factorizations is available using the Gram-Schmidt algorithm—see the discussion preceding Theorem 9 §4.5.

4.7.3 The Principal Axis Theorem

Recall that an $n \times n$ matrix A is said to be diagonalizable if $P^{-1}AP$ is a diagonal matrix for some invertible matrix P. By analogy, A is called **orthogonally diagonalizable** if an orthogonal matrix P can be found such that

$$P^{-1}AP = P^T AP \text{ is a diagonal matrix.}$$

4.7. ORTHOGONAL DIAGONALIZATION

The main reason for introducing orthogonal matrices is the fact that this condition turns out to characterize the symmetric matrices. This is one of the most useful results in linear algebra.

Theorem 7. Principal Axis Theorem. *The following are equivalent for an $n \times n$ matrix A:*

(1) *A is orthogonally diagonalizable.*

(2) *A has an orthonormal set of n eigenvectors.*

(3) *A is symmetric.*

Proof. (1)$\Leftrightarrow$(2). Let $P = [X_1 \cdots X_n]$ be an invertible $n \times n$ matrix with columns X_i. Then (1)$\Leftrightarrow$(2) follows from the following statements: P is orthogonal if and only if $\{X_1, \cdots, X_n\}$ is an orthonormal set in $\mathbb{R}^n$; and $P^{-1}AP$ is diagonal if and only if $\{X_1, \cdots, X_n\}$ is a basis of eigenvectors of A.

(1)$\Rightarrow$(3). Let $P^T A P = D$ where D is diagonal and $P^{-1} = P^T$. Then $A = PDP^T$ so, since D is symmetric, we have $A^T = P^{TT}D^T P^T = PDP^T = A$. This proves (3).

(3)$\Rightarrow$(1). Let A be an $n \times n$ symmetric matrix; we prove (1) by induction on n. If $n = 1$ then (1) is clear. In general, since A is symmetric let λ_1 be a real eigenvalue of A (Theorem 3 §2.5), and let $AX_1 = \lambda_1 X_1$ for some eigenvector X_1 in $\mathbb{R}^n$ which we may assume satisfies $\|X_1\| = 1$. Use the Gram-Schmidt Algorithm to find an orthonormal basis $\{X_1, X_2, \cdots, X_n\}$ of $\mathbb{R}^n$ containing this eigenvector X_1. Then $P_1 = [X_1 \; X_2 \; \cdots \; X_n]$ is an orthogonal matrix and

$$P_1^T A P_1 = P_1^{-1} A P_1 = \begin{bmatrix} \lambda_1 & B \\ 0 & A_1 \end{bmatrix}$$

in block form by Lemma 1. But $P_1^T A P_1$ is symmetric (because A is), so it follows that $B = 0$ and A_1 is symmetric of size $(n-1) \times (n-1)$. Hence, by induction, there exists an orthogonal $(n-1) \times (n-1)$ matrix Q such that $Q^T A_1 Q = D_1$ is diagonal. But then $P_2 = \begin{bmatrix} 1 & 0 \\ 0 & Q \end{bmatrix}$ is an orthogonal matrix, and

$$\begin{aligned}(P_1 P_2)^T A (P_1 P_2) &= P_2^T (P_1^T A P_1) P_2 \\ &= \begin{bmatrix} 1 & 0 \\ 0 & Q^T \end{bmatrix} \begin{bmatrix} \lambda_1 & 0 \\ 0 & A_1 \end{bmatrix} \begin{bmatrix} 1 & 0 \\ 0 & Q \end{bmatrix} \\ &= \begin{bmatrix} \lambda_1 & 0 \\ 0 & D_1 \end{bmatrix}\end{aligned}$$

is diagonal. Since $P_1 P_2$ is orthogonal (Corollary to Theorem 4), this proves (1). $\square$

A set of n orthonormal eigenvectors for a symmetric matrix A is is called a set of **principal axes** for A. The name comes from geometry, and will be clarified in Section 4.8 when we discuss quadratic forms.

Example 3. If $A = \begin{bmatrix} 1 & 0 & 2 \\ 0 & 1 & -1 \\ 2 & -1 & 5 \end{bmatrix}$, find an orthogonal matrix P such that $P^T A P$ is diagonal.

Solution. The procedure is exactly the same as for diagonalizing A. The characteristic polynomial of A is $c_A(x) = x(x-1)(x-6)$, so A has eigenvalues $\lambda_1 = 0$, $\lambda_2 = 1$ and $\lambda_3 = 6$, with corresponding eigenvectors $X_1 = [-2 \ 1 \ 1]^T$, $X_2 = [1 \ 2 \ 0]^T$, and $X_3 = [2 \ -1 \ 5]^T$. Surprisingly, these eigenvectors are orthogonal. Hence, after normalizing, we obtain a set $\{\frac{1}{\sqrt{6}} X_1, \frac{1}{\sqrt{5}} X_2, \frac{1}{\sqrt{30}} X_3\}$ of principal axes for A. These yield the orthogonal matrix

$$P = \begin{bmatrix} \frac{1}{\sqrt{6}} X_1 & \frac{1}{\sqrt{5}} X_2 & \frac{1}{\sqrt{30}} X_3 \end{bmatrix} = \frac{1}{\sqrt{30}} \begin{bmatrix} -2\sqrt{5} & \sqrt{6} & 2 \\ \sqrt{5} & 2\sqrt{6} & -1 \\ \sqrt{5} & 0 & 5 \end{bmatrix}.$$

This matrix P is a diagonalizing matrix for A (which happens to be orthogonal), and so $P^T A P = P^{-1} A P = diag(\lambda_1, \lambda_2, \lambda_3) = diag(0, 1, 6)$. $\square$

The fact that the eigenvectors in Example 3 were orthogonal is no coincidence. We know that they are independent because they correspond to distinct eigenvalues (Theorem 2); the fact that the matrix is *symmetric* forces them to be orthogonal. To see why, we need the following fact about symmetric matrices.

Lemma 3. *If A is a symmetric $n \times n$ matrix in $\mathbb{R}^n$ then $(AX) \bullet Y = X \bullet (AY)$ for all columns X and Y in $\mathbb{R}^n$.*

Proof. We have $A^T = A$ so $(AX) \bullet Y = (AX)^T Y = X^T A^T Y = X^T (AY) = X \bullet (AY)$. $\square$

This gives a strengthening of Theorem 2 for symmetric matrices.

Theorem 8. *If A is a symmetric matrix then eigenvectors of A corresponding to distinct eigenvalues are orthogonal.*

Proof. Suppose that $AX = \lambda X$ and $AY = \mu Y$ where $\lambda \neq \mu$ are distinct eigenvalues. Then Lemma 3 gives

$$\lambda (X \bullet Y) = (\lambda X) \bullet Y = (AX) \bullet Y = X \bullet (AY) = X \bullet (\mu Y) = \mu (X \bullet Y).$$

Hence $(\lambda - \mu)(X \bullet Y) = 0$, so $X \bullet Y = 0$ because $\lambda \neq \mu$. $\square$

With Theorem 8 in hand, the process for orthogonally diagonalizing a symmetric $n \times n$ matrix A is a straightforward extension of the diagonalization procedure in Section 2.3.

Orthogonal Diagonalization Algorithm. *To diagonalize an $n \times n$ symmetric matrix A:*

4.7. ORTHOGONAL DIAGONALIZATION

Step 1. *Find the distinct eigenvalues λ.*

Step 2. *Obtain a basis for each eigenspace $E_\lambda(A)$ as the basic solutions of $(\lambda I - A)X = 0$.*

Step 3. *Obtain an orthonormal basis of each $E_\lambda(A)$ using the Gram-Schmidt procedure if necessary.*

Step 4. *The set of all the eigenvectors in Step 3 is an orthonormal basis of $\mathbb{R}^n$ by Theorem 5.*

Step 5. *If P is the orthogonal matrix with the basis in Step 4 as its columns, then $P^T A P$ is diagonal.*

Example 4. *Orthogonally diagonalize the symmetric matrix*
$$A = \begin{bmatrix} 5 & -2 & 4 \\ -2 & 8 & 2 \\ 4 & 2 & 5 \end{bmatrix}.$$

Solution. The characteristic polynomial is $c_A(x) = x(x-9)^2$, so the eigenvalues are $\lambda_1 = 0$ and $\lambda_2 = 9$. Bases for the eigenspaces are

$$E_{\lambda_1}(A) = span\{[2\ 1\ -2]^T\} \quad \text{and} \quad E_{\lambda_2}(A) = span\{[1\ -2\ 0]^T, [1\ 0\ 1]^T\}.$$

Note that $[2\ 1\ -2]^T$ from $E_{\lambda_1}(A)$ is orthogonal to both vectors $[1\ -2\ 0]^T$ and $[1\ 0\ 1]^T$ in $E_{\lambda_2}(A)$, as Theorem 8 guarantees, but this basis of $E_{\lambda_2}(A)$ is not orthogonal. Applying the Gram-Schmidt algorithm yields an orthogonal basis: $E_{\lambda_2}(A) = span\{[1\ -2\ 0]^T, [4\ 2\ 5]^T\}$. Hence $\{[2\ 1\ -2]^T, [1\ -2\ 0]^T, [4\ 2\ 5]^T\}$ is an orthogonal basis of $\mathbb{R}^3$ consisting of eigenvectors of A, whence normalization gives the orthonormal basis (principal axes for A)

$$\{\tfrac{1}{3}[2\ 1\ -2]^T, \tfrac{1}{\sqrt{5}}[1\ -2\ 0]^T, \tfrac{1}{3\sqrt{5}}[4\ 2\ 5]^T\}.$$

Hence $P = \begin{bmatrix} \frac{2}{3} & \frac{1}{\sqrt{5}} & \frac{4}{3\sqrt{5}} \\ \frac{1}{3} & \frac{-2}{\sqrt{5}} & \frac{2}{3\sqrt{5}} \\ \frac{-2}{3} & 0 & \frac{5}{3\sqrt{5}} \end{bmatrix} = \frac{1}{3\sqrt{5}}\begin{bmatrix} 2\sqrt{5} & 3 & 4 \\ 1\sqrt{5} & -6 & 2 \\ -2\sqrt{5} & 0 & 5 \end{bmatrix}$ is an orthogonal matrix such that $P^T A P = P^{-1} A P = diag(0, 9, 9)$. $\square$

4.7.4 Triangulation

If we substitute "upper triangular" for "diagonal" in the Principal Axis Theorem, we can weaken the requirement that A is symmetric to asking only that all the eigenvalues are real. The proof is virtually the same.

Theorem 9. Triangulation Theorem. *If A is a square matrix with real eigenvalues, there exists an orthogonal matrix P such that $P^T A P$ is upper triangular.*

Proof. If A is an $n \times n$ matrix with real eigenvalues, the proof of (3)$\Rightarrow$(1) of Theorem 7 goes through except that B need not be zero and D_1 is upper triangular, not diagonal. We leave the details to the reader. $\square$

Unfortunately, Theorem 9 provides no systematic way to find the matrix P. An algorithm exists which is a refinement of the method for orthogonal diagonalization, but it is beyond the scope of this book. Even so, the theorem is useful as it stands and reveals important information about the determinant and trace of a matrix.

Here the **trace** trA of an $n \times n$ matrix A is defined to be the sum of the main diagonal elements of A, in other words:

$$\text{If } A = [a_{ij}] \text{ then } trA = a_{11} + a_{22} + \cdots + a_{nn}.$$

It is evident that $tr(A+B) = trA + trB$ and that $tr(cA) = c\,trA$ holds for all $n \times n$ matrices A and B, and all scalars c. The following fact is more surprising.

Lemma 4. *Let A and B be $n \times n$ matrices. Then:*

(1) $tr(AB) = tr(BA)$.

(2) *If P is an invertible matrix then $tr(P^{-1}AP) = trA$.*

Proof. Write $[a_{ij}]$ and $B = [b_{ij}]$. Then the (i,i)-entry of AB is $d_i = a_{i1}b_{1i} + a_{i2}b_{2i} + \cdots + a_{in}b_{ni} = \Sigma_j a_{ij}b_{ji}$. Hence

$$tr(AB) = d_1 + d_2 + \cdots + d_n = \Sigma_i d_i = \Sigma_i(\Sigma_j a_{ij}b_{ji}).$$

Similarly we have $tr(BA) = \Sigma_i(\Sigma_j b_{ij}a_{ji})$. Since these two double sums are the same, this proves (1). Then (2) follows from (1) because $tr(P^{-1}AP) = tr[P^{-1}(AP)] = tr[(AP)P^{-1}] = trA$. $\square$

Our reason for looking at Lemma 4 here is that it gives the first part of

Theorem 10. *Let A be an $n \times n$ matrix with real eigenvalues $\lambda_1, \lambda_2, \cdots, \lambda_n$ (possibly not all distinct). Then*

$$trA = \lambda_1 + \lambda_2 + \cdots + \lambda_n \quad \text{and} \quad detA = \lambda_1\lambda_2\cdots\lambda_n.$$

Proof. By Theorem 9, let P be a matrix (in fact orthogonal) such that $P^{-1}AP = U$ is upper triangular. Then $trA = trU$ (by Lemma 4) and $detA = detU$. The result now follows because U has the same eigenvalues as A (Theorem 6 §2.3), so they appear on the main diagonal of U. $\square$

4.7. ORTHOGONAL DIAGONALIZATION

Exercises

1. In each case either show that the statement is true or give an example showing that it is false.

   (a) Every eigenvalue of a matrix has a unit eigenvector (that is one of length 1).

   (b) If P has orthogonal columns and $det P = \pm 1$, then P is orthogonal.

   (c) Every orthogonal matrix is invertible.

   (d) Every orthogonal matrix is symmetric.

   (e) Every diagonal matrix is orthogonal.

   (f) If A is symmetric then $P^T AP$ is also symmetric for all matrices P.

2. By normalizing the columns, make each of the following matrices orthogonal.

   (a) $\begin{bmatrix} 2 & -2 \\ 1 & 4 \end{bmatrix}$ (b) $\begin{bmatrix} 3 & 4 \\ -4 & 3 \end{bmatrix}$

   (c) $\begin{bmatrix} cos\theta & -sin\theta & 0 \\ sin\theta & cos\theta & 0 \\ 0 & 0 & 3 \end{bmatrix}$ (d) $\begin{bmatrix} 1 & 0 & 2 \\ -1 & 1 & 1 \\ 1 & 1 & -1 \end{bmatrix}$

   (e) $\begin{bmatrix} -1 & 2 & 2 \\ 2 & -1 & 2 \\ 2 & 2 & -1 \end{bmatrix}$ (f) $\begin{bmatrix} 2 & 3 & -6 \\ 6 & 2 & 3 \\ -3 & 6 & 2 \end{bmatrix}$

3. If the first two rows of an orthogonal matrix are $[\frac{1}{3} \ \frac{2}{3} \ \frac{2}{3}]$ and $[\frac{2}{3} \ \frac{-2}{3} \ \frac{1}{3}]$, find all possible third rows.

4. If $\{X_1, X_2, \cdots, X_n\}$ is an orthonormal basis of $\mathbb{R}^n$, find all Y in $\mathbb{R}^n$ such that $\{Y, X_2, \cdots, X_n\}$ is an orthonormal basis of $\mathbb{R}^n$. [*Hint:* Expand Y in terms of $\{X_1, X_2, \cdots, X_n\}$ by the Expansion Theorem.]

5. For each matrix A, find an orthogonal matrix P such that $P^T AP$ is diagonal.

   (a) $A = \begin{bmatrix} 3 & 0 & 0 \\ 0 & 2 & 2 \\ 0 & 2 & 5 \end{bmatrix}$ (b) $A = \begin{bmatrix} 3 & 0 & 7 \\ 0 & 5 & 0 \\ 7 & 0 & 3 \end{bmatrix}$

   (c) $A = \begin{bmatrix} 3 & 3 & 0 \\ 3 & 3 & 0 \\ 0 & 0 & 6 \end{bmatrix}$ (d) $A = \begin{bmatrix} 1 & 3 & 0 \\ 3 & 13 & 2 \\ 0 & 2 & 1 \end{bmatrix}$

   (e) $A = \begin{bmatrix} 3 & 5 & 0 & 0 \\ 5 & 3 & 0 & 0 \\ 0 & 0 & 1 & 7 \\ 0 & 0 & 7 & 1 \end{bmatrix}$ (f) $A = \begin{bmatrix} 3 & 5 & -1 & 1 \\ 5 & 3 & 1 & -1 \\ -1 & 1 & 3 & 5 \\ 1 & -1 & 5 & 3 \end{bmatrix}$

6. If P is an orthogonal matrix and c is a scalar, show that cP is orthogonal if and only if $c = \pm 1$.

7. If P is a triangular, orthogonal matrix, show that P is diagonal and each diagonal entry is 1 or -1.

8. Show that an $n \times n$ matrix P is orthogonal if and only if $\|PX\| = \|X\|$ for every column X in $\mathbb{R}^n$. [*Hint*: If $\|PX\| = \|X\|$ replace X by $X + Y$ to show that $(PX) \bullet (PY) = X \bullet Y$ for all X and Y.]

9. An $n \times n$ matrix P is called a **permutation matrix** if the columns of P are just the columns of I_n in some order.

    (a) Show that every permutation matrix is orthogonal.

    (b) Show that the rows of a permutation matrix are the rows of I_n in some order.

10. Let A be an $m \times n$ matrix with columns $C_1, C_2, \cdots, C_n$. Explain why the (i, j)-entry of $A^T A$ is $C_i \bullet C_j$.

11. Show that the following are equivalent for an $m \times n$ matrix A:

    (i) A has orthogonal columns.

    (ii) $A = PD$ where P has orthonormal columns and D is diagonal and invertible.

    (iii) $A^T A$ is invertible and diagonal.

    [*Hint*: For (i)$\Rightarrow$(ii) take $D = diag(\|C_1\|, \|C_2\|, \cdots, \|C_n\|)$ where C_j is column j of A. Exercise 10 is useful for (iii)$\Rightarrow$(i).]

12. Show that the following are equivalent for a symmetric matrix A:

    (i) A is orthogonal.

    (ii) $A^2 = I$.

    (iii) The only eigenvalues of A are ± 1.

    [*Hint*: For (iii)$\Rightarrow$(ii) use the Principal Axis Theorem.]

13. If the columns $C_1, C_2, \cdots, C_n$ of the $n \times n$ matrix $A = [a_{ij}]$ are orthogonal, show that the (i, j)-entry of A^{-1} is $\frac{a_{ji}}{\|C_i\|^2}$. [*Hint*: Use Exercise 10 to show that $A^T A = D = diag(\|C_1\|^2, \|C_2\|^2, \cdots, \|C_n\|^2)$.]

14. If A is symmetric, show that $\lambda \geq 0$ for every eigenvalue of A if and only if $A = B^2$ for some symmetric matrix B. [*Hint* : If $P^T AP = D = diag(\lambda_1, \cdots, \lambda_n)$ where $P^{-1} = P^T$ and each $\lambda_i > 0$, take $B = P \, diag(\sqrt{\lambda_1}, \cdots, \sqrt{\lambda_n}) P^T$.]

15. Extend Theorem 8 as follows: Let λ and μ be distinct eigenvectors of a square matrix A, and hence of A^T. If X is an eigenvector of A corresponding to λ, and Y is an eigenvector of A^T corresponding to μ, show that X and Y are orthogonal.

4.7. ORTHOGONAL DIAGONALIZATION

16. Matrices A and B are called **orthogonally similar** (written $A \overset{o}{\sim} B$) if $B = P^T A P$ for some orthogonal matrix P. Hence the Principal Axis Theorem asserts that a matrix A is symmetric if and only if $A \overset{o}{\sim} D$ for some diagonal matrix D.

    (a) Show: (i) $A \overset{o}{\sim} A$ for all A; (ii) If $A \overset{o}{\sim} B$ then $B \overset{o}{\sim} A$; (iii) If $A \overset{o}{\sim} B$ and $B \overset{o}{\sim} C$ then $A \overset{o}{\sim} C$.

    (b) If $A \overset{o}{\sim} B$, show that $A^T \overset{o}{\sim} B^T$.

    (c) If A and B are invertible and $A \overset{o}{\sim} B$, show that $A^{-1} \overset{o}{\sim} B^{-1}$.

    (d) If $A \overset{o}{\sim} B$, show that A and B have the same eigenvalues. [*Hint*: Theorem 6 §2.3.]

    (e) If A and B are symmetric, show that $A \overset{o}{\sim} B$ if and only if A and B have the same eigenvalues.

17. Let A be an $n \times n$ matrix. If $(AX) \bullet Y = X \bullet (AY)$ holds for all columns X and Y in $\mathbb{R}^n$, show that A is symmetric. (Converse to Lemma 3.) [*Hint*: Show that $(AX) \bullet Y = X \bullet (A^T Y)$ holds for all columns X and Y in $\mathbb{R}^n$. If $X^T B Y = 0$ for all columns X and Y, show that $B = 0$ by taking X and Y to be various columns of I_n.]

18. A square matrix E is called a projection matrix if $E^2 = E = E^T$. (See Exercise 17 §4.6.)

    (a) If E is a projection matrix, show that $P = I - 2E$ is orthogonal and symmetric.

    (b) If P is an orthogonal, symmetric matrix, show that $E = \frac{1}{2}(I - P)$ is a projection matrix.

    (c) If Q has orthonormal columns, show that $E = QQ^T$ is a projection matrix.

19. Show that every 2×2 orthogonal matrix has the form $\begin{bmatrix} \cos\theta & -\sin\theta \\ \sin\theta & \cos\theta \end{bmatrix}$ or $\begin{bmatrix} \cos\theta & \sin\theta \\ \sin\theta & -\cos\theta \end{bmatrix}$ for some angle θ. [*Hint*: See the discussion preceding Theorem 7 §3.5.]

20. A square matrix A is said to be **nilpotent** if $A^k = 0$ for some $k \geq 1$.

    (a) If N is a square upper triangular matrix with 0's on the main diagonal, show that N is nilpotent.

    (b) If A is nilpotent, show that $P^T A P$ has the form in (a) for some orthogonal matrix P. [*Hint*: Theorem 9.]

    (c) If A is a square matrix with real eigenvalues, show that $A = S + N$ for some symmetric matrix S and some nilpotent matrix N. [*Hint*: Theorem 9 and (a).]

21. (a) If A is nilpotent, show that every eigenvalue of A is zero.

    (b) If every eigenvalue of A is zero, use Theorem 9 to show that A is nilpotent.

    (c) Show that the only nilpotent, diagonalizable matrix A is $A = 0$.

22. If A is a square matrix with real eigenvalues, show that $Q^T A Q$ is lower triangular for some orthogonal matrix Q.

23. If C is an $n \times n$ matrix with its nonzero columns orthogonal, show that $C = PD$ where P is orthogonal and D is diagonal with non-negative entries.

4.8 QUADRATIC FORMS[17]

The Principal Axis Theorem states that every symmetric matrix can be orthogonally diagonalized. This result has far reaching consequences, some of which are investigated in this section. We begin with quadratic forms which provide a geometrical view (and the name!) of the Principal Axis Theorem. This leads to the study of positive definite matrices, and to the Cholesky decomposition. Then we look constrained optimization where a quantity is maximized (or minimized) subject to a condition on the variables.

4.8.1 Quadratic Forms

The equation of a curve in the plane can often be made simpler by a change of coordinates, and the same is true for the equation of a surface in $\mathbb{R}^3$. In this section we show how to use the Principal Axis Theorem to accomplish this for quadratic curves in n variables.

Example 1. Consider the graph of the equation $x_1 x_2 = 1$ as shown in the diagram. The curve is clearly symmetric about the line $x_2 = x_1$. Hence we are going to look at the equation of this curve in terms of new variables y_1 and y_2, where the y_1-axis is the line $x_2 = x_1$. To do this, we obtain the y_1- and y_2-axes from the x_1- and x_2-axes by a counterclockwise rotation of $\frac{\pi}{4}$. Recall that the matrix of the rotation through an angle θ is $\begin{bmatrix} cos\theta & -sin\theta \\ sin\theta & cos\theta \end{bmatrix}$ by Theorem 3 §3.5. Hence, if the vector $[x_1 \ x_2]^T$ is rotated through $\frac{\pi}{4}$ the resulting vector $[y_1 \ y_2]^T$ is given by

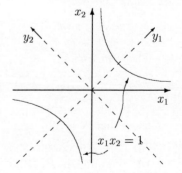

---

[17] This section can be omitted without loss of continuity.

4.8. QUADRATIC FORMS

$$\begin{bmatrix} y_1 \\ y_2 \end{bmatrix} = \begin{bmatrix} \cos(\frac{\pi}{4}) & -\sin(\frac{\pi}{4}) \\ \sin(\frac{\pi}{4}) & \cos(\frac{\pi}{4}) \end{bmatrix} \begin{bmatrix} x_1 \\ x_2 \end{bmatrix}$$

$$= \begin{bmatrix} \frac{1}{\sqrt{2}} & -\frac{1}{\sqrt{2}} \\ \frac{1}{\sqrt{2}} & \frac{1}{\sqrt{2}} \end{bmatrix} \begin{bmatrix} x_1 \\ x_2 \end{bmatrix}$$

$$= \frac{1}{\sqrt{2}} \begin{bmatrix} x_1 - x_2 \\ x_1 + x_2 \end{bmatrix}.$$

The matrix $\begin{bmatrix} \frac{1}{\sqrt{2}} & -\frac{1}{\sqrt{2}} \\ \frac{1}{\sqrt{2}} & \frac{1}{\sqrt{2}} \end{bmatrix}$ is orthogonal, so

$$\begin{bmatrix} x_1 \\ x_2 \end{bmatrix} = \begin{bmatrix} \frac{1}{\sqrt{2}} & -\frac{1}{\sqrt{2}} \\ \frac{1}{\sqrt{2}} & \frac{1}{\sqrt{2}} \end{bmatrix}^T \begin{bmatrix} y_1 \\ y_2 \end{bmatrix} = \frac{1}{\sqrt{2}} \begin{bmatrix} y_1 + y_2 \\ -y_1 + y_2 \end{bmatrix}.$$

Hence $x_1 = \frac{1}{\sqrt{2}}(y_1 + y_2)$ and $x_2 = \frac{1}{\sqrt{2}}(-y_1 + y_2)$, so the equation $x_1 x_2 = 1$ becomes $y_2^2 - y_1^2 = 2$ in terms of these new variables. In this form we recognize the equation as that of a hyperbola in the y_1-y_2-coordinate system. The new y_1- and y_2-axes are called the **principal axes** of the hyperbola. This is the source of the name Principal Axis Theorem. □

Note that we chose the new y_1- and y_2-axes in Example 1 by observing a symmetry of the graph of the original equation $x_1 x_2 = 1$. While this is still possible for equations in three variables (the graph is then a surface in $\mathbb{R}^3$), this type of geometric insight is not available when more than three variables are involved. The approach we adopt is as follows: The change of variables in Example 1 carried the equation $x_1 x_2 = 1$ to the equation $y_2^2 - y_1^2 = 2$ which has no cross term $y_1 y_2$. It is in this form that the Principal Axis Theorem applies to a much wider class of equations and actually *reveals* symmetries in their graphs.

An expression like $x_1^2 - x_2^2 + 2x_3^2 + 3x_1 x_2 - x_1 x_3$ is an example of a quadratic form in the variables x_1, x_2 and x_3. In general, a **quadratic form** q in the n variables $x_1, x_2, \cdots, x_n$ is a linear combination of the terms $x_1^2, x_2^2, \cdots, x_n^2$, $x_1 x_2, x_1 x_3, x_2 x_3, \cdots$:

$$q = r_1 x_1^2 + r_2 x_2^2 + \cdots + r_n x_n^2 + r_{12} x_1 x_2 + r_{13} x_1 x_3 + r_{23} x_2 x_3 + \cdots$$

where the coefficients r_i and r_{ij} are real numbers. Even though q is not a linear function of the x_i, it can be easily described using matrix multiplication. Indeed if we write $X = [x_1 \, x_2 \, \cdots \, x_n]^T$, q can be written in **matrix form**

$$q = X^T A X \text{ where } A \text{ is a } symmetric \, n \times n \text{ matrix.}$$

In fact, the main diagonal entries of the matrix A are the coefficients (possibly zero) of x_1^2, x_2^2 and x_n^2 in order and, if $i \neq j$, the (i,j)- and (j,i)-entries of A are both equal to half the coefficient of $x_i x_j$. Here is an example.

Example 2. *Write* $q = x_1^2 + 3x_3^2 + 2x_1 x_2 - x_1 x_3$ *in matrix form* $q = X^T A X$ *where A is symmetric.*

Solution. Applying the above rule, we get

$$q = [x_1 \ x_2 \ x_3] \begin{bmatrix} 1 & 1 & -\frac{1}{2} \\ 1 & 0 & 0 \\ -\frac{1}{2} & 0 & 3 \end{bmatrix} \begin{bmatrix} x_1 \\ x_2 \\ x_3 \end{bmatrix}.$$

The reader should verify that this really works. □

The reason for writing quadratic forms as in Example 2 is that the Principal Axis Theorem can be applied to the symmetric matrix A. Suppose that a quadratic form q is written in matrix form as

$$q = X^T A X$$

where A is a symmetric $n \times n$ matrix and $X = [x_1 \ x_2 \ \cdots \ x_n]^T$. We want to find new variables $Y = [y_1 \ y_2 \ \cdots \ y_n]^T$ such that, when q is written in terms of $y_1, y_2, \cdots, y_n$, there are no cross terms $y_i y_j$ where $i \neq j$. This amounts to asking that $q = Y^T D Y$ where D is a *diagonal* matrix. Since A is symmetric, the Principal Axis Theorem guarantees the existence of an orthogonal matrix P such that

$$P^T A P = D = \begin{bmatrix} \lambda_1 & 0 & \cdots & 0 \\ 0 & \lambda_2 & \cdots & 0 \\ \vdots & \vdots & \ddots & \vdots \\ 0 & 0 & \cdots & \lambda_n \end{bmatrix}$$

where the λ_i are the (real but not necessarily distinct) eigenvalues of the symmetric matrix A. Note that $P^{-1} = P^T$. Now define the new variables Y by

$$X = PY \text{ equivalently } Y = P^T X.$$

Then substitution in the equation $q = X^T A X$ gives

$$q = (PY)^T A (PY) = Y^T (P^T A P) Y = Y^T D Y = \lambda_1 y_1^2 + \lambda_2 y_2^2 + \cdots + \lambda_n y_n^2.$$

Hence this change of variables has produced the desired simplification of q (and we say that the quadratic form q has been **diagonalized**). The following theorem summarizes this discussion.

Theorem 1. *Let $q = X^T A X$ be a quadratic form where A is a symmetric $n \times n$ matrix and $X = [x_1 \ x_2 \ \cdots \ x_n]^T$. Let P be an orthogonal matrix such that $P^T A P = diag(\lambda_1, \lambda_2, \cdots, \lambda_n)$ where the λ_i are the (real) eigenvalues of A repeated according to their multiplicities. Define new variables $Y = [y_1 \ y_2 \ \cdots \ y_n]^T$ by*

$$X = PY \text{ equivalently } Y = P^T X.$$

Then

$$q = \lambda_1 y_1^2 + \lambda_2 y_2^2 + \cdots + \lambda_n y_n^2$$

4.8. QUADRATIC FORMS

in terms of these new variables y_i.

Example 3. The quadratic form $q = x_1 x_2$ in Example 1 has matrix form $q = X^T A X$ where $X = [x_1 \; x_2]^T$ and $A = \begin{bmatrix} 0 & \frac{1}{2} \\ \frac{1}{2} & 0 \end{bmatrix}$. The eigenvalues are $-\frac{1}{2}$ and $\frac{1}{2}$, with corresponding eigenvectors $\begin{bmatrix} 1 \\ -1 \end{bmatrix}$ and $\begin{bmatrix} 1 \\ 1 \end{bmatrix}$, respectively. They are (of course) orthogonal, and after normalizing the orthogonal matrix P is $P = \frac{1}{\sqrt{2}} \begin{bmatrix} 1 & 1 \\ -1 & 1 \end{bmatrix}$. Hence the new variables are $\begin{bmatrix} y_1 \\ y_2 \end{bmatrix} = Y = P^T X = \frac{1}{\sqrt{2}} \begin{bmatrix} 1 & -1 \\ 1 & 1 \end{bmatrix} \begin{bmatrix} x_1 \\ x_2 \end{bmatrix}$. Thus $y_1 = \frac{1}{\sqrt{2}}(x_1 - x_2)$ and $y_2 = \frac{1}{\sqrt{2}}(x_1 + x_2)$ as in Example 1.

Note that this "diagonalizing" change of variables is not unique. For example, if we take $[-1 \; 1]^T$ as the eigenvector corresponding to the eigenvalue $-\frac{1}{2}$, the resulting formulas are $y_1' = \frac{1}{\sqrt{2}}(-x_1 + x_2)$ and $y_2' = \frac{1}{\sqrt{2}}(x_2 + x_1)$. Then y_2' is the same as y_2, but y_1' is the negative of the preceding y_1. Geometrically, this amounts to pointing the y_1-axis in the opposite direction in Example 1. □

Example 4. *Diagonalize the quadratic form* $q = x_1^2 + 2x_2^2 + x_3^2 - 2x_1 x_2 - 2x_2 x_3$.

Solution. Here $q = X^T A X$ where $A = \begin{bmatrix} 1 & -1 & 0 \\ -1 & 2 & -1 \\ 0 & -1 & 1 \end{bmatrix}$ and $X = \begin{bmatrix} x_1 \\ x_2 \\ x_3 \end{bmatrix}$.

The eigenvalues of A are $\lambda_1 = 0$, $\lambda_2 = 1$ and $\lambda_3 = 3$, with (orthogonal) eigenvectors $G_1 = [1 \; 1 \; 1]^T$, $G_2 = [-1 \; 0 \; 1]^T$, and $G_3 = [1 \; -2 \; 1]^T$ respectively. Normalizing gives the principal axes $F_1 = \frac{1}{\sqrt{3}} G_1$, $F_2 = \frac{1}{\sqrt{2}} G_2$, and $F_3 = \frac{1}{\sqrt{6}} G_3$, and so we obtain an orthogonal matrix

$$P = [F_1 \; F_2 \; F_3] = \begin{bmatrix} \frac{1}{\sqrt{3}} & -\frac{1}{\sqrt{2}} & \frac{1}{\sqrt{6}} \\ \frac{1}{\sqrt{3}} & 0 & -\frac{2}{\sqrt{6}} \\ \frac{1}{\sqrt{3}} & \frac{1}{\sqrt{2}} & \frac{1}{\sqrt{6}} \end{bmatrix}$$

such that $P^T A P = diag(\lambda_1, \lambda_2, \lambda_3) = diag(0, 1, 3)$. Thus the new variables $[y_1 \; y_2 \; y_3]^T = Y = P^T X$ are

$$y_1 = \frac{1}{\sqrt{3}}(x_1 + x_2 + x_3), \; y_2 = \frac{1}{\sqrt{2}}(-x_1 + x_3), \; y_3 = \frac{1}{\sqrt{6}}(x_1 - 2x_2 + x_3).$$

If we compute $X = PY$ and substitute in q, we diagonalize the quadratic form in the sense that $q = \lambda_1 y_1^2 + \lambda_2 y_2^2 + \lambda_3 y_3^2 = y_2^2 + 3y_3^2$. □

These methods shed light on graphs of the form $q = 1$ where q is a quadratic form in two variables x_1 and x_2 (as in Example 1). The graph of the equation $rx_1^2 + sx_2^2 = 1$ is called an **ellipse** if $rs > 0$ and a **hyperbola** if $rs < 0$. Our theory guarantees that every equation $ax_1^2 + bx_1 x_2 + cx_2^2 = 1$ can be transformed into such a diagonal form by a change of coordinates. The next theorem shows that this can always be achieved by a rotation (as in Example 1), and it also gives a simple way to decide from the coefficients a, b and c whether it is an ellipse or a hyperbola.

Theorem 2. *Consider the quadratic form $q = ax_1^2 + bx_1x_2 + cx_2^2$.*

(1) *There is a counterclockwise rotation of the coordinate axes about the origin such that q has the form $q = ry_1^2 + sy_2^2$ in the new variables y_1 and y_2.*

(2) *The graph of the equation $ax_1^2 + bx_1x_2 + cx_2^2 = 1$ is an ellipse if $b^2 - 4ac < 0$ and it is a hyperbola if $b^2 - 4ac > 0$.*

Proof. The matrix of q is $A = \begin{bmatrix} a & \frac{1}{2}b \\ \frac{1}{2}b & c \end{bmatrix}$. Let λ_1 and λ_2 be the eigenvalues of A with orthonormal eigenvectors X_1 and X_2. Hence $P = [X_1 \ X_2]$ is an orthogonal matrix and $P^T A P = diag(\lambda_1, \lambda_2)$. We have $det P = \pm 1$, so by interchanging X_1 and X_2 if necessary, we can ensure that $det P = 1$ and hence (by Theorem 7 §3.5) that multiplication by P is a rotation of $\mathbb{R}^2$. Since the new coordinates y_1 and y_2 are given by $[y_1 \ y_2]^T = P^T [x_1 \ x_2]^T$, the new axes are obtained by a rotation (since P^T is also orthogonal and $det(P^T) = 1$). This proves (1). Furthermore, the fact that $P^T = P^{-1}$ gives

$$\lambda_1 \lambda_2 = det \begin{bmatrix} \lambda_1 & 0 \\ 0 & \lambda_2 \end{bmatrix} = det(P^T A P) = det A = -\tfrac{1}{4}(b^2 - 4ac).$$

Since $q = \lambda_1 y_1^2 + \lambda_2 y_2^2$ in the new coordinate system, the graph of the equation $q = 1$ is an ellipse if $\lambda_1 \lambda_2 > 0$ (that is $b^2 - 4ac < 0$), and it is a hyperbola if $\lambda_1 \lambda_2 < 0$ (that is $b^2 - 4ac > 0$). This proves (2). □

There are many ways to diagonalize a given quadratic form q. For example, consider $q = 3x_1^2 + 4x_1x_2 + 2x_2^2$. Then $q = X^T A X$ where $A = \begin{bmatrix} 3 & 2 \\ 2 & 2 \end{bmatrix}$. This matrix has eigenvalues $\lambda_1 = \frac{1}{2}(5 + \sqrt{17})$ and $\lambda_2 = \frac{1}{2}(5 - \sqrt{17})$, and so q can be written in the form $q = \lambda_1 y_1^2 + \lambda_2 y_2^2$ for appropriate new variables y_1 and y_2. On the other hand, q can be written as follows:

$$q = x_1^2 + 2(x_1 + x_2)^2 \quad \text{and} \quad q = 3(x_1 + \tfrac{2}{3}x_2)^2 + \tfrac{2}{3}x_2^2.$$

These formulas reveal two other changes of variables that diagonalize q:

$$\begin{array}{rl} y_1' = & x_1 \\ y_2' = & x_1 + x_2 \end{array} \quad \text{and} \quad \begin{array}{rl} y_1'' = & x_1 + \tfrac{2}{3}x_2 \\ y_2'' = & x_2 \end{array}.$$

The general problem of how such changes of variables are related has been studied in great detail; however this is beyond the scope of this book.

4.8.2 Positive Definite Matrices

If A is an $n \times n$ symmetric matrix, the quadratic form $q = X^T A X$ can be viewed as a function $q : \mathbb{R}^n \to \mathbb{R}$ given by $q(X) = X^T A X$. If $n = 2$ and $X = [x_1 \ x_2]^T$, then $q = q(x_1, x_2)$ is a function of two variables and we can look at the graph of q in $\mathbb{R}^3$. Three examples are shown in the diagram.

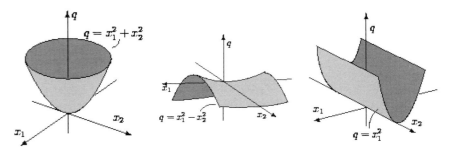

In this section we examine the important special case where the graph stays strictly above the x_1-x_2-plane except at the origin (as in the first example in the diagram). Moreover, we treat the general case where $q = q(x_1, x_2, \cdots, x_n)$ is a function of n variables.

If $q = q(X) = X^T A X$ is a quadratic form on $\mathbb{R}^n$, it is clear that $q(0) = 0$, and q is said to be a **positive definite form** if $q(X) > 0$ whenever $X \neq 0$. For example, it is clear that $q(x_1, x_2) = x_1^2 + 2x_2^2$ is positive definite; however it is sometimes less clear.

Example 5. If $A = \begin{bmatrix} 2 & -1 \\ -1 & 1 \end{bmatrix}$ and $X = \begin{bmatrix} x_1 \\ x_2 \end{bmatrix}$, show that the corresponding form $q = X^T A X = 2x_1^2 + x_2^2 - 2x_1 x_2$ is positive definite.

Solution. We can write $q = x_1^2 + (x_1 - x_2)^2$, so the only way that q can be zero is if $x_1 = 0$ and $x_1 - x_2 = 0$; that is if $X = 0$. □

The Principal Axis Theorem provides the following important characterization of these positive definite forms $q = X^T A X$ in terms of the (real) eigenvalues of the symmetric matrix A.

Theorem 3. *If A a symmetric matrix, the quadratic form $q(X) = X^T A X$ is positive definite if and only if every eigenvalue of A is positive.*

Proof. By the Principal Axis Theorem let P be an orthogonal matrix such that $P^T A P = D = diag(\lambda_1, \lambda_2, \cdots, \lambda_n)$ where the λ_i are the eigenvalues of A. Suppose $\lambda_i > 0$ for each i. If $X \neq 0$, write $Y = P^T X = [y_1 \ y_2 \ \cdots \ y_n]^T$. Then $Y \neq 0$ because P^T is invertible. Since $X = PY$, this means that

$$q(X) = X^T A X = Y^T (P^T A P) Y = Y^T D Y = \lambda_1 y_1^2 + \lambda_2 y_2^2 + \cdots + \lambda_n y_n^2 > 0$$

because each $\lambda_i > 0$ and some $y_i \neq 0$. Conversely, if $q(X) > 0$ for all $X \neq 0$, let $X_j = PE_j$ where E_j is column j of I_n. Then $0 < q(X_j) = X_j^T A X_j = E_j^T(P^T A P)E_j = E_j^T D E_j = \lambda_j$, as required. □

Motivated by Theorem 3, a square matrix is called **positive definite** if it is symmetric and all its eigenvalues are positive. Hence Theorem 3 asserts that a symmetric matrix A is positive definite if and only if the corresponding quadratic form $q(X) = X^T A X$ is a positive definite form. In particular, Example 5 shows that $\begin{bmatrix} 2 & -1 \\ -1 & 1 \end{bmatrix}$ is a positive definite matrix.

Corollary. *Every positive definite matrix A is invertible; indeed $\det A > 0$.*

Proof. Since A is symmetric, we have $P^T A P = D = diag(\lambda_1, \lambda_2, \cdots, \lambda_n)$ for some orthogonal matrix P. Hence $\det A = \det D = \lambda_1 \lambda_2 \cdots \lambda_n > 0$ because each $\lambda_i > 0$. □

These positive definite matrices arise frequently when optimization (maximum or minimum) problems are encountered, and so have applications throughout science and engineering. They are also used in factor analysis in statistics, and we will encounter them when discussing general inner products in Chapter 5.

Example 6. *If U is any invertible $n \times n$ matrix, show that $A = U^T U$ is positive definite.*

Solution. Let $q(X) = X^T A X$ be the quadratic form associated with A. If $X \neq 0$ in $\mathbb{R}^n$ then $q(X) = X^T A X = X^T(U^T U)X = (UX)^T(UX) = \|UX\|^2 > 0$ because $UX \neq 0$ (U is invertible). Hence Theorem 3 applies. □

It is remarkable that the converse to Example 6 is also true. In fact every positive definite matrix A can be factored as $A = U^T U$ where U is an upper triangular matrix with positive elements on the main diagonal. However, before verifying this, we introduce another concept that is central to any discussion of positive definite matrices.

If A is any $n \times n$ matrix, let $^{(r)}A$ denote the $r \times r$ submatrix of A in the upper left corner of A; that is $^{(r)}A$ is the matrix obtained from A by deleting the last $n - r$ rows and columns. The matrices $^{(1)}A$, $^{(2)}A$, $^{(3)}A, \cdots, ^{(n)}A = A$ are called the **principal submatrices** of A.

Example 7. If $A = \begin{bmatrix} 10 & 5 & 2 \\ 5 & 3 & 2 \\ 2 & 2 & 3 \end{bmatrix}$ then $^{(1)}A = [10]$, $^{(2)}A = \begin{bmatrix} 10 & 5 \\ 5 & 3 \end{bmatrix}$ and $^{(3)}A = A$.

Lemma 1. *If A is positive definite, so is each principal submatrix $^{(r)}A$ for $r = 1, 2, \cdots, n$.*

4.8. QUADRATIC FORMS

Proof. Write $A = \begin{bmatrix} {}^{(r)}A & P \\ Q & R \end{bmatrix}$ in block form. If $Y \neq 0$ in $\mathbb{R}^r$, write $X = \begin{bmatrix} Y \\ 0 \end{bmatrix}$ in $\mathbb{R}^n$. Then $X \neq 0$ so the fact that A is positive definite gives

$$0 < X^T A X = [Y^T \ 0] \begin{bmatrix} {}^{(r)}A & P \\ Q & R \end{bmatrix} \begin{bmatrix} Y \\ 0 \end{bmatrix} = Y^T \left({}^{(r)}A \right) Y.$$

This shows that ${}^{(r)}A$ is positive definite by Theorem 3.[18] □

If A is positive definite, Lemma 1 and the Corollary to Theorem 3 show that $det\left({}^{(r)}A \right) > 0$ for every r. This proves part of the following theorem which contains the converse to Example 6, and characterizes the positive definite matrices among the symmetric ones.

Theorem 4. *The following conditions are equivalent for a symmetric $n \times n$ matrix A:*

(1) *A is positive definite.*

(2) *$det\left({}^{(r)}A \right) > 0$ for each $r = 1, 2, \cdots, n$.*

(3) *$A = U^T U$ where U is an upper triangular matrix with positive entries on the main diagonal.*

*Furthermore, the factorization in (3) is unique (called the **Cholesky Factorization** of A).*

Proof. (3)$\Rightarrow$(1) is Example 6, and (1)$\Rightarrow$(2) follows from Lemma 1 and the Corollary to Theorem 3.

(2)$\Rightarrow$(3). Assume (2) and proceed by induction on n. If $n = 1$ then $A = [a]$ where $a > 0$ by (2), so take $U = [\sqrt{a}]$. If $n > 1$, write $B = {}^{(n-1)}A$. Then B is symmetric and satisfies (2) so, by induction, we have $B = U^T U$ as in (3) where U is of size $(n-1) \times (n-1)$. Then, as A is symmetric, it has block form $A = \begin{bmatrix} B & P \\ P^T & a \end{bmatrix}$ where P is a column in $\mathbb{R}^{n-1}$ and a is in $\mathbb{R}$. If we write $X = (U^T)^{-1} P$ and $b = a - X^T X$, block multiplication gives

$$A = \begin{bmatrix} U^T U & P \\ P^T & a \end{bmatrix} = \begin{bmatrix} U^T & 0 \\ X^T & 1 \end{bmatrix} \begin{bmatrix} U & X \\ 0 & b \end{bmatrix},$$

as the reader can verify. Taking determinants and applying Theorem 3 §2.2 gives $detA = det(U^T) detU \cdot b = b\,(detU)^2$. Hence $b > 0$ because $detA > 0$ by (2). But then the above factorization can be written $A = \begin{bmatrix} U^T & 0 \\ X^T & \sqrt{b} \end{bmatrix} \begin{bmatrix} U & X \\ 0 & \sqrt{b} \end{bmatrix}$. Since U has positive diagonal entries this is a Cholesky factorization of A. This proves (3).

---

[18] A similar argument shows that, if B is any matrix obtained from a positive definite matrix A by deleting certain rows and deleting the *same* columns, then B is also positive definite.

As to the uniqueness, suppose that $A = U^TU = U_1^TU_1$ are two Cholesky factorizations. Write $D = UU_1^{-1} = (U^T)^{-1}U_1^T$. Then D is upper triangular (because $D = UU_1^{-1}$) and lower triangular (because $D = (U^T)^{-1}U_1^T$) and so is a diagonal matrix. Thus $U = DU_1$ and $U_1 = DU$ (since $U_1^T = U^TD$), so it suffices to show that $D = I$. But eliminating U_1 gives $U = D^2U$, so $D^2 = I$ because U is invertible. Since the diagonal entries of D are positive (this is true of U and U_1) it follows that $D = I$. □

The remarkable thing is that the matrix U in the Cholesky factorization is easy to obtain from A using row operations. The key is that Step 1 of the following algorithm is possible for any positive definite matrix A. A proof of the algorithm is given following Example 8.

Algorithm for the Cholesky Factorization. *If A is a positive definite matrix, the Cholesky factorization $A = U^TU$ can be obtained as follows:*

Step 1 *Carry A to an upper triangular matrix U_1 with positive diagonal entries by row operations each of which adds a multiple of a row to a lower row.*

Step 2. *Obtain U from U_1 by dividing each row of U_1 by the square root of the diagonal entry in that row.*

Example 8. *Find the Cholesky factorization of* $A = \begin{bmatrix} 10 & 5 & 2 \\ 5 & 3 & 2 \\ 2 & 2 & 3 \end{bmatrix}$.

Solution. The matrix A is positive definite by Theorem 4 because $det^{(1)}A = 10 > 0$, $det^{(2)}A = 5 > 0$ and $det^{(3)}A = det\, A = 3 > 0$. Hence Step 1 of the Algorithm is carried out as follows:

$$A = \begin{bmatrix} 10 & 5 & 2 \\ 5 & 3 & 2 \\ 2 & 2 & 3 \end{bmatrix} \to \begin{bmatrix} 10 & 5 & 2 \\ 0 & \frac{1}{2} & 1 \\ 0 & 1 & \frac{13}{5} \end{bmatrix} \to \begin{bmatrix} 10 & 5 & 2 \\ 0 & \frac{1}{2} & 1 \\ 0 & 0 & \frac{3}{5} \end{bmatrix} = U_1.$$

Now carry out Step 2 on U_1 to obtain $U = \begin{bmatrix} \sqrt{10} & \frac{5}{\sqrt{10}} & \frac{2}{\sqrt{10}} \\ 0 & \frac{1}{\sqrt{2}} & \sqrt{2} \\ 0 & 0 & \frac{\sqrt{3}}{\sqrt{5}} \end{bmatrix}$. The reader can verify that $U^TU = A$. □

Proof of the Cholesky Algorithm. If A is positive definite, let $A = U^TU$ be the Cholesky factorization, and let $D = diag(d_1, \cdots, d_n)$ be the common diagonal of U and U^T. Then U^TD^{-1} is lower triangular with ones on the diagonal (call such matrices LT-1). Hence $L = (U^TD^{-1})^{-1}$ is also LT-1, and so $I_n \to L$ by a sequence of row operations each of which adds a multiple of a row to a lower row (verify; modify columns right to left). But then $A \to LA$ by the same

4.8. QUADRATIC FORMS

sequence of row operations (see the discussion preceding Theorem 1 §1.6). Since $LA = [D(U^T)^{-1}][U^TU] = DU$ is upper triangular with positive entries on the diagonal, this shows that Step 1 of the Algorithm is possible.

Turning to Step 2, let $A \to U_1$ as in Step 1 so that $U_1 = L_1A$ where L_1 is LT-1. Since A is symmetric, we get

$$L_1U_1^T = L_1(L_1A)^T = L_1A^TL_1^T = L_1AL_1^T = U_1L_1^T. \qquad (*)$$

Let $D_1 = diag(e_1, \cdots, e_n)$ denote the diagonal entries of U_1. Then $(*)$ gives $L_1(U_1^TD_1^{-1}) = U_1L_1^TD_1^{-1}$, and this is both upper triangular (right side) and LT-1 (left side), and so must equal I_n. In particular, $U_1^TD_1^{-1} = L_1^{-1}$. Now let $D_2 = diag(\sqrt{e_1}, \cdots, \sqrt{e_n})$, so that $D_2^2 = D_1$. If we write $U = D_2^{-1}U_1$, we have

$$U^TU = (U_1^TD_2^{-1})(D_2^{-1}U_1) = U_1^T(D_2^2)^{-1}U_1 = (U_1^TD_1^{-1})U_1 = (L_1^{-1})U_1 = A.$$

This proves Step 2 because $U = D_2^{-1}U_1$ is formed by multiplying each row of U_1 by the square root of its diagonal entry (verify). □

4.8.3 Constrained Optimization

It is a frequent occurrence in applications that a function $q = q(x_1, x_2, \cdots, x_n)$ of n variables, called an **objective function**, is to be made as large or as small as possible among all vectors $X = [x_1\ x_2\ \cdots\ x_n]^T$ lying in a certain region of $\mathbb{R}^n$ called the **feasible region**. A wide variety of objective functions q arise in practice; our primary concern here is to examine one important situation where q is a quadratic form. The next example gives some indication of how such problems arise.

Example 9. A politician proposes to spend x_1 dollars annually on health care and x_2 dollars annually on education. She is constrained in her spending by various budget pressures, and one model of this is that the expenditures x_1 and x_2 should satisfy a constraint like

$$5x_1^2 + 3x_2^2 \leq 15.$$

Since $x_i \geq 0$ for each i, the feasible region is the shaded area shown in the diagram. Any choice of feasible point $[x_1\ x_2]^T$ in this region will satisfy the budget constraints. However, these choices have different effects on voters, and the politician wants to choose $X = [x_1\ x_2]^T$ to maximize some measure $q = q(x_1, x_2)$ of voter satisfaction. Thus the assumption is that, for any value of c, all points on the graph of $q(x_1, x_2) = c$ have the same appeal to voters. Hence the goal is to find the largest value of c for which the graph of $q(x_1, x_2) = c$ contains a feasible point.

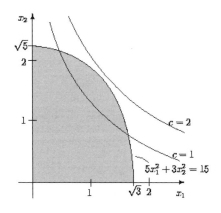

The choice of the function q depends upon many factors; we will show how to solve the problem for any quadratic form q (even with more than two variables). In the diagram the function q is given by

$$q(x_1, x_2) = x_1 x_2,$$

and the graphs of $q(x_1, x_2) = c$ are shown for $c = 1$ and $c = 2$. As c increases the graph of $q(x_1, x_2) = c$ moves up and to the right. From this it is clear that there will be a solution for some value of c between 1 and 2 (in fact the smallest value is $c = \frac{1}{2}\sqrt{15} = 1.94$ to two decimal places). □

The constraint $5x_1^2 + 3x_2^2 \leq 15$ in Example 9 can be put in a standard form. If we divide through by 15, it becomes $(\frac{x_1}{\sqrt{3}})^2 + (\frac{x_2}{\sqrt{5}})^2 \leq 1$. This suggests that we introduce new variables $Y = [y_1 \; y_2]^T$ where $y_1 = \frac{x_1}{\sqrt{3}}$ and $y_2 = \frac{x_2}{\sqrt{5}}$. Then the constraint becomes $\|Y\|^2 \leq 1$, equivalently $\|Y\| \leq 1$. In terms of these new variables, the objective function is $q = \sqrt{15} y_1 y_2$, and we want to maximize this subject to $\|Y\| \leq 1$. When this is done, the maximizing values of x_1 and x_2 are obtained from $x_1 = \sqrt{3} \, y_1$ and $x_2 = \sqrt{5} \, y_2$.

Hence, for constraints like that in Example 9, there is no real loss in generality in assuming that the constraint takes the form $\|X\| \leq 1$. In this case the Principal Axis Theorem solves the problem. Recall that a vector in $\mathbb{R}^n$ of length 1 is called a *unit vector*.

Theorem 5. *Consider the quadratic form $q = q(X) = X^T A X$ where A is an $n \times n$ symmetric matrix, and let λ_1 and λ_n denote the largest and smallest eigenvalues of A, respectively. Then*:

(1) $max\{q(X) \mid \|X\| \leq 1\} = \lambda_1$, *and* $q(F_1) = \lambda_1$, *where F_1 is any unit eigenvector corresponding to λ_1.*

(2) $min\{q(X) \mid \|X\| \leq 1\} = \lambda_n$, *and* $q(F_n) = \lambda_n$, *where F_n is any unit eigenvector corresponding to λ_n.*

Proof. Since A is symmetric, let the (real) eigenvalues λ_i of A be ordered as to size as follows: $\lambda_1 \geq \lambda_2 \geq \cdots \geq \lambda_n$. By the Principal Axis Theorem, let P be an orthogonal matrix such that $P^T A P = D = diag(\lambda_1, \lambda_2, \cdots, \lambda_n)$. Define $Y = P^T X$, equivalently $X = PY$, and observe that $\|Y\| = \|X\|$ because $\|Y\|^2 = Y^T Y = X^T (PP^T) X = X^T X = \|X\|^2$. If we write $Y = [y_1 \; y_2 \; \cdots \; y_n]^T$, then

$$\begin{aligned} q(X) &= q(PY) = (PY)^T A (PY) \\ &= Y^T (P^T A P) Y = Y^T D Y \\ &= \lambda_1 y_1^2 + \lambda_2 y_2^2 + \cdots + \lambda_n y_n^2. \end{aligned} \quad (**)$$

Now assume that $\|X\| \leq 1$. Since $\lambda_i \leq \lambda_1$ for each i, (**) gives

$$q(X) = \lambda_1 y_1^2 + \lambda_2 y_2^2 + \cdots + \lambda_n y_n^2 \leq \lambda_1 y_1^2 + \lambda_1 y_2^2 + \cdots + \lambda_1 y_n^2 = \lambda_1 \|Y\|^2 \leq \lambda_1$$

4.8. QUADRATIC FORMS

because $\|Y\| = \|X\| \leq 1$. This shows that $q(X)$ cannot exceed λ_1 when $\|X\| \leq 1$. To see that this maximum is actually achieved let F_n be a unit eigenvector corresponding to λ_1. Then

$$q(F_1) = F_1^T A F_1 = F_1^T (\lambda_1 F_1) = \lambda_1 (F_1^T F_1) = \lambda_1 \|F_1\|^2 = \lambda_1.$$

Hence λ_1 is the maximum value of $q(X)$ when $\|X\| \leq 1$, proving (1). The proof of (2) is analogous. □

The set of all vectors X in $\mathbb{R}^n$ such that $\|X\| \leq 1$ is called the **unit ball**. If $n = 2$ it is often called the unit disk and consists of the unit circle and its interior; if $n = 3$ it is the unit sphere and its interior. It is worth noting that the maximum value of a quadratic form $q(X)$ as X ranges *throughout* the unit ball is actually attained for a unit vector X on the *boundary* of the unit ball.

Theorem 5 is important for applications involving vibrations in areas as diverse as aerodynamics and particle physics, and the maximum and minimum values in the theorem are often found using advanced calculus to minimize the quadratic form on the unit ball. The algebraic approach using the Principal Axis Theorem gives a geometrical interpretation of the optimal values because they are eigenvalues.

Example 10. *Maximize and minimize the form $q(X) = 3x_1^2 + 14x_1x_2 + 3x_2^2$ subject to $\|X\| \leq 1$.*

Solution. The matrix of q is $A = \begin{bmatrix} 3 & 7 \\ 7 & 3 \end{bmatrix}$, with eigenvalues $\lambda_1 = 10$ and $\lambda_2 = -4$ and corresponding unit eigenvectors $F_1 = \frac{1}{\sqrt{2}}[1 \ 1]^T$ and $F_2 = \frac{1}{\sqrt{2}}[1 \ -1]^T$. Hence, among all unit vectors X in $\mathbb{R}^2$, $q(X)$ takes its maximal value 10 at $X = F_1$, and the minimum value of $q(X)$ is -4 when $X = F_2$. □

As noted above, the objective function in a constrained optimization problem need not be a quadratic form. We conclude with an example where the objective function is linear, and the feasible region is determined by linear constraints.

Example 11. A manufacturer makes x_1 units of product 1, and x_2 units of product 2, at a profit of $70 and $50 per unit respectively, and wants to choose x_1 and x_2 to maximize the total profit $p(x_1, x_2) = 70x_1 + 50x_2$. However x_1 and x_2 are not arbitrary; for example, $x_1 \geq 0$ and $x_2 \geq 0$. Other conditions also come into play.

Each unit of product 1 costs $1200 to produce and requires 2000 square feet of warehouse space; and a unit of product 2 requires $1300 to produce and requires 1100 square feet of space. If the total warehouse space is 11300 square feet, and if the total production budget is $8700, x_1 and x_2 must also satisfy the conditions

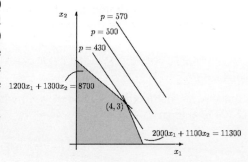

$$2000x_1 + 1100x_2 \leq 11300,$$
$$1200x_1 + 1300x_2 \leq 8700.$$

The feasible region in the plane satisfying these constraints (and $x_1 \geq 0$, $x_2 \geq 0$) is shaded in the diagram. If the profit equation $70x_1 + 50x_2 = p$ is plotted for various values of p, the resulting lines are parallel, with p increasing with distance from the origin. Hence the best choice occurs for the line $70x_1 + 50x_2 = 430$ that touches the shaded region at the point $(4, 3)$. So the profit p has a maximum of $p = 430$ for $x_1 = 4$ units and $x_2 = 3$ units. □

Example 11 is a simple case of the general **linear programming** problem[19] which arises in economic, management, network and scheduling applications. Here the objective function is a linear combination $q = a_1 x_1 + a_2 x_2 + \cdots + a_n x_n$ of the variables, and the feasible region consists of the vectors $X = [x_1 \; x_2 \; \cdots \; x_n]^T$ in $\mathbb{R}^n$ which satisfy a set of linear inequalities of the form $b_1 x_1 + b_2 x_2 + \cdots + b_n x_n \leq b$. There is a good method (an extension of the Gaussian algorithm) called the **simplex algorithm** for finding the maximum and minimum values of q when X ranges over such a feasible set. As Example 11 suggests, the optimal values turn out to be vertices of the feasible set. In particular, they are on the boundary of the feasible region, as is the case in Theorem 5.

4.8.4 Statistical Principal Component Analysis

Linear algebra is important in multivariate analysis in statistics, and we conclude with a very short look at one application of diagonalization in this area. A main feature of probability and statistics is the idea of a *random variable* X, that is a real valued function which takes its values according to a probability law (called its *distribution*). Random variables occur in a wide variety of contexts; examples include the number of meteors falling per square kilometer in a given region, the price of a share of Microsoft, or the duration of a long distance telephone call from a certain city.

The values of a random variable X are distributed about a central number μ, called the *mean* of X. The mean can be calculated from the distribution as the

---

[19] We do not pursue this here, and the reader is referred to "Linear Programming and Extensions" by N. Wu and R. Coppins, McGraw-Hill, 1981.

4.8. QUADRATIC FORMS

expectation $E(X) = \mu$ of the random variable X. Functions of a random variable are again random variables. In particular, $(X - \mu)^2$ is a random variable, and the *variance* of the random variable X, denoted $var(X)$, is defined to be the number

$$var(X) = E\{(X - \mu)^2\} \text{ where } \mu = E(X).$$

It is not difficult to see that $var(X) \geq 0$ for every random variable X. The number $\sigma = \sqrt{var(X)}$ is called the *standard deviation* of X, and is a measure of how much the values of X are spread about the mean μ of X. A main goal of statistical inference is finding reliable methods for estimating the mean and the standard deviation of a random variable X by sampling the values of X.

If two random variables X and Y are given, and their joint distribution is known, then functions of X and Y are also random variables. In particular, $X + Y$ and aX are random variables for any real number a, and we have

$$E(X + Y) = E(X) + E(Y) \quad \text{and} \quad E(aX) = aE(X).[20]$$

An important question is how much the random variables X and Y depend on each other. One measure of this is the *covariance* of X and Y, denoted $cov(X, Y)$, defined by

$$cov(X, Y) = E\{(X - \mu)(Y - \nu)\} \text{ where } \mu = E(X) \text{ and } \nu = E(Y).$$

Clearly, $cov(X, X) = var(X)$. If $cov(X, Y) = 0$ then X and Y have little relationship to each other and are said to be *uncorrelated*. [21]

Multivariate statistical analysis deals with a family $X_1, X_2, \cdots, X_n$ of random variables with means $\mu_i = E(X_i)$ and variances $\sigma_i^2 = var(X_i)$ for each i. We denote the covariance of X_i and X_j by $\sigma_{ij} = cov(X_i, X_j)$. Then the *covariance matrix* of the random variables $X_1, X_2, \cdots, X_n$ is defined to be the $n \times n$ matrix

$$\Sigma = [\sigma_{ij}]$$

whose (i, j)-entry is σ_{ij}. The matrix Σ is clearly symmetric; in fact it can be shown that Σ is **positive semidefinite** in the sense that $\lambda \geq 0$ for every eigenvalue λ of Σ. (In reality, Σ is positive definite in most cases of interest.) So suppose that the eigenvalues of Σ are $\lambda_1 \geq \lambda_2 \geq \cdots \geq \lambda_n \geq 0$, with corresponding eigenvectors $\bar{E}_1, \bar{E}_2, \cdots, \bar{E}_n$. [22] If $P = [\bar{E}_1 \ \bar{E}_2 \ \cdots \ \bar{E}_n]$, the Principal Axis Theorem shows that P is an orthogonal matrix and

$$P^T \Sigma P = diag(\lambda_1, \lambda_2, \cdots, \lambda_n).$$

---

[20] Because of this we say that $E(\ )$ is a linear transformation from the vector space of all random variables to the space of real numbers. In Section 5.3 we will have more to say about such general linear transformations.

[21] If X and Y are independent in the sense of probability theory, then they are uncorrelated; however, the converse is not true in general.

[22] In probability and statistics random variables are commonly denoted by upper case letters. Although it is in conflict with our usage of upper case letters for *matrices*, we will bow to convention and stay with the upper case notation, and denote column matrices of random variables as $\bar{X}$.

If we write $\bar{X} = [X_1\ X_2\ \cdots\ X_n]^T$, the procedure for diagonalizing a quadratic form (Theorem 1) gives new variables $\bar{Y} = [Y_1\ Y_2\ \cdots\ Y_n]^T$ defined by

$$\bar{Y} = P^T \bar{X}. \qquad (***)$$

These new random variables $Y_1, Y_2, \cdots, Y_n$ are called the **principal components** of the original random variables X_i, and are linear combinations of the X_i. Furthermore, it can be shown that

$$cov(Y_i, Y_j) = 0 \text{ if } i \neq j \quad \text{and} \quad var(Y_i) = \lambda_i \text{ for each } i.$$

Of course the principal components Y_i point along the principal axes of the quadratic form $q = \bar{X}^T \Sigma \bar{X}$.

The sum of the variances of a set of random variables is called the **total variance** of the variables, and determining the source of this total variance is one of the benefits of principal component analysis. The fact that the matrices Σ and $diag(\lambda_1, \lambda_2, \cdots, \lambda_n)$ are similar means that they have the same trace, that is

$$\sigma_{11} + \sigma_{22} + \cdots + \sigma_{nn} = \lambda_1 + \lambda_2 + \cdots + \lambda_n.$$

This means that the principal components Y_i have the same total variance as the original random variables X_i. Moreover, the fact that $\lambda_1 \geq \lambda_2 \geq \cdots \geq \lambda_n \geq 0$ means that most of this variance resides in the first few Y_i. In practice, statisticians find that studying these first few Y_i (and ignoring the rest) gives an accurate analysis of the total system variability. This results in substantial data reduction since often only a few Y_i suffice for all practical purposes. Furthermore, these Y_i are easily obtained as linear combinations of the X_i. Finally, the analysis of the principal components often reveals relationships among the X_i that were not previously suspected, and so results in interpretations that would not otherwise have been made.[23]

Exercises

1. In each case either show that the statement is true or give an example showing that it is false. Throughout, A denotes a square matrix.

    (a) Every positive definite matrix is orthogonally diagonalizable.

    (b) If all eigenvalues of A are positive then A is symmetric.

    (c) Every symmetric matrix with positive determinant is positive definite.

    (d) If A is positive definite and $X^T A X = 0$ where X is in $\mathbb{R}^n$ then $X = 0$.

    (e) If A is positive definite and $det A = 1$ then A is orthogonal.

---

[23] A more detailed analysis of principal components can be found for example in R.A. Johnson and D.W. Wichern, *Applied Multivariate Statistical Analysis* 4Ed., Prentice Hall, 1998.

4.8. QUADRATIC FORMS

(f) If U is invertible and all its eigenvalues are positive, then U is positive definite.

2. In each case find a symmetric matrix A such that $q = X^T B X$ takes the form $q = X^T A X$.

(a) $B = \begin{bmatrix} 2 & -1 & 1 \\ 1 & -1 & 3 \\ 0 & 2 & 1 \end{bmatrix}$ (b) $B = \begin{bmatrix} 2 & 2 & 7 \\ 4 & 1 & -2 \\ -1 & 0 & 5 \end{bmatrix}$

3. In each case find a change of variables that will diagonalize the quadratic form q, and express q in terms of these new variables.

(a) $q = x_1^2 + 4x_1x_2 + x_2^2$.

(b) $q = 2x_1^2 + 4x_1x_2 - x_2^2$.

(c) $q = x_1^2 + 4x_1x_2 + 3x_2^2$.

(d) $q = -x_1^2 + 4x_1x_2 + 3x_2^2$.

(e) $q = x_1^2 + 2x_2^2 + x_3^2 - 2x_1x_2 - 2x_2x_3$.

(f) $q = 5x_1^2 + 8x_2^2 + 5x_3^2 - 4x_1x_2 - 8x_1x_3 - 4x_2x_3$.

4. In each case decide if the graph of the equation is an ellipse or an hyperbola, and find a change of variables that transforms the equation to one with no cross terms.

(a) $3x_1^2 - 4x_1x_2 = 2$.

(b) $2x_1^2 + 4x_1x_2 + 5x_2^2 = 1$.

5. In each case find the maximum and minimum of $q(X) = X^T A X$ among all columns X such that $\|X\| \leq 1$.

(a) $q(X) = 3x_1^2 + 4x_1x_2 + x_2^2$, $X = [x_1 \; x_2]^T$.

(b) $q(X) = 2x_1^2 - 5x_1x_2 - 3x_2^2$, $X = [x_1 \; x_2]^T$.

(c) $q(X) = 3x_1^2 - x_2^2 - 2x_3^2 - 2x_1x_2 + 8x_2x_3$, $X = [x_1 \; x_2 \; x_3^T]$.

(d) $q(X) = x_1^2 - 3x_2^2 - x_3^2 - 4x_1x_3 + 4x_2x_3$, $X = [x_1 \; x_2 \; x_3]^T$.

6. In each case find the Cholesky factorization of the matrix A.

(a) $A = \begin{bmatrix} 2 & 3 \\ 3 & 5 \end{bmatrix}$ (b) $A = \begin{bmatrix} 3 & -2 \\ -2 & 2 \end{bmatrix}$

(c) $A = \begin{bmatrix} 10 & 4 & 3 \\ 4 & 2 & 2 \\ 3 & 2 & 3 \end{bmatrix}$ (d) $A = \begin{bmatrix} 5 & 3 & -2 \\ 3 & 2 & -1 \\ -2 & -1 & 4 \end{bmatrix}$

7. If A is positive definite, show that every diagonal entry of A is positive. [*Hint*: Consider $E_j^T A E_j$ where E_j is column j of the identity matrix.]

8. (a) If A is symmetric and invertible, show that A^2 is positive definite. [*Hint*: Show that $X^T A^2 X = \|AX\|^2$ for all X in $\mathbb{R}^n$.]

   (b) If $A = \begin{bmatrix} 3 & 2 \\ 2 & 1 \end{bmatrix}$, show that A^2 is positive definite but that A is not.

   (c) If A is positive definite, show that A^k is positive definite for all $k \geq 1$. [*Hint*: A is symmetric and invertible. Use induction on k.]

   (d) If A^k is positive definite where k is odd, show that A is positive definite.

9. If A is positive definite, show that A^{-1} is also positive definite. [*Hint*: Theorem 3.]

10. If A and B are positive definite $n \times n$ matrices, show that $A + B$ is positive definite.

11. If A and B are positive definite matrices (possibly different sizes), show that $\begin{bmatrix} A & 0 \\ 0 & B \end{bmatrix}$ is positive definite. [*Hint*: See the proof of Lemma 1.]

12. If A is an $n \times n$ positive definite matrix, and if U is an $n \times m$ matrix of rank m, show that $U^T A U$ is positive definite. [*Hint*: Theorem 3 §4.4.]

13. If A is positive definite, show that $A = B^2$ where B is positive definite. [*Hint*: Let $P^T A P = D = diag(\lambda_1, \cdots, \lambda_n)$, $\lambda_i > 0$, and consider $D_0 = diag(\sqrt{\lambda_1}, \cdots, \sqrt{\lambda_n})$.]

14. If A is positive definite, show that $A = CC^T$ for some matrix C with orthogonal columns. [*Hint*: Let $P^T A P = D = diag(\lambda_1, \cdots, \lambda_n)$, $\lambda_i > 0$. Write $D_0 = diag(\sqrt{\lambda_1}, \cdots, \sqrt{\lambda_n})$ and take $C = P D_0$.]

15. Call a square matrix **unit triangular** if it is triangular with 1's on the main diagonal.

    (a) Suppose that an invertible matrix A can be factored as $A = LDU$ where L is unit lower triangular, U is unit upper triangular, and D is diagonal with positive diagonal entries. Show that this factorization is unique: that is if $A = L_1 D_1 U_1$ is another such factorization, show that $L_1 = L$, $D_1 = D$ and $U_1 = U$. [*Hint*: If $L_1 D_1 U_1 = LDU$, scrutinize $L^{-1} L_1 D_1 = DUU_1^{-1}$.]

    (b) Show a matrix A is positive definite if and only if it has a factorization of the form $A = U^T DU$, which is a special form of (a). [*Hint*: Cholesky.]

16. Consider the equation $ax_1^2 + bx_1 x_2 + cx_2^2 = 1$ where $b \neq 0$. Write $X = [x_1 \ x_2]^T$, and let $Y = [y_1 \ y_2]^T$ denote the new variables obtained by rotating through an angle θ, so that (Theorem 3 §3.5) $Y = RX$ where $R = \begin{bmatrix} cos\theta & -sin\theta \\ sin\theta & cos\theta \end{bmatrix}$. If θ is an angle such that $cos\,2\theta = \frac{c-a}{\sqrt{b^2 + (a-c)^2}}$ and $sin\,2\theta = \frac{b}{\sqrt{b^2+(a-c)^2}}$, show that the resulting equation has no cross

term. [*Hint*: R is orthogonal, so $X = R^T Y$. Substitute into the equation $ax_1^2 + bx_1x_2 + cx_2^2 = 1$.]

17. Let $q = X^T A X$ be a quadratic form where A is an $n \times n$ symmetric matrix.

    (a) Show that A is uniquely determined by the function q, that is if $X^T B X = X^T A X$ holds for all columns X in $\mathbb{R}^n$, show that $B = A$. [*Hint*: If $C = B - A$ show that $X^T C X = 0$ for all X. Replace X by $X + Y$ to conclude that $X^Y C Y = 0$ for all X and Y. Then consider the columns of the identity matrix.]

    (b) If new variables Y are defined by $Y = U^{-1} X$ where U is an invertible matrix, show that q is given in terms of the new variables by $q = Y^T B Y$ where $B = U^T A U$.

    (c) Two square matrices A and B are said to be **congruent** (written $A \stackrel{c}{\sim} B$) if $B = U^T A U$ for some invertible matrix U. Show that:

    (i) $A \stackrel{c}{\sim} A$ for all A; If $A \stackrel{c}{\sim} B$ then $B \stackrel{c}{\sim} A$; and, If $A \stackrel{c}{\sim} B$ and $B \stackrel{c}{\sim} C$ then $A \stackrel{c}{\sim} C$.

    (ii) If $A \stackrel{c}{\sim} B$ then A is symmetric if and only if B is symmetric.

    (iii) A is symmetric if and only $A \stackrel{c}{\sim} D$ for some diagonal matrix D.

    (iv) If $A \stackrel{a}{\sim} B$ then $\operatorname{rank} A = \operatorname{rank} B$.

4.9 LINEAR TRANSFORMATIONS

In Section 3.5 we investigated 2×2 matrices by viewing them as transformations of the plane $\mathbb{R}^2$. Not only does this give a graphic interpretation of a matrix which is useful in analyzing the matrix, it also provides a way of studying certain geometrical properties of the plane (for example we found that rotations about the origin correspond to simple matrices depending on the angle of rotation). In this section we do the same thing for $n \times n$ matrices, and view them as linear transformations of $\mathbb{R}^n$.

4.9.1 Transformations $\mathbb{R}^n \to \mathbb{R}^m$

A transformation $T : \mathbb{R}^n \to \mathbb{R}^m$ is a rule[24] that associates with every vector X in $\mathbb{R}^n$ a uniquely determined vector $T(X)$ in $\mathbb{R}^m$. The situation can be visualized as in the diagram.

---

[24]Transformations $\mathbb{R} \to \mathbb{R}$ are usually called *functions* or *maps*. However the term "transformation" is commonly used in linear algebra.

To specify a transformation $T : \mathbb{R}^n \to \mathbb{R}^m$, *every* vector X in $\mathbb{R}^n$ must be assigned to a *uniquely determined* vector $T(X)$ in $\mathbb{R}^m$, called the **image** of X under T. This is referred to as specifying the **action** of the transformation T, or as **defining** T.

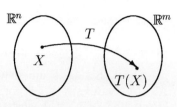

Most transformations $\mathbb{R}^n \to \mathbb{R}^m$ are of little interest in linear algebra; the important ones for us are those that preserve addition and scalar multiplication. More precisely, a transformation $T : \mathbb{R}^n \to \mathbb{R}^m$ is called a **linear transformation** if it satisfies the following two properties:

T1. $T(X + Y) = T(X) + T(Y)$ for all X and Y in $\mathbb{R}^n$.
       (T preserves addition)
T2. $T(aX) = a T(X)$ for all X in $\mathbb{R}^n$ and all scalars a.
       (T preserves scalar multiplication)

A linear transformation $\mathbb{R}^n \to \mathbb{R}^n$ is called a **linear operator** on $\mathbb{R}^n$.

Example 1. As we saw in Section 3.5, rotation about the origin, reflection about a line through the origin, and projection on a line through the origin are all examples of linear operators on $\mathbb{R}^2$.

Example 2. The transformation $T : \mathbb{R}^n \to \mathbb{R}^m$ given by $T(X) = 0$ for all X in $\mathbb{R}^n$ is a linear transformation because $0 + 0 = 0$. It is called the **zero transformation**, and is usually denoted $T = O$.

Example 3. The transformation $1_{\mathbb{R}^n} : \mathbb{R}^n \to \mathbb{R}^n$ defined by $1_{\mathbb{R}^n}(X) = X$ for all X in $\mathbb{R}^n$ is a linear transformation (verify), called the **identity operator** on $\mathbb{R}^n$.

Example 4. *If U is a subspace of $\mathbb{R}^n$, define the projection operator $T : \mathbb{R}^n \to \mathbb{R}^n$ by $T(X) = proj_U(X)$ for all X in $\mathbb{R}^n$. Show that T is a linear transformation.*

Solution. If $\{F_1, F_2, \cdots F_k\}$ is any orthogonal basis of U, we have

$$T(X) = \frac{X \bullet F_1}{\|F_1\|^2} F_1 + \frac{X \bullet F_2}{\|F_2\|^2} F_2 + \cdots + \frac{X \bullet F_k}{\|F_k\|^2} F_k$$

for each X in $\mathbb{R}^n$. As the reader can verify, the result follows because, for each i, we have $(X + Y) \bullet F_i = X \bullet F_i + Y \bullet F_i$ and $(aX) \bullet F_i = a(X \bullet F_i)$ for all X and Y in $\mathbb{R}^n$ and all scalars a.   □

Example 5. If Y is a fixed vector in $\mathbb{R}^n$, the transformation $T : \mathbb{R}^n \to \mathbb{R}^n$ defined by $T(X) = X + Y$ for all X in $\mathbb{R}^n$ is called Y-**translation**. Thus

4.9. LINEAR TRANSFORMATIONS

$T(0) = 0 + Y = Y$, so the map T is *not* a linear transformation if $Y \neq 0$. (If T were linear then $T(0) = 0$ by T2.)

We now come to the prototype example of a linear transformation $\mathbb{R}^n \to \mathbb{R}^m$.

Example 6. If A is any $m \times n$ matrix, define a transformation $T : \mathbb{R}^n \to \mathbb{R}^m$ by $T(X) = AX$ for all X in $\mathbb{R}^n$. This is a linear transformation for any choice of A, called the **matrix transformation** induced by A. Indeed, T1 holds because $T(X + Y) = A(X + Y) = AX + AY = T(X) + T(Y)$ for all X and Y in $\mathbb{R}^n$, and T2 can be similarly verified. □

It turns out that *every* linear transformation $\mathbb{R}^n \to \mathbb{R}^m$ is a matrix transformation. This fact depends in part on the following general properties of linear transformations: As well as preserving addition and scalar multiplication, they also preserve the zero vector, negatives, and linear combinations.

Theorem 1. *Let $T : \mathbb{R}^n \to \mathbb{R}^m$ be a linear transformation, let X and X_i denote vectors in $\mathbb{R}^n$, and let $a_1, a_2, \cdots, a_k$ denote scalars. Then:*

(1) $T(0) = 0.$

(2) $T(-X) = -T(X).$

(3) $T(a_1 X_1 + a_2 X_2 + \cdots + a_k X_k) = a_1 T(X_1) + a_2 T(X_2) + \cdots + a_k T(X_k).$

Proof. (1) and (2) follow from $T(aX) = aT(X)$ by taking $a = 0$ and $a = -1$, respectively. We prove (3) by induction on k. If $k = 1$ it is T2. If it holds for some $k \geq 1$ then

$$T(a_1 X_1 + a_2 X_2 + \cdots + a_k X_k + a_{k+1} X_{k+1})$$
$$= T(a_1 X_1 + a_2 X_2 + \cdots + a_k X_k) + T(a_{k+1} X_{k+1})$$
$$= [a_1 T(X_1) + a_2 T(X_2) + \cdots + a_k T(X_k)] + a_{k+1} T(X_{k+1})$$

where we used T1 at step 1, and we used T2 and the induction hypothesis at step 2. This completes the proof. □

Example 7. *Suppose $T : \mathbb{R}^4 \to \mathbb{R}^3$ is a linear transformation such that*

$$T\left([1\ 0\ -1\ 2]^T\right) = [2\ 1\ 0]^T \text{ and } T\left([1\ 1\ -2\ 0]^T\right) = [-1\ 2\ 4\ ]^T.$$

Compute $T\left([-2\ -5\ 7\ 6]^T\right).$

Solution. For convenience, write $X_1 = [1\ 0\ -1\ 2]^T$, $X_2 = [1\ 1\ -2\ 0]^T$ and $X = [-2\ -5\ 7\ 6]^T$. At first glance the problem seems impossible: We are given $T(X_1)$ and $T(X_2)$, but no way is apparent to find $T(X)$. However, since T is linear, part (3) of Theorem 1 shows that we can find $T(X)$ if X is *any* linear combination of X_1 and X_2. So we try to express X in the form $X = rX_1 + sX_2$.

Equating components gives two linear equation for r and s, and the solution is $X = 3X_1 - 5X_2$. Hence Theorem 1 gives

$$\begin{aligned} T(X) &= T(3X_1 - 5X_2) \\ &= 3T(X_1) - 5T(X_2) \\ &= 3[2\ 1\ 0]^T - 5[-1\ 2\ 4\ ]^T \\ &= [11\ -7\ -20]^T. \end{aligned}$$

This is what we wanted. □

Theorem 1 enables us to show that every linear transformation $\mathbb{R}^n \to \mathbb{R}^m$ is a matrix transformation. The proof requires a fact about matrix multiplication that will be used again later.

Lemma 1. *If A is an $m \times n$ matrix and $AX = 0$ for all X in $\mathbb{R}^n$, then $A = 0$.*

Proof. Write $I_n = [E_1\ E_2\ \cdots\ E_n]$ in terms of its columns. Then $AE_i = 0$ for each i by hypothesis, so

$$A = AI_n = A[E_1\ E_2\ \cdots\ E_n] = [AE_1\ AE_2\ \cdots\ AE_n] = [0\ 0\ \cdots\ 0] = 0.$$

□

Theorem 2. *Let $T : \mathbb{R}^n \to \mathbb{R}^m$ be a linear transformation. Then T is the matrix transformation induced by a uniquely determined $m \times n$ matrix A:*

$$T(X) = AX \text{ for all } X \text{ in } \mathbb{R}^n.$$

Furthermore, A is given in terms of its columns by

$$A = [T(E_1)\ T(E_2)\ \cdots\ T(E_n)]$$

where $\{E_1,\ E_2,\ \cdots\ E_n\}$ is the standard basis of $\mathbb{R}^n$. The matrix A is called the **standard matrix** *of T.*

Proof. Let $X = [x_1\ x_2\ \cdots\ x_n]^T$ in $\mathbb{R}^n$, so that $X = x_1 E_1 + x_2 E_2 + \cdots + x_n E_n$. Hence Theorem 1 gives

$$\begin{aligned} T(X) &= x_1 T(E_1) + x_2 T(E_2) + \cdots + x_n T(E_n) \\ &= [T(E_1)\ T(E_2)\ \cdots\ T(E_n)] \begin{bmatrix} x_1 \\ x_2 \\ \vdots \\ x_n \end{bmatrix} \\ &= AX \end{aligned}$$

for every X in $\mathbb{R}^n$, using block multiplication (Theorem 3 §1.4). If another $m \times n$ matrix B exists such that $T(X) = BX$ for all X in $\mathbb{R}^n$, then $AX = T(X) = BX$

4.9. LINEAR TRANSFORMATIONS

holds for all X. Thus $(A - B)X = 0$ for all columns X in $\mathbb{R}^n$, whence $A - B = 0$ by Lemma 1. □

In other words, every linear transformation $\mathbb{R}^n \to \mathbb{R}^m$ is the matrix transformation induced by its standard matrix. Because of Theorem 2, linear transformations $T : \mathbb{R}^n \to \mathbb{R}^m$ and matrix transformations $\mathbb{R}^n \to \mathbb{R}^m$ are different names for the same thing. Both interpretations have virtue depending on the situation.

Example 8. *If $U = span\{[2\ 1\ 2]^T, [1\ 2\ -2]^T\}$, find the standard matrix of the projection operator T onto U, that is $T(X) = proj_U(X)$ for all X in $\mathbb{R}^3$.*

Solution. Write $F_1 = [2\ 1\ 2]^T$, $F_2 = [1\ 2\ -2]^T$ and $X = [x_1\ x_2\ x_3]^T$. Since $\{F_1, F_2\}$ is an orthogonal basis of U, we have

$$
\begin{aligned}
proj_U(X) &= \frac{X \bullet F_1}{\|F_1\|^2} F_1 + \frac{X \bullet F_2}{\|F_2\|^2} F_2 \\
&= \frac{2x_1 + x_2 + 2x_3}{9} \begin{bmatrix} 2 \\ 1 \\ 2 \end{bmatrix} + \frac{x_1 + 2x_2 - 2x_3}{9} \begin{bmatrix} 1 \\ 2 \\ -2 \end{bmatrix} \\
&= \frac{1}{9} \begin{bmatrix} 5 & 4 & 2 \\ 4 & 5 & -2 \\ 2 & -2 & 8 \end{bmatrix} \begin{bmatrix} x_1 \\ x_2 \\ x_3 \end{bmatrix}.
\end{aligned}
$$

The standard matrix of T is now apparent. □

Let $T : \mathbb{R}^n \to \mathbb{R}^m$ and $S : \mathbb{R}^m \to \mathbb{R}^k$ be two transformations (not necessarily linear) which link together as follows: $\mathbb{R}^n \xrightarrow{T} \mathbb{R}^m \xrightarrow{S} \mathbb{R}^k$. As in Section 3.5, we define the **composite transformation** $S \circ T : \mathbb{R}^n \to \mathbb{R}^k$ by

$$(S \circ T)(X) = S[T(X)] \quad \text{for all } X \text{ in } \mathbb{R}^n.$$

Thus the action of $S \circ T$ can be described as "first T, then S". Now suppose that S and T are linear transformations with standard matrices A and B respectively. Then

$$(S \circ T)(X) = S[T(X)] = A[B(X)] = (AB)X$$

for all X in $\mathbb{R}^n$. Hence $S \circ T$ is the linear transformation with standard matrix AB. We record this for reference.

Theorem 3. *Let $\mathbb{R}^n \xrightarrow{T} \mathbb{R}^m \xrightarrow{S} \mathbb{R}^k$ be linear transformations. Then the composite $S \circ T : \mathbb{R}^n \to \mathbb{R}^k$ is also a linear transformation. Moreover, if S and T have standard matrices A and B respectively, then $S \circ T$ has standard matrix AB.*

As for functions in general, two linear transformations $S : \mathbb{R}^n \to \mathbb{R}^m$ and $T : \mathbb{R}^n \to \mathbb{R}^m$ are said to be **equal**, written $S = T$, if they have the same effect on every vector. In other words

$$S = T \quad \text{if and only if} \quad S(X) = T(X) \text{ for every vector } X \text{ in } \mathbb{R}^n.$$

We describe the requirement that $S(X) = T(X)$ for all X by saying that S and T have the same **action**. The next result asserts that a linear transformation $\mathbb{R}^n \to \mathbb{R}^m$ is determined by its effect on a spanning set of $\mathbb{R}^n$. The result is an easy consequence of Theorem 1.

Theorem 4. *Let $S : \mathbb{R}^n \to \mathbb{R}^m$ and $T : \mathbb{R}^n \to \mathbb{R}^m$ be two linear transformations, and suppose that $\mathbb{R}^n = span\{Y_1, Y_2, \cdots, Y_k\}$. If $S(Y_i) = T(Y_i)$ for each i, then $S = T$.*

Proof. Given X in $\mathbb{R}^n$, write $X = a_1 Y_1 + a_2 Y_2 + \cdots + a_k Y_k$ where each a_i is in $\mathbb{R}$. Then Theorem 1 gives

$$\begin{aligned} S(X) &= a_1 S(Y_1) + a_2 S(Y_2) + \cdots + a_k S(Y_k) \\ &= a_1 T(Y_1) + a_2 T(Y_2) + \cdots + a_k T(Y_k) \\ &= T(X). \end{aligned}$$

Since this holds for all X in $\mathbb{R}^n$, we have $S = T$. □

Theorem 4 can also be used to *construct* linear transformations. In fact, if a basis of $\mathbb{R}^n$ is given, there is a unique linear transformation $\mathbb{R}^n \to \mathbb{R}^m$ which has any desired effect on the basis vectors. More precisely, we have

Theorem 5. *Let $\mathcal{F} = \{F_1, F_2, \cdots, F_n\}$ be any basis of $\mathbb{R}^n$, and select any n vectors $\{Z_1, Z_2, \cdots, Z_n\}$ in $\mathbb{R}^m$, possibly not distinct. Then there is a unique linear transformation $T : \mathbb{R}^n \to \mathbb{R}^m$ such that*

$$T(F_i) = Z_i \quad \text{holds for each } i = 1, 2, \cdots, n.$$

In fact T can be defined as follows: If X is any vector in $\mathbb{R}^n$, express X as a linear combination of the vectors in the basis $\mathcal{F}$, say

$$X = a_1 F_1 + a_2 F_2 + \cdots + a_n F_n.$$

Then $T(X)$ is given by

$$T(X) = T(a_1 F_1 + a_2 F_2 + \cdots + a_n F_n) = a_1 Z_1 + a_2 Z_2 + \cdots + a_n Z_n. \quad (*)$$

Proof: If $T : \mathbb{R}^n \to \mathbb{R}^m$ is any linear transformation with the property that $T(F_i) = Z_i$ for each i, it is clear from Theorem 1 that (*) holds. Hence T is uniquely determined *if it exists*. The natural temptation is to define T by (*). That is, if X is any vector in $\mathbb{R}^n$, use the fact that $\mathcal{F}$ spans $\mathbb{R}^n$ to express $X = a_1 F_1 + a_2 F_2 + \cdots + a_n F_n$, and take

$$T(X) = a_1 Z_1 + a_2 Z_2 + \cdots + a_n Z_n.$$

However, there is one possible flaw in this definition: If $X = b_1 F_1 + b_2 F_2 + \cdots + b_n F_n$ is a *possibly different* linear combination for X, we would get

$$T(X) = b_1 Z_1 + b_2 Z_2 + \cdots + b_n Z_n,$$

4.9. LINEAR TRANSFORMATIONS

and this might be different from $a_1 Z_1 + a_2 Z_2 + \cdots + a_n Z_n$. However, this is where the assumption that $\mathcal{F}$ is also independent comes in: Since

$$a_1 F_1 + a_2 F_2 + \cdots + a_n F_n = b_1 F_1 + b_2 F_2 + \cdots + b_n F_n,$$

the independence of the F_i implies that $a_i = b_i$ for each i (Theorem 1 §4.2). Hence the definition (*) is unambiguous. It is now a routine matter to verify that T is linear, so the theorem is proved. □

Theorem 5 is used extensively as a way to define linear transformations by specifying their action on a basis. The definition in (*) is referred to as *extending T linearly* to $\mathbb{R}^n$.

Example 9. *Find a linear transformation $T : \mathbb{R}^3 \to \mathbb{R}^2$ such that $T\left([1\ 1\ 0]^T\right) = [3\ -2]^T$, $T\left([1\ 0\ 1]^T\right) = [2\ -1]^T$ and $T\left([0\ 1\ 1]^T\right) = [1\ 0]^T$.*

Solution. For convenience, write $F_1 = [1\ 1\ 0]^T$, $F_2 = [1\ 0\ 1]^T$ and $F_3 = [0\ 1\ 1]^T$. Then $\mathcal{F} = \{F_1, F_2, F_3\}$ is a basis of $\mathbb{R}^3$, so Theorem 5 applies. An arbitrary vector $X = [x_1\ x_2\ x_3]^T$ in $\mathbb{R}^3$ has the following $\mathcal{F}$-expansion:

$$X = \frac{x_1 + x_2 - x_3}{2} F_1 + \frac{x_1 - x_2 + x_3}{2} F_2 + \frac{-x_1 + x_2 + x_3}{2} F_3.$$

Hence

$$\begin{aligned} T(X) &= \tfrac{x_1+x_2-x_3}{2} T(F_1) + \tfrac{x_1-x_2+x_3}{2} T(F_2) + \tfrac{-x_1+x_2+x_3}{2} T(F_3) \\ &= \tfrac{x_1+x_2-x_3}{2} \begin{bmatrix} 3 \\ -2 \end{bmatrix} + \tfrac{x_1-x_2+x_3}{2} \begin{bmatrix} 2 \\ -1 \end{bmatrix} + \tfrac{-x_1+x_2+x_3}{2} \begin{bmatrix} 1 \\ 0 \end{bmatrix} \\ &= \tfrac{1}{2} \begin{bmatrix} 4x_1 + 2x_2 \\ -3x_1 - x_2 + x_3 \end{bmatrix}. \end{aligned}$$

Note that this shows that the standard matrix of T is $\tfrac{1}{2} \begin{bmatrix} 4 & 2 & 0 \\ -3 & -1 & 1 \end{bmatrix}$. □

4.9.2 Changing Coordinates

Consider the equation

$$x_1^2 + x_1 x_2 + x_2^2 = 3;$$

its graph is shown in the diagram. If the plane is rotated counterclockwise through $\frac{\pi}{4}$ radians, the variables x_1 and x_2 become

$$y_1 = \tfrac{1}{\sqrt{2}}(x_1 - x_2)$$

and

$$y_1 = \tfrac{1}{\sqrt{2}}(x_1 + x_2),$$

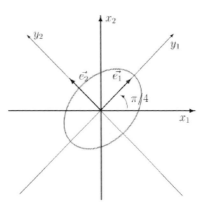

respectively. In terms of y_1 and y_2 the equation assumes a much simpler form: $y_1^2 + 3y_1^2 = 6$. The reason is that the curve is symmetric about the new y_1- and y_2-axes. Finding these axes amounts to finding the unit vectors $\vec{e}_1 = \frac{1}{\sqrt{2}}[1\ 1]^T$ and $\vec{e}_2 = \frac{1}{\sqrt{2}}[-1\ 1]^T$ along these axes,[25] that is finding a new basis $\{\vec{e}_1, \vec{e}_2\}$ of $\mathbb{R}^2$ that is compatible with the graph.

More generally, it often happens that one basis of $\mathbb{R}^n$ is more convenient than another. Changing the basis of $\mathbb{R}^n$ amounts to specifying a new "coordinate system" in $\mathbb{R}^n$, and we now investigate how the "coordinates" of a given vector change when the basis is changed. We begin by clarifying the relationship between bases and "coordinates".

Let $\mathcal{F} = \{F_1, F_2, \cdots, F_n\}$ be a basis of $\mathbb{R}^n$. Since $\mathcal{F}$ spans $\mathbb{R}^n$, every vector X in $\mathbb{R}^n$ is a linear combination $X = x_1 F_1 + x_2 F_2 + \cdots + x_n F_n$ where the x_i are real numbers. Moreover, since $\mathcal{F}$ is linearly independent, the numbers x_i are uniquely determined.[26] These numbers x_i are called the $\mathcal{F}$- **coordinates** of X, and the column

$$C_\mathcal{F}(X) = [x_1\ x_2\ \cdots\ x_n]^T$$

is called the $\mathcal{F}$-**coordinate vector** of X.

Example 10. If $\mathcal{E} = \{E_1, E_2, \cdots, E_n\}$ is the standard basis of $\mathbb{R}^n$, then $C_\mathcal{E}(X) = X$ for every X in $\mathbb{R}^n$. The reason is that if $X = [x_1\ x_2\ \cdots\ x_n]^T$ then $X = x_1 E_1 + x_2 E_2 + \cdots + x_n E_n$.

Example 11. If $\mathcal{F} = \{F_1, F_2, \cdots, F_n\}$ is an orthogonal basis of $\mathbb{R}^n$ then, for each X in $\mathbb{R}^n$,

$$C_\mathcal{F}(X) = \left[\ \frac{X \bullet F_1}{\|F_1\|^2}\ \ \frac{X \bullet F_2}{\|F_2\|^2}\ \ \cdots\ \ \frac{X \bullet F_n}{\|F_n\|^2}\ \right]^T$$

because $X = \frac{X \bullet F_1}{\|F_1\|^2} F_1 + \frac{X \bullet F_2}{\|F_2\|^2} F_2 + \cdots + \frac{X \bullet F_n}{\|F_n\|^2} F_n$ by the Expansion Theorem (Theorem 6 §4.5). This formula is particularly simple if $\mathcal{F}$ is an orthonormal basis since then all the denominators $\|F_i\|^2$ equal 1.

The following simple fact about coordinate vectors will be used several times.

Lemma 2. *If $\mathcal{F} = \{F_1, F_2, \cdots, F_n\}$ is a basis of $\mathbb{R}^n$, let $P = [F_1\ F_2\ \cdots\ F_n]$ be the matrix with the F_i as its columns. Then $C_\mathcal{F}(X) = P^{-1}X$ for all X in $\mathbb{R}^n$.*

Proof. Let $X = x_1 F_1 + x_2 F_2 + \cdots + x_n F_n$, so that $C_\mathcal{F}(X) = [x_1\ x_2\ \cdots\ x_n]^T$. Block multiplication gives:

---

[25]These vectors can be *discovered* using diagonalization (see Section 4.8): $\vec{e}_1$ and $\vec{e}_2$ are the principal axes of the quadratic form $q = x_1^2 + x_1 x_2 + x_2^2$.

[26]For the x_i to be unique, the vectors in $\mathcal{F}$ must always be listed in the same order—we then call $\mathcal{F}$ an *ordered* basis. Thus, when working with coordinates, we assume that the bases are ordered.

4.9. LINEAR TRANSFORMATIONS

$$PC_{\mathcal{F}}(X) = [F_1 \ F_2 \ \cdots \ F_n] \begin{bmatrix} x_1 \\ x_2 \\ \vdots \\ x_n \end{bmatrix} = x_1 F_1 + x_2 F_2 + \cdots + x_n F_n = X.$$

The lemma follows because P is invertible (its columns form a basis of $\mathbb{R}^n$). □

Note that Lemma 2 shows that $C_{\mathcal{F}} : \mathbb{R}^n \to \mathbb{R}^n$ is a linear transformation (with standard matrix P^{-1}) for any basis $\mathcal{F}$ of $\mathbb{R}^n$. Note further that if the basis $\mathcal{F}$ is orthonormal, then P is actually an orthogonal matrix (that is $P^{-1} = P^T$).

Now suppose that $\mathcal{F} = \{F_1, F_2, \cdots, F_n\}$ and $\mathcal{G} = \{G_1, G_2, \cdots, G_n\}$ are two bases of $\mathbb{R}^n$. We want to discover how the coordinate vectors $C_{\mathcal{F}}(X)$ and $C_{\mathcal{G}}(X)$ are related. By Lemma 2, we have

$$C_{\mathcal{F}}(X) = P^{-1}X \quad \text{and} \quad C_{\mathcal{G}}(X) = Q^{-1}X \quad \text{for all } X \text{ in } \mathbb{R}^n,$$

where $P = [F_1 \ F_2 \ \cdots \ F_n]$ and $Q = [G_1 \ G_2 \ \cdots \ G_n]$. Hence, for all X in $\mathbb{R}^n$, we have

$$\begin{aligned} C_{\mathcal{G}}(X) &= Q^{-1}X = Q^{-1}PC_{\mathcal{F}}(X) \\ &= Q^{-1}[F_1 \ F_2 \ \cdots \ F_n]C_{\mathcal{F}}(X) \\ &= [Q^{-1}F_1 \ Q^{-1}F_2 \ \cdots \ Q^{-1}F_n]C_{\mathcal{F}}(X) \\ &= [C_{\mathcal{G}}(F_1) \ C_{\mathcal{G}}(F_2) \ \cdots \ C_{\mathcal{G}}(F_n)]C_{\mathcal{F}}(X) \end{aligned}$$

With this in mind, if $\mathcal{F}$ and $\mathcal{G}$ are any two bases of $\mathbb{R}^n$ we define the matrix

$$P_{\mathcal{G} \leftarrow \mathcal{F}} = [C_{\mathcal{G}}(F_1) \ C_{\mathcal{G}}(F_2) \ \cdots \ C_{\mathcal{G}}(F_n)].$$

The matrix $P_{\mathcal{G} \leftarrow \mathcal{F}}$ is called the **change** matrix from $\mathcal{F}$ to $\mathcal{G}$. This matrix depends only on the two bases $\mathcal{F}$ and $\mathcal{G}$, and not on the vector X; on the other hand, the above calculation shows that

$$C_{\mathcal{G}}(X) = P_{\mathcal{G} \leftarrow \mathcal{F}} C_{\mathcal{F}}(X)$$

holds for *all* X in $\mathbb{R}^n$. This is the desired relationship between the $\mathcal{F}$- and $\mathcal{G}$-coordinates of X in $\mathbb{R}^n$, and we state it as the first part of the following theorem.

Theorem 6. *If $\mathcal{F} = \{F_1, F_2, \cdots, F_n\}$ and $\mathcal{G} = \{G_1, G_2, \cdots, G_n\}$ are (ordered) bases of $\mathbb{R}^n$ then*

$$C_{\mathcal{G}}(X) = P_{\mathcal{G} \leftarrow \mathcal{F}} C_{\mathcal{F}}(X) \quad \text{for all } X \text{ in } \mathbb{R}^n.$$

Moreover, if $\mathcal{G} = \{G_1, G_2, \cdots, G_n\}$ then the matrix $P_{\mathcal{G} \leftarrow \mathcal{F}}$ can be computed as follows:

$$[[G_1 \ G_2 \ \cdots \ G_n] \ [F_1 \ F_2 \ \cdots \ F_n]] \to [I_n \ P_{\mathcal{G} \leftarrow \mathcal{F}}].$$

In other words, carry the double matrix on the left to reduced row-echelon form, and $P_{\mathcal{G} \leftarrow \mathcal{F}}$ appears on the right.

Proof. It remains to verify the second part of the theorem. Write $P = [F_1 \ F_2 \ \cdots \ F_n]$ and $Q = [G_1 \ G_2 \ \cdots \ G_n]$ as before. Since P is invertible, we can carry $P \to I_n = P^{-1}P$ by row operations, so the same series of operations carries $Q \to P^{-1}Q$ by Theorem 1 §1.6. But

$$\begin{aligned} P^{-1}Q &= P^{-1}[G_1 \ G_2 \ \cdots \ G_n] = [P^{-1}G_1 \ P^{-1}G_2 \ \cdots \ P^{-1}G_n] \\ &= [C_{\mathcal{F}}(G_1) \ C_{\mathcal{F}}(G_2) \ \cdots \ C_{\mathcal{F}}(G_n)] \\ &= C_{\mathcal{G} \leftarrow \mathcal{F}} \end{aligned}$$

by Lemma 2. Hence doing the operations to the double matrix gives $[P \ Q] \to [I_n \ C_{\mathcal{G} \leftarrow \mathcal{F}}]$, as required. $\square$

These change matrices are all invertible; in fact we have the following result which explains the notation $P_{\mathcal{G} \leftarrow \mathcal{F}}$.

Corollary. Let $\mathcal{F}$, $\mathcal{G}$ and $\mathcal{H}$ be three ordered bases of $\mathbb{R}^n$. Then:

(1) $P_{\mathcal{F} \leftarrow \mathcal{F}} = I_n$.

(2) $P_{\mathcal{G} \leftarrow \mathcal{F}}$ is invertible, and $(P_{\mathcal{G} \leftarrow \mathcal{F}})^{-1} = P_{\mathcal{F} \leftarrow \mathcal{G}}$.

(3) $P_{\mathcal{H} \leftarrow \mathcal{G}} P_{\mathcal{G} \leftarrow \mathcal{F}} = P_{\mathcal{H} \leftarrow \mathcal{F}}$.

Proof. (1) is a routine computation, and (1) and (3) together imply (2) (verify). To prove (3), let X be a vector in $\mathbb{R}^n$ and apply Theorem 6 three times:

$$P_{\mathcal{H} \leftarrow \mathcal{G}} P_{\mathcal{G} \leftarrow \mathcal{F}} C_{\mathcal{F}}(X) = P_{\mathcal{H} \leftarrow \mathcal{G}} C_{\mathcal{G}}(X) = C_{\mathcal{H}}(X) = P_{\mathcal{H} \leftarrow \mathcal{F}} C_{\mathcal{F}}(X).$$

Since this holds for all X in $\mathbb{R}^n$, (3) follows from Lemmas 1 and 2. $\square$

Example 12. Find $P_{\mathcal{G} \leftarrow \mathcal{F}}$ where $\mathcal{F} = \{[2 \ -1]^T, [1 \ 5]^T\}$ and $\mathcal{G} = \{[1 \ 3]^T, [3 \ 4]^T\}$ are two bases of $\mathbb{R}^2$.

Solution. As in the last part of Theorem 6, we compute

$$\begin{bmatrix} 1 & 3 & 2 & 1 \\ 3 & 4 & -1 & 5 \end{bmatrix} \to \begin{bmatrix} 1 & 0 & -\frac{11}{5} & \frac{11}{5} \\ 0 & 1 & \frac{7}{5} & -\frac{2}{5} \end{bmatrix}, \text{ so } P_{\mathcal{G} \leftarrow \mathcal{F}} = \frac{1}{5}\begin{bmatrix} -11 & 11 \\ 7 & -2 \end{bmatrix}.$$

The reader can verify that the columns of $P_{\mathcal{G} \leftarrow \mathcal{F}}$ are indeed $C_{\mathcal{G}}([2 \ -1]^T)$ and $C_{\mathcal{G}}([1 \ 5]^T)$. $\square$

4.9.3 The $\mathcal{F}$-matrix of a Linear Operator

Suppose T is a rotation about the line L through the origin in $\mathbb{R}^3$. Finding the standard matrix of T is not convenient because the usual coordinate axes bear no relation to the line L. Everything is simpler if we use a new *intrinsic* set of axes, one pointing along the line L and the other two orthogonal to L (that

4.9. LINEAR TRANSFORMATIONS

is in the plane $L^\perp$). Choosing such a set of axes amounts to choosing a basis $\mathcal{F}$ for $\mathbb{R}^3$ with one vector in L and the other two vectors in $L^\perp$. The idea is to find an "$\mathcal{F}$-matrix" for the rotation that plays the same role relative to $\mathcal{F}$ that the standard matrix plays relative to the standard basis. It turns out that computing this $\mathcal{F}$-matrix is not too difficult because $\mathcal{F}$ is naturally aligned with the line L. Since the $\mathcal{F}$-matrix is intimately related to the basis $\mathcal{F}$, it is not surprising that the $\mathcal{F}$-coordinates $C_\mathcal{F}(X)$ will be used to describe each vector X.[27]

More precisely, if $T : \mathbb{R}^n \to \mathbb{R}^n$ is any linear operator, we look for a basis $\mathcal{F}$ of $\mathbb{R}^n$ and a matrix B such that

$$C_\mathcal{F}[T(X)] = B\, C_\mathcal{F}(X) \quad \text{for all } X \text{ in } \mathbb{R}^n. \qquad (**)$$

Such a matrix B is unique if it exists by Lemmas 1 and 2 (verify). Moreover, B describes the action of T because we can easily compute $T(X)$ from the $\mathcal{F}$-coordinates $C_\mathcal{F}[T(X)]$—see Example 14 below.

Furthermore, B always exists for any choice of basis $\mathcal{F} = \{F_1, F_2, \cdots, F_n\}$. Indeed, let A denote the standard matrix of T, and write $P = [F_1\ F_2\ \cdots\ F_n]^T$ as in Lemma 2. Then, using Lemma 2 twice, we have

$$C_\mathcal{F}(T(X)) = P^{-1}(T(X)) = P^{-1}(AX) = P^{-1}A(P\, C_\mathcal{F}(X)) = (P^{-1}AP)\, C_\mathcal{F}(X).$$

Hence $B = P^{-1}AP$ works in $(**)$. This matrix is called the $\mathcal{F}$-*matrix* of T and is denoted $M_\mathcal{F}(T)$. We have shown that $M_\mathcal{F}(T)$ is the unique matrix with the property that $C_\mathcal{F}[T(X)] = M_\mathcal{F}(T)\, C_\mathcal{F}(X)$ for all X in $\mathbb{R}^n$. This discussion proves most of the following theorem.

Theorem 7. *Let $T : \mathbb{R}^n \to \mathbb{R}^n$ be a linear operator, and let $\mathcal{F}$ be a basis of $\mathbb{R}^n$. Then a uniquely determined $n \times n$ matrix $M_\mathcal{F}(T)$ exists (called the $\mathcal{F}$-**matrix** of T) such that*

$$C_\mathcal{F}[T(X)] = M_\mathcal{F}(T)\, C_\mathcal{F}(X) \quad \text{for all } X \text{ in } \mathbb{R}^n.$$

Furthermore, if $\mathcal{F} = \{F_1, F_2, \cdots, F_n\}$ and we put $P = [F_1\ F_2\ \cdots\ F_n]$, we have:

(1) $M_\mathcal{F}(T) = P^{-1}AP$ *where A is the standard matrix of T.*

(2) $M_\mathcal{F}(T) = \begin{bmatrix} C_\mathcal{F}(T(F_1)) & C_\mathcal{F}(T(F_2)) & \cdots & C_\mathcal{F}(T(F_n)) \end{bmatrix}$.

Proof. Only (2) remains to be proved. By Lemma 2 we have $P\, C_\mathcal{F}(T(F_j)) = T(F_j) = AF_j$ for each j. Hence

$$\begin{aligned} AP &= A[F_1\ F_2\ \cdots\ F_n] \\ &= [\ AF_1\ \ AF_2\ \ \cdots\ \ AF_n\ ] \\ &= [\ P\, C_\mathcal{F}(T(F_1))\ \ P\, C_\mathcal{F}(T(F_2))\ \ \cdots\ \ P\, C_\mathcal{F}(T(F_n))\ ] \\ &= P\, [\ C_\mathcal{F}(T(F_1))\ \ C_\mathcal{F}(T(F_2))\ \ \cdots\ \ C_\mathcal{F}(T(F_n))\ ] \end{aligned}$$

---
[27] The coordinates relative to the standard basis $\mathcal{E}$ are given by $C_\mathcal{E}(X) = X$ for all X in $\mathbb{R}^n$.

Now (2) follows using (1). □

Example 13. If $\mathcal{E}$ is the standard basis of $\mathbb{R}^n$, then $M_{\mathcal{E}}(T)$ is the standard matrix of T for any linear operator T on $\mathbb{R}^n$. This follows from Example 10 and part (2) of Theorem 7.

If we know the $\mathcal{F}$-matrix of a linear operator $T : \mathbb{R}^n \to \mathbb{R}^n$ for some basis $\mathcal{F}$ of $\mathbb{R}^n$, it is a routine matter to use (the equation in) Theorem 7 to find the action of T. Here is an example.

Example 14. *Find the action of the operator* $T : \mathbb{R}^n \to \mathbb{R}^n$ *where* $M_{\mathcal{F}}(T) = \begin{bmatrix} 1 & 2 & 0 \\ 2 & -1 & 3 \\ 1 & 1 & 1 \end{bmatrix}$ *and* $\mathcal{F} = \{[1\ 0\ 0]^T, [0\ 1\ 1]^T, [0\ -1\ 1]^T\}$.

Solution. For convenience let $F_1 = [1\ 0\ 0]^T$, $F_2 = [0\ 1\ 1]^T$ and $F_3 = [0\ -1\ 1]^T$ denote the vectors in $\mathcal{F}$. By Theorem 7 we know that

$$C_{\mathcal{F}}[T(X)] = M_{\mathcal{F}}(T)\, C_{\mathcal{F}}(X)$$

for any $X = [x_1\ x_2\ x_3]^T$ in $\mathbb{R}^n$. The hard part is to compute $C_{\mathcal{F}}(X)$ since it is necessary to express the vector X as a linear combination of the vectors F_i in $\mathcal{F}$. In this case $\mathcal{F}$ is actually orthogonal, so the Expansion Theorem simplifies the work:

$$X = \frac{X \bullet F_1}{\|F_1\|^2} F_1 + \frac{X \bullet F_2}{\|F_2\|^2} F_2 + \frac{X \bullet F_3}{\|F_3\|^2} F_3$$

$$= \frac{x_1}{1} F_1 + \frac{x_2 + x_3}{2} F_2 + \frac{-x_2 + x_3}{2} F_3.$$

It follows that $C_{\mathcal{F}}(X) = \frac{1}{2}[2x_1\ x_2 + x_3\ x_3 - x_2]^T$ so

$$\begin{aligned}C_{\mathcal{F}}[T(X)] &= M_{\mathcal{F}}(T)\, C_{\mathcal{F}}(X) \\ &= \begin{bmatrix} 1 & 2 & 0 \\ 2 & -1 & 3 \\ 1 & 1 & 1 \end{bmatrix} \frac{1}{2} \begin{bmatrix} 2x_1 \\ x_2 + x_3 \\ x_3 - x_2 \end{bmatrix} \\ &= \begin{bmatrix} x_1 + x_2 + x_3 \\ 2x_1 - 2x_2 + x_3 \\ x_1 + x_3 \end{bmatrix}.\end{aligned}$$

Finally, this gives

$$\begin{aligned}T(X) &= (x_1 + x_2 + x_3)F_1 + (2x_1 - 2x_2 + x_3)F_2 + (x_1 + x_3)F_3 \\ &= \frac{1}{9}\begin{bmatrix} x_1 + x_2 + x_3 \\ x_1 - 2x_2 \\ 3x_1 - 2x_2 + 2x_3 \end{bmatrix}.\end{aligned}$$

This is what we wanted. □

Part (2) of Theorem 7 is very useful because it means that we can construct the $\mathcal{F}$-matrix of an operator T without knowing the standard matrix of T.

4.9. LINEAR TRANSFORMATIONS

More importantly, it means that we can often *choose* the basis $\mathcal{F}$ in such a way that the matrix $M_\mathcal{F}(T)$ has a relatively simple form. This is accomplished by choosing $\mathcal{F}$ so that it is naturally compatible with the action of the operator T. We describe this by saying that it is an **intrinsic** basis. This is a key technique in linear algebra; here are two examples.

Example 15. *Describe the rotations of $\mathbb{R}^3$ about a line L through the origin by finding an intrinsic basis $\mathcal{F}$ for $\mathbb{R}^3$ and computing $M_\mathcal{F}(T)$.*

Solution. Let T denote the rotation about L through the angle θ. Choose a unit vector $\vec{f}_1$ in L. Since $dim(L) = 1$, we have $dim(L^\perp) = 2$ by Theorem 3 §4.6 so let $\{\vec{f}_2, \vec{f}_3\}$ be an orthonormal basis of $L^\perp$. Then $\mathcal{F} = \{\vec{f}_1, \vec{f}_2, \vec{f}_3\}$ is an orthonormal basis of $\mathbb{R}^3$, and it remains to compute $M_\mathcal{F}(T)$. The action of T is described in the first diagram. We have

$$T(\vec{f}_1) = \vec{f}_1$$

because $\vec{f}_1$ is in L and T is rotation about L. Moreover, $T(\vec{f}_2)$ and $T(\vec{f}_3)$ lie in the plane $L^\perp$ and so are obtained by rotating $\vec{f}_1$ and $\vec{f}_2$ in $L^\perp$ through the angle θ.

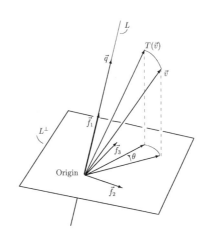

By the second diagram, the result is

$T(\vec{f}_2) = cos\theta \; \vec{f}_2 + sin\theta \; \vec{f}_3$ and
$T(\vec{f}_3) = -sin\theta \; \vec{f}_2 + cos\theta \; \vec{f}_3$.

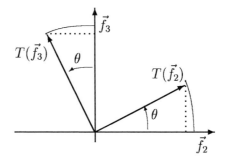

Hence Theorem 7 gives

$$C_\mathcal{F}[T] = \begin{bmatrix} C_\mathcal{F}(T(\vec{f}_1)) & C_\mathcal{F}(T(\vec{f}_2)) & C_\mathcal{F}(T(\vec{f}_3)) \end{bmatrix}$$
$$= \begin{bmatrix} 1 & 0 & 0 \\ 0 & cos\theta & -sin\theta \\ 0 & sin\theta & cos\theta \end{bmatrix}.$$

This describes the operator T. Note that $C_\mathcal{F}[T]$ is actually an orthogonal matrix here. We return to this in Theorem 12 below. □

Example 16. *Describe the reflection T of $\mathbb{R}^3$ in a plane U through the origin by finding an intrinsic basis $\mathcal{F}$ for $\mathbb{R}^3$ and computing $M_\mathcal{F}(T)$.*

Solution. We have $dim\, U = 2$ so $dim(U^\perp) = 1$ by Theorem 3 §4.6. So let $\vec{f}_1$ be a unit vector in $U^\perp$, and choose an orthonormal basis $\{\vec{f}_2, \vec{f}_3\}$ of U. Then $\mathcal{F} = \{\vec{f}_1, \vec{f}_2, \vec{f}_3\}$ is an orthonormal basis of $\mathbb{R}^3$. If T is reflection in U, it is clear that $T(\vec{f}_1) = -\vec{f}_1$ while $T(\vec{f}_2) = \vec{f}_2$ and $T(\vec{f}_3) = \vec{f}_3$. Hence part (2) of Theorem 8 gives

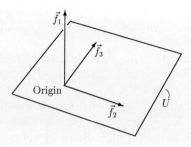

$$M_\mathcal{F}(T) = \begin{bmatrix} C_\mathcal{F}[T(\vec{f}_1)] & C_\mathcal{F}[T(\vec{f}_2)] & C_\mathcal{F}[T(\vec{f}_3)] \end{bmatrix}$$

$$= \begin{bmatrix} -1 & 0 & 0 \\ 0 & 1 & 0 \\ 0 & 0 & 1 \end{bmatrix}.$$

Thus our choice of the basis $\mathcal{F}$ has made $M_\mathcal{F}(T)$ into a diagonal matrix. □

4.9.4 Similarity

Recall that two $n \times n$ matrices A and B are said to be **similar** if $B = P^{-1}AP$ for some invertible matrix P. In this case we write $A \sim B$. The similarity relation $\sim$ enjoys the following basic properties (see Section 2.3):

1. $A \sim A$ for all square matrices A.

2. If $A \sim B$ then $B \sim A$.

3. If $A \sim B$ and $B \sim C$ then $A \sim C$.

Similar matrices enjoy many of the same features. For example, two similar matrices have the same determinant, trace, characteristic polynomial and eigenvectors, and if one of them is diagonalizable, so is the other.

The formula $M_\mathcal{F}(T) = P^{-1}AP$ in part (1) of Theorem 7 provides an important new characterization of when two matrices are similar. The proof requires one preliminary fact.

Lemma 3. *Let $\mathcal{F} = \{F_1, F_2, \cdots, F_n\}$ be any basis of $\mathbb{R}^n$, and write $P = [F_1 \ F_2 \ \cdots \ F_n]$. If A is an $n \times n$ matrix, define the operator $T_A : \mathbb{R}^n \to \mathbb{R}^n$ by $T_A(X) = AX$ for all X in $\mathbb{R}^n$. Then*

$$M_\mathcal{F}(T_A) = P^{-1}AP.$$

4.9. LINEAR TRANSFORMATIONS

Proof. The definition of the matrix $M_{\mathcal{F}}(T)$ of an operator T gives

$$\begin{aligned} M_{\mathcal{F}}(T_A) &= \begin{bmatrix} C_{\mathcal{F}}(T_A(F_1)) & C_{\mathcal{F}}(T_A(F_2)) & \cdots & C_{\mathcal{F}}(T_A(F_n)) \end{bmatrix} \\ &= \begin{bmatrix} C_{\mathcal{F}}(AF_1) & C_{\mathcal{F}}(AF_2) & \cdots & C_{\mathcal{F}}(AF_n) \end{bmatrix} \end{aligned}$$

But Lemma 2 gives $P C_{\mathcal{F}}(X) = X$ for every X in $\mathbb{R}^n$. Hence left multiplication by P gives

$$\begin{aligned} P M_{\mathcal{F}}(T_A) &= \begin{bmatrix} P C_{\mathcal{F}}(AF_1) & P C_{\mathcal{F}}(AF_2) & \cdots & P C_{\mathcal{F}}(AF_n) \end{bmatrix} \\ &= \begin{bmatrix} AF_1 & AF_2 & \cdots & AF_n \end{bmatrix} \\ &= AP. \end{aligned}$$

Now the lemma follows by left multiplication by P^{-1}. $\square$

Now we can give our characterization of similarity in terms of operators.

Theorem 8. *If A and B are $n \times n$ matrices then $A \sim B$ if and only if there exists an operator $T : \mathbb{R}^n \to \mathbb{R}^n$ and bases $\mathcal{F}$ and $\mathcal{G}$ of $\mathbb{R}^n$ such that $A = M_{\mathcal{F}}(T)$ and $B = M_{\mathcal{G}}(T)$.*

Proof. If such T, $\mathcal{F}$ and $\mathcal{G}$ exist, let A_0 denote the standard matrix of T. Then Theorem 7 shows that $M_{\mathcal{F}}(T) \sim A_0$ and $M_{\mathcal{G}}(T) \sim A_0$, so $M_{\mathcal{F}}(T) \sim M_{\mathcal{G}}(T)$ by the above properties of similarity.

Conversely, suppose $A \sim B$, say $B = P^{-1}AP$ where $P = [F_1 \; F_2 \; \cdots \; F_n]$ is invertible. If $\mathcal{F} = \{F_1, F_2, \cdots, F_n\}$ is the basis of $\mathbb{R}^n$ consisting of the columns of P, then $M_{\mathcal{F}}(T_A) = P^{-1}AP = B$ by Lemma 3. On the other hand, if $\mathcal{E}$ is the standard basis of $\mathbb{R}^n$ then $M_{\mathcal{E}}(T_A) = A$ (verify). This proves the theorem. $\square$

If A is a square matrix, define the **similarity class** $sim(A)$ of A to be the set of all matrices similar to A. Thus, for example, A is diagonalizable if and only if $sim(A)$ contains a diagonal matrix. In general, even for a non-diagonalizable matrix A, we can ask for the "nicest" matrix in $sim(A)$. Theorem 8 enters into the discussion by providing another description of $sim(A)$. It shows that, given an $n \times n$ matrix A, there exists an operator $T : \mathbb{R}^n \to \mathbb{R}^n$ such that $sim(A)$ is the set of all $\mathcal{F}$-matrices of T. More compactly,

$$sim(A) = \{M_{\mathcal{F}}(T) \mid \mathcal{F} \text{ is some basis of } \mathbb{R}^n\}.$$

Hence the task of finding the "nicest" matrix similar to A becomes: Given an operator $T : \mathbb{R}^n \to \mathbb{R}^n$, find a basis such that $M_{\mathcal{F}}(T)$ is as "nice" as possible. Because operators have a geometric interpretation, this gives new insights into similarity. We return to it in Chapter 5.

4.9.5 Isometries

To illustrate all these concepts, we conclude with a look at a very important class of operators on $\mathbb{R}^n$ called isometries that arise in geometry and in other applications. Using geometrical arguments, it was established in Section 3.5 that rotations about the origin and reflections in a line through the origin are all linear operators on $\mathbb{R}^2$. Similar geometric arguments work in $\mathbb{R}^3$, but we are going to give an algebraic proof that is valid in $\mathbb{R}^n$ for any n. The key observation is that rotations and reflections are distance preserving in the following sense. A transformation $T : \mathbb{R}^n \to \mathbb{R}^n$ (possibly not linear) is said to be **distance preserving** if

$$\|T(X) - T(Y)\| = \|X - Y\| \quad \text{for all } X \text{ and } Y \text{ in } \mathbb{R}^n, \quad (***)$$

that is the distance between $T(X)$ and $T(Y)$ is the same as the distance between X and Y for all vectors X and Y in $\mathbb{R}^n$. Distance preserving maps need not be linear: The Y-translations in Example 5 are distance preserving for every choice of Y (verify) but they are not linear if $Y \neq 0$. Surprisingly, the distance preserving transformations that fix the origin are actually linear.

Theorem 9. *Let $T : \mathbb{R}^n \to \mathbb{R}^n$ be a distance preserving transformation with the property that $T(0) = 0$. Then T is linear.*

Proof. Let X and Y denote vectors in $\mathbb{R}^n$. Observe first that the equation $\|T(X) - T(Y)\|^2 = \|X - Y\|^2$ (from $(***)$) becomes

$$\|T(X)\|^2 - 2T(X) \bullet T(Y) + \|T(Y)\|^2 = \|X\|^2 - 2X \bullet Y + \|Y\|^2.$$

But $\|T(X)\| = \|X\|$ for every X in $\mathbb{R}^n$ (take $Y = 0$ in $(***)$), so we obtain

$$T(X) \bullet T(Y) = X \bullet Y \quad \text{for all } X \text{ and } Y \text{ in } \mathbb{R}^n. \quad (****)$$

Now let $\{F_1, F_2, \cdots F_n\}$ be an orthogonal basis of $\mathbb{R}^n$. It follows from $(****)$ that $\{T(F_1), T(F_2), \cdots T(F_n)\}$ is also an orthogonal basis [$T(F_i) \neq 0$ for each i because $\|T(F_i)\| = \|F_i\| \neq 0$]. Hence, to prove that $T(aX) = aT(X)$, it suffices to show that $[T(aX) - aT(X)] \bullet T(F_i) = 0$ for each i (see Example 3 §4.5). But $(****)$ gives

$$\begin{aligned}[T(aX) - aT(X)] \bullet T(F_i) &= T(aX) \bullet T(F_i) - aT(X) \bullet T(F_i) \\ &= (aX) \bullet F_i - a(X \bullet F_i) \\ &= 0,\end{aligned}$$

as required. A similar argument shows that $T(X + Y) = T(X) + T(Y)$ for all X and Y, so T is linear. $\square$

A *linear* operator $T : \mathbb{R}^n \to \mathbb{R}^n$ which is distance preserving is called an **isometry**. This agrees with the definition in Section 3.5 for operators on $\mathbb{R}^2$. The first consequence of Theorem 9 is that, as for $\mathbb{R}^2$, rotations and reflections fixing the origin are all isometries.

Corollary 1. *In $\mathbb{R}^3$ the following are all isometries:*

4.9. LINEAR TRANSFORMATIONS

(1) *Rotations around a line through the origin.*

(2) *Reflections in a plane through the origin.*

Proof. It is clear geometrically that these transformations are all distance preserving, and they all fix the origin. Hence Theorem 9 shows that they are also linear. □

Isometries are distance preserving, and it is easy to verify that translations are also distance preserving. Moreover, the composite of two distance preserving transformations is again distance preserving (verify). The following Corollary shows that *all* distance preserving transformations arise in this way.

Corollary 2. *Every distance preserving transformation* $S : \mathbb{R}^n \to \mathbb{R}^n$ *is the composite of an isometry followed by a translation. More precisely, there exists a vector Y in $\mathbb{R}^n$ and an isometry T of $\mathbb{R}^n$ such that*

$$S(X) = T(X) + Y \text{ for every } X \text{ in } \mathbb{R}^n.$$

Proof. If $S : \mathbb{R}^n \to \mathbb{R}^n$ is distance preserving, write $S(0) = Y$ and define $T : \mathbb{R}^n \to \mathbb{R}^n$ by $T(X) = S(X) - Y$ for every X in $\mathbb{R}^n$. Then $T(0) = 0$ because $S(0) = Y$ and, for all X and X_1 in $\mathbb{R}^n$,

$$\begin{aligned} \|T(X) - T(X_1)\| &= \|(S(X) - Y) - (S(X_1) - Y)\| \\ &= \|S(X) - S(X_1)\| \\ &= \|X - X_1\| \end{aligned}$$

because S is distance preserving. Hence T is distance preserving, and so is an isometry by Theorem 9. □

We have defined isometries as the linear operators that preserve distance. The next theorem shows that they can also be characterized as the linear operators that preserve either length, or dot products, or orthonormal bases.

Theorem 10. *The following are equivalent for a linear operator* $T : \mathbb{R}^n \to \mathbb{R}^n$:

(1) T *is an isometry.*

(2) $\|T(X)\| = \|X\|$ *for all X in $\mathbb{R}^n$.*

(3) $T(X) \bullet T(Y) = X \bullet Y$ *for all X and Y in $\mathbb{R}^n$.*

(4) *If $\{F_1, F_2, \cdots, F_n\}$ is any orthonormal basis of $\mathbb{R}^n$, so also is $\{T(F_1), T(F_2), \cdots, T(F_n)\}$.*

(5) *There exists an orthonormal basis $\{F_1, F_2, \cdots, F_n\}$ of $\mathbb{R}^n$ such that $\{T(F_1), T(F_2), \cdots, T(F_n)\}$ is also an orthonormal basis.*

Proof. (1)⇒(2). If T is an isometry, then $\|T(X)\| = \|T(X) - T(0)\| = \|X - 0\| = \|X\|$ for all X.

(2)⇒(3). Since T is linear, (2) gives $\|T(X) - T(Y)\| = \|T(X - Y)\| = \|X - Y\|$ for all X and Y. This means that T preserves distance, so (3) follows from the proof of (****) in Theorem 9.

(3)⇒(4). This is clear when one remembers that $\|T(F_i)\|^2 = T(F_i) \bullet T(F_i) = F_i \bullet F_i = \|F_i\|^2 = 1$ for each i.

(4)⇒(5). This is clear since $\mathbb{R}^n$ *has* orthonormal bases.

(5)⇒(1). Given X in $\mathbb{R}^n$, write $X = x_1 F_1 + x_2 F_2 + \cdots + x_n F_n$ where the x_i are real. Then $T(X) = x_1 T(F_1) + x_2 T(F_2) + \cdots + x_n T(F_n)$, so Pythagoras' Theorem and (5) give $\|T(X)\|^2 = x_1^2 + x_2^2 + \cdots + x_n^2 = \|X\|^2$. It follows that $\|T(X)\| = \|X\|$ for each X. Since T is linear, this gives $\|T(X) - T(Y)\| = \|T(X - Y)\| = \|X - Y\|$ for all X and Y, proving (1). □

Condition (5) in Theorem 10 is a useful test that T is an isometry: Simply find an orthonormal basis $\{F_1, F_2, \cdots, F_n\}$ of $\mathbb{R}^n$ (possibly the standard basis) and check whether $\{T(F_1), T(F_2), \cdots, T(F_n)\}$ is also orthonormal.

Example 17. Let $\{F_1, F_2, F_3, F_4\}$ be an orthonormal basis of $\mathbb{R}^4$. Using Theorem 5, let $T : \mathbb{R}^4 \to \mathbb{R}^4$ be the linear operator that satisfies $T(F_1) = F_2$, $T(F_2) = -F_1$, $T(F_3) = F_4$ and $T(F_4) = -F_3$. Then this map T is an isometry by Theorem 10 because $\{T(F_1), T(F_2), T(F_3), T(F_4)\} = \{F_2, -F_1, F_4, -F_3\}$ is another orthonormal basis of $\mathbb{R}^4$. □

Isometries are closely related to orthogonal matrices as the next theorem shows.

Theorem 11. *Let $T : \mathbb{R}^n \to \mathbb{R}^n$ be a linear operator with standard matrix A. The following are equivalent*:

(1) *T is an isometry.*

(2) *A is an orthogonal matrix.*

(3) *$M_{\mathcal{F}}(T)$ is an orthogonal matrix for every orthonormal basis $\mathcal{F}$ of $\mathbb{R}^n$.*

(4) *$M_{\mathcal{F}}(T)$ is an orthogonal matrix for some orthonormal basis $\mathcal{F}$ of $\mathbb{R}^n$.*

Proof. (1)⇒(2). Write $A = [C_1 \ C_2 \ \cdots \ C_n]$ in terms of its columns. If E_i is column i of the identity matrix, then $C_i = AE_i$ for each i (because $A = AI$) and we have $C_i \bullet C_j = (AE_i) \bullet (AE_j) = T(E_i) \bullet T(E_j) = E_i \bullet E_j$ by Theorem 10. Hence $\{C_1, C_2, \cdots, C_n\}$ is an orthonormal set, and (2) follows.

(2)⇒(3). If $\mathcal{F} = \{F_1, F_2, \cdots, F_n\}$ is an orthonormal basis of $\mathbb{R}^n$, write $P = [F_1 \ F_2 \ \cdots \ F_n]$ —an orthogonal matrix. By Theorem 7, $M_{\mathcal{F}}(T) = P^{-1}AP$, and this is orthogonal being a product of orthogonal matrices by (2).

(3)⇒(4). This is clear.

4.9. LINEAR TRANSFORMATIONS

(4)$\Rightarrow$(1). If $\mathcal{F} = \{F_1, F_2, \cdots, F_n\}$ is as in (4) and $P = [F_1 \; F_2 \; \cdots \; F_n]$, then $M_\mathcal{F}(T) = P^{-1}AP$ by Theorem 7. But P is an orthogonal matrix, so $A = PM_\mathcal{F}(T)P^{-1}$ is also an orthogonal matrix. Hence, for X in $\mathbb{R}^n$, $\|T(X)\|^2 = \|AX\|^2 = (AX)^T(AX) = X^T A^T AX = X^T X = \|X\|^2$. It follows that $\|T(X)\| = \|X\|$ for all X in $\mathbb{R}^n$, so T is an isometry by Theorem 10. $\square$

It was shown in Section 3.5 that every isometry of $\mathbb{R}^2$ is either a rotation about the origin or a reflection in a line through the origin. Moreover, we have seen that rotations about a line through the origin are isometries of $\mathbb{R}^3$, as are reflections in a plane through the origin. It is a remarkable fact that every isometry of $\mathbb{R}^3$ is either a rotation, a reflection, or a composite of two of these.

Theorem 12. *If $T : \mathbb{R}^3 \to \mathbb{R}^3$ is an isometry then T is either a rotation about a line through the origin, a reflection in a plane through the origin, or else a composite of two of these where the line is in fact perpendicular to the plane.*

Proof. If A is the standard matrix of T, the characteristic polynomial of A has degree 3 (as A is 3×3) and so has a real root λ_1, an eigenvalue of A. Since A is orthogonal, $\lambda_1 = 1$ or $\lambda_1 = -1$ by Theorem 5 §4.7. Let $\vec{f_1}$ be a unit eigenvector corresponding to λ_1, and let $U = \mathbb{R}\vec{f_1}$ denote the line through the origin parallel to $\vec{f_1}$. Hence $U^\perp$ is the plane through the origin perpendicular to U.

Claim. If $\vec{v}$ is in $U^\perp$ then $T(\vec{v})$ is also in $U^\perp$.

Proof. Let $\vec{u}$ be any vector in U, so that $A\vec{u} = \lambda_1 \vec{u}$. If $\vec{v}$ is in $U^\perp$ then, since A is an orthogonal matrix by Theorem 11,

$$\lambda_1[(A\vec{v}) \bullet \vec{u}] = (A\vec{v}) \bullet (\lambda_1 \vec{u}) = (A\vec{v}) \bullet (A\vec{u}) = \vec{v}^T A^T A\vec{u} = \vec{v}^T \vec{u} = \vec{v} \bullet \vec{u} = 0$$

because $\vec{v}$ is in $U^\perp$ and $\vec{u}$ is in U. As $\lambda_1 \neq 0$, we have $(A\vec{v}) \bullet \vec{u} = \vec{0}$ for every $\vec{u}$ in U. Hence $T(\vec{v}) = A\vec{v}$ is in $U^\perp$, proving the Claim.

The Claim shows that T induces a transformation

$$T_0 : U^\perp \to U^\perp \quad \text{given by} \quad T_0(\vec{v}) = T(\vec{v})$$

for all $\vec{v}$ in $U^\perp$. This mapping T_0 is an isometry of the plane $U^\perp$ (since T is an isometry) and so is either a rotation or a reflection by Theorem 7 §3.5.[28] In each case, we can choose a convenient orthonormal basis $\{\vec{f_2}, \vec{f_3}\}$ of $U^\perp$ which will show that the theorem is true.

Case 1. T_0 is a rotation of $U^\perp$ through an angle θ. In this case choose an orthonormal basis $\{\vec{f_2}, \vec{f_3}\}$ of $U^\perp$ such that θ is counter-clockwise with respect to the axes in the $\vec{f_2}$ and $\vec{f_3}$ directions. Then, as in the solution of Example 15, we have $T(\vec{f_2}) = cos\theta \, \vec{f_2} + sin\theta \, \vec{f_3}$ and $T(\vec{f_3}) = -sin\theta \, \vec{f_2} + cos\theta \, \vec{f_3}$. Moreover

---

[28]Strictly speaking, Theorem 7 §3.5 refers to isometries of the plane $\mathbb{R}^2$. However, as we will see in Chapter 5, the important thing is that $U^\perp$ and $\mathbb{R}^2$ are both *planes*, and so have the same dimension.

$T(\vec{f_1}) = \lambda_1 \vec{f_1}$. Since $\mathcal{F} = \{\vec{f_1}, \vec{f_2}, \vec{f_3}\}$ is an orthonormal basis of $\mathbb{R}^3$, part (2) of Theorem 7 gives

$$M_{\mathcal{F}}(T) = \begin{bmatrix} \lambda_1 & 0 & 0 \\ 0 & \cos\theta & -\sin\theta \\ 0 & \sin\theta & \cos\theta \end{bmatrix}.$$

If $\lambda_1 = 1$ this is the rotation through θ around the line $\mathbb{R}\vec{f_1}$ by Example 15. If $\lambda_1 = -1$ then

$$M_{\mathcal{F}}(T) = \begin{bmatrix} -1 & 0 & 0 \\ 0 & \cos\theta & -\sin\theta \\ 0 & \sin\theta & \cos\theta \end{bmatrix} = \begin{bmatrix} 1 & 0 & 0 \\ 0 & \cos\theta & -\sin\theta \\ 0 & \sin\theta & \cos\theta \end{bmatrix} \begin{bmatrix} -1 & 0 & 0 \\ 0 & 1 & 0 \\ 0 & 0 & 1 \end{bmatrix}$$

which (using Examples 15 and 16) is reflection in the plane $U^{\perp}$, followed by rotation of the angle θ about the line U. Hence the theorem holds in this case.

Case 2. T_0 is the reflection of the plane $U^{\perp}$ in a line L_0 in $U^{\perp}$ through the origin. In this case choose unit vectors $\vec{g_2}$ and $\vec{g_3}$ in $U^{\perp}$ where $\vec{g_2}$ is parallel to L_0 and $\vec{g_3}$ is orthogonal to L_0. Hence $T(\vec{g_2}) = \vec{g_2}$ and $T(\vec{g_3}) = -\vec{g_3}$. If $\lambda_1 = 1$ we use the orthonormal basis $\mathcal{G} = \{\vec{g_3}, \vec{f_1}, \vec{g_2}\}$ of $\mathbb{R}^3$. Then Theorem 7 gives

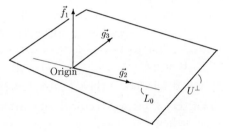

$$M_{\mathcal{G}}(T) = \begin{bmatrix} -1 & 0 & 0 \\ 0 & \lambda_1 & 0 \\ 0 & 0 & 1 \end{bmatrix} = \begin{bmatrix} -1 & 0 & 0 \\ 0 & 1 & 0 \\ 0 & 0 & 1 \end{bmatrix}$$

so T is reflection in the plane $span\{\vec{f_1}, \vec{g_2}\}$ (see Example 16). If $\lambda_1 = -1$ we use $\mathcal{G} = \{\vec{g_2}, \vec{g_3}, \vec{f_1}\}$. Hence

$$M_{\mathcal{G}}(T) = \begin{bmatrix} 1 & 0 & 0 \\ 0 & -1 & 0 \\ 0 & 0 & \lambda_1 \end{bmatrix} = \begin{bmatrix} 1 & 0 & 0 \\ 0 & -1 & 0 \\ 0 & 0 & -1 \end{bmatrix}$$

which is rotation of π around the line $\mathbb{R}\vec{g_2}$ by Example 15.

In fact we have proved that an isometry is either a rotation about a line through the origin, a reflection in a plane through the origin, or else the composite of a reflection in a plane through the origin followed by a rotation about a line through the origin perpendicular to the plane. □

4.9. LINEAR TRANSFORMATIONS

Exercises

1. In each case either show that the statement is true or give an example showing that it is false. Let $\{\vec{e}_1, \vec{e}_2\}$ denote the standard basis of $\mathbb{R}^2$.

   (a) If $T : \mathbb{R}^n \to \mathbb{R}^m$ is linear and $T(X) = 0$ for some X in $\mathbb{R}^n$, then $X = 0$.

   (b) There exists a linear transformation $T : \mathbb{R}^2 \to \mathbb{R}^3$ such that $T(\vec{e}_1) = [1 \ 1 \ 0]^T$ and $T(\vec{e}_2) = [0 \ 1 \ 1]^T$.

   (c) There exists a linear transformation $T : \mathbb{R}^2 \to \mathbb{R}^4$ such that $T(\vec{e}_1) = [1 \ -1 \ 3 \ -2]^T$ and $T(2\vec{e}_1) = [-1 \ 1 \ -3 \ 2]^T$.

   (d) If $T : \mathbb{R}^2 \to \mathbb{R}^3$ is linear and $\{X, Y\}$ is orthogonal in $\mathbb{R}^2$, then $\{T(X), T(Y)\}$ is orthogonal in $\mathbb{R}^3$.

   (e) If $T : \mathbb{R}^4 \to \mathbb{R}^2$ is defined by $T[x_1 \ x_2 \ x_3 \ x_4]^T = [x_4 \ x_3]^T$ then T is linear.

2. If Y and Z are vectors in $\mathbb{R}^n$, a vector X is said to be **between** Y and Z if $X = (1-f)Y + fZ$ for some number f such that $0 \le f \le 1$ [see Exercise 16 §3.1]. If $T : \mathbb{R}^n \to \mathbb{R}^m$ is a linear transformation and X is between Y and Z, show that $T(X)$ is between $T(Y)$ and $T(Z)$.

3. If a is a fixed real number, define $T : \mathbb{R}^n \to \mathbb{R}^n$ by $T(X) = aX$. Show that T is a linear operator (called a **scalar operator**). Find the standard matrix of T.

4. If $m < n$ define $T : \mathbb{R}^n \to \mathbb{R}^m$ by $T[x_1 \ x_2 \ \cdots \ x_n]^T = [x_1 \ x_2 \ \cdots \ x_m]^T$. Show that T is a linear transformation, and find its standard matrix.

5. If $m > n$ define $T : \mathbb{R}^n \to \mathbb{R}^m$ by
$$T[x_1 \ x_2 \ \cdots \ x_n]^T = [x_1 \ x_2 \ \cdots \ x_n \ 0 \ \cdots \ 0]^T.$$
Show that T is a linear transformation, and find its standard matrix.

6. (a) Find $T([x_1 \ x_2 \ x_3]^T)$ if $T : \mathbb{R}^3 \to \mathbb{R}^3$ is a linear transformation satisfying $T([1 \ 1 \ 0]^T) = [3 \ -1 \ 3]^T$, $T([0 \ 1 \ 1]^T) = [2 \ 1 \ 3]^T$ and $T([0 \ 0 \ 1]^T) = [-1 \ -2 \ 0]^T$.

   (b) Find $T([x_1 \ x_2 \ x_3 \ x_4]^T)$ if $T : \mathbb{R}^4 \to \mathbb{R}^2$ is a linear transformation satisfying $T([1 \ 1 \ 0 \ 0]^T) = [2 \ -1]^T$, $T([0 \ 1 \ 0 \ 0]^T) = [-2 \ 3]^T$, $T([0 \ 1 \ 0 \ 2]^T) = [1 \ 0]^T$ and $T([0 \ 0 \ 1 \ 0]^T) = [0 \ -1]^T$.

7. Let $\mathcal{F}$ denote a basis of $\mathbb{R}^3$, and let T be an operator on $\mathbb{R}^3$. In each case find $T([x_1 \ x_2 \ x_3]^T)$.

   (a) $\mathcal{F} = \left\{ \begin{bmatrix} 1 \\ 0 \\ 0 \end{bmatrix}, \begin{bmatrix} 1 \\ 1 \\ 0 \end{bmatrix}, \begin{bmatrix} 0 \\ 1 \\ 1 \end{bmatrix} \right\}$ and $M_\mathcal{F}(T) = \begin{bmatrix} 1 & 0 & -1 \\ 2 & 1 & 0 \\ -1 & 1 & 4 \end{bmatrix}$.

   (b) $\mathcal{F} = \left\{ \begin{bmatrix} 1 \\ 2 \\ 2 \end{bmatrix}, \begin{bmatrix} 2 \\ 1 \\ -2 \end{bmatrix}, \begin{bmatrix} 2 \\ -2 \\ 1 \end{bmatrix} \right\}$ and $M_\mathcal{F}(T) = \begin{bmatrix} 3 & -5 & 1 \\ 1 & 1 & 5 \\ -2 & 2 & 7 \end{bmatrix}$.

8. If $X \neq 0$ and Y are vectors in $\mathbb{R}^n$ and $\mathbb{R}^m$ respectively, show that there is a linear transformation $T : \mathbb{R}^n \to \mathbb{R}^m$ such that $T(X) = Y$. [*Hint*: Part (2) of Theorem 3 §4.3.]

9. A linear transformation $T : \mathbb{R}^n \to \mathbb{R}^1 = \mathbb{R}$ is called a **linear functional** on $\mathbb{R}^n$.

    (a) If X_0 is a fixed column in $\mathbb{R}^n$, show that $T(X) = X_0 \bullet X$ defines a linear functional T on $\mathbb{R}^n$.

    (b) Show that every linear functional on $\mathbb{R}^n$ arises as in (a) for some X_0 in $\mathbb{R}^n$.

10. (a) If X_0 is a fixed column in $\mathbb{R}^n$, show that $T(r) = rX_0$ defines a linear transformation $\mathbb{R} \to \mathbb{R}^n$.

    (b) Show that every linear transformation $\mathbb{R} \to \mathbb{R}^n$ arises as in (a) for some X_0 in $\mathbb{R}^n$.

11. Suppose that $\mathbb{R}^n = span\{Y_1, Y_2, \cdots, Y_k\}$. If $T : \mathbb{R}^n \to \mathbb{R}^m$ is a linear transformation and $T(Y_i) = 0$ for each i, show that $T = 0$ is the zero transformation.

12. If $T : \mathbb{R}^n \to \mathbb{R}^m$ is a linear transformation, define the **kernel** of T (denoted $kerT$) by $kerT = \{X \text{ in } \mathbb{R}^n \mid T(X) = 0\}$. Show that $kerT$ is a subspace of $\mathbb{R}^n$.

13. If $T : \mathbb{R}^n \to \mathbb{R}^m$ is a linear transformation, define the **image** of T (denoted imT) by $imT = \{T(X) \mid X \text{ in } \mathbb{R}^n\}$. Show that imT is a subspace of $\mathbb{R}^m$.

14. Let $T : \mathbb{R}^n \to \mathbb{R}^m$ be a linear transformation, and let $X_1, X_2, \cdots, X_k$ denote vectors in $\mathbb{R}^n$.

    (a) If $\{T(X_1), T(X_2), \cdots, T(X_k)\}$ is independent in $\mathbb{R}^m$, show that $\{X_1, X_2, \cdots, X_k\}$ is independent in $\mathbb{R}^n$.

    (b) If $\{X_1, X_2, \cdots, X_k\}$ spans $\mathbb{R}^n$, show that $\{T(X_1), T(X_2), \cdots, T(X_k)\}$ spans imT. (See the preceding exercise.)

15. Given bases $\mathcal{F} = \{[1\ 1]^T, [0\ 1]^T\}$ and $\mathcal{G} = \{[2\ 3]^T, [1\ 2]^T\}$ of $\mathbb{R}^2$, compute the change matrices $P_{\mathcal{G} \leftarrow \mathcal{F}}$ and $P_{\mathcal{F} \leftarrow \mathcal{G}}$. Verify that they are inverses of each other, and that $C_{\mathcal{G}}(X) = P_{\mathcal{G} \leftarrow \mathcal{F}} C_{\mathcal{F}}(X)$ holds for all X in $\mathbb{R}^2$.

16. In Theorem 9, prove that $T(X + Y) = T(X) + T(Y)$ for all X and Y in $\mathbb{R}^n$.

17. Show that the composite of two distance preserving isometries of $\mathbb{R}^n$ is again distance preserving.

18. If Y is a fixed vector in $\mathbb{R}^n$, let $Q_Y : \mathbb{R}^n \to \mathbb{R}^n$ denote translation by Y, that is $Q_Y(X) = X + Y$ for all X in $\mathbb{R}^n$. Show that $T \circ Q_Y = Q_{T(Y)} \circ T$ holds for every linear operator $T : \mathbb{R}^n \to \mathbb{R}^n$.

4.9. LINEAR TRANSFORMATIONS

19. Let U denote any subspace of $\mathbb{R}^n$ and write $P(X) = proj_U(X)$ for all X in $\mathbb{R}^n$. Define $T : \mathbb{R}^n \to \mathbb{R}^n$ by $T(X) = 2P(X) - X$ for all X in $\mathbb{R}^n$. We call T **reflection** in the subspace U by analogy with the case when $n = 3$ and U is a plane in $\mathbb{R}^n$ (see Theorem 1 §3.5). Show that T is an isometry of $\mathbb{R}^n$. [*Hint*: We have $T(X) = P(X) + [P(X) - X]$ and $X = P(X) + [X - P(X)]$. Use the Projection Theorem and Pythagoras' Theorem.]

20. If T is the reflection of $\mathbb{R}^3$ in a plane through the origin (see the preceding exercise), show that the standard matrix of T is diagonalizable.

21. If A is an $n \times n$ matrix, define $T_A : \mathbb{R}^n \to \mathbb{R}^n$ by $T_A(X) = AX$ for all X in $\mathbb{R}^n$.

    (a) If there exists a basis $\mathcal{F} = \{F_1, F_2, \cdots, F_n\}$ of $\mathbb{R}^n$ such that $M_\mathcal{F}(T_A)$ is diagonal, show that A is diagonalizable.

    (b) If A is diagonalizable, show that a basis $\mathcal{F} = \{F_1, F_2, \cdots, F_n\}$ of $\mathbb{R}^n$ exists such that $M_\mathcal{F}(T_A)$ is diagonal. [*Hint*: A has a basis of eigenvectors.]

22. Let $\mathcal{F}$ be a basis of $\mathbb{R}^n$. If A and B are $m \times n$ matrices and $A C_\mathcal{F}(X) = B C_\mathcal{F}(X)$ for all X in $\mathbb{R}^n$, show that $A = B$.

23. If $\mathcal{F}$ is a basis of $\mathbb{R}^n$, show that $M_\mathcal{F}(S \circ T) = M_\mathcal{F}(S) M_\mathcal{F}(T)$ for all linear operators S and T on $\mathbb{R}^n$. [*Hint*: For each X in $\mathbb{R}^n$, right multiply by $C_\mathcal{F}(X)$ and use Theorem 7 and the preceding exercise.]

24. Let T be a linear operator on $\mathbb{R}^n$ with standard matrix A. As in Section 3.5, we say that T is **invertible** if A is invertible, and define the **inverse operator** $T^{-1} : \mathbb{R}^n \to \mathbb{R}^n$ by $T^{-1}(X) = A^{-1}X$ for all X in $\mathbb{R}^n$.

    (a) If T is invertible, show that $T \circ T^{-1} = 1_{\mathbb{R}^n} = T^{-1} \circ T$.

    (b) If $\mathcal{F}$ is any basis of $\mathbb{R}^n$, show that $M_\mathcal{F}(T)$ is invertible and $M_\mathcal{F}(T^{-1}) = (M_\mathcal{F}(T))^{-1}$. [*Hint*: Preceding exercise.]

25. Show that the isometries of $\mathbb{R}^n$ enjoy the following properties:

    (1) $1_{\mathbb{R}^n}$ is an isometry.

    (2) If T is an isometry, so is T^{-1} (see the preceding exercise).

    (3) If S and T are isometries, then the composite $S \circ T$ is also an isometry.

    By virtue of these properties, the set of all isometries is called a *group* of operators.

26. (a) If T is an isometry of $\mathbb{R}^n$, show that $det(M_\mathcal{F}(T)) = \pm 1$ for any basis $\mathcal{F}$ of $\mathbb{R}^n$.

    (b) If T is an isometry of $\mathbb{R}^3$ with standard matrix A, show that: T is a rotation if and only if $det(M_\mathcal{F}(T)) = 1$ for any $\mathcal{F}$; and T is a reflection if and only if $det(M_\mathcal{F}(T)) = -1$ for any $\mathcal{F}$. [*Hint*: Theorem 12 and Exercise 23.]

27. Let $\mathcal{F} = \{F_1, \cdots, F_n\}$ and $\mathcal{G} = \{G_1, \cdots, G_n\}$ be bases of $\mathbb{R}^n$, and write $P_{\mathcal{G} \leftarrow \mathcal{F}} = P$. If $T : \mathbb{R}^n \to \mathbb{R}^n$ is a linear operator, show that $P M_{\mathcal{F}}(T) P^{-1} = M_{\mathcal{G}}(T)$. [*Hint*: Show that $P M_{\mathcal{F}}(T) = M_{\mathcal{G}}(T) P$ by right multiplying by $C_{\mathcal{F}}(X)$ for all X and applying Exercise 22.]

28. Extend Theorem 7 as follows: Let $T : \mathbb{R}^n \to \mathbb{R}^m$ be a linear transformation with standard matrix A. If $\mathcal{F} = \{F_1, \cdots, F_n\}$ and $\mathcal{G} = \{G_1, \cdots, G_m\}$ are bases of $\mathbb{R}^n$ and $\mathbb{R}^m$ respectively, write $P = [F_1 \cdots F_n]$ and $Q = [G_1 \cdots G_m]$.

    (a) Show that $C_{\mathcal{G}}(T(X)) = Q^{-1} A P C_{\mathcal{F}}(X)$ for all X in $\mathbb{R}^n$.

    (b) Show that $Q^{-1} A P = [D_{\mathcal{G}}(T(F_1)) \ D_{\mathcal{G}}(T(F_2)) \ \cdots \ D_{\mathcal{G}}(T(F_n))]$.

    [*Hint*: Proceed as in the proof of Theorem 7 and the paragraph preceding it.]

4.10 COMPLEX MATRICES[29]

Ever since we introduced eigenvalues, it has been clear that complex numbers play an important role in linear algebra. Even such a simple matrix as $A = \begin{bmatrix} 0 & -1 \\ 1 & 0 \end{bmatrix}$ has characteristic polynomial $x^2 + 1$, and so has nonreal eigenvalues i and $-i$. On the other hand, many of the applications of linear algebra involve only real numbers, so we have focused on $\mathbb{R}^n$ and on real matrices throughout this book. However, there is much to be gained by looking at $\mathbb{C}^n$ and at complex matrices, where $\mathbb{C}$ denotes the set of complex numbers. While a brief discussion of the complex numbers was given in Section 2.5, we give a more comprehensive discussion of complex matrices in this section, and extend several of our earlier results about real matrices. Only the basic properties of complex numbers will be needed—mainly complex arithmetic, conjugation and absolute value.

4.10.1 Complex Inner Products

Most of what we have done in Chapters 1 and 2 remains valid if the phrase "real number" is replaced by "complex number" wherever it occurs. The Gaussian algorithm works to solve systems of linear equations with complex coefficients, and the properties of matrix arithmetic and determinants remain the same. Furthermore, the proofs of theorems in the real case carry over unchanged to complex matrices. Consequently, we will not repeat those arguments again here.

The set of all n-tuples of complex numbers is denoted $\mathbb{C}^n$ and, as for $\mathbb{R}^n$, we denote them as columns (sometimes as rows) and refer to them as vectors. The addition and scalar multiplication of vectors in $\mathbb{C}^n$ is entirely analogous to $\mathbb{R}^n$. However, one difference arises when the length of a complex vector is considered. For example, using the formula from $\mathbb{R}^n$ on the vector $Z = [1 \ i]^T$ in $\mathbb{C}^2$, we find that Z has length $\sqrt{1^2 + i^2} = \sqrt{1 + (-1)} = 0$. Since it is important

---
[29] This section is not needed elsewhere in the book.

4.10. COMPLEX MATRICES

to retain the property that the length of a nonzero vector is positive, we see that the obvious extension of the length formula from $\mathbb{R}^n$ to $\mathbb{C}^n$ does not work. Fortunately, there is a satisfactory way to define the length of a vector in $\mathbb{C}^n$ which reduces to the usual length for vectors in $\mathbb{R}^n$. In fact, we can suitably extend the dot product.

Given $Z = [z_1 \ z_2 \ \cdots \ z_n]^T$ and $W = [w_1 \ w_2 \ \cdots \ w_n]^T$ in $\mathbb{C}^n$, define their **standard inner product** $\langle Z, W \rangle$ to be the complex number given by the formula

$$\langle Z, W \rangle = z_1 \bar{w}_1 + z_2 \bar{w}_2 + \cdots + z_n \bar{w}_n.$$

Here $\bar{w}$ denotes the conjugate of the complex number w. This reduces to the dot product on $\mathbb{R}^n$ since, if Z and W are actually in $\mathbb{R}^n$, then $\bar{w}_i = w_i$ for each i so $\langle Z, W \rangle = Z \bullet W$. Moreover, this leads to a satisfactory notion of length in $\mathbb{C}^n$. Indeed,

$$\langle Z, Z \rangle = z_1 \bar{z}_1 + z_2 \bar{z}_2 + \cdots + z_n \bar{z}_n = |z_1|^2 + |z_2|^2 + \cdots + |z_n|^2$$

where $|z|$ denotes the absolute value of the complex number z. Hence $\langle Z, Z \rangle > 0$ if $Z \neq 0$ so, as in $\mathbb{R}^n$, we define the **length** $\|Z\|$ of Z in $\mathbb{C}^n$ to be

$$\|Z\| = \sqrt{\langle Z, Z \rangle} = \sqrt{|z_1|^2 + |z_2|^2 + \cdots + |z_n|^2}.$$

Hence $\|Z\| = 0$ if and only if $Z = 0$ because $|z_1|^2 + |z_2|^2 + \cdots + |z_n|^2$ is a nonnegative real number which is zero if and only if every $z_i = 0$. This explains why the coefficients of W are conjugated in the definition of $\langle Z, W \rangle$, and it also proves (4) of the following theorem.

Theorem 1. Let Z, Z_1, W and W_1 denote vectors in $\mathbb{C}^n$, and let λ denote a complex number.

(1) $\langle Z + Z_1, W \rangle = \langle Z, W \rangle + \langle Z_1, W \rangle$ and $\langle Z, W + W_1 \rangle = \langle Z, W \rangle + \langle Z, W_1 \rangle$.

(2) $\langle \lambda Z, W \rangle = \lambda \langle Z, W \rangle$ and $\langle Z, \lambda W \rangle = \bar{\lambda} \langle Z, W \rangle$.

(3) $\langle Z, W \rangle = \overline{\langle W, Z \rangle}$.

(4) $\|Z\| \geq 0$ and $\|Z\| = 0$ if and only if $Z = 0$.

(5) $\|\lambda Z\| = |\lambda| \|Z\|$.

Proof. We verified (4) above, and we leave (1), (2) and (5) to the reader. To prove (3) let $Z = [z_1 \ z_2 \ \cdots \ z_n]^T$ and $W = [w_1 \ w_2 \ \cdots \ w_n]^T$. Since $\overline{(z+w)} = \bar{z} + \bar{w}$, $\overline{(zw)} = \bar{z}\bar{w}$ and $\overline{(\bar{z})} = z$ hold for all complex numbers z and w, we have

$$\begin{aligned}
\overline{\langle W, Z \rangle} &= \overline{w_1 \bar{z}_1 + \cdots + w_n \bar{z}_n} \\
&= \bar{w}_1 (\overline{\bar{z}_1}) + \cdots + \bar{w}_n (\overline{\bar{z}_n}) \\
&= \bar{w}_1 z_1 + \cdots + \bar{w}_n z_n \\
&= \langle Z, W \rangle.
\end{aligned}$$

This proves (3). □

The properties in Theorem 1 will be used frequently in what follows, usually without comment.

As for $\mathbb{R}^n$, a vector U in $\mathbb{C}^n$ is called a **unit vector** if $\|U\| = 1$. With this, property (5) in Theorem 1 shows that if $Z \neq 0$ in $\mathbb{C}^n$ then $\frac{1}{\|Z\|}Z$ is a unit vector.

Example 1. *If $Z = [i \ \ 1-i \ \ -3 \ \ 2+3i]^T$ and $W = [2-i \ \ 1 \ \ 1+i \ \ -i]^T$, compute: (1) $\|Z\|$; (2) A unit vector that is a positive scalar multiple of Z; (3) $\langle Z, W \rangle$.*

Solution. (1). $\|Z\|^2 = |i|^2 + |1-i|^2 + |-3|^2 + |2+3i|^2 = 1 + 2 + 9 + 13 = 25$, so $\|Z\| = 5$.

(2) Using (1), $U = \frac{1}{\|Z\|}Z = \frac{1}{5}Z$ does it.

(3). $\langle Z, W \rangle = (i)\overline{(2-i)} + (1-i)\overline{(1)} + (-3)\overline{(1+i)} + (2+3i)\overline{(-i)}$
$= i(2+i) + (1-i) - 3(1-i) + (2+3i)i$
$= -6 + 6i$. □

4.10.2 Complex Matrices

Matrix multiplication, addition, transposition and scalar multiplication (by complex scalars) is defined exactly as for real matrices, and the algebraic properties developed in Chapter 1 remain valid. However, the conjugation operation for complex numbers introduces another matrix operation. If $G = [g_{ij}]$ is a complex matrix, the **conjugate** $\bar{G}$ of G is defined to be the matrix

$$\bar{G} = [\bar{g}_{ij}]$$

obtained from G by conjugating every entry. It is a routine exercise to verify the following properties of matrix conjugation.

Theorem 2. *Let G and K denote complex matrices, and let λ be a complex number.*

(1) $\overline{G + K} = \bar{G} + \bar{K}$.

(2) $\overline{GK} = \bar{G}\bar{K}$.

(3) $\overline{(\lambda G)} = \bar{\lambda}\bar{G}$.

(4) $(\bar{G})^T = \overline{(G^T)}$.

(5) $\langle Z, W \rangle = Z^T \bar{W}$ *for all columns Z and W in $\mathbb{C}^n$.*

As for Theorem 1, these properties will be used repeatedly in what follows.

Matrix conjugation enables us to define the following notion which is essential in discussing complex analogues of the symmetric and orthogonal real

4.10. COMPLEX MATRICES

matrices. If G is a complex matrix, the **conjugate transpose** G^* of G is defined by

$$G^* = (\bar{G})^T = \overline{(G^T)}.$$

Observe that $A^* = A^T$ when A is a real matrix.

Example 2. If $G = \begin{bmatrix} -1 & 2-i & -2i \\ 3+4i & i & 7 \end{bmatrix}$ then $G^* = \begin{bmatrix} -1 & 3-4i \\ 2+i & -i \\ 2i & 7 \end{bmatrix}$. □

The following properties of the conjugate transpose follow easily from the definition of G^* and the rules governing transposition. As before these properties will be used several times below.

Theorem 3. *Let G and K denote complex matrices, and let λ be a complex number.*

(1) $(G^*)^* = G$.

(2) $(G + K)^* = G^* + K^*$.

(3) $(\lambda G)^* = \bar{\lambda} G^*$.

(4) $(GK)^* = K^* G^*$.

4.10.3 Hermitian and Unitary Matrices

If A is a real symmetric matrix, it is clear that $A^* = A$. The complex matrices that satisfy this condition turn out to be the most natural generalization of the symmetric matrices. A square complex matrix H is called **Hermitian**[30] if

$$H^* = H \text{ equivalently } H^T = \bar{H}.$$

Thus every real symmetric matrix is Hermitian. More generally, the Hermitian matrices are easy to recognize because the main diagonal entries must be real and the "reflection" of each nondiagonal entry in the main diagonal is the conjugate of that entry.

Example 3. $\begin{bmatrix} -1 & 2-i & -2i \\ 2+i & 7 & i \\ 2i & -i & 0 \end{bmatrix}$ is Hermitian, but $\begin{bmatrix} 0 & 3 & 1+i \\ 3 & 6 & 5i \\ 1-i & -5i & i \end{bmatrix}$

and $\begin{bmatrix} 2 & 1+i & 2-3i \\ 1-i & -4 & 5i \\ 3-2i & -5i & 3 \end{bmatrix}$ are not Hermitian. □

---

[30]Charles Hermite (1822-1901) was a French mathematician, who proved in 1855 that the eigenvalues of these matrices are all real (see Theorem 4). Hermite is best remembered as the first to show that the number e from calculus is not the root of any polynomial with integer coefficients.

These Hermitian matrices have an important characterization in terms of the standard inner product in $\mathbb{C}^n$.

Lemma 1. *An $n \times n$ complex matrix H is Hermitian if and only if*

$$\langle HZ, W \rangle = \langle Z, HW \rangle$$

for all columns Z and W in $\mathbb{C}^n$.

Proof. If H is Hermitian, we have $H^T = \bar{H}$. Hence, if Z and W are columns in $\mathbb{C}^n$, we have

$$\langle HZ, W \rangle = (HZ)^T \bar{W} = Z^T H^T \bar{W} = Z^T \bar{H} \bar{W} = Z^T \overline{(HW)} = \langle Z, HW \rangle.$$

Conversely, let E_j denote column j of I_n. If $H = [h_{ij}]$, some matrix arithmetic (and the hypothesis) gives

$$\begin{aligned}
\bar{h}_{ij} &= E_i^T \bar{H} E_j = E_i^T \overline{(HE_j)} = \langle E_i, HE_j \rangle \\
&= \langle HE_i, E_j \rangle = (HE_i)^T \bar{E}_j = E_i^T H^T E_j = h_{ji}
\end{aligned}$$

for all i and j. This shows that $\bar{H} = H^T$, as required. $\square$

Let G be an $n \times n$ complex matrix. As in the real case, a complex number λ is called an **eigenvalue** of G if

$$GZ = \lambda Z \text{ for some nonzero column } Z \neq 0 \text{ in } \mathbb{C}^n.$$

In this case Z is called an **eigenvector** of G corresponding to λ. The polynomial

$$c_G(x) = det(xI - G)$$

is called the **characteristic polynomial** of G. It has degree n as in the real case, but now it has complex coefficients. The proof of Theorem 2 §2.3 goes through to show that the eigenvalues of G are precisely the complex roots of $c_G(x)$. This reveals one reason why complex linear algebra is a more natural setting for the subject than its real counterpart: The so-called Fundamental Theorem of Algebra[31] asserts that $c_G(x)$ factors completely as follows:

$$c_G(x) = (x - \lambda_1)(x - \lambda_2) \cdots (x - \lambda_n)$$

where $\lambda_1, \lambda_2, \cdots \lambda_n$ are the eigenvalues of G, possibly with repetitions due to multiple roots.

Theorems 3 §2.5 and 8 §4.7 assert respectively that the eigenvalues of any symmetric matrix are real, and that eigenvectors corresponding to distinct eigenvalues are orthogonal. The next result shows that this remains true for any

---

[31] The Fundamental Theorem of Algebra actually shows that *every* polynomial $p(x)$ with complex coefficients has a complex root. This implies that $p(x)$ is a product of linear factors (see Appendix A.3). The theorem was first proved by Gauss in 1799 at the age of 22 years.

4.10. COMPLEX MATRICES

Hermitian matrix, and so confirms that the Hermitian matrices are the correct complex analogue of the symmetric real matrices. In the complex context, two columns Z and W in $\mathbb{C}^n$ are said to be **orthogonal** if $\langle Z, W \rangle = 0$.

Theorem 4. *Let H denote a Hermitian matrix.*

(1) *The eigenvalues of H are real.*

(2) *Eigenvectors of H corresponding to distinct eigenvalues are orthogonal.*

Proof. Let λ and μ be eigenvalues of H with eigenvectors Z and W respectively. Then $HZ = \lambda Z$ and $HW = \mu W$, so Lemma 1 and Theorem 1 give

$$\lambda \langle Z, W \rangle = \langle \lambda Z, W \rangle = \langle HZ, W \rangle = \langle Z, HW \rangle = \langle Z, \mu W \rangle = \bar{\mu} \langle Z, W \rangle. \qquad (*)$$

If $\lambda = \mu$ and $Z = W$, this reduces to $(\lambda - \bar{\lambda}) \|Z\|^2 = 0$, whence $\lambda = \bar{\lambda}$ and λ is real, proving (1). But then $\mu = \bar{\mu}$ too, so $(*)$ becomes $(\lambda - \mu) \langle Z, W \rangle = 0$. If $\lambda \neq \mu$, this implies $\langle Z, W \rangle = 0$, proving (2). $\square$

The main result about symmetric matrices is the Principal Axis Theorem which asserts that every symmetric real matrix A can be orthogonally diagonalized: that is there exists an orthogonal matrix P such that $P^T A P$ is diagonal. Here P is orthogonal if $P^{-1} = P^T$, and the next result identifies the complex analogues of these matrices. As in the real case, a set $\{Z_1, Z_2, \cdots, Z_n\}$ of columns in $\mathbb{C}^n$ is said to be **orthogonal** if $Z_i \neq 0$ for each i and $\langle Z_i, Z_j \rangle = 0$ whenever $i \neq j$. If in addition $\|Z_i\| = 1$ for each i, the set $\{Z_1, Z_2, \cdots, Z_n\}$ is called **orthonormal**.

Theorem 5. *The following conditions are equivalent for an $n \times n$ complex matrix U:*

(1) $U^{-1} = U^*$.

(2) *The columns of U are an orthonormal set in $\mathbb{C}^n$.*

(3) *The rows of U are an orthonormal set in $\mathbb{C}^n$.*

Proof. If C_j denotes column j of U, the (i, j)-entry of $U^T \bar{U}$ is $C_i^T \bar{C}_j = \langle C_i, C_j \rangle$. It follows that $U^T \bar{U} = I_n$ if and only if $\{C_1, \cdots, C_n\}$ is an orthonormal set. But $U^T \bar{U} = I_n$ if and only if $U^* U = I_n$, and (1)$\Leftrightarrow$(2) follows. The proof that (1)$\Leftrightarrow$(3) is similar. $\square$

A square complex matrix U is called **unitary** if it satisfies the conditions in Theorem 5. Thus a real matrix A is unitary if and only if it is orthogonal (because $A^* = A^T$). As in the real case, it is essential that the columns of U be ortho*normal* in Theorem 5.

Example 4. $\begin{bmatrix} 1+i & 1 \\ 1-i & i \end{bmatrix}$ has orthogonal columns, but the rows are not orthogonal. However, normalizing gives the unitary matrix $\begin{bmatrix} \frac{1}{2}(1+i) & \frac{1}{\sqrt{2}} \\ \frac{1}{2}(1-i) & \frac{i}{\sqrt{2}} \end{bmatrix} = \frac{1}{2}\begin{bmatrix} 1+i & \sqrt{2} \\ 1-i & \sqrt{2}i \end{bmatrix}$. □

4.10.4 Unitary Diagonalization

Given a real symmetric matrix A, there is an orthogonal matrix P such that $P^T A P$ is diagonal. This is the Principal Axis Theorem, and the diagonalization algorithm in Section 4.7 provides a systematic way to find P. In fact, the procedure works for any Hermitian matrix, except that the matrix P that we end up with is now unitary. Here is an example.

Example 5. *Given the Hermitian matrix* $H = \begin{bmatrix} 3 & 2-i \\ 2+i & 7 \end{bmatrix}$, *find a unitary matrix U such that U^*AU is diagonal.* □

Solution. The characteristic polynomial of H is $c_H(x) = det(xI - H) = (x-2)(x-8)$, so $\lambda_1 = 2$ and $\lambda_2 = 8$ are the eigenvalues of H (real as expected), with (orthogonal) eigenvectors $Z_1 = [2-i \ -1]^T$ and $Z_2 = [1 \ 2+i]^T$ respectively. Normalizing gives the unitary matrix $U = [\frac{1}{\sqrt{6}}Z_1 \ \frac{1}{\sqrt{6}}Z_2] = \frac{1}{\sqrt{6}}\begin{bmatrix} 2-i & 1 \\ -1 & 2+i \end{bmatrix}$, and the reader can verify that $U^*HU = \begin{bmatrix} \lambda_1 & 0 \\ 0 & \lambda_2 \end{bmatrix} = \begin{bmatrix} 2 & 0 \\ 0 & 8 \end{bmatrix}$ is diagonal. □

In Theorem 7 we are going to show that (as in Example 5) every Hermitian matrix H can be "unitarily diagonalized", that is U^*HU is diagonal for some unitary matrix U. However, with very little extra effort, we can obtain the following important theorem which says that every square complex matrix can be "unitarily triangularized".

Theorem 6. Schur's Theorem[32]. *If G is any $n \times n$ complex matrix, there exists a unitary matrix U such that*

$$U^*GU = T$$

where T is upper triangular. Moreover, the entries on the main diagonal of T are the eigenvalues $\lambda_1, \lambda_2, \cdots, \lambda_n$ of G, repeated according to their multiplicities.

---

[32]Issai Schur (1875-1941) was a German mathematician who did fundamental work in the theory of representations of groups as complex matrices.

4.10. COMPLEX MATRICES

Proof. We use induction on n. If $n = 1$, G is already upper triangular. If $n > 1$, assume the theorem is valid for complex matrices of size $(n-1) \times (n-1)$. Let λ_1 be an eigenvalue of G with unit eigenvector Z_1. By (the complex analogue of) Theorem 3 §4.3, Z_1 is part of a basis $\{Z_1, Z_2, \cdots, Z_n\}$ of $\mathbb{C}^n$ which, by (the analogue of) the Gram-Schmidt Algorithm, we may assume is orthonormal. Hence $U_1 = [Z_1 \ Z_2 \ \cdots \ Z_n]$ is a unitary matrix and the reader can verify that

$$U_1^* G U_1 = \begin{bmatrix} \lambda_1 & X_1 \\ 0 & G_1 \end{bmatrix}$$

in block form. By induction, there is an $(n-1) \times (n-1)$ unitary matrix V_1 such that $V_1^* G_1 V_1 = T_1$ is upper triangular. Hence $U_2 = \begin{bmatrix} 1 & 0 \\ 0 & V_1 \end{bmatrix}$ is unitary (verify), and so $U = U_1 U_2$ is also unitary (using Theorem 5). Moreover

$$\begin{aligned} U^* G U &= U_2^* (U_1^* G U_1) U_2 \\ &= \begin{bmatrix} 1 & 0 \\ 0 & V_1^* \end{bmatrix} \begin{bmatrix} \lambda_1 & X_1 \\ 0 & G_1 \end{bmatrix} \begin{bmatrix} 1 & 0 \\ 0 & V_1 \end{bmatrix} \\ &= \begin{bmatrix} \lambda_1 & X_1 V_1 \\ 0 & T_1 \end{bmatrix} \end{aligned}$$

is upper triangular, as required.

Finally, if $U^* G U = T$ as in the theorem, then T and G are similar and so have the same eigenvalues by (the analogue of) Theorem 6 §2.3. This proves the last sentence of the theorem. $\square$

Schur's theorem, together with the fact that traces and determinants of similar matrices are equal, gives

Corollary. *If $\lambda_1, \lambda_2, \cdots, \lambda_n$ are the eigenvalues (including multiplicities) of an $n \times n$ complex matrix G, then*

$$det G = \lambda_1 \lambda_2 \cdots \lambda_n \quad \text{and} \quad tr G = \lambda_1 + \lambda_2 + \cdots + \lambda_n \qquad \square$$

Note that if G is real the Corollary shows that the sum and product of the (possibly non-real) eigenvalues of G are actually real. This also follows from the fact that if λ is a complex root of any real polynomial then $\bar{\lambda}$ is also a root.

Schur's Theorem asserts that every square complex matrix can be "unitarily triangularized". This does not hold for "unitarily diagonalized". In fact, if $G = \begin{bmatrix} 1 & 1 \\ 0 & 1 \end{bmatrix}$ then $U^{-1} G U$ *cannot* be diagonal for *any* invertible complex matrix (verify). However all is well for Hermitian matrices, and Schur's Theorem gives the following important extension of the Principal Axis Theorem.

Theorem 7. Spectral Theorem. *If H is any Hermitian matrix, there exists a unitary matrix U such that U^*HU is diagonal.*

Proof. By Schur's Theorem, let $U^*HU = T$ be upper triangular, where U is unitary. Then $T^* = U^*H^*U^{**} = U^*HU = T$, so T is also lower triangular. Hence T is actually diagonal, as required. □

The Principal Axis Theorem asserts that a real, square matrix is symmetric if and only if it can be orthogonally diagonalized. The analogue for complex matrices would say that a matrix is Hermitian if and only if it can be unitarily diagonalized. Half of this is true by the Spectral Theorem, but there exist unitarily diagonalizable matrices that are not Hermitian. Here is an example.

Example 6. *Show that the non-Hermitian matrix $G = \begin{bmatrix} 0 & -1 \\ 1 & 0 \end{bmatrix}$ is unitarily diagonalizable.* □

Solution. The characteristic polynomial is $c_G(x) = x^2 + 1$, so the eigenvalues are $\lambda_1 = i$ and $\lambda_2 = -i$ with corresponding eigenvectors $Z_1 = [-1 \ \ i]^T$ and $Z_2 = [i \ \ -1]^T$. Normalizing gives the unitary matrix $U = [\frac{1}{\sqrt{2}}Z_1 \ \ \frac{1}{\sqrt{2}}Z_2]^T = \frac{1}{\sqrt{2}}\begin{bmatrix} -1 & i \\ i & -1 \end{bmatrix}$ such that $U^*GU = \begin{bmatrix} i & 0 \\ 0 & -i \end{bmatrix}$. □

We conclude with a simple characterization of which square complex matrices are unitarily diagonalizable. A complex matrix N is called **normal** if $NN^* = N^*N$. Hermitian and unitary matrices are easily seen to be normal, as is the matrix G in Example 6. In fact we have

Theorem 8. *An $n \times n$ matrix is unitarily diagonalizable if and only if it is normal.*

Proof. If N is unitarily diagonalizable, let $U^*NU = D$ be diagonal where U is unitary. Since $D^*D = DD^*$ (both D and D^* are diagonal), we obtain $U^*(N^*N)U = U^*(NN^*)U$, and $N^*N = NN^*$ follows by cancellation. Thus N is normal.

Conversely, assume that N is normal, that is $N^*N = NN^*$. By Schur's Theorem, let $U^*NU = T$ where T is upper triangular and U is unitary. We show that T is diagonal by induction on n. If $n = 1$ the matrix T is already diagonal. If $n > 1$, observe that

$$TT^* = U^*(NN^*)U = U^*(N^*N)U = T^*T.$$

If $T = [t_{ij}]$ then, equating $(1,1)$-entries in $TT^* = T^*T$ gives

$$|t_{11}|^2 + |t_{12}|^2 + \cdots + |t_{1n}|^2 = |t_{11}|^2$$

4.10. COMPLEX MATRICES

because T is upper triangular. It follows that $t_{12} = \cdots = t_{1n} = 0$, so T has the block form $T = \begin{bmatrix} t_{11} & 0 \\ 0 & T_1 \end{bmatrix}$. Then $TT^* = T^*T$ implies that $T_1 T_1^* = T_1^* T_1$, so T_1 is diagonal by induction. Thus T is diagonal, and the proof is complete. $\square$

We note in passing that the real normal 2×2 matrices are just the symmetric matrices together with the matrices of the form $\begin{bmatrix} a & -b \\ b & a \end{bmatrix}$.

Exercises

1. In each case either show that the statement is true or give an example showing that it is false.
   (a) Every invertible Hermitian matrix is unitary.
   (b) Every unitary matrix is Hermitian.
   (c) The diagonal entries of a Hermitian matrix are all real.

2. In each case compute the length of Z and W, and determine if they are orthogonal.
   (a) $Z = [-2 \ \ 1+i \ \ 4i]^T$ and $W = [1-i \ \ 2i \ \ 1]^T$.
   (b) $Z = [2+i \ \ 3i \ \ -1-i]^T$ and $W = [2 \ \ 2-i \ \ -i]^T$.
   (c) $Z = [2-i \ \ -1 \ \ 1+3i \ \ 2i]^T$ and $W = [-1 \ \ 2i \ \ 1+i \ \ i]^T$.
   (d) $Z = [1+2i \ \ -5 \ \ 2-i \ \ -2i]^T$ and $W = [2-i \ \ 1-i \ \ -3i \ \ 3+i]^T$.

3. In each case determine whether the given matrix is Hermitian, unitary or normal.
   (a) $Z = \begin{bmatrix} 2 & 3 \\ -3 & 2 \end{bmatrix}$.   (b) $Z = \begin{bmatrix} 1 & -i \\ i & 2 \end{bmatrix}$.
   (c) $Z = \begin{bmatrix} 1 & 1+i \\ 1+i & i \end{bmatrix}$.   (d) $Z = \frac{1}{\sqrt{2}} \begin{bmatrix} 1 & -1 \\ 1 & 1 \end{bmatrix}$.

4. In each case find a unitary matrix U such that U^*ZU is diagonal.
   (a) $Z = \begin{bmatrix} 1 & i \\ -i & 1 \end{bmatrix}$.   (b) $Z = \begin{bmatrix} 4 & 3-i \\ 3+i & 1 \end{bmatrix}$.
   (c) $Z = \begin{bmatrix} 1 & 0 & 1+i \\ 0 & 2 & 0 \\ 1-i & 0 & 0 \end{bmatrix}$.   (d) $Z = \begin{bmatrix} 1 & 0 & 0 \\ 0 & 1 & 1+i \\ 0 & 1-i & 2 \end{bmatrix}$.

5. If $A = \begin{bmatrix} 1 & 1 \\ 0 & 1 \end{bmatrix}$, show that $U^{-1}AU$ is not diagonal for any invertible complex matrix U.

6. If $A = \begin{bmatrix} 0 & -1 \\ 1 & 0 \end{bmatrix}$, show that $U^{-1}AU$ is not upper triangular for any invertible *real* matrix U.

7. If H and H_1 are Hermitian, show that HH_1 is Hermitian if and only if $HH_1 = H_1H$.

8. (a) Show that every unitary matrix is normal.

    (b) Show that every Hermitian matrix is normal.

9. (a) Describe the diagonal matrices that are Hermitian.

    (b) Describe the diagonal matrices that are unitary.

10. (a) Show that Z^*Z is Hermitian for any complex matrix Z (possible not square).

    (b) If $ZZ^* = I$, show that $H = Z^*Z$ satisfies $H^2 = H$.

11. If A is a real, normal 2×2 matrix, show that either A is symmetric or $A = \begin{bmatrix} a & -b \\ b & a \end{bmatrix}$ for some a and b.

12. Show that $\langle ZX, Y \rangle = \langle X, Z^*Y \rangle$ for all X and Y in $\mathbb{C}^n$, and all $n \times n$ complex matrices Z.

13. If Z is any $n \times n$ complex matrix, show that both $Z + Z^*$ and ZZ^* are Hermitian.

14. If Z is an invertible complex matrix, show that Z^* is also invertible, and $(Z^*)^{-1} = (Z^{-1})^*$.

15. Let U denote a unitary matrix.

    (a) Show that U^* is also unitary.

    (b) Show that U is invertible and that U^{-1} is also unitary.

    (c) Show that $\|UX\| = \|X\|$ for every column X in $\mathbb{C}^n$.

    (d) Show that $|\lambda| = 1$ for every eigenvalue λ of U.

16. If G is both Hermitian and unitary, show that every eigenvalue of G is 1 or -1.

17. Show that the product of two unitary matrices is again unitary.

18. If N is normal, show that zN is also normal for any complex number z.

19. If Z is any complex $n \times n$ matrix, show that U^*ZU is *lower* triangular for some unitary matrix U.

4.11. SINGULAR VALUE DECOMPOSITION

20. (a) Show that every complex matrix Z can be uniquely written as $Z = A + iB$ where A and B are real matrices.

    (b) If $Z = A + iB$ as in (a), show that Z is Hermitian if and only if A is symmetric and B is skew-symmetric ($B^T = -B$).

21. A complex matrix S is called **skew-Hermitian** if $S^* = -S$.

    (a) Show that $Z - Z^*$ is skew-Hermitian for every complex matrix Z.

    (b) If S is skew-Hermitian, show that S^2 and iS are Hermitian.

    (c) If S is skew-Hermitian, show that the eigenvalues of S are pure imaginary ($i\lambda$ for real λ).

    (d) Show that every complex matrix Z can be uniquely written as $Z = H + S$ where H is Hermitian and S is skew-Hermitian.

22. A complex matrix Z is called *nilpotent* if $Z^k = 0$ for some $k \geq 1$.

    (a) If Z is nilpotent, show that the only eigenvalue of Z is $\lambda = 0$.

    (b) If Z is $n \times n$, conclude that $c_Z(x) = x^n$.

    (c) If $Z \neq 0$, show that Z is not diagonalizable.

23. If H is Hermitian, show that all the coefficients of $c_H(x)$ are real numbers. [*Hint*: Theorem 4.]

24. (a) Show that $det\,\bar{Z} = \overline{det\,Z}$ for every $n \times n$ complex matrix Z. [*Hint*: Use induction on n and apply the Laplace expansion.]

    (b) If Z is Hermitian, show that $det(Z^*) = det\,Z$. [*Hint*: Theorem 4 and the Corollary to Schur's Theorem.]

4.11 SINGULAR VALUE DECOMPOSITION[33]

In this section we present an analogue of the Principal Axis Theorem for nonsquare matrices, a result which has been described[34] as "a highlight of linear algebra". The Principal Axis Theorem asserts that every symmetric matrix A can be written as a product $A = PDP^T$ where P is orthogonal and D is diagonal. Surprisingly, this result implies that, if A is *any* matrix (possibly not square), there exist orthogonal matrices P and Q such that $A = PDQ^T$ where D is a suitable $m \times n$ diagonal matrix (entries off the main diagonal are zero). This is called a singular value decomposition of A and is one of the most useful matrix factorizations in linear algebra. The decomposition is used throughout applied linear algebra, some examples being statistical pattern recognition, data

---

[33] This section is not needed elsewhere in this book.
[34] Gilbert Strang, Introduction to Linear Algebra, Wellesley–Cambridge Press, 1993, page 325.

analysis, artificial intelligence and data compression.[35] Moreover excellent algorithms exist for finding the decompostion, so it is important in numerical analysis.

4.11.1 The Singular Value Decomposition

Let A be an $m \times n$ matrix of rank r. A **singular value decomposition** (**SVD**) for A is a factorization
$$A = PDQ^T$$
where $P = [P_1\ P_2\ \cdots\ P_m]$ and $Q = [Q_1\ Q_2\ \cdots\ Q_n]$ are orthogonal matrices of sizes $m \times m$ and $n \times n$ respectively, and $D = diag(d_1, d_2, \cdots)$ is an $m \times n$ diagonal matrix where $d_1 \geq d_2 \geq \cdots \geq 0$. We are going to show in this section that every matrix A has an SVD, that D is uniquely determined by A, and that the columns P_i and Q_i are orthonormal eigenvectors of the symmetric matrices AA^T and A^TA, respectively. The properties of A are closely related to the properties of these symmetric matrices (for example see Theorem 5 §4.6). This is somewhat surprising because symmetric matrices are much easier to deal with than matrices in general, even square matrices. For example, the eigenvalues of any symmetric matrix are all real; in fact the eigenvalues of A^TA are real and *nonnegative* as we shall see.

If A is any real $m \times n$ matrix, let $\lambda_1, \cdots, \lambda_n$ denote the (real) eigenvalues of the $n \times n$ symmetric matrix A^TA and, for convenience, choose the notation so that
$$\lambda_1 \geq \lambda_2 \geq \lambda_3 \geq \cdots \geq \lambda_n.$$
By the Principal Axis Theorem, let $\{Q_1, \cdots, Q_n\}$ be an orthonormal basis of $\mathbb{R}^n$ where, for each i, Q_i is an eigenvector of A^TA corresponding to λ_i. Observe first that for all i and j
$$AQ_i \bullet AQ_j = (AQ_i)^T AQ_j = Q_i^T(A^TAQ_j) = Q_i^T(\lambda_j Q_j) = \lambda_j(Q_i \bullet Q_j) \quad (*)$$
Since $\|Q_i\| = 1$ for each i, taking $j = i$ in (*) shows that
$$\lambda_i = \|AQ_i\|^2 \quad \text{for each} \quad i = 1, 2, \cdots, n.$$
In particular, $\lambda_i \geq 0$ for each i. The numbers
$$\sigma_i = \sqrt{\lambda_i} \quad \text{for} \quad i = 1, 2, \cdots, n$$
are called the **singular values** of the matrix A. Thus
$$\sigma_1 \geq \sigma_2 \geq \cdots \geq \sigma_n \geq 0$$
and
$$\|AQ_i\|^2 = \sigma_i^2 \quad \text{for each } i. \qquad (**)$$

---

[35]For more on this, see "Linear Algebra with Applications" by J.T. Scheick, McGraw-Hill, 1997.

4.11. SINGULAR VALUE DECOMPOSITION

The following lemma is the first indication of the close relationship between these singular values and the structure of the matrix A. Recall that $colA = imA = \{AX \mid X \text{ in } \mathbb{R}^n\}$ by Theorem 5 §4.4.

Lemma 1. *Let A be any $m \times n$ matrix and suppose that $rankA = r$.*

(1) *The matrix A has exactly r nonzero singular values (counting multiplicities).*

(2) *$\{AQ_1, AQ_2, \cdots, AQ_r\}$ is an orthogonal basis of $colA = imA$.*

Proof. Suppose that s of the σ_i are nonzero, so $\sigma_1 \geq \sigma_2 \geq \cdots \geq \sigma_s > 0$ and $\sigma_{s+1} = \cdots = \sigma_n = 0$. Write $\mathcal{F} = \{AQ_1, AQ_2, \cdots, AQ_s\}$. Then $\mathcal{F} \subseteq imA = col(A)$, and $\mathcal{F}$ is an orthogonal set by (*) and (**). Since $dim(colA) = r$, both (1) and (2) follow if we can show that $\mathcal{F}$ spans imA. But if X is any vector in $\mathbb{R}^n$, write $X = r_1Q_1 + r_2Q_2 + \cdots + r_nQ_n$ where each r_i is in $\mathbb{R}$. Then (**) gives

$$\begin{aligned} AX &= r_1AQ_1 + r_2AQ_2 + \cdots + r_sAQ_s + \cdots + r_nAQ_n \\ &= r_1AQ_1 + r_2AQ_2 + \cdots + r_sAQ_s \in span\,\mathcal{F}. \end{aligned}$$

Since $imA = \{AX \mid X \text{ in } \mathbb{R}^n\}$, this shows that $\mathcal{F}$ spans imA, and so proves the lemma. $\square$

If $\sigma_1 \geq \cdots \geq \sigma_r$ are the nonzero singular values of the $m \times n$ matrix A, then $r = rankA$ by Lemma 1, and we define

$$\Sigma = \begin{bmatrix} diag(\sigma_1, \cdots, \sigma_r) & 0 \\ 0 & 0 \end{bmatrix}.$$

Thus Σ is the $m \times n$ block matrix with the $r \times r$ diagonal matrix $diag(\sigma_1, \cdots, \sigma_r)$ in the upper left corner and all other entries zero (some zero blocks may be absent if $r = m$ or $r = n$). In particular, Σ is an $m \times n$ diagonal matrix.

Now observe that (**) gives $\|AQ_i\| = \sigma_i \neq 0$ for each $i = 1, 2, \cdots, r$. Hence we define the unit vectors

$$P_i = \frac{1}{\sigma_i} AQ_i \quad \text{for each } i = 1, 2, \cdots, r \qquad (***)$$

in $\mathbb{R}^m$. Then Lemma 1 shows that $\{P_1, P_2, \cdots, P_r\}$ is an orthonormal set in $\mathbb{R}^m$, so extend it to an orthonormal basis $\{P_1, P_2, \cdots, P_r, P_{r+1}, \cdots, P_m\}$ of $\mathbb{R}^m$. Then

$$P = [P_1\ P_2\ \cdots\ P_m] \quad \text{and} \quad Q = [Q_1\ Q_2\ \cdots\ Q_n]$$

are orthogonal matrices and we have

Theorem 1. Singular Value Decomposition. *Let A be an $m \times n$ matrix of rank r with nonzero singular values $\sigma_1 \geq \cdots \geq \sigma_r$. If we define Σ, P and Q as above, then P and Q are orthogonal matrices, Σ is a diagonal matrix, and A factors as follows:*

$$A = P\Sigma Q^T.$$

Proof. We compute

$$P\Sigma = [P_1\ P_2\ \cdots\ P_m] \begin{bmatrix} \sigma_1 & 0 & \cdots & 0 & 0 & \cdots & 0 \\ 0 & \sigma_2 & \cdots & 0 & 0 & \cdots & 0 \\ \vdots & \vdots & \ddots & \vdots & \vdots & & \vdots \\ 0 & 0 & \cdots & \sigma_r & 0 & \cdots & 0 \\ 0 & 0 & \cdots & 0 & 0 & \cdots & 0 \\ \vdots & \vdots & & \vdots & \vdots & \ddots & \vdots \\ 0 & 0 & \cdots & 0 & 0 & \cdots & 0 \end{bmatrix}$$

$$= [\sigma_1 P_1\ \sigma_2 P_2\ \cdots\ \sigma_r P_r\ 0\ \cdots\ 0].$$

On the other hand, $Q = [Q_1\ Q_2\ \cdots\ Q_n]$ and we obtain

$$\begin{aligned} AQ &= A[Q_1\ Q_2\ \cdots\ Q_n] \\ &= [AQ_1\ AQ_2\ \cdots\ AQ_n] \\ &= [\sigma_1 P_1\ \sigma_2 P_2\ \cdots\ \sigma_r P_r\ 0\ \cdots\ 0] \end{aligned}$$

by (***) because $AQ_i = 0$ if $i > r$ by (**). It follows that $P\Sigma = AQ$, whence $A = P\Sigma Q^T$ as required. $\square$

The discussion leading up to Theorem 1 provides a routine method for finding a SVD for a matrix A which is useful when A is not too large.

Example 1. *Find a singular value decomposition of* $A = \begin{bmatrix} 1 & 2 & 0 \\ 2 & 1 & 0 \end{bmatrix}$.

Solution. $A^T A = \begin{bmatrix} 5 & 4 & 0 \\ 4 & 5 & 0 \\ 0 & 0 & 0 \end{bmatrix}$ has eigenvalues $\lambda_1 = 9$, $\lambda_2 = 1$ and $\lambda_3 = 0$ with orthonormal eigenvectors $Q_1 = \frac{1}{\sqrt{2}}[1\ 1\ 0]^T$, $Q_2 = \frac{1}{\sqrt{2}}[-1\ 1\ 0]^T$ and $Q_3 = [0\ 0\ 1]^T$ respectively. Thus the singular values are $\sigma_1 = 3$, $\sigma_2 = 1$ and $\sigma_3 = 0$. Then the formula $P_i = \frac{1}{\sigma_i} AQ_i$ gives $P_1 = \frac{1}{\sqrt{2}}[1\ 1]^T$ and $P_2 = \frac{1}{\sqrt{2}}[1\ -1]^T$. Hence we have

$$P = [P_1\ P_2] = \tfrac{1}{\sqrt{2}} \begin{bmatrix} 1 & 1 \\ 1 & -1 \end{bmatrix},\ \Sigma = \begin{bmatrix} 3 & 0 & 0 \\ 0 & 1 & 0 \end{bmatrix}$$

and

$$Q = \tfrac{1}{\sqrt{2}} \begin{bmatrix} 1 & -1 & 0 \\ 1 & 1 & 0 \\ 0 & 0 & \sqrt{2} \end{bmatrix}.$$

This gives a singular value decomposition $A = P\Sigma Q^T$, as the reader can verify. $\square$

Remark 1. For large matrices, the method in Example 1 is *not* the way the singular value decomposition is found in practice. One reason is that small singular values may be lost in the process due to roundoff error. There are efficient, iterative algorithms that accurately produce the singular values and the matrices P and Q.

4.11. SINGULAR VALUE DECOMPOSITION

Remark 2. In solving a system $AX = B$ of equations, roundoff error can be a serious problem. The singular value decomposition is useful in estimating just how sensitive the system is to error. If $\sigma_1 \geq \sigma_2 \geq \cdots \geq \sigma_r$ are the nonzero singular values, the ratio
$$c = \frac{\sigma_1}{\sigma_r}$$
is called the **condition number** of the matrix A. It turns out that larger values of c correspond to a greater danger of roundoff error.

The singular value decomposition of $A = P\Sigma Q^T$ is not unique. In the above construction the columns Q_i can be any orthonormal basis of eigenvectors of $A^T A$ and, even though $\{P_1, \cdots, P_r\}$ are given by (***), the only restriction on the columns $P_{r+1}, \cdots, P_m$ is that $\{P_1, \cdots, P_r, P_{r+1}, \cdots, P_m\}$ should be orthonormal. However the matrix Σ is uniquely determined by A. This is part of the following theorem which also gives some constraints on the matrices P and Q.

Theorem 2. Uniqueness of the Singular Value Decomposition. *Let A be an $m \times n$ matrix of rank r, with nonzero singular values $\sigma_1 \geq \sigma_2 \geq \cdots \geq \sigma_r$. Suppose that A can be factored as*
$$A = PSQ^T$$
where $P = [P_1 \ P_2 \ \cdots \ P_m]$ and $Q = [Q_1 \ Q_2 \ \cdots \ Q_n]$ are orthogonal matrices, and $S = diag(s_1, s_2, \cdots)$ is an $m \times n$ diagonal matrix with $s_1 \geq s_2 \geq \cdots \geq 0$. Then:

(1) $S = \Sigma$.

(2) $P_i = \frac{1}{\sigma_i}AQ_i$ for $1 \leq i \leq r$, $AQ_i = 0$ for $i > r$ and $(A^T A)Q_i = \sigma_i^2 Q_i$ for $1 \leq i \leq n$.

(3) $\{P_1, P_2, \cdots, P_m\}$ and $\{Q_1, Q_2, \cdots, Q_n\}$ are (orthonormal) sets of eigenvectors of AA^T and $A^T A$ respectively.

Proof. We have $A^T A = (QS^T P^T)(PSQ^T) = Q(S^T S)Q^T$. Hence the diagonal matrix $S^T S = diag(s_1^2, s_2^2, \cdots)$ is similar to $A^T A$ and so has the same eigenvalues $\sigma_1^2 \geq \sigma_2^2 \geq \cdots$. Since $s_1 \geq s_2 \geq \cdots \geq 0$, it follows that $s_i = \sigma_i$ for each i so $S = \Sigma$, proving (1). But then $P\Sigma = PS = AQ$ so (as in the proof of Theorem 1)
$$[\sigma_1 P_1 \ \sigma_2 P_2 \ \cdots \ \sigma_r P_r \ 0 \ \cdots \ 0] = [AQ_1 \ AQ_2 \ \cdots \ AQ_n]$$
Comparing columns gives the first two assertions in (2). But also $(A^T A)Q = Q(S^T S) = Q(\Sigma^T \Sigma) = Q\, diag(\sigma_1^2, \sigma_2^2, \cdots)$ and therefore it follows (by comparing columns) that $(A^T A)Q_i = \sigma_i^2 Q_i$ for each i. This completes the proof of (2) and shows that each Q_i is an eigenvector of $A^T A$. Similarly $(AA^T)P = P(\Sigma\Sigma^T)$, so the columns of P are eigenvectors of AA^T. This proves (3). □

4.11.2 The Fundamental Subspaces

A lot of important information about a matrix can be obtained from a singular value decomposition. The next result, important in its own right, will be needed.

Lemma 2. *If A is an $m \times n$ matrix then $null(A^T) = (colA)^\perp$ in $\mathbb{R}^m$.*

Proof. We use the fact (Theorem 5 §4.4) that $colA = imA = \{AX \mid X \text{ in } \mathbb{R}^n\}$. If Y is in $\mathbb{R}^m$, we have

$$\begin{aligned} A^T Y = 0 &\Leftrightarrow X \bullet (A^T Y) = 0 \quad \text{for all } X \text{ in } \mathbb{R}^n \\ &\Leftrightarrow X^T(A^T Y) = 0 \quad \text{for all } X \text{ in } \mathbb{R}^n \\ &\Leftrightarrow (AX) \bullet Y = 0 \quad \text{for all } X \text{ in } \mathbb{R}^n. \end{aligned}$$

It follows that Y is in $null(A^T)$ if and only if Y is in $(imA)^\perp$, as required. □

With this we can see that the singular value decomposition of a matrix A provides orthonormal bases of four important subspaces associated with A, sometimes called the **fundamental subspaces** of A.[36]

Theorem 3. *Let A be an $m \times n$ matrix of rank r, and let $A = P\Sigma Q^T$ be a singular value decomposition of A where $P = [P_1 \; P_2 \; \cdots \; P_m]$ and $Q = [Q_1 \; Q_2 \; \cdots \; Q_n]$ are orthogonal matrices. Then:*

(1) $\{P_1, P_2, \cdots, P_r\}$ *is an orthonormal basis of* $colA = imA$.

(2) $\{P_{r+1}, P_{r+2}, \cdots, P_m\}$ *is an orthonormal basis of* $null(A^T)$.

(3) $\{Q_1^T, \cdots, Q_r^T\}$ *is an orthonormal basis of* $rowA$.

(4) $\{Q_{r+1}, \cdots, Q_n\}$ *is an orthonormal basis of* $nullA$.

Proof. (1). By Theorem 2, $P_i = \frac{1}{\sigma_i} AQ_i = A(\frac{1}{\sigma_i} Q_i) \in imA = colA$ for each i. Hence (1) follows because $\{P_1, P_2, \cdots, P_r\}$ is orthonormal and $dim(colA) = r$.
(2). By (1) and Lemma 2, we have

$$null(A^T) = (colA)^\perp = [span\{P_1, \cdots, P_r\}]^\perp = span\{P_{r+1}, \cdots, P_m\}$$

because $\{P_1, \cdots, P_r, P_{r+1}, \cdots, P_m\}$ is an orthogonal basis of $\mathbb{R}^m$.
(3). If $1 \leq i \leq r$, then, since $A^T A Q_i = \sigma_i^2 Q_i$ by Theorem 2, we have $Q_i = A^T(\frac{1}{\sigma_i^2} AQ_i) \in im(A^T) = col(A^T)$. Now let $R_1, R_2, \cdots, R_m$ denote the rows of A. Then $R_1^T, R_2^T, \cdots, R_m^T$ are the columns of A^T, so there exist r_k in $\mathbb{R}$ such that $Q_i = \Sigma_{k=1}^m r_k R_k^T$. Thus $Q_i^T = \Sigma_{k=1}^m r_k R_k$ is in $rowA$ for $1 \leq i \leq r$. Since $\{Q_1^T, \cdots, Q_r^T\}$ is independent, (3) follows because $dim(rowA) = r$.

---

[36] A detailed discussion of these subspaces can be found in G. Strang, "Introduction to Linear Algebra" Wellesley Cambridge Press, 1993.

4.11. SINGULAR VALUE DECOMPOSITION

(4). We have $AQ_i = 0$ if $i > r$ by Theorem 2. Thus $\{Q_{r+1}, \cdots, Q_n\} \subseteq null A$, and we are done because $dim(null A) = n - r$ by Theorem 6 §4.4. □

A singular value decomposition $A = P\Sigma Q^T$ of an $m \times n$ matrix A gives a way to find a type of "inverse" for A. Let $\sigma_1 \geq \sigma_2 \geq \cdots \geq \sigma_r$ be the singular values of A, and write $D = diag(\sigma_1, \sigma_2, \cdots, \sigma_r)$. Then

$$\Sigma = \begin{bmatrix} D & 0 \\ 0 & 0 \end{bmatrix}_{m \times n} \text{ as before, and we write } \Sigma^+ = \begin{bmatrix} D^{-1} & 0 \\ 0 & 0 \end{bmatrix}_{n \times m}$$

in block form (where the subscript gives the size of the matrix). Then define the **pseudo-inverse** (or **Moore-Penrose inverse**) A^+ of A by

$$A^+ = Q\Sigma^+ P^T.$$

If A is $n \times n$ and invertible then $A^+ = A^{-1}$ is the usual inverse of A. Indeed in this case $r = n$ and Σ is invertible with $\Sigma^{-1} = \Sigma^+$, so

$$A^{-1} = (P\Sigma Q^T)^{-1} = Q\Sigma^{-1}P^T = Q\Sigma^+ P^T = A^+.$$

In fact, for an arbitrary $m \times n$ matrix A, it is readily verified by block multiplication that

$$AA^+ = P \begin{bmatrix} I_r & 0 \\ 0 & 0 \end{bmatrix}_{m \times m} P^T \quad \text{and} \quad A^+ A = Q \begin{bmatrix} I_r & 0 \\ 0 & 0 \end{bmatrix}_{n \times n} Q^T.$$

If A is invertible then $r = n$ and it follows that $AA^+ = I$ and $A^+ A = I$.

4.11.3 The Polar Decomposition

Recall that a matrix A is called positive definite if it is symmetric and $\lambda > 0$ for every eigenvalue λ of A. There is an important natural generalization of this notion. The proof of Theorem 3 §4.8 goes through to show that the following conditions are equivalent for a square matrix A:

(1) *A is symmetric and $\lambda \geq 0$ for every eigenvalue λ of A.*

(2) *A is symmetric and $q(X) \geq 0$ for every X in $\mathbb{R}^n$, where $q(X) = X^T AX$ is the quadratic form determined by A.*

A square matrix G is called **positive semidefinite** if it satisfies these conditions. The next example shows that these matrices occur in abundance.

Example 3. *If A is any $m \times n$ matrix, show that $A^T A$ is positive semidefinite.*

Solution. If λ is any eigenvalue of $A^T A$, let X be a corresponding eigenvector. Then $(A^T A)X = \lambda X$, so

$$\|AX\|^2 = (AX)^T AX = X^T(A^T AX) = X^T(\lambda X) = \lambda(X^T X) = \lambda \|X\|^2.$$

But $\|X\|^2 \neq 0$ because $X \neq 0$, so $\lambda = \frac{\|AX\|^2}{\|X\|^2} \geq 0$. □

Every positive semidefinite matrix G arises as in Example 3; in fact G can be written in the form $G = C^T C$ for some square matrix C with its nonzero rows orthogonal.

Theorem 4. *The following conditions are equivalent for an $n \times n$ matrix G:*

(1) G *is positive semidefinite.*

(2) $G = C^T C$ *where C is $n \times n$ and the nonzero rows of C are orthogonal.*

(3) $G = A^T A$ *for some $m \times n$ matrix A.*

Proof. (2)⇒(3) is clear, and (3)⇒(1) follows from Example 3.
(1)⇒(2). Given (1), let $\lambda_1, \cdots, \lambda_n$ be the (real) eigenvalues of G, so that $\lambda_i \geq 0$ for each i by (1). Since G is symmetric, the Principal Axis Theorem provides an orthogonal matrix P such that $P^T G P = D = diag(\lambda_1, \cdots, \lambda_n)$. Put $D_0 = diag(\sqrt{\lambda_1}, \cdots, \sqrt{\lambda_n})$ so that $D_0^2 = D$. If we take $C = D_0 P^T$ then row I of C is $\sqrt{\lambda_i}$ times row i of P^T for each i. Since P^T is an orthogonal matrix, this shows that the nonzero rows of C are orthogonal. Finally, a routine matrix calculation shows that $G = C^T C$, which proves (2). □

The following theorem is one reason for the importance of these positive semidefinite matrices. The proof requires the singular value decomposition.

Theorem 5. Polar Decomposition. *If A is any $n \times n$ matrix then $A = GR$ where G is positive semidefinite and R is orthogonal.*

Proof. Let $A = P \Sigma Q^T$ be a singular value decomposition of A. By inserting $I_n = P^T P$ in this product, we obtain another factorization

$$A = P \Sigma Q^T = P \Sigma (P^T P) Q^T = (P \Sigma P^T)(P Q^T).$$

Hence $A = GR$ where $G = P \Sigma P^T$ is positive definite (because it is similar to Σ and so has the same eigenvalues) and $R = PQ^T$ is orthogonal (because it is a product of orthogonal matrices). This is what we wanted. □

The proof of Theorem 5 gives formulas for the matrices G and R in the polar decomposition of A in terms of any singular value decomposition of A.

Example 4. *Find the polar decomposition of the matrix $A = \begin{bmatrix} 1 & 1 \\ 2 & 2 \end{bmatrix}$.*

Solution. We begin by finding the singular value decomposition $A = P \Sigma Q^T$. The eigenvalues of $A^T A$ are $\lambda_1 = 10$ and $\lambda_2 = 0$, with orthonormal eigenvectors $Q_1 = \frac{1}{\sqrt{2}} \begin{bmatrix} 1 \\ 1 \end{bmatrix}$ and $Q_2 = \frac{1}{\sqrt{2}} \begin{bmatrix} 1 \\ -1 \end{bmatrix}$. Thus $Q = \frac{1}{\sqrt{2}} \begin{bmatrix} 1 & 1 \\ 1 & -1 \end{bmatrix}$. The singular

4.11. SINGULAR VALUE DECOMPOSITION

values of A are $\sigma_1 = \sqrt{10}$ and $\sigma_2 = 0$, so we have $P_1 = \frac{1}{\sigma_1} AQ_1 = \frac{1}{\sqrt{5}} \begin{bmatrix} 1 \\ 2 \end{bmatrix}$. Since $\sigma_2 = 0$ we can choose any P_2 such that $\{P_1, P_2\}$ is an orthonormal basis of $\mathbb{R}^2$, say $P_2 = \frac{1}{\sqrt{5}} \begin{bmatrix} 2 \\ -1 \end{bmatrix}$. Hence $P = \frac{1}{\sqrt{5}} \begin{bmatrix} 1 & 2 \\ 2 & -1 \end{bmatrix}$. Finally $\Sigma = \begin{bmatrix} \sigma_1 & 0 \\ 0 & \sigma_2 \end{bmatrix} = \begin{bmatrix} \sqrt{10} & 0 \\ 0 & 0 \end{bmatrix}$, and the resulting singular value decomposition of A is $A = P\Sigma Q^T$ as is easily verified.

Finally, as in the proof of Theorem 5, we have

$$G = P\Sigma P^T = \frac{\sqrt{10}}{5} \begin{bmatrix} 1 & 2 \\ 2 & 4 \end{bmatrix} \quad \text{and} \quad R = PQ^T = \frac{1}{\sqrt{10}} \begin{bmatrix} 3 & -1 \\ 1 & 3 \end{bmatrix}.$$

Then G is positive semidefinite and R is orthogonal, and the reader can verify that $A = GR$. □

The name "polar decomposition" comes from the polar form for complex numbers. Recall (Section 2.5) that it is customary to identify the plane $\mathbb{R}^2$ with the set $\mathbb{C}$ of all complex numbers. This means that the complex number $z = a + ib$ is identified with the vector $\vec{z} = \begin{bmatrix} a \\ b \end{bmatrix}$ in $\mathbb{R}^2$. This being done, multiplication by z is a linear transformation $T : \mathbb{R}^2 \to \mathbb{R}^2$ given by $T(\vec{v}) = \vec{z}\vec{v}$ for all $\vec{v}$ in $\mathbb{R}^2$. Here the product $\vec{z}\vec{v}$ comes from complex multiplication. More precisely, if we write $\vec{v} = \begin{bmatrix} x \\ y \end{bmatrix} = x + iy$ then

$$\begin{aligned} \vec{z}\vec{v} &= (a+ib)(x+iy) = (ax - by) + i(bx + ay) \\ &= \begin{bmatrix} ax - by \\ bx + ay \end{bmatrix} = \begin{bmatrix} a & -b \\ b & a \end{bmatrix} \begin{bmatrix} x \\ y \end{bmatrix} \\ &= A\vec{v} \end{aligned}$$

where $A = \begin{bmatrix} a & -b \\ b & a \end{bmatrix}$. Hence multiplication by $z = a + ib$ is matrix multiplication by A.

Now recall that every complex number $z = a + ib$ has the *polar form* $z = re^{i\theta}$ where $r = \sqrt{a^2 + b^2} \geq 0$ and $e^{i\theta} = cos\theta + i\,sin\theta$ where θ is the angle given by $cos\theta = \frac{a}{r}$ and $sin\theta = \frac{b}{r}$. Thus multiplication by $z = re^{i\theta}$ is first multiplication by $e^{i\theta}$ followed by multiplication by r, that is multiplication by the orthogonal matrix $R = \begin{bmatrix} cos\theta & -sin\theta \\ sin\theta & cos\theta \end{bmatrix}$ followed by multiplication by the positive semidefinite matrix $G = \begin{bmatrix} r & 0 \\ 0 & r \end{bmatrix}$. Thus the polar form of z becomes

$$A = GR,$$

which explains the name *polar decomposition* of A. Note that G is actually positive definite in this case if $z \neq 0$.

Exercises

1. Suppose that A has singular values $\sigma_1 \geq \sigma_2 \geq \cdots \geq \sigma_r$. Find the singular values of kA where k is a real number.

2. Show that A and A^T have the same singular values. [*Hint* : Theorem 2.]

3. If $A = P\Sigma Q^T$ is a singular value decomposition of an invertible matrix A, find a singular value decomposition of A^{-1}.

4. If A is a square matrix, show that $|det A|$ is the product of the singular values of A.

5. Let A be an $m \times n$ matrix and let P and Q denote orthogonal matrices of sizes $m \times m$ and $n \times n$ respectively. Show that PA and AQ each have the same singular values as A.

6. If A denotes an $m \times n$ matrix, show that $rank A = n$ if and only if all the singular values of A are nonzero.

7. Let $A = P\Sigma Q^T$ be a singular value decomposition of an $m \times n$ matrix A. If $P_1, P_2, \cdots, P_m$ are the columns of P and $Q_1, Q_2, \cdots, Q_n$ are the columns of Q, show that $A = \sigma_1 P_1 Q_1^T + \sigma_2 P_2 Q_2^T + \cdots + \sigma_r P_r Q_r^T$ where the σ_i are the singular values of A.

8. Find a singular value decomposition of an $n \times n$ invertible matrix A as follows: Apply the Principal Axis Theorem to the symmetric matrix $A^T A$ to find an orthogonal matrix Q such that $Q^T(A^T A)Q = diag(\lambda_1, \cdots, \lambda_n)$ where $\lambda_1 \geq \cdots \geq \lambda_n$ are the eigenvalues of $A^T A$. If $D = diag(\sqrt{\lambda_1}, \cdots, \sqrt{\lambda_n})$ and $P = AQD^{-1}$, show that $A = PDQ^T$ and that this is a singular value decomposition of A.

9. If A is any $m \times n$ matrix show that:

   (a) AA^+ and A^+A are both symmetric matrices.

   (b) $AA^+A = A$ and $A^+AA^+ = A^+$.

10. If A is a positive definite matrix, show that the eigenvalues of A are the singular values of A.

11. Show that the product of two positive semidefinite matrices need not be positive semidefinite.

12. If G is positive semidefinite and orthogonal, show that $G = I$. [*Hint*: Theorem 5 §4.7.]

13. If G is an invertible positive semidefinite matrix, show that G^{-1} is also positive semidefinite.

14. If G is positive semidefinite, show that kG is also positive semidefinite for all real numbers $k > 0$.

4.11. SINGULAR VALUE DECOMPOSITION

15. If G is positive semidefinite, show that $U^T G U$ is also positive semidefinite for all invertible matrices U.

16. Show that polar forms are not unique by computing GP and GQ where

$$G = \begin{bmatrix} 1 & 1 \\ 1 & 1 \end{bmatrix}, \quad P = \frac{1}{\sqrt{2}} \begin{bmatrix} -1 & 1 \\ 1 & 1 \end{bmatrix} \quad \text{and} \quad Q = \frac{1}{\sqrt{2}} \begin{bmatrix} 1 & 1 \\ -1 & 1 \end{bmatrix}.$$

17. If A is any square matrix, show that it has a polar factorization of the form $A = RG$ where R is orthogonal and G is positive semidefinite.

18. Let B be any symmetric $n \times n$ matrix with (real) eigenvalues $\lambda_1 \geq \cdots \geq \lambda_n$. If X is any nonzero column in $\mathbb{R}^n$, define the **Rayleigh quotient** $R(X) = \frac{X^T B X}{\|X\|^2}$.

    (a) If X is an eigenvector of B corresponding to λ_i, show that $R(X) = \lambda_i$.

    (b) Show that $\lambda_1 \geq R(X) \geq \lambda_n$ for all $X \neq 0$ in $\mathbb{R}^n$.

    (c) Conclude that $\lambda_1 = max\{R(X) \mid X \neq 0 \text{ in } \mathbb{R}^n\}$ and $\lambda_n = min\{R(X) \mid X \neq 0 \text{ in } \mathbb{R}^n\}$.

    [*Hint*: For (b), since B is symmetric let $\{Q_1, \cdots, Q_n\}$ be an orthonormal basis of $\mathbb{R}^n$ where Q_i is an eigenvector corresponding to λ_i for each i. Given X in $\mathbb{R}^n$, write $X = r_1 Q_1 + \cdots + r_n Q_n$, so that $BX = r_1 \lambda_1 Q_1 + \cdots + r_n \lambda_n Q_n$. Use the fact that the Q_i are orthonormal to compute $\|X\|^2 = X \bullet X$ and $X^T B X = X \bullet (BX)$, and exploit the assumption that $\lambda_1 \geq \lambda_i \geq \lambda_n$ for all i.]

19. Show that the largest eigenvalue λ_1 of $A^T A$ is given by

$$\lambda_1 = max \left\{ \frac{\|AX\|^2}{\|X\|^2} \;\middle|\; X \neq 0 \text{ in } \mathbb{R}^n \right\}.$$

    [*Hint*: Apply the preceding exercise with $B = A^T A$.]

Chapter 5

VECTOR SPACES

5.1 EXAMPLES AND BASIC PROPERTIES

The study of $\mathbb{R}^n$ in Chapter 4 has revealed how useful concepts like subspace, basis and dimension can be. Surprisingly these concepts depend on ten simple properties of $\mathbb{R}^n$ (called axioms). In other words, these ten axioms are all that is needed to define these concepts and prove their essential properties. The reason that this is important is that there are examples other than $\mathbb{R}^n$ in which which these axioms are valid, and so notions like dimension can be developed in these new contexts. An abstract vector space is a "super" example in which the *only* thing we know is that the axioms hold. Surprisingly, even in this full generality, we can still define concepts (like subspace and dimension), and prove theorems about them (such as the fundamental theorem). Hence the concepts make sense, and theorems are valid, in *all* the examples because each example is a specific realization of the general space. This leads to many more applications. Furthermore, clarifying the notion of an abstract vector space gives us a useful new perspective on the specific examples, and points to unforeseen notions such as vector spaces of infinite dimension. This chapter is an introduction to these abstract vector spaces.

5.1.1 Vector Spaces

A **vector space** consists of a nonempty set V of elements (called **vectors**) that can be added and multiplied by a number[1] (called a **scalar**), and for which certain properties hold (called **axioms**). We assume that any two vectors $\mathbf{v}$ and $\mathbf{w}$ in V can be added and that the sum, denoted $\mathbf{v} + \mathbf{w}$, is again a vector in V. Similarly, we assume that any vector $\mathbf{v}$ in V can be multiplied by any scalar a, and that the product, denoted $a\mathbf{v}$, is again in V. These operations are called

---

[1] For us the scalars will be real numbers. However, they could be complex numbers and, more generally, the scalars could be drawn from an algebraic system called a **field**. Another example is the set $\mathbb{Q}$ of all rational numbers.

vector addition and **scalar multiplication** respectively, and the following axioms are assumed to hold:

Axioms for vector addition:

A1 *If $\mathbf{v}$ and $\mathbf{w}$ are in V, then $\mathbf{v} + \mathbf{w}$ is in V.*

A2 $\mathbf{v} + \mathbf{w} = \mathbf{w} + \mathbf{v}$ *for all $\mathbf{v}$ and $\mathbf{w}$ in V.*

A3 $\mathbf{u} + (\mathbf{v} + \mathbf{w}) = (\mathbf{u} + \mathbf{v}) + \mathbf{w}$ *for all $\mathbf{u}$, $\mathbf{v}$ and $\mathbf{w}$ in V.*

A4 *An element $\mathbf{0}$ in V exists such that $\mathbf{v} + \mathbf{0} = \mathbf{v}$ for all $\mathbf{v}$ in V.*

A5 *For each $\mathbf{v}$ in V an element $-\mathbf{v}$ in V exists such that and $\mathbf{v} + (-\mathbf{v}) = \mathbf{0}$.*

Axioms for scalar multiplication:

S1 *If $\mathbf{v}$ is in V then $a\mathbf{v}$ is in V for all scalars a.*

S2 $a(\mathbf{v} + \mathbf{w}) = a\mathbf{v} + a\mathbf{w}$ *for all $\mathbf{v}$ and $\mathbf{w}$ in V and all scalars a.*

S3 $(a + b)\mathbf{v} = a\mathbf{v} + b\mathbf{v}$ *for all $\mathbf{v}$ in V and all scalars a and b.*

S4 $a(b\mathbf{v}) = (ab)\mathbf{v}$ *for all $\mathbf{v}$ in V and all scalars a and b.*

S5 $1\mathbf{v} = \mathbf{v}$ *for all $\mathbf{v}$ in V.*

We describe Axioms A1 and S1 by saying that V is **closed** under vector addition and scalar multiplication, respectively. As we shall see, the property in Axiom A4 uniquely determines the vector $\mathbf{0}$, and $\mathbf{0}$ is called the **zero vector** of the vector space V. Similarly the vector $-\mathbf{v}$ in Axiom A5 is uniquely determined by $\mathbf{v}$, and is called the **negative** of $\mathbf{v}$. Note that combining Axiom A2 and A4 shows that $\mathbf{0} + \mathbf{v} = \mathbf{v}$ for all $\mathbf{v}$ in V. Similarly $(-\mathbf{v}) + \mathbf{v} = \mathbf{0}$ using Axioms A2 and A5.

Thus, to specify a vector space V, we must do several things. First we must say precisely what V is, that is we must specify the set of vectors we are considering. Next we must define how to add two of our vectors (to get a sum in V), and how to multiply any vector by any scalar (to get a scalar product in V). Finally, we must verify that the axioms hold. Here are several examples, starting with the central concept of Chapter 4.

Example 1. For each $n \geq 1$, $\mathbb{R}^n$ is a vector space using the usual addition and scalar multiplication. Here the vectors are the n-tuples, (which we continue to write as columns). The zero vector of $\mathbb{R}^n$ is the usual zero column, and the negative of a vector is the usual negative. Note that $\mathbb{R}^1 = \mathbb{R}$ is the set of real numbers with the usual operations. □

Example 2. If $m \geq 1$ and $n \geq 1$, the set $\mathbb{M}_{m,n}$ of all $m \times n$ matrices is a vector space using matrix addition and scalar multiplication. The set $\mathbb{M}_{m,n}$ is clearly closed under these operations, the zero vector is the zero matrix $0 = 0_{m,n}$ of size

5.1. EXAMPLES AND BASIC PROPERTIES

$m \times n$, and the negative of a matrix A in $\mathbb{M}_{m,n}$ is $-A$, the usual negative. The other axioms are basic properties of matrix addition and scalar multiplication. □

Example 3. The set $\mathbb{P}$ of all polynomials is a vector space. Here a **polynomial** $p(x)$ is an expression

$$p(x) = a_0 + a_1 x + a_2 x^2 + \cdots + a_n x^n$$

where x is a variable called an **indeterminant**, and $a_0, a_1, a_2, \cdots, a_n$ are scalars (possibly zero), called the **coefficients** of the polynomial; a_0 is called the **constant coefficient** of $p(x)$. If the coefficients are all zero, the polynomial is called the **zero polynomial** and is denoted simply as 0. Two polynomials are **equal** if they are identical, that is if corresponding coefficients are all equal. In particular,

$$a_0 + a_1 x + a_2 x^2 + \cdots + a_n x^n = 0$$

means that $a_0 = a_1 = a_2 = \cdots = a_n = 0$. (This is the reason for calling x an *indeterminate*.)

Two polynomials are added in the obvious way by adding corresponding coefficients, and the scalar product of a polynomial by a scalar a is obtained by multiplying every coefficient by a. For example,

$$\begin{aligned}(2 - 5x - 3x^3) + (1 + 3x + 7x^2) &= 3 - 2x + 7x^2 - 3x^3 \\ 4(-1 + 3x - 5x^3) &= -4 + 12x - 20x^3.\end{aligned}$$

It is a routine verification that these operations make $\mathbb{P}$ into a vector space; the zero vector is the zero polynomial, and the negative of a polynomial $p(x)$ is the polynomial $(-1)p(x)$ obtained from $p(x)$ by negating every coefficient.[2] □

If $p(x)$ is a nonzero polynomial, the highest power of x in $p(x)$ with nonzero coefficient is called the **degree** of $p(x)$, and is denoted as $deg\,p(x)$. The coefficient of this highest power is called the **leading coefficient** of $p(x)$. The degree of the zero polynomial is not defined. Some examples:

$$\begin{aligned}deg(3 - x) &= 1 \quad &\text{with leading coefficient } -1 \\ deg(2 + 7x^2) &= 2 \quad &\text{with leading coefficient } 7 \\ deg(4) &= 0 \quad &\text{with leading coefficient } 4\end{aligned}$$

The notion of degree leads to the following example.

Example 4. If $n \geq 0$ the set $\mathbb{P}_n = \{a_0 + a_1 x + a_2 x^2 + \cdots + a_n x^n \mid a_i \text{ in } \mathbb{R}\}$ consists of all polynomials of degree at most n, together with the zero polynomial. It is easy to see that $\mathbb{P}_n$ is a vector space using the addition and scalar multiplication of $\mathbb{P}$. □

---

[2] More information on polynomials can be found in Appendix A.3.

The next example is a vector space that arises frequently in analysis (the theoretical side of calculus).

Example 5. If D is a nonempty set of real numbers, let $\mathbb{F}[D]$ denote the set of all real valued functions defined on D. The set D is called the **domain** of the functions; usually D will be either $\mathbb{R}$ or an interval

$$[a, b] = \{x \text{ in } \mathbb{R} \mid a \leq x \leq b\}.$$

Two such functions f and g in $\mathbb{F}[D]$ are **equal** (written $f = g$) if they have equal values for all x in D, that is if

$$f(x) = g(x) \quad \text{for all } x \text{ in } D,$$

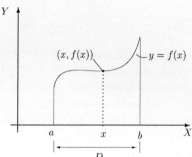

and we describe this situation by saying that f and g have the same **action**.

Given a scalar a and functions f and g, the action of the sum $f + g$ and scalar product af are defined as in calculus:

$$(f + g)(x) = f(x) + g(x) \quad \text{and} \quad (af)(x) = a\,f(x) \quad \text{for all } x \text{ in } D.$$

Thus the value of the function $f + g$ at each x is the sum of the values of f and g at x, and the value of af at x is a times the value of f at x. For this reason, these operations are called **pointwise** addition and scalar multiplication. They are the operations familiar from elementary algebra and calculus.

For example, if $f(x) = \sin x$ for all x and $g(x) = 1 + x$ for all x, then the sum $f + g$ is given by

$$(f + g)(x) = f(x) + g(x)$$
$$= (1 + x) + \sin x$$

for all x. The graph of the equation $y = (f + g)(x)$ is shown in the diagram together with $y = f(x)$ and $y = g(x)$.

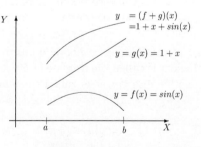

These pointwise operations make $\mathbb{F}[D]$ into a vector space. The zero vector in $\mathbb{F}[D]$ is the constant function f_0 that is identically zero for all values of x, that is $f_0(x) = 0$ for every x in D. The negative of a function f is the function $-f$ defined by $(-f)(x) = -f(x)$ for all x in D. The verification that axiom A4 holds in $\mathbb{F}[D]$ is as follows: If f is any function in $\mathbb{F}[D]$ we compute

$$(f + f_0)(x) = f(x) + f_0(x) = f(x) + 0 = f(x) \quad \text{for all } x \text{ in } D.$$

This means that $f + f_0 = f$ so, as this is valid for every f in $\mathbb{F}[D]$, it shows that axiom A4 holds in $\mathbb{F}[D]$. The reader should verify that the other axioms hold. $\square$

5.1. EXAMPLES AND BASIC PROPERTIES

Example 6. A set $\{\mathbf{0}\}$ with one element becomes a vector space if we define $\mathbf{0} + \mathbf{0} = \mathbf{0}$ and $a\mathbf{0} = \mathbf{0}$ for all scalars a. The axioms are easily verified, and $\{\mathbf{0}\}$ is called the **zero space**. □

While other examples will occur later, these suffice to illustrate the diversity of the vector space concept. We now turn to the business of proving theorems in the setting of an abstract vector space V (and which will then hold in *any* of these examples). At the outset, all we know about V is that the axioms are valid. While this does not seem to be much to go on, a surprising number of facts about V can be deduced from the axioms. The idea is to build up a body of information, one theorem at a time, and to use theorems already proved to help deduce new results. Our first result is very useful.

Theorem 1. Cancellation. *If* $\mathbf{w} + \mathbf{v} = \mathbf{u} + \mathbf{v}$ *in a vector space* V, *then* $\mathbf{w} = \mathbf{u}$.

Proof. If $\mathbf{v}$, $\mathbf{w}$ and $\mathbf{u}$ were numbers, we would simply subtract $\mathbf{v}$ from both sides to obtain $\mathbf{w} = \mathbf{u}$. We accomplish this by adding $-\mathbf{v}$ to both sides (using Axiom A5 to guarantee that $-\mathbf{v}$ is available). The result is

$$(\mathbf{w} + \mathbf{v}) + (-\mathbf{v}) = (\mathbf{u} + \mathbf{v}) + (-\mathbf{v}).$$

Using Axiom A3, this becomes $\mathbf{w} + (\mathbf{v} + (-\mathbf{v})) = \mathbf{u} + (\mathbf{v} + (-\mathbf{v}))$, and this is $\mathbf{w} + \mathbf{0} = \mathbf{u} + \mathbf{0}$ by Axiom A5 again. Hence $\mathbf{w} = \mathbf{u}$ by Axiom A4. This is what we wanted. □

In particular, Theorem 1 shows that there is only one zero vector in any vector space V. Indeed, if $\mathbf{0}'$ has the property in Axiom A4, then $\mathbf{v} + \mathbf{0} = \mathbf{v} = \mathbf{v} + \mathbf{0}'$ for all $\mathbf{v}$ in V, so $\mathbf{0} = \mathbf{0}'$ by cancellation. Moreover, a similar argument shows that there is only one negative of each vector (verify). Hence we speak of *the* zero vector in a vector space, and of *the* negative of any given vector.

The technique in the proof of Theorem 1 was to mimic the process of subtracting $\mathbf{v}$ from both sides of the equation by adding $-\mathbf{v}$ to both sides. If we do this to the equation

$$\mathbf{x} + \mathbf{v} = \mathbf{u}$$

the result is $\mathbf{x} = \mathbf{u} + (-\mathbf{v})$ as the reader can verify. Hence, by analogy with arithmetic, we define the **difference** $\mathbf{u} - \mathbf{v}$ of two vectors $\mathbf{u}$ and $\mathbf{v}$ by

$$\mathbf{u} - \mathbf{v} = \mathbf{u} + (-\mathbf{v}).$$

And, as in arithmetic, we say that $\mathbf{u} - \mathbf{v}$ is the result of having **subtracted** $\mathbf{v}$ from $\mathbf{u}$. Incidentally, we have shown that $\mathbf{x} = \mathbf{u} - \mathbf{v}$ is the unique solution to the equation $\mathbf{x} + \mathbf{v} = \mathbf{u}$.

The next theorem gives some basic facts about an arbitrary vector space V that will be used extensively later.

Theorem 2. *Let* $\mathbf{v}$ *denote a vector in a vector space* V, *and let* a *denote a scalar* .

(1) $0\mathbf{v} = \mathbf{0}$.

(2) $a\mathbf{0} = \mathbf{0}$.

(3) *If $a\mathbf{v} = \mathbf{0}$ then either $a = 0$ or $\mathbf{v} = \mathbf{0}$.*

(4) $(-1)\mathbf{v} = -\mathbf{v}$.

(5) $(-a)\mathbf{v} = -(a\mathbf{v}) = a(-\mathbf{v})$.

Proof. (1). By Axioms S3 and A4 we have $0\mathbf{v}+0\mathbf{v} = (0+0)\mathbf{v} = 0\mathbf{v} = 0\mathbf{v} + \mathbf{0}$. Now (1) follows by cancellation.

(2). Here $a\mathbf{0}+a\mathbf{0} = a(\mathbf{0}+\mathbf{0}) = a\mathbf{0} = a\mathbf{0}+\mathbf{0}$ by Axioms S2 and A4, so again cancellation does it.

(3). Let $a\mathbf{v} = \mathbf{0}$. If $a = 0$ we are done. If $a \neq 0$ then multiply the equation $a\mathbf{v} = \mathbf{0}$ on both sides by $\frac{1}{a}$. Then (2) and Axioms S4 and S5 give $\mathbf{v} = 1\mathbf{v} = (\frac{1}{a}a)\mathbf{v} = \frac{1}{a}(a\mathbf{v}) = \frac{1}{a}\mathbf{0} = \mathbf{0}$.

(4). Using (1) with Axioms S5, S3, A2 and A5, we have $(-1)\mathbf{v} + \mathbf{v} = (-1)\mathbf{v}+1\mathbf{v} = (-1+1)\mathbf{v} = 0\mathbf{v} = \mathbf{0} = -\mathbf{v}+\mathbf{v}$. Now cancellation applies.

(5). For the first equation, we use (4) and Axiom S4: $(-a)\mathbf{v} = ((-1)a)\mathbf{v} = (-1)(a\mathbf{v}) = -(a\mathbf{v})$. The other equation is verified similarly, and the proof is left to the reader. $\square$

Observe that parts (1), (2) and (3) of Theorem 2 can be combined in the following statement: If a is a scalar and $\mathbf{v}$ is a vector, then

$$a\mathbf{v} = \mathbf{0} \quad \text{if and only if either } a = 0 \text{ or } \mathbf{v} = \mathbf{0}.$$

This will be used several times in what follows.

Axioms S2 and S3 extend as follows. If $a, a_1, a_2, \cdots, a_n$ are scalars and $\mathbf{v}, \mathbf{v}_1, \mathbf{v}_2, \cdots, \mathbf{v}_n$ are vectors, we have[3]

$$a(\mathbf{v}_1 + \mathbf{v}_2 + \cdots + \mathbf{v}_n) = a\mathbf{v}_1 + a\mathbf{v}_2 + \cdots + a\mathbf{v}_n$$
$$(a_1 + a_2 + \cdots + a_n)\mathbf{v} = a_1\mathbf{v} + a_2\mathbf{v} + \cdots + a_n\mathbf{v}$$

These facts, together with the axioms, Theorem 2, and the definition of subtraction, enable us to manipulate and simplify sums of scalar multiples of vectors by collecting like terms, expanding and taking out common factors. This has already been discussed for matrices and vectors in $\mathbb{R}^n$, and manipulations in an arbitrary vector space are carried out in a similar way. For example,

$$3(\mathbf{v} - 2\mathbf{w}) - 4(\mathbf{v} + 2\mathbf{u}) + 2(3\mathbf{w} + 4\mathbf{u}) = 3\mathbf{v} - 6\mathbf{w} - 4\mathbf{v} - 8\mathbf{u} + 6\mathbf{w} + 8\mathbf{u} = -\mathbf{v}.$$

Computations like this are used in every discussion of vector spaces.

---

[3] It is a consequence of Axiom A3 that we can omit parentheses when writing a sum $\mathbf{v}_1 + \mathbf{v}_2 + \cdots + \mathbf{v}_n$.

5.1.2 Subspaces

If V is a vector space, a nonempty subset $U \subseteq V$ is called a **subspace** of V if U is itself a vector space using the addition and scalar multiplication of V. Thus $\mathbb{P}_n$ is a subspace of $\mathbb{P}$ and, if we regard polynomials as real valued functions defined on $\mathbb{R}$, $\mathbb{P}$ is a subspace of $\mathbb{F}[\mathbb{R}]$. Note that if U and W are subspaces of some vector space V, and if $U \subseteq W$, then U is a subspace of W.

Using Theorem 2, we can give a simple test for when a subset of a vector space is a subspace.

Theorem 3. Subspace Test. *A subset U of a vector space V is a subspace if and only if it satisfies the following three conditions:*

(1) **0** *is in U.*

(2) *If $\mathbf{u}$ is in U then $a\mathbf{u}$ is in U for all scalars a.*
  (U *is* **closed under scalar multiplication.**)

(3) *If $\mathbf{u}$ and $\mathbf{u}_1$ are in U, then $\mathbf{u} + \mathbf{u}_1$ is in U.*
  (U *is* **closed under addition.**)

Proof. If U is a subspace then (2) and (3) hold by Axioms S1 and A1 (applied to U), and if $\mathbf{u}$ is in U then Theorem 2 gives $\mathbf{0} = 0\mathbf{u}$, so $\mathbf{0}$ is in U by (2), and (1) holds.[4]

Conversely, suppose that (1), (2) and (3) hold for U. Then Axioms A1 and S1 hold by (2) and (3), and Axioms A2, A3, S2, S3, S4 and S5 all hold in U because they hold in V. Next, Axiom A4 holds in U because the zero vector of V is actually in U by (1), and so serves as the zero of U. Finally, if $\mathbf{u}$ is in U then its negative $-\mathbf{u}$ in the space V is actually in U by (1) because $-\mathbf{u} = (-1)\mathbf{u}$. This proves Axiom A5 for U. □

Note that the proof of Theorem 3 shows that if U is a subspace of V, then U and V share the same zero vector, and the negative of a vector in U is the same as its negative in V.

Subspaces of $\mathbb{R}^n$ were defined in Section 4.1 as subsets satisfying the conditions in Theorem 3. This gives many more examples of vector spaces.

Example 7. All the subspaces of $\mathbb{R}^n$ discussed Chapter 4 are vector spaces in their own right. For example, if A is any matrix then each of $null A$, $im A = col A$ and $row A$ is a vector space. Another example: $U^\perp$ is a vector space if U is any subspace of $\mathbb{R}^n$.

Example 8. If V is any vector space, then $\{\mathbf{0}\}$ and V are subspaces of V. The space $\{\mathbf{0}\}$ is called the **zero subspace** of V, and any subspace other than $\{\mathbf{0}\}$ and V is called a **proper subspace** of V.

---

[4]This shows that condition (1) follows from (2) provided U contains at least one vector. We prefer to retain (1) in Theorem 3 because checking it is usually easy and, if it fails, U *cannot* be a subspace.

Example 9. *If $\mathbf{v}$ is a vector in a vector space V, show that $\mathbb{R}\mathbf{v}$ is a subspace of V where $\mathbb{R}\mathbf{v} = \{r\mathbf{v} \mid r \text{ in } \mathbb{R}\}$.*

Solution. First, $\mathbf{0} = 0\mathbf{v}$ shows that $\mathbf{0}$ is in $\mathbb{R}\mathbf{v}$. Next, given $\mathbf{u} = r\mathbf{v}$ and $\mathbf{u}_1 = r_1\mathbf{v}$ in $\mathbb{R}\mathbf{v}$, we have

$$a\mathbf{u} = a(r\mathbf{v}) = (ar)\mathbf{v} \quad \text{and} \quad \mathbf{u} + \mathbf{u}_1 = r\mathbf{v} + r_1\mathbf{v} = (r + r_1)\mathbf{v}$$

for each scalar a. Hence $a\mathbf{u}$ and $\mathbf{u} + \mathbf{u}_1$ are both in $\mathbb{R}\mathbf{v}$, so the Subspace Test applies. □

Example 10. *Show that the set $U = \{A \text{ in } \mathbb{M}_{n,n} \mid A \text{ is symmetric}\}$ is a subspace of $\mathbb{M}_{n,n}$.*

Solution. Recall that A is symmetric if $A^T = A$. Hence the zero matrix 0 is in U because $0^T = 0$. If A and B are in U, then $A^T = A$ and $B^T = B$, so $(A+B)^T = A^T + B^T = A + B$ and $(aA)^T = a\,A^T = aA$ for all scalars a. Hence $A + B$ and aA are in U, so U is a subspace by the Subspace Test. □

Example 11. *If λ is a fixed real number, let $U = \{p(x) \text{ in } \mathbb{P} \mid p(\lambda) = 0\}$ be the set of all polynomials which have λ as a root. Show that U is a subspace of $\mathbb{P}$.*

Solution. Clearly the zero polynomial lies in U. If $p(x)$ and $q(x)$ lie in U then $(p+q)(\lambda) = p(\lambda) + q(\lambda) = 0 + 0 = 0$, so $p + q$ is in U. Similarly ap is in U for all scalars a (verify), and the Subspace Test applies. □

Example 12. *Show that the set of differentiable[5] functions is a subspace of $\mathbb{F}[\mathbb{R}]$.*

Solution. The zero function is differentiable (its derivative is again the zero function), and it is a theorem of calculus that sums and scalar multiples of differentiable functions are again differentiable. □

Let $\mathbf{v}_1, \mathbf{v}_2, \cdots, \mathbf{v}_n$ be vectors in a vector space V. As in $\mathbb{R}^n$, if $a_1, a_2, \cdots, a_n$ are scalars, a sum $a_1\mathbf{v}_1 + a_2\mathbf{v}_2 + \cdots + a_n\mathbf{v}_n$ is called a **linear combination** of the $\mathbf{v}_i$. The set of all such linear combinations is called the **span** of the $\mathbf{v}_i$ and is denoted

$$span\{\mathbf{v}_1, \mathbf{v}_2, \cdots, \mathbf{v}_n\} = \{r_1\mathbf{v}_1 + r_2\mathbf{v}_2 + \cdots + r_n\mathbf{v}_n \mid r_i \text{ in } \mathbb{R}\}.$$

In particular, $span\{\mathbf{v}\} = \{r\mathbf{v} \mid r \text{ in } \mathbb{R}\} = \mathbb{R}\mathbf{v}$ is the subspace in Example 9. These notions coincide with those in Chapter 4, and the proof of Theorem 1 §4.1 goes through verbatim to give

Theorem 4. *Let $U = span\{\mathbf{v}_1, \mathbf{v}_2, \cdots, \mathbf{v}_n\}$ in a vector space V.*

---

[5] Readers not familiar with calculus may omit this example with no loss of continuity.

5.1. EXAMPLES AND BASIC PROPERTIES

(1) U is a subspace of V containing each of the vectors $\mathbf{v}_1, \mathbf{v}_2, \cdots, \mathbf{v}_n$.

(2) U is the "smallest" subspace containing these vectors in the sense that any subspace of V that contains each of $\mathbf{v}_1, \mathbf{v}_2, \cdots, \mathbf{v}_n$ must already contain U.

Example 13. *If* $\mathbf{u}$, $\mathbf{v}$ *and* $\mathbf{w}$ *are vectors, show that* $span\{\mathbf{u}, \mathbf{v}, \mathbf{w}\} = span\{\mathbf{u}+\mathbf{v}, \mathbf{v}+\mathbf{w}, \mathbf{u}+\mathbf{w}\}$.

Solution. For convenience, write $U = span\{\mathbf{u}, \mathbf{v}, \mathbf{w}\}$ and $W = span\{\mathbf{u}+\mathbf{v}, \mathbf{v}+\mathbf{w}, \mathbf{u}+\mathbf{w}\}$. We have $W \subseteq U$ by Theorem 4 because each of $\mathbf{u}+\mathbf{v}$, $\mathbf{v}+\mathbf{w}$ and $\mathbf{u}+\mathbf{w}$ is in U. On the other hand,

$$\mathbf{u} = \tfrac{1}{2}(\mathbf{u}+\mathbf{v}) - \tfrac{1}{2}(\mathbf{v}+\mathbf{w}) + \tfrac{1}{2}(\mathbf{u}+\mathbf{w})$$

shows that $\mathbf{u}$ is in W. Similarly $\mathbf{v}$ and $\mathbf{w}$ are in W, so $U = span\{\mathbf{u}, \mathbf{v}, \mathbf{w}\} \subseteq W$, again by Theorem 4. □

Example 14. *Show that* $\mathbb{P}_3 = span\{2x + x^3, x^2 - x, 3x - 2, 5\}$.

Solution. For convenience, write $U = span\{2x + x^3, x^2 - x, 3x - 2, 5\}$. Then $U \subseteq \mathbb{P}_3$ is clear. Since $\mathbb{P}_3 = span\{1, x, x^2, x^3\}$, it suffices by Theorem 4 to show that each of 1, x, x^2 and x^3 is in U.

First $1 = \tfrac{1}{5} \cdot 5$ is in U because 5 is in U.

Then $x = \tfrac{1}{3}[(3x - 2) + 2 \cdot 1]$ is in U because both $3x - 2$ and 1 are in U.

Next, $x^2 = (x^2 - x) + x$ is in U because $x^2 - x$ and x are both in U.

Finally $x^3 = (2x + x^3) - 2 \cdot x$ is in U because $2x + x^3$ and x are in U.

This is what we wanted. □

Example 15. *Show that* $\mathbb{P}_n = span\{1, x, x^2, \cdots, x^n\}$.

Solution. We have $\mathbb{P}_n = \{a_0 + a_1 x + a_2 x^2 + \cdots + a_n x^n \mid a_i \text{ in } \mathbb{R}\}$ by Example 4. □

Every 2×2 matrix A in $\mathbb{M}_{2,2}$ can be written as follows:

$$A = \begin{bmatrix} a_{11} & a_{12} \\ a_{21} & a_{22} \end{bmatrix}$$
$$= a_{11} \begin{bmatrix} 1 & 0 \\ 0 & 0 \end{bmatrix} + a_{12} \begin{bmatrix} 0 & 1 \\ 0 & 0 \end{bmatrix} + a_{21} \begin{bmatrix} 0 & 0 \\ 1 & 0 \end{bmatrix} + a_{22} \begin{bmatrix} 0 & 0 \\ 0 & 1 \end{bmatrix}.$$

This reveals a useful spanning set for $\mathbb{M}_{2,2}$, and a similar decomposition works in $\mathbb{M}_{m,n}$. The $m \times n$ matrix with (i, j)-entry 1 and the other entries 0 is called the (i, j)-**matrix unit**, and is denoted E_{ij}. If $A = [a_{ij}]$ is a matrix in $\mathbb{M}_{m,n}$, it

is clear that $A = \Sigma_{i,j} a_{ij} E_{ij}$ where the sum is taken over all i and j such that $1 \leq i \leq m$ and $1 \leq j \leq n$. Hence $\mathbb{M}_{m,n}$ is spanned by these matrix units. For example
$$\mathbb{M}_{2,3} = span\{E_{11}, E_{12}, E_{13}, E_{21}, E_{22}, E_{23}\}.$$
In general there are mn matrix units in $\mathbb{M}_{m,n}$, and we have

Example 16. $\mathbb{M}_{m,n} = span\{E_{ij} \mid 1 \leq i \leq m \text{ and } 1 \leq j \leq n\}$ where the E_{ij} are the $m \times n$ matrix units.

Example 17. Suppose an object is suspended by a coil spring. If the object is pulled down from its rest position and released, it will begin to oscillate. It is known that the displacement y about the rest position is given in terms of the time t by a function of the following form:

$$y = a\,cos(\omega t) + b\,sin(\omega t)$$

where ω is a constant depending on the spring and the mass of the object. Thus the set of all functions describing such motions is the subspace of $\mathbb{F}[\mathbb{R}]$ spanned by $cos(\omega t)$ and $sin(\omega t)$.

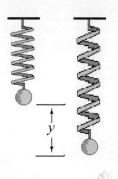

Exercises

The problems that involve calculus concepts can be omitted by students with no calculus background.

1. In each case either show that the statement is true or give an example showing that it is false.

   (a) Every vector space contains at least one vector.

   (b) Every nonzero vector space contains an infinite number of vectors.

   (c) $\{1 - x,\ 2 + x - x^2\}$ spans $\mathbb{P}_2$.

   (d) Working in $\mathbb{F}(\mathbb{R})$, 1 is in $span\{cos^2 x,\ sin^2 x\}$.

   (e) Working in $\mathbb{F}(\mathbb{R})$, x is in $span\{cos^2 x,\ sin^2 x\}$.

   (f) The set $\mathbb{Z} = \{0, \pm 1, \pm 2, \pm 3, \cdots\}$ is a vector space with the usual operations.

   (g) The set $\mathbb{Q}$ of all rational numbers (fractions) is a vector space with the usual operations.

   (h) Every nonzero vector space V contains a proper subspace (Example 8).

5.1. EXAMPLES AND BASIC PROPERTIES

2. Let V denote the set of all 1×3 matrices, and define addition on V as for $\mathbb{M}_{1,3}$. For each of the following definitions of scalar multiplication (denoted $\cdot$ in each case), decide if V is a vector space.

   (a) $a \cdot [x\ y\ z] = [ax\ ay\ z]$. (b) $a \cdot [x\ y\ z] = [0\ 0\ 0]$.
   (c) $a \cdot [x\ y\ z] = [ax\ ay\ 0]$. (d) $a \cdot [x\ y\ z] = [-ax\ -ay\ -az]$.
   (e) $a \cdot [x\ y\ z] = [x\ y\ z]$. (f) $a \cdot [x\ y\ z] = [ay\ ax\ az]$.

3. In each case decide if V is a vector space with the given operations. Support your answer.

   (a) V is the set of all non-negative real numbers; usual addition and multiplication.

   (b) V is the set $\mathbb{C}$ of complex numbers; usual addition and multiplication by a real number.

   (c) $V = \{\mathbf{0}\}$, where $\mathbf{0} + \mathbf{0} = \mathbf{0}$ and $a \cdot \mathbf{0} = \mathbf{0}$ for all real numbers a.

   (d) $V = \{[x\ 1\ z] \mid x\ \text{and}\ z\ \text{real}\}$; operations of $\mathbb{M}_{1,3}$.

   (e) $V = \{[x\ y] \mid x\ \text{and}\ y\ \text{real},\ x^2 = y^2\}$, operations of $\mathbb{M}_{1,2}$.

   (f) V is the set of all 2×2 matrices B; addition of $\mathbb{M}_{2,2}$ and $a \cdot B = |a|\,B$.

   (g) V is the set of all 2×2 matrices B; addition of $\mathbb{M}_{2,2}$ and $a \cdot B = aB^T$.

   (h) V is the set of functions $\mathbb{R} \to R$; pointwise addition and $(a \cdot f)(x) = f(ax)$ for all x.

4. Let $V = \{[x\ y] \mid x\ \text{and}\ y\ \text{real}\}$. Show that V is a vector space with the following operations:

   $$[x\ y] \dotplus [x_1\ y_1] = [x + x_1\ y + y_1 + 1] \text{ and } a \cdot [x\ y] = [ax\ ay + a - 1].$$

5. Let $V = \{v \text{ in } \mathbb{R} \mid v > 0\}$. Show that V is a vector space if the vector addition is ordinary multiplication, and scalar multiplication is defined by $a \cdot v = v^a$.

6. In each case determine if U is a subspace of $\mathbb{P}_3$. Support your answer.

   (a) $U = \{p(x) \mid p(1) = 1\}$
   (b) $U = \{p(x) \mid p(x) = 0 \text{ or } \deg p(x) = 3\}$
   (c) $U = \{x\,p(x) \mid p(x) \text{ in } \mathbb{P}_3\}$
   (d) $U = \{xp(x) + (1-x)q(x) \mid p(x) \text{ and } q(x) \text{ in } \mathbb{P}_2\}$
   (e) $U = \{xp(x) \mid p(x) \text{ in } \mathbb{P}_2\}$
   (f) $U = \{p(x) \text{ in } \mathbb{P}_3 \mid p(x) \text{ has constant coefficient } 0\}$

7. In each case determine if U is a subspace of $\mathbb{M}_{2,2}$. Support your answer.

   (a) $U = \left\{ \begin{bmatrix} a & b \\ 0 & c \end{bmatrix} \mid a, b, c \text{ in } \mathbb{R} \right\}$

(b) $U = \left\{ \begin{bmatrix} a & b \\ c & d \end{bmatrix} \mid a+c = b+d \right\}$

(c) $U = \{A \mid A = -A^T\}$

(d) $U = \{A \mid AB = BA\}$, where B is a fixed matrix in $\mathbb{M}_{2,2}$

(e) $U = \{A \mid det A = 0\}$

(f) $U = \{A \mid A^2 = A\}$

(g) $U = \{BA_0 \mid B \text{ in } \mathbb{M}_{2,3}\}$ where A_0 is a fixed 3×2 matrix.

8. In each case determine if U is a subspace of $\mathbb{F}[0,1]$. Support your answer.

   (a) $U = \{f \mid f(0) = 1\}$

   (b) $U = \{f \mid f(0) = 0\}$

   (c) $U = \{f \mid f(0) = f(1)\}$

   (d) $U = \{f \mid f(x) \geq 0 \text{ for all } x \text{ in } [0,1]\}$

   (e) $U = \{f \mid f(x) = f(y) \text{ for all } x \text{ and } y \text{ in } [0,1]\}$.

   (f) $U = \{f \mid f(x+y) = f(x) + f(y) \text{ for all } x \text{ and } y \text{ in } [0,1]\}$.

   (g) $U = \{f \mid f'(\frac{1}{2}) = 0$, where f' denotes the derivative of $f.\}$

   (h) $U = \{f \mid \int_0^1 f(x)dx = 0\}$

9. In each case determine whether $\mathbf{v}_1$ or $\mathbf{v}_2$ lies in $span\{\mathbf{u}, \mathbf{w}\}$.

   (a) $\mathbf{u} = x^2 - 2$, $\mathbf{w} = x^2 + x$; $\mathbf{v}_1 = x^2 + 1$, $\mathbf{v}_2 = x^2 - 2x - 6$.

   (b) $\mathbf{u} = x^3 - 2x + 1$, $\mathbf{w} = x^2 + x - 2$; $\mathbf{v}_1 = x^3 + 2x^2 - 3$, $\mathbf{v}_2 = x^3$.

   (c) $\mathbf{u} = \begin{bmatrix} 1 & -1 \\ 0 & 1 \end{bmatrix}$, $\mathbf{w} = \begin{bmatrix} 0 & 1 \\ 1 & -1 \end{bmatrix}$; $\mathbf{v}_1 = \begin{bmatrix} 1 & 0 \\ 1 & 0 \end{bmatrix}$, $\mathbf{v}_2 = \begin{bmatrix} 1 & 0 \\ 0 & 1 \end{bmatrix}$.

   (d) $\mathbf{u} = \begin{bmatrix} 1 & 0 \\ 0 & 1 \end{bmatrix}$, $\mathbf{w} = \begin{bmatrix} 0 & -1 \\ -1 & 1 \end{bmatrix}$; $\mathbf{v}_1 = \begin{bmatrix} 1 & 0 \\ 0 & 0 \end{bmatrix}$, $\mathbf{v}_2 = \begin{bmatrix} 5 & 3 \\ 3 & 2 \end{bmatrix}$.

   (e) $\mathbf{u} = sin^2 x$, $\mathbf{w} = cos^2 x$; $\mathbf{v}_1 = sinx$, $\mathbf{v}_2 = cos(2x)$. (Work in $\mathbb{F}[\mathbb{R}]$.)

   (f) $\mathbf{u} = sin^2 x$, $\mathbf{w} = cos^2 x$; $\mathbf{v}_1 = 2$, $\mathbf{v}_2 = x$. (Work in $\mathbb{F}[R]$.)

10. (a) Does $\left\{ \begin{bmatrix} 1 & 1 \\ 0 & 0 \end{bmatrix}, \begin{bmatrix} 1 & 0 \\ 1 & 0 \end{bmatrix}, \begin{bmatrix} 0 & 0 \\ 1 & 1 \end{bmatrix}, \begin{bmatrix} 0 & 1 \\ 0 & 1 \end{bmatrix} \right\}$ span $\mathbb{M}_{2,2}$? Support your answer.

    (b) Does $\{1+x, x+x^2, x^2+x^3, x^3+1\}$ span $\mathbb{P}_3$? Support your answer.

11. In each case find *all* vectors $\mathbf{x}$, $\mathbf{y}$, and $\mathbf{z}$ that satisfy the given equations.

    (a) $\begin{aligned} \mathbf{x} - \mathbf{y} + 2\mathbf{z} &= 0 \\ 2\mathbf{x} + 3\mathbf{y} - \mathbf{z} &= 0 \\ -\mathbf{x} + 6\mathbf{y} - 7\mathbf{z} &= 0 \end{aligned}$

    (b) $\begin{aligned} \mathbf{x} + \mathbf{y} - 2\mathbf{z} &= 0 \\ -\mathbf{x} - 3\mathbf{y} + 2\mathbf{z} &= 0 \\ -\mathbf{x} - 7\mathbf{y} + 2\mathbf{z} &= 0 \end{aligned}$

5.1. EXAMPLES AND BASIC PROPERTIES

12. If Y_0 is a fixed column in $\mathbb{R}^n$, let $U = \{AY_0 \mid A \text{ in } \mathbb{M}_{m,n}\}$.
    (a) Show that U is a subspace of $\mathbb{R}^m$.
    (b) Show that $U = \mathbb{R}^m$ if $Y_0 \neq 0$.

13. (Requires calculus) Show that the set $\mathbf{C}[a,b]$ of all continuous functions defined on the interval $[a,b]$ is a subspace of $\mathbb{F}[a,b]$.

14. (a) Show that the zero element of a vector space is unique. That is, if $\mathbf{z} + \mathbf{v} = \mathbf{v}$ holds for all $\mathbf{v}$, show that necessarily $\mathbf{z} = \mathbf{0}$.
    (b) Show that the negative of a vector $\mathbf{v}$ is uniquely determined by $\mathbf{v}$. That is, if $\mathbf{v} + \mathbf{v}' = \mathbf{0}$, show that necessarily $\mathbf{v}' = -\mathbf{v}$.

15. Show that $\mathbf{x} = \mathbf{u} - \mathbf{v}$ is the *unique* solution to the equation $\mathbf{x} + \mathbf{v} = \mathbf{u}$. State all axioms used.

16. Prove that $a(-\mathbf{v}) = (-a)\mathbf{v}$ for all scalars a and vectors $\mathbf{v}$, and so complete (5) of Theorem 2.

17. Show that $-\mathbf{0} = \mathbf{0}$ in every vector space V. Which vectors $\mathbf{v}$ satisfy $-\mathbf{v} = \mathbf{v}$?

18. If $\mathbf{v} \neq \mathbf{0}$ in a vector space V, show that $\mathbb{R}\mathbf{v}$ has only two subspaces, $\{\mathbf{0}\}$ and $\mathbb{R}\mathbf{v}$.

19. Given $\mathbf{v}$ in a vector space V, show that $\mathbb{R}\mathbf{v} = span\{a\mathbf{v}\}$ for all $a \neq 0$ in $\mathbb{R}$.

20. Let D denote a set of real numbers. If a is a real number, define the **constant function** $f_a : D \to \mathbb{R}$ by $f_a(x) = a$ for all x in D.
    (a) Show that the set $U = \{f_a \mid a \text{ in } \mathbb{R}\}$ of all constant functions is a subspace of $\mathbb{F}[D]$.
    (b) Show further that $U = span\{f_1\} = \mathbb{R}f_1$.
    (c) Show that the constant function f_0 is the zero vector in $\mathbb{F}[D]$ (see Example 5).

21. Suppose $a\mathbf{v} = b\mathbf{v}$ in a vector space.
    (a) If $a \neq b$, show that $\mathbf{v} = \mathbf{0}$.
    (b) If $\mathbf{v} \neq \mathbf{0}$, show that $a = b$.

22. Suppose $a\mathbf{v} = a\mathbf{w}$ in a vector space.
    (a) If $\mathbf{v} \neq \mathbf{w}$, show that $a = 0$.
    (b) If $a \neq 0$ show that $\mathbf{v} = \mathbf{w}$.

23. Let U be a subspace of a vector space V.
    (a) If $a\mathbf{v}$ is in U where $a \neq 0$, show that $\mathbf{v}$ is in U.
    (b) If $\mathbf{v}$ and $\mathbf{w} + \mathbf{v}$ are in U, show that $\mathbf{w}$ is in U.

24. (a) Show that $\mathbb{M}_{2,2} = span\{\begin{bmatrix} 1 & 0 \\ 0 & 0 \end{bmatrix}, \begin{bmatrix} 1 & 0 \\ 0 & 1 \end{bmatrix}, \begin{bmatrix} 0 & 1 \\ 1 & 0 \end{bmatrix}, \begin{bmatrix} 0 & 1 \\ 0 & 0 \end{bmatrix}\}$.

    (b) Show that $\mathbb{P}_3 = span\{1 + x^3, x + x^2, 1 + x + x^2, 2x - x^2\}$.

25. If $\mathbb{M}_{n,n} = span\{A_1, A_2, \cdots, A_k\}$, show that $\mathbb{M}_{n,n} = span\{A_1^T, A_2^T, \cdots, A_k^T\}$.

26. If $A_1, A_2, \cdots, A_k$ are $m \times n$ matrices and $A_1 Y = A_2 Y = \cdots = A_k Y = 0$ for some $Y \neq 0$ in $\mathbb{R}^n$, show that $\{A_1, A_2, \cdots, A_k\}$ cannot span $\mathbb{M}_{m,n}$.

27. Let X and Y be nonempty sets of vectors in a vector space V. If $X \subseteq Y$, show that $span\{X\} \subseteq span\{Y\}$.

28. Show that no finite set of polynomials can span $\mathbb{P}$.

29. Let U be a *nonempty* subset in a vector space V.

    (a) Show that U is a subspace of V if and only if $\mathbf{u} + a\mathbf{u}_1$ is in U for all $\mathbf{u}$ and $\mathbf{u}_1$ in U and all a in $\mathbb{R}$.

    (b) Show that U is a subspace of V if and only if U is closed under addition and scalar multiplication.

30. By calculating $(1+1)(\mathbf{v}+\mathbf{w})$ two ways, show that axiom A2 follows from the other axioms.

31. Use induction on n to prove each of the following in an arbitrary vector space V.

    (a) $a(\mathbf{v}_1 + \mathbf{v}_2 + \cdots + \mathbf{v}_n) = a\mathbf{v}_1 + a\mathbf{v}_2 + \cdots + a\mathbf{v}_n$ for all a in $\mathbb{R}$ and $\mathbf{v}_i$ in V.

    (b) $(a_1 + a_2 + \cdots + a_n)\mathbf{v} = a_1\mathbf{v} + a_2\mathbf{v} + \cdots + a_n\mathbf{v}$ for all a_i in $\mathbb{R}$ and $\mathbf{v}$ in V.

5.2 INDEPENDENCE AND DIMENSION

The deepest results about vector spaces stem from the idea of dimension, and this in turn relies on the notion of linear independence. For $\mathbb{R}^n$ these concepts were discussed in Sections 4.2 and 4.3; in this section, they are extended to an arbitrary vector space V in a natural way and general versions of the basic theorems are derived, usually with similar proofs.

5.2.1 Independence and the Fundamental Theorem

Let $\{\mathbf{v}_1, \mathbf{v}_2, \cdots, \mathbf{v}_n\}$ be a set of vectors in a vector space V. A linear combination $r_1\mathbf{v}_1 + r_2\mathbf{v}_2 + \cdots + r_n\mathbf{v}_n$ of these vectors is said to **vanish** if it equals **0**. The **trivial** linear combination

$$0\mathbf{v}_1 + 0\mathbf{v}_2 + \cdots + 0\mathbf{v}_n$$

5.2. INDEPENDENCE AND DIMENSION

(with every coefficient zero) certainly vanishes. As in $\mathbb{R}^n$, the set $\{\mathbf{v}_1, \mathbf{v}_2, \cdots, \mathbf{v}_n\}$ is called **linearly independent** (or simply **independent**[6]) if the *only* linear combination that vanishes is the trivial one. Thus, to check if a set $\{\mathbf{v}_1, \mathbf{v}_2, \cdots, \mathbf{v}_n\}$ is independent, the procedure (as in $\mathbb{R}^n$) is as follows:

Step 1. *Set a linear combination equal to zero* $r_1\mathbf{v}_1 + r_2\mathbf{v}_2 + \cdots + r_n\mathbf{v}_n = \mathbf{0}$.

Step 2. *Show somehow that this implies that all the coefficients are zero:* $r_i = 0$ *for each i.*

Of course, if some nontrivial linear combination vanishes (that is if Step 2 is impossible) then the vectors are *not* independent.

Example 1. Show that $\{1 + x^2, \, 2 - x + x^2, \, 1 - x^2\}$ is independent in $\mathbb{P}$.

Solution. Suppose $r_1(1+x^2) + r_2(2-x+x^2) + r_3(1-x^2) = 0$ in $\mathbb{P}$. Collecting powers of x gives

$$(r_1 + 2r_2 + r_3) + (-r_2)x + (r_1 + r_2 - r_3)x^2 = 0.$$

The definition of equality in $\mathbb{P}$ means that every coefficient is zero:

$$r_1 + 2r_2 + r_3 = 0, \quad -r_2 = 0 \quad \text{and} \quad r_1 + r_2 - r_3 = 0.$$

It is a routine task to verify that the only solution of these equations is the trivial one $r_1 = r_2 = r_3 = 0$. This proves independence. □

The next example gives a condition for independence in $\mathbb{P}$ which will be referred to several times below.

Example 2. Let $p_1(x), p_2(x), \cdots, p_k(x)$ be nonzero polynomials in $\mathbb{P}$ of distinct degrees. Show that $\{p_1(x), p_2(x), \cdots, p_k(x)\}$ is linearly independent.

Solution. There is no loss of generality in assuming that $\deg p_1(x) < \deg p_2(x) < \cdots < \deg p_k(x)$. Let $\deg p_k(x) = m_k$, and let the leading coefficient of $p_k(x)$ be a_k. Suppose that

$$r_1 p_1(x) + r_2 p_2(x) + \cdots + r_k p_k(x) = 0 \text{ where each } r_i \text{ is in } \mathbb{R}. \quad (*)$$

Then $r_k a_k = 0$ because it is the coefficient of x^{m_k} on the left side of $(*)$. Since $a_k \neq 0$ this means that $r_k = 0$, so $(*)$ becomes $r_1 p_1(x) + r_2 p_2(x) + \cdots + r_{k-1} p_{k-1}(x) = 0$. Now repeat the argument to conclude that $r_{k-1} = 0$. Continue to show that $r_i = 0$ for each i. □

Example 3. Show that $\{\sin x, \cos x\}$ is independent in the space $\mathbb{F}[0, 2\pi]$ of all real valued functions on the interval $[0, 2\pi]$.

---

[6]The adjective "linear" is sometimes needed because there are other notions of independence in mathematics. However, none of these occur in these notes.

Solution. Suppose that a linear combination vanishes in $\mathbb{F}[0, 2\pi]$. This means that
$$r_1 \sin(x) + r_2 \cos(x) = 0.$$
By the definition of equality in $\mathbb{F}[0, 2\pi]$, this must hold for *all* values of x in the interval $[0, 2\pi]$. Taking $x = 0$ yields $r_2 = 0$ because $\sin 0 = 0$ and $\cos 0 = 1$. Similarly $r_1 = 0$ follows by taking $x = \frac{\pi}{2}$ (because $\sin(\frac{\pi}{2}) = 1$ and $\cos(\frac{\pi}{2}) = 0$). □

Example 4. *If $\{\mathbf{u}, \mathbf{v}, \mathbf{w}\}$ is independent, show that $\{\mathbf{u} + \mathbf{v}, \mathbf{v} + \mathbf{w}, \mathbf{u} + \mathbf{w}\}$ is also independent.*

Solution. Suppose a linear combination vanishes: $r_1(\mathbf{u} + \mathbf{v}) + r_2(\mathbf{v} + \mathbf{w}) + r_3(\mathbf{u} + \mathbf{w}) = \mathbf{0}$. Collecting terms in $\mathbf{u}$, $\mathbf{v}$ and $\mathbf{w}$, this becomes
$$(r_1 + r_3)\mathbf{u} + (r_1 + r_2)\mathbf{v} + (r_2 + r_3)\mathbf{w} = \mathbf{0}.$$
Since $\{\mathbf{u}, \mathbf{v}, \mathbf{w}\}$ is independent, this must be the trivial combination, that is all the coefficients must vanish:
$$r_1 + r_3 = 0, \ r_1 + r_2 = 0, \ r_2 + r_3 = 0.$$
Since this system of equations has only the trivial solution $r_1 = r_2 = r_3 = 0$, we have shown that $\{\mathbf{u} + \mathbf{v}, \mathbf{v} + \mathbf{w}, \mathbf{u} + \mathbf{w}\}$ is independent. □

Example 5. *If $\mathbf{v} \neq \mathbf{0}$ in a vector space V, show that $\{\mathbf{v}\}$ is linearly independent.*

Solution. Here a vanishing linear combination has the form $r\mathbf{v} = \mathbf{0}$. Since $\mathbf{v} \neq \mathbf{0}$, this means $r = 0$ by Theorem 2 §5.1. □

Example 6. *Show that no independent set can contain the zero vector.*

Solution. If the zero vector is in an independent set, consider the linear combination where the zero vector has coefficient one and every other coefficient is zero. This combination vanishes and is nontrivial, contrary to the independence of the set. □

Remarkably, the Fundamental Theorem from Chapter 4 (for $\mathbb{R}^n$) holds in every abstract vector space. In fact, both proofs of Theorem 1 §4.3 go through as written. Because of its importance we give the general version of the second proof.

Theorem 1. Fundamental Theorem. *Suppose that V is a vector space and $V = \text{span}\{\mathbf{v}_1, \mathbf{v}_2, \cdots, \mathbf{v}_n\}$. If $\{\mathbf{u}_1, \mathbf{u}_2, \cdots, \mathbf{u}_m\}$ is any linearly independent subset of V, then $m \leq n$.*

Proof. We assume that $m > n$ and show that this leads to a contradiction. As $V = \text{span}\{\mathbf{v}_1, \mathbf{v}_2, \cdots, \mathbf{v}_n\}$, write $\mathbf{u}_1 = r_1\mathbf{v}_1 + r_2\mathbf{v}_2 + \cdots + r_n\mathbf{v}_n$ where the r_i are

5.2. INDEPENDENCE AND DIMENSION

in $\mathbb{R}$. Then some $r_i \neq 0$ because $\mathbf{u}_1 \neq 0$ (by Example 6). By relabeling the $\mathbf{v}_i$ if necessary, assume that $r_1 \neq 0$. Then $V = span\{\mathbf{u}_1, \mathbf{v}_2, \cdots, \mathbf{v}_n\}$ by Theorem 4 §5.1 (verify). Now use this to write $\mathbf{u}_2 = t_1\mathbf{u}_1 + r_2\mathbf{v}_2 + \cdots + r_n\mathbf{v}_n$. Then some $r_i \neq 0$ because $\{\mathbf{u}_1, \mathbf{u}_2\}$ is independent (verify), say $r_2 \neq 0$ after possibly relabeling the $\mathbf{v}_i$. Then $V = span\{\mathbf{u}_1, \mathbf{u}_2, \mathbf{v}_3, \cdots, \mathbf{v}_n\}$, again using Theorem 4 §5.1. Since we are assuming that $m > n$, this procedure continues until all the vectors $\mathbf{v}_1, \mathbf{v}_2, \cdots, \mathbf{v}_n$ are replaced by $\mathbf{u}_1, \mathbf{u}_2, \cdots, \mathbf{u}_n$ and we have $V = span\{\mathbf{u}_1, \mathbf{u}_2, \cdots, \mathbf{u}_n\}$. But then $\mathbf{u}_{n+1}$ is a linear combination of $\mathbf{u}_1, \mathbf{u}_2, \cdots, \mathbf{u}_n$, contrary to the independence of the $\mathbf{u}_i$. This contradiction completes the proof.[7]
□

As for $\mathbb{R}^n$, a finite set of vectors $\{\mathbf{b}_1, \mathbf{b}_2, \cdots, \mathbf{b}_n\}$ in a vector space V is called a **basis** of V if it is independent and $V = span\{\mathbf{b}_1, \mathbf{b}_2, \cdots, \mathbf{b}_n\}$. Then we have immediately

Theorem 2. Invariance Theorem. *If $\{\mathbf{b}_1, \mathbf{b}_2, \cdots, \mathbf{b}_n\}$ and $\{\mathbf{d}_1, \mathbf{d}_2, \cdots, \mathbf{d}_m\}$ are two bases of a vector space V, then $n = m$.*

Proof. We have $m \leq n$ by the Fundamental Theorem because $V = span\{\mathbf{b}_1, \mathbf{b}_2, \cdots, \mathbf{b}_n\}$ and $\{\mathbf{d}_1, \mathbf{d}_2, \cdots, \mathbf{d}_m\}$ is independent. Now interchange the roles of the $\mathbf{b}_i$ and $\mathbf{d}_j$ to get $n \leq m$. □

With this we can define the notion of dimension in *any* vector space that has a basis. In fact, we define the **dimension** of V to be the number of vectors in any basis of V, and denote this number by $dim V$. The Invariance Theorem shows that this is well defined in the sense that we get the same dimension no matter which basis we choose. Of course this agrees with the dimension we used in Chapter 4 for subspaces of $\mathbb{R}^n$. In particular, $dim(\mathbb{R}^n) = n$ as in Section 4.3 because the standard basis contains n vectors.

The zero space $\{\mathbf{0}\}$ deserves special mention. We define its dimension to be zero:
$$dim\{\mathbf{0}\} = 0.$$
This is equivalent to regarding $\{\mathbf{0}\}$ as having an empty basis, and so is consistent with the fact (Example 6) that the zero vector cannot belong to any independent set. Hence the statement

$$dim V = n \quad \text{if and only if} \quad V \text{ has a basis of } n \text{ vectors}$$

is valid even if $n = 0$.

Example 7. $dim(\mathbb{P}_n) = n + 1$ because $\mathcal{B} = \{1, x, x^2, \cdots, x^n\}$ is a basis, called the **standard basis** of $\mathbb{P}_n$. Indeed, $\mathbb{P}_n = span\,\mathcal{B}$ by Example 15 §5.1, and $r_0 +$

---
[7]In addition to showing that $m \leq n$, this argument proves that m of the (spanning) vectors $\mathbf{v}_i$ can be replaced by the (independent) vectors $\mathbf{u}_1, \mathbf{u}_2, \cdots, \mathbf{u}_m$. In this form the result is called the **Steinitz Exchange Lemma**.

$r_1 x + \cdots + r_n x^n = 0$ implies each $r_i = 0$ by equating corresponding coefficients (see Example 3 §5.1).

Example 8. $dim(\mathbb{M}_{m,n}) = mn$ because the set $\mathcal{B}$ of mn matrix units in $\mathbb{M}_{m,n}$ is a basis (it spans $\mathbb{M}_{m,n}$ by Example 16 §5.1, and the reader can verify that it is independent). We call $\mathcal{B}$ the **standard basis** of $\mathbb{M}_{m,n}$.

Example 9. The set $\mathbb{C}$ of complex numbers is a vector space using complex addition and multiplication by a real number. The set $\{1, i\}$ is a basis. Indeed, $\mathbb{C} = span\{1, i\}$ because every complex number z has the form $z = a + bi$ where a and b are real, and $\{1, i\}$ is independent because the only way $a + bi = 0$ can happen is if $a = 0$ and $b = 0$ (see Section 2.5). Hence $dim(\mathbb{C}) = 2$.

The next example illustrates the power of the Fundamental Theorem.

Example 10. *If A is any $n \times n$ matrix, show that there is a nonzero polynomial $f(x)$ of degree at most n^2 such that $f(A) = 0$.*

Solution. The $n^2 + 1$ matrices $I, A, A^2, \cdots, A^{n^2}$ cannot be independent as vectors in $\mathbb{M}_{n,n}$ by the Fundamental Theorem because $dim(\mathbb{M}_{n,n}) = n^2$. Hence a nontrivial linear combination vanishes; that is there exist scalars $r_0, r_1, r_2, \cdots, r_{n^2}$, not all zero such that $r_0 I + r_1 A + r_2 A^2 + \cdots + r_{n^2} A^{n^2} = 0$. So take $f(x) = r_0 + r_1 x + r_2 x^2 + \cdots + r_{n^2} x^{n^2}$. □

In fact we can do better. The characteristic polynomial $c_A(x)$ has degree n and satisfies $c_A(A) = 0$. This is the Cayley-Hamilton Theorem, and we prove it in Section 6. However, it is remarkable that we can deduce that $f(A) = 0$ for *some* nonzero polynomial $f(x)$ using only the Fundamental Theorem.

Lest we begin thinking that *every* vector space has a basis, we have

Example 11. *Show that the vector space $\mathbb{P}$ does not have a basis.*

Solution 1. If $\{p_1(x), p_2(x), \cdots, p_n(x)\}$ is a basis of $\mathbb{P}$, then let x^m be the highest power of x appearing in any of the polynomials $p_i(x)$. Then it is clear that every $p_i(x)$ is in $\mathbb{P}_m$, and hence that $\mathbb{P} \subseteq \mathbb{P}_m$. But this is impossible because x^{m+1} is in $\mathbb{P}$ but not in $\mathbb{P}_m$.

Solution 2. If $dim \mathbb{P} = n$ then $\mathbb{P}$ is spanned by n vectors, so the existence of $n + 1$ independent vectors $\{1, x, x^2, \cdots, x^n\}$ in $\mathbb{P}$ contradicts the Fundamental Theorem. □

We say that a vector space V is **infinite dimensional** if it does not have a (finite) basis. Hence Example 11 shows that $\mathbb{P}$ is infinite dimensional. There is

5.2. INDEPENDENCE AND DIMENSION

a way [8] to define what we mean by an infinite basis of a vector space (the set $\{1, x, x^2, x^3, \cdots, x^n, \cdots\}$ qualifies in $\mathbb{P}$) and a version of the Invariance Theorem holds. But that is beyond the scope of this book. Instead, we focus on the vector spaces which do have a finite basis. In fact, we can weaken this requirement as follows.

A vector space V is called **finite dimensional** if it has a finite spanning set, that is if
$$V = span\{\mathbf{v}_1, \mathbf{v}_2, \cdots, \mathbf{v}_n\}$$
for some finite set $\{\mathbf{v}_1, \mathbf{v}_2, \cdots, \mathbf{v}_n\}$ of vectors in V. We are going to show that such a space V *has* a basis (and so $dim V \leq n$ by the Fundamental Theorem). To do this, we need two lemmas that were proved for $\mathbb{R}^n$ in Sections 4.2 and 4.3. The first of these shows how to enlarge a known independent set, and the proof is identical to that of Lemma 1 §4.3.

Lemma 1. Independent Lemma. *Suppose that $\{\mathbf{v}_1, \mathbf{v}_2, \cdots, \mathbf{v}_n\}$ is a linearly independent set of vectors in a vector space V. If $\mathbf{v}$ is any vector in V that is not in $span\{\mathbf{v}_1, \mathbf{v}_2, \cdots, \mathbf{v}_n\}$, then $\{\mathbf{v}, \mathbf{v}_1, \mathbf{v}_2, \cdots, \mathbf{v}_n\}$ is also linearly independent.*

As in Section 4.2, call a set of vectors **linearly dependent** (or simply **dependent**) if it is not linearly independent, that is if some nontrivial linear combination vanishes. Then the proof of Theorem 3 §4.2 goes through as written to prove

Lemma 2. Dependent Lemma. *A set $\{\mathbf{v}_1, \mathbf{v}_2, \cdots, \mathbf{v}_n\}$ of vectors in a vector space V is linearly dependent if and only if one of the $\mathbf{v}_i$ is a linear combination of the rest.*

These lemmas enable us to prove the following theorem which reveals two important ways to show that bases exist in a finite dimensional vector space. Both are used frequently.

Theorem 3. *Let $V \neq \{\mathbf{0}\}$ be a finite dimensional vector space, say V is spanned by n vectors.*

(1) *V has a basis and $dim V \leq n$.*

(2) *Every linearly independent subset of V is part of a basis.*

(3) *Every finite spanning set of V contains a basis.*

Proof. We first prove (2), then (1) and (3).

---

[8] If V is a vector space, a (possibly infinite) set $\mathcal{B}$ of vectors in V is called **independent** if every finite subset of $\mathcal{B}$ is linearly independent, and $\mathcal{B}$ is said to **span** V if every vector in V is a linear combination of (finitely many) vectors in $\mathcal{B}$. The set $\mathcal{B}$ is called a **basis** of V if it is both independent and spans V.

(2). Let $\{\mathbf{v}_1, \mathbf{v}_2, \cdots, \mathbf{v}_k\}$ be independent[9] in V. If $span\{\mathbf{v}_1, \mathbf{v}_2, \cdots, \mathbf{v}_k\} = V$ we are done. Otherwise, choose $\mathbf{v}_{k+1}$ in V which is not in $span\{\mathbf{v}_1, \mathbf{v}_2, \cdots, \mathbf{v}_k\}$. Then $\{\mathbf{v}_1, \mathbf{v}_2, \cdots, \mathbf{v}_k, \mathbf{v}_{k+1}\}$ is independent by Lemma 1, and we are done if $span\{\mathbf{v}_1, \mathbf{v}_2, \cdots, \mathbf{v}_k, \mathbf{v}_{k+1}\} = V$. If not repeat the process to create larger and larger independent sets in V. Since V is finite dimensional, this cannot continue indefinitely by the Fundamental Theorem, so we must reach a basis of V at some stage.

(1). Since $V \neq \{\mathbf{0}\}$ let $\mathbf{v}_1$ be a nonzero vector in V. Then $\{\mathbf{v}_1\}$ is independent, so V has a basis by (2). Finally $dim V \leq n$ by the Fundamental Theorem because V is spanned by n vectors.

(3). Let $V = span\{\mathbf{v}_1, \mathbf{v}_2, \cdots, \mathbf{v}_n\}$. If $\{\mathbf{v}_1, \mathbf{v}_2, \cdots, \mathbf{v}_n\}$ is independent, we are done. Otherwise one of the $\mathbf{v}_i$ is in the span of the rest by Lemma 2; relabeling if necessary let $\mathbf{v}_1$ lie in $span\{\mathbf{v}_2, \cdots, \mathbf{v}_n\}$. Then $V = span\{\mathbf{v}_2, \cdots, \mathbf{v}_n\}$ by Theorem 4 §5.1, so we are done if $\{\mathbf{v}_2, \cdots, \mathbf{v}_n\}$ is independent. If not, repeat the process. If we encounter a basis at some stage, we are done. Otherwise we ultimately reach $V = span\{\mathbf{v}_n\} = \mathbb{R}\mathbf{v}_n$. But then $\{\mathbf{v}_n\}$ is the desired basis because $\mathbf{v}_n \neq \mathbf{0}$ (as $V \neq \{\mathbf{0}\}$). □

Theorem 3 shows that a vector space is finite dimensional if and only if it has a finite basis. It also gives the following useful test for when a subset is a basis.

Theorem 4. *Let V be a vector space with $dim V = n$. If $\mathcal{B}$ is a set of exactly n vectors of V, then*

$$\mathcal{B} \text{ is independent if and only if } \mathcal{B} \text{ spans } V.$$

In either case, $\mathcal{B}$ is a basis of V.

Proof. If $\mathcal{B}$ is independent but does not span V, then $\mathcal{B}$ is part if a basis of more than n vectors by Theorem 3, contrary to Theorem 2. Similarly, if $\mathcal{B}$ spans V but is not independent, then $\mathcal{B}$ contains a basis of fewer than n vectors by Theorem 3, again contrary to Theorem 2. □

Theorem 4 is a very useful "labor saving device" for a vector space V of dimension n, since it eliminates the need to verify one or the other of independence and spanning when showing that a set $\mathcal{B}$ of n vectors is a basis of V. Here is an example illustrating how it is used.

Example 12. *Show that $\mathcal{B} = \{1, (x-a), (x-a)^2, \cdots, (x-a)^n\}$ is a basis of $\mathbb{P}_n$ for every scalar a.*

Solution. The set $\mathcal{B}$ is independent by Example 2 because it consists of polynomials of distinct degrees. Hence $\mathcal{B}$ is a basis of $\mathbb{P}_n$ by Theorem 4 because $dim(\mathbb{P}_n) = n + 1$. □

---

[9]Because V is finite dimensional, the independence forces it to be a *finite* subset by the Fundamental Theorem.

5.2. INDEPENDENCE AND DIMENSION

It follows from Example 12 that every polynomial $p(x)$ of degree at most n can be uniquely written in the form

$$p(x) = r_0 + r_1(x-a) + r_2(x-a)^2 + \cdots + r_n(x-a)^n$$

where the coefficients r_i are in $\mathbb{R}$. This can be proved other ways, but this argument illustrates once more the power of these methods.

We conclude with another "labor saving" result, this time for reducing the work in showing that a subspace of V is all of V.

Theorem 5. *Let U be a subspace of a vector space V, and assume $dim V - n$.*

(1) *U has a basis and $dim U \leq n$.*

(2) *If $dim U = n$ then $U = V$.*

(3) *Every basis of U is part of a basis of V.*

Proof. (1). If $U = \{\mathbf{0}\}$ then it has an empty basis and (1) holds because $dim U = 0 \leq n$. Otherwise, let $\mathbf{u}_1 \neq \mathbf{0}$ be a vector in U. If $span\{\mathbf{u}_1\} = U$ we are done. If not, the construction in the proof of (2) of Theorem 3 either gives a basis of U or creates arbitrarily large independent subsets of V. Hence (1) follows because the latter possibility contradicts the Fundamental Theorem.

(2). If $dim U = n$ then any basis $\{\mathbf{u}_1, \mathbf{u}_2, \cdots, \mathbf{u}_n\}$ of U is a basis of V by Theorem 4. But then $V = span\{\mathbf{u}_1, \mathbf{u}_2, \cdots, \mathbf{u}_n\} = U$, proving (2).

(3). This follows from Theorem 3. $\square$

Example 13. *If a is a real number, let $W = \{p(x) \text{ in } \mathbb{P}_n \mid p(a) = 0\}$ denote the set of all polynomials in $\mathbb{P}_n$ with a as a root. Show that $\{(x-a), (x-a)^2, \cdots, (x-a)^n\}$ is a basis of W.*

Solution. Observe first that $(x-a)^k$ is in W for each k, and that $\{(x-a), (x-a)^2, \cdots, (x-a)^n\}$ is linearly independent by Example 2. For convenience, write

$$U = span\{(x-a), (x-a)^2, \cdots, (x-a)^n\}.$$

Then $U \subseteq W \subseteq \mathbb{P}_n$, $dim U = n$ and $dim(\mathbb{P}_n) = n+1$. Hence $n \leq dim W \leq n+1$ by Theorem 5 so, since $dim W$ is an integer, either $dim W = n$ or $dim W = n+1$. If $dim W = n+1 = dim(\mathbb{P}_n)$ then $W = \mathbb{P}_n$ by Theorem 4, a contradiction (for example, 1 is in $\mathbb{P}_n$ but not in W). So $dim W = n = dim U$. But then the fact that U is a subspace of W gives $W = U$, again by Theorem 5. This is what we wanted. $\square$

Example 14. *(Requires calculus)* Let f' and $f'' = (f')'$ denote the first and second derivatives of a function f in $\mathbb{F}[\mathbb{R}]$. Then solutions f to the differential equation

$$f'' + \omega^2 f = 0 \text{ where } \omega \text{ is a real constant} \qquad (**)$$

are called **simple harmonic motions** because they describe oscillations like that of a pendulum or a weight on a spring (see Example 17 §5.1). The set H of all these solutions is a subspace of $\mathbb{F}[\mathbb{R}]$. In fact, if f and g are in H, the property $(f+g)'' = f'' + g''$ of differentiation gives

$$\begin{aligned}(f+g)'' + \omega(f+g) &= (f''+g'') + (\omega f + \omega g) \\ &= (f'' + \omega f) + (g'' + \omega g) \\ &= 0 + 0 = 0,\end{aligned}$$

so $f + g$ is in H. Similarly, rf is in H for all r in $\mathbb{R}$ because $(rf)'' = rf''$. If f_1 and f_2 are defined by $f_1(x) = sin(\omega x)$ and $f_2(x) = cos(\omega x)$, it is easy to see that f_1 and f_2 both lie in H, and that $\{f_1, f_2\}$ is independent (verify). Remarkably, it is a theorem of calculus that H is two dimensional, so $\{f_1, f_2\}$ is actually a basis of H by Theorem 4. Hence every solution f to (**) is given by

$$f(x) = a\,sin(\omega x) + b\,cos(\omega x) \text{ for all } x$$

where a and b are uniquely determined constants. □

Exercises

1. In each case either show that the statement is true or give an example showing that it is false.

    (a) If $\{\mathbf{u}, \mathbf{v}\}$ and $\{\mathbf{w}, \mathbf{z}\}$ are both independent, then $\{\mathbf{u}, \mathbf{v}, \mathbf{w}, \mathbf{z}\}$ is independent.

    (b) If one of $\{\mathbf{u}, \mathbf{v}\}$ and $\{\mathbf{w}, \mathbf{z}\}$ is dependent, then $\{\mathbf{u}, \mathbf{v}, \mathbf{w}, \mathbf{z}\}$ is dependent.

    (c) $\{\mathbf{u}, \mathbf{u}\}$ is never independent.

    (d) If $\{\mathbf{u}, \mathbf{v}, \mathbf{w}\}$ is independent then $a\mathbf{u} + b\mathbf{v} + c\mathbf{w} = \mathbf{0}$ for some scalars a, b and c, not all zero.

    (e) If $\{\mathbf{u}, \mathbf{v}\}$ is independent, so is $\{\mathbf{u}, \mathbf{u} + \mathbf{v}\}$.

    (f) If $\{\mathbf{u}, \mathbf{v}\}$ is independent, so is $\{\mathbf{u}, \mathbf{v}, \mathbf{u} + \mathbf{v}\}$.

    (g) If $\{\mathbf{u}, \mathbf{v}, \mathbf{w}\}$ is independent, so is $\{\mathbf{u}, \mathbf{v}\}$.

    (h) If $\{\mathbf{u}, \mathbf{v}, \mathbf{w}\}$ is dependent, so is $\{\mathbf{u}, \mathbf{v}\}$.

    (i) $\mathbb{P}_2$ has a basis of polynomials $p(x)$ such that $p(0) = 0$.

    (j) $\mathbb{P}_2$ has a basis of polynomials $p(x)$ such that $p(0) = 1$.

    (k) Every basis of $\mathbb{M}_{2,2}$ contains an invertible matrix.

    (l) No basis of $\mathbb{M}_{2,2}$ contains a matrix A such that $A^2 = 0$.

    (m) No basis of $\mathbb{M}_{2,2}$ consists of matrices A such that $A\begin{bmatrix}1 & 0\end{bmatrix}^T = 0$.

5.2. INDEPENDENCE AND DIMENSION

2. In each case determine if the set of vectors is independent in V.

   (a) $V = \mathbb{P}_2$; $\{1+x, 2-x^2, x^2+2x\}$

   (b) $V = \mathbb{P}_2$; $\{2+x, 1+x^2, 1+x+x^2\}$

   (c) $V = \mathbb{M}_{2,2}$; $\{\begin{bmatrix} 1 & 0 \\ 0 & 0 \end{bmatrix}, \begin{bmatrix} 0 & 1 \\ 1 & 0 \end{bmatrix}, \begin{bmatrix} 0 & 1 \\ 1 & 1 \end{bmatrix}, \begin{bmatrix} 1 & 0 \\ 0 & -1 \end{bmatrix}\}$

   (d) $V = \mathbb{M}_{2,2}$; $\{\begin{bmatrix} 1 & 1 \\ 0 & 0 \end{bmatrix}, \begin{bmatrix} 1 & 0 \\ 1 & 0 \end{bmatrix}, \begin{bmatrix} 0 & 0 \\ 1 & 1 \end{bmatrix}, \begin{bmatrix} 0 & 1 \\ 0 & 1 \end{bmatrix}\}$

   (e) $V = \mathbb{F}[0,1]$; $\{\frac{1}{x^2-4}, \frac{1}{x^2+3x+2}, \frac{1}{x^2-x-2}\}$

   (f) $V = \mathbb{F}[1,2]$; $\{\frac{1}{x}, \frac{1}{x^2}, \frac{1}{x^3}\}$

   (g) $V = \mathbb{F}[0, 2\pi]$; $\{1, \cos^2 x, \sin^2 x\}$

   (h) $V = \mathbb{F}[0, 2\pi]$; $\{x, \cos^2 x, \sin^2 x\}$

3. In each case find a basis and calculate the dimension of the subspace U of V.

   (a) $V = \mathbb{P}_2$; $U = span\{1 - 2x, 3x - x^2, 3 - 2x^2\}$

   (b) $V = \mathbb{P}_2$; $U = \{p(x) \mid p(1) = 0\}$

   (c) $V = \mathbb{P}_2$; $U = \{p(x) \mid p(x) = p(-x)\}$

   (d) $V = \mathbb{P}_2$; $U = span\{x - 2, x^2 + 2x, 4 - x^2\}$

   (e) $V = \mathbb{M}_{2,2}$; $U = span\left\{\begin{bmatrix} 1 & 2 \\ -1 & 0 \end{bmatrix}, \begin{bmatrix} 2 & 0 \\ 1 & 3 \end{bmatrix}, \begin{bmatrix} 1 & -6 \\ 5 & 6 \end{bmatrix}\right\}$

   (f) $V = \mathbb{M}_{2,2}$; $U = \{A \mid AB = BA\}$ where $B = \begin{bmatrix} 1 & 1 \\ 0 & 0 \end{bmatrix}$

   (g) $V = \mathbb{M}_{2,2}$; $U = \{A \mid AB = CA\}$, $B = \begin{bmatrix} 1 & 0 \\ 0 & 0 \end{bmatrix}$, $C = \begin{bmatrix} 0 & 1 \\ 1 & 0 \end{bmatrix}$

   (h) $V = \mathbb{M}_{2,2}$; $U = \{A \mid \text{The columns of } A \text{ have equal sums}\}$

   (i) $V = \mathbb{M}_{3,3}$; $U = \{A \mid A^T = -A\}$

   (j) $V = \mathbb{R}^4$; $U = \{X \mid X^T A = 0\}$ where $A = \begin{bmatrix} 1 & -1 & 0 & 2 \\ -2 & 3 & -1 & 4 \end{bmatrix}^T$

   (k) $V = \mathbb{R}^4$; $U = \{[a\ b\ c\ d]^T \mid a + b + c + d = 0\}$

   (m) $V = span\{\mathbf{u}, \mathbf{v}, \mathbf{w}, \mathbf{z}\}$ where $\{\mathbf{u}, \mathbf{v}, \mathbf{w}, \mathbf{z}\}$ is independent;
   $U = span\{\mathbf{u}+\mathbf{v}, \mathbf{v}+\mathbf{w}, \mathbf{w}+\mathbf{z}, \mathbf{z}+\mathbf{u}\}$

4. Find a basis of $\mathbb{P}_3$ consisting of polynomials whose coefficients sum to 2. What if the coefficients sum to 0?

5. (a) Show that any (nonempty) subset of an independent set is independent.

   (b) Show that any set of vectors that contains a dependent set is itself dependent.

6. Consider the space $\mathbb{C}$ of complex numbers, and let z denote an element of $\mathbb{C}$.

   (a) If z is not real, show that $\{z, z^2\}$ is a basis of $\mathbb{C}$.

   (b) If z is not real or pure imaginary, show that $\{z, \bar{z}\}$ is a basis of $\mathbb{C}$.

7. Show that if $\mathcal{B} = \{p_0(x), p_1(x), p_2(x), \cdots, p_n(x)\}$ are polynomials in $\mathbb{P}_n$ of distinct degrees, then $\mathcal{B}$ is a basis of $\mathbb{P}_n$.

8. If V is finite dimensional, show that every (possibly infinite) spanning set $\mathcal{S}$ contains a basis. [*Hint*: Show first that $\mathcal{S}$ contains a finite spanning set.]

9. Suppose that A is an $n \times n$ matrix such that $A^3 = 0$ but $A^2 \neq 0$. Show that $\{I, A, A^2\}$ is independent in $\mathbb{M}_{n,n}$. Generalize.

10. Let A and B be nonzero matrices in $\mathbb{M}_{n,n}$. If A is symmetric and B is skew-symmetric (that is $B^T = -B$), show that $\{A, B\}$ is independent.

11. Let f and g be functions in $\mathbb{F}[0, 1]$ such that $f(0) = 0$ and $g(0) = 1$, and $f(1) = 1$ and $g(1) = 0$. Show that $\{f, g\}$ is independent.

12. Let $\{A_1, A_2, \cdots, A_k\}$ be a set of matrices in $\mathbb{M}_{m,n}$, and let P and Q be invertible matrices of sizes $m \times m$ and $n \times n$, respectively.

    (a) If $\{A_1, A_2, \cdots, A_k\}$ is independent, show that $\{PA_1Q, PA_2Q, \cdots, PA_kQ\}$ is independent.

    (b) If $\{A_1, A_2, \cdots, A_k\}$ spans $\mathbb{M}_{m,n}$, show that $\{PA_1Q, PA_2Q, \cdots, PA_kQ\}$ spans $\mathbb{M}_{m,n}$.

13. Let A be an $m \times n$ matrix, and let $\{B_1, B_2, \cdots, B_k\}$ be independent columns in $\mathbb{R}^n$. If $X_1, X_2, \cdots, X_k$ are solutions to the equations $AX_i = B_i$ for each $i = 1, 2, \cdots, k$, show that $\{X_1, X_2, \cdots, X_k\}$ is independent in $\mathbb{R}^n$.

14. Let $p(x)$ and $q(x)$ be polynomials of degree at least 1. If $\{p(x), q(x)\}$ is independent, show that $\{p(x), q(x), p(x)q(x)\}$ is also independent.

15. Call a function f in $\mathbb{F}[a, b]$ a **constant function** if there exists a constant c such that $f(x) = c$ for all x. Show that the set of all constant functions is a subspace of $\mathbb{F}[a, b]$ of dimension 1.

16. Show that $\{a + bx, a_1 + b_1 x\}$ is a basis of P_1 if and only if $\{[a\ b]^T, [a_1\ b_1]^T\}$ is a basis of $\mathbb{R}^2$.

17. Find the dimension of the subspace $U = span\{1,\ cos^2\theta,\ cos(2\theta)\}$ of $\mathbb{F}[-\pi, \pi]$. Support your answer.

18. Find a basis of $\mathbb{M}_{2,2}$ consisting of matrices A with the property that $A^2 = A$.

5.2. INDEPENDENCE AND DIMENSION

19. If A is a 2×2 matrix and $U = \{B \text{ in } \mathbb{M}_{2,2} \mid AB = BA\}$, show that $dim U \geq 2$. [Hint: I and A are in U. Consider separately the case that $A = aI$ for some a in $\mathbb{R}$.]

20. A polynomial $p(x)$ is called **even** if $p(x) = p(-x)$, and $p(x)$ is called **odd** if $p(x) = -p(-x)$. Let E_n and O_n denote the sets of even and odd polynomials in $\mathbb{P}_n$.

    (a) Show E_n is a subspace of $\mathbb{P}_n$, and find its dimension.

    (b) Show O_n is a subspace of $\mathbb{P}_n$, and find its dimension.

21. If $\mathbf{v} \neq \mathbf{0}$ in V, show that the only subspaces of $\mathbb{R}\mathbf{v}$ are $\{\mathbf{0}\}$ and $\mathbb{R}\mathbf{v}$.

22. If $dim V = 2$ and U is any subspace of V other than $\{\mathbf{0}\}$ and V, show that $U = \mathbb{R}\mathbf{v}$ for some nonzero vector $\mathbf{v}$ in V. What does this say about the plane $\mathbb{R}^2$?

23. If U is a subspace of $\mathbb{P}_1$, show that $U = \{\mathbf{0}\}$, $U = \mathbb{P}_1$, $U = \mathbb{R}$ or $U = \mathbb{R}(a+x)$ for some a in $\mathbb{R}$.

24. If $\{\mathbf{u}, \mathbf{v}, \mathbf{w}, \mathbf{z}\}$ is independent, which of the following are independent? Support your answer.

    (a) $\{\mathbf{u} - \mathbf{v}, \mathbf{v} - \mathbf{w}, \mathbf{w} - \mathbf{u}\}$

    (b) $\{\mathbf{u} + \mathbf{v}, \mathbf{v} + \mathbf{w}, \mathbf{w} + \mathbf{u}\}$

    (c) $\{\mathbf{u} - \mathbf{v}, \mathbf{v} - \mathbf{w}, \mathbf{w} - \mathbf{z}, \mathbf{z} - \mathbf{u}\}$

    (d) $\{\mathbf{u} + \mathbf{v}, \mathbf{v} + \mathbf{w}, \mathbf{w} + \mathbf{u}, \mathbf{z} + \mathbf{u}\}$

25. Show that $\{\mathbf{u}, \mathbf{v}\}$ is dependent if and only if one of $\mathbf{u}$ and $\mathbf{v}$ is a scalar multiple of the other.

26. If $\mathbf{v} \neq \mathbf{0}$ is a vector in a vector space V, show that there exists a basis of V containing $\mathbf{v}$.

27. If $\{\mathbf{v}_1, \mathbf{v}_2, \mathbf{v}_3, \mathbf{v}_4\}$ is independent in V, show that $\{\mathbf{v}_1, \mathbf{v}_2 + a\mathbf{v}_1, \mathbf{v}_3 + c\mathbf{v}_2 + d\mathbf{v}_1, \mathbf{v}_4 + e\mathbf{v}_3 + f\mathbf{v}_2 + g\mathbf{v}_1\}$ is also independent. Generalize.

28. If $\{\mathbf{v}_1, \mathbf{v}_2, \cdots, \mathbf{v}_k\}$ is independent in V, and if $\mathbf{w}$ is not in $span\{\mathbf{v}_1, \mathbf{v}_2, \cdots, \mathbf{v}_k\}$, show that the set $\{\mathbf{w} + \mathbf{v}_1, \mathbf{w} + \mathbf{v}_2, \cdots, \mathbf{w} + \mathbf{v}_k\}$ is also independent.

29. Suppose that $\{\mathbf{v}_1, \mathbf{v}_2, \cdots, \mathbf{v}_k\}$ is a minimal spanning set for a vector space V, that is no spanning set of V contains fewer than k vectors. Show that $\{\mathbf{v}_1, \mathbf{v}_2, \cdots, \mathbf{v}_k\}$ is a basis of V.

30. Suppose that $\{\mathbf{v}_1, \mathbf{v}_2, \cdots, \mathbf{v}_k\}$ is a maximal independent set in a vector space V, that is no independent set in V contains more than k vectors. Show that $\{\mathbf{v}_1, \mathbf{v}_2, \cdots, \mathbf{v}_k\}$ is a basis of V.

31. Show that $\mathbb{F}[0,1]$ is infinite dimensional.

32. Let $U \subseteq W$ be subspaces of V, and suppose that $dim U = k$ and $dim W = m$ where $k < m$. If l is any integer such that $k \leq l \leq m$, show that there is a subspace X such that $U \subseteq X \subseteq W$ and $dim X = l$. [Hint: Theorem 3.]

33. Let $\mathbb{S}$ denote the set of all infinite sequences $(a_0, a_1, a_2, \cdots)$ of real numbers a_i. Define equality, addition and scalar multiplication on $\mathbb{S}$ as follows:
    $(a_0, a_1, a_2, \cdots) = (b_0, b_1, b_2, \cdots)$ if and only if $a_i = b_i$ for all i.
    $(a_0, a_1, a_2, \cdots) + (b_0, b_1, b_2, \cdots) = (a_0 + b_0, a_1 + b_1, a_2 + b_2, \cdots)$.
    $r(a_0, a_1, a_2, \cdots) = (ra_0, ra_1, ra_2, \cdots)$ for all real numbers r.

    (a) Show that $\mathbb{S}$ is a vector space and that $\mathbb{S}$ is infinite dimensional.

    (b) Show that $P = \{(a_0, a_1, a_2, \cdots) \mid a_i = 0$ for all but a finitely many $i\}$ is a subspace of $\mathbb{S}$. Is $dim P$ infinite? Support your answer.

    (c) Show that the set $C = \{(a, a, a, \cdots) \mid a$ in $\mathbb{R}\}$ is a subspace of $\mathbb{S}$. What is $dim C$?

    (d) Show that the convergent sequences are an infinite dimensional subspace of $\mathbb{S}$ (requires calculus).

34. Let $\mathbb{S}$ be the space of sequences in the preceding exercise, and let a and b be real numbers.

    (a) Show that $U = \{(x_0, x_1, x_2, \cdots) \mid x_{k+2} = ax_k + bx_{k+1}$ for all $k \geq 0\}$ is a subspace of $\mathbb{S}$.

    (b) Show that the space in (a) has dimension 2. [*Hint*: $\{\mathbf{e}, \mathbf{f}\}$ is a basis where $\mathbf{e}$ is the sequence starting with $x_0 = 1$ and $x_1 = 0$, and $\mathbf{f}$ is the sequence starting with $x_0 = 0$ and $x_1 = 1$. If $\mathbf{v} = (v_0, v_1, \cdots)$ is in $\mathbb{S}$, use induction to show that $\mathbf{v} = v_0\mathbf{e} + v_1\mathbf{f}$.]

35. If U and W are subspaces of V, define their **intersection** $U \cap W$ and **sum** $U + W$ by

    $$U \cap W = \{\mathbf{v} \text{ in } V \mid \mathbf{v} \text{ is in both } U \text{ and } W\}$$

    and

    $$U + W = \{\mathbf{u} + \mathbf{w} \mid \mathbf{u} \text{ is in } U \text{ and } \mathbf{w} \text{ is in } W\}$$

    (a) Show that $U \cap W$ and $U + W$ are subspaces of V.

    (b) If U and W have bases $\{\mathbf{u}_1, \mathbf{u}_2\}$ and $\{\mathbf{v}_1, \mathbf{v}_2, \mathbf{v}_3\}$ respectively, and if $U \cap W = \{\mathbf{0}\}$, show that $\{\mathbf{u}_1, \mathbf{u}_2, \mathbf{v}_1, \mathbf{v}_2, \mathbf{v}_3\}$ is independent.

    (c) If U and W are both finite dimensional, show that $U + W$ is also finite dimensional (even though V may not be finite dimensional). [*Hint*: If $\mathcal{C}$ and $\mathcal{D}$ are (finite) bases of U and W respectively show that $\mathcal{C} \cup \mathcal{D} = \{\mathbf{v}$ in $V \mid \mathbf{v}$ is in $\mathcal{C}$ or $\mathbf{v}$ is in $\mathcal{D}\}$ spans $U + W$.]

    (d) If V is finite dimensional, show that $dim(U + W) = dim U + dim W - dim(U \cap W)$. [*Hint*: If $\mathcal{B}$ is a basis of $U \cap W$, let (by Theorem 3) $\mathcal{C}$ and $\mathcal{D}$ be bases of U and W respectively each containing $\mathcal{B}$. Show that $\mathcal{C} \cup \mathcal{D}$ is a basis of $U + W$.]

5.3 LINEAR TRANSFORMATIONS

In Section 3.5 we studied transformations of $\mathbb{R}^2$ such as rotations, reflections and projections, and we found that they are all given by multiplication by certain 2×2 matrices. The key idea that connects a geometrical transformation $T : \mathbb{R}^2 \to \mathbb{R}^2$ like rotation with a matrix is that T is *linear*. That is, T preserves addition and scalar multiplication in the sense that

$$T(\vec{v} + \vec{w}) = T(\vec{v}) + T(\vec{w}) \quad \text{and} \quad T(r\,\vec{v}) = r\,T(\vec{v}) \qquad (*)$$

for all vectors $\vec{v}$ and $\vec{w}$ in $\mathbb{R}^2$ and all scalars r. In Section 4.10 we applied this to linear transformations $T : \mathbb{R}^n \to \mathbb{R}^m$ and found that these too are given by matrix multiplication. This connection between the abstract concept of linearity and the more concrete, computational notion of matrix multiplication is very useful (one example is our analysis of the isometries of $\mathbb{R}^3$).

The reason for recalling all this here is that the conditions in (*) make sense much more generally, and lead to the general idea of a linear transformation $T : V \to W$ where V and W are arbitrary vector spaces. This is the main theme of this section (and the next). It reveals a new perspective on matrices and provides a whole new way to view similarity. It also increases the applicability of linear algebra since, once one knows what they are, linear transformations appear everywhere.

5.3.1 Linear Transformations

If V and W are two vector spaces, a transformation[10] $T : V \to W$ is a rule that associates with every vector **v** in V a uniquely determined vector $T(\mathbf{v})$ in W. A transformation $T : V \to W$ is called a **linear transformation** if it preserves addition and scalar multiplication in the sense that it has the following properties:

T1 $T(\mathbf{v} + \mathbf{v}_1) = T(\mathbf{v}) + T(\mathbf{v}_1)$ *for all vectors* **v** *and* $\mathbf{v}_1$ *in* V.

T2 $T(r\mathbf{v}_1) = r\,T(\mathbf{v})$ *for all vectors* **v** *in* V *and all scalars* r.

A linear transformation $T : V \to V$ is called a **linear operator** on the vector space V.

The action of a linear transformation $T : V \to W$ can be visualized as in the diagram. The vector space V is called the **domain** of T, and W is the **codomain** of T. If **v** is a vector in V, the vector $T(\mathbf{v})$ is called the **image** of **v** under T.

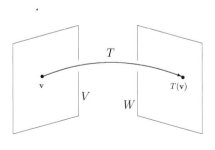

---

[10]Another word for transformation is "function", but the word "transformation" is well established in linear algebra.

Example 1. Every $m \times n$ matrix A induces a linear transformation $T : \mathbb{R}^n \to \mathbb{R}^m$ given by
$$T(X) = AX \text{ for all } X \text{ in } \mathbb{R}^n.$$
In this case Axioms T1 and T2 are simple properties of matrix multiplication. Moreover, *every* linear transformation $\mathbb{R}^n \to \mathbb{R}^m$ is induced by some $m \times n$ matrix A (Theorem 2 §4.9). □

Example 2. Transposition is a linear transformation. That is, the map
$$T : \mathbb{M}_{m,n} \to \mathbb{M}_{n,m} \text{ given by } T(A) = A^T \text{ for all } A \text{ in } \mathbb{M}_{m,n}$$
is T is a linear transformation. Now Axioms T1 and T2 are basic properties of transposition:
$$T(A + B) = (A + B)^T = A^T + B^T = T(A) + T(B)$$
and
$$T(rA) = (rA)^T = r A^T = rT(A)$$
for all A and B in $\mathbb{M}_{m,n}$ and all r in $\mathbb{R}$. □

Example 3. If a is a fixed real number, define a transformation $E_a : \mathbb{P}_n \to \mathbb{R}$ by $E_a[p(x)] = p(a)$ for all polynomials $p(x)$ in $\mathbb{P}_n$. Then E_a is linear. Indeed, Axioms T1 and T2 are routine observations:
$$E_a[p(x) + q(x)] = p(a) + q(a) = E_a[p(x)] + E_a[q(x)]$$
and
$$E_a[r\, p(x)] = r\, p(a) = r\, E_a[p(x)]$$
for all $p(x)$ and $q(x)$ in $\mathbb{P}_n$ and all r in $\mathbb{R}$. The transformation E_a is called **evaluation** at a. □

The next example is important in analysis.

Example 4. Differentiation is a linear transformation. Let V denote the subspace of $\mathbb{F}[\mathbb{R}]$ consisting of all differentiable real valued functions defined on $\mathbb{R}$ (see Example 12 §5.1). If we denote the derivative of a function f by f', define
$$T : V \to \mathbb{F}[\mathbb{R}] \text{ by } T(f) = f'$$
Then T is a linear transformation. In fact Axioms T1 and T2 read $(f+g)' = f' + g'$ and $(rf)' = rf'$ for all r in $\mathbb{R}$, and these are basic facts about differentiation. □

Example 5. If V and W are vector spaces, the following are linear transformations.

Identity operator on V

5.3. LINEAR TRANSFORMATIONS

$$1_V : V \to V \text{ given by } 1_V(\mathbf{v}) = \mathbf{v} \text{ for all } \mathbf{v} \text{ in } V.$$

Zero transformation $V \to W$

$$0 : V \to W \text{ given by } 0(\mathbf{v}) = \mathbf{0} \text{ for all } \mathbf{v} \text{ in } V.$$

For a in $\mathbb{R}$, the **scalar operator on** V

$$a : V \to V \text{ given by } a(\mathbf{v}) = a\mathbf{v} \text{ for all } \mathbf{v} \text{ in } V.$$

In addition to preserving sums and scalar products, linear transformations preserve the zero vector, negatives and linear combinations. These useful properties are recorded in the next result.

Theorem 1. *Let $T : V \to W$ be a linear transformation. Then:*

(1) $T(\mathbf{0}) = \mathbf{0}$.[11]

(2) $T(-\mathbf{v}) = -T(\mathbf{v})$ *for all* $\mathbf{v}$ *in* V.

(3) $T(r_1\mathbf{v}_1 + r_2\mathbf{v}_2 + \cdots + r_k\mathbf{v}_k) = r_1 T(\mathbf{v}_1) + r_2 T(\mathbf{v}_2) + \cdots + r_k T(\mathbf{v}_k)$ *for all r_i in $\mathbb{R}$ and all $\mathbf{v}_i$ in V.*

Proof. (1). $T(\mathbf{0}) = T(0\mathbf{v}) = 0\, T(\mathbf{v}) = \mathbf{0}$ for any $\mathbf{v}$ in V by Axiom T2.

(2). $T(-\mathbf{v}) = T((-1)\mathbf{v}) = (-1)T(\mathbf{v}) = -T(\mathbf{v})$, again by Axiom T2.

(3). If $k = 1$ this is Axiom T2. If $k > 1$ we use induction on k. If it holds for k vectors then

$$\begin{aligned} T(r_1\mathbf{v}_1 + r_2\mathbf{v}_2 + \cdots &+ r_k\mathbf{v}_k + r_{k+1}\mathbf{v}_{k+1}) \\ &= T(r_1\mathbf{v}_1 + r_2\mathbf{v}_2 + \cdots + r_k\mathbf{v}_k) + T(r_{k+1}\mathbf{v}_{k+1}) \\ &= r_1 T(\mathbf{v}_1) + r_2 T(\mathbf{v}_2) + \cdots + r_k T(\mathbf{v}_k) + r_{k+1} T(\mathbf{v}_{k+1}) \end{aligned}$$

by the induction assumption and Axioms T1 and T2. Hence it holds for $k+1$ vectors. $\square$

The next two examples illustrate the importance of the fact that linear transformations preserve linear combinations (property (3) in Theorem 1).

Example 6. *Suppose that $T : V \to W$ is a linear transformation. If $T(\mathbf{v} + 2\mathbf{v}_1) = \mathbf{w}$ and $T(3\mathbf{v} - 5\mathbf{v}_1) = \mathbf{w}_1$, find $T(\mathbf{v})$ and $T(\mathbf{v}_1)$ in terms of $\mathbf{w}$ and $\mathbf{w}_1$.*

Solution. Because T is linear, the given equations imply that

$$\begin{aligned} T(\mathbf{v}) + 2\, T(\mathbf{v}_1) &= \mathbf{w} \\ 3\, T(\mathbf{v}) - 5\, T(\mathbf{v}_1) &= \mathbf{w}_1 \end{aligned}.$$

---

[11]Note that $\mathbf{0}$ is playing two roles here. On the left side it is the zero vector in V, and it is the zero vector in W on the right side. This is a common abuse of notation, and is harmless once everyone knows it is going on.

If we subtract 3 times the first equation from the second we get $-11T(\mathbf{v}_1) = \mathbf{w}_1 - 3\mathbf{w}$. Thus $T(\mathbf{v}_1) = \frac{3}{11}\mathbf{w} - \frac{1}{11}\mathbf{w}_1$. Substitution in the first equation gives $T(\mathbf{v}) = \frac{5}{11}\mathbf{w} + \frac{2}{11}\mathbf{w}_1$. □

Example 7. *Suppose that $T(3 - 5x) = 2$ and $T(1 - x + 2x^2) = 1 + x$ where $T : \mathbb{P}_2 \to \mathbb{P}_2$ is a linear transformation. Find $T(6 - 11x - 3x^2)$.*

Solution. This seems difficult until we realize that (using Theorem 1) we can find $T[p(x)]$ for any polynomial $p(x)$ that is a linear combination of $3 - 5x$ and $1 - x + 2x^2$. So we look for scalars r and s such that

$$6 - 11x - 3x^2 = r(3 - 5x) + s(1 - x + 2x^2).$$

Equating coefficients gives linear equations for r and s : $6 = 3r+s$, $-11 = -5r-s$ and $-3 = 2s$. These equations have the unique solution $r = \frac{5}{2}$ and $s = -\frac{3}{2}$. Then

$$\begin{aligned} T(6 - 11x - 3x^2) &= rT(3 - 5x) + sT(1 - x + 2x^2) \\ &= \tfrac{5}{2}T(3 - 5x) - \tfrac{3}{2}T(1 - x + 2x^2) \\ &= \tfrac{5}{2}2 - \tfrac{3}{2}(1 + x) \\ &= \tfrac{7}{2} - \tfrac{3}{2}x. \end{aligned}$$

This is what we wanted. □

If $T : V \to W$ is a linear transformation and $T(\mathbf{v}_1), T(\mathbf{v}_2), \cdots, T(\mathbf{v}_k)$, are known, property (3) in Theorem 1 shows that $T(r_1\mathbf{v}_1 + r_2\mathbf{v}_2 + \cdots + r_k\mathbf{v}_k)$ can be computed for any choice of the scalars r_i. In particular, if $V = span\{\mathbf{v}_1, \mathbf{v}_2, \cdots, \mathbf{v}_k\}$ then $T(\mathbf{v})$ is determined for every $\mathbf{v}$ in V by the choice of $T(\mathbf{v}_1)$, $T(\mathbf{v}_2), \cdots, T(\mathbf{v}_k)$ in W. The next theorem states this differently. Recall that two transformations $S : V \to W$ and $T : V \to W$ are **equal** (written $S = T$) if $S(\mathbf{v}) = T(\mathbf{v})$ for all $\mathbf{v}$ in V.

Theorem 2. *Let $S : V \to W$ and $T : V \to W$ be two linear transformations, and assume that $V = span\{\mathbf{v}_1, \mathbf{v}_2, \cdots, \mathbf{v}_k\}$. If $S(\mathbf{v}_i) = T(\mathbf{v}_i)$ for each i, then $S = T$.*

Proof. Given $\mathbf{v}$ in V, write $\mathbf{v} = r_1\mathbf{v}_1 + r_2\mathbf{v}_2 + \cdots + r_k\mathbf{v}_k$ where r_i is in $\mathbb{R}$ for each i. Since $S(\mathbf{v}_i) = T(\mathbf{v}_i)$ for each i, Theorem 1 gives

$$\begin{aligned} S(\mathbf{v}) &= r_1S(\mathbf{v}_1) + r_2S(\mathbf{v}_2) + \cdots + r_kS(\mathbf{v}_k) \\ &= r_1T(\mathbf{v}_1) + r_2T(\mathbf{v}_2) + \cdots + r_kT(\mathbf{v}_k) \\ &= T(\mathbf{v}). \end{aligned}$$

Since $\mathbf{v}$ was arbitrary in V, this shows that $S = T$. □

Thus a linear transformation $T : V \to W$ is determined by its effect on a spanning set of V. If the spanning set is actually a basis, we can say more. The following lemma will be needed.

5.3. LINEAR TRANSFORMATIONS

Lemma 1. *If $\{\mathbf{b}_1, \mathbf{b}_2, \cdots, \mathbf{b}_n\}$ is a basis of a vector space V, every vector $\mathbf{v}$ in V can be uniquely represented as a linear combination*

$$\mathbf{v} = r_1\mathbf{b}_1 + r_2\mathbf{b}_2 + \cdots + r_n\mathbf{b}_n \text{ where the } r_i \text{ are in } \mathbb{R}.$$

That is if $\mathbf{v} = s_1\mathbf{b}_1 + s_2\mathbf{b}_2 + \cdots + s_n\mathbf{b}_n$ is another such representation, s_i in $\mathbb{R}$, then $s_i = r_i$ for each i.

Proof. Since $V = span\{\mathbf{b}_1, \mathbf{b}_2, \cdots, \mathbf{b}_n\}$, the vector $\mathbf{v}$ has at least one such representation; if

$$r_1\mathbf{b}_1 + r_2\mathbf{b}_2 + \cdots + r_n\mathbf{b}_n = \mathbf{v} = s_1\mathbf{b}_1 + s_2\mathbf{b}_2 + \cdots + s_n\mathbf{b}_n$$

are two such representations then $(r_1-s_1)\mathbf{b}_1 + (r_2-s_2)\mathbf{b}_2 + \cdots + (r_n-s_n)\mathbf{b}_n = \mathbf{0}$. Hence $r_i - s_i = 0$ for each i because $\{\mathbf{b}_1, \mathbf{b}_2, \cdots, \mathbf{b}_n\}$ is linearly independent. □

Theorem 3. *Let $\{\mathbf{b}_1, \mathbf{b}_2, \cdots, \mathbf{b}_n\}$ be a basis of a vector space V, and let $\{\mathbf{w}_1, \mathbf{w}_2, \cdots, \mathbf{w}_n\}$ be arbitrary vectors in W (possibly not distinct). Then there exists a unique linear transformation $T : V \to W$ such that*

$$T(\mathbf{b}_i) = \mathbf{w}_i \text{ for each } i = 1, 2, \cdots, n.$$

Furthermore, the action of T is given as follows: If

$$\mathbf{v} = r_1\mathbf{b}_1 + r_2\mathbf{b}_2 + \cdots + r_n\mathbf{b}_n$$

is in V where each r_i is in $\mathbb{R}$, then

$$T(\mathbf{v}) = T(r_1\mathbf{b}_1 + r_2\mathbf{b}_2 + \cdots + r_n\mathbf{b}_n) = r_1\mathbf{w}_1 + r_2\mathbf{w}_2 + \cdots + r_n\mathbf{w}_n. \quad (**)$$

Proof. If T exists it is unique by Theorem 2 because $\{\mathbf{b}_1, \mathbf{b}_2, \cdots, \mathbf{b}_n\}$ spans V. But if we write $\mathbf{v} = r_1\mathbf{b}_1 + r_2\mathbf{b}_2 + \cdots + r_n\mathbf{b}_n$, r_i in $\mathbb{R}$, then the r_i are uniquely determined by $\mathbf{v}$ and so $T(\mathbf{v}) = r_1\mathbf{w}_1 + r_2\mathbf{w}_2 + \cdots + r_n\mathbf{w}_n$ is also uniquely determined by $\mathbf{v}$. Hence the formula $(**)$ defines a transformation $T : V \to W$ which clearly satisfies $T(\mathbf{b}_i) = \mathbf{w}_i$ for each i. The verification that this transformation T is linear is left to the reader. □

Example 8. *Find a linear transformation $T : \mathbb{P}_2 \to \mathbb{P}_4$ such that*

$$T(1) = x^4, \ T(1+x) = 1+x^3, \text{ and } T(1+x^2) = 1-x^2.$$

Solution. The set $\{1, \ 1+x, \ 1+x^2\} \subseteq \mathbb{P}_2$ is independent because the polynomials have distinct degrees, and so it is a basis of $\mathbb{P}_2$ because $deg(\mathbb{P}_2) = 3$. By comparing coefficients, the expansion of an arbitrary polynomial in $\mathbb{P}_2$ in terms of this basis is

$$r_0 + r_1 x + r_2 x^2 = (r_0 - r_1 - r_2) \cdot 1 + r_1(1+x) + r_2(1+x^2).$$

Hence the required linear transformation is given by

$$\begin{aligned} T(r_0 + r_1 x + r_2 x^2) &= (r_0 - r_1 - r_2) \cdot T(1) + r_1 \cdot T(1+x) + r_2 \cdot T(1+x^2) \\ &= (r_0 - r_1 - r_2)x^4 + r_1(1+x^3) + r_2(1-x^2) \\ &= (r_1 + r_2) - r_2 x^2 + r_1 x^3 + (r_0 - r_1 - r_2)x^4. \end{aligned}$$

$\square$

In practice, linear transformations $T : V \to W$ are usually described by simply giving the action of T on some convenient basis of V; a complete description of $T(\mathbf{v})$ for every vector $\mathbf{v}$ in V as in Example 8 is usually not provided (or needed).

5.3.2 Kernel and Image

Every linear transformation $T : V \to W$ gives rise to two subspaces, called the kernel and image of T, which encode a lot of information about T. The **kernel** of T, denoted $ker T$, is the subspace of V consisting of all vectors that T carries to $\mathbf{0}$; more formally,

$$ker T = \{\mathbf{v} \text{ in } V \mid T(\mathbf{v}) = \mathbf{0}\}.$$

The **image** of T, denoted $im T$, is the subspace of W consisting of all images under T of vectors in V:

$$im T = \{T(\mathbf{v}) \mid \mathbf{v} \text{ in } V\}.$$

We begin by verifying that $ker T$ and $im T$ are indeed subspaces of V and W.

Lemma 2. *Let $T : V \to W$ be a linear transformation. Then $ker T$ is a subspace of V and $im T$ is a subspace of W.*

Proof. The fact that $T(\mathbf{0}) = \mathbf{0}$ shows that both $ker T$ and $im T$ contain the zero vector. If $\mathbf{v}_1$ and $\mathbf{v}_2$ are both in $ker T$ then $T(\mathbf{v}_1) = \mathbf{0}$ and $T(\mathbf{v}_2) = \mathbf{0}$, so

$$T(\mathbf{v}_1 + \mathbf{v}_2) = T(\mathbf{v}_1) + T(\mathbf{v}_2) = \mathbf{0} + \mathbf{0} = \mathbf{0} \quad \text{and} \quad T(r\mathbf{v}_1) = rT(\mathbf{v}_1) = r\mathbf{0} = \mathbf{0}$$

for all r in $\mathbb{R}$. Hence $\mathbf{v}_1 + \mathbf{v}_2$ and $r\mathbf{v}_1$ are both in $ker T$, which shows that $ker T$ is a subspace of V.

If $\mathbf{w}_1$ and $\mathbf{w}_2$ are both in $im T$, write $\mathbf{w}_1 = T(\mathbf{v}_1)$ and $\mathbf{w}_2 = T(\mathbf{v}_2)$ where $\mathbf{v}_1$ and $\mathbf{v}_2$ are in V. Then

$$\mathbf{w}_1 + \mathbf{w}_2 = T(\mathbf{v}_1) + T(\mathbf{v}_2) = T(\mathbf{v}_1 + \mathbf{v}_2) \quad \text{and} \quad r\mathbf{w}_1 = rT(\mathbf{v}_1) = T(r\mathbf{v}_1)$$

show that $\mathbf{w}_1 + \mathbf{w}_2$ and $r\mathbf{w}_1$ are both in $im T$ for all r in $\mathbb{R}$. Hence $im T$ is a subspace of W.
$\square$

The subspaces $ker T$ and $im T$, with their relationship to T, are depicted in the diagram.

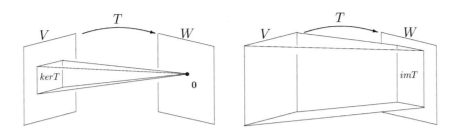

Familiar subspaces of $\mathbb{R}^n$ and $\mathbb{M}_{n,n}$ often occur in a natural way as images or kernels of linear transformations. Here are three examples.

Example 9. If A is an $m \times m$ matrix, let $T : \mathbb{R}^n \to \mathbb{R}^m$ be the matrix transformation given by $T(X) = AX$ for all X in $\mathbb{R}^n$. The subspaces $null A$ and $im A$ arise as follows:

$$ker T = \{X \mid AX = 0\} = null A \quad \text{and} \quad im T = \{AX \mid X \text{ in } \mathbb{R}^n\} = im A.$$

Example 10. If U is a subspace of $\mathbb{R}^n$, let $T : \mathbb{R}^n \to \mathbb{R}^n$ be the projection on U, that is define $T(X) = proj_U(X)$ for all X in $\mathbb{R}^n$. Show that

$$im T = U \quad \text{and} \quad ker T = U^\perp.$$

Solution. Note first that T is a linear transformation by Example 4 §4.9. The Projection Theorem (Theorem 2 §4.6) asserts that

$$T(X) \text{ is in } U \quad \text{and} \quad X - T(X) \text{ is in } U^\perp \text{ for all } X \text{ in } \mathbb{R}^n. \qquad (***)$$

Since $T(X)$ is in U for all X, we have $im T \subseteq U$. On the other hand, if X is in U, then (***) shows that $X - T(X)$ is in $U \cap U^\perp = \{0\}$ for all X, so $X = T(X)$ is in $im T$. This proves that $im T = U$.

Similarly, (***) shows that $X - T(X)$ is in $U^\perp$ for all X, so $ker T \subseteq U^\perp$. But if X is in $U^\perp$, then (***) shows that $T(X) = X - (X - T(X))$ is in $U \cap U^\perp = \{0\}$, and it follows that X is in $ker T$. This shows that $ker T = U^\perp$. □

Recall that a matrix S is called **skew-symmetric** if $S^T = -S$.

Example 11. Let $T : \mathbb{M}_{n,n} \to \mathbb{M}_{n,n}$ be defined by $T(A) = A - A^T$ for all matrices A in $\mathbb{M}_{n,n}$. Show that

$$ker T = \{A \text{ in } \mathbb{M}_{n,n} \mid A \text{ is symmetric}\}$$

and

$$im T = \{S \text{ in } \mathbb{M}_{n,n} \mid S \text{ is skew-symmetric}\}.$$

Solution. Clearly, $ker T = \{A \mid A^T = A\}$ is the set of symmetric matrices. On the other hand, $T(A) = A - A^T$ is skew-symmetric for all A (verify), and every skew-symmetric matrix S has the form $S = A - A^T$ where $A = \frac{1}{2}S$ (verify). This proves the second assertion. □

Example 12. As in Example 3, if a is a number, let $E_a : \mathbb{P}_n \to \mathbb{R}$ be evaluation at a, that is $E_a[p(x)] = p(a)$ for all polynomials $p(x)$ in $\mathbb{P}_n$. Then

$$ker(E_a) = \{p(x) \mid p(a) = 0\}$$

consists of the polynomials having a as a root. On the other hand,

$$im(E_a) = \{p(a) \mid p(x) \text{ in } \mathbb{P}_n\} = \mathbb{R}$$

because every real number b equals $p(a)$ for some polynomial $p(x)$ (for example $p(x) = b$, the constant polynomial). □

Example 13.[12] Let $T : \mathbb{P}_n \to \mathbb{P}_n$ be the differentiation operator: $T[p(x)] = p'(x)$ for all $p(x)$ in $\mathbb{P}_n$. Then T is linear (see Example 4) and

$$kerT = \{p(x) \mid p'(x) = 0\} \text{ consists of the constant polynomials}$$

because they are the only ones whose derivative is identically zero. On the other hand,

$$imT = \{p'(x) \mid p(x) \text{ in } \mathbb{P}_n\} = \mathbb{P}_{n-1}.$$

Indeed, $imT \subseteq \mathbb{P}_{n-1}$ because $p'(x)$ is always 0 or of degree one less than $p(x)$, and $\mathbb{P}_{n-1} \subseteq imT$ because every polynomial in $\mathbb{P}_{n-1}$ has degree at most $n-1$ and so is the derivative of some polynomial $p(x)$ in $\mathbb{P}_n$. □

The kernel and image of a linear transformation T are closely related to two other properties of T. If $T : V \to W$ is a linear transformation, we say that:

T is **onto** if $imT = W$.
T is **one-to-one** if $T(\mathbf{v}_1) = T(\mathbf{v}_2)$ implies that $\mathbf{v}_1 = \mathbf{v}_2$.

Thus $T : V \to W$ is onto if every vector $\mathbf{w}$ in W is the image of some vector in V. Furthermore, T is one-to-one if no vector $\mathbf{w}$ in W is the image of two distinct vectors in V.

Example 14. The linear transformation $T : \mathbb{R}^2 \to \mathbb{R}$ given by $T([r\ s]^T) = r$ is onto but not one-to-one. In fact, T is onto because every r in $\mathbb{R}$ is the image of some vector in $\mathbb{R}^2$ (say $T([r\ 0]^T) = r$). But T is not one-to-one. For example, $T([1\ 0]^T) = T([1\ 1]^T)$ but $[1\ 0]^T \neq [1\ 1]^T$.

Example 15. The linear transformation $T : \mathbb{R} \to \mathbb{R}^2$ given by $T(r) = [r\ 0]^T$ is one-to-one but not onto. Indeed, T is one-to-one because $T(r_1) = T(r_2)$ means $[r_1\ 0]^T = [r_2\ 0]^T$, whence $r_1 = r_2$. But T is not onto because, for example, the vector $[0\ 1]^T$ in $\mathbb{R}^2$ is not the image under T of any vector r in $\mathbb{R}$.

---

[12] Requires polynomial calculus.

5.3. LINEAR TRANSFORMATIONS

The onto transformations $T : V \to W$ are the ones for which the subspace imT of W is as *large* as possible (namely $imT = W$). By contrast, the next result shows that the one-to-one transformations are those for which the subspace $kerT$ of V is as *small* as possible in V (namely $kerT = \{\mathbf{0}\}$). This is a useful test for when T is one-to-one.

Theorem 4. *Let $T : V \to W$ be a linear transformation. Then*

$$T \text{ is one-to-one if and only if } kerT = \{\mathbf{0}\}.$$

Proof. If T is one-to-one, let $\mathbf{v}$ be a vector in $kerT$, that is $T(\mathbf{v}) = \mathbf{0}$. Hence $T(\mathbf{v}) = T(\mathbf{0})$ by Theorem 1, so $\mathbf{v} = \mathbf{0}$ by the one-to-one condition. It follows that $kerT = \{\mathbf{0}\}$. Conversely, assume that $kerT = \{\mathbf{0}\}$. To show that T is one-to-one, let $T(\mathbf{v}_1) = T(\mathbf{v}_2)$; we must show that $\mathbf{v}_1 = \mathbf{v}_2$. But the linearity of T gives $T(\mathbf{v}_1 - \mathbf{v}_2) = T(\mathbf{v}_1) - T(\mathbf{v}_2) = \mathbf{0}$, so $\mathbf{v}_1 - \mathbf{v}_2$ is in $kerT = \{\mathbf{0}\}$. Thus $\mathbf{v}_1 = \mathbf{v}_2$, as required. □

Example 16. If A is an $m \times n$ matrix, consider the matrix transformation $T : \mathbb{R}^n \to \mathbb{R}^m$ given by $T(X) = AX$ for all X in $\mathbb{R}^n$. Theorem 3 §4.4 asserts that $rankA = n$ if and only if the system $AX = 0$ has only the trivial solution $X = 0$. In our present language, this is:

$$T \text{ is one-to-one if and only if } rankA = n.$$

Similarly, Theorem 4 §4.4 shows that $rankA = m$ if and only if the system $AX = B$ has a solution for every column B in $\mathbb{R}^m$. This gives:

$$T \text{ is onto if and only if } rankA = m.$$
□

5.3.3 The Dimension Theorem

If A is an $m \times n$ matrix, the corollary of Theorem 6 §4.4 asserts that

$$n = dim(imA) + dim(nullA).$$

This equation is important in matrix theory because $nullA$ is the set of all solutions X to the system $AX = 0$ of linear equations, and $dim(imA) = dim(colA) = rankA$ (see Theorem 5 § 4.4). If $T : \mathbb{R}^n \to \mathbb{R}^m$ is given by $T(X) = AX$ for all X in $\mathbb{R}^n$, then Example 9 shows that $imA = imT$ and $nullA = kerT$. Consequently, the following theorem is a profound extension of the above equation.

Theorem 5. Dimension Theorem. *Let $T : V \to W$ be a linear transformation. If both $kerT$ and imT are finite dimensional, then V is also finite dimensional and*

$$dimV = dim(imT) + dim(kerT).$$

Proof. Let $\{\mathbf{b}_1, \mathbf{b}_2, \cdots, \mathbf{b}_k\}$ be a basis of $kerT$ and, since $imT = \{T(\mathbf{v}) \mid \mathbf{v}$ in $V\}$, let $\{T(\mathbf{d}_1), T(\mathbf{d}_2), \cdots, T(\mathbf{d}_r)\}$ be a basis of imT where each $\mathbf{d}_j$ is in V. Then $dim(kerT) = k$ and $dim(imT) = r$, so it suffices to show that

$$\mathcal{B} = \{\mathbf{b}_1, \mathbf{b}_2, \cdots, \mathbf{b}_k, \mathbf{d}_1, \mathbf{d}_2, \cdots, \mathbf{d}_r\}$$

is a basis of V.

$\mathcal{B}$ *spans* V. If $\mathbf{v}$ is in V, then $T(\mathbf{v})$ is in imT, so write $T(\mathbf{v}) = s_1 T(\mathbf{d}_1) + s_2 T(\mathbf{d}_2) + \cdots + s_r T(\mathbf{d}_r)$ where each s_j is in $\mathbb{R}$. Then $T(\mathbf{v}) = T(s_1 \mathbf{d}_1 + s_2 \mathbf{d}_2 + \cdots + s_r \mathbf{d}_r)$ by Theorem 1, and so

$$T[\mathbf{v} - (s_1 \mathbf{d}_1 + s_2 \mathbf{d}_2 + \cdots + s_r \mathbf{d}_r)] = T(\mathbf{v}) - T(s_1 \mathbf{d}_1 + s_2 \mathbf{d}_2 + \cdots + s_r \mathbf{d}_r) = \mathbf{0}.$$

Hence $\mathbf{v} - (s_1 \mathbf{d}_1 + s_2 \mathbf{d}_2 + \cdots + s_r \mathbf{d}_r)$ lies in $kerT$, and so is a linear combination of $\{\mathbf{b}_1, \mathbf{b}_2, \cdots, \mathbf{b}_k\}$. It follows that $\mathbf{v}$ is in $span\mathcal{B}$, as required.

$\mathcal{B}$ *is linearly independent.* Suppose that t_i and s_j in $\mathbb{R}$ exist such that

$$t_1 \mathbf{b}_1 + t_2 \mathbf{b}_2 + \cdots + t_k \mathbf{b}_k + s_1 \mathbf{d}_1 + s_2 \mathbf{d}_2 + \cdots + s_r \mathbf{d}_r = \mathbf{0}. \qquad (****)$$

Since each $\mathbf{b}_i$ is in $kerT$, applying T gives $s_1 T(\mathbf{d}_1) + s_2 T(\mathbf{d}_2) + \cdots + s_r T(\mathbf{d}_r) = \mathbf{0}$. It follows that each $s_j = 0$, whence (****) becomes $t_1 \mathbf{b}_1 + t_2 \mathbf{b}_2 + \cdots + t_k \mathbf{b}_k = \mathbf{0}$. Hence each $t_i = 0$, and we are done. $\square$

As noted above, the Dimension Theorem provides another proof of the important fact that the dimension of the space of solutions of a system $AX = 0$ equals $n - r$ where A is an $n \times m$ matrix of rank r. Another illustration: The result that $dimU + dimU^\perp = n$ for all subspaces U of $\mathbb{R}^n$ (Theorem 3 §4.6) follows immediately from the Dimension Theorem and Example 10.

The following consequence of the Dimension Theorem will be referred to later.

Corollary. Let $T : V \to W$ be a linear transformation. If $dimV = dimW$ is finite, then T is one-to-one if and only if T is onto.

Proof. The Dimension Theorem gives $dim(imT) + dim(kerT) = dimV$. If T is one-to-one then $kerT = \{\mathbf{0}\}$ by Theorem 4, so $dim(kerT) = 0$. Hence $dim(imT) = dimV = dimW$, whence $imT = W$ by Theorem 5 §5.2, and T is onto. Conversely, if T is onto, then $imT = W$ so $dim(imT) = dimW = dimV$. It follows that $dim(kerT) = 0$, so that $kerT = \{\mathbf{0}\}$ and T is one-to-one. $\square$

If $T : V \to W$ is a linear transformation where $dimV = n$ is finite, and if either of $dim(kerT)$ or $dim(imT)$ is known, then the Dimension Theorem provides the other. This can be useful because one of these dimensions can be easier to compute than the other. Applications of this often depends on *defining* an appropriate linear transformation. This is illustrated in the following example which gives a different proof of an important result from Section 1.5.

5.3. LINEAR TRANSFORMATIONS

Example 17. If $BC = I$ where B and C are $n \times n$ matrices, show that also $CB = I$.

Solution. Define $T : \mathbb{M}_{n,n} \to \mathbb{M}_{n,n}$ by $T(A) = CA$ for all A in $\mathbb{M}_{n,n}$. Then T is linear (verify). If A is in $kerT$, then $0 = T(A) = CA$, so $A = IA = BCA = B0 = 0$. This shows that $kerT = \{0\}$, so that T is one-to-one. But then T is onto by the Corollary to the Dimension Theorem. In particular $T(D) = I$ for some matrix D, that is $CD = I$. Hence $D = ID = BCD = BI = B$. Thus $CB = CD = I$, as asserted. □

Exercises

Throughout these exercises, V always denotes a vector space.

1. True or False?

    (a) $T : \mathbb{R} \to \mathbb{R}$ is a linear transformation where $T(r) = r^2$ for all r in $\mathbb{R}$.

    (b) $T : \mathbb{R} \to \mathbb{R}$ is a linear transformation where $T(r) = |r|$ for all r in $\mathbb{R}$.

    (c) $T(X) = AX$ defines a linear transformation $\mathbb{R}^5 \to \mathbb{R}^6$ for every 5×6 matrix A.

    (d) If $T : V \to W$ is a linear transformation and $T(\mathbf{v}_1) = T(\mathbf{v}_2)$, then $\mathbf{v}_1 - \mathbf{v}_1$ is in $kerT$.

    (e) If $T : V \to W$ is a linear transformation and $T(\mathbf{v}) = \mathbf{0}$ for some $\mathbf{v} \neq \mathbf{0}$, then $T = 0$ is the zero transformation.

    (f) If $T : V \to W$ is linear and $kerT = V$ then $W = 0$.

    (g) If $T : V \to W$ is linear and $imT = 0$ then $T = 0$ is the zero transformation.

    (h) If $T : \mathbb{R}^n \to \mathbb{R}^n$ is defined by $T(X) = proj_{\mathbb{R}^n}(X)$ for all X in $\mathbb{R}^n$, then $T = 1_{\mathbb{R}^n}$.

    (i) A linear transformation $\mathbb{R}^n \to \mathbb{R}^m$ cannot be both one-to-one and onto.

    (j) If $T : V \to W$ is one-to-one and $dimV = dimW = n$, then $imT = W$.

    (k) Every linear transformation $T : \mathbb{R}^6 \to \mathbb{R}^5$ is onto.

    (l) There exists a linear transformation T such that $T(\mathbf{v}_1)$ is in $\mathbb{F}[0,1]$ and $T(\mathbf{v}_2)$ is in $\mathbb{R}^3$.

2. Let $T : V \to W$ denote a linear transformation. In each case either prove the statement or give an example showing that it is false.

    (a) If $dimV = 6$, $dimW = 4$ and $dim(kerT) = 2$ then T is onto.

    (b) If $dimV = 4$ and $dimW = 3$ then T is one-to-one.

    (c) If $\{\mathbf{e}_1, \mathbf{e}_2, \mathbf{e}_3, \mathbf{e}_4\}$ is a basis of V and $T(\mathbf{e}_2) = \mathbf{0} = T(\mathbf{e}_4)$ then $dim(imT) \leq 2$.

(d) If $dim(kerT) \leq dimW$ then $dimW \geq \frac{1}{2}dimV$.

(e) If T is one-to-one then $dimV \leq dimW$.

(f) If $dimV \leq dimW$ then T is one-to-one.

(g) If T is onto then $dimV \geq dimW$.

(h) If $dimV \geq dimW$ then T is onto.

3. Show that the following are linear transformations.

    (a) $T : \mathbb{C} \to \mathbb{C}$ where $T(z) = \bar{z}$, where $\bar{z}$ denotes the conjugate of the complex number z.

    (b) $T : \mathbb{M}_{m,n} \to \mathbb{M}_{k,l}$ where $T(A) = PAQ$ where P and Q are fixed $k \times m$ and $n \times l$ matrices.

    (c) $T : \mathbb{P}_n \to \mathbb{R}$ where $T(a_0 + a_1 x + \cdots + a_n x^n) = a_n$.

    (d) $T : \mathbb{R}^n \to \mathbb{R}$ where $T(X) = Y \bullet X$, where Y is a fixed vector in $\mathbb{R}^n$.

    (e) $T : \mathbb{P}_n \to \mathbb{P}_n$ where $T[p(x)] = p(x+1)$ for all polynomials $p(x)$ in $\mathbb{P}_n$.

    (f) $T : V \to \mathbb{R}$ where $T(r_1 \mathbf{e}_1 + \cdots + r_n \mathbf{e}_n) = r_1$, where $\{\mathbf{e}_1, \cdots, \mathbf{e}_n\}$ is a fixed basis of V.

    (g) $T : \mathbb{M}_{n,n} \to \mathbb{R}$ where $T(A) = trA$, where trA denotes the trace of the matrix A.

    (h) $T : \mathbb{M}_{m,n} \to \mathbb{R}^m$ where $T(A)$ is column 2 of A.

    (i) $T : \mathbb{R} \to \mathbb{F}[D]$, where $T(a) = f_a$ is the constant function defined by $f_a(x) = a$ for all x in D. (See Exercise 20 §5.1.)

4. Decide whether the following transformations $T : \mathbb{M}_{n,n} \to \mathbb{R}$ are linear. Support your answer.

    (a) $T : \mathbb{M}_{n,n} \to \mathbb{R}$ where $T(A) = detA$

    (b) $T : \mathbb{M}_{n,n} \to \mathbb{R}$ where $T(A) = rankA$

5. In each case find a linear transformation T with the given properties, and compute $T(\mathbf{v})$.

    (a) $T : \mathbb{P}_2 \to \mathbb{P}_4$, $T(1) = x^4$, $T(x + x^2) = 1$, $T(x - x^2) = x + x^3$; $\mathbf{v} = a + bx + cx^2$

    (b) $T : \mathbb{M}_{2,2} \to \mathbb{R}^2$, $T \begin{bmatrix} 1 & 0 \\ 0 & 1 \end{bmatrix} = \begin{bmatrix} 1 \\ 2 \end{bmatrix}$, $T \begin{bmatrix} 0 & 1 \\ 0 & 0 \end{bmatrix} = \begin{bmatrix} 0 \\ 1 \end{bmatrix}$, $T \begin{bmatrix} 0 & 0 \\ 1 & 0 \end{bmatrix} = \begin{bmatrix} 1 \\ 0 \end{bmatrix}$, $T \begin{bmatrix} 0 & 0 \\ 0 & 1 \end{bmatrix} = \begin{bmatrix} 0 \\ 0 \end{bmatrix}$; $\mathbf{v} = \begin{bmatrix} a & b \\ c & d \end{bmatrix}$

6. If $T : V \to V$ is linear and $\mathbf{e}$ and $\mathbf{f}$ are in V, find $T(3\mathbf{e} + \mathbf{f})$ and $T(\mathbf{f})$ in terms of $\mathbf{e}$ and $\mathbf{f}$ if $T(\mathbf{e} - \mathbf{f}) = 2\mathbf{e} - \mathbf{f}$ and $T(2\mathbf{f} - \mathbf{e}) = \mathbf{e} + \mathbf{f}$.

7. Let $T : V \to V$ be a linear transformation, and let $\{\mathbf{b}_1, \mathbf{b}_2, \cdots, \mathbf{b}_n\}$ be a basis of V.

(a) If $T(\mathbf{b}_i) = \mathbf{b}_i$ for each i, show that $T = 1_V$ is the identity transformation.

(b) If $T(\mathbf{b}_i) = \mathbf{0}$ for each i, show that $T = 0$ is the zero transformation.

8. Let $T : V \to W$ be a linear transformation. Give a careful proof of the fact that $T(\mathbf{v} - \mathbf{v}_1) = T(\mathbf{v}) - T(\mathbf{v}_1)$ for all $\mathbf{v}$ and $\mathbf{v}_1$ in V. Cite every axiom or theorem used.

9. Describe all linear transformations $T : \mathbb{R} \to V$. [Hint: $r = r1$ for all r in $\mathbb{R}$.]

10. A linear transformation $T : V \to \mathbb{R}$ is called a **linear functional**.

    (a) Let $T : V \to \mathbb{R}$ be a linear functional where $dim V = n$. If $\mathcal{B}$ is a basis of V, show that there exists a vector Y in $\mathbb{R}^n$ such that $T(\mathbf{v}) = Y \bullet C_\mathcal{B}(\mathbf{v})$ for all in V. Compare with Exercise 9 §4.9.

    (b) If V is the set of all integrable functions in $\mathbb{F}[0,1]$, show that $T(f) = \int_0^1 f(t)dt$ defines a linear functional $\mathbb{F}[0,1] \to \mathbb{R}$.

11. Let V be a finite dimensional vector space. Given $\mathbf{v} \neq \mathbf{0}$ in V and $\mathbf{w}$ in a space W, show that there exists a linear transformation $T : V \to W$ such that $T(\mathbf{v}) = \mathbf{w}$. What happens if $\mathbf{v} = \mathbf{0}$?

12. Show that the following are equivalent for a linear transformation $T : V \to W$.

    (a) $ker T = V$. (b) $im T = \{\mathbf{0}\}$. (c) $T = 0$.

13. Let $T : V \to W$ be a linear transformation, let $\mathcal{B} = \{\mathbf{b}_1, \mathbf{b}_2, \cdots, \mathbf{b}_n\}$ be vectors in V, and write $\mathcal{C} = \{T(\mathbf{b}_1), T(\mathbf{b}_2), \cdots, T(\mathbf{b}_n)\}$.

    (a) If $\mathcal{C}$ is independent, show that $\mathcal{B}$ is independent.

    (b) If T is one-to-one and $\mathcal{B}$ is independent, show that $\mathcal{C}$ is independent.

    (c) If $W = span \mathcal{C}$ show that T is onto.

    (d) If $V = span \mathcal{B}$, show that $im T = span \mathcal{C}$.

14. Given vectors $\mathbf{v}_1, \mathbf{v}_2, \cdots, \mathbf{v}_n$ in a vector space V, define $T : \mathbb{R}^n \to V$ by $T[r_1 \; r_2 \; \cdots \; r_n]^T = r_1 \mathbf{v}_1 + r_2 \mathbf{v}_2 + \cdots + r_n \mathbf{v}_n$.

    (a) Show that T is one-to-one if and only if $\{\mathbf{v}_1, \mathbf{v}_2, \cdots, \mathbf{v}_n\}$ is linearly independent.

    (b) Show that T is onto if and only if $\{\mathbf{v}_1, \mathbf{v}_2, \cdots, \mathbf{v}_n\}$ spans V.

15. Let U be a subspace of a finite dimensional vector space V.

    (a) Show that $U = ker T$ for some linear operator $T : V \to V$. [Hint: Theorem 3.]

    (b) Show that $U = im S$ for some linear operator $S : V \to V$. [Hint: Theorem 3.]

16. Let $T : \mathbb{M}_{n,n} \to \mathbb{R}$ be the trace map: $T(A) = trA$ for all A in $\mathbb{M}_{n,n}$. Show that $dim(kerT) = n^2 - 1$.

17. If A is an $n \times n$ matrix, let $U = \{B \text{ in } \mathbb{M}_{m,n} \mid BA = 0\}$ and let $W = \{BA \mid B \text{ in } \mathbb{M}_{m,n}\}$. Use the Dimension Theorem to show that U and W are both subspaces of $\mathbb{M}_{m,n}$, and that $dimU + dimW = mn$.

18. Use the Dimension Theorem to prove Theorem 1 §1.3: A homogeneous system of linear equations that has more variables than equations must have nontrivial solutions.

19. Let U and W denote the spaces of even and odd polynomials in $\mathbb{P}_n$ (see Exercise 20 §5.2). Use the Dimension Theorem to show that $dimU + dimW = n + 1$. [*Hint*: Consider $T : \mathbb{P}_n \to \mathbb{P}_n$ where $T[p(x)] = p(x) - p(-x)$.]

20. Let U and W denote the spaces of symmetric and skew-symmetric $n \times n$ matrices, respectively. Show that $dimU + dimW = n^2$. [*Hint* : Example 11.]

21. Consider the subspace $W = \{p(x) \text{ in } \mathbb{P}_n \mid p(a) = 0\}$ of $\mathbb{P}_n$. Use the Dimension Theorem to prove that $\{(x-a), (x-a)^2, (x-a)^3, \cdots, (x-a)^n\}$ is a basis of W. [*Hint*: Use the evaluation map $E_a : \mathbb{P}_n \to \mathbb{R}$, Example 2 §5.2 and Theorem 4 §5.2.]

22. Use the Dimension Theorem to show that every matrix A in $\mathbb{M}_{n,n}$ has the form $A = B^T - 3B$ for some B in $\mathbb{M}_{n,n}$.

23. Use the Dimension Theorem to show that every polynomial $p(x)$ in $\mathbb{P}_{n-1}$ can be written in the form $p(x) = q(x+1) - q(x)$ for some polynomial $q(x)$ in $\mathbb{P}_n$.

24. Show that differentiation is the only linear transformation $\mathbb{P}_n \to \mathbb{P}_n$ that satisfies $T(x^k) = kx^{k-1}$ for all $k = 0, 1, 2, \cdots, n$.

25. If a and b are real numbers, define $T_{a,b} : \mathbb{C} \to \mathbb{C}$ by $T_{a,b}(r + si) = ar + bsi$.

    (a) Show that $T_{a,b}$ is linear and $T_{a,b}(\bar{z}) = \overline{T_{a,b}(z)}$ for all z in $\mathbb{C}$. Here $\bar{z}$ denotes conjugation.

    (b) If $T : \mathbb{C} \to \mathbb{C}$ is linear and satisfies $T(\bar{z}) = \overline{T(z)}$ for all z in $\mathbb{C}$, show that $T = T_{a,b}$ for some a and b. [*Hint*: Show that $T(1)$ is real and $T(i)$ is pure imaginary.]

26. If a is in $\mathbb{R}$, consider the evaluation transformation $E_a : \mathbb{P}_n \to \mathbb{R}$ where $E_a[f(x)] = f(a)$.

    (a) Show that $E_a(x^k) = [E_a(x)]^k$ for each $k = 0, 1, 2, \cdots, n$. (Note that $x^0 = 1$.)

    (b) If $T : \mathbb{P}_n \to \mathbb{R}$ is linear and satisfies $T(x^k) = [T(x)]^k$ for each $k = 0, 1, 2, \cdots, n$, show that $T = E_a$ for some a in $\mathbb{R}$. [*Hint*: If $T = E_a$ for some a, what is $T(x)$?]

27. If $T : \mathbb{M}_{n,n} \to \mathbb{R}$ is linear and satisfies $T(AB) = T(BA)$ for all A and B in $\mathbb{M}_{n,n}$, show that there exists a constant k such that $T(A) = k\,tr(A)$ for all A in $\mathbb{M}_{n,n}$, where tr denotes the trace map (See Lemma 4 §4.7). [*Hint*: If E_{ij} denote the matrix units in $\mathbb{M}_{n,n}$ (see Example 16 §5.1) and take $k = T(E_{11})$. Use the fact that

$$E_{ij}E_{km} = \begin{cases} 0 & \text{if } j \neq k \\ E_{im} & \text{if } j = k \end{cases}$$

28. Let V and W denote finite dimensional vector spaces. Using Theorems 3 and 5:

    (a) Show that $dim V \leq dim W$ if and only if there exists a one-to-one linear transformation $T : V \to W$.

    (b) Show that $dim V \geq dim W$ if and only if there exists an onto linear transformation $T : V \to W$.

5.4 ISOMORPHISMS AND MATRICES

Two vector spaces can appear to be quite different but, on closer examination, be the same underlying space expressed with different symbols. In this case we say that the spaces are isomorphic. This concept is investigated in this section, and is described in two ways using linear transformations: one using the one-to-one and onto properties, and the other using composition. The notion of composition leads naturally to a way of introducing the matrix of any linear transformation T, and using it to describe the action of T.

5.4.1 Isomorphisms

If V and W are vector spaces, a linear transformation $T : V \to W$ is called an **isomorphism** if it is both one-to-one and onto. If an isomorphism $V \to W$ exists, we say that V and W are **isomorphic**[13] vector spaces and write

$$V \cong W.$$

Example 1. $V \cong V$ for every vector space V because $1_V : V \to V$ is an isomorphism.

Example 2. $\mathbb{M}_{m,n} \cong \mathbb{M}_{n,m}$ because transposition $T : \mathbb{M}_{m,n} \to \mathbb{M}_{n,m}$ is an isomorphism.

An isomorphism $T : V \to W$ induces a *pairing*

$$\mathbf{v} \longleftrightarrow T(\mathbf{v})$$

---

[13]The word comes from two Greek words: *iso* meaning "same" and *morphos* meaning "form".

between the vectors **v** in V and the vectors $T(\mathbf{v})$ in W. Every vector in each space is paired with exactly one vector in the other space. Moreover, this pairing preserves vector addition and scalar multiplication, and hence preserves all vector space properties. The change of notation $\mathbf{v} \longmapsto T(\mathbf{v})$ changes the vector space V into W. Hence, roughly speaking, isomorphic spaces are the same except for notation. The following example illustrates this.

Example 3. We have $\mathbb{R}^2 \cong \mathbb{C}$ because, for example, the map $T : \mathbb{R}^2 \to \mathbb{C}$ given by $T([x\ y]^T) = x + iy$ is an isomorphism (verify).

The pairing T turns the Euclidean plane $\mathbb{R}^2$ into the complex plane $\mathbb{C}$. $\mathbb{C}$ is the same vector space as $\mathbb{R}^2$, but with a different interpretation: The ordered pair $[x\ y]^T$ is now viewed as the complex number $x + iy$. We obtain one from the other by changing symbols. □

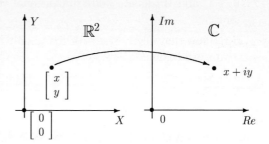

The following theorem is surprising because it shows that, up to isomorphism, there is *only one* n-dimensional vector space.

Theorem 1. *If V has dimension n, then $V \cong \mathbb{R}^n$.*

Proof. Choose a basis $\mathcal{B} = \{\mathbf{b}_1, \mathbf{b}_2, \cdots, \mathbf{b}_n\}$ of V and define $T : \mathbb{R}^n \to V$ by $T([r_1, r_2, \cdots, r_n]^T) = r_1\mathbf{b}_1 + r_2\mathbf{b}_2 + \cdots + r_n\mathbf{b}_n$. Then T is one-to-one because $\mathcal{B}$ is independent, and T is onto because $\mathcal{B}$ spans V (verify). Thus T is an isomorphism, so $V \cong \mathbb{R}^n$. □

Because of Theorem 1, it seems at first glance that the study of an abstract n-dimensional vector space V is a waste of time: Since V is just $\mathbb{R}^n$ except for a change of symbols, simply study $\mathbb{R}^n$ and be done with it! But the meaning is in the interpretation. Viewing $\mathbb{R}^2$ as $\mathbb{C}$ casts a whole new light on the plane — for example the geometrical interpretation of complex multiplication adds new life to $\mathbb{R}^2$ and provides an important new way to study rotations in the plane. Similarly, one verifies that $\mathbb{R}^{n+1} \cong \mathbb{P}_n$, and interpreting vectors in $\mathbb{R}^{n+1}$ as polynomials greatly enriches the structure of $\mathbb{R}^{n+1}$ because polynomials can be multiplied and composed. Another example is $\mathbb{R}^{n^2} \cong \mathbb{M}_{n,n}$ (verify) with the attendant view of vectors in $\mathbb{R}^{n^2}$ as $n \times n$ matrices.

On the other hand, revealing the vector space nature of $\mathbb{C}$, $\mathbb{P}_n$, and $\mathbb{M}_{n,n}$ allows general vector space theorems and techniques to be used to study complex numbers, polynomials and matrices. These are powerful methods and yield new information in all three cases.

We know that every linear transformation preserves linear combinations of vectors; the isomorphisms are the ones preserving bases.

5.4. ISOMORPHISMS AND MATRICES

Theorem 2. *If V and W are finite dimensional vector spaces, the following are equivalent for a linear transformation $T : V \to W$:*

(1) *T is an isomorphism.*

(2) *If $\{\mathbf{b}_1, \mathbf{b}_2, \cdots, \mathbf{b}_n\}$ is any basis of V then $\{T(\mathbf{b}_1), T(\mathbf{b}_2), \cdots, T(\mathbf{b}_n)\}$ is a basis of W.*

(3) *There exists a basis $\{\mathbf{b}_1, \mathbf{b}_2, \cdots, \mathbf{b}_n\}$ of V such that $\{T(\mathbf{b}_1), T(\mathbf{b}_2), \cdots, T(\mathbf{b}_n)\}$ is a basis of W.*

Proof. (1)$\Rightarrow$(2). If $\{\mathbf{b}_1, \mathbf{b}_2, \cdots, \mathbf{b}_n\}$ is any basis of V, then $\{T(\mathbf{b}_1), T(\mathbf{b}_2), \cdots, T(\mathbf{b}_n)\}$ is independent because T is one-to-one, and it spans W because T is onto. The verifications are left to the reader.

(2)$\Rightarrow$(3). This follows because V *has* a finite basis (being finite dimensional).

(3)$\Rightarrow$(1). Given a basis $\{\mathbf{b}_1, \mathbf{b}_2, \cdots, \mathbf{b}_n\}$ as in (3), T is one-to-one because $\{T(\mathbf{b}_1), T(\mathbf{b}_2), \cdots, T(\mathbf{b}_n)\}$ is independent and T is onto because it spans W. Again the details are left to the reader. □

Condition (3) in Theorem 2 is a useful test for when a linear transformation is actually an isomorphism, and it can be used to *construct* isomorphisms. The next example provides an illustration.

Example 4. *If V is the space of all 2×2 symmetric matrices, find an isomorphism $T : \mathbb{P}_2 \to V$ such that $T(x) = I$.*

Solution. Write $\mathbf{d}_1 = I = \begin{bmatrix} 1 & 0 \\ 0 & 1 \end{bmatrix}$, $\mathbf{d}_2 = \begin{bmatrix} 1 & 0 \\ 0 & 0 \end{bmatrix}$ and $\mathbf{d}_3 = \begin{bmatrix} 0 & 1 \\ 1 & 0 \end{bmatrix}$. Then $\mathcal{B} = \{1, x, x^2\}$ and $\mathcal{D} = \{\mathbf{d}_1, \mathbf{d}_2, \mathbf{d}_3\}$ are bases of $\mathbb{P}_2$ and V respectively. By Theorem 3 §5.3 there is a linear transformation $T : \mathbb{P}_2 \to V$ such that $T(1) = \mathbf{d}_2$, $T(x) = \mathbf{d}_1 = I$ and $T(x^2) = \mathbf{d}_3$. Since T clearly takes the basis $\mathcal{B}$ to the basis $\mathcal{D}$, it is an isomorphism by Theorem 2. [14] □

Theorem 3. *Let V and W denote finite dimensional vector spaces.*

(1) *$V \cong W$ if and only if $dimV = dimW$.*

(2) *If $dimV = dimW$, a linear transformation $T : V \to W$ is an isomorphism if and only if it is either one-to-one or onto.*

Proof. Part (2) restates the Corollary to the Dimension Theorem (Theorem 5 §5.3). To prove (1), suppose $V \cong W$, let $T : V \to W$ be an isomorphism, and choose a basis $\{\mathbf{b}_1, \mathbf{b}_2, \cdots, \mathbf{b}_n\}$ of V. Then $\{T(\mathbf{b}_1), T(\mathbf{b}_2), \cdots, T(\mathbf{b}_n)\}$ is a basis

---

[14] Of course we can extend T linearly to get the formula $T(a+bx+cx^2) = a\mathbf{d}_1 + b\mathbf{d}_2 + c\mathbf{d}_3 = \begin{bmatrix} a+b & c \\ c & a \end{bmatrix}$. However such detail is usually unnecessary and it is customary to simply specify T on a basis.

of W by Theorem 2, so $dim W = n = dim V$. Conversely, if $dim V = n = dim W$ let $\{\mathbf{b}_1, \mathbf{b}_2, \cdots, \mathbf{b}_n\}$ and $\{\mathbf{d}_1, \mathbf{d}_2, \cdots, \mathbf{d}_n\}$ be bases of V and W respectively. By Theorem 3 §5.3 let $T : V \to W$ be the linear transformation such that $T(\mathbf{b}_i) = \mathbf{d}_i$ for all i. Then T is an isomorphism by Theorem 2, so $V \cong W$. □

Condition (1) in Theorem 3 shows that the isomorphic relation $\cong$ has the following three properties. If U, V and W denote finite dimensional vector spaces, then:

(1) $V \cong V$.

(2) If $V \cong W$ then $W \cong V$.

(3) If $V \cong W$ and $W \cong U$ then $V \cong U$.

Because of these properties, $\cong$ is called an **equivalence relation**, or simply an **equivalence**.[15]

5.4.2 Composition

There is another way to describe isomorphisms based on the idea of composing linear transformations. Suppose that $V \xrightarrow{T} W \xrightarrow{S} U$ are two linear transformations. As in Section 4.9 we define their **composite** $S \circ T : V \to U$ by

$$(S \circ T)(\mathbf{v}) = S[T(\mathbf{v})] \text{ for all } \mathbf{v} \text{ in } V.$$

The action of $S \circ T$ is best described by saying "first T then S" (note the order), and is illustrated in the diagram. If $T : V \to V$, is an operator, we denote $T \circ T = T^2$, $T \circ T \circ T = T^3$, etc.

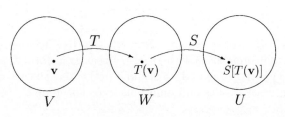

Example 5. If A and B are $n \times n$ matrices, define $S : \mathbb{R}^n \to \mathbb{R}^n$ and $T : \mathbb{R}^n \to \mathbb{R}^n$ by $S(X) = AX$ and $T(X) = BX$ for all X in $\mathbb{R}^n$. Then $(S \circ T)(X) = (AB)X$ for all X in $\mathbb{R}^n$, that is $S \circ T$ is multiplication by the product AB of the matrices for S and T.[16] Similarly, $T \circ S$ is multiplication by the matrix BA. In particular, it can happen that $S \circ T \neq T \circ S$ — simply choose A and B such that $AB \neq BA$.

The basic properties of composition are listed in the following theorem.

Theorem 4. *Let $V \xrightarrow{T} W \xrightarrow{S} U \xrightarrow{R} Z$ denote linear transformations. Then:*

---

[15] Other examples are the similarity relation $\sim$ for matrices and the equality relation $=$.
[16] We extend this in Theorem 10 below.

5.4. ISOMORPHISMS AND MATRICES

(1) $S \circ T$ is a linear transformation.

(2) $T \circ 1_V = T = 1_W \circ T$.

(3) $(R \circ S) \circ T = R \circ (S \circ T)$.

(4) If S and T are both one-to-one (respectively onto), so is $S \circ T$.

Proof. Let $\mathbf{v}$ and $\mathbf{w}$ be vectors in V. Then the linearity of S and T gives

$$\begin{aligned}(S \circ T)(\mathbf{v} + \mathbf{w}) &= S[T(\mathbf{v} + \mathbf{w})] \\ &= S[T(\mathbf{v}) + T(\mathbf{w})] \\ &= S[T(\mathbf{v})] + S[T(\mathbf{w})] \\ &= (S \circ T)(\mathbf{v}) + (S \circ T)(\mathbf{w}).\end{aligned}$$

Hence $S \circ T$ preserves addition; the proof that $(S \circ T)(r\mathbf{v}) = r(S \circ T)(\mathbf{v})$ for all r in $\mathbb{R}$ is similar and is left to the reader. This proves (1). Now, for any vector $\mathbf{v}$ in V, compute

$$\begin{aligned}[(R \circ S) \circ T](\mathbf{v}) &= (R \circ S)[T(\mathbf{v})] \\ &= R\{S[T(\mathbf{v})]\} \\ &= R\{(S \circ T)(\mathbf{v})\} \\ &= [R \circ (S \circ T)](\mathbf{v}).\end{aligned}$$

This proves (3). The verification of (2) is similar and is left to the reader, as is the proof of (4). □

Because of property (3) in Theorem 4 we write the transformation $(R \circ S) \circ T = R \circ (S \circ T)$ simply as $R \circ S \circ T$. The proof shows that the action of $R \circ S \circ T$ is first T, then S, and finally R, in the order of the arrows $V \xrightarrow{T} W \xrightarrow{S} U \xrightarrow{R} Z$.

We are now ready to characterize isomorphisms in terms of composition.

Theorem 5. Let V and W be finite dimensional vector spaces. The following conditions are equivalent for a linear transformation $T : V \to W$:

(1) T is an isomorphism.

(2) There exists a linear transformation $S : W \to V$ such that $S \circ T = 1_V$ and $T \circ S = 1_W$.

Moreover, if these conditions are satisfied the transformation S is uniquely determined by T.

Proof. (1)$\Rightarrow$(2). Assume that T is an isomorphism. Choose a basis $\{\mathbf{b}_1, \mathbf{b}_2, \cdots, \mathbf{b}_n\}$ of V and write $\mathbf{d}_i = T(\mathbf{b}_i)$ for each i. Then $\{\mathbf{d}_1, \mathbf{d}_2, \cdots, \mathbf{d}_n\}$ is a basis of W by Theorem 2, so there exists a linear transformation $S : W \to V$ such that $S(\mathbf{d}_i) = \mathbf{b}_i$ for each i (by Theorem 3 §5.3). But then $(S \circ T)(\mathbf{b}_i) = S[T(\mathbf{b}_i)] = S(\mathbf{d}_i) = \mathbf{b}_i = 1_V(\mathbf{b}_i)$ for each i. This means that $S \circ T = 1_V$ by

Theorem 2 §5.3 because $\{\mathbf{b}_1, \mathbf{b}_2, \cdots, \mathbf{b}_n\}$ spans V. Similarly $T \circ S = 1_W$ because $(T \circ S)(\mathbf{d}_i) = 1_W(\mathbf{d}_i)$ for each i. This proves (2).

(2)⇒(1). Suppose S exists as in (2). If $\mathbf{w}$ is in W then $\mathbf{w} = 1_W(\mathbf{w}) = (T \circ S)(\mathbf{w}) = T[S(\mathbf{w})]$. Hence $\mathbf{w}$ is in imT, and it follows that T is onto. On the other hand, if $T(\mathbf{v}) = \mathbf{0}$ where $\mathbf{v}$ is in V, then $\mathbf{v} = 1_V(\mathbf{v}) = (S \circ T)(\mathbf{v}) = S[T(\mathbf{v})] = S(\mathbf{0}) = \mathbf{0}$. This shows that $ker T = \{\mathbf{0}\}$, and so T is one-to-one. Hence (1) is proved.

Finally, suppose that $S_1 : W \to V$ is another transformation satisfying the conditions in (2): $S_1 \circ T = 1_V$ and $T \circ S_1 = 1_W$. Then Theorem 4 gives $S_1 = S_1 \circ 1_W = S_1 \circ (T \circ S) = (S_1 \circ T) \circ S = 1_V \circ S = S$. This proves the uniqueness in the last sentence of the Theorem. □

If $T : V \to W$ is an isomorphism, the unique linear transformation $S : W \to V$ that satisfies $S \circ T = 1_V$ and $T \circ S = 1_W$ is called the **inverse** of T and is denoted T^{-1} (as for matrices). Hence, if it exists, the transformation $T^{-1} : W \to V$ is uniquely determined by the equations

$$T^{-1} \circ T = 1_V \quad \text{and} \quad T \circ T^{-1} = 1_W.$$

These equations mean that T and T^{-1} are related by the **fundamental identities**:

$$T^{-1}[T(\mathbf{v})] = \mathbf{v} \quad \text{for all} \quad \mathbf{v} \text{ in } V, \quad \text{and} \quad T[T^{-1}(\mathbf{w})] = \mathbf{w} \quad \text{for all } \mathbf{w} \text{ in } W.$$

In other words each of T and T^{-1} "reverses" the action of the other. Among other things, this gives a way to find the action of T^{-1} if the effect of T on a basis is known. Here is an example.

Example 6. Let $\vec{e}_1 = [1\ 0\ 0]^T$, $\vec{e}_2 = [0\ 1\ 0]^T$ and $\vec{e}_3 = [0\ 0\ 1]^T$ denote the standard basis of $\mathbb{R}^3$. Consider the transformation $T : \mathbb{R}^3 \to \mathbb{P}_2$ given by

$$T(\vec{e}_1) = 1, \quad T(\vec{e}_2) = x - 1 \quad \text{and} \quad T(\vec{e}_3) = x^2 - 1.$$

Show that T is an isomorphism and find the action of T^{-1}.

Solution. Since $\{T(\vec{e}_1), T(\vec{e}_2), T(\vec{e}_3)\} = \{1, x-1, x^2-1\}$ is a basis of $\mathbb{P}_2$, it follows from Theorem 2 that T is an isomorphism. Hence $T^{-1} : \mathbb{P}_2 \to \mathbb{R}^3$ exists by Theorem 5. Moreover, since T^{-1} reverses the action of T we have

$$T^{-1}(1) = \vec{e}_1, \quad T^{-1}(x-1) = \vec{e}_2 \quad \text{and} \quad T^{-1}(x^2-1) = \vec{e}_3.$$

This defines T^{-1} because $\{1, x-1, x^2-1\}$ is a basis of $\mathbb{P}_2$. (If a formula is needed, we extend linearly to get $T^{-1}(a + bx + cx^2) = [a+b+c\ b\ c]^T$.) □

The following example illustrates that it is sometimes possible to show that a linear transformation $T : V \to W$ is an isomorphism by explicitly constructing the inverse $T^{-1} : W \to V$.

5.4. ISOMORPHISMS AND MATRICES

Example 7. *Consider the transpose transformation $T : \mathbb{M}_{n,n} \to \mathbb{M}_{n,n}$ where $T(A) = A^T$ for all A in $\mathbb{M}_{n,n}$. Show that T is an isomorphism and find T^{-1}.*

Solution. Rather than show that T is one-to-one and onto (which is not difficult!), we utilize the fact that $(A^T)^T = A$ for every matrix A in $\mathbb{M}_{n,n}$. In terms of T, this asserts that $(T \circ T)(A) = T[T(A)] = [A^T]^T = A = 1_{\mathbb{M}_{n,n}}(A)$ for all A in $\mathbb{M}_{n,n}$. This means that $T \circ T = 1_{\mathbb{M}_{n,n}}$, and hence that T is invertible and $T^{-1} = T$. $\square$

By contrast, we can sometimes show that a linear transformation is an isomorphism even though it is difficult to give an explicit formula for the inverse.

Example 8. *Define $T : \mathbb{P}_n \to \mathbb{R}^{n+1}$ by $T[p(x)] = [p(0)\ p(1)\ p(2)\ \cdots\ p(n)]^T$ for all $p(x)$ in $\mathbb{P}_n$. Show that T is an isomorphism.*

Solution. The verification that T is linear is left to the reader (see Example 3 §5.3). If $T[p(x)] = 0$ then $p(k) = 0$ for each $k = 0, 1, 2, \cdots, n$, so the polynomial $p(x)$ has $n+1$ distinct roots. But this implies that $p(x) = 0$ is the zero polynomial because no nonzero polynomial of degree at most n can have more than n roots (see Appendix A.3). Hence T is one-to-one. It follows by Theorem 3 that T is an isomorphism because $dim(\mathbb{P}_n) = n + 1 = dim(\mathbb{R}^{n+1})$. In particular, the inverse transformation $T^{-1} : \mathbb{R}^{n+1} \to \mathbb{P}_n$ exists, but we have not given an explicit formula for it. To do so requires some ingenuity. $\square$

The following theorem contains some basic properties of inverses of linear transformations that are the analogs of the same properties for matrices (see Theorem 3 §1.5). The proof uses Theorems 4 and 5, and is left to the reader.

Theorem 6. *Let $V \xrightarrow{T} W \xrightarrow{S} U$ be linear transformations. Then:*

(1) $1_V : V \to V$ *is an isomorphism, and* $(1_V)^{-1} = 1_V$.

(2) *If T is an isomorphism, so is T^{-1}, and* $(T^{-1})^{-1} = T$.

(3) *If S and T are both isomorphisms, so is $S \circ T$, and* $(S \circ T)^{-1} = T^{-1} \circ S^{-1}$.

5.4.3 Coordinates

In Section 4.9 we saw that every linear transformation $S : \mathbb{R}^n \to \mathbb{R}^m$ is given by $S(X) = AX$ for some $m \times n$ matrix A. In much the same way we can use matrix multiplication to describe the action of *any* linear transformation $T : V \to W$. However matrices must act on $\mathbb{R}^n$, while T acts on the abstract space V. So we must first find a way to convert vectors in V or W into columns. As in Section 4.9, this is done by using coordinates.

Let $\mathcal{B} = \{\mathbf{b}_1, \mathbf{b}_2, \cdots, \mathbf{b}_n\}$ be a basis of a vector space V. Hence every vector $\mathbf{v}$ in V is represented as a linear combination $\mathbf{v} = r_1\mathbf{b}_1 + r_2\mathbf{b}_2 + \cdots + r_n\mathbf{b}_n$

where the numbers r_i in $\mathbb{R}$ are uniquely[17] determined by $\mathbf{v}$. Hence, as in Section 4.9, we define the $\mathcal{B}$-**coordinate vector** $C_{\mathcal{B}}(\mathbf{v})$ of the vector $\mathbf{v}$ as follows:

$$C_{\mathcal{B}}(\mathbf{v}) = [r_1\ r_2\ \cdots\ r_n]^T \text{ when } \mathbf{v} = r_1\mathbf{b}_1 + r_2\mathbf{b}_2 + \cdots + r_n\mathbf{b}_n.$$

The numbers r_i are called the $\mathcal{B}$-**coordinates** of the vector $\mathbf{v}$.

Example 9. This is illustrated in the diagram where $\vec{e}_1$ and $\vec{e}_2$ are the standard basis vectors of $\mathbb{R}^2$, which give rise to the usual X_1- and X_2-axes. Thus a vector $\vec{v} = [4\ 3]^T$ has the usual coordinates $x_1 = 4$ and $x_2 = 3$.

However, another basis $\mathcal{B} = \{\mathbf{b}_1, \mathbf{b}_2\}$ is also depicted in the diagram, giving rise to the B_1- and B_2-axes. The vector $\vec{v}$ is represented as $\vec{v} = 3\mathbf{b}_1 + 2\mathbf{b}_2$, so $C_{\mathcal{B}}(\vec{v}) = [3\ 2]^T$. □

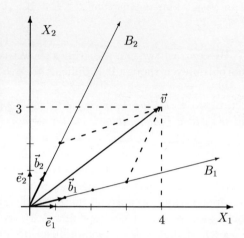

Example 10. If $\mathcal{E}$ is the standard basis of $\mathbb{R}^n$ (consisting of the columns of I_n), then $C_{\mathcal{E}}(X) = X$ for all columns X in $\mathbb{R}^n$.

Example 11. If $\mathcal{B} = \{1, x, x^2, \cdots, x^n\}$ in $\mathbb{P}_n$ then $C_{\mathcal{B}}(a_0 + a_1 x + \cdots + a_n x^n) = [a_0\ a_1\ \cdots\ a_n]^T$.

Example 12. If $\mathcal{B} = \left\{ \begin{bmatrix} 1 & 0 \\ 0 & 0 \end{bmatrix}, \begin{bmatrix} 0 & 1 \\ 0 & 0 \end{bmatrix}, \begin{bmatrix} 0 & 0 \\ 1 & 0 \end{bmatrix}, \begin{bmatrix} 0 & 0 \\ 0 & 1 \end{bmatrix} \right\}$ in $\mathbb{M}_{2,2}$ then $C_{\mathcal{B}} \begin{bmatrix} a & b \\ c & d \end{bmatrix} = [a\ b\ c\ d]^T$.

If $\mathcal{B}$ is a basis of the n-dimensional vector space V, the formation of $\mathcal{B}$-coordinates defines a transformation

$$C_{\mathcal{B}} : V \to \mathbb{R}^n.$$

In fact, $C_{\mathcal{B}}$ is an isomorphism (see Theorem 7 below). Moreover, as in Section 4.9, there is a simple way of relating the coordinates of a vector in V with respect to two different bases $\mathcal{B}$ and $\mathcal{D}$ of V. If $\mathcal{B} = \{\mathbf{b}_1, \mathbf{b}_2, \cdots, \mathbf{b}_n\}$, we define the **change matrix** $P_{\mathcal{D} \leftarrow \mathcal{B}}$ from $\mathcal{B}$ to $\mathcal{D}$ to be the matrix

$$P_{\mathcal{D} \leftarrow \mathcal{B}} = [C_{\mathcal{D}}(\mathbf{b}_1)\ C_{\mathcal{D}}(\mathbf{b}_2)\ \cdots\ C_{\mathcal{D}}(\mathbf{b}_n)]$$

---

[17]This requires that the basis B is *ordered*, that is the order in which the basis vectors are listed is taken into account. As in Section 4.9, when working with coordinates we assume (usually without saying so) that the bases are ordered.

5.4. ISOMORPHISMS AND MATRICES

with $C_\mathcal{D}(\mathbf{b}_1), C_\mathcal{D}(\mathbf{b}_2), \cdots, C_\mathcal{D}(\mathbf{b}_n)$ as its columns. When $V = \mathbb{R}^n$ this reduces to the notion discussed in Section 4.9. The basic properties of these concepts are collected in the following theorem.

Theorem 7. *Let $\mathcal{B}$ and $\mathcal{D}$ be bases of the n-dimensional vector space V. Then:*

(1) $C_\mathcal{B} : V \to \mathbb{R}^n$ *is an isomorphism.*

(2) $C_\mathcal{D}(\mathbf{v}) = P_{\mathcal{D} \leftarrow \mathcal{B}} \, C_\mathcal{B}(\mathbf{v})$ *for all vectors $\mathbf{v}$ in V.*

Proof. Write $\mathcal{B}$ out explicitly as $\mathcal{B} = \{\mathbf{b}_1, \mathbf{b}_2, \cdots, \mathbf{b}_n\}$.

(1). Let $\mathbf{v} = r_1 \mathbf{b}_1 + r_2 \mathbf{b}_2 + \cdots + r_n \mathbf{b}_n$ and $\mathbf{w} = s_1 \mathbf{b}_1 + s_2 \mathbf{b}_2 + \cdots + s_n \mathbf{b}_n$ be vectors in V, so that $C_\mathcal{B}(\mathbf{v}) = [r_1 \; r_2 \; \cdots \; r_n]^T$ and $C_\mathcal{B}(\mathbf{w}) = [s_1 \; s_2 \; \cdots \; s_n]^T$. We have $\mathbf{v} + \mathbf{w} = (r_1 + s_1)\mathbf{b}_1 + (r_2 + s_2)\mathbf{b}_2 + \cdots + (r_n + s_n)\mathbf{b}_n$, so

$$\begin{aligned} C_\mathcal{B}(\mathbf{v} + \mathbf{w}) &= [r_1 + s_1 \; r_2 + s_2 \; \cdots \; r_n + s_n]^T \\ &= [r_1 \; r_2 \; \cdots \; r_n]^T + [s_1 \; s_2 \; \cdots \; s_n]^T \\ &= C_\mathcal{B}(\mathbf{v}) + C_\mathcal{B}(\mathbf{w}). \end{aligned}$$

Similarly one verifies that $C_\mathcal{B}(r\mathbf{v}) = r \, C_\mathcal{B}(\mathbf{v})$ for all r in $\mathbb{R}$. Hence $C_\mathcal{B}$ is linear; it is onto by its definition, and it is one-to-one because $\mathcal{B}$ is independent. This proves (1).

(2). Given any vector $\mathbf{v}$ in V, write $\mathbf{v} = r_1 \mathbf{b}_1 + r_2 \mathbf{b}_2 + \cdots + r_n \mathbf{b}_n$ as before. Since $C_\mathcal{B}$ is linear by (1), block multiplication gives

$$\begin{aligned} P_{\mathcal{D} \leftarrow \mathcal{B}} \, C_\mathcal{B}(\mathbf{v}) &= [C_\mathcal{D}(\mathbf{b}_1) \; C_\mathcal{D}(\mathbf{b}_2) \; \cdots \; C_\mathcal{D}(\mathbf{b}_n)] \begin{bmatrix} r_1 \\ r_2 \\ \vdots \\ r_n \end{bmatrix} \\ &= r_1 C_\mathcal{D}(\mathbf{b}_1) + r_2 C_\mathcal{D}(\mathbf{b}_2) + \cdots + r_n C_\mathcal{D}(\mathbf{b}_n) \\ &= C_\mathcal{D}(r_1 \mathbf{b}_1 + r_2 \mathbf{b}_2 + \cdots + r_n \mathbf{b}_n) \\ &= C_\mathcal{D}(\mathbf{v}). \end{aligned}$$

This is what we wanted. $\square$

Example 13. If $\mathcal{B} = \{1, x, x^2\}$ and $\mathcal{D} = \{1 + x, x + x^2, x^2 + 1\}$ in $\mathbb{P}_2$, we have

$$P_{\mathcal{D} \leftarrow \mathcal{B}} = [C_\mathcal{D}(1) \; C_\mathcal{D}(x) \; C_\mathcal{D}(x^2)] = \tfrac{1}{2} \begin{bmatrix} 1 & 1 & -1 \\ -1 & 1 & 1 \\ 1 & -1 & 1 \end{bmatrix}.$$

(For example, the first column comes from $1 = \tfrac{1}{2}\{(1+x) - (x+x^2) + (x^2+1)\}$.)

Hence
$$C_\mathcal{D}(a + bx + cx^2) = P_{\mathcal{D} \leftarrow \mathcal{B}} \, C_\mathcal{B}(a + bx + cx^2)$$
$$= \tfrac{1}{2} \begin{bmatrix} 1 & 1 & -1 \\ -1 & 1 & 1 \\ 1 & -1 & 1 \end{bmatrix} \begin{bmatrix} a \\ b \\ c \end{bmatrix}$$
$$= \tfrac{1}{2} \begin{bmatrix} a + b - c \\ -a + b + c \\ a - b + c \end{bmatrix}$$

for all a, b and c in $\mathbb{R}$. As a check, this gives the expansion of $a + bx + cx^2$ in terms of the basis $\mathcal{D}$:

$$a + bx + cx^2 =$$
$$\tfrac{1}{2}(a + b - c)(1 + x) + \tfrac{1}{2}(-a + b + c)(x + x^2) + \tfrac{1}{2}(a - b + c)(x^2 + 1).$$

The reader can verify that this is indeed the case. $\square$

The following useful result will be used several times below.

Lemma 1. *Let $\mathcal{B}$ be a basis of the n-dimensional vector space V. If A and B are $m \times n$ matrices such that $A \, C_\mathcal{B}(\mathbf{v}) = B \, C_\mathcal{B}(\mathbf{v})$ for all $\mathbf{v}$ in V, then $A = B$.*

Proof. Let $A \, C_\mathcal{B}(\mathbf{v}) = B \, C_\mathcal{B}(\mathbf{v})$ for all $\mathbf{v}$ in V. If we write $\mathcal{B} = \{\mathbf{b}_1, \mathbf{b}_2, \cdots, \mathbf{b}_n\}$, then $C_\mathcal{B}(\mathbf{b}_j)$ is column j of the identity matrix (verify). It follows that $A \, C_\mathcal{B}(\mathbf{b}_j)$ is column j of A, so the equation $A \, C_\mathcal{B}(\mathbf{b}_j) = B \, C_\mathcal{B}(\mathbf{b}_j)$ shows that column j of A equals column j of B. Since this holds for each j, it shows that $A = B$. $\square$

These change matrices $P_{\mathcal{D} \leftarrow \mathcal{B}}$ are all invertible. In fact, we have

Theorem 8. *Let $\mathcal{B}$, $\mathcal{D}$, and $\mathcal{E}$ denote three bases of a vector space V. Then:*

(1) $P_{\mathcal{B} \leftarrow \mathcal{B}} = I$.

(2) $P_{\mathcal{D} \leftarrow \mathcal{B}}$ *is invertible and* $(P_{\mathcal{D} \leftarrow \mathcal{B}})^{-1} = P_{\mathcal{B} \leftarrow \mathcal{D}}$.

(3) $P_{\mathcal{E} \leftarrow \mathcal{D}} P_{\mathcal{D} \leftarrow \mathcal{B}} = P_{\mathcal{E} \leftarrow \mathcal{B}}$.

Proof. (1) is a routine verification, and (2) follows from (1) and (3). Finally (3) follows by right multiplying the left side by $C_\mathcal{B}(\mathbf{v})$, applying condition (2) in Theorem 7 three times, and using Lemma 1. $\square$

Property (3) in Theorem 8 explains the notation $P_{\mathcal{D} \leftarrow \mathcal{B}}$.

5.4.4 The Matrix of a Linear Transformation

The results in Theorems 4 and 6 about composition of linear transformations bear a striking resemblance to the analogous properties of matrix multiplication. Furthermore, we showed in Theorem 2 §4.9 that every linear transformation $\mathbb{R}^n \to \mathbb{R}^m$ is given by multiplication by an $m \times n$ matrix. Hence it will come as no surprise that, for any linear transformation T, there is a way to associate a matrix to T which can be used to describe the action of T.

So let $T : V \to W$ be a linear transformation where $dim V = n$ and $dim W = m$. We are interested in computing $T(\mathbf{v})$ for any vector $\mathbf{v}$ in V using matrix multiplication. This can be accomplished as follows. Let $\mathcal{B}$ and $\mathcal{D}$ be bases of V and W respectively. The idea is to compute $T(\mathbf{v})$ by first finding its $\mathcal{D}$-coordinate vector $C_{\mathcal{D}}[T(\mathbf{v})]$. This is a column in $\mathbb{R}^m$, and it turns out that there exists an $m \times n$ matrix $M_{\mathcal{D}\mathcal{B}}(T)$ such that

$$C_{\mathcal{D}}[T(\mathbf{v})] = M_{\mathcal{D}\mathcal{B}}(T)\, C_{\mathcal{B}}(\mathbf{v}) \quad \text{for all } \mathbf{v} \text{ in } V.$$

The matrix $M_{\mathcal{D}\mathcal{B}}(T)$ depends only on T and the bases $\mathcal{B}$ and $\mathcal{D}$, and is called the $\mathcal{D}\mathcal{B}$-**matrix** of T. Moreover, it is easy to compute. Indeed, if the basis $\mathcal{B}$ is listed as $\mathcal{B} = \{\mathbf{b}_1, \mathbf{b}_2, \cdots, \mathbf{b}_n\}$ then

$$M_{\mathcal{D}\mathcal{B}}(T) = [C_{\mathcal{D}}[T(\mathbf{b}_1)]\ \ C_{\mathcal{D}}[T(\mathbf{b}_2)]\ \cdots\ C_{\mathcal{D}}[T(\mathbf{b}_n)]] \tag{*}$$

is the matrix with the coordinate vectors

$$C_{\mathcal{D}}[T(\mathbf{b}_1)],\ C_{\mathcal{D}}[T(\mathbf{b}_2)],\ \cdots,\ C_{\mathcal{D}}[T(\mathbf{b}_n)]$$

as its columns. The following theorem summarizes this discussion.

Theorem 9. *Let $T : V \to W$ be a linear transformation where $dim V = n$ and $dim W = m$, and let $\mathcal{B}$ and $\mathcal{D}$ be bases of V and W respectively. Then*

$$C_{\mathcal{D}}[T(\mathbf{v})] = M_{\mathcal{D}\mathcal{B}}(T)\, C_{\mathcal{B}}(\mathbf{v}) \quad \text{for all } \mathbf{v} \text{ in } V$$

where $M_{\mathcal{D}\mathcal{B}}(T) = [C_{\mathcal{D}}[T(\mathbf{b}_1)]\ \ C_{\mathcal{D}}[T(\mathbf{b}_2)]\ \cdots\ C_{\mathcal{D}}[T(\mathbf{b}_n)]]$ is the $m \times n$ $\mathcal{D}\mathcal{B}$-matrix of T.

Proof. Let $\mathcal{B} = \{\mathbf{b}_1, \mathbf{b}_2, \cdots, \mathbf{b}_n\}$ and write $\mathbf{v} = r_1\mathbf{b}_1 + r_2\mathbf{b}_2 + \cdots + r_n\mathbf{b}_n$, so that $C_{\mathcal{B}}(\mathbf{v}) = [r_1\ r_2\ \cdots\ r_n]^T$. Since both $C_{\mathcal{D}}$ and T are linear, we have

$$\begin{aligned}
M_{\mathcal{D}\mathcal{B}}(T)\, C_{\mathcal{B}}(\mathbf{v}) &= [C_{\mathcal{D}}[T(\mathbf{b}_1)]\ \ C_{\mathcal{D}}[T(\mathbf{b}_2)]\ \cdots\ C_{\mathcal{D}}[T(\mathbf{b}_n)]] \begin{bmatrix} r_1 \\ r_2 \\ \vdots \\ r_n \end{bmatrix} \\
&= r_1 C_{\mathcal{D}}[T(\mathbf{b}_1)] + r_2 C_{\mathcal{D}}[T(\mathbf{b}_2)] + \cdots + r_n C_{\mathcal{D}}[T(\mathbf{b}_n)] \\
&= C_{\mathcal{D}}[T(r_1\mathbf{b}_1 + r_2\mathbf{b}_2 + \cdots + r_n\mathbf{b}_n)] \\
&= C_{\mathcal{D}}[T(\mathbf{v})].
\end{aligned}$$

using block multiplication. This is what we wanted. □

It is instructive to interpret Theorem 9 in terms of composition. Let $T : V \to W$ be linear where $\dim V = n$ and $\dim W = m$, and let $\mathcal{B} = \{\mathbf{b}_1, \mathbf{b}_2, \cdots, \mathbf{b}_n\}$ and $\mathcal{D}$ be bases of V and W respectively. For convenience, write $M = M_{\mathcal{DB}}(T)$ and let $T_M : \mathbb{R}^n \to \mathbb{R}^m$ be the matrix transformation given by $T_M(X) = MX$ for all X in $\mathbb{R}^n$. The diagram expresses the relationship between the transformations T and T_M and the two coordinate isomorphisms $C_\mathcal{B}$ and $C_\mathcal{D}$.

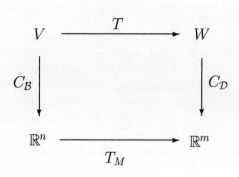

Then the conclusion in Theorem 9 is that $T_M[C_\mathcal{B}(\mathbf{v})] = C_\mathcal{D}[T(\mathbf{v})]$ holds for all $\mathbf{v}$ in V, in other words,

$$T_M \circ C_\mathcal{B} = C_\mathcal{D} \circ T \quad \text{equivalently} \quad T = C_\mathcal{D}^{-1} \circ T_M \circ C_\mathcal{B}.$$

This is often expressed by saying that the diagram is *commutative*: That is, starting with a vector $\mathbf{v}$ in V, the effect on $\mathbf{v}$ of applying the transformations in the diagram is the same no matter which route is taken to $\mathbb{R}^m$.

Example 14. Let $\mathcal{B}$ and $\mathcal{D}$ be bases of V and W respectively. Then the identity operator $1_V : V \to V$ has matrix $M_{\mathcal{BB}}(1_V) = I$, and the zero transformation $0 : V \to W$ has matrix $M_{\mathcal{DB}}(0) = 0$.

The next theorem shows that the construction of the matrix of a linear transformation preserves composition in the sense that composing linear transformations means multiplying their matrices.

Theorem 10. *Let $V \xrightarrow{T} W \xrightarrow{S} U$ be linear transformations, and let $\mathcal{B}$, $\mathcal{D}$ and $\mathcal{G}$ denote bases of V, W and U respectively. Then we have the composite $S \circ T : V \to U$ and the matrices are related as follows:*

$$M_{\mathcal{GB}}(S \circ T) = M_{\mathcal{GD}}(S)\, M_{\mathcal{DB}}(T).$$

Proof. Given $\mathbf{v}$ in V, right multiply the right side by $C_\mathcal{B}(\mathbf{v})$ and use Theorem 9 three times:

$$\begin{aligned} M_{\mathcal{GD}}(S)\, M_{\mathcal{DB}}(T)\, C_\mathcal{B}(\mathbf{v}) &= M_{\mathcal{GD}}(S)\, C_\mathcal{D}[T(\mathbf{v})] \\ &= C_\mathcal{G}\{S[T(\mathbf{v})]\} \\ &= C_\mathcal{G}\{(S \circ T)(\mathbf{v})\} \\ &= M_{\mathcal{GB}}(S \circ T) C_\mathcal{B}(\mathbf{v}). \end{aligned}$$

5.4. ISOMORPHISMS AND MATRICES

The result now follows because we can "cancel" $C_\mathcal{B}(\mathbf{v})$ by Lemma 1. $\square$

We can now characterize isomorphisms in terms of their matrices.

Theorem 11. *The following are equivalent for a linear transformation* $T : V \to W$ *where* $\dim V = n = \dim W$.

(1) T *is an isomorphism.*

(2) $M_{\mathcal{DB}}(T)$ *is invertible for all bases* $\mathcal{B}$ *and* $\mathcal{D}$ *of* V *and* W *respectively.*

(3) *There exist bases* $\mathcal{B}$ *and* $\mathcal{D}$ *of* V *and* W *respectively such that* $M_{\mathcal{DB}}(T)$ *is invertible.*

When this is the case, $M_{\mathcal{DB}}(T)^{-1} = M_{\mathcal{BD}}(T^{-1})$.

Proof. (1)$\Rightarrow$(2). Given (1), apply Theorem 10 to $T^{-1} \circ T = 1_V$ to get $M_{\mathcal{BD}}(T^{-1}) M_{\mathcal{DB}}(T) = M_{\mathcal{BB}}(1_V) = I$. Similarly $T \circ T^{-1} = 1_W$ gives the reverse equation $M_{\mathcal{DB}}(T) M_{\mathcal{BD}}(T^{-1}) = I$. This proves (2) and the last sentence.

(2)$\Rightarrow$(3). This is clear since V and W *have* bases (they are finite dimensional).

(3)$\Rightarrow$(1). Assume (3) and let $\mathcal{B} = \{\mathbf{b}_1, \mathbf{b}_2, \cdots, \mathbf{b}_n\}$ and $\mathcal{D}$ be bases of V and W, respectively. Then $M_{\mathcal{DB}}(T)$ is invertible by (3), so its columns $\{C_\mathcal{D}[T(\mathbf{b}_1)], C_\mathcal{D}[T(\mathbf{b}_2)], \cdots, C_\mathcal{D}[T(\mathbf{b}_n)]\}$ form a basis of $\mathbb{R}^n$. Since $C_\mathcal{D}^{-1} : \mathbb{R}^n \to W$ is an isomorphism, it follows from Theorem 2 that $\{T(\mathbf{b}_1), T(\mathbf{b}_2), \cdots, T(\mathbf{b}_n)\}$ is a basis of W. Hence T is an isomorphism, again by Theorem 2. $\square$

We conclude this section with a fact about rank: If V and W are finite dimensional vector spaces, the matrices of a linear transformation $T : V \to W$ all have the *same* rank. This follows from the fact (see below) that

$$\text{rank}\, M_{\mathcal{BD}}(T) = \dim(\text{im}\, T) \qquad (**)$$

for all bases $\mathcal{B}$ and $\mathcal{D}$ of V and W, respectively ($\text{im}\, T = \{T(\mathbf{v}) \mid \mathbf{v} \text{ in } V\}$ does not depend on $\mathcal{B}$ or $\mathcal{D}$). To prove (**), write $M_{\mathcal{DB}}(T) = A$ for simplicity, and recall that $\text{rank}\, A = \dim(\text{col}\, A)$. Hence it suffices to show that $\text{im}\, T \cong \text{col}\, A$ and, to this end, define

$$S : \text{im}\, T \to \text{col}\, A \text{ by } S(T(\mathbf{v})) = C_\mathcal{D}(T(\mathbf{v})) \text{ for all } T(\mathbf{v}) \text{ in } \text{im}\, T.$$

This makes sense because $C_\mathcal{D}(T(\mathbf{v})) = A\, C_\mathcal{B}(\mathbf{v})$ for all $\mathbf{v}$ by Theorem 9, and $A\, C_\mathcal{B}(\mathbf{v})$ is in $\text{col}\, A$ because $\text{col}\, A = \{AX \mid X \text{ in } \mathbb{R}^m\}$. Moreover S is linear and one-to-one because this is true of $C_\mathcal{B}$, and it is onto because every X in $\mathbb{R}^m$ has the form $X = C_\mathcal{B}(\mathbf{v})$ ($C_\mathcal{B}$ is onto). Hence S is an isomorphism, so (**) is proved.

Because of (**), it is natural to define the **rank** of T by

$$\text{rank}\, T = \dim(\text{im}\, T).$$

Hence we have proved

Theorem 12. *If $T : V \to W$ is a linear transformation where V and W are finite dimensional, then*

$$rank\, T = rank(M_{\mathcal{BD}}(T))$$

for all bases $\mathcal{B}$ and $\mathcal{D}$ of V and W, respectively.

If $T : V \to W$ is a linear transformation of rank r, the next example exhibits a choice of bases for V and W that makes the matrix of T as nice as possible.

Example 15. *Let $T : V \to W$ be a linear transformation where $dim V = n$ and $dim W = m$. Show that there exist bases $\mathcal{B}$ and $\mathcal{D}$ of V and W such that*

$$M_{\mathcal{DB}}(T) = \begin{bmatrix} I_r & 0 \\ 0 & 0 \end{bmatrix} \quad \text{where } r = rank\, T.$$

Solution. By extending a basis of $ker\, T$, let

$$\mathcal{B} = \{\mathbf{b}_1, \mathbf{b}_2, \cdots, \mathbf{b}_r, \mathbf{b}_{r+1}, \cdots, \mathbf{b}_n\}$$

be a basis of V such that $\{\mathbf{b}_{r+1}, \cdots, \mathbf{b}_n\}$ is a basis of $ker\, T$. One verifies that $\{T(\mathbf{b}_1), T(\mathbf{b}_2), \cdots, T(\mathbf{b}_r)\}$ is a basis of $im\, T$ (so $rank\, T = r$), and we can extend it to a basis

$$\mathcal{D} = \{T(\mathbf{b}_1), T(\mathbf{b}_2), \cdots, T(\mathbf{b}_r), \mathbf{d}_{r+1}, \cdots, \mathbf{d}_m\}$$

of W. Hence

$$
\begin{aligned}
M_{\mathcal{DB}}(T) &= [C_{\mathcal{D}}(T(\mathbf{b}_1)) \cdots C_{\mathcal{D}}(T(\mathbf{b}_r)) \; C_{\mathcal{D}}(T(\mathbf{b}_{r+1})) \cdots C_{\mathcal{D}}(T(\mathbf{b}_n))] \\
&= \begin{bmatrix} I_r & 0 \\ 0 & 0 \end{bmatrix}
\end{aligned}
$$

in block form, as required. $\square$

Exercises

1. In each case either show that the statement is true or give an example showing that it is false. Assume that T is linear throughout.

    (a) If $T : V \to V$ satisfies $T^2 = T$ then T is an isomorphism.

    (b) If $T : V \to V$ satisfies $im\, T \subseteq ker\, T$ then $T^2 = 0$.

    (c) If $T : \mathbb{M}_{n,n} \to \mathbb{M}_{n,n}$ and $T(U) = U$ for some invertible matrix U, then T is an isomorphism.

    (d) If $T : V \to W$ is an isomorphism then $T^{-1}(T(\mathbf{w})) = \mathbf{w}$ for every vector $\mathbf{w}$ in W.

5.4. ISOMORPHISMS AND MATRICES

(e) Every isomorphism $T : \mathbb{M}_{n,n} \to \mathbb{M}_{n,n}$ satisfies $T(I) = I$.

(f) If $dim V = 4$ and $dim W = 5$, it is not possible to have an isomorphism $V \to W$.

(g) Let $T : V \to W$ be linear and let $\mathcal{B} = \{\mathbf{b}_1, \mathbf{b}_2\}$ and $\mathcal{D} = \{\mathbf{d}_1, \mathbf{d}_2, \mathbf{d}_3\}$ be bases of V and W, respectively. If $T(\mathbf{b}_2) = 5\mathbf{d}_1 + 7\mathbf{d}_2 - 2\mathbf{d}_3$, then column 2 of $M_{\mathcal{DB}}(T)$ is $[5\ 7\ -2]^T$.

(h) If $T : \mathbb{M}_{2,3} \to \mathbb{M}_{3,2}$ is defined by $T(A) = A^T$ for each A in $\mathbb{M}_{2,3}$, then $T^{-1} = T$.

2. In each case show that T is an isomorphism and give a formula for the action of T^{-1}.

   (a) $T : \mathbb{R}^3 \to \mathbb{P}_2$ with $T([a\ b\ c]^T) = (a+b) + (b+c)x + (c+a)x^2$.

   (b) $T : \mathbb{C} \to \mathbb{C}$ with $T(z) = \bar{z}$.

   (c) $T : \mathbb{M}_{mn} \to \mathbb{M}_{mn}$ with $T(A) = UAV$, where U and V are fixed invertible matrices.

   (d) $T : \mathbb{P}_1 \to \mathbb{R}^2$ with $T[p(x)] = [p(0)\ p(1)]^T$.

   (e) $T : \mathbb{P}_n \to \mathbb{P}_n$ with $T[p(x)] = p(3-x)$. [Hint: Consider T^2.]

   (f) $T : V \to V$ with $T(\mathbf{v}) = k\mathbf{v}$, where $k \neq 0$ is a fixed real number.

   (g) $T : \mathbb{P}_1 \to \mathbb{P}_1$ with $T(a + bx) = (b - a) - ax$, [Hint: Consider T^3.]

   (h) $T : \mathbb{R}^2 \to \mathbb{R}^2$ with $T([x\ y]^T) = [ky - x\ y]^T$, where k is a real number.

3. If $T : \mathbb{R}^4 \to \mathbb{R}^4$ is defined by $T([x\ y\ z\ w]^T) = [-y\ x-y\ z\ -w]^T$, show that $T^6 = 1_{\mathbb{R}^4}$, and so determine T^{-1}.

4. If V is the vector space in Exercise 5 §5.1, find an isomorphism $\mathbb{R} \to V$.

5. If $T : V \to V$ is linear, show that $T^2 = 1_V$ if and only if T is an isomorphism and $T^{-1} = T$.

6. In each case find the coordinates of $\mathbf{v}$ with respect to the basis $\mathcal{B}$ of V.

   (a) $V = \mathbb{P}_2$, $\mathbf{v} = x^2 - 3x + 4$, $\mathcal{B} = \{x + 2,\ x^2 - 1,\ 1\}$.

   (b) $V = \mathbb{M}_{2,2}$, $\mathbf{v} = \begin{bmatrix} 2 & 1 \\ 0 & -1 \end{bmatrix}$,
   $\mathcal{B} = \left\{ \begin{bmatrix} 1 & 1 \\ 0 & 0 \end{bmatrix}, \begin{bmatrix} 1 & 0 \\ 1 & 0 \end{bmatrix}, \begin{bmatrix} 1 & 0 \\ 0 & 1 \end{bmatrix}, \begin{bmatrix} 0 & 0 \\ 0 & 1 \end{bmatrix} \right\}$.

7. In each case find the matrix $M_{\mathcal{DB}}(T)$.

   (a) $T : \mathbb{P}_3 \to \mathbb{P}_2$ where $T(a + bx + cx^2) = b + cx$, $\mathcal{B} = \{1, x, x^2\}$ and $\mathcal{D} = \{1, x\}$.

   (b) $T : \mathbb{M}_{2,2} \to \mathbb{M}_{2,2}$ where $T(A) = A^T$,
   $\mathcal{B} = \mathcal{D} = \left\{ \begin{bmatrix} 1 & 0 \\ 0 & 0 \end{bmatrix}, \begin{bmatrix} 0 & 1 \\ 0 & 0 \end{bmatrix}, \begin{bmatrix} 0 & 0 \\ 1 & 0 \end{bmatrix}, \begin{bmatrix} 0 & 0 \\ 0 & 1 \end{bmatrix} \right\}$.

   (c) $T : \mathbb{M}_{2,2} \to \mathbb{R}$ where $T(A) = trA$,
   $\mathcal{B} = \left\{ \begin{bmatrix} 1 & 0 \\ 0 & 0 \end{bmatrix}, \begin{bmatrix} 0 & 1 \\ 0 & 0 \end{bmatrix}, \begin{bmatrix} 0 & 0 \\ 1 & 0 \end{bmatrix}, \begin{bmatrix} 0 & 0 \\ 0 & 1 \end{bmatrix} \right\}$, $\mathcal{D} = \{1\}$.

8. In each case find the action of T given the matrix $M_{\mathcal{DB}}(T)$.

   (a) $T : \mathbb{P}_2 \to \mathbb{R}^2$, $M_{\mathcal{DB}}(T) = \begin{bmatrix} 1 & -1 & 0 \\ 2 & 0 & 1 \end{bmatrix}$, $\mathcal{B} = \{1, x, x^2\}$, $\mathcal{D} = \{[1\ 1]^T, [1\ 0]^T\}$.

   (b) $T : \mathbb{R}^3 \to \mathbb{M}_{2,2}$, $M_{\mathcal{DB}}(T) = \begin{bmatrix} 1 & 0 & 0 \\ -1 & 2 & 1 \\ 0 & 1 & 2 \\ 1 & 0 & 0 \end{bmatrix}$, where $\mathcal{B}$ is the standard basis of $\mathbb{R}^3$ and $\mathcal{D} = \left\{ \begin{bmatrix} 1 & 0 \\ 0 & 0 \end{bmatrix}, \begin{bmatrix} 0 & 1 \\ 0 & 0 \end{bmatrix}, \begin{bmatrix} 0 & 0 \\ 1 & 0 \end{bmatrix}, \begin{bmatrix} 0 & 0 \\ 0 & 1 \end{bmatrix} \right\}$.

9. Let $D : \mathbb{P}_3 \to \mathbb{P}_2$ be the differentiation map: $D[p(x)] = p'(x)$. Find $M_{\mathcal{DB}}(T)$ where $\mathcal{B} = \{1, x, x^2, x^3\}$ and $\mathcal{D} = \{1, x, x^2\}$, and use it to compute $D(a + bx + cx^2 + dx^3)$.

10. If $T : V \to \mathbb{R}^m$ and let $\mathcal{B} = \{\mathbf{b}_1, \mathbf{b}_2, \cdots, \mathbf{b}_n\}$ be a basis of V. If $\mathcal{E}$ is the standard basis of $\mathbb{R}^m$, show that $M_{\mathcal{EB}}(T) = [T(\mathbf{b}_1)\ T(\mathbf{b}_2)\ \cdots\ T(\mathbf{b}_n)]$.

11. Let $T : V \to W$ be an isomorphism. If $\mathcal{B} = \{\mathbf{b}_1, \mathbf{b}_2, \cdots, \mathbf{b}_n\}$ is a basis of V, Theorem 2 asserts that $\mathcal{D} = \{T(\mathbf{b}_1), T(\mathbf{b}_2), \cdots, T(\mathbf{b}_n)\}$ is a basis of W. Show that $M_{\mathcal{DB}}(T) = I_n$.

12. Let U be any $n \times n$ invertible matrix. If $\mathcal{D} = \{\mathbf{d}_1, \mathbf{d}_2, \cdots, \mathbf{d}_n\}$ where $\mathbf{d}_j$ is column j of U, show that $M_{\mathcal{ED}}(1_{\mathbb{R}^n}) = U$ where $\mathcal{E}$ is the standard basis of $\mathbb{R}^n$.

13. Let $V \xrightarrow{T} W \xrightarrow{S} U$ be linear transformations.

    (a) If S and T are both onto, prove that $S \circ T$ is onto.

    (b) If S and T are both one-to-one, prove that $S \circ T$ is one-to-one.

14. Let $V \xrightarrow{T} W \xrightarrow{S} U$ be linear transformations.

    (a) If T is onto and $S \circ T = S_1 \circ T$ where $S_1 : W \to U$, show that $S = S_1$.

    (b) If S is one-to-one and $S \circ T = S \circ T_1$ where $T_1 : V \to W$, show that $T = T_1$.

15. Let $V \xrightarrow{T} W \xrightarrow{S} U$ be linear transformations.

    (a) If $S \circ T$ is one-to-one, show that T is one-to-one.

    (b) If $S \circ T$ is onto, show that S is onto.

16. Let $V \xrightarrow{T} W \xrightarrow{S} V$ be linear transformations such that $S \circ T = 1_V$. If $\dim V = \dim W$ is finite, show that also $T \circ S = 1_W$ (so T is an isomorphism and $S = T^{-1}$).

17. Let $\mathbb{S}$ be the space of all sequences $(a_0, a_1, a_2, \cdots)$ of real numbers with componentwise operations (See Exercise 33 §5.2). Define $T : \mathbb{S} \to \mathbb{S}$ and

$R : \mathbb{S} \to \mathbb{S}$ by $T(a_0, a_1, a_2, \cdots) = (a_1, a_2, a_3, \cdots)$ and $R(a_0, a_1, a_2, \cdots) = (0, a_0, a_1, a_2, a_3, \cdots)$.

(a) Show that T and R are linear transformations.

(b) Show that $T \circ R = 1_{\mathbb{S}}$ but $R \circ T \neq 1_{\mathbb{S}}$.

(c) Show that T is onto but not one-to-one, and that R is one-to-one but not onto.

18. Let $T : V \to W$ be a linear transformation where $dim V$ and $dim W$ are finite. If T is one-to-one, show that there exists a linear transformation $S : W \to V$ such that $S \circ T = 1_V$. [Hint : If $\mathcal{B} = \{\mathbf{b}_1, \mathbf{b}_2, \cdots, \mathbf{b}_n\}$ is a basis of V, show that $\{T(\mathbf{b}_1), T(\mathbf{b}_2), \cdots, T(\mathbf{b}_n)\}$ is independent.]

19. Let $T : V \to W$ be a linear transformation where $dim V$ and $dim W$ are finite. If T is onto, show that there exists a linear transformation $S : W \to V$ such that $T \circ S = 1_W$. [Hint : By Theorem 3 §5.2. find a basis $\{\mathbf{b}_1, \mathbf{b}_2, \cdots, \mathbf{b}_m, \mathbf{b}_{m+1}, \cdots, \mathbf{b}_n\}$ of V such that $\{\mathbf{b}_{m+1}, \cdots, \mathbf{b}_n\}$ is a basis of $ker T$. Show that $\{T(\mathbf{b}_1), T(\mathbf{b}_2), \cdots, T(\mathbf{b}_m)\}$ is independent and (as T is onto) spans W.]

20. (a) If U is a fixed $n \times n$ matrix, define $S_U : \mathbb{M}_{n,n} \to \mathbb{M}_{n,n}$ by $S_U(A) = UA$ for all A in $\mathbb{M}_{n,n}$.

(a) Show that S_U is linear.

(b) If U is invertible, show that S_U is an isomorphism.

(c) Is every linear transformation $\mathbb{M}_{n,n} \to \mathbb{M}_{n,n}$ equal to S_U for some U? Defend your answer.

21. If A is an $m \times n$ matrix, let $T_A : \mathbb{R}^n \to \mathbb{R}^m$ be defined by $T_A(X) = AX$ for all X in $\mathbb{R}^n$.

(a) Show that $T_A \circ T_B = T_{AB}$ and $T_{I_n} = 1_{\mathbb{R}^n}$.

(b) If A is invertible and $n \times n$, show that T_A is invertible and $(T_A)^{-1} = T_{A^{-1}}$.

(c) If $T_A = T_B$, show that $A = B$.

(d) If A, B and C are three matrices, show that $A(BC) = (AB)C$ by considering $T_A \circ (T_B \circ T_C) = (T_A \circ T_B) \circ T_C$. [Hint: (a) and (c).]

(e) If $\mathcal{E}$ and $\mathcal{F}$ are the standard bases of $\mathbb{R}^n$ and $\mathbb{R}^m$ respectively, show that $M_{\mathcal{F}\mathcal{E}}(T_A) = A$.

22. Let $T : V \to W$ and $T_1 : V \to W$ be linear transformations. If $M_{\mathcal{D}\mathcal{B}}(T) = M_{\mathcal{D}\mathcal{B}}(T_1)$ for some bases $\mathcal{B}$ and $\mathcal{D}$ of V and W, show that $T = T_1$.

23. Prove Theorem 6. [Hint: Mimic the proof of Theorem 3 §1.5.]

24. Show that $M_{\mathcal{D}'\mathcal{B}'}(T) P_{\mathcal{B}' \leftarrow \mathcal{B}} = P_{\mathcal{D}' \leftarrow \mathcal{D}} M_{\mathcal{D}\mathcal{B}}(T)$ where $T : V \to W$ is a linear transformation, $\mathcal{B}$ and $\mathcal{B}'$ are bases of V, and $\mathcal{D}$ and $\mathcal{D}'$ are bases of W. [Hint: Right multiply by $C_{\mathcal{B}}(\mathbf{v})$ for arbitrary $\mathbf{v}$ in V and use Lemma 1 and Theorems 7 and 9.]

25. Let $T : V \to W$ be a linear transformation, and suppose that $S \circ T = 1_V$ and $T \circ S = 1_W$ for some transformation $S : W \to V$ (not assumed to be linear). Show that S must be linear.

26. Let m and n be fixed integers. If A and B are $m \times n$ matrices, define $A \overset{r}{\sim} B$ to mean that $B = UA$ for some invertible $m \times m$ matrix U.

    (a) Show that $\overset{r}{\sim}$ is an equivalence relation on the set $\mathbb{M}_{m,n}$ of all $m \times n$ matrices. (See the discussion following Theorem 3)

    (b) Show that $A \overset{r}{\sim} B$ if and only if $A \to B$ by row operations. [Hint: Section 1.6.]

5.5 LINEAR OPERATORS AND SIMILARITY

While the study of linear transformations from one vector space to another is important, the central problem of linear algebra is to understand the structure of an operator from a vector space to itself. If $T : V \to V$ is any linear operator where $dim V = n$, it is possible (see Example 15 §5.4) to choose bases $\mathcal{B}$ and $\mathcal{D}$ of V such that the matrix of T has a very simple form indeed: $M_{\mathcal{D}\mathcal{B}}(T) = \begin{bmatrix} I_r & 0 \\ 0 & 0 \end{bmatrix}$ where $r = rank T$. Consequently nothing can be learned about T from examining the matrices $M_{\mathcal{D}\mathcal{B}}(T)$ where $\mathcal{B}$ and $\mathcal{D}$ can be chosen arbitrarily. But if we insist that $\mathcal{B}$ and $\mathcal{D}$ are equal, the situation changes and our central problem amounts to finding a basis $\mathcal{B}$ such that $M_{\mathcal{B}\mathcal{B}}(T)$ is as simple as possible. The complete solution to this task is beyond the scope of this book, but we outline the main ideas in the following two sections.

5.5.1 The $\mathcal{B}$-Matrix of an Operator

Let $T : V \to V$ be a linear operator where $dim V = n$. If $\mathcal{B}$ is a basis of V, we write
$$M_{\mathcal{B}}(T) = M_{\mathcal{B}\mathcal{B}}(T)$$
for simplicity, and call this the $\mathcal{B}$-**matrix** of of T. Hence if $\mathcal{B}$ is listed (in a fixed order as) as $\mathcal{B} = \{\mathbf{b}_1, \mathbf{b}_2, \cdots, \mathbf{b}_n\}$ then
$$M_{\mathcal{B}}(T) = [C_{\mathcal{B}}(T(\mathbf{b}_1)) \ C_{\mathcal{B}}(T(\mathbf{b}_2)) \ \cdots \ C_{\mathcal{B}}(T(\mathbf{b}_n))]$$
as in Section 5.4, where $C_{\mathcal{B}}(\mathbf{v})$ denotes the coordinate vector of $\mathbf{v}$ with respect to the basis $\mathcal{B}$. Note that this agrees with the notation in Section 4.9 when $V = \mathbb{R}^n$.

Example 1. If $w = a + bi$ is a fixed complex number, define $T : \mathbb{C} \to \mathbb{C}$ by $T(z) = wz$ for all z in $\mathbb{C}$. Then T is linear (verify) and, if $\mathcal{B} = \{1, i\}$, we have
$$M_{\mathcal{B}}(T) = [C_{\mathcal{B}}(w) \ C_{\mathcal{B}}(wi)] = \begin{bmatrix} a & -b \\ b & a \end{bmatrix}.$$

5.5. LINEAR OPERATORS AND SIMILARITY

If $w \neq 0$, then $|w| = \sqrt{a^2 + b^2} \neq 0$ and it is instructive to write this as

$$M_{\mathcal{B}}(T) = \sqrt{a^2 + b^2} \begin{bmatrix} \frac{a}{\sqrt{a^2+b^2}} & \frac{-b}{\sqrt{a^2+b^2}} \\ \frac{b}{\sqrt{a^2+b^2}} & \frac{a}{\sqrt{a^2+b^2}} \end{bmatrix}$$

$$= |w| \begin{bmatrix} \cos\theta & -\sin\theta \\ \sin\theta & \cos\theta \end{bmatrix}$$

where θ is the argument of w. Hence, if we identify the complex number

$z = x + iy$ with the point $\begin{bmatrix} x \\ y \end{bmatrix}$,

multiplication by w has the effect of a rotation through the angle θ followed by expansion by $|w|$. This is depicted in the diagram. □

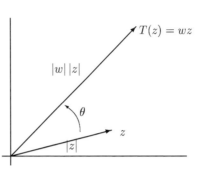

The following theorem collects several results from Section 5.4 for reference (specialized to operators).

Theorem 1. Let $T : V \to V$ be a linear operator where $\dim V = n$. Then:

(1) $C_{\mathcal{B}}(T(\mathbf{v})) = M_{\mathcal{B}}(T) \, C_{\mathcal{B}}(\mathbf{v})$ for all $\mathbf{v}$ in V.

(2) If $S : V \to V$ is another operator, $M_{\mathcal{B}}(S \circ T) = M_{\mathcal{B}}(S) \, M_{\mathcal{B}}(T)$.

(3) T is an isomorphism if and only if $M_{\mathcal{B}}(T)$ is invertible for some (equivalently every) basis $\mathcal{B}$ of V.

(4) If T is invertible then $M_{\mathcal{B}}(T^{-1}) = [M_{\mathcal{B}}(T)]^{-1}$.

Example 2. If A is an $n \times n$ matrix, define $T_A : \mathbb{R}^n \to \mathbb{R}^n$ by $T_A(X) = AX$ for all X in $\mathbb{R}^n$. If $\mathcal{E}$ is the standard basis of $\mathbb{R}^n$, show that

$$M_{\mathcal{E}}(T_A) = A.$$

Solution. We have $\mathcal{E} = \{E_1, E_2, \cdots, E_n\}$ where the E_j are the columns of I_n. Since $C_{\mathcal{E}}(X) = X$ for all X in $\mathbb{R}^n$, we have

$$\begin{aligned} M_{\mathcal{E}}(T_A) &= [C_{\mathcal{E}}(AE_1) \ C_{\mathcal{E}}(AE_2) \ \cdots \ C_{\mathcal{E}}(AE_n)] \\ &= [AE_1 \ AE_2 \ \cdots \ AE_n] \\ &= A[E_1 \ E_2 \ \cdots \ E_n] \\ &= AI_n = A \end{aligned}$$

as required. □

Example 3. Let $D : \mathbb{P}_2 \to \mathbb{P}_2$ be the differentiation operator defined by $D(p(x)) = p'(x)$ for all polynomials $p(x)$ in $\mathbb{P}_2$. If $\mathcal{B} = \{1, x, x^2\}$, we have

$$\begin{aligned} M_\mathcal{B}(D) &= [C_\mathcal{B}(D(1)) \ C_\mathcal{B}(D(x)) \ C_\mathcal{B}(D(x^2))] \\ &= [C_\mathcal{B}(0) \ C_\mathcal{B}(1) \ C_\mathcal{B}(2x)] \\ &= \begin{bmatrix} 0 & 1 & 0 \\ 0 & 0 & 2 \\ 0 & 0 & 0 \end{bmatrix}. \end{aligned}$$

Hence, if $p(x) = a + bx + cx^2$, we have

$$C_\mathcal{B}\{D(p(x))\} = M_\mathcal{B}(D)\, C_\mathcal{B}(p(x)) = \begin{bmatrix} 0 & 1 & 0 \\ 0 & 0 & 2 \\ 0 & 0 & 0 \end{bmatrix} \begin{bmatrix} a \\ b \\ c \end{bmatrix} = \begin{bmatrix} b \\ 2c \\ 0 \end{bmatrix}.$$

Thus Theorem 1 gives $D(p(x)) = b + (2c)x$, as it should. Note that $M_\mathcal{B}(D)$ is not invertible here, so D is not invertible. Of course this can also be seen from the fact that $\ker D \neq 0$ (for example, $D(1) = 0$). $\square$

Example 4. Let $p_0(x), p_1(x), \cdots, p_n(x)$ be polynomials in $\mathbb{P}_n$ with $\deg p_k(x) = k$ for each k. If $T : \mathbb{P}_n \to \mathbb{P}_n$ is defined by $T(x^k) = p_k(x)$ for each k, find $M_\mathcal{B}(T)$ where $\mathcal{B} = \{1, x, \cdots, x^n\}$, and conclude that $\mathcal{D} = \{p_0(x), p_1(x), \cdots, p_n(x)\}$ is a basis of $\mathbb{P}_n$. (See Example 2 §5.2).

Solution. The highest power of x occurring in $p_k(x)$ is x^k (because $\deg p_k(x) = k$). Hence $M_\mathcal{B}(T)$ has the form

$$\begin{aligned} M_\mathcal{B}(T) &= [C_\mathcal{B}(p_0(x)) \ C_\mathcal{B}(p_1(x)) \ C_\mathcal{B}(p_2(x)) \ \cdots \ C_\mathcal{B}(p_n(x))] \\ &= \begin{bmatrix} a_0 & * & * & \cdots & * \\ 0 & a_1 & * & \cdots & * \\ 0 & 0 & a_2 & \cdots & * \\ \vdots & \vdots & \vdots & & \vdots \\ 0 & 0 & 0 & \cdots & a_n \end{bmatrix} \end{aligned}$$

where $a_k \neq 0$ is the leading coefficient of $p_k(x)$ for each k. It follows that $M_\mathcal{B}(T)$ is invertible (its determinant is $a_0 a_1 a_2 \cdots a_n \neq 0$), so T is an isomorphism by Theorem 1. Hence $\mathcal{D}$ is a basis of $\mathbb{P}_n$ by Theorem 2 §5.4. $\square$

The next example shows how a property of matrices can be deduced by first verifying the same property of operators, and then using Theorem 1. This technique is important because properties of operators are often easier to visualize.

Example 5. Let $T : V \to V$ be an operator on the n-dimensional space V.

(1) *Show that an isomorphism $S : V \to V$ exists such that $T \circ S \circ T = T$.*

(2) *If A is an $n \times n$ matrix, show that $AUA = A$ for some invertible matrix U.*

5.5. LINEAR OPERATORS AND SIMILARITY

Solution. (1). By extending a basis of $kerT$, choose a basis $\mathcal{B} = \{\mathbf{b}_1, \mathbf{b}_2, \cdots, \mathbf{b}_r, \mathbf{b}_{r+1}, \cdots, \mathbf{b}_n\}$ of V where $\{\mathbf{b}_{r+1}, \cdots, \mathbf{b}_n\}$ is a basis of $kerT$. Then $\{T(\mathbf{b}_1), T(\mathbf{b}_2), \cdots, T(\mathbf{b}_r)\}$ is independent (verify) so extend it to a basis $\mathcal{D} = \{T(\mathbf{b}_1), T(\mathbf{b}_2), \cdots, T(\mathbf{b}_r), \mathbf{d}_{r+1}, \cdots, \mathbf{d}_n\}$ of V. We use this basis $\mathcal{D}$ to define $S : V \to V$ as follows:

$$S(T(\mathbf{b}_i)) = \mathbf{b}_i \text{ for } 1 \leq i \leq r \quad \text{and} \quad S(\mathbf{d}_j) = \mathbf{b}_j \text{ for } r < j \leq n.$$

Then S is an isomorphism because it carries the basis $\mathcal{D}$ to the basis $\mathcal{B}$. To see that $T \circ S \circ T = T$, we verify that both $T \circ S \circ T$ and T have the same effect on every vector in the basis $\mathcal{B}$. In fact:

$$(T \circ S \circ T)(\mathbf{b}_i) = T(S(T(\mathbf{b}_i))) = \begin{cases} T(\mathbf{b}_i) & \text{for } i \leq r \text{ by the definition of } S, \\ T(\mathbf{b}_i) & \text{for } i > r \text{ because both equal } \mathbf{0}. \end{cases}$$

(2). Given A, define $T = T_A : \mathbb{R}^n \to \mathbb{R}^n$ as in Example 2, and construct an isomorphism $S : \mathbb{R}^n \to \mathbb{R}^n$ as in (1). If $\mathcal{E}$ is the standard basis of $\mathbb{R}^n$, then $M_{\mathcal{E}}(T) = M_{\mathcal{E}}(T_A) = A$ by Example 2, and $U = M_{\mathcal{E}}(S)$ is invertible by Theorem 1. Finally

$$AUA = M_{\mathcal{E}}(T) M_{\mathcal{E}}(S) M_{\mathcal{E}}(T) = M_{\mathcal{E}}(T \circ S \circ T) = M_{\mathcal{E}}(T) = A$$

again by Theorem 1. This is what we wanted. □

5.5.2 Change of Basis

If $T : V \to V$ is an operator, it is critical to understand how the matrix $M_{\mathcal{B}}(T)$ changes when the basis $\mathcal{B}$ is changed. To do so, recall that if $\mathcal{B} = \{\mathbf{b}_1, \mathbf{b}_2, \cdots, \mathbf{b}_n\}$ and $\mathcal{D}$ are two bases of V the change matrix $P_{\mathcal{D} \leftarrow \mathcal{B}}$ is defined by

$$P_{\mathcal{D} \leftarrow \mathcal{B}} = [C_{\mathcal{D}}(\mathbf{b}_1) \ C_{\mathcal{D}}(\mathbf{b}_2) \ \cdots \ C_{\mathcal{D}}(\mathbf{b}_n)],$$

and that changes of coordinates are given (in Theorem 7 §5.4) by

$$C_{\mathcal{D}}(\mathbf{v}) = P_{\mathcal{D} \leftarrow \mathcal{B}} C_{\mathcal{B}}(\mathbf{v}) \quad \text{for all } \mathbf{v} \text{ in } V.$$

We now show that these change matrices are also closely related to the way the matrix of the operator T changes. More precisely, we claim that

$$P_{\mathcal{B} \leftarrow \mathcal{D}} M_{\mathcal{D}}(T) = M_{\mathcal{B}}(T) P_{\mathcal{B} \leftarrow \mathcal{D}}. \tag{*}$$

To see this, right multiply the left side by $C_{\mathcal{D}}(\mathbf{v})$ and use Theorem 1:

$$\begin{aligned} P_{\mathcal{B} \leftarrow \mathcal{D}} M_{\mathcal{D}}(T) C_{\mathcal{D}}(\mathbf{v}) &= P_{\mathcal{B} \leftarrow \mathcal{D}} C_{\mathcal{D}}[T(\mathbf{v})] \\ &= C_{\mathcal{B}}[T(\mathbf{v})] \\ &= M_{\mathcal{B}}(T) C_{\mathcal{B}}(\mathbf{v}) \\ &= M_{\mathcal{B}}(T) P_{\mathcal{B} \leftarrow \mathcal{D}} C_{\mathcal{D}}(\mathbf{v}). \end{aligned}$$

Since this holds for all vectors $\mathbf{v}$, (*) follows because $C_\mathcal{D}(\mathbf{v})$ can be "cancelled" by Lemma 1 §5.4. This proves

Theorem 2. *Let $T : V \to V$ be a linear operator where $\dim V = n$. If $\mathcal{B}$ and $\mathcal{D}$ are two ordered bases of V, the matrices*

$$M_\mathcal{B}(T) \quad \text{and} \quad M_\mathcal{D}(T) \quad \text{are similar.}$$

More precisely, if $P = P_{\mathcal{B} \leftarrow \mathcal{D}}$ then

$$M_\mathcal{D}(T) = P^{-1} M_\mathcal{B}(T) P.$$

Note that $(P_{\mathcal{B} \leftarrow \mathcal{D}})^{-1} = P_{\mathcal{D} \leftarrow \mathcal{B}}$ by Theorem 8 §5.4. Hence the formula in Theorem 2 takes the easily remembered form $M_\mathcal{D}(T) = P_{\mathcal{D} \leftarrow \mathcal{B}} M_\mathcal{B}(T) P_{\mathcal{B} \leftarrow \mathcal{D}}$.

Example 6. Define $T : \mathbb{P}_1 \to \mathbb{P}_1$ by $T(a+bx) = (a+b) - (2a+b)x$. If $\mathcal{B} = \{1, x\}$, then

$$M_\mathcal{B}(T) = [C_\mathcal{B}(T(1)) \ C_\mathcal{B}(T(x))] = [C_\mathcal{B}(1 - 2x) \ C_\mathcal{B}(1 - x)] = \begin{bmatrix} 1 & 1 \\ -2 & -1 \end{bmatrix}.$$

Now let $\mathcal{D} = \{x - 1, x\}$ and compute

$$M_\mathcal{D}(T) = [C_\mathcal{D}(T(x-1)) \ C_\mathcal{D}(T(x))] = [C_\mathcal{D}(x) \ C_\mathcal{D}(1-x)] = \begin{bmatrix} 0 & -1 \\ 1 & 0 \end{bmatrix}.$$

Thus the matrices $M_\mathcal{B}(T)$ and $M_\mathcal{D}(T)$ are similar by Theorem 2. In fact, we compute

$$P = P_{\mathcal{B} \leftarrow \mathcal{D}} = [C_\mathcal{B}(x-1) \ C_\mathcal{B}(x)] = \begin{bmatrix} -1 & 0 \\ 1 & 1 \end{bmatrix},$$

so Theorem 2 gives $P^{-1} M_\mathcal{B}(T) P = M_\mathcal{D}(T)$ as the reader can verify. $\square$

Of course we knew in Theorem 8 §4.9 that *every* pair of similar $n \times n$ matrices A and B arises as in Theorem 2 as the matrices of some operator $T : \mathbb{R}^n \to \mathbb{R}^n$ with respect to two bases of $\mathbb{R}^n$. We restate the result here for reference.

Lemma 1. *If A and B are similar $n \times n$ matrices, say $B = P^{-1}AP$, then*

$$A = M_\mathcal{E}(T_A) \quad \text{and} \quad B = M_\mathcal{F}(T_A)$$

where $T_A : \mathbb{R}^n \to \mathbb{R}^n$ is defined by $T_A(X) = AX$ for all X in $\mathbb{R}^n$, $\mathcal{E}$ is the standard basis of $\mathbb{R}^n$ and $\mathcal{F} = \{F_1, F_2, \cdots, F_n\}$ consists of the columns F_j of P.

Proof. We have $A = M_\mathcal{E}(T_A)$ by Example 2, and $B = M_\mathcal{F}(T_A)$ by Lemma 3 §4.9. $\square$

5.5. LINEAR OPERATORS AND SIMILARITY

We saw in Theorem 12 §5.4 that the rank of every matrix of a linear operator T equals $dim(imT)$, and we called this number the **rank** of T, denoting it as $rankT$. Hence we have

$$rankT = rank(M_\mathcal{B}(T)) \quad \text{for every basis } \mathcal{B} \text{ of } V.$$

The fact that this is independent of the choice of basis can also be seen from Theorem 2 because similar matrices have the same rank (Corollary 4 of Theorem 2 §4.4). When viewed in this way, it is clear that Theorem 2 allows us to define the determinant, the trace and the characteristic polynomial of any linear operator in the same way. The reason is that similar matrices also have the same determinant, the same trace and the same characteristic polynomial (Theorem 6 §2.3 and Lemma 4 §4.7).

Thus, for example, Theorem 2 shows that if $T : V \to V$ is a linear operator (where $dimV$ is finite), then $det[M_\mathcal{B}(T)] = det[M_\mathcal{D}(T)]$ for all bases $\mathcal{B}$ and $\mathcal{D}$ of V. We call this common value the determinant of T; more formally, we define the **determinant** $detT$ of an operator $T : V \to V$ by

$$detT = det[M_\mathcal{B}(T)] \quad \text{for } any \text{ basis } \mathcal{B} \text{ of } V.$$

This makes sense because it is independent of the choice of the basis $\mathcal{B}$ and so depends only on the operator T. In the same way, similar matrices have the same trace and the same characteristic polynomial. Hence we can define the **trace** trT of T and the **characteristic polynomial** $c_T(x)$ of T by

$$trT = tr[M_\mathcal{B}(T)] \quad \text{for any basis } \mathcal{B} \text{ of } V.$$

$$c_T(x) = c_{M_\mathcal{B}(T)}(x) \quad \text{for any basis } \mathcal{B} \text{ of } V.$$

In this way, properties of determinants, traces and characteristic polynomials that are proved for matrices actually hold for operators. For example, we have $det(AB) = detA \, detB$ and $tr(AB) = tr(BA)$ for all square matrices A and B. These equations translate to the fact that

$$det(S \circ T) = detS \, detT \quad \text{and} \quad tr(S \circ T) = tr(T \circ S)$$

hold for any operators S and T on the same space V. The verifications of these properties (and others) are left as exercises for the reader.

5.5.3 Diagonalization

Possibly the most important application of the concept of similarity is its use when discussing diagonalization. Recall that an $n \times n$ matrix A is said to be *diagonalizable* if it is similar to a diagonal matrix D. By Theorem 2 this happens if there exists an operator $T : V \to V$ and bases $\mathcal{B}$ and $\mathcal{D}$ of V such that $A = M_\mathcal{B}(T)$ and $D = M_\mathcal{D}(T)$. This, in turn, focuses our attention on how to find a basis $\mathcal{D}$ of V for which $M_\mathcal{D}(T)$ is diagonal.

To this end, list the basis $\mathcal{D}$ as $\mathcal{D} = \{\mathbf{d}_1, \mathbf{d}_2, \cdots, \mathbf{d}_n\}$ and assume that $M_\mathcal{D}(T) = diag(\lambda_1, \lambda_2, \cdots, \lambda_n)$ where the λ_i are real numbers. Then

$$[C_\mathcal{D}(T(\mathbf{d}_i))\ C_\mathcal{D}(T(\mathbf{d}_i))\ \cdots\ C_\mathcal{D}(T(\mathbf{d}_i))] = M_\mathcal{D}(T) = diag(\lambda_1, \lambda_2, \cdots, \lambda_n).$$

Comparing columns shows that $C_\mathcal{D}(T(\mathbf{d}_i)) = [0\ \cdots\ \lambda_i\ \cdots 0]^T$, and hence that

$$T(\mathbf{d}_i) = \lambda_i \mathbf{d}_i \quad \text{for each } i = 1, 2, \cdots, n.$$

This prompts the following definitions: As for matrices, we say that a number λ is an **eigenvalue** of the operator T if

$$T(\mathbf{v}) = \lambda \mathbf{v} \quad \text{for some vector } \mathbf{v} \neq \mathbf{0},$$

and in this case we say that $\mathbf{v}$ is an **eigenvector** of T corresponding to λ. Finally, the subspace

$$E_\lambda(T) = \{\mathbf{v} \text{ in } V \mid T(\mathbf{v}) = \lambda \mathbf{v}\}$$

of V is called the **eigenspace** of T corresponding to λ. All this terminology is consistent with our earlier usage for matrices because, as the reader can verify, the eigenvalues and eigenvectors of an $n \times n$ matrix A are the same as the eigenvalues and eigenvectors of the operator $T_A : \mathbb{R}^n \to \mathbb{R}^n$, where $T_A(X) = AX$ for all X in $\mathbb{R}^n$.

We can now characterize when the $\mathcal{D}$-matrix of an operator is diagonal.

Theorem 3. *The following conditions are equivalent for an operator $T : V \to V$ where $dim V = n$:*

(1) *$M_\mathcal{D}(T)$ is diagonal for some basis $\mathcal{D}$.*

(2) *There exists a basis of V consisting of eigenvectors of T.*

Moreover, if $\mathcal{D} = \{\mathbf{d}_1, \mathbf{d}_2, \cdots, \mathbf{d}_n\}$ is a basis of eigenvectors and $T(\mathbf{d}_i) = \lambda_i \mathbf{d}_i$ for each i, then

$$M_\mathcal{D}(T) = diag(\lambda_1, \lambda_2, \cdots, \lambda_n).$$

Proof. The preceding discussion proves that (1)$\Rightarrow$(2). Conversely, if $\mathcal{D}$ and the λ_i are as in the theorem, we compute $M_\mathcal{D}(T)$ in terms of its columns:

$$\begin{aligned} M_\mathcal{D}(T) &= [C_\mathcal{D}(T(\mathbf{d}_1))\ C_\mathcal{D}(T(\mathbf{d}_2))\ \cdots\ C_\mathcal{D}(T(\mathbf{d}_n))] \\ &= [C_\mathcal{D}(\lambda_1 \mathbf{d}_1)\ C_\mathcal{D}(\lambda_2 \mathbf{d}_2)\ \cdots\ C_\mathcal{D}(\lambda_n \mathbf{d}_n)] \\ &= diag(\lambda_1, \lambda_2, \cdots, \lambda_n). \end{aligned}$$

This proves that (2)$\Rightarrow$(1), and also proves the last sentence in the theorem. $\square$

A linear operator $T : V \to V$ is called **diagonalizable** if it satisfies the conditions in Theorem 3.

5.5. LINEAR OPERATORS AND SIMILARITY

Example 7. Let $R_\theta : \mathbb{R}^2 \to \mathbb{R}^2$ denote rotation about the origin through the angle $\theta \neq 0$, (a linear operator by Theorem 3 §3.5). Show geometrically and algebraically that R_θ is not diagonalizable.

Solution. If $\vec{v} \neq \vec{0}$ is a vector in $\mathbb{R}^2$, then $R_\theta(\vec{v})$ is the result of rotating $\vec{v}$ through the angle $\theta \neq 0$, and so cannot equal $\lambda \vec{v}$ for any real number λ (see the diagram). Hence R_θ has no eigenvectors, and so is not diagonalizable.

For an algebraic argument, recall that the matrix of R_θ is $\begin{bmatrix} \cos\theta & -\sin\theta \\ \sin\theta & \cos\theta \end{bmatrix}$ with respect to the standard basis (Theorem 3 § 3.5), so the characteristic polynomial is $c_{R_\theta}(x) = x^2 + 1$. This means that R_θ has no real eigenvalues, and so is not diagonalizable. □

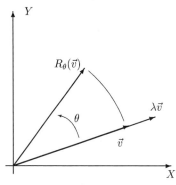

Before characterizing the diagonalizable operators, we need the next result which reveals the close connection between the eigenvalues, eigenvectors and eigenspaces of T and those of each of its matrices.

Lemma 2. Let $T : V \to V$ be a linear operator where $\dim V = n$, let $\mathcal{B}$ be a basis of V and let λ be a real number. Then:

(1) $C_\mathcal{B} : E_\lambda(T) \to E_\lambda(M_\mathcal{B}(T))$ is an isomorphism.

(2) The eigenvalues λ of T are precisely the eigenvalues of the matrix $M_\mathcal{B}(T)$. Moreover, if $\mathbf{v}$ is in V and we write $\mathbf{v} = C_\mathcal{B}(X)$ where X is in $\mathbb{R}^n$, then

$$\mathbf{v} \text{ is in } E_\lambda(T) \quad \text{if and only if} \quad X \text{ is in } E_\lambda(M_\mathcal{B}(T)).$$

Proof. For convenience, write $A = M_\mathcal{B}(T)$, so that $C_\mathcal{B}(T(\mathbf{v})) = AC_\mathcal{B}(\mathbf{v})$ for all $\mathbf{v}$ in V. If $T(\mathbf{v}) = \lambda \mathbf{v}$ where $\mathbf{v} \neq \mathbf{0}$, then the linearity of $C_\mathcal{B}$ gives

$$AC_\mathcal{B}(\mathbf{v}) = C_\mathcal{B}(T(\mathbf{v})) = C_\mathcal{B}(\lambda \mathbf{v}) = \lambda C_\mathcal{B}(\mathbf{v}),$$

so λ is an eigenvalue of A and $C_\mathcal{B}(\mathbf{v})$ is a corresponding eigenvector. This shows that every eigenvalue λ of T is also an eigenvalue of A, and also that $C_\mathcal{B} : E_\lambda(T) \to E_\lambda(A)$ is a linear transformation (one-to-one). On the other hand, if $AX = \lambda X$ where $X \neq 0$, write $X = C_\mathcal{B}(\mathbf{v})$ where $\mathbf{v} \neq \mathbf{0}$ is in V. Then

$$C_\mathcal{B}(T(\mathbf{v})) = AC_\mathcal{B}(\mathbf{v}) = AX = \lambda X = \lambda C_\mathcal{B}(\mathbf{v}) = C_\mathcal{B}(\lambda \mathbf{v}).$$

Hence $T(\mathbf{v}) = \lambda \mathbf{v}$ because $C_\mathcal{B}$ is one-to-one. This shows that every eigenvalue λ of A is also an eigenvalue of T, and that $C_\mathcal{B} : E_\lambda(T) \to E_\lambda(A)$ is onto. This proves all our assertions. □

Example 8. Let $\mathcal{B} = \{\mathbf{v}, \mathbf{w}\}$ be a basis of V, and define $T : V \to V$ by $T(\mathbf{v}) = 2\mathbf{v}$ and $T(\mathbf{w}) = -3\mathbf{v} + 5\mathbf{w}$. Find the matrix $M_{\mathcal{B}}(T)$ and use it to find the eigenvectors of T.

Solution. We have $M_{\mathcal{B}}(T) = [C_{\mathcal{B}}(T(\mathbf{v})) \ C_{\mathcal{B}}(T(\mathbf{w}))] = \begin{bmatrix} 2 & -3 \\ 0 & 5 \end{bmatrix}$. Hence the eigenvalues of $M_{\mathcal{B}}(T)$ are $\lambda_1 = 2$ and $\lambda_2 = 5$, with corresponding eigenvectors $X_1 = \begin{bmatrix} 1 \\ 0 \end{bmatrix}$ and $X_2 = \begin{bmatrix} -1 \\ 1 \end{bmatrix}$. Since $X_1 = C_{\mathcal{B}}(\mathbf{v})$ and $X_2 = C_{\mathcal{B}}(\mathbf{w} - \mathbf{v})$, it follows by Lemma 2 that $\mathbf{v}_1 = \mathbf{v}$ is an eigenvector of T corresponding to λ_1 (as is clear from the definition of T) and $\mathbf{v}_2 = \mathbf{w} - \mathbf{v}$ is an eigenvector of T corresponding to λ_2. Of course, we could have discovered these eigenvectors $\mathbf{v}_1$ and $\mathbf{v}_2$ for T directly, and then used them to find X_1 and X_2 using Lemma 2. □

Lemma 2 enables us to characterize when an operator is diagonalizable in terms of its matrices. The result is not surprising.

Theorem 4. *If $T : V \to V$ is a linear operator where $\dim V = n$, the following are equivalent:*

(1) *T is diagonalizable.*

(2) *$M_{\mathcal{B}}(T)$ is a diagonalizable matrix for every basis $\mathcal{B}$ of V.*

(3) *$M_{\mathcal{B}}(T)$ is a diagonalizable matrix for some basis $\mathcal{B}$ of V.*

Moreover, we can pass back and forth between the eigenvector bases of V and those of $M_{\mathcal{B}}(T)$ using the coordinate isomorphism $C_{\mathcal{B}} : V \to \mathbb{R}^n$.

Proof. As before, write $A = M_{\mathcal{B}}(T)$ for convenience.

(1)⇒(2). If T is diagonalizable, there exists a basis $\mathcal{D}$ of V such that $M_{\mathcal{D}}(T)$ is diagonal. Since A is similar to $M_{\mathcal{D}}(T)$ by Theorem 2, it follows that A is diagonalizable.

(2)⇒(3). This is clear.

(3)⇒(1). Given (3), A is a diagonalizable matrix, so let $\{X_1, X_2, \cdots, X_n\}$ be a basis of $\mathbb{R}^n$ consisting of eigenvectors of A. If $X_j = C_{\mathcal{B}}(\mathbf{b}_j)$ for each j, then $\mathcal{D} = \{\mathbf{b}_1, \mathbf{b}_2, \cdots, \mathbf{b}_n\}$ is a basis of V (because $C_{\mathcal{B}}^{-1}$ is an isomorphism and $\mathbf{b}_j = C_{\mathcal{B}}^{-1}(X_j)$ for each j). Since $\mathcal{D}$ consists of eigenvectors of T by Lemma 2, this proves (1).

Finally the last sentence follows from Lemma 2. □

Example 9. Let $T : \mathbb{M}_{2,2} \to \mathbb{M}_{2,2}$ be defined by $T(A) = A^T$ for each matrix A in $\mathbb{M}_{2,2}$. Determine if T is diagonalizable and, if so, find a basis of $\mathbb{M}_{2,2}$ consisting of eigenvectors of T.

Solution. Consider the standard basis $\mathcal{B} = \{E_{11}, E_{12}, E_{21}, E_{22}\}$ of $\mathbb{M}_{2,2}$ where E_{ij} is the matrix unit with 1 in the (i, j)-position and zeros elsewhere. Then

$$T(E_{11}) = E_{11}, \ T(E_{12}) = E_{21}, \ T(E_{21}) = E_{12}, \text{ and } T(E_{22}) = E_{22}$$

5.5. LINEAR OPERATORS AND SIMILARITY

as is easily verified. Hence

$$M_{\mathcal{B}}(T) = \begin{bmatrix} 1 & 0 & 0 & 0 \\ 0 & 0 & 1 & 0 \\ 0 & 1 & 0 & 0 \\ 0 & 0 & 0 & 1 \end{bmatrix}$$

so the characteristic polynomial of T is $c_T(x) = c_{M_{\mathcal{B}}(T)}(x) = (x-1)^3(x+1)$. Thus $M_{\mathcal{B}}(T)$ (and hence T) has eigenvalues $\lambda_1 = 1$ of multiplicity 3 and $\lambda_2 = -1$ of multiplicity 1. Moreover, corresponding eigenvectors of $M_{\mathcal{B}}(T)$ are

$$X_1 = \begin{bmatrix} 1 \\ 0 \\ 0 \\ 0 \end{bmatrix}, Y_1 = \begin{bmatrix} 0 \\ 1 \\ 1 \\ 0 \end{bmatrix} \text{ and } Z_1 = \begin{bmatrix} 0 \\ 0 \\ 0 \\ 1 \end{bmatrix} \text{ for } \lambda_1, \text{ and } X_2 = \begin{bmatrix} 0 \\ -1 \\ 1 \\ 0 \end{bmatrix} \text{ for } \lambda_2.$$

Hence $M_{\mathcal{B}}(T)$ is diagonalizable by Theorem 3 §4.7, so T is diagonalizable by Theorem 4. In addition, we have

$$X_1 = C_{\mathcal{B}} \begin{bmatrix} 1 & 0 \\ 0 & 0 \end{bmatrix}, \quad Y_1 = C_{\mathcal{B}} \begin{bmatrix} 0 & 1 \\ 1 & 0 \end{bmatrix},$$
$$Z_1 = C_{\mathcal{B}} \begin{bmatrix} 0 & 0 \\ 0 & 1 \end{bmatrix}, \quad X_2 = C_{\mathcal{B}} \begin{bmatrix} 0 & -1 \\ 1 & 0 \end{bmatrix}.$$

Hence Lemma 2 shows that $\mathcal{D} = \left\{ \begin{bmatrix} 1 & 0 \\ 0 & 0 \end{bmatrix}, \begin{bmatrix} 0 & 1 \\ 1 & 0 \end{bmatrix}, \begin{bmatrix} 0 & 0 \\ 0 & 1 \end{bmatrix}, \begin{bmatrix} 0 & -1 \\ 1 & 0 \end{bmatrix} \right\}$ is a basis of eigenvectors of T (showing again that T is diagonalizable). □

Furthermore, if $T : \mathbb{M}_{2,2} \to \mathbb{M}_{2,2}$ is the transposition operator in Example 9, observe that the eigenspaces of T are

$$E_{\lambda_1}(T) = \{A \mid A^T = A\} \quad \text{and} \quad E_{\lambda_2}(T) = \{A \mid A^T = -A\},$$

and so consist of all symmetric and all skew-symmetric matrices, respectively. Furthermore, if we separate out the vectors in $\mathcal{D}$ corresponding to different eigenvalues, we obtain bases for the eigenspaces:

$$E_{\lambda_1}(T) = \text{span}\left\{ \begin{bmatrix} 1 & 0 \\ 0 & 0 \end{bmatrix}, \begin{bmatrix} 0 & 1 \\ 1 & 0 \end{bmatrix}, \begin{bmatrix} 0 & 0 \\ 0 & 1 \end{bmatrix} \right\}$$
$$= \left\{ \begin{bmatrix} a & b \\ b & c \end{bmatrix} \mid a, b \text{ and } c \text{ in } \mathbb{R} \right\},$$
$$E_{\lambda_2}(T) = \text{span}\left\{ \begin{bmatrix} 0 & -1 \\ 1 & 0 \end{bmatrix} \right\}$$
$$= \left\{ \begin{bmatrix} 0 & -d \\ d & 0 \end{bmatrix} \mid d \text{ in } \mathbb{R} \right\}.$$

This phenomenon holds more generally, and depends only upon the fact that $T^2 = 1_{\mathbb{M}_{2,2}}$. To see this, we need the techniques developed in the next section (this result is Exercise 26 §5.6).

Exercises

1. In each case either show that the statement is true or give an example showing that it is false. Throughout, A and B denote $n \times n$ matrices and S and T are operators $V \to V$.

   (a) If $det A = det B$ then A and B are similar.

   (b) If A is similar to B then kA is similar to kB for any scalar k.

   (c) If $T = R^{-1} \circ S \circ R$ for some isomorphism $R : V \to V$, then S and T have the same eigenvalues.

   (d) If A is diagonalizable and $A = M_\mathcal{B}(T)$ for some basis $\mathcal{B}$, then V has a basis of T-eigenvectors.

   (e) If $\mathbf{v}$ and $\mathbf{w}$ are distinct eigenvectors of T, then $\{\mathbf{v}, \mathbf{w}\}$ is independent.

   (f) If $A = M_\mathcal{B}(T)$ and $B = M_\mathcal{B}(S)$ for some basis $\mathcal{B}$, then A and B are similar.

   (g) If $\mathbf{v}$ is an eigenvector of T, the same is true of $k\mathbf{v}$ for all scalars $k \neq 0$.

   (h) If $\mathbf{v}$ is an eigenvector of T and $\mathbf{w}$ is in $ker T$, then $\mathbf{v} + \mathbf{w}$ is an eigenvector of T.

2. Let $T : \mathbb{P}_2 \to \mathbb{P}_2$ be given by $T(a + bx + cx^2) = (2a - b) + (a + b + c)x + (c - a)x^2$.

   (a) Find $M_\mathcal{B}(T)$ where $\mathcal{B} = \{1, x, x^2\}$.

   (b) Is T invertible? Defend your answer.

3. Let $T : \mathbb{M}_{2,2} \to \mathbb{M}_{2,2}$ be given by $T \begin{bmatrix} a & b \\ c & d \end{bmatrix} = \begin{bmatrix} a+b+c & 2c-a \\ b+d & c-d \end{bmatrix}$.

   (a) Find $M_\mathcal{B}(T)$ where $\mathcal{B} = \{E_{11}, E_{12}, E_{21}, E_{22}\}$ is the standard basis of matrix units.

   (b) Is T invertible? Defend your answer.

4. If $T : \mathbb{P}_1 \to \mathbb{P}_1$ has matrix $M_\mathcal{B}(T) = \begin{bmatrix} 1 & -2 \\ -2 & 3 \end{bmatrix}$ where $\mathcal{B} = \{1, x\}$, find $T^{-1}(2x - 3)$.

5. If $T : \mathbb{M}_{2,2} \to \mathbb{M}_{2,2}$ has matrix

$$M_\mathcal{B}(T) = \begin{bmatrix} 1 & 0 & 1 & 1 \\ 3 & 1 & -1 & 0 \\ 0 & 0 & -1 & 0 \\ 2 & 0 & 0 & 1 \end{bmatrix}$$

   where $\mathcal{B} = \{E_{11}, E_{12}, E_{21}, E_{22}\}$ is the standard basis of matrix units, find $T^{-1} \begin{bmatrix} 2 & 3 \\ -1 & 1 \end{bmatrix}$.

6. If B is a fixed 2×2 matrix, define $T : \mathbb{M}_{2,2} \to \mathbb{M}_{2,2}$ by $T(A) = BA$ for all A in $\mathbb{M}_{2,2}$. If $B = \begin{bmatrix} a & b \\ c & d \end{bmatrix}$, find a basis $\mathcal{B}$ of $\mathbb{M}_{2,2}$ such that $M_\mathcal{B}(T) = \begin{bmatrix} aI_2 & bI_2 \\ cI_2 & dI_2 \end{bmatrix}$.

7. (a) Find the eigenvalues and eigenvectors of the matrix $A = \begin{bmatrix} 1 & 2 \\ 3 & 2 \end{bmatrix}$.

   (b) If $T : \mathbb{R}^2 \to \mathbb{R}^2$ is defined by $T(X) = AX$ for all X in $\mathbb{R}^2$, use (a) to find a basis of eigenvectors of T.

   (c) If $T : \mathbb{P}_1 \to \mathbb{P}_1$ is defined by $T(a + bx) = (a + 2b) + (3a + 2b)x$, use (a) to find a basis of eigenvectors of T.

8. Let $D : \mathbb{P}_3 \to \mathbb{P}_3$ be the differentiation operator given by $D(p(x)) = p'(x)$.

   (a) Find the matrix of D with respect to the basis $\mathcal{B} = \{1, x, x^2, x^3\}$.

   (b) Is D an isomorphism? Justify your answer.

   (c) Show that $D^4 = 0$ is the zero operator.

9. Let $\mathcal{B} = \{\mathbf{b}_1, \mathbf{b}_2, \mathbf{b}_3\}$ be a basis of V, and let $\mathbf{d}_1 = 2\mathbf{b}_1 - \mathbf{b}_2$, $\mathbf{d}_2 = \mathbf{b}_2 - \mathbf{b}_3$ and $\mathbf{d}_3 = \mathbf{b}_1 + \mathbf{b}_2 + \mathbf{b}_3$.

   (a) Show that $\mathcal{D} = \{\mathbf{d}_1, \mathbf{d}_2, \mathbf{d}_3\}$ is a basis of V.

   (b) Find the change matrix $P_{\mathcal{B} \leftarrow \mathcal{D}}$.

10. Let $\mathcal{B} = \{\mathbf{b}_1, \mathbf{b}_2, \mathbf{b}_3\}$ be a basis of V. If $T : V \to V$ is an operator satisfying $T(\mathbf{b}_1) = \mathbf{b}_1 + \mathbf{b}_2$, $T(\mathbf{b}_2) = \mathbf{b}_2 + \mathbf{b}_3$ and $T(\mathbf{b}_3) = \mathbf{b}_1$, find:

    (a) $M_\mathcal{B}(T)$.

    (b) $T^{-1}(\mathbf{b}_2)$.

11. Let $\mathcal{B} = \{\mathbf{b}_1, \mathbf{b}_2, \mathbf{b}_3\}$ be a basis of V. If $T : V \to V$ is an operator with matrix $M_\mathcal{B}(T) = \begin{bmatrix} 1 & 0 & 3 \\ -2 & 1 & 4 \\ 0 & -1 & 1 \end{bmatrix}$, find $T(2\mathbf{b}_2 - 5\mathbf{b}_3)$.

12. Let $\mathcal{B}$ be a basis of V where $\dim V = n$. If A is $n \times n$, show that $A = M_\mathcal{B}(T)$ for some operator $T : V \to V$.

13. Describe the operator $T : V \to V$ if it has the property that $V = E_\lambda(T)$ for some real number λ.

14. Let $\mathcal{B} = \{\mathbf{v}, \mathbf{w}\}$ be a basis of V, and let $T : V \to V$ be defined by $T(\mathbf{v}) = \mathbf{v}$ and $T(\mathbf{w}) = \mathbf{v} - \mathbf{w}$.

    (a) Find a basis of eigenvectors of T.

    (b) Find the matrix $M_\mathcal{B}(T)$ and diagonalize it.

15. Let $T : \mathbb{P}_n \to \mathbb{P}_n$ be defined by $T(p(x)) = p(x) + x\, p'(x)$ where p' indicated the derivative. Show that T is an isomorphism by considering $M_\mathcal{B}(T)$ where $\mathcal{B} = \{1, x, x^2, \cdots, x^n\}$.

16. If k is any number, define $T : \mathbb{M}_{2,2} \to \mathbb{M}_{2,2}$ by $T(A) = A + kA^T$. Show that T is an isomorphism if $k \neq \pm 1$ by considering $M_\mathcal{B}(T)$ where $\mathcal{B} = \left\{ \begin{bmatrix} 1 & 0 \\ 0 & 0 \end{bmatrix}, \begin{bmatrix} 0 & 0 \\ 0 & 1 \end{bmatrix}, \begin{bmatrix} 0 & 1 \\ 1 & 0 \end{bmatrix}, \begin{bmatrix} 0 & -1 \\ 1 & 0 \end{bmatrix}, \right\}$.

17. Let $Q : \mathbb{R}^2 \to \mathbb{R}^2$ denote reflection in some line L through the origin.

    (a) Show geometrically that any direction vector $\vec{d}$ for L is an eigenvector of Q.

    (b) Use a geometrical argument to find eigenvectors independent of $\vec{d}$.

    (c) Show that Q is diagonalizable.

18. Let S and T be linear operators $V \to V$ where $dim V$ is finite. Show that $det(S \circ T) = det S \, det T$.

19. Let S and T be linear operators $V \to V$ where $dim V$ is finite. Show that $tr(S \circ T) = tr(T \circ S)$. [Hint: Lemma 4 §4.7.]

20. Let S and T be two operators on a vector space V of dimension n.

    (a) If $T = R \circ S$ for some isomorphism $R : V \to V$, show that $ker S = ker T$.

    (b) If $ker S = ker T$ show that $T = R \circ S$ for some isomorphism $R : V \to V$. [Hint: Let $\mathcal{B} = \{\mathbf{b}_1, \cdots, \mathbf{b}_r, \mathbf{b}_{r+1}, \cdots, \mathbf{b}_n\}$ be a basis of V such that $\{\mathbf{b}_{r+1}, \cdots, \mathbf{b}_n\}$ is a basis of $ker S = ker T$. Then show that there exist bases of V of the form $\mathcal{D} = \{S(\mathbf{b}_1), \cdots, S(\mathbf{b}_r), \mathbf{d}_{r+1}, \cdots, \mathbf{d}_n\}$ and $\mathcal{F} = \{T(\mathbf{b}_1), \cdots, T(\mathbf{b}_r), \mathbf{f}_{r+1}, \cdots, \mathbf{f}_n\}$. Define R on the basis $\mathcal{D}$ by taking $R[S(\mathbf{b}_i)] = T(\mathbf{b}_i)$ for $1 \leq i \leq r$ and $R(\mathbf{d}_j) = \mathbf{f}_j$ for $r < j \leq n$.]

    (c) If A and B are $n \times n$ matrices, show that $null A = null B$ if and only if $B = CA$ for some invertible matrix C. [Hint: See Example 2.]

5.6 INVARIANT SUBSPACES

As noted earlier, the central problem in linear algebra is to find a way to discover the "simplest" matrix for a linear operator $T : V \to V$. Often the way to do this is to regard T as an operator on subspaces of V of smaller dimension, and then find a way to build up the matrix of T on V from smaller matrices. The notion of a T-invariant subspace of V is a basic tool in this endeavor.

5.6. INVARIANT SUBSPACES

5.6.1 Invariant Subspaces

Suppose $T : V \to V$ is a linear operator. A subspace U of V is called **T-invariant** if

$T(\mathbf{u})$ is in U for every vector $\mathbf{u}$ in U.

If we write $T(U) = \{T(\mathbf{u}) \mid \mathbf{u}$ in $U\}$, this condition can be compactly expressed as

$$T(U) \subseteq U.$$

This is shown in the diagram.

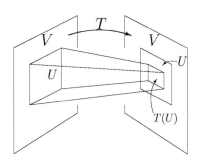

Example 1. $\{\mathbf{0}\}$ and V are T-invariant subspaces of V for every operator $T : V \to V$.

Example 2. If λ is an eigenvalue of the operator $T : V \to V$, show that the eigenspace $E_\lambda(T)$ is a T-invariant subspace of V.

Solution. If $\mathbf{u}$ is in $E_\lambda(T)$, then $T(\mathbf{u}) = \lambda\mathbf{u}$ is in $E_\lambda(T)$ because $E_\lambda(T)$ is a subspace. $\square$

Example 3. Let V be a 2-dimensional vector space with basis $\{\mathbf{v}, \mathbf{w}\}$, and define $T : V \to V$ by taking $T(\mathbf{v}) = -\mathbf{w}$ and $T(\mathbf{w}) = \mathbf{v}$. Show that V has no T-invariant subspace other than $\{\mathbf{0}\}$ and V.

Solution. Suppose that a T-invariant subspace U exists other than $\{\mathbf{0}\}$ or V; we show that this leads to a contradiction. Since $dim V = 2$, we have $dim U = 1$ so $U = \mathbb{R}\mathbf{u}$ for some vector $\mathbf{u} \neq \mathbf{0}$ in V. Since $T(\mathbf{u})$ is in U, we have $T(\mathbf{u}) = \lambda\mathbf{u}$ for some real number λ. On the other hand, $\mathbf{u}$ is in $V = span\{\mathbf{v}, \mathbf{w}\}$, so we can write $\mathbf{u} = a\mathbf{v} + b\mathbf{w}$ where a and b are in $\mathbb{R}$. Hence the equation $\lambda\mathbf{u} = T(\mathbf{u})$ becomes

$$\lambda a\mathbf{v} + \lambda b\mathbf{w} = \lambda \mathbf{u} = T(\mathbf{u}) = T(a\mathbf{v} + b\mathbf{w}) = aT(\mathbf{v}) + bT(\mathbf{w}) = -a\mathbf{w} + b\mathbf{v}.$$

Since $\{\mathbf{v}, \mathbf{w}\}$ is independent, it follows that $\lambda a = b$ and $\lambda b = -a$. But then $\lambda^2 a = \lambda(\lambda a) = \lambda b = -a$, so that $(\lambda^2 + 1)a = 0$. Hence $a = 0$ because λ is real. Similarly $\lambda^2 b = -b$ so $b = 0$, whence $\mathbf{u} = a\mathbf{v} + b\mathbf{w} = \mathbf{0}$, a contradiction. $\square$

We now give a useful test for when a subspace U is T-invariant: Simply check that T carries every vector in some (usually finite) spanning set for U back into U.

Lemma 1. If $T : V \to V$ is an operator and $U = span\{\mathbf{u}_1, \mathbf{u}_2, \cdots, \mathbf{u}_m\}$ is a subspace of V, then U is T-invariant if and only if $T(\mathbf{u}_i)$ is in U for each $i = 1, 2, \cdots, m$.

Proof. If U is T-invariant then certainly $T(\mathbf{u}_i)$ is in U for each i. Conversely, assume that each $T(\mathbf{u}_i)$ is in U, and let $\mathbf{u}$ be any vector in U. Then $\mathbf{u} = r_1\mathbf{u}_1 + r_2\mathbf{u}_2 + \cdots + r_m\mathbf{u}_m$ for some r_i in $\mathbb{R}$, so $T(\mathbf{u}) = r_1T(\mathbf{u}_1) + r_2T(\mathbf{u}_2) + \cdots + r_mT(\mathbf{u}_m)$ is also in U because U is a subspace. Thus U is T-invariant. $\square$

Example 4. Write $B = \begin{bmatrix} 1 & -1 \\ 1 & 0 \end{bmatrix}$, define $T : \mathbb{M}_{2,2} \to \mathbb{M}_{2,2}$ by $T(A) = BA$ for all A in $\mathbb{M}_{2,2}$, and consider the subspace $U = \text{span}\{I, B\}$ of $\mathbb{M}_{2,2}$. Show that U is a T-invariant subspace of $\mathbb{M}_{2,2}$ which is not an eigenspace of T.

Solution. We have $T(I) = B$ and $T(B) = B^2$, so it is enough (by Lemma 1) to show that B^2 is in U. This follows from the fact that $B^2 = B - I$ (verify). (This can be seen as follows: The characteristic polynomial of B is $c_B(x) = x^2 - x + 1$, and the Cayley-Hamilton Theorem (Theorem 7 below) asserts that $c_B(B) = 0$.) To see that U is not an eigenspace of T, simply observe that $T(B) = B^2 = B - I \neq \lambda B$ for any number λ (otherwise $(1 - \lambda)B = I$, a contradiction). $\square$

We now come to the principal reason for the importance of T-invariant subspaces. If $T : V \to V$ is an operator and U is a T-invariant subspace of V, the fact that $T(\mathbf{u})$ is in U for each vector $\mathbf{u}$ in U means that

$$T : U \to U \text{ is an operator } \textit{on the vector space } U,$$

called the **restriction** of T to U. This is very important for two reasons: First, the dimension of U will in general be smaller than the dimension of V so it will be easier to find a "simple" matrix for $T : U \to U$ (possibly by induction on the dimension). Second, if a basis of U is extended to a basis of V, the corresponding matrix of $T : V \to V$ takes a simplified, block upper triangular form. This is the content of the following theorem.

Theorem 1. Let $T : V \to V$ be a linear operator where $\dim V = n$, and let U be a T-invariant subspace of V. If $\mathcal{B}_0 = \{\mathbf{u}_1, \mathbf{u}_2, \cdots, \mathbf{u}_m\}$ is any basis of U, extend it to a basis

$$\mathcal{B} = \{\mathbf{u}_1, \mathbf{u}_2, \cdots, \mathbf{u}_m, \mathbf{v}_{m+1}, \cdots, \mathbf{v}_n\}$$

of V in any way at all. Then

$$M_\mathcal{B}(T) = \begin{bmatrix} M_{\mathcal{B}_0}(T) & Y \\ 0 & A \end{bmatrix}$$

in block triangular form, where $M_{\mathcal{B}_0}(T)$ is the $\mathcal{B}_0$-matrix of the restriction of T to U.

Proof. First write the $n \times n$ matrix $M_\mathcal{B}(T)$ in terms of its columns:

$M_\mathcal{B}(T)$
$= [C_\mathcal{B}(T(\mathbf{u}_1)) \ C_\mathcal{B}(T(\mathbf{u}_2)) \ \cdots \ C_\mathcal{B}(T(\mathbf{u}_m)) \ C_\mathcal{B}(T(\mathbf{v}_{m+1})) \ \cdots \ C_\mathcal{B}(T(\mathbf{v}_n))].$

5.6. INVARIANT SUBSPACES

Since $\mathbf{u}_i$ is in U for each $i = 1, 2, \cdots, m$, the fact that U is T-invariant means $T(\mathbf{u}_i)$ is in U. Hence the $\mathcal{B}$-expansion of $T(\mathbf{u}_i)$ has the form

$$T(\mathbf{u}_i) = r_1\mathbf{u}_1 + r_2\mathbf{u}_2 + \cdots + r_m\mathbf{u}_m + 0\mathbf{v}_{m+1} \cdots + 0\mathbf{v}_n$$

where $r_1, r_2, \cdots, r_m$ are real numbers. It follows that

$$C_\mathcal{B}(T(\mathbf{u}_i)) = [r_1 \ r_2 \ \cdots \ r_m \ 0 \ 0 \ \cdots \ 0]^T \text{ and } C_{\mathcal{B}_0}(T(\mathbf{u}_i)) = [r_1 \ r_2 \ \cdots \ r_m]^T.$$

This means that, for each i, the column $C_\mathcal{B}(T(\mathbf{u}_i))$ in $M_\mathcal{B}(T)$ consists of the column $C_{\mathcal{B}_0}(T(\mathbf{u}_i))$ on top and zeros below. Since our hypotheses give us no information about the remaining columns $C_\mathcal{B}(T(\mathbf{v}_k))$ in $M_\mathcal{B}(T)$, this shows that $M_\mathcal{B}(T)$ has the triangular form given in the Theorem. □

Example 5. *In Example 4 we considered the operator* $T : \mathbb{M}_{2,2} \to \mathbb{M}_{2,2}$ *defined by* $T(A) = BA$ *for all* A *in* $\mathbb{M}_{2,2}$, *where* $B = \begin{bmatrix} 1 & -1 \\ 1 & 0 \end{bmatrix}$, *and we showed that* $U = \mathrm{span}\,\{I, B\}$ *is a T-invariant subspace of* $\mathbb{M}_{2,2}$. *Complete* $\mathcal{B}_o = \{I, B\}$ *to a basis* $\mathcal{B}$ *of* $\mathbb{M}_{2,2}$, *and so illustrate Theorem 1.*

Solution. $\mathcal{B}_o$ is independent because $B \neq kI$ for any real number k. One possible completion (verify) of $\mathcal{B}_o$ to a basis of $\mathbb{M}_{2,2}$ is $\mathcal{B} = \{I, B, E_{11}, E_{12}\}$ where $E_{11} = \begin{bmatrix} 1 & 0 \\ 0 & 0 \end{bmatrix}$ and $E_{12} = \begin{bmatrix} 0 & 1 \\ 0 & 0 \end{bmatrix}$. Since $B^2 = B - I$ (see Example 4), we see that

$$\begin{array}{ll}
T(I) = BI = B & \text{so } C_\mathcal{B}(T(I)) = [0 \ 1 \ 0 \ 0]^T. \\
T(B) = B^2 = B - I & \text{so } C_\mathcal{B}(T(B)) = [-1 \ 1 \ 0 \ 0]^T. \\
T(E_{11}) = BE_{11} = B + E_{12} & \text{so } C_\mathcal{B}(T(E_{11})) = [0 \ 1 \ 0 \ 1]^T. \\
T(E_{12}) = BE_{12} = I - E_{11} + E_{12} & \text{so } C_\mathcal{B}(T(E_{12})) = [1 \ 0 \ -1 \ 1]^T.
\end{array}$$

Hence the $\mathcal{B}$-matrix of T is

$$M_\mathcal{B}(T) = [C_\mathcal{B}(T(I)) \ C_\mathcal{B}(T(B)) \ C_\mathcal{B}(T(E_{11})) \ C_\mathcal{B}(T(E_{12}))]$$
$$= \begin{bmatrix} 0 & -1 & 0 & 1 \\ 1 & 1 & 1 & 0 \\ 0 & 0 & 0 & -1 \\ 0 & 0 & 1 & 1 \end{bmatrix}.$$

This has the block upper triangular form expected from Theorem 1. □

Theorem 1 has an important consequence that will be used later.

Theorem 2. *Let* $T : V \to V$ *be an operator where* $\dim V = n$, *let* U *be a T-invariant subspace of* V, *and let* $q(x)$ *denote the characteristic polynomial of the restriction of* T *to* U. *Then*

$$c_T(x) = p(x)\,q(x) \text{ for some polynomial } p(x).$$

Proof. Let $dim U = m$. Choose a basis $\mathcal{B}_0$ of U, so that $q(x) = det[xI_m - M_{\mathcal{B}_0}(T)]$. Extend $\mathcal{B}_0$ to a basis $\mathcal{B}$ of V. Then $M_\mathcal{B}(T)$ has the block form $M_\mathcal{B}(T) = \begin{bmatrix} M_{\mathcal{B}_0}(T) & Y \\ 0 & A \end{bmatrix}$ by Theorem 1 for some $(n-m) \times (n-m)$ matrix A. Now a routine matrix computation (using Theorem 3 §2.2) gives

$$\begin{aligned} c_T(x) &= c_{M_\mathcal{B}(T)}(x) \\ &= det[xI_n - M_\mathcal{B}(T)] \\ &= det \begin{bmatrix} xI_m - M_{\mathcal{B}_0}(T) & -Y \\ 0 & xI_{n-m} - A \end{bmatrix} \\ &= det[xI_m - M_{\mathcal{B}_0}(T)] \, det[xI_{n-m} - A] \\ &= q(x) \, c_A(x). \end{aligned}$$

Theorem 2 follows with $p(x) = c_A(x)$. $\square$

If $T : V \to V$ is a linear operator, the triangular matrix

$$M_\mathcal{B}(T) = \begin{bmatrix} M_{\mathcal{B}_0}(T) & Y \\ 0 & A \end{bmatrix}$$

in Theorem 1 results from choosing the basis $\mathcal{B}$ so that the first few vectors are a basis $\mathcal{B}_0$ of some T-invariant subspace of V. If we can choose the remaining vectors in $\mathcal{B}$ so that they also span a T-invariant subspace, we can make $Y = 0$ and so get a block diagonal matrix for T. Describing how this is done requires the idea of a direct sum decomposition of V, and we digress slightly to discuss this concept.

5.6.2 Direct Sums

Let V be any vector space. If U and W are subspaces of V, define their **sum** $U + W$ and **intersection** $U \cap W$ as follows:

$$U + W = \{\mathbf{u} + \mathbf{w} \mid \mathbf{u} \text{ in } U \text{ and } \mathbf{w} \text{ in } W\}$$

and

$$U \cap W = \{\mathbf{v} \text{ in } V \mid \mathbf{v} \text{ is in both } U \text{ and } W\}.$$

Thus $U + W$ consists of all sums of a vector in U and a vector in W, while $U \cap W$ consists of all vectors in both U and W. The routine proof of the following lemma is left to the reader.

Lemma 2. *If U and W are subspaces of V, then $U + W$ and $U \cap W$ are both subspaces of V.*

The most important pairs of subspaces U and W of a vector space V turn out to be those for which

$$U + W = V \quad \text{and} \quad U \cap W = \{\mathbf{0}\}.$$

5.6. INVARIANT SUBSPACES

When this is the case we say that V is the **direct sum** of U and W, we indicate this by writing
$$V = U \oplus W,$$
and we call each of U and W a **complement** of the other in V. A subspace U can have more than one complement.

Example 6. If $U = \mathbb{R}[1\ 0]$ in $V = \mathbb{M}_{1,2}$, show that each of $W_1 = \mathbb{R}[0\ 1]$, $W_2 = \mathbb{R}[1\ 1]$ and $W_3 = \mathbb{R}[1\ -1]$ are complements of U in $\mathbb{R}^2$.

Solution. We show that $V = U \oplus W_2$, and leave the other two cases to the reader. To see that $V = U + W_2$, observe that
$$[x\ y] = (x - y)[1\ 0] + y[1\ 1] \text{ is in } U + W_2 \text{ for all } x \text{ and } y.$$

Now suppose that $[x\ y]$ is in both $U = \mathbb{R}[1\ 0]$ and $W_2 = \mathbb{R}[1\ 1]$. Then $[x\ y] = r[1\ 0] = s[1\ 1]$ for some r and s in $\mathbb{R}$, so comparing second entries gives $0 = s$. Thus $[x\ y] = s[1\ 1] = 0$, and we have proved that $U \cap W_2 = 0$. Hence $V = U \oplus W_2$. □

Direct sum decompositions occur frequently as the following examples show.

Example 7. Consider the subspaces $U = \{A \text{ in } \mathbb{M}_{n,n} \mid A^T = A\}$ and $W = \{A \text{ in } \mathbb{M}_{n,n} \mid A^T = -A\}$ of $\mathbb{M}_{n,n}$ consisting of all symmetric and all skew-symmetric matrices, respectively. Show that $\mathbb{M}_{n,n} = U \oplus W$.

Solution. Here the fact that $U \cap W = \{\mathbf{0}\}$ is the observation that only the zero matrix is both symmetric and skew-symmetric (verify). To see that $U + W = \mathbb{M}_{n,n}$, let A be any matrix in $\mathbb{M}_{n,n}$ and consider the decomposition $A = \frac{1}{2}(A + A^T) + \frac{1}{2}(A - A^T)$. It is a routine exercise to verify that $\frac{1}{2}(A + A^T)$ is symmetric (and so is in U) and $\frac{1}{2}(A - A^T)$ is skew-symmetric (and so is in W). Hence $\mathbb{M}_{n,n} = U + W$. □

Example 8. Consider the space $\mathbb{F}[a, b]$ of all real valued functions defined on the interval $[a, b]$. A function f in $\mathbb{F}[a, b]$ is called **even** if $f(-x) = f(x)$ for all x in $[a, b]$, and f is called **odd** if $f(-x) = -f(x)$ for all x in $[a, b]$. If U consists of all even functions and W consists of all odd functions, show that $\mathbb{F}[a, b] = U \oplus W$.

Solution. We leave to the reader the verification that U and W are both subspaces of $\mathbb{F}[a, b]$. If f is in $U \cap W$ then f is both even and odd, so $f(-x) = f(x)$ and $f(-x) = -f(x)$ both hold for all x in $[a, b]$. It follows that $f(x) = 0$ for all x; in other words $f = f_0$ is the zero vector in $\mathbb{F}[a, b]$ (see Example 5 §5.1). Hence $U \cap W = \{f_0\}$ is the zero subspace of $\mathbb{F}[a, b]$.

Next, if f is any function in $\mathbb{F}[a, b]$, define $g : [a, b] \to \mathbb{R}$ and $h : [a, b] \to \mathbb{R}$ by
$$g(x) = \tfrac{1}{2}\{f(x) + f(-x)\} \quad \text{and} \quad h(x) = \tfrac{1}{2}\{f(x) - f(-x)\}$$

for all x in $[a, b]$. Then g and h both lie in $\mathbb{F}[a, b]$, and the reader can verify that g is even and h is odd. Moreover $g(x) + h(x) = f(x)$ holds for all x in $[a, b]$, which means that $f = g + h$. Hence $\mathbb{F}[a, b] = U + W$. □

Example 9. If U denotes any subspace of $\mathbb{R}^n$, show that

$$\mathbb{R}^n = U \oplus U^\perp$$

where $U^\perp = \{X \text{ in } \mathbb{R}^n \mid X \bullet Y = 0 \text{ for all } Y \text{ in } U\}$ is the orthogonal complement of U.

Solution. If X is in both U and $U^\perp$ then $0 = X \bullet X = \|X\|^2$, so $X = 0$. Hence $U \cap U^\perp = \{0\}$. To see that $\mathbb{R}^n = U + U^\perp$, let X be any vector in $\mathbb{R}^n$ and write $P = proj_U(X)$. Then $X = P + (X - P)$, and the Projection Theorem (Theorem 2 §4.6) asserts that P is in U and $(X - P)$ is in $U^\perp$. It follows that $\mathbb{R}^n = U + U^\perp$. □

Clearly, $V = U + W$ if and only if every vector **v** in V has the form $\mathbf{v} = \mathbf{u} + \mathbf{w}$ for some **u** in U and **w** in W. We are interested in the situation where this representation is *unique*, that is where condition (1) in the following lemma holds.

Lemma 3. *The following conditions are equivalent for subspaces U and W of a vector space V:*

(1) *If $\mathbf{u} + \mathbf{w} = \mathbf{u}_1 + \mathbf{w}_1$ with $\mathbf{u}, \mathbf{u}_1$ in U and $\mathbf{w}, \mathbf{w}_1$ in W, then $\mathbf{u} = \mathbf{u}_1$ and $\mathbf{w} = \mathbf{w}_1$.*

(2) *If $\mathbf{u} + \mathbf{w} = \mathbf{0}$ where $\mathbf{u}$ is in U and $\mathbf{w}$ is in W, then $\mathbf{u} = \mathbf{0}$ and $\mathbf{w} = \mathbf{0}$.*

(3) $U \cap W = \{\mathbf{0}\}$.

Proof. (1)⇒(2). If $\mathbf{u} + \mathbf{w} = \mathbf{0}$, (2) follows by taking $\mathbf{u}_1 = \mathbf{0}$ and $\mathbf{w}_1 = \mathbf{0}$ in (1).

(2)⇒(3). If **v** is in $U \cap W$ then $\mathbf{v} + (-\mathbf{v}) = \mathbf{0}$ so $\mathbf{v} = \mathbf{0}$ by (2) because **v** is in U and $-\mathbf{v}$ is in W.

(3)⇒(1). Suppose that $\mathbf{u} + \mathbf{w} = \mathbf{u}_1 + \mathbf{w}_1$ as in (1). Then $\mathbf{u} - \mathbf{u}_1 = \mathbf{w}_1 - \mathbf{w}$, and this vector is in both U (left side) and W (right side). Hence $\mathbf{u} - \mathbf{u}_1 = \mathbf{0}$ and $\mathbf{w}_1 - \mathbf{w} = \mathbf{0}$ by (3), proving (1). □

We can now give some important characterizations of direct sum decompositions.

Theorem 3. *If U and W are subspaces of V where $\dim V$ is finite, the following are equivalent:*

(1) $V = U \oplus W$.

5.6. INVARIANT SUBSPACES

(2) *Every vector* $\mathbf{v}$ *in* V *has a unique representation in the form* $\mathbf{v} = \mathbf{u} + \mathbf{w}$ *with* $\mathbf{u}$ *in* U *and* $\mathbf{w}$ *in* W.

(3) *If* $\mathcal{B}_1 = \{\mathbf{u}_1, \mathbf{u}_2, \cdots, \mathbf{u}_m\}$ *and* $\mathcal{B}_2 = \{\mathbf{w}_1, \mathbf{w}_2, \cdots, \mathbf{w}_k\}$ *are any bases of* U *and* W *respectively, then* $\mathcal{B} = \{\mathbf{u}_1, \mathbf{u}_2, \cdots, \mathbf{u}_m, \mathbf{w}_1, \mathbf{w}_2, \cdots, \mathbf{w}_k\}$ *is a basis of* V.

(4) *There exist bases* $\mathcal{B}_1 = \{\mathbf{u}_1, \mathbf{u}_2, \cdots, \mathbf{u}_m\}$ *and* $\mathcal{B}_2 = \{\mathbf{w}_1, \mathbf{w}_2, \cdots, \mathbf{w}_k\}$ *of* U *and* W *respectively such that* $\mathcal{B} = \{\mathbf{u}_1, \mathbf{u}_2, \cdots, \mathbf{u}_m, \mathbf{w}_1, \mathbf{w}_2, \cdots, \mathbf{w}_k\}$ *is a basis of* V.

Proof. (1)$\Rightarrow$(2). Since $V = U + W$, every vector $\mathbf{v}$ in V has the form $\mathbf{v} = \mathbf{u} + \mathbf{w}$ for some vectors $\mathbf{u}$ in U and $\mathbf{w}$ in W. This representation is unique by Lemma 3 because $U \cap W = \{\mathbf{0}\}$.

(2)$\Rightarrow$(3). If $\mathbf{v}$ is in V, use (2) to write $\mathbf{v} = \mathbf{u} + \mathbf{w}$ where $\mathbf{u}$ is in U and $\mathbf{w}$ is in W. Since $\mathbf{u}$ and $\mathbf{w}$ are in $span\{\mathcal{B}_1\}$ and $span\{\mathcal{B}_2\}$ respectively, it follows that $\mathbf{v} = \mathbf{u} + \mathbf{w}$ is in $span\{\mathcal{B}\}$. Hence $V = span\{\mathcal{B}\}$.

To see that $\mathcal{B}$ is independent, suppose numbers s_i and t_j exist such that

$$s_1\mathbf{u}_1 + s_2\mathbf{u}_2 + \cdots + s_m\mathbf{u}_m + t_1\mathbf{w}_1 + t_2\mathbf{w}_2 + \cdots + t_k\mathbf{w}_k = \mathbf{0}.$$

If we write $\mathbf{u} = s_1\mathbf{u}_1 + s_2\mathbf{u}_2 + \cdots + s_m\mathbf{u}_m$ and $\mathbf{w} = t_1\mathbf{w}_1 + t_2\mathbf{w}_2 + \cdots + t_k\mathbf{w}_k$, we have $\mathbf{u} + \mathbf{w} = \mathbf{0}$ with $\mathbf{u}$ in U and $\mathbf{w}$ in W. The uniqueness in (2) implies that $\mathbf{u} = \mathbf{0}$ and $\mathbf{w} = \mathbf{0}$, and then the independence of $\mathcal{B}_1$ and $\mathcal{B}_2$ forces $s_i = 0$ for each i and $t_j = 0$ for each j. Hence $\mathcal{B}$ is independent.

(3)$\Rightarrow$(4). This is is clear since U and W have bases.

(4)$\Rightarrow$(1). Given (4), every vector $\mathbf{v}$ in V is in $span\{\mathcal{B}\}$, and so there exist s_i and t_j in $\mathbb{R}$ such that

$$\mathbf{v} = s_1\mathbf{u}_1 + s_2\mathbf{u}_2 + \cdots + s_m\mathbf{u}_m + t_1\mathbf{w}_1 + t_2\mathbf{w}_2 + \cdots + t_k\mathbf{w}_k.$$

Hence $\mathbf{v} = \mathbf{u} + \mathbf{w}$ where $\mathbf{u} = s_1\mathbf{u}_1 + s_2\mathbf{u}_2 + \cdots + s_m\mathbf{u}_m$ is in U and $\mathbf{w} = t_1\mathbf{w}_1 + t_2\mathbf{w}_2 + \cdots + t_k\mathbf{w}_k$ is in W. It follows that $V = U + W$. If $\mathbf{v}$ is in $U \cap W$, (4) gives $\mathbf{v} = s_1\mathbf{u}_1 + s_2\mathbf{u}_2 + \cdots + s_m\mathbf{u}_m$ and $\mathbf{v} = t_1\mathbf{w}_1 + t_2\mathbf{w}_2 + \cdots + t_k\mathbf{w}_k$ with s_i and t_j in $\mathbb{R}$. Then the independence of $\mathcal{B}$ implies that $s_i = 0 = t_j$ for all i and j, so $\mathbf{v} = \mathbf{0}$. This proves that $U \cap W = \{\mathbf{0}\}$, and so proves (1). $\square$

If $\mathcal{B} = \{\mathbf{u}_1, \mathbf{u}_2, \cdots, \mathbf{u}_m, \mathbf{w}_1, \mathbf{w}_2, \cdots, \mathbf{w}_k\}$ is any basis of a vector space V, then condition (4) of Theorem 3 shows that $V = U \oplus W$ where we have $U = span\{\mathbf{u}_1, \mathbf{u}_2, \cdots, \mathbf{u}_m\}$ and $W = span\{\mathbf{w}_1, \mathbf{w}_2, \cdots, \mathbf{w}_k\}$. Moreover, every direct sum decomposition of V arises in this way by partitioning some basis $\mathcal{B}$ of V into two parts.

The following useful result is an immediate consequence of Condition (4) in Theorem 3.

Theorem 4. *If* V *is a finite dimensional vector space and* $V = U \oplus W$ *where* U *and* W *are subspaces of* V, *then* $dim V = dim U + dim W$.

5.6.3 Reducible Operators

Returning to T-invariant subspaces, we can now prove the promised refinement of Theorem 1 wherein we outline how to find a basis such that the matrix of the operator T is block diagonal.

Theorem 5. *Let $T : V \to V$ be a linear operator where $\dim V = n$, let*

$$V = U \oplus W \text{ where both } U \text{ and } W \text{ are } T\text{-invariant,}$$

and let $\mathcal{B}_1 = \{\mathbf{u}_1, \mathbf{u}_2, \cdots, \mathbf{u}_m\}$ and $\mathcal{B}_2 = \{\mathbf{w}_1, \mathbf{w}_2, \cdots, \mathbf{w}_k\}$ be bases of U and W respectively. Then $\mathcal{B} = \{\mathbf{u}_1, \mathbf{u}_2, \cdots, \mathbf{u}_m, \mathbf{w}_1, \mathbf{w}_2, \cdots, \mathbf{w}_k\}$ is a basis of V and $M_\mathcal{B}(T)$ has the block diagonal form

$$M_\mathcal{B}(T) = \begin{bmatrix} M_{\mathcal{B}_1}(T) & 0 \\ 0 & M_{\mathcal{B}_2}(T) \end{bmatrix}$$

where $M_{\mathcal{B}_1}(T)$ and $M_{\mathcal{B}_2}(T)$ are the matrices of the restrictions of T to U and W respectively.

Proof. Theorem 4 shows that $\mathcal{B}$ is a basis of V, and Theorem 1 shows that the columns in $M_\mathcal{B}(T)$ corresponding to $\mathcal{B}_1$ have the required form because U is T-invariant. Since W is also T-invariant, a similar argument works for the remaining columns of $M_\mathcal{B}(T)$. $\square$

An operator $T : V \to V$ is called **reducible** if a direct sum decomposition $V = U \oplus W$ can be found such that both U and W are T-invariant.

Example 10. *In Example 4 we considered the operator $T : \mathbb{M}_{2,2} \to \mathbb{M}_{2,2}$ defined by $T(A) = BA$ for all A in $\mathbb{M}_{2,2}$, where $B = \begin{bmatrix} 1 & -1 \\ 1 & 0 \end{bmatrix}$. Show that T is reducible.*

Solution. In Example 4 we showed that $U = \text{span}\{I, B\}$ is a T-invariant subspace of $\mathbb{M}_{2,2}$ with basis $\mathcal{B}_1 = \{I, B\}$. In Example 5 we completed $\mathcal{B}_1$ to a basis $\{I, B, E_{11}, E_{12}\}$ and found that $\mathbb{M}_{2,2} = U \oplus W_1$ where $W_1 = \text{span}\{E_{11}, E_{12}\}$. The problem is that W_1 is *not* T-invariant.

However a different completion of the basis $\mathcal{B}_1$ *does* produce a T-invariant complement of U. If we write $\mathcal{B}_2 = \{E_{11}, E_{21}\}$, we find that $W = \text{span}\{\mathcal{B}_2\}$ is T-invariant by Lemma 1 because $T(E_{11}) = E_{11} + E_{21}$ and $T(E_{21}) = -E_{11}$. Moreover, $\mathcal{B} = \{I, B, E_{11}, E_{21}\}$ is a basis of V (verify) so $V = U \oplus W$. Hence T is reducible. In fact,

$$M_\mathcal{B}(T) = \begin{bmatrix} M_{\mathcal{B}_1}(T) & 0 \\ 0 & M_{\mathcal{B}_2}(T) \end{bmatrix}$$

where $M_{\mathcal{B}_1}(T) = \begin{bmatrix} 0 & -1 \\ 1 & 1 \end{bmatrix}$ because $T(I) = B$ and $T(B) = B - I$, and

5.6. INVARIANT SUBSPACES

$$M_{\mathcal{B}_2}(T) = \begin{bmatrix} 1 & -1 \\ 1 & 0 \end{bmatrix} \text{ because } T(E_{11}) = E_{11} + E_{21} \text{ and } T(E_{21}) = -E_{11}. \quad \square$$

The operator T in Example 10 has a natural T-invariant subspace U, but the first complement W_1 of U we considered was not T-invariant. However we succeeded in finding a T-invariant complement W of U, so T is in fact reducible. Sometimes, however, an operator can fail to be reducible even though it has an invariant subspace; the problem is that *no* invariant complement exists. Here is an example.

Example 11. *Let $\{\mathbf{u}, \mathbf{v}\}$ be a basis of V and define $T : V \to V$ by $T(\mathbf{u}) = \mathbf{u}$ and $T(\mathbf{v}) = \mathbf{u} + \mathbf{v}$. Show that $U = \mathbb{R}\mathbf{u}$ is T-invariant but has no T-invariant complement in V.*

Solution. U is T-invariant by Lemma 1 because $T(\mathbf{u}) = \mathbf{u}$. Suppose that $V = U \oplus W$ where W is also T-invariant. Then $dim W = 1$ by Theorem 4 (because $dim V = 2$), say $W = \mathbb{R}\mathbf{w}$. Write $\mathbf{w} = a\mathbf{u} + b\mathbf{v}$ where a and b are in $\mathbb{R}$. Then $b \neq 0$ because otherwise $\mathbf{w} = a\mathbf{u}$ is in $U \cap W = \{\mathbf{0}\}$. Since W is T-invariant, there is a real number λ such that $T(\mathbf{w}) = \lambda \mathbf{w}$. Since $\mathbf{w} = a\mathbf{u} + b\mathbf{v}$, the fact that T is linear gives

$$\lambda a \mathbf{u} + \lambda b \mathbf{v} = \lambda \mathbf{w} = T(\mathbf{w}) = aT(\mathbf{u}) + bT(\mathbf{v}) = a\mathbf{u} + b(\mathbf{u} + \mathbf{v}) = (a+b)\mathbf{u} + b\mathbf{v}.$$

Since $\{\mathbf{u}, \mathbf{v}\}$ is independent, this gives $\lambda a = a + b$ and $\lambda b = b$. Hence $\lambda = 1$ (because $b \neq 0$) so that $a + b = \lambda a = a$. This gives $b = 0$, a contradiction. $\square$

Sometimes, a direct sum decomposition can be found into two eigenspaces of an operator. This not only shows that the operator is reducible, but diagonalizable as well. The next example is an illustration.

Example 12. *An operator $T : V \to V$ is called an **idempotent** if $T^2 = T$.*

(1) *If T is an idempotent, show that T is diagonalizable by showing that V is the direct sum of its eigenspaces.*

(2) *If $A = A^2$ is any idempotent matrix, show that A is similar to $\begin{bmatrix} 0 & 0 \\ 0 & I_m \end{bmatrix}$ for some integer m (so A is diagonalizable).*

Solution. (1) If λ is an eigenvalue of T, we first show that either $\lambda = 0$ or $\lambda = 1$. Indeed, if $T(\mathbf{v}) = \lambda \mathbf{v}$ where $\mathbf{v} \neq \mathbf{0}$, then

$$\lambda \mathbf{v} = T(\mathbf{v}) = T^2(\mathbf{v}) = T(T(\mathbf{v})) = T(\lambda \mathbf{v}) = \lambda T(\mathbf{v}) = \lambda^2 \mathbf{v}.$$

Hence $\lambda = \lambda^2$ because $\mathbf{v} \neq \mathbf{0}$, so $\lambda = 0$ or $\lambda = 1$. Hence there are two eigenspaces:

$$E_0(T) = \{\mathbf{v} \text{ in } V \mid T(\mathbf{v}) = \mathbf{0}\} \quad \text{and} \quad E_1(T) = \{\mathbf{v} \text{ in } V \mid T(\mathbf{v}) = \mathbf{v}\}.$$

We claim that $V = E_0(T) \oplus E_1(T)$. It is clear that $E_0(T) \cap E_1(T) = \{\mathbf{0}\}$. Moreover $V = E_0(T) + E_1(T)$ because any vector $\mathbf{v}$ in V can be written in the form $\mathbf{v} = (\mathbf{v} - T(\mathbf{v})) + T(\mathbf{v})$ and (since $T^2 = T$) it is easy to verify that $\mathbf{v} - T(\mathbf{v})$ is in $E_0(T)$ and that $T(\mathbf{v})$ is in $E_1(T)$. Hence

$$V = E_0(T) \oplus E_1(T)$$

and both $E_0(T)$ and $E_1(T)$ are T-invariant (by Example 2). Moreover, if $\mathcal{B}_0$ and $\mathcal{B}_1$ are bases of $E_0(T)$ and $E_1(T)$ respectively, the matrices of the restriction of T to $E_0(T)$ and $E_1(T)$ are $M_{\mathcal{B}_0}(T) = 0$ and $M_{\mathcal{B}_1}(T) = I_m$ where $m = dim(E_1(T))$. Thus Theorem 5 provides a basis $\mathcal{B}$ of V such that

$$M_\mathcal{B}(T) = \begin{bmatrix} M_{\mathcal{B}_0}(T) & 0 \\ 0 & M_{\mathcal{B}_1}(T) \end{bmatrix} = \begin{bmatrix} 0 & 0 \\ 0 & I_m \end{bmatrix}.$$

In particular, T is diagonalizable.

(2) If A is $n \times n$ and $A^2 = A$, define $T_A : \mathbb{R}^n \to \mathbb{R}^n$ by $T_A(X) = AX$ for all X in $\mathbb{R}^n$. Then $T_A^2 = T_A$ (verify) so, by part (1), there exists a basis $\mathcal{B}$ such that

$$M_\mathcal{B}(T_A) = \begin{bmatrix} 0 & 0 \\ 0 & I_m \end{bmatrix} \text{ where } m = dim(E_1(A)).$$

Since $A = M_\mathcal{E}(T_A)$ where $\mathcal{E}$ is the standard basis of $\mathbb{R}^n$, we have $A \sim M_\mathcal{B}(T_A)$ by Theorem 2 §5.5, as required. $\square$

If V is a vector space, we always have the trivial direct decomposition

$$V = \{\mathbf{0}\} \oplus V = V \oplus \{\mathbf{0}\}.$$

We call all other direct decompositions $V = U \oplus W$ **proper**. Thus $V = U \oplus W$ is proper if $U \neq V$ and $W \neq V$ (equivalently $U \neq \{\mathbf{0}\}$ and $W \neq \{\mathbf{0}\}$). If $T : V \to V$ is an operator, the vector space V is called T**-irreducible** if no proper direct decomposition $V = U \oplus W$ can be found where both U and W are T-invariant. This certainly happens if V has no T-invariant subspaces except $\{\mathbf{0}\}$ and V, and such a space is called T**-simple**.[18] If $dim V = 1$ then V is T-simple for every operator $T : V \to V$ (verify), and Example 3 describes a 2-dimensional, T-simple space. However, even though every T-simple space is T-irreducible, it is possible to have a T-irreducible space that is not T-simple.

Example 13. *As in Example 11, let $\mathcal{B} = \{\mathbf{u}, \mathbf{v}\}$ be a basis of V and define $T : V \to V$ by $T(\mathbf{u}) = \mathbf{u}$ and $T(\mathbf{v}) = \mathbf{u} + \mathbf{v}$. Show that V is a T-irreducible space that is not T-simple.*

Proof. The subspace $\mathbb{R}\mathbf{u}$ is T-invariant by Lemma 1, so V is not T-simple because $dim V = 2$. In Example 11 we showed that $\mathbb{R}\mathbf{u}$ has no T-complement in V; we now show that V is in fact irreducible. Suppose on the contrary that T

---
[18] T-simple spaces are sometimes call **completely T-irreducible**, especially in older books, but we prefer the term T-simple.

5.6. INVARIANT SUBSPACES

is reducible, say $V = Z \oplus W$ where Z and W are proper, T-invariant subspaces of V. Since $dim V = 2$, the fact that Z and W are proper means that $dim Z = 1$ and $dim W = 1$, say $Z = \mathbb{R}\mathbf{z}$ and $W = \mathbb{R}\mathbf{w}$ where $\mathbf{z} \neq \mathbf{0}$ and $\mathbf{w} \neq \mathbf{0}$. Moreover, since Z and W are T-invariant it follows that $\mathbf{z}$ and $\mathbf{w}$ are both eigenvectors of T (verify). But $\{\mathbf{z}, \mathbf{w}\}$ is independent because $\mathbb{R}\mathbf{z} \cap \mathbb{R}\mathbf{w} = \{\mathbf{0}\}$, and so is a basis of eigenvectors of T. This means that T is diagonalizable, and hence that $M_\mathcal{B}(T) = \begin{bmatrix} 1 & 1 \\ 0 & 1 \end{bmatrix}$ is a diagonalizable matrix, a contradiction by Example 7 §2.3. So T is irreducible. $\square$

5.6.4 The Cayley-Hamilton Theorem

We conclude this section by using invariant subspaces to prove a classical theorem of linear algebra, the Cayley-Hamilton Theorem, which asserts that every matrix satisfies its characteristic polynomial. The following example provides an illustration.

Example 14. If $A = \begin{bmatrix} 1 & -1 \\ 2 & 3 \end{bmatrix}$ then the characteristic polynomial is $c_A(x) = det(xI - A) = x^2 - 4x + 5$. If we substitute A for x in $c_A(x)$ the result is $c_A(A) = A^2 - 4A + 5I$.[19] The Cayley-Hamilton Theorem asserts that in fact $c_A(A) = 0$; indeed

$$c_A(A) = A^2 - 4A + 5I = \begin{bmatrix} -1 & -4 \\ 8 & 7 \end{bmatrix} - \begin{bmatrix} 4 & -4 \\ 8 & 12 \end{bmatrix} + \begin{bmatrix} 5 & 0 \\ 0 & 5 \end{bmatrix} = \begin{bmatrix} 0 & 0 \\ 0 & 0 \end{bmatrix},$$

as expected. $\square$

Our approach to the Cayley-Hamilton Theorem, is to first prove it for operators by showing that every operator $T : V \to V$ satisfies its characteristic polynomial. This clearly entails defining what is meant by $p(T)$ where $p(x)$ is any polynomial. This, in turn, means that we must define sums and scalar multiples of operators.

Suppose that $S : V \to V$ and $T : V \to V$ are operators and that a is a scalar. We define new operators $S + T : V \to V$ and $aT : V \to V$ as follows:

$$(S + T)(\mathbf{v}) = S(\mathbf{v}) + T(\mathbf{v}) \quad \text{for all } \mathbf{v} \text{ in } V$$

and

$$(aT)(\mathbf{v}) = a\,T(\mathbf{v}) \quad \text{for all } \mathbf{v} \text{ in } V.$$

It is a routine matter to verify that $S + T$ and aT are both linear; they are called the **sum** of S and T, and the **scalar product** of T by a, respectively. If we recall that the power T^k means the composition of T with itself k times, we can define $p(T)$ for every polynomial $p(x)$. Indeed, if

$$p(x) = a_0 + a_1 x + a_2 x^2 + a_3 x^3 + \cdots + a_m x^m \text{ where each } a_i \text{ is in } \mathbb{R},$$

---

[19] Note that we are thinking of 5 in $c_A(x)$ as $5x^0$, and then taking $5A^0 = 5I$ in $c_A(A)$.

we define the operator $p(T) : V \to V$ by

$$p(T) = a_0 + a_1 T + a_2 T^2 + a_3 T^3 + \cdots + a_m T^m.$$

Thus, for example, if $p(x) = 3 - 5x + 7x^3$, the action of $p(T)$ on a vector $\mathbf{v}$ is

$$p(T)(\mathbf{v}) = (3 - 5T + 7T^3)(\mathbf{v}) = 3\mathbf{v} - 5T(\mathbf{v}) + 7T^3(\mathbf{v}) \text{ for all } \mathbf{v} \text{ in } V.$$

Of course $T^3(\mathbf{v})$ means $T(T(T(\mathbf{v})))$ for each vector $\mathbf{v}$.

Now suppose that $\mathbf{v}$ is any nonzero vector in V where $dim V$ is finite, and consider the subsets

$$\{\mathbf{v}\}, \; \{\mathbf{v}, \; T(\mathbf{v})\}, \; \{\mathbf{v}, \; T(\mathbf{v}), \; T^2(\mathbf{v})\}, \cdots.$$

The first of these sets is independent because $\mathbf{v} \neq \mathbf{0}$. But they cannot all be independent because $dim V$ is finite, so one of them in dependent. Let $m \geq 1$ be the smallest integer such that $\{\mathbf{v}, T(\mathbf{v}), T^2(\mathbf{v}), \cdots, T^m(\mathbf{v})\}$ is dependent. Then

$$\mathcal{B}_0 = \{\mathbf{v}, \; T(\mathbf{v}), \; T^2(\mathbf{v}), \cdots, T^{m-1}(\mathbf{v})\}$$

is independent. (If $m = 1$ this is taken to mean that $\mathcal{B}_0 = \{\mathbf{v}\}$.)[20] Having found $\mathcal{B}_0$, write

$$U = span \mathcal{B}_0 = span\{\mathbf{v}, \; T(\mathbf{v}), \; T^2(\mathbf{v}), \cdots, T^{m-1}(\mathbf{v})\}. \tag{*}$$

so that $dim(U) = m$ and $\mathcal{B}_0$ is a basis of U.

Since $\{\mathbf{v}, T(\mathbf{v}), T^2(\mathbf{v}), \cdots, T^{m-1}(\mathbf{v}), T^m(\mathbf{v})\}$ is dependent, Lemma 1 §5.2 implies that $T^m(\mathbf{v})$ is in U. This has two consequences: First we can write

$$T^m(\mathbf{v}) = -r_0 \mathbf{v} - r_1 T(\mathbf{v}) - r_2 T^2(\mathbf{v}) - \cdots - r_{m-1} T^{m-1}(\mathbf{v}) \tag{**}$$

where each r_i is in $\mathbb{R}$. (The reason for the minus signs will be apparent shortly.) The second consequence of the fact that $T^m(\mathbf{v})$ is in U is that it shows (using Lemma 1) that U is a T-invariant subspace of V, called the T-**cyclic subspace** generated by $\mathbf{v}$. Hence $T : U \to U$ is an operator on U, and (*) and (**) show that the $\mathcal{B}_0$-matrix of T is

$$M_{\mathcal{B}_0}(T) = \begin{bmatrix} 0 & 0 & \cdots & 0 & -r_0 \\ 1 & 0 & \cdots & 0 & -r_1 \\ 0 & 1 & \cdots & 0 & -r_2 \\ \vdots & \vdots & & \vdots & \vdots \\ 0 & 0 & \cdots & 1 & -r_{m-1} \end{bmatrix}. \tag{***}$$

If we let $q(x)$ denote the characteristic polynomial of the restriction of T to U, we have

$$\begin{aligned} q(x) &= c_{M_{\mathcal{B}_0}(T)}(x) = det[xI - M_{\mathcal{B}_0}(T)] \\ &= r_0 + r_1 x + r_2 x^2 + \cdots + r_{m-1} x^{m-1} + x^m \end{aligned} \tag{****}$$

---

[20] Note that $\{\mathbf{v}, T(\mathbf{v})\}$ is dependent if and only if $\mathbf{v}$ is an eigenvector of T (verify).

5.6. INVARIANT SUBSPACES

as the reader can verify.[21] For this reason, the matrix in (***) is called the **companion matrix** of the polynomial $q(x) = r_0 + r_1 x + r_2 x^2 + \cdots + r_{m-1} x^{m-1} + x^m$.

We can now prove the operator version of the Cayley-Hamilton Theorem.

Theorem 6. *If $T : V \to V$ is a linear operator where $\dim V = n$, then T satisfies its characteristic polynomial, that is*
$$c_T(T) = 0.$$

Proof. We are to prove that $c_T(T)(\mathbf{v}) = \mathbf{0}$ for every vector $\mathbf{v}$ in V. This is clear if $\mathbf{v} = \mathbf{0}$ because $c_T(T)$ is linear. If $\mathbf{v} \neq \mathbf{0}$ we use it to construct U, $\mathcal{B}_0$ and $q(x)$ as above. If we complete $\mathcal{B}_0$ to a basis of V in any way at all, then Theorem 2 asserts that
$$c_T(x) = p(x)\, q(x) \text{ for some polynomial } p(x).$$
Hence $c_T(T)(\mathbf{v}) = p(T)\,[q(T)(\mathbf{v})]$ so it suffices to show that $q(T)(\mathbf{v}) = \mathbf{0}$. But (****) and (**) give
$$q(T)(\mathbf{v}) = r_0 \mathbf{v} + r_1 T(\mathbf{v}) + r_2 T^2(\mathbf{v}) + \cdots + r_{m-1} T^{m-1}(\mathbf{v}) + T^m(\mathbf{v}) = \mathbf{0}.$$
This completes the proof. $\square$

The Cayley-Hamilton Theorem is the matrix version of Theorem 6. We need one more observation about matrices of operators: If $S : V \to V$ and $T : V \to V$ are operators on a vector space V and $\mathcal{B}$ is any basis of V, then
$$\begin{aligned} M_\mathcal{B}(S+T) &= M_\mathcal{B}(S) + M_\mathcal{B}(T) \\ M_\mathcal{B}(aT) &= a\, M_\mathcal{B}(T) \text{ for any scalar } a \\ M_\mathcal{B}(S \circ T) &= M_\mathcal{B}(S) M_\mathcal{B}(S) \end{aligned}$$

The last of these is part of Theorem 1 §5.5, and the routine verifications of the first two are left as an exercise.

Theorem 7. Cayley-Hamilton Theorem.[22] *Every square matrix A satisfies its characteristic polynomial, that is*
$$c_A(A) = 0.$$

Proof. Define $T_A : \mathbb{R}^n \to \mathbb{R}^n$ by $T_A(X) = AX$ for all X in $\mathbb{R}^n$. Then $c_{T_A}(T_A) = 0$ by Theorem 6. If $\mathcal{E}$ is the standard basis of $\mathbb{R}^n$, then
$$M_\mathcal{E}(T_A) = A \quad \text{and} \quad c_{T_A}(x) = c_{M_\mathcal{E}(T_A)}(x) = c_A(x).$$

---

[21] This is the reason for the negative signs in (**).

[22] This theorem is named after the English mathematician Arthur Cayley (1821-1895), one of the pioneers in matrix theory, and William Rowan Hamilton (1805-1865), an Irish mathematician famous for his work on physical dynamics. The theorem was established for a special class of matrices by Hamilton in 1853, and Cayley announced the general version five years later.

Now write $c_A(x) = r_0 + r_1 x + r_2 x^2 + \cdots + r_n x^n$, and compute

$$\begin{aligned}
c_A(A) &= r_0 I + r_1 A + r_2 A^2 + \cdots + r_n A^n \\
&= r_0 M_{\mathcal{E}}(1_{\mathbb{R}^n}) + r_1 M_{\mathcal{E}}(T_A) + r_2 M_{\mathcal{E}}(T_A)^2 + \cdots + r_n M_{\mathcal{E}}(T_A)^n \\
&= M_{\mathcal{E}}(r_0 1_{\mathbb{R}^n} + r_1 T_A + r_2 T_A^2 + \cdots + r_n T_A^n) \\
&= M_{\mathcal{E}}(c_{T_A}(T_A)) \\
&= M_{\mathcal{E}}(0) \\
&= 0.
\end{aligned}$$

This is what we wanted. □

Exercises

Throughout these exercises let V denote a vector space and let U and W denote subspaces of V.

1. In each case either show that the statement is true or give an example showing that it is false. Throughout, let $T : V \to V$ denote an operator.

   (a) $\ker T$ is T-invariant.

   (b) $\mathrm{im}\, T$ is T-invariant.

   (c) It can happen that V has no T-invariant subspace.

   (d) $U+W$ contains both U and W, and is contained in every such subspace.

   (e) $U \cap W$ is contained in both U and W, and contains every such subspace.

   (f) If T is an isomorphism, the only T-invariant subspaces of V are V and $\{\mathbf{0}\}$.

   (g) If $\mathbf{v}$ is an eigenvector of T then $\mathbb{R}\mathbf{v}$ is T-invariant.

   (h) If $\mathbb{R}\mathbf{v}$ is T-invariant where $\mathbf{v} \neq \mathbf{0}$, then $\mathbf{v}$ is an eigenvector of T.

2. Let V have basis $\mathcal{B} = \{\mathbf{b}_1, \mathbf{b}_2, \mathbf{b}_3, \mathbf{b}_4\}$, and define $T : V \to V$ by

   $$T(\mathbf{b}_1) = \mathbf{b}_1, \quad T(\mathbf{b}_2) = \mathbf{b}_2 + 2\mathbf{b}_4, \quad T(\mathbf{b}_3) = \mathbf{b}_1, \quad T(\mathbf{b}_4) = \mathbf{0}.$$

   Show that $T^2 = T$, find a basis $\mathcal{B}_o$ of $E_1(T)$, extend it to a basis of V, and illustrate Theorem 1.

3. Let $T : \mathbb{P}_2 \to \mathbb{P}_2$ be defined by $T(a + bx + cx^2) = (a - 2b - c) - (a + 3b + c)x - (a + 4b)x^2$.

   (a) Show that $U = \mathrm{span}\{1, x + x^2\}$ is T-invariant.

   (b) Use (a) to find a basis $\mathcal{B}$ of such that $M(T)$ is block upper triangular.

   (c) Use (b) to compute $c_T(x)$.

4. (a) Show that every subspace of $E_\lambda(T)$ is T-invariant.

   (b) Describe the action of the restriction of T to $E_\lambda(T)$.

5.6. INVARIANT SUBSPACES

5. If $T : V \to V$ is an operator and U is a subspace of V, show that $T(U)$ is a subspace of V, T-invariant if U is T-invariant.

6. In each case show that $V = U \oplus W$.

   (a) $V = \mathbb{M}_{2,2}$, $U = \left\{ \begin{bmatrix} a & b \\ a & b \end{bmatrix} \mid a, b \text{ in } \mathbb{R} \right\}$ and $W = \left\{ \begin{bmatrix} c & c \\ d & -d \end{bmatrix} \mid a, b \text{ in } \mathbb{R} \right\}$.

   (b) $V = \mathbb{P}_3$, $U = \{a + b(x - x^2) \mid a, b \text{ in } \mathbb{R}\}$ and $W = \{c(1 + x) + dx^3 \mid c, d \text{ in } \mathbb{R}\}$.

   (c) $V = \mathbb{R}^4$, $U = \{[a \ -b \ b \ a]^T \mid a, b \text{ in } \mathbb{R}\}$ and $W = \{[c \ d \ c \ d]^T \mid c, d \text{ in } \mathbb{R}\}$.

7. Define $T : \mathbb{P}_3 \to \mathbb{P}_3$ by $T(a + bx + cx^2 + dx^3) = (2a + b) + (-5a - 2b)x + (c + 2d)x^2 - (c + d)x^3$ for all a, b, c and d. Show that $\mathbb{P}_3$ has two 2-dimensional T-invariant subspaces U and W such that $\mathbb{P}_3 = U \oplus W$, but that T has no real eigenvalue. [Hint: First find $M_\mathcal{B}(T)$ where $\mathcal{B} = \{1, x, x^2, x^3\}$.]

8. Let $T : V \to V$ be an operator.

   (a) Show that every subspace of V is T-invariant if and only every nonzero vector $\mathbf{v}$ in V is an eigenvector of T.

   (b) Show that λ exists such that $T(\mathbf{v}) = \lambda \mathbf{v}$ for every vector $\mathbf{v}$ in V.

9. Let V be a finite dimensional vector space. Show that V and $\{\mathbf{0}\}$ are the only subspaces of V that are T-invariant for every operator $T : V \to V$. [Hint: Theorem 3 §5.3.]

10. If $V = U \oplus W$ and $V = U \oplus W_1$ where V is finite dimensional, show that $dim W = dim W_1$.

11. Show that $U \cap W = \{\mathbf{0}\}$ if and only if $\{\mathbf{u}, \mathbf{w}\}$ is independent for all nonzero vectors $\mathbf{u}$ in U and $\mathbf{w}$ in W.

12. If $U \cap W = \{\mathbf{0}\}$ and $dim U + dim W = dim V$, show that $V = U \oplus W$.

13. Suppose that $V = U \oplus W$ where both U and W are T-invariant. If the restrictions of T to U and W are both diagonalizable, show that T is diagonalizable.

14. Let $T : V \to V$ be a linear operator where $dim V = n$.

    (a) Show that $E_\lambda(T) \cap E_\mu(T) = \{\mathbf{0}\}$ if $\lambda \neq \mu$.

    (b) If $\lambda \neq \mu$ are the only eigenvalues of an operator T, and if $dim(E_\lambda(T)) + dim(E_\mu(T)) = n$, show that T is diagonalizable.

    (c) If $\lambda \neq \mu$ are the only eigenvalues of T and $dim V = 2$, show that T is diagonalizable.

    (d) Give an example where T has exactly two distinct eigenvalues but is not diagonalizable.

15. Let $T : V \to V$ be an operator. If V is T-simple and $T \neq 0$, show that T is an isomorphism. (**Schur'sLemma.**) [*Hint:* Consider $kerT$ and imT.]

16. Let S and T be linear operators $V \to V$, and assume that $S \circ T = T \circ S$.

    (a) If λ is an eigenvalue of S, show that $E_\lambda(S)$ is T-invariant.

    (b) Show that imS and $kerS$ are both T-invariant.

    (c) If U is an T-invariant subspace of V, show that $S(U)$ is also T-invariant.

17. Let $T : V \to V$ be an operator, and let U be the T-cyclic subspace generated by $\mathbf{v} \neq \mathbf{0}$ in V (see the discussion preceding Theorem 6). If W is any T-invariant subspace that contains $\mathbf{v}$, show that $U \subseteq W$. [Thus U is the "smallest" T-invariant subspace containing $\mathbf{v}$.]

18. Determine if the following are true or false. Justify your answer.

    (a) If $dimV = 3$, and $V = U \oplus W$ where both U and W are T-invariant, then T is diagonalizable.

    (b) If $dimV = 2$, and $V = U \oplus W$ where both U and W are T-invariant, then T is diagonalizable.

19. If $T : V \to V$ is an operator and $dimV = 1$, show that V is T-simple.

20. Let $T : V \to V$ be an operator. If $U = \mathbb{R}\mathbf{u}$ is a 1-dimensional subspace of V, show that U is T-invariant if and only if $\mathbf{u}$ is an eigenvector of T.

21. (a) If $dimV > 1$ and V is T-simple, show that T has no real eigenvalue.

    (b) Let $A = \begin{bmatrix} B & 0 \\ 0 & C \end{bmatrix}$ where $B = \begin{bmatrix} 2 & 1 \\ -5 & -2 \end{bmatrix}$ and $C = \begin{bmatrix} -1 & 1 \\ -2 & 1 \end{bmatrix}$. Show the converse of (a) is false as follows: Consider the operator $T_A : \mathbb{R}^4 \to \mathbb{R}^4$ such that $T_A(X) = AX$ for all X in $\mathbb{R}^4$. Show that T_A has no real eigenvalue, but that $\mathbb{R}^4 = U \oplus W$ where U and W are both 2-dimensional and T_A-invariant.

22. Let $T : V \to V$ be an operator where $dimV = 2$. Show that V is T-reducible if and only if T is diagonalizable.

23. Let $T : V \to V$ be an operator with real eigenvalue λ. If $\mathcal{B}_0$ is any basis of $E_\lambda(T)$, show that $M_{\mathcal{B}_0}(T) = \lambda I$.

24. If $T : V \to V$ is an operator and U and W are T-invariant subspaces of V, show that $U \cap W$ and $U + W$ are also T-invariant.

25. (a) If $V = U \oplus W$, define $T : V \to V$ as follows: Given $\mathbf{v}$ in V, write it as $\mathbf{v} = \mathbf{u} + \mathbf{w}$ with $\mathbf{u}$ in U and $\mathbf{w}$ in W, and take $T(\mathbf{v}) = \mathbf{u}$. Show that T is an operator, $T^2 = T$, $imT = U$ and $kerT = W$.

    (b) Show that every operator $T : V \to V$ that satisfies $T^2 = T$ arises as in (a) for some direct decomposition $V = U \oplus W$. [*Hint*: Try $U = imT$ and $W = kerT$.]

26. (a) An operator $T : V \to V$ is called an **involution** if $T^2 = 1_V$. (Examples: conjugation on $\mathbb{C}$; transposition on $\mathbb{M}_{n,n}$; $T(p(x)) = p(1-x)$ on $\mathbb{P}_n$.) If T is an involution show that T is diagonalizable by showing that V is the direct sum of the eigenspaces of T. [*Hint*: Example 12.]

    (b) If $A^2 = I$ where A is a square matrix, show that A is similar to a matrix of the block form $\begin{bmatrix} I_m & 0 \\ 0 & -I_k \end{bmatrix}$. [*Hint*: Example 12.]

27. Let $S : V \to V$ and $T : V \to V$ be operators, let $\mathcal{B}$ be a basis of V, and let a be a scalar.

    (a) Show that $S + T$ and aT are operators.

    (b) Show that $M_\mathcal{B}(S+T) = M_\mathcal{B}(S) + M_\mathcal{B}(T)$ and $M_\mathcal{B}(aT) = aM_\mathcal{B}(T)$

5.7 GENERAL INNER PRODUCTS

The importance of the dot product in $\mathbb{R}^n$ would be difficult to exaggerate. It leads to the idea of an orthogonal basis and hence to a number of results of both theoretical and practical interest, including the Projection Theorem and approximation, the Principal Axis Theorem, and the Principal Value Decomposition. In this section we generalize the dot product to the idea of an inner product on an arbitrary vector space, and so see that the techniques we developed for $\mathbb{R}^n$ in Chapter 4 using the dot product will generalize to a much wider context.

5.7.1 Inner Products

An **inner product** $\langle\,,\,\rangle$ on a vector space V is a function that assigns a real number $\langle \mathbf{v}, \mathbf{w} \rangle$ to every pair of vectors $\mathbf{v}$ and $\mathbf{w}$ in V in such a way that the following axioms are satisfied for all vectors $\mathbf{u}$, $\mathbf{v}$, and $\mathbf{w}$ in V and all scalars r:

P1 $\langle \mathbf{v}, \mathbf{w} \rangle = \langle \mathbf{w}, \mathbf{v} \rangle$.

P2 $\langle \mathbf{u}, \mathbf{v} + \mathbf{w} \rangle = \langle \mathbf{u}, \mathbf{v} \rangle + \langle \mathbf{u}, \mathbf{w} \rangle$.

P3 $\langle r\mathbf{v}, \mathbf{w} \rangle = r \langle \mathbf{v}, \mathbf{w} \rangle$.

P4 $\langle \mathbf{v}, \mathbf{v} \rangle > 0$ whenever $\mathbf{v} \neq \mathbf{0}$.

A vector space endowed with an inner product is called an **inner product space**. Note that $\langle \mathbf{v}, r\mathbf{w} \rangle = r \langle \mathbf{v}, \mathbf{w} \rangle$ is an immediate consequence of Axioms P1 and P3, and that $\langle \mathbf{v} + \mathbf{w}, \mathbf{u} \rangle = \langle \mathbf{v}, \mathbf{u} \rangle + \langle \mathbf{w}, \mathbf{u} \rangle$ follows from Axioms P1 and P2.

Before giving examples, we clarify Axiom P4. Axiom P2 gives $\langle \mathbf{v}, \mathbf{0} \rangle = \langle \mathbf{v}, \mathbf{0} + \mathbf{0} \rangle = \langle \mathbf{v}, \mathbf{0} \rangle + \langle \mathbf{v}, \mathbf{0} \rangle$, so

$$\langle \mathbf{v}, \mathbf{0} \rangle = 0 \text{ for every vector } \mathbf{v} \text{ in } V.$$

With Axiom P1, this proves the first part of the following theorem and gives some useful refinements of Axiom P4 (parts (2) and (3) in the theorem).

Theorem 1. *If* $\mathbf{v}$ *is any vector in an inner product space, then:*

(1) $\langle \mathbf{0}, \mathbf{v} \rangle = 0 = \langle \mathbf{v}, \mathbf{0} \rangle$.

(2) $\langle \mathbf{v}, \mathbf{v} \rangle \geq 0$ *for all vectors* $\mathbf{v}$.

(3) $\langle \mathbf{v}, \mathbf{v} \rangle = 0$ *if and only if* $\mathbf{v} = \mathbf{0}$.

Example 1. The dot product $\langle X, Y \rangle = X \bullet Y$ is an inner product on $\mathbb{R}^n$ (see Theorem 1 §4.5).

Example 2. Define $\langle \, , \, \rangle$ on $\mathbb{M}_{n,n}$ by $\langle A, B \rangle = tr(AB^T)$, where trA denotes the trace of the square matrix A. Then Axioms P1, P2 and P3 follow from the fact that $tr(\ )$ is a linear transformation which satisfies $tr(A^T) = trA$ for all A. If $R_1, \cdots, R_n$ are the rows of A then the (i,j)-entry of AA^T is $R_i \bullet R_j$ (verify), so

$$\langle A, A \rangle = tr(AA^T) = R_1 \bullet R_1 + R_2 \bullet R_2 + \cdots + R_n \bullet R_n$$
$$= \|R_1\|^2 + \|R_2\|^2 + \cdots + \|R_n\|^2.$$

Hence if $A \neq 0$ then some $R_i \neq 0$, so $\|R_i\|^2 > 0$ and $\langle A, A \rangle > 0$. This is Axiom P4. We note in passing that $\langle A, A \rangle$ is the sum of the squares of all n^2 entries of the $n \times n$ matrix A. □

Example 3. Any $n+1$ distinct real numbers $a_0, a_1, a_2, \cdots, a_n$ determine an inner product $\langle \, , \, \rangle$ on $\mathbb{P}_n$ as follows: Define

$$\langle p(x), q(x) \rangle = p(a_0)q(a_0) + \cdots + p(a_n)q(a_n)$$

for all polynomials $p(x)$ and $q(x)$ in $\mathbb{P}_n$. Axioms P1, P2 and P3 are routine verifications. If $p(x) \neq 0$ then $p(a_i) \neq 0$ for some i because nonzero polynomials of degree at most n can have at most n distinct roots (see Appendix A.3). But then $\langle p(x), p(x) \rangle = p(a_0)^2 + p(a_1)^2 + \cdots + p(a_n)^2 > 0$, proving Axiom P4. □

The next example involves calculus.

Example 4. The set $\mathbf{C}[a, b]$ of all continuous real valued functions on the interval $[a, b]$ is a subspace of $\mathbb{F}[a, b]$. Define $\langle \, , \, \rangle$ on $\mathbf{C}[a, b]$ by

5.7. GENERAL INNER PRODUCTS

$$\langle f, g \rangle = \int_a^b f(x)g(x)dx.$$

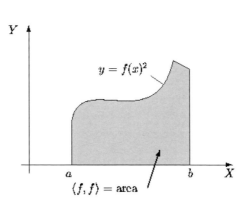

Again, all the axioms except P4 are routine properties of integrals. But if $\langle f, f \rangle = 0$ then $\int_a^b f(x)^2 dx = 0$. Since $f(x)^2$ is continuous and $f(x)^2 \geq 0$ for all x in $[a, b]$, it is a theorem of calculus (see the diagram) that $f(x)^2 = 0$ for all x. Hence $f = 0$ is the zero function, proving that Axiom P4 holds. □

Because of axioms P2 and P3, every inner product is linear in the first variable, that is

$$\langle r_1 \mathbf{v}_1 + r_2 \mathbf{v}_2 + \cdots + r_m \mathbf{v}_m, \mathbf{w} \rangle = r_1 \langle \mathbf{v}_1, \mathbf{w} \rangle + r_2 \langle \mathbf{v}_2, \mathbf{w} \rangle + \cdots + r_m \langle \mathbf{v}_m, \mathbf{w} \rangle$$

for all vectors $\mathbf{v}_i$ and $\mathbf{w}$. Of course linearity in the second variable now follows from Axiom P1. Hence computations with an inner product are carried out much as for the dot product in $\mathbb{R}^n$.

In fact the linearity enables us to easily describe *all* inner products $\langle \,,\, \rangle$ on $\mathbb{R}^n$. Let $\mathcal{E} = \{E_1, E_2, \cdots, E_n\}$ denote the standard basis of $\mathbb{R}^n$, and consider the $n \times n$ matrix $P = [p_{ij}]$ where

$$p_{ij} = \langle E_i, E_j \rangle \quad \text{for all } i \text{ and } j.$$

If $X = [x_1 \, x_2 \, \cdots \, x_n]^T$ and $Y = [y_1 \, y_2 \, \cdots \, y_n]^T$ are any two vectors in $\mathbb{R}^n$, then $X = \Sigma_i x_i E_i$ and $Y = \Sigma_j y_j E_j$, so the linearity of $\langle \,,\, \rangle$ gives $\langle X, Y \rangle$ as a matrix product:

$$\langle X, Y \rangle = \langle \Sigma_i x_i E_i \,,\, \Sigma_j y_j E_j \rangle = \Sigma_{i,j} x_i y_j \langle E_i, E_j \rangle = \Sigma_{i,j} x_i p_{ij} y_j = X^T PY.$$

In particular $X^T PX = \langle X, X \rangle > 0$ whenever $X \neq 0$ by P4. Since P is symmetric (by Axiom P1), this shows that P is a positive definite matrix (see Section 4.8). On the other hand, if P is any positive definite matrix and we define $\langle X, Y \rangle = X^T PY$ for all X and Y in $\mathbb{R}^n$, it is a routine matter to verify that $\langle \,,\, \rangle$ is an inner product on $\mathbb{R}^n$. Hence we have

Theorem 2. *Every positive definite matrix P gives rise to an inner product $\langle \,,\, \rangle$ on $\mathbb{R}^n$ given by*

$$\langle X, Y \rangle = X^T PY \quad \text{for all } X \text{ and } Y \text{ in } \mathbb{R}^n.$$

Moreover, every inner product on $\mathbb{R}^n$ arises in this way from some positive definite matrix P.

Note that the dot product on $\mathbb{R}^n$ corresponds to $P = I$ in Theorem 2.

Theorem 2 can also be stated in the language of quadratic forms in Section 4.8: If $\langle \, , \, \rangle$ is an inner product on $\mathbb{R}^n$, then $q(X) = \langle X, X \rangle$ is a positive definite quadratic form in n variables, and every such quadratic form arises in this way from an inner product.

Example 5. Define $\langle \, , \, \rangle$ on $\mathbb{R}^2$ by
$$\langle [x_1, x_2]^T, [y_1, y_2]^T \rangle = x_1 y_1 - 2x_1 y_2 - 2x_2 y_1 + 5x_2 y_2.$$

Hence $P = \begin{bmatrix} 1 & -2 \\ -2 & 5 \end{bmatrix}$ in Theorem 2. Since P is a positive definite matrix by Theorem 4 §4.8, it follows that $\langle \, , \, \rangle$ is an inner product. Note that Axioms P1, P2 and P3 hold for *any* symmetric matrix P; Axiom P4 follows directly in this case because $\langle [x_1, x_2]^T, [x_1, x_2]^T \rangle = x_1^2 - 4x_1 x_2 + 5x_2^2 = (x_1 - 2x_2)^2 + x_2^2$. □

5.7.2 Norms and Orthogonality

We showed in Theorem 1 that $\langle \mathbf{v}, \mathbf{v} \rangle \geq 0$ holds for all vectors $\mathbf{v}$ in any inner product space V. Hence, as in $\mathbb{R}^n$, we can define the **norm**[23] of $\mathbf{v}$ by
$$\|\mathbf{v}\| = \sqrt{\langle \mathbf{v}, \mathbf{v} \rangle}.$$

A vector of norm 1 is called a **unit vector**.

Example 6. *Compute $\|p(x)\|$ if $p(x) = 2 - x - x^2$ where $\mathbb{P}_2$ has the norm in Example 3 determined by the numbers -1, 1 and 2.*

Solution. $\|p(x)\|^2 = \langle p(x), p(x) \rangle = p(-1)^2 + p(1)^2 + p(2)^2 = 2^2 + 0^2 + (-4)^2 = 20$, so $\|p(x)\| = 2\sqrt{5}$. □

The proof of Theorem 2 §4.5 and its Corollary goes through as written to prove the following fundamental inequalities about norms.

Theorem 3. *Let V be an inner product space.*

(1) **Cauchy-Schwarz Inequality.** $|\langle \mathbf{v}, \mathbf{w} \rangle| \leq \|\mathbf{v}\| \, \|\mathbf{w}\|$ *for all $\mathbf{v}$ and $\mathbf{w}$ in V.*

(2) **Triangle Inequality.** $\|\mathbf{v} + \mathbf{w}\| \leq \|\mathbf{v}\| + \|\mathbf{w}\|$ *for all $\mathbf{v}$ and $\mathbf{w}$ in V.*

Again motivated by the situation in $\mathbb{R}^n$, we say that two vectors $\mathbf{v}$ and $\mathbf{w}$ are **orthogonal** if
$$\langle \mathbf{v}, \mathbf{w} \rangle = 0$$

---
[23]The term *length* is used for $\mathbb{R}^n$, reflecting the geometric interpretation of $\mathbb{R}^2$ and $\mathbb{R}^3$, but *norm* is used more commonly for spaces of polynomials or functions.

5.7. GENERAL INNER PRODUCTS

and we say that a set $\{\mathbf{e}_0, \mathbf{e}_1, \cdots, \mathbf{e}_m\}$ of vectors is an **orthogonal set**[24] if

$$\mathbf{e}_i \neq \mathbf{0} \text{ for each } i, \text{ and } \langle \mathbf{e}_i, \mathbf{e}_j \rangle = 0 \text{ for all } i \neq j.$$

An orthogonal set consisting of unit vectors is called an **orthonormal** set. If $\{\mathbf{e}_0, \mathbf{e}_1, \cdots, \mathbf{e}_m\}$ is an orthogonal set of vectors, we can **normalize** it to obtain the orthonormal set $\{\frac{1}{\|\mathbf{e}_0\|}\mathbf{e}_0, \frac{1}{\|\mathbf{e}_1\|}\mathbf{e}_1, \cdots, \frac{1}{\|\mathbf{e}_m\|}\mathbf{e}_m\}$.

Example 7. *Show that $\{\sin x, \cos x\}$ is orthogonal in the space $\mathbf{C}[-\pi, \pi]$ of all continuous real valued functions on the interval $[-\pi, \pi]$.*

Solution. We have $\langle \sin x, \cos x \rangle = \int_{-\pi}^{\pi} \sin x \cos x \, dx = -\frac{1}{4}[\cos(2x)]_{-\pi}^{\pi} = 0.$ □

The following theorem includes Theorems 4 and 5 in Section 4.5 in the case of the dot product in $\mathbb{R}^n$, and the same proofs go through.

Theorem 4. *Let $\{\mathbf{e}_0, \mathbf{e}_1, \cdots, \mathbf{e}_m\}$ be an orthogonal set in an inner product space V.*

(1) *$\{\mathbf{e}_0, \mathbf{e}_1, \cdots, \mathbf{e}_m\}$ is independent.*

(2) Pythagoras' Theorem.
$$\|\mathbf{e}_0 + \mathbf{e}_1 + \cdots + \mathbf{e}_m\|^2 = \|\mathbf{e}_0\|^2 + \|\mathbf{e}_1\|^2 + \cdots + \|\mathbf{e}_m\|^2.$$

Part (1) in Theorem 4 suggests the concept of an orthogonal basis for an arbitrary inner product space V, and the general version of the expansion theorem (Theorem 6 in Section 4.5) remains valid.

Theorem 5. Expansion Theorem. *Let $\{\mathbf{e}_1, \mathbf{e}_2, \cdots, \mathbf{e}_n\}$ be an orthogonal basis of an inner product space V. Then*

$$\mathbf{v} = \frac{\langle \mathbf{v}, \mathbf{e}_0 \rangle}{\|\mathbf{e}_0\|^2}\mathbf{e}_0 + \frac{\langle \mathbf{v}, \mathbf{e}_1 \rangle}{\|\mathbf{e}_1\|^2}\mathbf{e}_1 + \cdots + \frac{\langle \mathbf{v}, \mathbf{e}_n \rangle}{\|\mathbf{e}_n\|^2}\mathbf{e}_n \quad \text{for every vector } \mathbf{v} \text{ in } V.$$

We are going to make use of this general version of the Expansion Theorem in several situations. The first application requires an important orthogonal basis of $\mathbb{P}_n$.

Example 8. Lagrange polynomials.[25] Consider the inner product on $\mathbb{P}_n$ determined (see Example 3) by $n+1$ distinct real numbers, $a_0, a_1, a_2, \cdots, a_n$:

$$\langle p(x), q(x) \rangle = p(a_0)q(a_0) + p(a_1)q(a_1) + \cdots + p(a_n)q(a_n).$$

---

[24] We are labeling from 0 here in anticipation of the main application to infinite orthogonal sets, where this is customary.

[25] Joseph Louis Lagrange (1736-1813) was one of the great mathematicians and made important contributions to many parts of mathematics. He is also remembered for his skill at exposition, and his *Méchanique Analytique* was described by Hamilton as a "scientific poem".

Define the Lagrange polynomials $\delta_0(x), \delta_1(x), \delta_2(x), \cdots, \delta_n(x)$ by

$$\delta_k(x) = \frac{\Pi_{i \neq k}(x - a_i)}{\Pi_{i \neq k}(a_k - a_i)}$$

where $\Pi_{i \neq k}(x - a_i)$ means the product of all the terms $(x - a_i)$ with $(x - a_k)$ omitted. Note that the denominator of $\delta_k(x)$ is nonzero because the a_i are distinct. Clearly each $\delta_k(x)$ has degree n, and we have

$$\delta_k(a_k) = 1 \quad \text{and} \quad \delta_k(a_i) = 0 \quad \text{if } i \neq k. \qquad (*)$$

If $p(x)$ is any polynomial $\mathbb{P}_n$, this shows that

$$\langle p(x), \delta_k(x) \rangle = p(a_k) \quad \text{for each } k.$$

In particular, if we take $p(x) = \delta_i(x)$ it follows that

$$\mathcal{D} = \{\delta_0(x), \delta_1(x), \delta_2(x), \cdots, \delta_n(x)\}$$

is an orthonormal basis of $\mathbb{P}_n$. Hence the Expansion Theorem gives the following simple formula for the expansion of any polynomial $p(x)$ in terms of the basis $\mathcal{D}$:

$$p(x) = p(a_0)\,\delta_0(x) + p(a_1)\,\delta_1(x) + \cdots + p(a_n)\,\delta_n(x).$$

More generally, let f be any function in $\mathbb{F}[a, b]$, where $[a, b]$ is an interval containing all the a_i. Then (*) shows that the polynomial

$$p_f(x) = f(a_0)\,\delta_0(x) + f(a_1)\,\delta_1(x) + \cdots + f(a_n)\,\delta_n(x) \qquad (**)$$

agrees with f on the $n+1$ numbers a_i, that is

$$f(a_i) = p_f(a_i) \quad \text{holds for each } i = 1, 2, \cdots, n.$$

The polynomial $p_f(x)$ is called the **Lagrange interpolation polynomial** for f, and is the unique polynomial in $\mathbb{P}_n$ that coincides with f at each a_i. $\square$

The Lagrange polynomials have many applications, one of which we explore in the following example.

Example 9. Numerical integration. Suppose that we want to approximate $\int_a^b f(x)dx$. The idea is to choose $n+1$ distinct numbers $a_0, a_1, a_2, \cdots, a_n$ in the interval $[a, b]$, let $\delta_0(x), \delta_1(x), \delta_2(x), \cdots, \delta_n(x)$ be the corresponding Lagrange polynomials, and approximate f by the Lagrange interpolation polynomial p_f given in (**). Then we obtain the approximation (called a **Gaussian quadrature formula**)

$$\int_a^b f(x)dx \approx \int_a^b p_f(x)dx = f(a_0)\,d_0 + f(a_1)\,d_1 + \cdots + f(a_n)\,d_n$$

5.7. GENERAL INNER PRODUCTS

where $\approx$ means "approximately equal", and $d_i = \int_a^b p_f(x)\delta_i(x)dx$ for each i. This approximation is exact by Example 8 if f is a polynomial of degree at most n and, with careful choice of the a_i, is an important tool in numerical analysis. □

The techniques for finding orthogonal sets in an arbitrary inner product space V are entirely analogous to those in $\mathbb{R}^n$. The following lemma shows how to enlarge any orthogonal set; the proof of Lemma 1 §4.5 goes through as written.

Lemma 1. Orthogonal Lemma. *Let $\{\mathbf{e}_0, \mathbf{e}_1, \cdots, \mathbf{e}_m\}$ be an orthogonal set in an inner product space V. Given $\mathbf{v}$ in V, write*

$$\mathbf{e}_{m+1} = \mathbf{v} - \left(\frac{\langle \mathbf{v}, \mathbf{e}_0 \rangle}{\|\mathbf{e}_0\|^2} \mathbf{e}_0 + \frac{\langle \mathbf{v}, \mathbf{e}_1 \rangle}{\|\mathbf{e}_1\|^2} \mathbf{e}_1 + \cdots + \frac{\langle \mathbf{v}, \mathbf{e}_m \rangle}{\|\mathbf{e}_m\|^2} \mathbf{e}_m \right).$$

Then $\langle \mathbf{e}_{m+1}, \mathbf{e}_i \rangle = 0$ for each $i = 1, 2, \cdots, m$, and if $\mathbf{v}$ is not in span$\{\mathbf{e}_0, \mathbf{e}_1, \cdots, \mathbf{e}_m\}$ then $\{\mathbf{e}_0, \mathbf{e}_1, \cdots, \mathbf{e}_m, \mathbf{e}_{m+1}\}$ is an orthogonal set.

As in $\mathbb{R}^n$, the Orthogonal Lemma leads to the general version of the Gram-Schmidt Algorithm for obtaining an orthogonal basis of a subspace from any convenient basis. Again the proof of Theorem 8 §4.5 goes through.

Theorem 6. Gram-Schmidt Algorithm. *If $\{\mathbf{b}_0, \mathbf{b}_1, \cdots, \mathbf{b}_m\}$ is any basis of a subspace U of an inner product space V, construct $\mathbf{e}_0, \mathbf{e}_1, \cdots, \mathbf{e}_m$ in U successively as follows:*

$$\begin{aligned}
\mathbf{e}_0 &= \mathbf{b}_0 \\
\mathbf{e}_1 &= \mathbf{b}_1 - \frac{\langle \mathbf{b}_1, \mathbf{e}_0 \rangle}{\|\mathbf{e}_0\|^2} \mathbf{e}_0 \\
\mathbf{e}_2 &= \mathbf{b}_2 - \frac{\langle \mathbf{b}_2, \mathbf{e}_0 \rangle}{\|\mathbf{e}_0\|^2} \mathbf{e}_0 - \frac{\langle \mathbf{b}_2, \mathbf{e}_1 \rangle}{\|\mathbf{e}_1\|^2} \mathbf{e}_1 \\
&\vdots \\
\mathbf{e}_k &= \mathbf{b}_k - \frac{\langle \mathbf{b}_k, \mathbf{e}_0 \rangle}{\|\mathbf{e}_0\|^2} \mathbf{e}_0 - \frac{\langle \mathbf{b}_k, \mathbf{e}_1 \rangle}{\|\mathbf{e}_1\|^2} \mathbf{e}_1 - \cdots - \frac{\langle \mathbf{b}_k, \mathbf{e}_{k-1} \rangle}{\|\mathbf{e}_{k-1}\|^2} \mathbf{e}_{k-1} \\
&\vdots
\end{aligned}$$

for each $k = 1, 2, 3, \cdots, m$. Then:

(1) $\{\mathbf{e}_0, \mathbf{e}_1, \cdots, \mathbf{e}_m\}$ *is an orthogonal basis of U.*

(2) *span$\{\mathbf{e}_0, \mathbf{e}_1, \cdots, \mathbf{e}_k\}$ = span$\{\mathbf{b}_0, \mathbf{b}_1, \cdots, \mathbf{b}_k\}$ for each $k = 0, 1, \cdots, m$.*

Example 10. *If $V = \mathbb{P}_3$ with the inner product $\langle p, q \rangle = \int_{-1}^{1} p(x)q(x)dx$, apply the Gram-Schmidt algorithm to the basis $\{1, x, x^2, x^3\}$ of $\mathbb{P}_3$ to obtain the orthogonal basis $\{1, x, \frac{1}{3}(3x^2 - 1), \frac{1}{5}(5x^3 - 3x)\}$.*

Solution. By the Gram-Schmidt algorithm, we obtain polynomials $q_0, q_1, \cdots$ one after the other.[26] To begin with, we have $q_0(x) = 1$. The algorithm continues:

$$\begin{aligned}
q_1(x) &= x - \frac{\langle x, q_0 \rangle}{\|q_0\|^2} q_0 = x - \frac{0}{2} q_1 = x, \\
q_2(x) &= x^2 - \frac{\langle x^2, q_0 \rangle}{\|q_0\|^2} q_0 - \frac{\langle x^2, q_1 \rangle}{\|q_1\|^2} q_1 \\
&= x^2 - \frac{2/3}{2} q_0 - 0 q_1 = x^2 - \tfrac{1}{3} = \tfrac{1}{3}(3x^2 - 1), \\
q_3(x) &= x^3 - \frac{\langle x^3, q_0 \rangle}{\|q_0\|^2} q_0 - \frac{\langle x^3, q_1 \rangle}{\|q_1\|^2} q_1 - \frac{\langle x^3, q_2 \rangle}{\|q_2\|^2} q_2 \\
&= x^3 - 0 q_0 - \frac{2/5}{2/3} q_1 - 0 q_2 \\
&= x^3 - \tfrac{3}{5} x = \tfrac{1}{5}(5x^3 - 3x). \quad \square
\end{aligned}$$

The polynomials in Example 10 all have leading coefficient 1. In applications (for example differential equations) it is customary to take scalar multiples $p(x)$ of these polynomials such that $p(1) = 1$. The result is the orthogonal basis

$$\{1, x, \tfrac{1}{2}(3x^2 - 1), \tfrac{1}{2}(5x^3 - 3x)\} \quad \text{of } \mathbb{P}_3.$$

These are the first four **Legendre polynomials**, and we will return to them below.

Let U be a subspace of an arbitrary inner product space V. As in $\mathbb{R}^n$, define the **orthogonal complement** $U^\perp$ of U in V by

$$U^\perp = \{\mathbf{v} \text{ in } V \mid \langle \mathbf{u}, \mathbf{v} \rangle = 0 \text{ for all } \mathbf{u} \text{ in } U\}.$$

This is a subspace of V even if V is not finite-dimensional, and we have

Theorem 7. *If U is a finite dimensional subspace of an arbitrary inner product space V then $V = U \oplus U^\perp$.*

Proof. If $\mathbf{v}$ is in $U \cap U^\perp$ then $\|\mathbf{v}\|^2 = \langle \mathbf{v}, \mathbf{v} \rangle = 0$, so $\mathbf{v} = \mathbf{0}$. Hence $U \cap U^\perp = \{\mathbf{0}\}$. Now choose an orthogonal basis $\{\mathbf{e}_0, \mathbf{e}_1, \cdots, \mathbf{e}_m\}$ of U. If $\mathbf{v}$ is any vector in V, the Orthogonal Lemma shows that $\mathbf{v} - (r_0 \mathbf{e}_0 + r_1 \mathbf{e}_1 + \cdots + r_m \mathbf{e}_m)$ is in $U^\perp$ for certain real numbers r_i. It follows that $V = U + U^\perp$. $\square$

If U is a subspace of $\mathbb{R}^n$, we showed that $U^{\perp\perp} = U$ (Theorem 3 §4.6). However, lest the reader get the idea that *all* properties of orthogonal complements in $\mathbb{R}^n$ extend to the general case, we now give an example where $U^{\perp\perp} \neq U$.

Example 11. *Consider $\mathbb{P}$ with the inner product $\langle p, q \rangle = \int_0^1 p(x) q(x) dx$, and let U denote the subspace of all polynomials with zero constant term. Show that $U^\perp = \{0\}$, so $U^{\perp\perp} = \mathbb{P} \neq U$.*

---

[26] For simplicity of notation, we occasionally write a polynomial $q(x)$ simply as q. This amounts to thinking of q as a function in $\mathbb{F}(\mathbb{R})$.

5.7. GENERAL INNER PRODUCTS

Solution. Observe first that $U = \{x\, q(x) \mid q(x) \text{ in } \mathbb{P}\}$. Suppose that $p(x)$ is in $U^\perp$. Then $\langle p(x), x\, q(x)\rangle = 0$ for all $q(x)$ in $\mathbb{P}$, so

$$\langle x\, p(x), q(x)\rangle = \int_0^1 x\, p(x) q(x) dx = \int_0^1 p(x)(x\, q(x)) dx = \langle p(x), x\, q(x)\rangle = 0$$

for all $q(x)$ in $\mathbb{P}$. Taking $q(x) = x\, p(x)$ gives $\|x\, p(x)\|^2 = 0$, whence $x\, p(x) = 0$. This implies that $p(x) = 0$, and hence that $U^\perp = \{0\}$, as asserted. □

It is no coincidence that the subspace U is infinite dimensional in Example 10; we show below (Theorem 9) that $U^{\perp\perp} = U$ holds for all finite dimensional subspaces of any inner product space.

5.7.3 Projections and Approximation

Let $\mathbf{v}$ be a vector in an inner product space V, and let U be a finite dimensional subspace of V with orthogonal basis $\{\mathbf{e}_0, \mathbf{e}_1, \cdots, \mathbf{e}_m\}$. As in Section 4.6, define the **projection** of $\mathbf{v}$ on U by

$$proj_U(\mathbf{v}) = \frac{\langle \mathbf{v}, \mathbf{e}_0\rangle}{\|\mathbf{e}_0\|^2}\mathbf{e}_0 + \frac{\langle \mathbf{v}, \mathbf{e}_1\rangle}{\|\mathbf{e}_1\|^2}\mathbf{e}_1 + \cdots + \frac{\langle \mathbf{v}, \mathbf{e}_m\rangle}{\|\mathbf{e}_m\|^2}\mathbf{e}_m. \qquad (***)$$

As the notation implies, $proj_U(\mathbf{v})$ depends on U and $\mathbf{v}$ only, and not on the choice of orthogonal basis $\{\mathbf{e}_0, \mathbf{e}_1, \cdots, \mathbf{e}_m\}$ of U (the proof of Lemma 1 §4.6 goes through). The basic properties of projections are given in the following result which extends the Projection Theorem for $\mathbb{R}^n$ (Theorem 2 §4.6) to the case where V is an arbitrary inner product which need not be finite dimensional.

Theorem 8. Projection Theorem. *Let U be a finite dimensional subspace of an arbitrary inner product space V. Then:*

(1) *$proj_U(\mathbf{v})$ is in U and $\mathbf{v} - proj_U(\mathbf{v})$ is in $U^\perp$ for every vector $\mathbf{v}$ in V.*

(2) *$proj_U : V \to V$ is a linear transformation with image U and kernel $U^\perp$.*

Proof. (1). That $proj_U(\mathbf{v})$ is in U for every vector $\mathbf{v}$ follows from $(***)$ because each $\mathbf{e}_i$ is in U, and $\mathbf{v} - proj_U(\mathbf{v})$ is in $U^\perp$ because it is orthogonal to each $\mathbf{e}_i$ by the Orthogonal Lemma.

(2). The linearity of $proj_U : V \to V$ follows from Axioms P2 and P3 (verify). We have $im(proj_U) \subseteq U$ by (1); and $U \subseteq im(proj_U)$ because $\mathbf{u} = proj_U(\mathbf{u})$ by the Expansion Theorem for every $\mathbf{u}$ in U. Hence $U = im(proj_U)$. Turning to the kernel, if $\mathbf{v}$ is in $ker(proj_U)$ then $\mathbf{v} = \mathbf{v} - proj_U(\mathbf{v})$ is in $U^\perp$ by the Projection Theorem. Thus $ker(proj_U) \subseteq U^\perp$. On the other hand, if $\mathbf{v}$ is in $U^\perp$ then $(***)$ shows that $proj_U(\mathbf{v}) = \mathbf{0}$, so $U^\perp \subseteq ker(proj_U)$. Hence $U^\perp = ker(proj_U)$. □

The projection theorem has an important consequence which we record in the following theorem.

Theorem 9. *If U is a finite dimensional subspace in an arbitrary inner product space V, then $U^{\perp\perp} = U$.*

Proof. The inclusion $U \subseteq U^{\perp\perp}$ always holds (verify). If $\mathbf{v}$ is in $U^{\perp\perp}$, write $\mathbf{p} = proj_U(\mathbf{v})$. Then $\mathbf{p}$ is in U by the Projection Theorem so $\mathbf{p}$ is in $U^{\perp\perp}$. Hence $\mathbf{v} - \mathbf{p}$ is in $U^{\perp\perp}$. But $\mathbf{v} - \mathbf{p}$ is in $U^{\perp}$ by the Projection Theorem, and it follows that $\mathbf{v} = \mathbf{p}$. Thus $\mathbf{v}$ is in U, proving that $U^{\perp\perp} \subseteq U$. This is what we wanted. □

Here is an example outlining how the Projection Theorem can be applied to time series analysis in statistics.

Example 12. Statistical Trend Analysis. Let f be a function of time t, say some measure of stock market value, or the degree of global warming, and assume that f is known only at times $t_0, t_1, \cdots, t_m$. One constantly arising question is whether the data values $f(t_0), f(t_1), \cdots, f(t_m)$ exhibit a "linear trend", in which case we might use the least squares line $f(t) = a_0 + a_1 t$ to estimate other values of $f(t)$. If there is a "quadratic trend" we would fit a quadratic $f(t) = a_0 + a_1 t + a_2 t^2$, and so on for a "cubic trend", etc. The difficulty in deciding which trend is appropriate is that, when a polynomial is fitted to the data, the coefficients may not be "statistically independent".

To remedy this, one procedure is to use the inner product on $\mathbb{P}_m$ (from Example 3) given by

$$\langle p, q \rangle = p(t_0)q(t_0) + p(t_1)q(t_1) + \cdots + p(t_m)q(t_m),$$

and apply the Gram-Schmidt Algorithm to the basis $\{1, x, x^2, \cdots, x^m\}$ of $\mathbb{P}_m$ to obtain an orthogonal basis $\{q_0, q_1, \cdots, q_m\}$. Let p_f be the Lagrange interpolation polynomial for f in Example 8 (that is the unique polynomial in $\mathbb{P}_m$ such that $p_f(t_i) = f(t_i)$ for each i). Assuming (as is usually the case) that we are only interested in trends of degree no more than 4, we compute the projection of p_f onto $\mathbb{P}_4$:

$$proj_{\mathbb{P}_4}(p_f) = b_0 q_0 + b_1 q_1 + b_2 q_2 + b_3 q_3 + b_4 q_4.$$

The coefficients $b_i = \frac{\langle p_f, q_i \rangle}{\|q_i\|^2}$ are called the **trend coefficients** for f. They are easily computed, statistically independent (for most data), and b_1 measures the linear trend, b_2 measures the quadratic trend, etc. More information on this can be found in books on linear models in statistics.[27] □

We now turn to an extension of the Approximation Theorem for $\mathbb{R}^n$ (Section 4.6) which is valid for inner product spaces that may not be finite dimensional.

Theorem 10. Approximation Theorem. *Let U be a finite dimensional subspace of an arbitrary inner product space V. If $\mathbf{v}$ is any vector in V, then $proj_U(\mathbf{v})$ is the vector in U that is closest to $\mathbf{v}$ in the sense that*

$$\|\mathbf{v} - proj_U(\mathbf{v})\| < \|\mathbf{v} - \mathbf{u}\| \quad \text{for all } \mathbf{u} \text{ in } U, \quad \mathbf{u} \neq proj_U(\mathbf{v}).$$

---
[27] For example, I. Guttman, *Linear Models, An Introduction*, Wiley, 1982.

5.7. GENERAL INNER PRODUCTS

Proof. For convenience write $\mathbf{p} = proj_U(\mathbf{v})$. Given $\mathbf{u}$ in U, we can write $\mathbf{v} - \mathbf{u} = (\mathbf{v} - \mathbf{p}) + (\mathbf{p} - \mathbf{u})$ where $\mathbf{v} - \mathbf{p}$ is in $U^\perp$ and $\mathbf{p} - \mathbf{u}$ is in U. Hence Pythagoras' Theorem gives $\|\mathbf{v} - \mathbf{u}\|^2 = \|\mathbf{v} - \mathbf{p}\|^2 + \|\mathbf{p} - \mathbf{u}\|^2$. This does it because $\|\mathbf{p} - \mathbf{u}\| > 0$ if $\mathbf{u} \neq \mathbf{p}$. □

In Chapter 4 we used the $\mathbb{R}^n$-version of the Approximation Theorem to find all best approximations to a solution of an inconsistent system of linear equations, and to find least squares approximating polynomials for finite sets of data pairs. Both these applications required the finite dimensionality of the underlying vector space $\mathbb{R}^n$. However, much can be said when the underlying space is infinite dimensional.

An infinite set $\{\mathbf{e}_0, \mathbf{e}_1, \mathbf{e}_2, \cdots, \mathbf{e}_n, \cdots\}$ in an inner product space V is called **orthogonal** if

$$\mathbf{e}_i \neq 0 \text{ for each } i \quad \text{and} \quad \langle \mathbf{e}_i, \mathbf{e}_i \rangle = 0 \text{ whenever } i \neq j.$$

Such orthogonal sets can be found by an extension of the Gram-Schmidt Algorithm.

We need the following idea. An infinite set $\{\mathbf{b}_0, \mathbf{b}_1, \mathbf{b}_2, \cdots, \mathbf{b}_m, \cdots\}$ of vectors in an inner product space is called (linearly) **independent** if $\{\mathbf{b}_0, \mathbf{b}_1, \mathbf{b}_2, \cdots, \mathbf{b}_m\}$ is independent for each m. One example is $\{1, x, x^2, \cdots, x^n, \cdots\}$ in $\mathbb{P}$.

Example 13. Suppose that $\{\mathbf{b}_0, \mathbf{b}_1, \mathbf{b}_2, \cdots, \mathbf{b}_m, \cdots\}$ is an infinite independent set of vectors in an inner product space. Apply the Gram-Schmidt process to the vectors $\mathbf{b}_i$ as follows: Take

$$\mathbf{e}_0 = \mathbf{b}_0$$

and, having determined $\mathbf{e}_0, \mathbf{e}_1, \mathbf{e}_2, \cdots, \mathbf{e}_{m-1}$ for some $m \geq 1$, take

$$\mathbf{e}_m = \frac{\langle \mathbf{b}_m, \mathbf{e}_0 \rangle}{\|\mathbf{e}_0\|^2}\mathbf{e}_0 + \frac{\langle \mathbf{b}_m, \mathbf{e}_1 \rangle}{\|\mathbf{e}_1\|^2}\mathbf{e}_1 + \cdots + \frac{\langle \mathbf{b}_{m-1}, \mathbf{e}_{m-1} \rangle}{\|\mathbf{e}_{m-1}\|^2}\mathbf{e}_{m-1}.$$

This produces an infinite orthogonal set $\{\mathbf{e}_0, \mathbf{e}_1, \mathbf{e}_2, \cdots, \mathbf{e}_m, \cdots\}$ by Theorem 6. In fact we have $span\{\mathbf{e}_0, \mathbf{e}_1, \mathbf{e}_2, \cdots, \mathbf{e}_m\} = span\{\mathbf{b}_0, \mathbf{b}_1, \mathbf{b}_2, \cdots, \mathbf{b}_m\}$ for each m. □

Now assume that an infinite orthogonal set

$$\mathcal{E} = \{\mathbf{e}_0, \mathbf{e}_1, \mathbf{e}_2, \cdots, \mathbf{e}_n, \cdots\}$$

is given in an inner product space V. We use it to define subspaces U_m as follows:

$$U_m = span\{\mathbf{e}_0, \mathbf{e}_1, \mathbf{e}_2, \cdots, \mathbf{e}_m\} \quad \text{for each } m \geq 1.$$

Clearly $U_0 \subseteq U_1 \subseteq U_2 \subseteq \cdots \subseteq U_m \subseteq \cdots$, and $dim(U_m) = m + 1$ for each m because $\{\mathbf{e}_0, \mathbf{e}_1, \mathbf{e}_2, \cdots, \mathbf{e}_m\}$ is an (orthogonal) basis of U_m. Given $\mathbf{v}$ in V and $m \geq 1$, define

$$\mathbf{v}_m = proj_{U_m}(\mathbf{v}) = \frac{\langle \mathbf{v}, \mathbf{e}_0 \rangle}{\|\mathbf{e}_0\|^2}\mathbf{e}_0 + \frac{\langle \mathbf{v}, \mathbf{e}_1 \rangle}{\|\mathbf{e}_1\|^2}\mathbf{e}_1 + \cdots + \frac{\langle \mathbf{v}, \mathbf{e}_m \rangle}{\|\mathbf{e}_m\|^2}\mathbf{e}_m. \quad (****)$$

The vector $\mathbf{v}_m$ is in U_m for each m, and is called the mth $\mathcal{E}$-**approximation** of $\mathbf{v}$. The reason for the name will be apparent soon.

Recall that Theorem 10 with $U = U_{m+1}$ gives $\|\mathbf{v} - \mathbf{v}_{m+1}\| \leq \|\mathbf{v} - \mathbf{u}\|$ for each vector $\mathbf{u}$ in U_{m+1}. But $\mathbf{v}_m$ is in U_{m+1} because $U_m \subseteq U_{m+1}$, so we have

$$\|\mathbf{v} - \mathbf{v}_{m+1}\| \leq \|\mathbf{v} - \mathbf{v}_m\| \quad \text{for each } m = 0, 1, 2, \cdots$$

Using this we obtain

$$\|\mathbf{v} - \mathbf{v}_0\| \geq \|\mathbf{v} - \mathbf{v}_1\| \geq \|\mathbf{v} - \mathbf{v}_2\| \geq \cdots \geq \|\mathbf{v} - \mathbf{v}_m\| \geq \cdots.$$

Thus the vectors $\mathbf{v}_0, \mathbf{v}_1, \mathbf{v}_2, \cdots$ are better and better approximations to $\mathbf{v}$ (this is the reason for calling these $\mathbf{v}_i$ the $\mathcal{E}$-approximations of $\mathbf{v}$). Furthermore, each vector $\mathbf{v}_m$ is easily computable from $\mathbf{v}$ and the $\mathbf{e}_i$ because of (****).

Hence the following is a general method in approximation theory: Given a vector $\mathbf{v}$ in an inner product space V, find an orthogonal set $\{\mathbf{e}_0, \mathbf{e}_1, \mathbf{e}_2, \cdots, \mathbf{e}_n, \cdots\}$ in V such that the vectors $\mathbf{v}_m$ approach $\mathbf{v}$, that is

$$\lim_{n \to \infty} \|\mathbf{v} - \mathbf{v}_m\| = 0.$$

Then $\mathbf{v}_m$ is a suitable approximation to $\mathbf{v}$ if m is large enough.[28]

Of course this may not be the case for some particular infinite orthogonal set of vectors $\mathbf{e}_i$. But there are several situations where it does happen, many in spaces of functions, and the results are important tools for a variety of applications in science and engineering. A complete discussion of this topic is well beyond the scope of this book, and we content ourselves with two examples, one where the $\mathbf{e}_i$ are trigonometric functions and the other where the $\mathbf{e}_i$ are orthogonal polynomials.

5.7.4 Fourier Approximation

Consider the space $\mathbf{C}[-\pi, \pi]$ of all continuous functions on the interval $[-\pi, \pi]$, endowed with the inner product

$$\langle f, g \rangle = \int_{-\pi}^{\pi} f(x)g(x)dx.$$

Then standard integration techniques (using trigonometric identities) show that

$$\mathcal{F} = \{1,\ cos\,x,\ sin\,x,$$
$$cos\,(2x),\ sin(2x),\ cos(3x),\ sin(3x), \cdots,\ cos(mx),\ sin(mx), \cdots\}$$

is an orthogonal set in $\mathbf{C}[-\pi, \pi]$, and that the norms of these functions are given by

$$\|1\|^2 = 2\pi \quad \text{and} \quad \|cos(mx)\|^2 = \|sin(mx)\|^2 = \pi \text{ for each } m \geq 1.$$

---

[28] We say that a sequence of numbers $a_1, a_2, a_3, \cdots$ approaches a number a, and write $\lim_{n \to \infty} a_n = a$, if the numbers a_i become and remain very close to a as n increases. More precisely: We ask that for any small number $\varepsilon > 0$ there should exist an integer N such that $|a_n - a| < \varepsilon$ for all $n \geq N$. This notion is fundamental in calculus.

5.7. GENERAL INNER PRODUCTS

Slightly modifying our earlier notation, we write

$$\mathcal{F}_m = \text{span}\{1,\ \cos x,\ \sin x,$$
$$\cos(2x),\ \sin(2x),\ \cos(3x),\ \sin(3x),\cdots,\cos(mx),\ \sin(mx)\}.$$

If f is any function in $\mathbf{C}[-\pi,\pi]$, the mth $\mathcal{F}$-approximation of f (called the mth **Fourier approximation** [29]) is the function f_m given (as in Equation (****) above) by

$$f_m(x) = a_0 + a_1 \cos x + b_1 \sin x$$
$$+ a_2 \cos(2x) + b_2 \sin(2x) + \cdots + a_m \cos(mx) + b_m \sin(mx)$$

for all x in $[-\pi,\pi]$. Here the a_i and b_i are called the **Fourier coefficients** of f, and are given by

$$a_0 = \frac{\langle f,1 \rangle}{\|1\|^2} = \frac{1}{2\pi}\int_{-\pi}^{\pi} f(x)\,dx,$$

$$a_k = \frac{\langle f, \cos(kx) \rangle}{\|\cos(kx)\|^2} = \frac{1}{\pi}\int_{-\pi}^{\pi} f(x)\cos(kx)\,dx \quad \text{for } k=1,2,3,\cdots,$$

$$b_k = \frac{\langle f, \sin(kx) \rangle}{\|\sin(kx)\|^2} = \frac{1}{\pi}\int_{-\pi}^{\pi} f(x)\sin(kx)\,dx \quad \text{for } k=1,2,3,\cdots.$$

Calculating these coefficients can be tedious (integration by parts is often useful), but one simplification is worth noting. Recall (see Example 8 §5.6) that

$$\int_{-\pi}^{\pi} f(x)\,dx = 0 \text{ if } f \text{ is an odd function}$$

and

$$\int_{-\pi}^{\pi} f(x)\,dx = 2\int_{0}^{\pi} f(x)\,dx \text{ if } f \text{ is even.}$$

This is useful in calculating Fourier coefficients because $\cos(kx)$ is even and $\sin(kx)$ is odd for every $k \geq 1$, and because the product of an even function and an odd function is odd while the product of two odd (or two even) functions is even. In particular, we have

$$a_k = 0 \text{ for all } k \text{ if } f \text{ is odd}, \quad \text{and} \quad b_k = 0 \text{ for all } k \text{ if } f \text{ is even}.$$

This simplifies the work in the following example.

Example 14. Let f be the absolute value function defined by $f(x) = |x|$ for all x in the interval $[-\pi,\pi]$. Then f is even so $b_k = 0$ for all k, and we have

$$a_0 = \frac{\pi}{2}$$

and

$$a_k = \begin{cases} 0 & \text{if } k \geq 1 \text{ is even} \\ -\frac{4}{\pi k^2} & \text{if } k \geq 1 \text{ is odd} \end{cases}.$$

---
[29] Named for J.B.J. Fourier (1768-1830), a French mathematician who used these techniques in 1822 to investigate heat conduction in solids.

Hence the fifth Fourier approximation of f is

$$f_5(x) = \tfrac{\pi}{2} - \tfrac{4}{\pi}\{\cos x + \tfrac{1}{9}\cos(3x) + \tfrac{1}{25}\cos(5x)\}.$$

The graphs of f and f_5 are plotted in the diagram. □

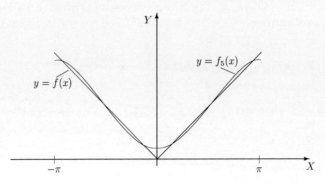

The reason for the importance of Fourier approximations is the following fundamental theorem.[30]

Theorem 11. *Let f be any function in $\mathbf{C}[-\pi, \pi]$ such that $f(-\pi) = f(\pi)$. Then*

$$\|f - f_m\| \text{ approaches } 0.$$

In fact, we have more:

$$f_m(x) \text{ approaches } f(x) \text{ for all } x \text{ in the interval } [-\pi, \pi].$$

This is an important result in the theory of Fourier approximation, and shows that f has a representation as an infinite series

$$f(x) = a_0 + a_1 \cos x + b_1 \sin x + a_2 \cos(2x) + b_2 \sin(2x) + \cdots$$

called the **Fourier series** of f. A full discussion of this is beyond the scope of this book.

Example 15. *Expand $f(x) = x^2$ on the interval $[-\pi, \pi]$ in a Fourier series, and so obtain a series expansion of $\tfrac{\pi^2}{6}$.*

Solution. Here f is an even function so all the Fourier sine coefficients b_k are zero. As to the cosine coefficients:

$$a_0 = \tfrac{1}{2\pi}\int_{-\pi}^{\pi} x^2 dx = \tfrac{\pi^2}{3} \quad \text{and} \quad a_k = \tfrac{1}{\pi}\int_{-\pi}^{\pi} x^2 \cos(kx)dx = \tfrac{4}{k^2}(-1)^k \text{ for } k \geq 1,$$

---

[30]See for example R.V. Churchill and J.W. Brown, *Fourier Series and Boundary Value Problems*, 4 Ed., McGraw-Hill, 1987.

5.7. GENERAL INNER PRODUCTS

where the second integral requires integration by parts twice, and uses the fact that $cos(k\pi) = (-1)^k$ for every integer k. Hence the Fourier series for x^2 is

$$x^2 = \frac{\pi^2}{3} - 4\left[\cos x - \frac{1}{2^2}\cos(2x) + \frac{1}{3^2}\cos(3x) - \frac{1}{4^2}\cos(4x) + \cdots\right]$$

for $-\pi \leq x \leq \pi$. In particular, taking $x = \pi$ gives an infinite series for $\frac{\pi^2}{6}$:

$$\frac{\pi^2}{6} = 1 + \frac{1}{2^2} + \frac{1}{3^2} + \frac{1}{4^2} + \cdots.$$

Many other such formulas can be proved using Theorem 11. $\square$

Finally, we observe that the functions $cos(kx)$ and $sin(kx)$ are easily generated as an electric voltage (if x is time), so the Fourier approximations to f can be obtained in electronic form. Hence Fourier series play a fundamental role in electrical engineering. In fact, they are an effective tool for solving many of the partial differential equations that arise throughout science and engineering.

5.7.5 Legendre Polynomials

Consider the inner product space $\mathbb{P}$ with the inner product

$$\langle p, q \rangle = \int_{-1}^{1} p(x)q(x)dx.$$

If we apply the Gram-Schmidt algorithm to the infinite independent set $\{1, x, x^2, \cdots x^m, \cdots\}$ in $\mathbb{P}$, we obtain an infinite orthogonal set $\{q_0, q_1, q_2, \cdots, q_m, \cdots\}$ in $\mathbb{P}$ given recursively as follows:

$$q_0 = 1,$$
$$q_m = x^m - \frac{\langle x^m, q_0 \rangle}{\|q_0\|^2}q_0 - \frac{\langle x^m, q_1 \rangle}{\|q_1\|^2}q_1 - \cdots - \frac{\langle x^m, q_{m-1} \rangle}{\|q_{m-1}\|^2}q_{m-1} \quad \text{for } m \geq 1.$$

We investigated these polynomials in Example 10, and calculated the first few: $q_0 = 1$, $q_1 = x$, $q_2 = \frac{1}{3}(3x^2 - 1)$ and $q_3 = \frac{1}{5}(5x^3 - 3x)$. It is easy to see (by induction) that q_m has degree m for each m and has leading coefficient 1. In particular, $\{q_0, q_1, q_2, \cdots, q_m\}$ is an orthogonal basis of $\mathbb{P}_m$ for each m because $dim(\mathbb{P}_m) = m + 1$.

Surprisingly, there is a length 2 recurrence for the q_m that is much easier to use than the Gram-Schmidt formula. To derive it, consider the polynomial xq_m. It has degree $m + 1$ so it lies in $\mathbb{P}_{m+1}$, and hence the Expansion Theorem gives

$$xq_m = r_0q_0 + r_1q_1 + r_2q_2 + \cdots + r_{m+1}q_{m+1} \quad \text{where } r_i = \frac{\langle xq_m, q_i \rangle}{\|q_i\|^2} \quad \text{for each } i.$$

But $\langle xq_m, q_i \rangle = \langle q_m, xq_i \rangle$ (verify), and $\langle q_m, xq_i \rangle = 0$ for $i \leq m - 2$ because then xq_i is in $\mathbb{P}_{m-1}$. It follows that $r_i = 0$ for all $i \leq m - 2$, so

$$xq_m = r_{m-1}q_{m-1} + r_mq_m + r_{m+1}q_{m+1}.$$

Moreover, since $deg\, q_k = k$ for each k, comparing the coefficients of x^{m+1} on each side of this equation gives $r_{m+1} = 1$. Hence we obtain the following length 2 recurrence for the polynomials q_i:

$$q_{m+1} = \left\{ x - \frac{\langle xq_m, q_m \rangle}{\|q_m\|^2} \right\} q_m - \frac{\langle xq_m, q_{m-1} \rangle}{\|q_{m-1}\|^2} q_{m-1} \quad \text{for } m \geq 1.$$

Since we know that $q_0 = 1$ and $q_1 = x$, this formula gives every q_m one after the other. In particular, we find that $q_4 = \frac{1}{35}(35x^4 - 30x^2 + 3)$ and $q_5 = \frac{1}{63}(63x^5 - 70x^3 + 15x)$.

However, there is an even simpler recurrence, but for the scalar multiples p_m of the q_m which have the property that

$$p_m(1) = 1 \quad \text{for all } m.$$

These are the **Legendre polynomials**[31] $p_0(x), p_1(x), p_2(x), \cdots p_m(x), \cdots$ and they are defined by

$$p_m(x) = \frac{1}{q_m(1)} q_m(x) \quad \text{for each } m = 1, 2, \cdots.$$

Using the above expressions for the q_i, the first six Legendre polynomials are

$$\begin{aligned}
p_0 &= 1 \\
p_1 &= x \\
p_2 &= \tfrac{1}{2}(3x^2 - 1) \\
p_3 &= \tfrac{1}{2}(5x^3 - 3x) \\
p_4 &= \tfrac{1}{8}(35x^4 - 30x^2 + 3) \\
p_5 &= \tfrac{1}{8}(63x^5 - 70x^3 + 15x).
\end{aligned}$$

The Legendre polynomials form an orthogonal set (because they are scalar multiples of the q_m) and find use in a wide variety of applications.

Moreover, an argument much like the one above for the q_m shows that the polynomials p_m also satisfy a length 2 recurrence. However it is much simpler than the recurrence for the q_m and we record the result in the following theorem.

Theorem 12. *The Legendre polynomials satisfy the recurrence*

$$p_{m+1} = \frac{2m+1}{m+1} x p_m - \frac{m}{m+1} p_{m-1} \quad \text{for all } m \geq 1.$$

$\square$

This gives an easy way to write the polynomials p_m down, beginning with $p_0 = 1$ and $p_1 = x$.

---

[31] A.M. Legendre (1752-1833) was a French mathematician, known for his work in number theory and differential equations.

5.7. GENERAL INNER PRODUCTS

The proof of Theorem 12 is beyond the scope of this book. It is usually derived from the fact that the Legendre polynomials first arose as solutions to differential equations. Indeed, $y = p_m(x)$ is a solution to **Legendre's differential equation**

$$(1 - x^2)\frac{d^2 y}{dx^2} - 2x\frac{dy}{dx} + m(m+1)y = 0,$$

and many treatments of these polynomials begin with this fact. Another observation is that these polynomials are the coefficients in the power series expansion of the function $(1 - 2xz + z^2)^{-\frac{1}{2}}$ regarded as a function of z, that is

$$(1 - 2xz + z^2)^{-\frac{1}{2}} = p_0(x) + p_1(x)z + p_2(x)z^2 + \cdots + p_m(x)z^m + \cdots$$

when $|z| \leq 1$. Because of this the function $(1-2xz+z^2)^{-\frac{1}{2}}$ is called a **generating function** for the Legendre polynomials $p_m(x)$, and it leads quickly to many of their properties. One elegant example is **Rodrigue's formula:**

$$p_m(x) = \frac{1}{2^m m!}\frac{d^m}{dx^m}(x^2 - 1)^m \text{ for each } m \geq 1.$$

All these facts, and many others, can be found in books on special functions [32] where other families of orthogonal polynomials are discussed.

Exercises

Throughout these exercises, V denotes an inner product space with inner product $\langle \, , \, \rangle$.

1. In each case determine which of Axioms P1-P4 fail.
   (a) $V = \mathbb{R}^2, \langle [x_1 \ x_2]^T, [y_1 \ y_2]^T \rangle = x_1 x_2 y_1 y_2$.
   (b) $V = \mathbb{R}^3, \langle [x_1 \ x_2 \ x_3]^T, [y_1 \ y_2 \ y_3]^T \rangle = x_1 y_1 - x_2 y_2 + x_3 y_3$.
   (c) $V = \mathbb{C}, \langle z, w \rangle = re(z\bar{w})$ where $re(z)$ denotes the real part of the complex number z.
   (d) $V = \mathbb{P}_3, \langle p(x), q(x) \rangle = p(1)q(1)$.
   (e) $V = \mathbb{M}_{2,2}, \langle A, B \rangle = det(AB^T)$.
   (f) $V = \mathbb{F}[0, 1], \langle f, g \rangle = f(0)g(1) + f(1)g(0)$.

2. Deduce that $\langle \mathbf{v}, \mathbf{0} \rangle = 0$ for all $\mathbf{v}$ in V space using only Axioms P1 and P3.

3. Show that $\mathbf{v}$ and $\mathbf{w}$ are orthogonal in V if and only if $\|\mathbf{v} + \mathbf{w}\| = \|\mathbf{v} - \mathbf{w}\|$.

4. (a) Show that $\langle \mathbf{v}, \mathbf{w} \rangle = \frac{1}{4}\{\|\mathbf{v} + \mathbf{w}\|^2 - \|\mathbf{v} - \mathbf{w}\|^2\}$ for all $\mathbf{v}$ and $\mathbf{w}$ in V.
   (b) If two inner products on V have the same norm functions, show that they are equal.

---
[32] See for example, N.N. Lebedev, *Special Functions and their Applications*, Dover, 1972.

5. If $V = span\{\mathbf{v}_1, \mathbf{v}_2, \cdots, \mathbf{v}_n\}$ and $\langle \mathbf{v}, \mathbf{v}_i \rangle = \langle \mathbf{w}, \mathbf{v}_i \rangle$ for each i, show that $\mathbf{v} = \mathbf{w}$.

6. Let $\mathbb{D}_n$ denote the space of all functions $f : \{1, 2, \cdots, n\} \to \mathbb{R}$ with pointwise addition and scalar multiplication (see Example 5 §5.1). For each $k = 1, 2, \cdots, n$, define δ_k in $\mathbb{D}_n$ by $\delta_k(i) = \begin{cases} 1 & \text{if } i = k \\ 0 & \text{if } i \neq k \end{cases}$.

   (a) If $\langle f, g \rangle = f(1)g(1) + f(2)g(2) + \cdots + f(n)g(n)$, show that $\langle\,,\,\rangle$ is an inner product on $\mathbb{D}_n$.

   (b) Show that $\{\delta_1, \delta_2, \cdots, \delta_n\}$ is an orthonormal basis of $\mathbb{D}_n$.

   (c) Show that $f = f(1)\delta_1 + f(2)\delta_2 + \cdots + f(n)\delta_n$ for every f in $\mathbb{D}_n$.

7. Show that every inner product on $\mathbb{R}^n$ has the form $\langle X, Y \rangle = (UX) \bullet (UY)$ for some invertible upper triangular matrix U with positive diagonal entries. [*Hint*: Theorem 4 §4.8.]

8. If V is any finite dimensional vector space and $\mathcal{B}$ is any basis of V, show that $\langle\,,\,\rangle$ is an inner product on V if we define $\langle \mathbf{v}, \mathbf{w} \rangle = C_\mathcal{B}(\mathbf{v}) \bullet C_\mathcal{B}(\mathbf{w})$ for all $\mathbf{v}$ and $\mathbf{w}$ in V.

9. If P is any $n \times n$ matrix, define $\langle\,,\,\rangle$ on $\mathbb{R}^n$ by $\langle X, Y \rangle = X^T P Y$ for all X and Y in $\mathbb{R}^n$.

   (a) Show that $\langle\,,\,\rangle$ satisfies Axioms P2 and P3, and that Axiom P1 holds if P is symmetric.

   (b) Show that if Axiom P1 holds then P is necessarily symmetric. [*Hint*: Show that $X^T(P - P^T)Y = 0$ for all X and Y in $\mathbb{R}^n$.]

10. In each case use the Gram-Schmidt process to convert the basis $\{1, x, x^2\}$ of $\mathbb{P}_2$ into an orthogonal basis of $\mathbb{P}_2$.

    (a) $\langle p, q \rangle = p(0)q(0) + p(1)q(1) + p(2)q(2)$.

    (b) $\langle p, q \rangle = \int_0^1 p(x)q(x)dx$.

11. Using the inner product $\langle A, B \rangle = tr(AB^T)$ on $M_{2,2}$, find the matrix in $U = span\{I, J\}$ closest to $A = \begin{bmatrix} 1 & -1 \\ 2 & 3 \end{bmatrix}$, where $I = \begin{bmatrix} 1 & 0 \\ 0 & 1 \end{bmatrix}$ and $J = \begin{bmatrix} 1 & 1 \\ 1 & 1 \end{bmatrix}$.

12. If $\mathcal{B} = \{\mathbf{b}_1, \mathbf{b}_2, \cdots, \mathbf{b}_n\}$ is any orthonormal basis of V, show that $\langle \mathbf{v}, \mathbf{w} \rangle = C_\mathcal{B}(\mathbf{v}) \bullet C_\mathcal{B}(\mathbf{w})$ for all $\mathbf{v}$ and $\mathbf{w}$ in V.

13. If $\mathcal{B} = \{\mathbf{b}_1, \mathbf{b}_2, \cdots, \mathbf{b}_n\}$ is any basis of V, show that there exists a positive definite matrix P such that $\langle \mathbf{v}, \mathbf{w} \rangle = C_\mathcal{B}(\mathbf{v})^T P C_\mathcal{B}(\mathbf{w})$ for all $\mathbf{v}$ and $\mathbf{w}$ in V. [*Hint*: Theorem 2.]

14. If $\{\mathbf{b}_1, \mathbf{b}_2, \cdots, \mathbf{b}_n\}$ is an orthonormal basis of V and $\mathbf{v} = r_1\mathbf{b}_1 + r_2\mathbf{b}_2 + \cdots + r_n\mathbf{b}_n$, show that $\|\mathbf{v}\|^2 = r_1^2 + r_2^2 + \cdots + r_n^2$. (**Parseval's Formula.**)

5.7. GENERAL INNER PRODUCTS

15. If $\{\mathbf{b}_1, \mathbf{b}_2, \cdots, \mathbf{b}_n\}$ is an orthogonal basis of V, show that

$$\langle \mathbf{v}, \mathbf{w} \rangle = \frac{\langle \mathbf{v}, \mathbf{b}_1 \rangle \langle \mathbf{w}, \mathbf{b}_1 \rangle}{\|\mathbf{b}_1\|^2} + \cdots + \frac{\langle \mathbf{v}, \mathbf{b}_n \rangle \langle \mathbf{w}, \mathbf{b}_n \rangle}{\|\mathbf{b}_n\|^2}$$

    for all $\mathbf{v}$ and $\mathbf{w}$ in V. [Hint: Theorem 5.]

16. Show that $(U + W)^\perp = U^\perp \cap W^\perp$ for all subspaces U and W of V.

17. (a) Show that $U^\perp + W^\perp \subseteq (U \cap W)^\perp$ for all subspaces U and W of V.

    (b) If $dim V = n$, show that $dim(U^\perp) = n - dim U$ for every subspace U of V.

    (c) If $dim V = n$, show that $(U \cap W)^\perp = U^\perp + W^\perp$. [Hint: Use (b), and Exercise 25 §4.3 (twice) to show that $dim(U^\perp + W^\perp) = dim((U \cap W)^\perp)$. Then apply (a).]

18. Let $U_1 \subseteq U_2$ be finite dimensional subspaces of an inner product space V. If $\mathbf{v}$ is any vector in V, show that $\|\mathbf{v} - proj_{U_2}(\mathbf{v})\| \leq \|\mathbf{v} - proj_{U_1}(\mathbf{v})\|$.

19. Let $W_1 \subseteq W_2 \subseteq \cdots$ be nonzero subspaces of an inner product space V such that $W_m \neq W_{m+1}$ for each m. Choose $\mathbf{b}_1 \neq \mathbf{0}$ in W_1 and, for $m \geq 2$, choose $\mathbf{b}_m$ in W_m but not in W_{m-1}. Show that $\{\mathbf{b}_1, \mathbf{b}_2, \mathbf{b}_3, \cdots\}$ is an infinite independent set.

20. Consider the space $\mathbf{C}[a, b]$ with the inner product $\langle f, g \rangle = \int_a^b f(x)g(x)dx$. Show that $\langle fh, g \rangle = \langle f, hg \rangle$ for all f, g and h in $\mathbf{C}[a, b]$.

21. (a) Show that $\frac{\pi^2}{8} = 1 + \frac{1}{3^2} + \frac{1}{5^2} + \cdots$. [Hint: Example 14 and Theorem 11.]

    (b) Show that $\frac{\pi^2}{12} = 1 - \frac{1}{2^2} + \frac{1}{3^2} - \frac{1}{4^2} + \frac{1}{5^2} - \cdots$. [Hint: Example 15 and Theorem 11.]

22. If V is a vector space, a function $\| \| : V \to \mathbb{R}$ is called a **norm** if: (1) $\|\mathbf{v}\| \geq 0$ for all in V; (2) $\|\mathbf{v}\| = 0$ if and only if $\mathbf{v} = \mathbf{0}$; (3) $\|r\mathbf{v}\| = |r| \|\mathbf{v}\|$ for all r in $\mathbb{R}$ and all $\mathbf{v}$ in V; and (4) $\|\mathbf{v} + \mathbf{w}\| \leq \|\mathbf{v}\| + \|\mathbf{w}\|$ for all $\mathbf{v}$ and $\mathbf{w}$ in V.

    (a) Show that $\|\mathbf{v}\| = \sqrt{\langle \mathbf{v}, \mathbf{v} \rangle}$ is a norm in this sense for any inner product $\langle , \rangle$ on V.

    (b) Show that $\|[x\ y]^T\| = |x| + |y|$ is a norm on $\mathbb{R}^2$ which does not arise from any inner product. [Hint: Theorem 2.]

23. Show that $\{1, \cos x, \cos(2x), \cos(3x), \cdots\}$ is an orthogonal set in $C[0, \pi]$ using the inner product $\langle f, g \rangle = \int_0^\pi f(x)g(x)dx$. [Hint: $\cos x \cos y = \frac{1}{2}[\cos(x+y) + \cos(x-y)]$.]

24. Compute the Legendre polynomials p_6 and p_7.

25. Show that the Legendre polynomial p_m is an even function if m is even, and an odd function if m is odd. [*Hint*: Theorem 12 and induction.]

26. Show that $p_m(-1) = (-1)^m$ for each $m = 0, 1, 2, \cdots$.

27. Let a_m denote the leading coefficient of the Legendre polynomial p_m.
    (a) Show that $a_{m+1} = \frac{2m+1}{m+1} a_m$ for each $m \geq 1$. [*Hint*: Theorem 12.]
    (b) Deduce that $a_m = \frac{(2m-1)(2m-3)\cdots(3)(1)}{m!}$ for each $m \geq 1$.

APPENDIX

A.1 Basic Trigonometry

The traditional degree measure for angles is not convenient for mathematics or science. A much better measure, called radian measure, is given geometrically as follows. Suppose a standard Cartesian coordinate system is given in the plane. The circle with radius 1, centered at the origin, is called the **unit circle**. An angle θ is said to be in **standard position** if it is measured counterclockwise from the positive X-axis as in Figure 1. The **radian measure** of θ is the length of the arc on the unit circle determined by the angle. Since the circumference of the unit circle is 2π, we have the basic relationship between degree and radian measures:

$$2\pi = 360°$$

Other important equivalences follow:

$$180° = \pi, \quad 90° = \frac{\pi}{2}, \quad 60° = \frac{\pi}{3}, \quad 45° = \frac{\pi}{4}, \quad 30° = \frac{\pi}{6}$$

Note that negative angles are measured *clockwise* from the positive X-axis, and their radian measure is the negative of the length of the corresponding arc on the unit circle. For example, $-\frac{\pi}{3}$ radians is the angle $60°$ measured clockwise from the positive X-axis (shown in Figure 3). Unless otherwise mentioned, all angles in this book are given in radians.[33]

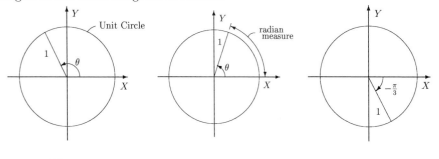

Figure 1 Figure 2 Figure 3

---

[33] A good way to visualize radian measure is to think of the θ-axis as wrapped around the unit circle with the origin at the point $P(1, 0)$. The positive θ-axis is wrapped counterclockwise, and the negative θ-axis is wrapped clockwise.

The word "trigonometry" comes from the Greek word *trigonon* for triangle, and so means "measurement of triangles". The trigonometric functions *cosine* and *sine* were invented for this purpose. For **acute** angles (between 0 and $\frac{\pi}{2}$) these functions are based upon the fundamental **similarity property** of triangles:[34]

> *If two triangles are similar (have the same angles) the ratios of the lengths of corresponding sides are equal.*

We only need this fact for right triangles, that is triangles in which one angle is a **right angle** $\frac{\pi}{2}$.

Now suppose that θ is an acute angle in a right triangle as in Figure 4. The side opposite the right angle is called the hypotenuse, and the other two sides are opposite and adjacent to the angle θ. The **sine** and **cosine** functions $sin\,\theta$ and $cos\,\theta$ are defined as ratios of the lengths of these sides:

$$sin\,\theta = \frac{a}{h} = \frac{\text{opposite}}{\text{hypotenuse}} \qquad cos\,\theta = \frac{b}{h} = \frac{\text{adjacent}}{\text{hypotenuse}}$$

These ratios do not depend on the size of the right triangle (by the similarity property) and so depend only on the angle θ. This fact makes them very useful for practical applications of geometry such as surveying.

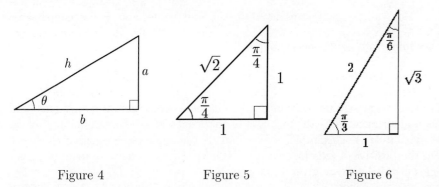

Figure 4   Figure 5   Figure 6

The numerical values of $sin\,\theta$ and $cos\,\theta$ can be obtained from any scientific calculator, but these values are only approximations (albeit to several decimal places). The following example gives three special triangles for which the exact ratios are known.

Example 1. Find $sin\,\theta$ and $cos\,\theta$ for $\theta = \frac{\pi}{4} = 45°$, $\theta = \frac{\pi}{6} = 30°$ and $\theta = \frac{\pi}{3} = 60°$.

Solution. The triangle in Figure 5 is half the square of side 1, so both acute angles equal $\frac{\pi}{4}$. The hypotenuse has length $\sqrt{2}$ by Pythagoras' theorem, so $cos\,\frac{\pi}{4} = \frac{1}{\sqrt{2}}$ and $sin\,\frac{\pi}{4} = \frac{1}{\sqrt{2}}$.

---
[34] Pythagoras' theorem also follows from the similarity property (see Theorem 4, §3.1).

A.1 BASIC TRIGONOMETRY

The triangle in Figure 6 is half the equilateral triangle of side 2, and so the acute angles are $\frac{\pi}{6}$ and $\frac{\pi}{3}$. The hypotenuse has length 2 and the short side has length 1, so the third side has length $\sqrt{3}$ by Pythagoras' theorem. Hence $cos\frac{\pi}{3} = \frac{1}{2} = sin\frac{\pi}{6}$, and $sin\frac{\pi}{3} = \frac{\sqrt{3}}{2} = cos\frac{\pi}{6}$. □

While the trigonometric functions of an acute angle are useful for many practical purposes, it soon becomes apparent that it is necessary to extend these functions to arbitrary angles. That is, we must first make clear what we mean by $sin\,\theta$ and $cos\,\theta$ where θ is a possible non-acute angle.

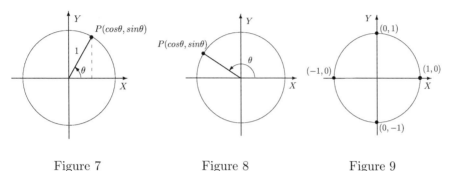

Figure 7   Figure 8   Figure 9

Consider an acute angle θ in standard position as in Figure 7. The angle θ determines a unique point P on the unit circle, and P has coordinates $P(cos\,\theta, sin\,\theta)$ because the unit circle has radius 1. This gives a simple way to define $cos\,\theta$ and $sin\,\theta$ for any angle θ :

> Given an angle θ in standard position, $cos\,\theta$ and $sin\,\theta$ are defined to be the X- and Y-coordinates of the point P on the unit circle determined by θ. Thus
> $$P = P(cos\,\theta, sin\,\theta).$$

This is illustrated in Figure 8, and makes it easy to find the cosine and sine of basic angles.

Example 2. *Find the cosine and sine of the following angles*: $\theta = 0$, $\theta = \frac{\pi}{2}$, $\theta = \pi$, $\theta = \frac{3\pi}{2}$, *and* $\theta = 2\pi$.

Solution. The coordinates of the points on the unit circle corresponding to these angles are plotted in Figure 9. Hence

$$\begin{array}{lllll} cos0 = 1 & cos\frac{\pi}{2} = 0 & cos\pi = -1 & cos\frac{3\pi}{2} = 0 & cos2\pi = 1 \\ sin0 = 0 & sin\frac{\pi}{2} = 1 & sin\pi = 0 & sin\frac{3\pi}{2} = -1 & sin2\pi = 0 \end{array}$$

Notice that the angles $\theta = 0$ and $\theta = 2\pi$ have the same cosine and sine because they determine the same point on the unit circle. □

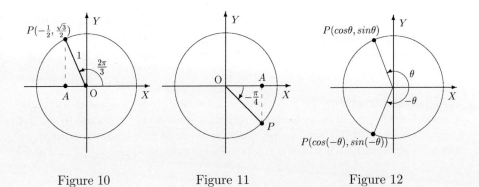

Figure 10   Figure 11   Figure 12

Example 3. *Find $\cos\theta$ and $\sin\theta$ for the angles $\theta = \frac{2\pi}{3}$ and $\theta = -\frac{\pi}{4}$.*

Solution. The angle $\theta = \frac{2\pi}{3}$ is given in standard position in Figure 10. The triangle AOP has angles $\frac{\pi}{3}$ and $\frac{\pi}{6}$ at vertices O and P respectively. Since the hypotenuse has length 1 (it is in the unit circle), the sides AO and AP have lengths $\frac{1}{2}$ and $\frac{\sqrt{3}}{2}$ respectively. Hence $P = P(-\frac{1}{2}, \frac{\sqrt{3}}{2})$, so $\cos\frac{2\pi}{3} = -\frac{1}{2}$ and $\sin\frac{2\pi}{2} = \frac{\sqrt{3}}{2}$.

Turning to $\theta = -\frac{\pi}{4}$, the angle $-\frac{\pi}{4}$ is shown in standard position in Figure 11. Now the triangle AOP has angles $\frac{\pi}{4}$ at both vertices O and P so, as the hypotenuse has length 1, sides AO and PA both have length $\frac{1}{\sqrt{2}}$. Thus $P = P(\frac{1}{\sqrt{2}}, -\frac{1}{\sqrt{2}})$, and so $\cos(-\frac{\pi}{4}) = \frac{1}{\sqrt{2}}$ and $\sin(-\frac{\pi}{4}) = -\frac{1}{\sqrt{2}}$. □

The observation in Example 2 that the angles $\theta = 0$ and $\theta = 2\pi$ have the same cosine and sine is a special case of a more general fact: If θ is any angle, then adding any multiple of 2π to θ (even a negative multiple) does not change the point on the unit circle because 2π corresponds to a full rotation. This is an important property of the cosine and sine functions.

Theorem 1. *If θ is any angle then*

$$\begin{aligned}\cos(\theta + 2k\pi) &= \cos\theta \\ \sin(\theta + 2k\pi) &= \sin\theta\end{aligned} \qquad \text{for all } k = 0, \pm 1, \pm 2, \cdots.$$

Because of Theorem 1, the cosine and sine functions are said to be **periodic functions** with **period** 2π.

The geometrical relationship between $\cos\theta$ and $\sin\theta$ leads to many identities for these functions. For example, the points $P(\cos\theta, \sin\theta)$ and $P(\cos(-\theta), \sin(-\theta))$ corresponding to θ and $-\theta$ are reflections of each other in the X-axis (see Figure 12). This gives the following important formulas:

$$\begin{aligned}\cos(-\theta) &= \cos\theta \\ \sin(-\theta) &= -\sin\theta\end{aligned} \qquad \text{for all angles } \theta.$$

Moreover the equation of the unit circle is $x^2 + y^2 = 1$, so the fact that $P(\cos\theta, \sin\theta)$ lies on the unit circle gives the following basic relationship between these functions:

Theorem 2. $cos^2\theta + sin^2\theta = 1$ *for all angles θ.*

This is the fundamental identity of trigonometry.

A.2 Induction

Suppose you are given the following sequence of equations:

$$\begin{aligned} 1 &= 1 \\ 1+3 &= 4 \\ 1+3+5 &= 9 \\ 1+3+5+7 &= 16 \\ 1+3+5+7+9 &= 25 \\ &\vdots \end{aligned}$$

There is a clear pattern: If $k = 1, 2, 3, 4$ or 5, the sum of the first k odd integers equals k^2. It is intriguing to speculate whether this is true for *every* integer $k \geq 1$. At first glance this seems to be a difficult question since the answer entails checking infinitely many cases $k = 1, 2, 3, \cdots$. But a clever idea makes it easy to prove that it is indeed true for every k.

Since the sum of the first k odd integers can be written $1+2+3+\cdots+(2k-1)$, the conjecture takes the form

$$1 + 2 + 3 + \cdots + (2k - 1) = k^2 \qquad \text{for every integer } k \geq 1.$$

For convenience, think of the assertion that $1 + 2 + 3 + \cdots + (2k-1) = k^2$ as a statement p_k about the integer k. Then we have an infinite sequence of statements

$$p_1, p_2, p_3, \cdots, p_k, \cdots,$$

and we want to verify that p_k is true for every $k \geq 1$. The idea is as follows: First show that p_1 is true, then show that if p_1 is true then p_2 must be true, then show that the truth of p_2 forces the truth of p_3, then that p_3 implies p_4, and so on indefinitely. In general we show that if p_k is true then p_{k+1} is also true. If we can do this then p_k must be true for every $k \geq 1$.

In the above example, p_k is the assertion that $1+2+3+\cdots+(2k-1) = k^2$. If this holds for some k, we want to show that p_{k+1} must also be true. In this case, the statement p_{k+1} reads

$$1 + 2 + 3 + \cdots + (2k - 1) + (2k + 1) = (k + 1)^2$$

by replacing k by $k+1$ in p_k. If we assume that p_k is true, the left side becomes

$$1 + 2 + 3 + \cdots + (2k - 1) + (2k + 1) = k^2 + (2k + 1) = (k + 1)^2,$$

so p_{k+1} is true, as required. Hence p_k is indeed true for each k.

There is a compact way to formulate this method. If p and q are statements, the assertion "*If p is true then q is true*" is expressed by saying that p **implies** q, and is written as
$$p \Rightarrow q.$$
Hence the above method can be given as follows:

Principle of Mathematical Induction. *Suppose a sequence of statements $p_1, p_2, p_3, \cdots, p_k, \cdots$ is given such that:*

1. *p_1 is true.*
2. *$p_k \Rightarrow p_{k+1}$ for every $k \geq 1$.*

Then p_k is true for every $k \geq 1$.

This is usually taken as one of the axioms for the positive integers, and is one of the most useful techniques in all of mathematics.[35] The following examples give a small indication of its range of application.

Example 1. *Show that $1 + 2 + 3 + \cdots + k = \frac{1}{2}k(k+1)$ for all $k \geq 1$.*

Solution. Let p_k be the statement that $1 + 2 + 3 + \cdots + k = \frac{1}{2}k(k+1)$. Then p_1 is true (it reads $1 = \frac{1}{2}1(2)$. To prove that $p_k \Rightarrow p_{k+1}$, we assume that p_k is true for some value of $k \geq 1$, and verify that p_{k+1} is also true: $1 + 2 + 3 + \cdots + (k+1) = \frac{1}{2}(k+1)(k+2)$. But the left side is
$$1 + 2 + 3 + \cdots + k + (k+1) = \tfrac{1}{2}k(k+1) + (k+1) = \tfrac{1}{2}(k+1)(k+2)$$
using p_k. Hence $p_k \Rightarrow p_{k+1}$, as required. □

Example 2. *If $x \neq 1$ is a number, show that*
$$1 + x + x^2 + \cdots + x^k = \frac{x^{k+1} - 1}{x - 1} \quad \text{for all} \quad k \geq 1.$$

Solution. Let p_k denote the statement $1 + x + x^2 + \cdots + x^k = \frac{x^{k+1}-1}{x-1}$. If $k = 1$ then p_1 reads $1 + x = \frac{x^2-1}{x-1}$ which is true because $x^2 - 1 = (x+1)(x-1)$. If we assume that p_k is true for some $k \geq 1$ then we must prove $p_{k+1} : 1 + x + x^2 + \cdots + x^{k+1} = \frac{x^{k+2}-1}{x-1}$. But the left side of p_{k+1} reads

$$\begin{aligned} 1 + x + x^2 + \cdots + x^k + x^{k+1} &= \tfrac{x^{k+1}-1}{x-1} + x^{k+1} = \tfrac{(x^{k+1}-1) + x^{k+1}(x-1)}{x-1} \\ &= \tfrac{x^{k+2}-1}{x-1}. \end{aligned}$$

using p_k. This is p_{k+1}, so we have proved that $p_k \Rightarrow p_{k+1}$. □

---

[35] A graphic description of the principal of induction can be given as follows: View the statements p_k as a series of dominos lined up in such a way that if p_k falls over it knocks p_{k+1} over. Then if we push p_1 over, every domino will fall.

A.3 POLYNOMIALS

When using the principal of induction we verify that p_1 is true and that $p_k \Rightarrow p_{k+1}$ for each $k \geq 1$, and so conclude that p_k is true for all $k \geq 1$. There is nothing special about 1 here; we can start the induction at any fixed integer m. This extended principle is as follows: Suppose that

1. p_m is true.

2. $p_k \Rightarrow p_{k+1}$ for each $k \geq m$.

Then p_k is true for all $k \geq m$.

Here is an example where this extended principle is needed. Recall that the number $k!$ (called "k-factorial") denotes the product of all the numbers from k to 1:

$$k! = k(k-1)(k-2)(k-3)\cdots 3 \cdot 2 \cdot 1.$$

Hence $2! = 2$, $3! = 6$, $4! = 24$, etc.

Example 3. *Show that $2^k < k!$ for all $k \geq 4$.*

Solution. If we let p_k denote the statement $2^k < k!$ for each $k \geq 1$, then p_1, p_2 and p_3 are all false as is easily verified. But p_4 is true because $2^4 = 16 < 24 = 4!$. If we assume that p_k is true for some $k \geq 4$, then $2^k < k!$ and we compute:

$$2^{k+1} = 2 \cdot 2^k < 2 \cdot k! < (k+1)k! = (k+1)!$$

because $2 < k+1$. This completes the induction. □

A.3 Polynomials

Expressions like $3 - 5x$, $1 + 3x - 2x^2$ are examples of polynomials. In general, a **polynomial** is an expression of the form

$$f(x) = a_0 + a_1 x + a_2 x^2 + \cdots + a_n x^n$$

where the a_i are numbers, called the **coefficients** of the polynomial, and x is a variable called an **indeterminate**. The number a_0 is called the **constant** term of the polynomial. The polynomial with every coefficient zero is called the **zero polynomial**, and is denoted simply as 0.

If $f(x) \neq 0$, the coefficient of the highest power of x appearing in $f(x)$ is called the **leading term** of $f(x)$, and the highest power itself is called the **degree** of the polynomial and is denoted $\deg f(x)$. Hence

$-1 + 5x + 3x^2$ has constant term -1, leading term 3, and degree 2,
7 has constant term 7, leading term 7, and degree 0,
$6x - 3x^3 + x^4 - x^5$ has constant term 0, leading term -1, and degree 5.

Two polynomials $f(x)$ and $g(x)$ are called **equal** if every coefficient of $f(x)$ is the same as the corresponding coefficient of $g(x)$. More precisely, if

$$f(x) = a_0 + a_1 x + a_2 x^2 + \cdots \quad \text{and} \quad g(x) = b_0 + b_1 x + b_2 x^2 + \cdots$$

are polynomials, then

$$f(x) = g(x) \quad \text{if and only if} \quad a_0 = b_0,\ a_1 = b_1,\ a_2 = b_2, \cdots.$$

In particular, this means that

$$f(x) = 0 \text{ is the zero polynomial} \quad \text{if and only if} \quad a_0 = 0, a_1 = 0, a_2 = 0, \cdots.$$

This is the reason for calling x an indeterminate.

Let $f(x)$ and $g(x)$ denote nonzero polynomials of degrees n and m respectively, say

$$f(x) = a_0 + a_1 x + a_2 x^2 + \cdots + a_n x^n \quad \text{and} \quad g(x) = b_0 + b_1 x + b_2 x^2 + \cdots + b_m x^m$$

where $a_n \neq 0$ and $b_m \neq 0$. If these expressions are multiplied, the result is

$$f(x)\,g(x) = a_0 b_0 + (a_0 b_1 + a_1 b_0)x + +(a_0 b_2 + a_1 b_1 + a_2 b_0)x + \cdots + a_n b_m x^{n+m}.$$

Since a_n and b_m are nonzero numbers, their product $a_n b_m \neq 0$ and we have

Theorem 1. *If $f(x)$ and $g(x)$ are nonzero polynomials of degrees n and m respectively, their product $f(x)\,g(x)$ is also nonzero and*

$$deg[f(x)\,g(x)] = n + m.$$

Example 1. $(2 - x + 3x^2)(3 + x^2 - 5x^3) = 6 - 3x + 11x^2 - 11x^3 + 8x^4 - 15x^5$.

If $f(x)$ is any polynomial, the next theorem shows that $f(x) - f(a)$ is a multiple of the polynomial $x - a$. In fact we have the

Remainder Theorem. *If $f(x)$ is a polynomial of degree n and a is any number, then there exists a polynomial $q(x)$ such that*

$$f(x) = (x - a)q(x) + f(a)$$

where either $q(x) = 0$ or $deg\, q(x) \leq n - 1$.

Proof. Write $f(x) = a_0 + a_1 x + a_2 x^2 + \cdots + a_n x^n$ where the a_i are numbers, so that $f(a) = a_0 + a_1 a + a_2 a^2 + \cdots + a_n a^n$. If these expressions are subtracted, the constant terms cancel and we obtain

$$f(x) - f(a) = a_1(x - a) + a_2(x^2 - a^2) + \cdots + a_n(x^n - a^n).$$

A.3 POLYNOMIALS

Hence it suffices to show that, for each $k \geq 1$, $x^k - a^k = (x-a)p(x)$ for some polynomial $p(x)$ of degree $k-1$. This is clear if $k = 1$. If it holds for some value k, the fact that

$$x^{k+1} - a^{k+1} = (x-a)x^k + a(x^k - a^k)$$

shows that it holds for $k+1$. Hence the proof is complete by induction. $\square$

There is a systematic procedure for finding the polynomial $q(x)$ in the Remainder theorem. It is illustrated below for $f(x) = x^3 - 3x^2 + x - 1$ and $a = 2$. The polynomial $q(x)$ is generated on the top line one term at a time as follows: First x^2 is chosen because $x^2(x-2)$ has the same x^3-term as $f(x)$, and this is subtracted from $f(x)$ to leave a "remainder" of $-x^2 + x - 1$. Next, the second term on top is $-x$ because $-x(x-2)$ has the same x^2-term, and this is subtracted to leave $-x - 1$. Finally the third term on top is -1, and the process ends with a "remainder" of -3.

$$\begin{array}{r|rrrrr}
 & x^2 & - & x & - & 1 \\ \hline
x-2 \,) & x^3 & - & 3x^2 & + x & - 1 \\
 & x^3 & - & 2x^2 & & \\ \hline
 & & - & x^2 & + x & - 1 \\
 & & - & x^2 & + 2x & \\ \hline
 & & & & - x & - 1 \\
 & & & & - x & + 2 \\ \hline
 & & & & & - 3
\end{array}$$

Hence $x^3 - 3x^2 + x - 1 = (x-2)(x^2 - x - 1) + (-3)$. Hence the final remainder is $-3 = f(2)$ as is easily verified. This procedure is called the **division algorithm**. [36]

A real number a is called a **root** of the polynomial $f(x)$ if $f(a) = 0$. Hence for example, 1 is a root of $f(x) = 2 - x + 3x^2 - 4x^3$, but -1 is not a root because $f(-1) = 10 \neq 0$. If $f(x)$ is a multiple of $x - a$ we say that $x - a$ is a **factor** of $f(x)$. Hence the Remainder Theorem shows immediately that if a is a root of $f(x)$ then $x - a$ is a factor of $f(x)$. But the converse is also true: If $x - a$ is a factor of $f(x)$, say $f(x) = (x-a)q(x)$, then $f(a) = (a-a)q(a) = 0$. This proves the

Factor Theorem. *If $f(x)$ is a polynomial and a is a number, then $x - a$ is a factor of $f(x)$ if and only if a is a root of $f(x)$.*

Example 2. If $f(x) = x^3 - 2x^2 - 6x + 4$, then $f(-2) = 0$, so $x - (-2) = x + 2$ is a factor of $f(x)$. In fact, the division algorithm gives $f(x) = (x+2)(x^2 - 4x + 2)$. $\square$

---

[36] This procedure can be used to divide $f(x)$ by *any* nonzero polynomial $d(x)$ in place of $x - a$; the remainder then is a polynomial that is either zero or of degree less than the degree of $d(x)$.

Consider the polynomial $f(x) = x^3 - 3x + 2$. Then 1 is clearly a root of $f(x)$, and the division algorithm gives $f(x) = (x-1)(x^2 + x - 2)$. But 1 is also a root of $x^2 + x - 2$; in fact $x^2 + x - 2 = (x-1)(x+2)$. Hence

$$f(x) = (x-1)^2(x+2)$$

and we say that the root 1 has **multiplicity** 2.

Note that every nonzero constant polynomial $f(x) = b \neq 0$ has *no* roots. However there do exist non-constant polynomials with no roots. For example, if $g(x) = x^2 + 1$ then $g(a) = a^2 + 1 \geq 1$ for every real number a, so a is not a root. However the *complex* number i is a root of $g(x)$, and we return to this below.

Now suppose that $f(x)$ is any nonzero polynomial. We claim that it can be factored in the following form:

$$f(x) = (x - a_1)(x - a_2) \cdots (x - a_m)g(x)$$

where $a_1, a_2, \cdots, a_m$ are the roots of $f(x)$ and $g(x)$ has no root (where the a_i may have repetitions, and may not appear at all (if $f(x)$ has no real root).

The above calculation shows that the polynomial $f(x) = x^3 - 3x + 2$ has roots 1 and -2, with 1 of multiplicity two. Counting the root -2 once, we say that $f(x)$ has three roots counting multiplicities. The next theorem shows that no polynomial can have more roots than its degree even if multiplicities are counted.

Theorem 2. *If $f(x)$ is a nonzero polynomial of degree n, then $f(x)$ has at most n roots counting multiplicities.*

Proof. If $n = 0$ then $f(x)$ is a constant and has no roots. So the theorem is true if $n = 0$. (It also holds for $n = 1$ because, if $f(x) = a + bx$ where $b \neq 0$, then the only root is $-\frac{a}{b}$.) In general, suppose that the theorem holds for some value of $n \geq 0$, and let $f(x)$ have degree $n + 1$. We must show that $f(x)$ has at most $n+1$ roots counting multiplicities. This is certainly true if $f(x)$ has no root. On the other hand, if a is a root of $f(x)$ the factor theorem shows that $f(x) = (x-a)q(x)$ for some polynomial $q(x)$, and $q(x)$ has degree n by Theorem 1. By induction, $q(x)$ has at most n roots. But if b is any root of $f(x)$ then

$$(b - a)q(b) = f(b) = 0$$

so either $b = a$ or b is a root of $q(x)$. It follows that $f(x)$ has at most n roots. This completes the induction and so proves Theorem 2. □

As we have seen, a polynomial may have *no* root, for example $f(x) = x^2 + 1$. Of course $f(x)$ has complex roots i and $-i$, where i is the complex number such that $i^2 = -1$. But Theorem 2 even holds for complex roots: the number of complex roots (counting multiplicities) cannot exceed the degree of the polynomial. Moreover, the Fundamental Theorem of Algebra asserts that the only nonzero polynomials with no complex root are the constant polynomials.

SELECTED HINTS AND SOLUTIONS

References to Explorations and Lessons refer to the ILAW web tutorial.

Chapter 1

LINEAR EQUATIONS AND MATRICES

1.1 Matrices

1. Put each table in matrix form, and use matrix operations (including matrix transpose if necessary, making sure that corresponding entries have the same meaning) to perform the required operations. [ILAW: Try Explorations 1.1.1 to 1.1.4 to learn how to handle large matrices.]

2. An $m \times n$ matrix is a rectangular array of m rows and n columns; you just have to remember that the first number is for rows, the second for columns.

3. Many operations on data tables as in Exercise 1 require the use of these basic operations. Can you make up more examples? Do you think more operations than those covered are needed?

4. (a) $\begin{bmatrix} -118 & -9 & -645 \\ 920 & 236 & 72 \\ -965 & 495 & 1849 \end{bmatrix}$. [ILAW: Verify with Explorations 1.1.1-1.1.4.]

5. (a) Use Exploration 1.1.4 to help you formulate and test a conjecture.

6. (b) False as $A + (-A) = 0$ for any matrix A.

7. For (a), the bottom right corner gives $d = 1$, then the bottom left corner gives $c = d = 1$, then the top right corner gives $b = c = 1$, and finally the top left corner gives $a = b - c = 1 - 1 = 0$.

8. (a) Basic algebraic manipulations give

$$A = \frac{1}{2}\left(\begin{bmatrix} 1 & 4 \\ 0 & 7 \\ -2 & 0 \end{bmatrix} + \begin{bmatrix} 2 & 0 \\ -3 & 4 \\ 0 & 8 \end{bmatrix}\right) = \begin{bmatrix} \frac{3}{2} & 2 \\ -\frac{3}{2} & \frac{11}{2} \\ -1 & 4 \end{bmatrix}.$$

9. Algebraic manipulations give $3(5A - 3B) + 6(B - 4A) + 3(2A + B) = 15A - 9B + 6B - 24A + 6A + 3B = -3A$. Remember that the operations of addition, subtraction, and scalar multiplication of matrices are very similar to those of ordinary numbers. The transpose operation does not exist for numbers, and matrix multiplication is coming later.

10. (a) The diagonal of $A + B$ is the diagonal of A plus the diagonal of B.

11. (a) If $A = [a_{ij}]$, then the (i,j)-entry of $(c+d)A$ is $(c+d)a_{ij} = ca_{ij} + da_{ij}$. On the other hand, the (i,j)-entry of cA is ca_{ij} and the (i,j)-entry of dA is da_{ij}, so the (i,j)-entry of $cA + dA$ is $ca_{ij} + da_{ij}$. Thus $(c+d)A$ and $cA + dA$ have the same (i,j)-entries for each i and j. This means that $(c+d)A = cA + dA$.

516   Selected Hints and Solutions

1.2 Linear Equations

1. (a) $X = [-s+t+1 \quad s+t+2 \quad s \quad t]^T$ means that $x_1 = -s+t+1$, $x_2 = s+t+2$, $x_3 = s$ and $x_4 = t$. To verify that X is a solution for any values of s and t, replace in both equations; the first one gives:

$$\begin{aligned} x_1 - 2x_2 + 3x_3 + x_4 &= (-s+t+1) - 2(s+t+2) + 3s + t \\ &= (-s - 2s + 3s) + (t - 2t + t) + (1 - 4) \\ &= -3. \end{aligned}$$

    This verifies the first equation. The second equation is verified similarly.

2. Remember that $X = [2 \quad -3]$ is considered to be *one* solution of the equation $2x - y = 7$, although it contains two values $x = 2$ and $y = -3$.
    Which of the following is a system of linear equations, and what is the difference in solving them?

$$\begin{array}{rcl} x - y &=& 1 \\ x + y &=& 17 \end{array} \quad \text{and} \quad \begin{array}{rcl} x^2 - y &=& 2 \\ x^2 + y &=& 7 \end{array}$$

    [ILAW: Use Exploration 1.2.1 to write various linear equations and test solutions.]

3. Make sure that you understand the Gaussian Algorithm. [ILAW: Exploration 1.2.3 has a built-in *Suggested Step* which follows the Gaussian Algorithm.]

4. Form the augmented matrix of each system, transform it to reduced row-echelon form from which you can read the solutions. [ILAW: Use Exploration 1.2.2 or 1.2.3.]
    The reduced row-echelon form of the augmented matrix in part (f) is

$$\begin{bmatrix} 1 & 0 & -\frac{13}{11} & 0 \\ 0 & 1 & -\frac{36}{11} & 0 \\ 0 & 0 & 0 & 1 \end{bmatrix}.$$

    The last row shows that the system has no solution.

5. Form the augmented matrix, and perform row operations to transform the matrix into a form where you can read the solutions. Is your procedure close to the Gaussian elimination?

6. The rank of a matrix is the number of leading 1's in its reduced row-echelon form. If the system has no solution then there is nothing much more that we can say. But if there are solutions, the number of parameters is the number of variables minus the number of leading 1's.

7. It is explicitly mentioned that the system has a solution, so we only need to compute the number of parameters. But there are 7 variables and the rank cannot be more than 5, so we must have at least 2 parameters.

8. Try all possibilities and different methods to solve the equation. The problem comes down to whether a and b are zero or not, right?

9. Solving these types of problems means that you have a good grasp of the techniques to solve systems of linear equations. The best method is to transform the augmented matrix of each system to reduced row-echelon form, and split into cases depending on the values of a, b and c only when necessary.

Selected Hints and Solutions   517

For (c), the augmented matrix is $\begin{bmatrix} 1 & a & 1 \\ b & 2 & 5 \end{bmatrix}$. Adding $-b$ times Row 1 to Row 2 (valid for *any* b) gives $\begin{bmatrix} 1 & a & 1 \\ 0 & 2-ab & 5-b \end{bmatrix}$. We can see that if $2 - ab \neq 0$, there will be a unique solution. Now if $2 - ab = 0$, we get $\begin{bmatrix} 1 & a & 1 \\ 0 & 0 & 5-b \end{bmatrix}$. Therefore if $5-b = 0$ (and therefore $b = 5$ and $a = 2/b = 2/5$) there are infinitely many solutions. If finally $5 - b \neq 0$ (and therefore $b \neq 5$ but $a = 2/b$), there will be no solution.

10. A proof that this is correct will come in Chapter 4. Notice that the rank of a matrix can be seen from any of its row-echelon forms. [ILAW: Use Exploration 1.2.4 to formulate and test a conjecture.]

11. You can find the rank as soon as the matrix has been transformed to an upper triangular form, meaning that all entries below the main diagonal are 0.

12. At the four intersections, the traffic flow coming in must be equal to the flow going out, giving the system of linear equations:

$$\begin{aligned} f_4 + 70 &= f_1 \\ f_1 &= 60 + f_2 \\ f_2 + 50 &= f_3 \\ f_3 &= 60 + f_4 \end{aligned}$$

The general solution must use only non negative integers since each variable represents a traffic flow. Change the data or create your own downtown streets, and analyze the traffic patterns.

13. Take the data and express the requirements as a system of linear equations. How do you set up the variables? The solutions must again be taken as non negative. Must they be integers?

14. (c) False: $R = \begin{bmatrix} 1 & 0 & -1 & 2 \\ 0 & 1 & -1 & 3 \end{bmatrix}$.

(h) True. The reduced row-echelon form for A contains all the leading ones in the same form for C.

[ILAW: Use appropriate Explorations to test the statements and find counterexamples if necessary.]

1.3 Homogeneous Systems

1. Notice that only the coefficient matrix is given. Transform the augmented matrices to reduced row-echelon form as usual [ILAW: Use Exploration 1.3.1], and read the solutions. For (d), the general solution is

$$X = \begin{bmatrix} 2t - 3s \\ t \\ -5s \\ s \end{bmatrix} = t \begin{bmatrix} 2 \\ 1 \\ 0 \\ 0 \end{bmatrix} + s \begin{bmatrix} -3 \\ 0 \\ -5 \\ 1 \end{bmatrix}.$$

The column matrices $X_1 = [2 \ 1 \ 0 \ 0]^T$ and $X_2 = [-3 \ 0 \ -5 \ 1]^T$ are called the basic solutions.

2. To balance (a), we let

$$\begin{aligned} v &= \text{the number of ammonia molecules present,} \\ w &= \text{the number of copper oxide molecules present,} \\ x &= \text{the number of nitrogen molecules present,} \\ y &= \text{the number of copper molecules present,} \\ z &= \text{the number of water molecules present.} \end{aligned}$$

The requirements can now be written as:

$$\begin{aligned} v &= 2x & \text{(To balance N)} \\ 3v &= 2z & \text{(To balance H)} \\ w &= y & \text{(To balance Cu)}, \\ w &= z & \text{(To balance O)}, \end{aligned}$$

The general solution is $X = [\frac{2}{3}s \ \ s \ \ \frac{1}{3}s \ \ s \ \ s]^T$. A possible particular solution using $s = 3$ gives

$$2NH_3 + 3CuO \rightarrow N_2 + 3Cu + 3H_2O.$$

3. In a homogeneous system we don't have to worry about having no solution; the important question is whether we get only the trivial solution or else infinitely many.

4. (b) Here $n = 6$ and $r \leq 4$, so $n - r \geq 2$. [ILAW: Use Explorations 1.3.1 or 1.3.2 to formulate and test conjectures.]

5. Can you think of a system of three equations in two variables that has nontrivial solutions? [ILAW: Explorations 1.3.1 or 1.3.2.]

6. You need to understand the role of the number of variables versus the number of equations for homogeneous systems. [ILAW: Explorations 1.3.1 or 1.3.2.]

7. For (g), the system has no nontrivial solutions only when no parameters are introduced, and therefore when $m = n$. [ILAW: Explorations 1.3.1 or 1.3.2.]

8. Does the last column play any role in the rank of the augmented matrix? [ILAW: Explorations 1.3.1 or 1.3.2.]

1.4 Matrix Multiplication

1. (a) $\begin{bmatrix} -6 & -9 \\ -14 & 14 \end{bmatrix}$ [ILAW: Exploration 1.4.1.]

2. Make sure you know when a product cannot be done. (c) $\begin{bmatrix} 7 \\ 19 \end{bmatrix}$

   (h) Not defined: A is 2×3, D is 2×1. [ILAW: Exploration 1.4.1.]

3. Now you have no choice but to write down explicitly how a matrix multiplication is done, and when it cannot be done.

4. Check the possible sizes of A and B.

5. This is similar to the previous problem, check the possible sizes of A and B.

6. Check the possible sizes of A, B, and C.

Selected Hints and Solutions

7. (b) $-3AB + 2AC + A^2B$. [ILAW: Use $A^2 = AA$ and $B^2 = BB$ in Exploration 1.4.2.]

8. Because matrices generally do not commute (that is $AB \neq BA$ in general), algebraic manipulations vary from the corresponding manipulations for ordinary numbers. For matrices, $(A+B)^2 = A^2 + AB + BA + B^2$, so $(A+B)^2 = A^2 + 2AB + B^2$ exactly when $AB = BA$.

9. Compute A^2 by making use of the two given identities, for example $A^2 = AA = (AB)A$.

10. This problem tests your general knowledge of matrix multiplication. [ILAW: Use Exploration 1.4.1.]
    (a) True. If row i of A is zero, then row i of AB is zero.
    (g) False. $A = \begin{bmatrix} 1 & 0 \\ 0 & 0 \end{bmatrix}$, $B = \begin{bmatrix} 0 & 0 \\ 0 & 1 \end{bmatrix}$.

    Use appropriate Explorations to help you test some statements. For (j), use 1.4.1 to formulate and test a conjecture for 2×2 matrices.

11. $(AA^T)^T = (A^T)^T A^T = AA^T$, and therefore AA^T is symmetric by definition.

12. Use the fact that $AB = (AB)^T$ by assumption.

13. The general solution of a system $AX = B$ is given by *one* particular solution X_0 plus the general solution of the associated homogeneous system $AX = 0$. Carefully look at the general solution of both systems $AX = B$ and $AX = 0$. [ILAW: Exploration 1.2.3 and 1.3.2.]
    (b) The general solution is

    $$X = \begin{bmatrix} -2 \\ 2 \\ 0 \end{bmatrix} + s \begin{bmatrix} 1 \\ -3 \\ 1 \end{bmatrix} = X_0 + sX_1,$$

    where X_0 is a particular solution of $AX = B$, and sX_1 is the general solution of $AX = 0$.

14. As you are now an expert on the material in Sections 2 and 3, you can come up with a good numerical example. [ILAW: Explorations 1.2.2 and 1.3.2.]

15. Under these conditions, what will the reduced row-echelon form of the augmented matrix of the system $AX = 0$ look like? Then what will the reduced row-echelon form of the augmented matrix of the system $AX = B$ look like?

16. Again, what can the reduced row-echelon form of the augmented matrix look like?

17. If there was such a B, then multiply the matrix equation $AX = 0$ on the left by B and see what happens. Why multiply on the left and not on the right? Does it make a difference?

18. Do some hand calculations to help you formulate a conjecture. [ILAW: Exploration 1.4.1.]

19. Try some 3×3 examples formulate a conjecture. [ILAW: Exploration 1.4.1.]

20. In this case, $A^n = A$ for each odd $n = 1, 3, 5, \cdots$

21. Substitute in the equation. [ILAW: Exploration 1.4.1.]

22. Things happen with matrix multiplication that don't happen with ordinary number multiplication. For (b), look at the entries of a 2×2 symmetric matrix A, and consider the top left entry of the matrix A^2; what can it be?

23. Make sure you understand how block multiplication is done, and why it can be useful for large matrices. [ILAW: Exploration 1.4.4.]

24. Use Theorem 4, but notice that you will have to come up with a slightly different pattern as the matrix in part (a) has its block of 0's on the top right. You can quickly see a pattern if you do a few calculations. [ILAW: Exploration 1.4.4.]
    (a) If $A = \begin{bmatrix} B & 0 \\ X & C \end{bmatrix}$, then $A^n = \begin{bmatrix} B & 0 \\ X & 2^{n-1}C \end{bmatrix}$.

25. This is an algebraic question similar to Exercise 22. For (b), you get $\begin{bmatrix} I & 0 \\ 0 & I \end{bmatrix}$, which is simply the identity matrix (of the appropriate size).

26. Do the block multiplication a few times and make a conjecture.

27. Make sure you understand the translation between a graph and its adjacency matrix of 0's and 1's, and why the entries of the powers of the adjacency matrix give the corresponding number of paths between vertices. [ILAW: Exploration 1.4.5.]

28. Write $I_n = [C_1 \; C_2 \; \cdots \; C_n]$ where C_j is column j of I_n, and compute $B = BI_n$ to see what happens. [ILAW: Exploration 1.4.1.]

29. A must commute with $\begin{bmatrix} 0 & 1 \\ 0 & 0 \end{bmatrix}$ and $\begin{bmatrix} 0 & 0 \\ 1 & 0 \end{bmatrix}$; see what that means.

30. Look at the top left entry and bottom right entry of AA^T.

31. (b) If $E^2 = E$, then $(I - E)^2 = (I - E)(I - E) = I^2 - IE - EI + E^2 = I - 2E + E = I - E$.

32. Do the product and see if you can choose x, y, z and w to recapture the matrix A. Why must $a \neq 0$?

33. Calculate successive powers A^n and formulate a conjecture. [ILAW: Exploration 1.4.1.]

1.5 Matrix Inverses

1. $(A^{-1})^{-1}$ is by definition a matrix B such that $A^{-1}B = I = BA^{-1}$, so $B = A$.

2. Make sure you understand the procedure. [ILAW: Exploration 1.5.1 and 1.5.3.]
    (d) The inverse is $\begin{bmatrix} 0 & -1 & 0 \\ 1 & 1 & -1 \\ 2 & 2 & -3 \end{bmatrix}$.

3. To find A^{-1}, you form the augmented matrix $[A : I]$, and you must transform A to I; otherwise there is no inverse. You could go first to a row-echelon form and then come back to each column and put 0's above every leading 1 (try it a few times). A variation is to go straight for the reduced row-echelon form, creating the leading 1's and immediately placing 0's above and below, and then moving to the next column. Any other ideas? [ILAW: Exploration 1.5.3.]

Selected Hints and Solutions

4. This is like Exercise 1. Find A^{-1}, cut and paste it back to find its inverse and see what you get. Of course you should get back A itself, that is $(A^{-1})^{-1} = A$. [ILAW: Exploration 1.5.1.]

5. First do a bit of algebra to isolate A. For (d), take the transpose on each side to get $((A^{-1} - 3I)^T)^T = 5 \begin{bmatrix} 1 & 2 \\ 3 & 4 \end{bmatrix}^T$, and therefore $(A^{-1} - 3I) = 5 \begin{bmatrix} 1 & 3 \\ 2 & 4 \end{bmatrix}$. So $A^{-1} = 3I + 5 \begin{bmatrix} 1 & 3 \\ 2 & 4 \end{bmatrix} = \begin{bmatrix} 8 & 15 \\ 10 & 23 \end{bmatrix}$. Finally $A = (A^{-1})^{-1} = \begin{bmatrix} 8 & 15 \\ 10 & 23 \end{bmatrix}^{-1} = \frac{1}{34} \begin{bmatrix} 23 & -15 \\ -10 & 8 \end{bmatrix}$.

6. It is important that you understand the use of matrix inverses to solve a system $AX = B$. Multiply on the left by A^{-1} to get $A^{-1}AX = A^{-1}B$, so the unique solution is $X = A^{-1}B$. Hence compute A^{-1} and multiply to get the solution $X = A^{-1}B$.

7. Use the strategy of the previous exercise. Each of the systems can be written as $AX = B$ and hopefully the coefficient matrix A is invertible, so $X = A^{-1}B$ will be the unique solution. For (d), the coefficient matrix is $A = \begin{bmatrix} 3 & 5 & 0 \\ 1 & 2 & 1 \\ 3 & 7 & 1 \end{bmatrix}$ and $A^{-1} = \begin{bmatrix} 1 & 1 & -1 \\ -0.4 & -0.6 & 0.6 \\ -0.2 & 1.2 & -0.2 \end{bmatrix}$, so

$$X = A^{-1}B = \begin{bmatrix} 1 & 1 & -1 \\ -0.4 & -0.6 & 0.6 \\ -0.2 & 1.2 & -0.2 \end{bmatrix} \begin{bmatrix} -5 \\ 1 \\ 0 \end{bmatrix} = \begin{bmatrix} -4 \\ 1.4 \\ 2.2 \end{bmatrix}.$$

8. This is worth repeating. If A is invertible, then $A^{-1}B$ is a solution of $AX = B$ so we have at least one solution. But if X is any solution of $AX = B$, then $A^{-1}(AX) = A^{-1}B$ giving $IX = A^{-1}B$, and so $X = A^{-1}B$ is the only solution. If you change B, you may get a different solution to $AX = B$, but there is a unique solution for each B.

9. The point here is that you don't need to restrict yourself to column matrices for B in $AX = B$, or to having only X to the right of A. For (b), multiply $XA = \begin{bmatrix} 2 & 3 & -1 \\ -1 & 0 & 5 \end{bmatrix}$ on the right by the given A^{-1} to get $(XA)A^{-1} = \begin{bmatrix} 2 & 3 & -1 \\ -1 & 0 & 5 \end{bmatrix} \begin{bmatrix} 1 & 0 & 2 \\ 1 & 2 & 1 \\ 3 & 5 & 3 \end{bmatrix}$, so $X = \begin{bmatrix} 2 & 1 & 4 \\ 14 & 25 & 13 \end{bmatrix}$.

10. Repeat a similar argument to Exercise 8.

11. The matrix C plays a role similar to A^{-1}, although it may not be a square matrix. But it is good enough for left multiplication and cancelling A in $AX = B$.

12. There are a few ways to show this. One is to show that no 2×2 matrix C can be found so that $CA = I$, as the second column will always be 0 for any C. Another is to show that the reduced row-echelon form of A cannot be I, again because of the second column. Another is to create a system of linear equations $AX = B$ without a unique solution.

13. (a) False: $A = \begin{bmatrix} 0 & 1 \\ 0 & 0 \end{bmatrix}$.
    (c) True. It implies that $A(\frac{1}{3}A^2) = I = (\frac{1}{3}A^2)A$. Alternatively, it shows $(detA)^3 = 3^n$ where A is $n \times n$, so $detA \neq 0$.
    (e) [ILAW: Use Explorations 1.1.2 and 1.5.1.]

14. Is U invertible?

15. If $A^2 - A - 2I = 0$, verify that $A^2 - A = 2I$, and so $A[\frac{1}{2}(A - I)] = I$. This means that $A^{-1} = \frac{1}{2}(A - I)$.

16. If $A^2 = 0$, verify that $(I - A)(I + A) = I$.

17. This is a summary of many properties of matrix inverses, and so is a good review of the section. Part (f) uses Corollary 2 to Theorem 5, but is not obvious at first. [ILAW: Exploration 1.5.5.]

18. Conjugates will return in Chapter 2.
    (a) [ILAW: Exploration 1.4.1.]
    (b) If $P^{-1}AP$ has an inverse C, so that $CP^{-1}AP = I = P^{-1}APC$, then
    $$A(PCP^{-1}) = (PP^{-1})A(PCP^{-1}) = P(P^{-1}APC)P^{-1} = PIP^{-1} = I,$$
    and therefore $A^{-1} = PCP^{-1}$. Alternatively, if $B = P^{-1}AP$ is an invertible matrix, then
    $$PBP^{-1} = (PP^{-1})A(PP^{-1}) = IAI = A.$$
    So $A = PBP^{-1}$ is also invertible since P, B, and P^{-1} are invertible.

19. If $P(A+X)Q = B$ and both P and Q are invertible, then $P^{-1}P(A+X)QQ^{-1} = P^{-1}BQ^{-1}$ (notice the order), so $A+X = P^{-1}BQ^{-1}$ and finally $X = P^{-1}BQ^{-1} - A$.

20. (a) If $AC = CA$, first left multiply by $(A)^{-1}$.

21. For (a), check any product AB.

22. Use $\sin^2\theta + \cos^2\theta = 1$.

23. For (b) with A^{-1} in block form, use block multiplication and see what you get.

24. Use A^{-1} by multiplying on the left.

25. For (b), use A^{-1} and B^{-1} by appropriate multiplication.

26. If one of them is invertible, use its inverse with an appropriate multiplication.

27. You can't use B^{-1} before you have shown it exists!

28. For (a) write $C = (AB)^{-1}$. Then $A(BC) = (AB)(AB)^{-1} = I$ so A is invertible by Corollary 1 of Theorem 5. Similarly for B.

29. Just do the algebraic calculations. [ILAW: Exploration 1.4.1.]

30. $(cA)(\frac{1}{c}A^{-1}) = (c\frac{1}{c})(AA^{-1}) = 1I = I$, and similarly the other way around.

1.6 Elementary Matrices

1. You need to find which row operation has been performed on the identity matrix to yield the given matrix E. In (d), the operation was "add -3 times row 3 to row 1"; the inverse operation would be "add $+3$ times row 3 to row 1", as performing this operation on E would give the identity matrix back.

Selected Hints and Solutions 523

2. Start with A and see which operation yields B. [ILAW: Exploration 1.6.1.]

3. Perform two row operations one after the other. [ILAW: Exploration 1.6.1.]

4. Elementary matrices are used to perform elementary row operations by matrix multiplication.

5. A matrix A is invertible exactly if its reduced row-echelon form is the identity matrix, therefore transform A to I, and represent each row operation by the corresponding elementary matrix. [ILAW: Exploration 1.6.2.]
   (b) We get

$$\begin{bmatrix} 1 & 0 \\ 0 & 1 \end{bmatrix} = \begin{bmatrix} 1 & 0.5 \\ 0 & 1 \end{bmatrix} \begin{bmatrix} 1 & 0 \\ 0 & -2 \end{bmatrix} \begin{bmatrix} 1 & 0 \\ -1 & 1 \end{bmatrix} \begin{bmatrix} 0.5 & 0 \\ 0 & 1 \end{bmatrix} \begin{bmatrix} 2 & -1 \\ 1 & -1 \end{bmatrix},$$

and therefore

$$\begin{bmatrix} 2 & -1 \\ 1 & -1 \end{bmatrix} = \left(\begin{bmatrix} 1 & 0.5 \\ 0 & 1 \end{bmatrix} \begin{bmatrix} 1 & 0 \\ 0 & -2 \end{bmatrix} \begin{bmatrix} 1 & 0 \\ -1 & 1 \end{bmatrix} \begin{bmatrix} 0.5 & 0 \\ 0 & 1 \end{bmatrix} \right)^{-1}$$

$$= \begin{bmatrix} 0.5 & 0 \\ 0 & 1 \end{bmatrix}^{-1} \begin{bmatrix} 1 & 0 \\ -1 & 1 \end{bmatrix}^{-1} \begin{bmatrix} 1 & 0 \\ 0 & -2 \end{bmatrix}^{-1} \begin{bmatrix} 1 & 0.5 \\ 0 & 1 \end{bmatrix}^{-1}$$

$$= \begin{bmatrix} 2 & 0 \\ 0 & 1 \end{bmatrix} \begin{bmatrix} 1 & 0 \\ 1 & 1 \end{bmatrix} \begin{bmatrix} 1 & 0 \\ 0 & -0.5 \end{bmatrix} \begin{bmatrix} 1 & -0.5 \\ 0 & 1 \end{bmatrix}.$$

6. Use Theorem 1. [ILAW: Using Exploration 1.6.2.]

   (d) $U = \frac{1}{6} \begin{bmatrix} 5 & 1 & 0 \\ -1 & 1 & 0 \\ -16 & -2 & 6 \end{bmatrix}$.

7. Since you are not required to have zero entries above leading 1's in a row-echelon form, there are many possibilities. [ILAW: Using Exploration 1.6.3.]

8. Find row operations carrying A to B, and use Theorem 1. [ILAW: Using Exploration 1.6.2.]

9. Elementary matrices are the *building blocks* of invertible matrices; not only is every elementary matrix invertible, but any invertible matrix can be written as a product of elementary matrices.

10. Having F as an elementary matrix on the right as in $B = AF$ means that operations are done on *columns* instead of rows. We don't really have this setup, but $B = AF$ if and only if $B^T = (AF)^T = F^T A^T$, so you can transform A^T to B^T in the usual way, giving the transpose of the desired elementary matrix. See also the discussion following Example 2.

11. You should check it for each of the three kinds of elementary matrices.

12. (a) Elementary matrices are invertible, so $B = EA$ if and only if $E^{-1}B = A$. So is E^{-1} an elementary matrix? For (b), is adding 2 times row 1 to three times row 4 really an elementary row operation?

13. For (a), we get

$$UAV = \begin{bmatrix} -1 & 1 & 0 \\ -2 & 1 & 0 \\ 2 & -1 & 1 \end{bmatrix} \begin{bmatrix} 1 & -1 & 2 & 1 \\ 2 & -1 & 0 & 3 \\ 0 & 1 & -4 & 1 \end{bmatrix} \begin{bmatrix} 1 & 0 & 2 & -2 \\ 0 & 1 & 4 & -1 \\ 0 & 0 & 1 & 0 \\ 0 & 0 & 0 & 1 \end{bmatrix}$$

$$= \begin{bmatrix} 1 & 0 & 0 & 0 \\ 0 & 1 & 0 & 0 \\ 0 & 0 & 0 & 0 \end{bmatrix}$$

[ILAW: Use Exploration 1.6.2 to find a matrix U such that $R = UA$ is the reduced row-echelon form of A, and then use the transpose to find a matrix V^T such that $S = V^T R^T$ is in reduced row-echelon form. Then U and V are as desired, and $UAV = \begin{bmatrix} I_r & 0 \\ 0 & 0 \end{bmatrix}$ where r is the rank of A.]

14. If the sequence of elementary row operations corresponds to the invertible matrix U (that is U is the product of the corresponding elementary matrices), then $P = UA$ and $Q = UI = U$, so $P = QA$.

15. If a matrix A is carried to 0, then $0 = UA$ for some invertible matrix U, and therefore $0 = U^{-1}0 = U^{-1}(UA) = A$.

16. We already discussed that invertible matrices can be written as products of elementary matrices.

17. If R is the reduced row-echelon form of A, then we know that $R = WA$ for some invertible matrix W, so $A = W^{-1}R$ and you can take $U = W^{-1}$. For (b), use the fact that the reduced row-echelon form is unique.

18. Retrace the argument leading to Theorem 1 if column operations are done instead of row operations. Use the remark following Example 2. V is the product of the corresponding elementary *column* operations.

19. (c) $A \stackrel{r}{\sim} B$ means $A \to B$ by row operations, and $B \stackrel{r}{\sim} C$ means $B \to C$ by row operations. Hence $A \to C$ by the row operations $A \to B$ followed by the row operations $B \to C$. Alternatively, $B = UA$ and $C = VB$ by (a) where U and V are invertible. Hence $C = VB = V(UA) = (VU)A$, so $A \stackrel{r}{\sim} C$ because VU is invertible.
(e) Show that both A and B can be carried to the same row-echelon matrix.

1.7 LU-Factorization

1. To solve $LUX = B$, solve $LY = B$ and then $UX = Y$. [ILAW: Exploration 1.7.1.]
For (b), the unique solution to $LY = B$ is $Y = [2 \quad 3\frac{2}{3} \quad -3 \quad -21\frac{2}{3}]^T$, but the system $UX = Y$ has no solution. Therefore the original system $AX = B$ has no solution.

Selected Hints and Solutions 525

2. Make sure you understand the LU-algorithm process. [ILAW: Exploration 1.7.2.] For (f),

$$\begin{bmatrix} 1 & -3 & 2 & 0 \\ 2 & -5 & -3 & 1 \\ 0 & -1 & 3 & 7 \\ -1 & 3 & 2 & 5 \end{bmatrix} = \begin{bmatrix} 1 & 0 & 0 & 0 \\ 2 & 1 & 0 & 0 \\ 0 & 1 & 10 & 0 \\ -1 & 0 & 4 & 2.6 \end{bmatrix} \begin{bmatrix} 1 & -3 & 2 & 0 \\ 0 & 1 & -7 & 1 \\ 0 & 0 & 1 & 0.6 \\ 0 & 0 & 0 & 1 \end{bmatrix}$$

3. If A can be transformed to a row-echelon form $U = PA$ without using row interchanges, then $A = P^{-1}U = LU$ where L is lower triangular and U is upper triangular. So one method is simply to do this and invert P.

4. If the transformation of a matrix A to any row-echelon form requires row interchanges, then we can find a permutation matrix P such that PA does not require row interchanges. For (b), one such P is

$$P = \begin{bmatrix} 0 & 0 & 0 & 1 \\ 0 & 0 & 1 & 0 \\ 0 & 1 & 0 & 0 \\ 1 & 0 & 0 & 0 \end{bmatrix}$$

One LU-factorization of PA is

$$PA = \begin{bmatrix} 0 & 0 & 0 & 1 \\ 0 & 0 & 1 & 0 \\ 0 & 1 & 0 & 0 \\ 1 & 0 & 0 & 0 \end{bmatrix} \begin{bmatrix} 0 & 0 & 0 & 3 \\ 0 & 0 & 2 & -4 \\ 0 & -1 & 0 & 5 \\ 1 & 3 & -2 & 3 \end{bmatrix}$$

$$= \begin{bmatrix} 1 & 0 & 0 & 0 \\ 0 & -1 & 0 & 0 \\ 0 & 0 & 2 & 0 \\ 0 & 0 & 0 & 3 \end{bmatrix} \begin{bmatrix} 1 & 3 & -2 & 3 \\ 0 & 1 & 0 & -5 \\ 0 & 0 & 1 & -2 \\ 0 & 0 & 0 & 1 \end{bmatrix} = LU$$

Although the upper triangular matrix is in row-echelon form, this need not be the case.

5. For (c), a permutation matrix is product of elementary matrices, and is therefore invertible.

6. If $A = LU$ is an LU-factorization, consider the form of L and U and use the fact that L is invertible. [ILAW: Explorations 1.7.1 and 1.7.2.]

7. Why not use A itself as L? Any candidate for U?

8. Try first on a 2×2 matrix and attempt to interchange the two rows without using the row interchange operation. This does not solve our problem with LU-factorization as you will see that it requires adding a multiple of a row to a row above it. [ILAW: Explorations 1.6.1.]

9. This is a similar idea.

10. A permutation matrix P can be written as product of elementary matrices, so P^{-1} is the product of the inverse of these elementary matrices in reverse order. Check that they are of the right kind.

11. Show first that U is invertible with 1's on the main diagonal. If $A = L_1U_1$ is another LU-factorization, show that $L_1^{-1}L = U_1U^{-1}$ and observe that this is both upper and lower triangular.

12. Once you have an LU-factorization $A = LU$, there are quite a few things you can do keeping L lower triangular and U upper triangular. For example find an invertible diagonal matrix D such that $LD = L'$ is unit lower triangular, so $A = LU = L'(D^{-1}U)$.

1.8 Markov Chains

1. Since there is only rain and shine, there is enough data given to fill in the required Markov chain setup with initial condition $S_0 = \begin{bmatrix} 0.8 \\ 0.2 \end{bmatrix}$ and transition matrix $P = \begin{bmatrix} 0.8 & 0.4 \\ 0.2 & 0.6 \end{bmatrix}$. We can now calculate, rounding off to 2 decimals,

$$S_1 = PS_0 = \begin{bmatrix} 0.72 \\ 0.28 \end{bmatrix}, S_2 = PS_1 = \begin{bmatrix} 0.69 \\ 0.31 \end{bmatrix}, S_3 = PS_2 = \begin{bmatrix} 0.68 \\ 0.32 \end{bmatrix}, \ldots$$

On the other hand, the long term prediction is the state vector S such that $S = PS$, or equivalently $(I - P)S = 0$. The solution of this system of linear equations, where the sum of the entries of S must add to 1, is

$$S = \begin{bmatrix} 2/3 \\ 1/3 \end{bmatrix} \approx \begin{bmatrix} 0.67 \\ 0.33 \end{bmatrix}.$$

It will take until S_5 to equal S with 2 decimals accuracy. [ILAW: Exploration 1.8.1.]

2. We are asking for the state vector S such that $PS = S$. Remember that the entries of S must add to 1 as each entry denotes the probability of being in that state. The procedure is to simply solve the homogeneous system $(I - P)S = 0$, and choosing the solution that adds to 1. [ILAW: Exploration 1.8.2.]
   For (c), the probability it is in state 2 after 3 transitions is $\frac{5}{8}$.

3. You form the graph with n vertices if P is $n \times n$, and draw an edge from v_j to v_i if the (i,j) entry of P is positive. Then P is regular if there is an integer n such that you can go from any vertex in the graph to any other using exactly n edges. Using the graph notions we learned in 1.4.5, you can replace each positive entry of P by 1 to obtain the adjacency matrix of the graph, and the verification that P is regular can be done by finding a large enough power of the adjacency matrix so that all entries are positive. For (d), the matrix is regular, and already the second power of P (or the adjacency matrix) has all entries positive. [ILAW: Exploration 1.4.5.]

4. The transition matrix, with the first column corresponding to beef soup, second to chicken soup, and third to vegetable soup, is

$$P = \begin{bmatrix} 0 & 2/3 & 2/3 \\ 1/2 & 0 & 1/3 \\ 1/2 & 1/3 & 0 \end{bmatrix}.$$

Note that each column of P sums to 1; this is useful in determining P from the data.

Selected Hints and Solutions   527

For (a), $S_0 = \begin{bmatrix} 1 \\ 0 \\ 0 \end{bmatrix}$, $S_1 = \begin{bmatrix} 0 \\ 1/2 \\ 1/2 \end{bmatrix}$, and $S_2 = \begin{bmatrix} 2/3 \\ 1/6 \\ 1/6 \end{bmatrix}$, so the probability of eating beef soup 2 days later is $2/3$.

5. This is a 4 state Markov chain, and if columns correspond in order to regions R_1, R_2, R_3, and R_4, then the transition matrix is

$$P = \begin{bmatrix} 1/2 & 1/6 & 1/6 & 1/4 \\ 0 & 1/2 & 1/6 & 1/8 \\ 0 & 1/6 & 1/2 & 1/8 \\ 1/2 & 1/6 & 1/6 & 1/2 \end{bmatrix}.$$

You should make sure that you understand how to create such a transition matrix from the data. [ILAW: Exploration 1.8.2.]

Question (a) asks for the steady state S such that $PS = S$ (if it exists), because in the long term, the state vectors S_n are very close to S. We get $S = \begin{bmatrix} 0.3 \\ 0.15 \\ 0.15 \\ 0.4 \end{bmatrix}$, giving the corresponding proportion of time the pack hunts in each region. For (b), you need to calculate S_3 as S_0 corresponds to Monday.

[ILAW: You can use Exploration 1.8.2 to simulate what will happen on Thursday if the pack hunts in region R_1 on Monday. Go to the simulation option, and run 3 steps since we are asking to simulate S_3, Monday to Tuesday, Tuesday to Wednesday, and Wednesday to Thursday. Press "Run", and watch what happens. Starting from region R_1 (S_0), the computer randomly selects a new region according to the probabilities of the transition matrix, and takes a count in the matrix N_3 of the region it ends up in and calculates the visited frequencies in M_3. If you let it run long enough, M_3 should get very close to S_3, so such a simulation could be used to approximate S_3.]

6. This is a 3 state Markov chain. The only difficulty is to create the transition matrix, but I am leaving this one to you. Once you think you have it right, go back and read the problem slowly to verify that each entry of the matrix corresponds to the given data.

7. Very similar.

8. This is an application to social classes. Since each transition is measured in 1 generation, you can estimate how long it would take to reach a steady state.

9. There are only two states here, on time or late.

10. Again only two states, Election and No election.

11. This is a very interesting example where the transition matrix

$$P = \begin{bmatrix} 1 & \frac{1}{2} & 0 & 0 & 0 \\ 0 & 0 & \frac{1}{2} & 0 & 0 \\ 0 & \frac{1}{2} & 0 & \frac{1}{2} & 0 \\ 0 & 0 & \frac{1}{2} & 0 & 0 \\ 0 & 0 & 0 & \frac{1}{2} & 1 \end{bmatrix}$$

is not regular; you can verify this either by showing that no power of the matrix has all entries positive, or by considering the associated graph and finding two vertices from which it is impossible to go from one to the other along the same number of edges.

In any case, you can calculate the successive state vectors $S_{n+1} = PS_n$, using S_3 to answer part (b), but also look at S_n for large values of n to make a prediction on the steady state. Even though the transition matrix is not regular for some Markov chains, it is still possible that a steady state vector exists. The condition that P is regular is *sufficient* for the chain to have a steady state, but not *necessary*. Using this steady state vector, make some calculations as to who would profit in the long run in a casino setting. Modify the game to get different outcomes.

12. For (b), assume the (i,i)-entry is 1, and consider the graph and the arrows in and out of state i.

13. For (b), use part (a).
Part (c) is interesting, and shows that a steady state vector always exists for stochastic matrices. The problem is that you may never get close to it by successive iterations of the form $S_{n+1} = PS_n$. To see that such an S exists, we must find a nontrivial solution to $PS = S$, or $(I - P)S = 0$. But the rows add up to 0, so the last row is a combination of the previous ones and the rank of $(I - P)$ is at most $n - 1$, giving infinitely many solutions.

14. Here $P^3 = I$ so $I = P^3 = P^6 = \cdots$, $P = P^4 = P^7 = \cdots$, and $P^2 = P^5 = P^8 = \cdots$. Hence P^n has zero entries for every n, so the chain is not regular.
The graph, shown in the diagram, shows $1 \to 1$ only by $3, 6, 9, \cdots$ transitions, while $1 \to 2$ only by $1, 4, 7, \cdots$ transitions. Thus no number k of transitions carry both $1 \to 1$ and $1 \to 2$. On the other hand, $PS = S$ where $S = \left[\begin{smallmatrix}\frac{1}{3} & \frac{1}{3} & \frac{1}{3}\end{smallmatrix}\right]^T$, so S is a fixed probability vector for the chain. If we start with $S_0 = [1\ 0\ 0]^T$ then $S_0 = S_3 = S_6 = \cdots$, $S_1 = S_4 = S_7 = \cdots$, and $S_2 = S_5 = S_8 = \cdots$. Thus the entries of S_n are all 0's and 1's, and so do not get closer and closer to $\frac{1}{3}$.

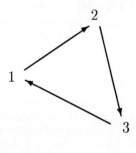

Chapter 2

DETERMINANTS AND DIAGONALIZATION

2.1 The Laplace Expansion

1. $C_{12}(A) = (-1)^{1+2}det(A_{12}) = -1det\begin{bmatrix} 2 & -7 \\ -3 & 4 \end{bmatrix} = -1(8 - 21) = 13$.

2. The lower right entry of an $n \times n$ matrix A is the (n,n) entry, so the sign is $(-1)^{n+n} = (-1)^{2n} = +1$.

3. For (d), $det\begin{bmatrix} a+1 & a \\ a & a-1 \end{bmatrix} = (a+1)(a-1) - a^2 = -1$ for any value of a!

Selected Hints and Solutions

For (k), you can use Exploration 2.1.1 to evaluate the Laplace expansion along the $4th$ row (for example) to get:

$$\det \begin{bmatrix} 1 & 3 & -1 & 1 \\ -1 & 1 & -3 & 1 \\ 5 & 2 & 8 & 2 \\ 2 & 4 & 0 & -1 \end{bmatrix} =$$

$$2(-1)\det \begin{bmatrix} 3 & -1 & 1 \\ 1 & -3 & 1 \\ 2 & 8 & 2 \end{bmatrix} + 4(|1)\det \begin{bmatrix} 1 & -1 & 1 \\ -1 & -3 & 1 \\ 5 & 8 & 2 \end{bmatrix}$$

$$+0(-1)\det \begin{bmatrix} 1 & 3 & 1 \\ -1 & 1 & 1 \\ 5 & 2 & 2 \end{bmatrix} + (-1)(+1)\det \begin{bmatrix} 1 & 3 & -1 \\ -1 & 1 & -3 \\ 5 & 2 & 8 \end{bmatrix}$$

$$= (2)(-1)(-28) + (4)(1)(-14) + (0)(-1)(14) + (-1)(1)(0) = 0$$

Try along another row or column. Try it by hand to see if how much work is involved and whether you can get it right.

4. You should be able to explain to someone how to evaluate a determinant by the Laplace expansion, what are the cofactors and the signs of each matrix entry. Teaching or helping someone is actually a very good way to learn the material.

5. What happens if you evaluate the determinant using the Laplace expansion along that row (or column) of zeros? Check using Exploration 2.1.1.

6. The determinant of a matrix is unchanged if a multiple of a row (or column) is added to a different row (or column). This can simplify hand calculations a bit. For example, in (b), we could do:

$$\det B = \det \begin{bmatrix} -35 & 5 & 7 \\ -10 & -11 & 2 \\ 66 & 13 & -13 \end{bmatrix}$$

$$= \det \begin{bmatrix} -5 & 38 & 1 \\ -10 & -11 & 2 \\ 66 & 13 & -13 \end{bmatrix} \quad \text{(Subtract -3(Row 2) from Row 1)}$$

$$= \det \begin{bmatrix} -5 & 38 & 1 \\ 0 & -87 & 0 \\ 66 & 13 & -13 \end{bmatrix} \quad \text{(Subtract 2(Row 1) fromo Row 2)}$$

$$= -87 \det \begin{bmatrix} -5 & 1 \\ 66 & -13 \end{bmatrix} \quad \text{(Expand along Row 2)}$$

$$= -87(-1)$$

$$= 87.$$

[ILAW: Use Exploration 2.1.2 to verify your computation.]

7. Evaluate some determinants as indicated, and use the results to formulate a conjecture. We will get back to this in the next section. [ILAW: Exploration 2.1.1.]

8. Test with some matrices and give you some intuition into the problems. [ILAW: Exploration 2.1.1.]
   (b) False: Try $A = I_2$.
   (d) True - by Theorem 3.

9. The determinant of a lower or upper triangular matrix is simply the product of its diagonal entries. To prove this, first verify it for 2×2 matrices. Now to proceed by induction, assume that this is true for all $n \times n$ matrices. Given an $(n+1) \times (n+1)$ upper triangular matrix, evaluate its determinant along the first column and see what you get.

10. $det A = 126$. Use Theorem 2 to help understand what should happen, and then verify it directly. For (c), since B is obtained by multiplying row 2 of A by 3, then the determinant of B should be 3 times the determinant of A. [ILAW: Use Exploration 2.1.1.]

11. Since B is obtained from A by multiplying row 1 by 5, we get $det B = 5 det A$. So $det A = \frac{1}{5} det B = 2$. You should get comfortable with this kind problem before you try the next one.

12. Transform each matrix to row-echelon form for example (which is upper triangular). The determinant of an upper triangular matrix is simply the product of the diagonal entries, and the determinant of the original matrix can be traced back using Theorem 2.

    For (b), we can transform A to row-echelon form (with 2 decimal approximation)
    $$R = \begin{bmatrix} 1 & -2 & 0.66 \\ 0 & 1 & 0.35 \\ 0 & 0 & 1 \end{bmatrix} \text{ by}$$

    $$A \xrightarrow{\frac{1}{3}\text{Row 1}} A_1 \xrightarrow{\text{Row 2} - 7\text{Row 1}} A_2 \xrightarrow{\text{Row 3} - 7\text{Row 1}} A_3 \xrightarrow{\frac{1}{13}\text{Row 2}} A_4$$
    $$\xrightarrow{\frac{1}{19}\text{Row 3}} A_5 \xrightarrow{\text{Row 3} - \text{Row 2}} A_6 \xrightarrow{\frac{13}{9}\text{Row 3}} R$$

    Therefore
    $$\begin{aligned} det A &= 3det(A_1) = 3det(A_2) = 3det(A_3) = 39det(A_4) \\ &= 741det(A_5) = 741det(A_6) = 513det(R) = 513. \end{aligned}$$

13. For (b), using row operations, we get
    $$det \begin{bmatrix} 1 & -x & -x \\ x & -2 & -x \\ x & x & -3 \end{bmatrix} = det \begin{bmatrix} 1 & -x & -x \\ 0 & x^2 - 2 & x^2 - x \\ 0 & x^2 + x & x^2 - 3 \end{bmatrix}$$
    $$= 1\left[(x^2 - 2)(x^2 - 3) - (x^2 - x)(x^2 + x)\right]$$
    $$= (x^4 - 5x^2 + 6) - (x^4 - x^2)$$
    $$= -4x^2 + 6$$

    Therefore $det A = 0$ if $x^2 = 3/2$, that is $x = \pm\sqrt{3/2}$.

14. $det \begin{bmatrix} cos\theta & -sin\theta \\ sin\theta & cos\theta \end{bmatrix} = cos^2\theta + sin^2\theta = 1$ for every angle θ.

15. Since A is 3×3, $det(2A) = 2^3 det A = 8 det A$ (take 2 out of each row). We are given that $det(2A) = 6$, so this means that $8 det A = 6$, that is $det A = \frac{6}{8} = \frac{3}{4}$. If A is 4×4, then $det(2A) = 2^4 det A$, and the calculation proceeds. In general, if A is $n \times n$, $det(kA) = k^n det A$ for any scalar k.

Selected Hints and Solutions 531

16. If A is $n \times n$, then $-A$ is obtained by multiplying *each* row of A by -1, so $det(-A) = (-1)^n det A$ by Theorem 3. Now argue that $det(-A) = det A$ if n is even, or else if $det A = 0$.

17. In both cases $det A = ad - bc = det A^T$.

18. Use Theorem 5. The second argument uses (twice) the incorrect assertion that $det(X + Y) = det X + det Y$.

19. Using row and column operations,

$$det \begin{bmatrix} 0 & 1 & 1 & 1 \\ 1 & 0 & x & x \\ 1 & x & 0 & x \\ 1 & x & x & 0 \end{bmatrix} = det \begin{bmatrix} 0 & 1 & 1 & 1 \\ 1 & 0 & x & x \\ 0 & x & -x & 0 \\ 0 & x & 0 & -x \end{bmatrix}$$

$$= 1(-1) det \begin{bmatrix} 1 & 1 & 1 \\ x & -x & 0 \\ x & 0 & -x \end{bmatrix}$$

$$= (-1) det \begin{bmatrix} 1 & 0 & 0 \\ x & -2x & -x \\ x & -x & -2x \end{bmatrix}$$

$$= (-1)(4x^2 - x^2) = -3x^2.$$

20. Use row operations to simplify the first column as in Gaussian Elimination, and then perform Laplace expansion along that column. Use the fact that $y^2 - x^2 = (y - x)(y + x)$ to get a common factor in row 2 (similarly in row 3).

21. You could perform Laplace expansion directly along the first row. A option is to first add x times column 2 to column 1, and then expand along row 1. In the resulting 3×3 matrix, add x^2 times column 2 to column 1. Watch the polynomial "grow" in the lower left corner! Now you see how it works in general.

2.2 Determinant and Inverses

1. Create some matrices and do the computations until you understand what the product formula $det(AB) = det(A) det(B)$ really means.

2. The product formula $det(AB) = det(A) det(B)$ works for any square matrices A and B of the same size, so use it for $B = A^{-1}$ and explain what happens.

3. Create some matrices and do the computations until you understand what the formula $det(A^{-1}) = 1/det(A)$ really means.

4. Use the product formula; for (c), $A^2 = 3A$ implies that $det(A^2) = det(3A)$, so $det(A) det(A) = 3^n det(A)$. We could have $det(A) = 0$, but otherwise we conclude after division that $det(A) = 3^n$.

5. Use the fact that the product formula implies that $det(A^k) = (det A)^k$.

6. For (b):

$$det[B^T A^{-1} B^{-1} C A^2 (C^{-1})^T]$$
$$= det(B^T) det(A^{-1}) det(B^{-1}) det(C) det(A^2) det((C^{-1})^T)$$
$$= det(B)(1/det(A))(1/det(B)) det(C) det(A) det(A)(1/det(C))$$
$$= det(A).$$

7. We use that A is 3×3 to show that $det(2A^{-1}) = 2^3 det(A^{-1}) = 8/det(A)$, and therefore $det(A) = 8/5$.

8. For (b), $adj\,A = \begin{bmatrix} -8 & 2 & 6 \\ -2 & -2 & 4 \\ 10 & 10 & 0 \end{bmatrix}$ and $det\,A = 20$.

9. The adjoint is the *transpose* of the cofactor matrix.

10. Test some matrices and gain intuition into each problem. [ILAW: Explorations 2.2.1 and 2.2.2.]
    For (b), $adj(A)$ always exists, it is the transpose of the cofactor matrix.

11. Since $A^{-1} = \frac{1}{det(A)} adj\,A$, and $adj\,A$ is the transpose of the cofactor matrix, then the (2,3)-entry of A^{-1} must the (3,2)-cofactor divided by the determinant, right?

12. This is just to make sure you haven't forgotten the product formula for determinants.

13. You can compute $det(A)$ directly using the Laplace expansion, and you must show that it cannot be 0 (by Theorem 2).

14. For (b), add column 2 to column 3 to get

$$det\,A = det \begin{bmatrix} 1 & -c & 0 \\ 1 & 1 & 0 \\ c & -c & 1-c \end{bmatrix} = (1-c)(1+c),$$

so A has an inverse exactly when $det(A) \neq 0$, that is $c \neq \pm 1$.
For $c \neq \pm 1$,

$$A^{-1} = \tfrac{1}{det(A)} adj\,A = \tfrac{1}{det(A)} \begin{bmatrix} 1-c & -1-c & -2c \\ c-c^2 & 1-c^2 & c-c^2 \\ 0 & 1+c & 1+c \end{bmatrix}^T$$

$$= \tfrac{1}{det(A)} \begin{bmatrix} 1-c & c-c^2 & 0 \\ -1-c & 1-c^2 & 1+c \\ -2c & c-c^2 & 1+c \end{bmatrix}.$$

15. For (b),

$$y = \frac{det(A_2(B))}{det(A)} = \frac{\begin{bmatrix} 3 & 5 & 4 \\ 5 & 8 & 1 \\ -2 & -3 & 7 \end{bmatrix}}{\begin{bmatrix} 3 & -2 & 4 \\ 5 & 3 & 1 \\ -2 & 6 & 7 \end{bmatrix}} = \frac{-4}{263} \approx -0.0152.$$

16. For (b), use Theorem 3 twice:

$$det \begin{bmatrix} A & 0 & 0 \\ X & B & 0 \\ Z & X & C \end{bmatrix} = det\,A\, det \begin{bmatrix} B & 0 \\ X & C \end{bmatrix} = det\,A(det\,B\, det\,C) = -6.$$

17. Just partition the matrix in order to use Theorem 3.

18. Apply the product theorem to the Hint.

Selected Hints and Solutions

19. For (b), if A is 3×3 and $A^2 = -I$, then $det(A^2) = det(-I)$, so $(det A)^2 = (-1)^3 = -1$, which is impossible because $det A$ is a real number.
20. Write $det A = d$ for convenience, so that $A \, adj A = dI$. Take determinants and apply the product theorem. We get $det[A \, adj(A)] = det[dI]$, and thus $d \, det[adj(A)] = d^n$, and therefore $det(adj(A)) = d^{n-1}$.
21. Proceed as in the last problem.
22. Use $A^{-1} = \frac{1}{det(A)} adj A$.
23. For (b), $A = AI = AB^{-1}B = UB$, where $U = AB^{-1}$ is invertible.

2.3 Diagonalization and Eigenvalues

1. Compute A^{20} directly one power at a time. You can make it a bit faster by computing $AA = A^2$, then cut and paste to compute $A^2A^2 = A^4$, then $A^4A^4 = A^8$, then $A^8A^8 = A^{16}$, and finally $A^{16}A^4 = A^{20}$. [ILAW: Exploration 1.4.1.] For (b), we get $A = PDP^{-1}$, and therefore $A^{20} = PD^{20}P^{-1}$, right? But D^{20} is easily done because it is diagonal. [ILAW: Exploration 2.3.1.]]

2. For (b), $c_A(x) = (x-3)(x+2)$, and therefore the eigenvalues are $\lambda = 3, -2$, with eigenvectors $X_1 = \begin{bmatrix} 1 \\ -1 \end{bmatrix}$ and $X_2 = \begin{bmatrix} 1 \\ 4 \end{bmatrix}$. A diagonalizing matrix is $P = \begin{bmatrix} 1 & 1 \\ -1 & 4 \end{bmatrix}$.

   For (d), $c_A(x) = (x-1)(x-2)(x-3)$, and therefore the eigenvalues are $\lambda = 1, 2, 3$, with eigenvectors $X_1 = \begin{bmatrix} 1 \\ 1 \\ 1 \end{bmatrix}$, $X_2 = \begin{bmatrix} 2 \\ 1 \\ 1 \end{bmatrix}$, and $X_3 = \begin{bmatrix} 1 \\ 1 \\ 2 \end{bmatrix}$. A diagonalizing matrix is $P = \begin{bmatrix} 1 & 2 & 1 \\ 1 & 1 & 1 \\ 1 & 1 & 2 \end{bmatrix}$.

3. Make sure you clearly understand each step of the process; you should be able to describe it in your own words.

4. Read carefully the discussion before Theorem 3; the equation $P^{-1}AP = D$ implies that the diagonal entries of D are the eigenvalues of A, and that the columns of P are corresponding eigenvectors. There is no other way to diagonalize a matrix except to shuffle the order of the eigenvalues and corresponding eigenvectors.

5. For (b), $c_A(x) = (x-1)^2$ so $\lambda = 1$ is an eigenvalue of multiplicity 2. But $\lambda I - A = 1I - A \begin{bmatrix} 0 & -c \\ 0 & 0 \end{bmatrix}$, so the solutions of $(\lambda I - A)X = 0$ is a one-parameter family $t \begin{bmatrix} 0 \\ 1 \end{bmatrix}$ if $c \neq 0$. So A is not diagonalizable by Theorem 5.

6. For (b), the characteristic polynomial is $c_A(x) = (x+1)^2(x-3)$, so $\lambda_1 = -1$ is an eigenvalue of multiplicity 2. But $(-1)I - A = \begin{bmatrix} -3 & -2 & 1 \\ -1 & -2 & 0 \\ -1 & 2 & 1 \end{bmatrix}$, so $((-1)I - A)X = 0$ has general solution $X = t \begin{bmatrix} 2 \\ -1 \\ 4 \end{bmatrix}$, and there is only one

basic eigenvector $X_1 = \begin{bmatrix} 2 \\ -1 \\ 4 \end{bmatrix}$ corresponding to this eigenvalue. Thus A is not diagonalizable by Theorem 5.

7. For (a) and (e), we have already seen counterexamples in previous exercises. [ILAW: Explorations 2.2.1, 2.3.2 and 2.3.3.]

8. Let $AX = \lambda X$, $X \neq 0$. Then $A_1 X = (A - \alpha I)X = AX - \alpha X = \lambda X - \alpha X = (\lambda - \alpha)X$. Same eigenvectors.

9. The eigenvalues are $\cos\theta \pm i\sin\theta$.

10. We have $AX = \lambda X$ and $AY = \lambda Y$, so compute $A(X+Y)$ and $A(aX)$.

11. We have $AX = \lambda X$ for some $X \neq 0$. For (b), we then have $A^2 X = A(AX) = A(\lambda X) = \lambda(AX) = \lambda(\lambda X) = \lambda^2 X$, so λ^2 is an eigenvalue of A^2.

12. For (b), if μ is an eigenvalue of A^{-1}, then part (a) shows that $\lambda = \frac{1}{\mu}$ is an eigenvalue of $(A^{-1})^{-1} = A$, so $\mu = \frac{1}{\lambda}$ is of the desired form.

13. For (b),

$$DE = \begin{pmatrix} \lambda_1 & & \\ & \ddots & \\ & & \lambda_n \end{pmatrix} \begin{pmatrix} \mu_1 & & \\ & \ddots & \\ & & \mu_n \end{pmatrix} = \begin{pmatrix} \lambda_1\mu_1 & & \\ & \ddots & \\ & & \lambda_n\mu_1 \end{pmatrix}.$$

14. (b) Let $P^{-1}AP = D$ be diagonal. Then $D = D^T = P^T A^T (P^{-1})^T = P^T A^T (P^T)^{-1} = Q^{-1} A^T Q$ where $Q = (P^T)^{-1}$. Since Q is invertible, this shows that A^T is diagonalizable.

15. Let $P^{-1}AP = D = diag(\lambda_1, \lambda_2, \cdots, \lambda_n)$, then $A = PDP^{-1}$, so compute $det A$.

16. (a) If $AX = \lambda X$ for some column $X \neq 0$, show that $A^2 X = \lambda^2 X$. What is A^2 here?
    (b) Proceed as in Example 10.

17. (a) If $AX = \lambda X$ for some column $X \neq 0$, show that $\lambda X = \lambda^2 X$.
    (b) Proceed as in Example 10.

18. For (b) $A = I^{-1}AI$.
    For (d), if $B = P^{-1}AP$ and $C = A^{-1}BQ$, then $C = (PQ)^{-1}A(PQ)$.
    For (g), if $B = P^{-1}AP$ then $B^k = (P^{-1}AP)^k = P^{-1}A^k P$, using Theorem 1.

19. Compute the characteristic polynomial of the matrix A and verify that it has the required form. Then use the quadratic formula to compute the roots.

20. (a) Compare off-diagonal entries in $CD = DC$.
    (b) If $P^{-1}AP = D$ is diagonal, let $C = P^{-1}BP$, and apply (a).

Selected Hints and Solutions   535

2-4 Linear Dynamical Systems

1. The growth matrix is $A = \begin{bmatrix} \frac{1}{2} & \frac{1}{3} \\ 2 & 0 \end{bmatrix}$. [ILAW: Explorations 2.4.1 and 2.4.2.]

   For (a), the initial population is $V_0 = \begin{bmatrix} 100 \\ 60 \end{bmatrix}$. We can calculate by hand

   $$V_1 = AV_0 = \begin{bmatrix} 70 \\ 200 \end{bmatrix}$$

   $$V_2 = AV_1 \approx \begin{bmatrix} 101.67 \\ 140 \end{bmatrix}$$

   $$V_3 = AV_2 \approx \begin{bmatrix} 97.5 \\ 203.33 \end{bmatrix} \cdots$$

   For (b), we need to look at the eigenvalues. The characteristic polynomial of A is

   $$c_A(x) = \det \begin{bmatrix} x - \frac{1}{2} & -\frac{1}{3} \\ -2 & x \end{bmatrix} = x^2 - \frac{1}{2}x - \frac{2}{3}.$$

   By the quadratic formula, the roots are

   $$\lambda = \frac{1}{2}\left[\frac{1}{2} \pm \sqrt{\frac{1}{4} + \frac{8}{3}}\right] = \frac{1}{4}\left[1 \pm \sqrt{\frac{35}{3}}\right],$$

   so the eigenvalues are $\lambda_1 \approx 1.104$ and $\lambda_2 \approx -.604$. Since $\lambda_1 > 1$, we know already that the population will become very large as k increases, no matter what the initial population was, answering (b) and (c). The eigenvectors corresponding to λ_1 and λ_2 are $X_1 \approx \begin{bmatrix} 1.104 \\ 2 \end{bmatrix}$ and $X_2 \approx \begin{bmatrix} -.604 \\ 2 \end{bmatrix}$, so the diagonalizing matrix P is $P = \begin{bmatrix} 1.104 & -.604 \\ 2 & 2 \end{bmatrix}$. Since we are assuming that $V_0 = \begin{bmatrix} 100 \\ 60 \end{bmatrix}$, the coefficients b_1 and b_2 are given by

   $$\begin{bmatrix} b_1 \\ b_2 \end{bmatrix} = P^{-1}V_0 \approx \begin{bmatrix} .5855 & .1768 \\ -.5855 & .3232 \end{bmatrix} \begin{bmatrix} 100 \\ 60 \end{bmatrix} = \begin{bmatrix} 69.16 \\ -39.16 \end{bmatrix}.$$

   Therefore, the population can be approximated by:

   $$\begin{bmatrix} a_k \\ j_k \end{bmatrix} = V_k \approx b_1 \lambda_1^k X_1 = 69.16 \, (1.104)^k \begin{bmatrix} 1.104 \\ 2 \end{bmatrix}.$$

   Comparing entries gives

   $$a_k \approx 76.35 \, (1.104)^k \quad \text{and} \quad j_k \approx 138.32 \, (1.104)^k$$

   for sufficiently large k. This gives $a_k + j_k \approx 214.67 \, (1.104)^k$ as the total female population, and also shows that the breakdown of the population into adults and juveniles stabilizes with approximately $\frac{76.35}{214.67} = 0.356 = 35.6\%$ adults and $\frac{138.32}{214.67} = .644 = 64.4\%$ juveniles.

2. This analysis is very similar, but in this case the dominant eigenvalue is less than 1 and therefore the population is heading for extinction. [ILAW: Explorations 2.4.1 and 2.4.2.]

3. The population will stabilize if the dominant eigenvalue is 1.

4. This is a slightly more general analysis, and each case depends on whether the dominant eigenvalue is 1, less than 1, or larger than 1.

5. If we let $V_k = \begin{bmatrix} h_k \\ m_k \end{bmatrix}$, and $A = \begin{bmatrix} .55 & .005 \\ -18 & 1.2 \end{bmatrix}$, then the matrix recurrence takes the form

$$V_{k+1} = AV_k \text{ where } V_0 = \begin{bmatrix} 30 \\ 2000 \end{bmatrix}$$

Find the characteristic polynomial of the matrix A, and verify that the eigenvalues are $\lambda_1 = 1$ and $\lambda_2 = 0.75$, with corresponding eigenvectors $X_1 = \begin{bmatrix} 1 \\ 90 \end{bmatrix}$ and $X_2 = \begin{bmatrix} 1 \\ 40 \end{bmatrix}$. Since 1 is the dominant eigenvalue, we know that an equilibrium will be reached. This could also have been verified by the condition $\gamma\delta = (1-p)(q-1)$. An exact formula for V_k can be obtained as

$$V_k = b_1\lambda_1^k X_1 + b_2\lambda_2^k X_2 = 16(1)^k \begin{bmatrix} 1 \\ 90 \end{bmatrix} + 14(0.75)^k \begin{bmatrix} 1 \\ 40 \end{bmatrix}$$

and using the dominant eigenvalue approximated by

$$V_k \approx b_1\lambda_1^k X_1 = 16(1^k) \begin{bmatrix} 1 \\ 90 \end{bmatrix} = \begin{bmatrix} 16 \\ 1440 \end{bmatrix}$$

giving the equilibrium populations.

6. Starting with $V_0 = \begin{bmatrix} a \\ b \end{bmatrix}$, calculate V_1, V_2, V_3 and V_4.

2.5 Complex Eigenvalues

1. (f). $\frac{3-5i}{7-i} = \frac{3-5i}{7-i} \frac{7+i}{7+i} = \frac{21+3i-35i+5}{49+1} = \frac{13-16i}{25}$.
   (j). Compute $\left|\frac{z}{|z|}\right|$ for a few different values of z. Make a conjecture and verify it.
   [ILAW: Exploration 2.5.1.]

2. (b). $(2+xi)^2 = (4-x^2) + 4xi$. If this is equal to 4, then $4xi$ must be 0, and hence $x = 0$. By chance $x = 0$ works fine. Note that the equation $(2+xi)^2 = 6$ for example would have no real solution x.

3. For (b), the equation reduces to $z = \frac{1+5i}{1+2i} = \frac{11+3i}{5}$.
   (d). $z = 1 + i$.

4. $z = \pm i$.

5. (b) If $z = a + bi$, then $z - \bar{z} = 2bi$ is a complex number with no real part, that is pure imaginary.

6. For (b), $zw = (aa' - bb') + (ab' + ba')i$, and therefore $\overline{zw} = (aa' - bb') - (ab' + ba')i$. On the other hand, $\bar{z}\bar{w} = (a-bi)(a'-b'i) = (aa'-bb') + (-ab'-ba')i = (aa'-bb') - (ab'+ba')i$. This verifies C2.

Selected Hints and Solutions

7. In each case, test various hypothesis with concrete examples to help you gain intuition in these questions.
   For (b), think about what happens with i^2.
   For (d), consider $z = e^{i\theta}$.
   [ILAW: Exploration 2.5.1.]

8. The roots are given by the quadratic formula. Observe that (d) and (e) have the same roots, right?
   [ILAW: Exploration 2.5.1.]

9. Complex roots of real quadratics come in pairs, λ and $\bar{\lambda}$. For (d), the polynomial is therefore $(x-\lambda)(x-\bar{\lambda}) = x^2 - 2\operatorname{re}\lambda + |\lambda|^2 = x^2 - 4x + 29$. Verify your answer.
   [ILAW: Exploration 2.5.2.]

10. (b). The augmented matrix is $\begin{bmatrix} 1 & 1+i & -i & 2 \\ 1+i & 0 & 1-i & 3+i \\ -1 & 1-i & i & -1+i \end{bmatrix}$ and the reduced row-echelon form $\begin{bmatrix} 1 & 0 & -i & 2-i \\ 0 & 1 & 0 & 0.5+0.5i \\ 0 & 0 & 0 & 0 \end{bmatrix}$. The solutions are therefore

$$X = \begin{bmatrix} 2-i+is \\ 0.5+0.5i \\ s \end{bmatrix}$$ for any *complex* value of the parameter s!
   [ILAW: Exploration 1.2.3.]

11. (b). $\frac{1}{2i}\begin{bmatrix} 3+i & -1+i \\ -2-i & 1 \end{bmatrix} = \frac{1}{2}\begin{bmatrix} 1-3i & 1+i \\ -1+2i & -i \end{bmatrix}$.
    [ILAW: Exploration 1.5.3.]

12. (d). The eigenvalues are $\lambda_1 = 1$, $\lambda_2 = i$ and $\lambda_3 = -i$ with corresponding eigenvectors

$$X_1 = \begin{bmatrix} 1 \\ 1 \\ 0 \end{bmatrix}, \quad X_2 = \begin{bmatrix} 2 \\ 1 \\ 2-i \end{bmatrix}, \quad \text{and} \quad X_3 = \begin{bmatrix} 2 \\ 1 \\ 2+i \end{bmatrix}.$$

   [ILAW: Exploration 2.3.3.]

13. For (c), one such matrix is $P = \begin{bmatrix} i & -i \\ 1 & 1 \end{bmatrix}$.

14. Look at each entry and use the similar properties for complex numbers, $\overline{zw} = \bar{z}\bar{w}$, and $\overline{z+w} = \bar{z}+\bar{w}$.
    [ILAW: Exploration 1.4.1.]

15. If $AX = \lambda X$ for some complex $X \neq 0$, show that $A\bar{X} = \bar{\lambda}\bar{X}$ where $\bar{X} \neq 0$.

16. Use C1, C2 and C4 to show that $f(\bar{\lambda}) = \overline{f(\lambda)}$.

17. If $z = a+bi$, then $\bar{z} = a-bi$, so compute both sides.

18. If $z = a+bi$, then the requirement that $|z| = 1$ is equivalent to $a^2 + b^2 = 1$. Therefore these z's are exactly those complex numbers on the unit circle in the complex plane.

19. Use $z = a+bi$ and $w = c+di$, and compute both sides. Alternatively, apply C2 to $|\frac{z}{w} \times w| = |z|$.

20. (b). $2e^{-\pi i/2} = 2e^{3\pi i/2}$.
    (d). $2\sqrt{3}e^{7\pi i/6}$.

21. (b). $\frac{1}{2} - \frac{\sqrt{3}}{2}i$.
    (d). $1 - i$.

22. (b). $-\frac{1}{32}(1 + \sqrt{3}i)$.
    (d). $-i$.

23. (b). $1, \frac{1}{2}(-1 + \sqrt{3}i), \frac{1}{2}(-1 - \sqrt{3}i)$.

24. (b). $\sqrt{2}e^{\pi i/6}, \sqrt{2}e^{4\pi i/6}, \sqrt{2}e^{7\pi i/6}, \sqrt{2}e^{10\pi i/6}$.

25. (b). DeMoivre's Theorem gives $(e^{i\theta})^3 = e^{3\theta i}$.

26. Use $e^{i\theta} = cos\theta + isin\theta$.

2.6 Linear Recurrences

1. This is a two-place recurrence, and we can express it as $V_{k+1} = AV_k$, where $A = \begin{bmatrix} 0 & 1 \\ 6 & 1 \end{bmatrix}$ and $V_k = \begin{bmatrix} x_k \\ x_{k+1} \end{bmatrix}$. A diagonalizing matrix for A is $P = \begin{bmatrix} 1 & 1 \\ 3 & -2 \end{bmatrix}$. For (b), this gives the formula $V_k = \frac{3}{5}3^k \begin{bmatrix} 1 \\ 3 \end{bmatrix} + \frac{2}{5}(-2)^k \begin{bmatrix} 1 \\ -2 \end{bmatrix}$ giving $x_k = \frac{1}{5}\left[3^{k+1} - (-2)^{k+1}\right]$, $k \geq 0$.
   [ILAW: Exploration 2.6.1.]

2. For (b), $x_k = \frac{1}{3}(4 - (-2)^k)$.

3. We now have a three place relation. For (b), it can be expressed by $V_{k+1} = AV_k$, where $A = \begin{bmatrix} 0 & 1 & 0 \\ 0 & 0 & 1 \\ 2 & 1 & -2 \end{bmatrix}$ and $V_k = \begin{bmatrix} x_k \\ x_{k+1} \\ x_{k+2} \end{bmatrix}$. The eigenvalues of A are ± 1 and 2, and the formula (with 2 decimal approximation)

$$V_k = \frac{1}{2}(-1)^k \begin{bmatrix} 1 \\ -1 \\ 1 \end{bmatrix} + (0)(-2)^k \begin{bmatrix} 1 \\ -2 \\ 4 \end{bmatrix} + \frac{1}{2}(1)^k \begin{bmatrix} 1 \\ 1 \\ 1 \end{bmatrix}.$$

By looking at the top entries, we get $x_k = \frac{1}{2}(-1)^k + \frac{1}{2} = \frac{1}{2}((-1)^k + 1)$. The value of x_k therefore only depends on whether k is odd or even, giving 1 for k even, and 0 otherwise.

4. These are very interesting recurrences where complex numbers are used to (diagonalize the matrix and) compute a recurrence of ordinary integers. For (b), we obtained the formula (with 2 decimals approximation)

$$V_k \approx (0.11)(-0.69)^k \begin{bmatrix} 2.06 \\ -1.43 \\ 1 \end{bmatrix}$$

$$+(0.44 - 1.03i)(1.34 - 1.02i)^k \begin{bmatrix} 0.09 + 0.33i \\ 0.46 + 0.35i \\ 1 \end{bmatrix}$$

$$+(0.44 + 1.03i)(1.34 + 1.02i)^k \begin{bmatrix} 0.09 - 0.33i \\ 0.46 - 0.35i \\ 1 \end{bmatrix}$$

Although complex numbers are involved in the formula, notice that the result is always an integer!

5. For (a), check that $AX_i = \lambda_i X_i$ directly, and that λ_i is a root of the characteristic polynomial.

6. Note that the x_k are all real even though the formula involves non real numbers!

7. In Exercises 7-14, you must set up the recurrence for the numbers in question, and then solve it. In each case, enough information is provided to do this. In Exercise 7, $s_1 = 1$ (just one step) and $s_2 = 2$ (11 and 2). The general case can be argued as follows: One reaches the last step by either climbing the last step alone, or the last two steps at once. This gives the recurrence $s_{k+2} = s_k + s_{k+1}$.

8. Such a word either ends with a "b", or else the second last letter is a "b".

9. Look at the preceding problem.

10. Show that $x_{k+2} = 2x_{k+1} + 2x_k$ by considering how many stacks have a red, blue or gold chip on top.

11. Let a_k and b_k denote the number of α- and β-particles at time $t = k$ seconds. Find a_{k+1} and b_{k+1} in terms of a_k and b_k.

12. Can you tell when was the best year?

13. Consider the cases $r = 1$ and $r \neq 1$ separately. If $r \neq 1$ you will need the identity $1 + r + r^2 + \cdots + r^{n-1} = \frac{1-r^n}{1-r}$ for $n \geq 1$.

14. This is an extension of the methods presented in this section. For (b), we let $y_k = x_k + 5/6$, and obtain the recurrence $y_{k+2} = y_{k+1} + 6y_k$, $y_0 = y_1 = 11/6$. We get for example $y_{10} = 65704 + \frac{5}{6}$, and therefore $x_{10} = 65704$.

2.7 Polynomial Interpolation

1. These are straightforward polynomial interpolations.
   [ILAW: Explorations 2.7.1 and 1.2.3.]

   For (c), there are 4 data points and we therefore need an interpolating polynomial of degree $4 - 1 = 3$. Writing $p(x) = ax^3 + bx^2 + cx + d$, the requirements that p passes through the 4 points can be written in the following system of linear equations:

$$\begin{array}{rrrrl} a0^3 & +b0^2 & +c0 & +d & = 1 \\ a1^3 & +b1^2 & +c1 & +d & = 1 \\ a(-1)^3 & +b(-1)^2 & +c(-1) & +d & = 4 \\ a2^3 & +b2^2 & +c2 & +d & = 5 \end{array}$$

   The unique solution gives $p(x) = \frac{1}{6}x^3 + \frac{3}{2}x^2 - \frac{5}{3}x + 1$, and therefore we approximate the value at $x = \frac{1}{2}$ by $p(\frac{1}{2}) = \frac{9}{16}$.

2. For (a), the interpolating values for weeks 4 and 10 are \$4.1 and \$28.49 respectively. Adding the data point for week 4, the new interpolation gives a value of \$ -0.91, answering (b). This is certainly very disconcerting, and we must conclude that polynomial interpolation must be used with care. It is best used to interpolate values very close to known ones, thus avoiding the disparity shown here.

3. For (a), you must interpolate for example through the data points $(0, 40)$, $(30, 7.5)$, and $(35, 5)$.

   For (b), this fourth condition will require a polynomial of degree 3 of the form $p(x) = ax^3 + bx^2 + cx + d$. The 4 requirements can be written as $p(0) = 40$, $p(30) = 7.5$, $p(35) = 5$, and $p'(35) = 0$, where $p'(x) = 3ax^2 + 2bx + c$ is the first derivative of $p(x)$. The unique solution is (with 2 decimal approximation)
   $$p(x) \approx 0.0024x^3 - 0.14x^2 + 0.92x + 40$$

4. For (a), we obtain (with 2 decimals approximation):
   $$s_1 = -0.21x^3 + 0x^2 + 2.1x + 1 \quad \text{and} \quad s_2 = -0.21x^3 + 0x^2 + 2.21x + 1$$
   For (b), we obtain (with loads scaled to 1000 kg)
   $$s_1 = s_2(x) = \frac{7}{60}x^3 - \frac{7}{20}x^2 + \frac{5}{6}x + \frac{9}{10}$$
   giving $s(2.2) = 2.282$, slightly smaller than the result obtained from polynomial interpolation.

5. As in Example 2, the determinant of the Vandermonde matrix corresponding to the numbers $x_1, x_2, \cdots, x_n$ is the product of the form $(x_i - x_j)$ for all possible $j < i$. Therefore, if A is invertible, then its determinant is nonzero, and hence all x_i's must be distinct.

2.8 Systems of Differential Equations

1. (b) $c_1 \begin{bmatrix} 1 \\ 1 \end{bmatrix} e^{4x} + c_2 \begin{bmatrix} 5 \\ -1 \end{bmatrix} e^{-2x}$.

   (d) $c_1 \begin{bmatrix} -8 \\ 10 \\ 7 \end{bmatrix} e^{-x} + c_2 \begin{bmatrix} 1 \\ -2 \\ 1 \end{bmatrix} 2^{2x} + c_3 \begin{bmatrix} 1 \\ 0 \\ 1 \end{bmatrix} e^{4x}$.

2. Extend the construction following Example 2.

Selected Hints and Solutions 541

Chapter 3
VECTOR GEOMETRY

3.1 Geometric Vectors

1. For the 2-dimensional XY-plane, a point is specified by a first coordinate on the X-axis and the second on the Y-axis, and therefore the order is important. For (c), the coordinates of the terminal point of $\overrightarrow{PQ}$ in standard position are $(-18, -4)$, giving -18 as its X-component and -4 as its Y-component, and its matrix form as $\overrightarrow{PQ} = \begin{bmatrix} -18 \\ -4 \end{bmatrix}$. The length of $\overrightarrow{PQ}$, which is exactly the distance from P to Q, is $\left\lVert \overrightarrow{PQ} \right\rVert = \sqrt{(-18)^2 + (-4)^2} = \sqrt{340}$.
[ILAW: Explorations 3.1.1 to 3.1.6.]

2. We have $\vec{p} = \begin{bmatrix} 3 \\ -2 \end{bmatrix}$ and $\vec{q} = \begin{bmatrix} -1 \\ 6 \end{bmatrix}$, and therefore

$$\vec{p} + \vec{q} = \begin{bmatrix} 3 \\ -2 \end{bmatrix} + \begin{bmatrix} -1 \\ 6 \end{bmatrix} = \begin{bmatrix} 2 \\ 4 \end{bmatrix}$$

and
$$\lVert \vec{p} + \vec{q} \rVert = \sqrt{2^2 + 4^2} = \sqrt{20} = 2\sqrt{5}.$$

3. For (c), $\overrightarrow{PQ} = [-2 \; 5 \; 12]^T$, and $\left\lVert \overrightarrow{PQ} \right\rVert = \sqrt{(-2)^2 + 5^2 + 12^2} = \sqrt{173}$.

4. For (b), write $P(x, y, z)$, so that $\overrightarrow{PQ} = [-x \; -1-y \; 3-z]^T$. To have $\overrightarrow{PQ} = [3 \; 0 \; -2]^T$, we must therefore have $x = -3$, $y = -1$, and $z = 5$.

5. For (d), $\frac{1}{2}(\vec{u} + \vec{v} + 2\vec{w}) = [-4 \; 3/2 \; 2]^T$, and therefore $\left\lVert \frac{1}{2}(\vec{u} + \vec{v} + 2\vec{w}) \right\rVert = \sqrt{(-4)^2 + (3/2)^2 + 2^2} = \frac{1}{2}\sqrt{89}$.

6. (b). Since $\lVert a\vec{v} \rVert = |a| \lVert \vec{v} \rVert$ by Theorem 1, we have $\left\lVert \frac{1}{\lVert \vec{v} \rVert} \vec{v} \right\rVert = \frac{1}{\lVert \vec{v} \rVert} \lVert \vec{v} \rVert = 1$ for any nonzero vector $\vec{v}$.

7. Since vector operations are just matrix operations, they obey the same properties covered in Chapter 1. For (b), $5(\vec{u}-\vec{v}-2\vec{w}) + 3(2\vec{v}+3\vec{w}-3\vec{u}) - 2(3\vec{u}+\vec{v}+8\vec{w}) = (5-9-6)\vec{u} + (-5+6-2)\vec{v} + (-10+9-16)\vec{w} = -10\vec{u} - \vec{v} - 17\vec{w}$.

8. Since vector operations are those of matrix operations, we can do a bit of algebra to solve for $\vec{x}$. For (c), $2\vec{x} - 3\vec{v} = \lVert \vec{u} \rVert^2 \vec{u} + 4 \lVert \vec{u} \rVert^2 \vec{x}$, and therefore $(2 - 4\lVert \vec{u} \rVert^2)\vec{x} = 3\vec{v} + \lVert \vec{u} \rVert^2 \vec{u}$, that is $\vec{x} = \frac{1}{(2-4\lVert \vec{u} \rVert^2)}[3\vec{v} + \lVert \vec{u} \rVert^2 \vec{u}] =$, giving finally $\vec{x} = \frac{-1}{150}[-73 \; 108 \; -184]^T$.

9. This kind of question will be with us again in later chapters. It is asking whether the vector $\vec{x}$ can be written as a *linear combination* of the vectors $\vec{u}$, $\vec{v}$, and $\vec{w}$. For (c), we are asked to find a, b and c such that

$$[9 \; 3 \; 0]^T = a[2 \; -1 \; 3]^T + b[0 \; 2 \; -3]^T + c[6 \; 1 \; 3]^T$$

which is identical to solving the system of linear equations

$$\begin{array}{rrrcl} 2a & +0b & +6c & = & 9 \\ -a & +2b & +c & = & 3 \\ 3a & -3b & +3c & = & 0 \end{array}$$

By the methods of Chapter 1, we see that there is no solution in this case. [ILAW: Exploration 1.2.3.]

10. Two vectors are parallel if they have same or opposite direction, but an easy way to verify this is to check whether the two vectors are multiples of each other. For (c), if $\vec{v} = a\vec{u}$, then the first component gives $a = -4$, but the second component gives $a = +4$; this contradiction shows that the two vectors cannot be parallel.

11. Since $\vec{u}$ is parallel to $\vec{v}$, we must have $\vec{u} = a\vec{v}$ for some number a. Therefore $\|\vec{u}\| = |a|\|\vec{v}\|$, and $\|\vec{u}\| = 3\|\vec{v}\|$ translates into $|a|\|\vec{v}\| = 3\|\vec{v}\|$. Since $\vec{v} \neq \vec{0}$, we conclude that $a = \pm 3$ and the only possibilities are $\vec{u} = \pm 3\vec{v}$.

12. Since $\overrightarrow{PQ} = a\vec{d}$ for some scalar a, we have $5 = \left\|\overrightarrow{PQ}\right\| = |a|\|\vec{d}\|$. For (b), this gives $a = \pm 5/\sqrt{13}$. Writing $Q(x, y, z)$, we must have $[x - 1 \ y - 2 \ z + 5]^T = \pm \frac{5}{\sqrt{13}}[0 \ -3 \ 2]^T$, giving $Q(1, 2 - \frac{15}{\sqrt{13}}, -5 + \frac{10}{\sqrt{13}})$, or $Q(1, 2 + \frac{15}{\sqrt{13}}, -5 - \frac{10}{\sqrt{13}})$.

13. Writing $S(x, y, z)$, we must have for example $\overrightarrow{PS} = \overrightarrow{QR}$. For (b), this gives $[x - 2 \ y \ z + 1]^T = [3 + 2 \ -1 - 4 \ 0 - 1]^T$, and finally $S(7, -5, -2)$.

14. If $M(x, y, z)$ is the midpoint between P and Q, then $\overrightarrow{PM} = \frac{1}{2}\overrightarrow{PQ}$. For (b), this gives $[x-1 \ y-2 \ z-3]^T = \frac{1}{2}[-3-1 \ 5-2 \ -3-3]^T$, and finally $M(-1, 7/2, 0)$.

15. For (a), if T is the point a third of the way from P to Q, then $\overrightarrow{PT} = \frac{1}{3}\overrightarrow{PQ}$, giving $T(\frac{2}{3}, \frac{-1}{3}, \frac{11}{3})$.

16. If A is the point a fraction r of the way from P to Q, then $\overrightarrow{PA} = r\overrightarrow{PQ}$. Using position vectors $\vec{a}, \vec{p}$, and $\vec{q}$ for the points A, P and Q respectively, this translates into $\vec{a} - \vec{p} = r(\vec{q} - \vec{p})$, giving $\vec{a} = (1 - r)\vec{p} + r\vec{q}$.

17. (d). False. Try $\vec{w} = -\vec{v}$.
    (f). False. The direction of $t\vec{v}$ is the same as $\vec{v}$ only when t is positive, and opposite when t is negative.
    (g). False. There is a subtle case when $\vec{w} = \vec{0}$ and $\vec{v} \neq \vec{0}$, because here $\vec{v}$ is parallel to $\vec{v} = \vec{v} + \vec{w}$, but no vector is parallel to $\vec{0}$ by definition.
    [ILAW: Explorations 3.1.1 to 3.1.6.]

18. For (b), $\overrightarrow{EF} = \overrightarrow{EB} + \overrightarrow{BF} = \frac{1}{2}\overrightarrow{AB} + \frac{1}{2}\overrightarrow{BC} = \frac{1}{2}(\overrightarrow{AB} + \overrightarrow{BC}) = \frac{1}{2}\overrightarrow{AC}$. This shows that $\overrightarrow{EF}$ is parallel to $\overrightarrow{AC}$ and half as long.

19. First $\overrightarrow{AE} = \overrightarrow{BD}$ by the symmetry of the hexagon. Hence $\overrightarrow{AB} + \overrightarrow{AE} = \overrightarrow{AB} + \overrightarrow{BD} = \overrightarrow{AD}$. Similarly, $\overrightarrow{AC} + \overrightarrow{AF} = \overrightarrow{AD}$.

20. $\mathcal{C}$ consists of exactly the points P in the plane at a distance r from its center, here the origin, and therefore $\|\vec{p}\| = r$.

21. Proceed similarly to the previous exercise.

3.2 Dot Product and Projections

1. For (f), $\vec{v} \bullet \vec{w} = (3)(4) + (-1)(2) + (5)(-2) = 0$.
   [ILAW: Exploration 3.2.1.]

2. For (d), $\vec{v} \bullet \vec{v} = (1)(1) + (-1)(-1) + (1)(1) = 3$, and $\|\vec{v}\| = \sqrt{\vec{v} \bullet \vec{v}} = \sqrt{3}$.

3. We know that the angle between two vectors is acute, a right angle, or obtuse, depending on whether their dot product is respectively positive, zero, or negative. For (h), $\vec{v} \bullet \vec{w} = -43$, and therefore the angle between the two vectors is obtuse, larger than $\pi/2$ and no more than π radians.

Selected Hints and Solutions   543

4. Use the formula $\cos\theta = \frac{\vec{v}\bullet\vec{w}}{\|\vec{v}\|\|\vec{w}\|}$.
   For (d), we get $\cos\theta = 0.4749$, and thus $\theta \approx 0.34\pi$. This is close to $60°$.
   [ILAW: Exploration 3.2.2.]

5. The easiest method we have to guarantee two vectors to be orthogonal is to ensure that their dot product is zero. Therefore $\vec{u}$ perpendicular to $\vec{v}$ translates into $2a - b + 3c = 0$, and $\vec{u}$ perpendicular to $\vec{w}$ translates into $a + b - 2c = 0$. Solving this homogeneous system of linear equations will produce all possible values of a, b, and c as required.
   [ILAW: Exploration 1.3.1.]

6. The angle at vertex A is the angle between the two vectors $\overrightarrow{AB}$ and $\overrightarrow{AC}$, and similarly for each vertex. You need to go around and find which vertex has a right angle.

7. The line through A and B is parallel to the vector $\overrightarrow{AB}$, and similarly for C and D. Therefore we only need to verify that the vectors $\overrightarrow{AB}$ and $\overrightarrow{CD}$ are perpendicular.

8. As in the last exercise, use the vectors $\overrightarrow{AB}$ and $\overrightarrow{CD}$. For (d), $\overrightarrow{AB} = [10\ -5\ -3]^T$ and $\overrightarrow{CD} = [1\ -1\ 5]^T$, giving $\overrightarrow{AB}\bullet\overrightarrow{CD} = 0$. The two vectors, and hence the two lines, are perpendicular. For (f), the lines are parallel because $\overrightarrow{CD} = (-2)\overrightarrow{AB}$ (see Theorem 3 §3.1).

9. For (d), $proj_{\vec{w}}\vec{v} = \frac{\vec{v}\bullet\vec{w}}{\|\vec{w}\|^2}\vec{w} = \frac{38}{57}[5\ -4\ 4]^T$.
   [ILAW: Exploration 3.2.3.]

10. The required vector $\vec{v}_1$ is the projection of $\vec{v}$ on $\vec{w}$, and $\vec{v}_2 = \vec{v} - \vec{v}_1$. For (d), $\vec{v}_1 = \frac{1}{3}[-1\ -4\ 1]^T$, and $\vec{v}_2 = \frac{1}{3}[10\ -2\ 2]^T$.

11. Draw a diagram. We can compute both $\overrightarrow{AC}$ and $\overrightarrow{AB} = 2\overrightarrow{AM}$. Now $\overrightarrow{BD} = \overrightarrow{BA} + \overrightarrow{AC} + \overrightarrow{CD} = -\overrightarrow{AB} + \overrightarrow{AC} - \overrightarrow{AB}$.

12. Since $\overrightarrow{AC} = \overrightarrow{AM} + \overrightarrow{MC} = \overrightarrow{AM} + \frac{1}{2}\overrightarrow{BC}$, we can then find C. From this we can also obtain B.

13. As in Example 5 (see Figure 7) let $\vec{v}$ and $\vec{w}$ be vectors along adjacent sides of the parallelogram. Use the fact that the diagonals are orthogonal to show that $\|\vec{v}\| = \|\vec{w}\|$.

14. Express $\overrightarrow{AC}$ and $\overrightarrow{BC}$ in terms of $\vec{v} = \overrightarrow{OA}$ and $\vec{w} = \overrightarrow{OC}$ where O is the center.

15. Place the solid with one vertex at the origin as in the diagram.

16. For (e), how you can get a vanishing projection.
    [ILAW: Explorations 3.2.1 to 3.2.3.]

17. Write $\vec{v} = [x_1\ y_1\ z_1]^T$ and $\vec{w} = [x_2\ y_2\ z_2]^T$. For (5), $(a\vec{v})\bullet\vec{w} = (ax_1)x_2 + (ay_1)y_2 + (az_1)z_2 = a(\vec{v}\bullet\vec{w})$. Similarly, $\vec{v}\bullet(a\vec{w}) = a(\vec{v}\bullet\vec{w})$.

18. Use Example 2 and the assumption that the vectors are perpendicular.

19. $(\vec{v}+\vec{w})\bullet(\vec{v}-\vec{w}) = \vec{v}\bullet\vec{v} - \vec{v}\bullet\vec{w} + \vec{w}\bullet\vec{v} - \vec{w}\bullet\vec{w} = \vec{v}\bullet\vec{v} - \vec{w}\bullet\vec{w} = \|\vec{v}\|^2 - \|\vec{w}\|^2$.

20. As in Example 2, verify $\|\vec{v}+\vec{w}\|^2 = \|\vec{v}\|^2 + 2(\vec{v}\bullet\vec{w}) + \|\vec{w}\|^2$ and $\|\vec{v}-\vec{w}\|^2 = \|\vec{v}\|^2 - 2(\vec{v}\bullet\vec{w}) + \|\vec{w}\|^2$.

21. Compute $(\vec{u}+\vec{v}+\vec{w})\bullet(\vec{u}+\vec{v}+\vec{w})$ using the orthogonality assumption.

22. For (b), $proj_{\vec{d}}(a\vec{v}) = \frac{a\vec{v}\bullet\vec{d}}{\|\vec{d}\|^2}\vec{d} = a(\frac{\vec{v}\bullet\vec{d}}{\|\vec{d}\|^2})\vec{d} = a\ proj_{\vec{d}}(\vec{v})$.

23. Use $\vec{i} = [1\ 0\ 0]^T$ as a vector parallel to the X-axis, so that α is the angle between the vectors $\vec{i}$ and $\vec{v}$.

24. For (a), $\vec{d}$ is the vector from the origin to the point $P(1, m)$ on the line. For (b), use Theorem 3.

25. Express $\vec{d'}$ as a multiple of $\vec{d}$ and use (1) of Theorem 4.

26. For (a), use the fact that $|\cos\theta| \leq 1$ for any angle θ. For (b), think about when $\cos\theta = 1$.

3.3 Lines and Planes

1. A point $P(x, y, z)$ is given by its coordinates x, y, and z. The position vector of that point P is the vector $\vec{p}$ with its tail at the origin and tip at P. Therefore
$$\vec{p} = \begin{bmatrix} x \\ y \\ z \end{bmatrix},$$
and now x, y, and z are called the components of $\vec{p}$.
[ILAW: Exploration 3.3.1.]

2. In each case, we need to find a direction vector $\vec{d}$ for the line together with a point P on the line; this is what you need to write the scalar equations of the line. [ILAW: Exploration 3.3.2.]
For (a), P is given and you can use $\vec{d} = [0\ -1\ 3]^T$; the scalar equations become

$$\begin{array}{rcrl}
x & = & 4 & \\
y & = & -3 & -t \\
z & = & 5 & +3t
\end{array}$$

For (k), the point Q on the line such that $\overrightarrow{PQ}$ is perpendicular to the line corresponds to the point on the line closest to P, so you can use Exploration 3.3.2. We find $Q(2.8, -1.6, -3)$, and therefore we can use $\vec{d} = \overrightarrow{QP} = [3.2\ 1.6\ 0]^T$ as a direction vector for the required line. We get the scalar equations:

$$\begin{array}{rcrl}
x & = & 6 & +3.2t \\
y & = & 0 & +1.6t \\
z & = & -3 & -6t
\end{array}$$

3. Write the scalar equations of the line through A and B using $\overrightarrow{AB}$ as a direction vector. Then verify whether the points P and Q satisfy the equations. Q does and P doesn't.

4. One way to do this is to write the scalar equations of the line through A and B, and then verify whether the third point C satisfies the equations.

5. These lines can go through any point $P(x_0, y_0, z_0)$, but must have a direction vector parallel to the X-axis, for example $\vec{d} = [1\ 0\ 0]^T$. The scalar equations must then be of the form:

$$\begin{array}{rcrl}
x & = & x_0 & +t \\
y & = & y_0 & \\
z & = & z_0 &
\end{array}$$

Selected Hints and Solutions

6. The line through A and B has scalar equations:

$$\begin{aligned} x &= 1 + t \\ y &= -1 + t \\ z &= 2 - t \end{aligned}$$

and therefore every point C on the line has the form $C(1+t, -1+t, 2-t)$ for some t. Now $\left\|\overrightarrow{AC}\right\|^2 = t^2 + t^2 + t^2 = 3t^2$, and $\left\|\overrightarrow{BC}\right\|^2 = 3 - 6t + 3t^2$. Therefore $\left\|\overrightarrow{AC}\right\| = 2\left\|\overrightarrow{BC}\right\|$ implies that $t = 2$ or $t = 2/3$, giving the two points $C(3, 1, 0)$ and $C(5/3, -1/3, 4/3)$.

7. The points of intersection are those points $P(x, y, z)$ satisfying both equations. For (d), use s as a parameter for the first line, and P must satisfy $x = 3s = 1 - 2t$, $y = 1 - s = 3 + t$, and $z = 2 - s = 5 + t$. The first two equations have unique solution $s = 5$, $t = -7$. The last equation however is not satisfied with these values, and there is therefore no point of intersection between these two lines.

8. For (b), the point Q closest to P is $Q(\frac{-3}{13}, \frac{15}{13}, 4)$, and the shortest distance is $\frac{1}{13}\sqrt{1846} \approx 3.305$.
   [ILAW: Exploration 3.2.2.]

9. We move on to planes. A point is on the plane exactly if its coordinates satisfy the equation; this is precisely what the equation means. So P does and Q doesn't.

10. In each case, we need a point on the plane and a normal vector for the plane. For (b), we are given a point P on the plane and we can use the same normal $\vec{n} = [1\ -2\ 3]^T$ as the given plane since the two are parallel. Hence the equation is $x - 2y + 3z = k$ for some k by Theorem 1. The fact that $P(1, -2, 4)$ is in the plane gives $k = 17$. The equation is thus $x - 2y + 3z = 17$.
    [ILAW: Exploration 3.3.3.]
    For (e), the plane must contain the midpoint $P(2, -4, 8)$ between A and B, and has normal $\overrightarrow{AB} = [2\ -2\ 2]^T$. An equation is therefore $2x - 2y + 2z = 28$, or equivalently $x - y + z = 14$.

11. The point $O(0, 0, 0)$ must satisfy the equation.

12. If a plane is perpendicular to the Z-axis, then the vector $[0\ 0\ 1]^T$ can be used as a normal. Hence the equations of all such planes must be of the form $0x + 0y + z = k$, or $z = k$ for some k.

13. The line lies in the given plane if every point of the line lies in the plane. For (b), any point P on the line has the form $P(-1, 3 + 2t, 1 + 5t)$ for some t, and for each of those points, $x - y + z = -1 - 3 - 2t + 1 + 5t = -3 + 3t$. This is equal to 1 only for $t = \frac{4}{3}$, and therefore the line intersects the plane in only one point, so does not entirely lie in the plane.

14. This is the same technique as in the previous exercise. The line in part (a) has no chance since its direction vector is not perpendicular to the normal of the plane.

15. For (b), any point P in the line has the form $P(5, -3 + 2t, 1 - 3t)$ for some t, and for each of those points, $2x + 4y - 3z = 10 - 12 + 8t - 3 + 9t = -5 + 17t$. This equals 1 only for $t = \frac{6}{17}$, giving the point of intersection $P(5, -\frac{39}{17}, -\frac{1}{17})$.

16. This will happen exactly when the three points are colinear, right?

546   Selected Hints and Solutions

17. Use Exploration 3.3.3 to help you visualize the question and support your computation. For (d), choose any point P_0 in the plane and write $\vec{v} = \overrightarrow{P_0P}$. Then the shortest distance is $\|proj_{\vec{n}}(\vec{v})\| = \frac{|\vec{v}\bullet\vec{n}|}{\|\vec{n}\|} = \frac{35}{\sqrt{14}} \approx 9.35$. The point Q on the plane closest to P is $Q(-9.5, -2.5, 4)$.

18. For (a), there is a unique solution $X = [1\frac{1}{3} \; -\frac{1}{3} \; -\frac{1}{3}]^T$, so this is the solution closest to P. For (b), the solution set is a line with scalar equations

$$\begin{bmatrix} x \\ y \\ z \end{bmatrix} = \begin{bmatrix} 3-4t \\ 2-t \\ t \end{bmatrix}.$$

The point on that line closest to P is $Q(\frac{-5}{9}, \frac{10}{9}, 1)$.
[ILAW: Exploration 3.3.2.]
For (c), the given point is a solution itself, and is therefore as close as you can get. For (d), the solution set is a plane, and you can use Exploration 3-3-3 to help you find the point on that plane closest to the given point P.

19. For (d), what are the solutions of an equation $ax + by + cz = k$ and any nonzero multiple of it? For (h), how would you define a line being parallel to a plane?

20. Draw a diagram and convince yourself that this is true geometrically. You can also write the equation of the line through two points A and B, the general equation of a plane containing A and B, and verify algebraically that any point on the line satisfies the equation of the plane.

21. The scalar equations are:

$$\begin{aligned} x &= x_0 &+ at \\ y &= y_0 &+ bt \\ z &= z_0 &+ ct \end{aligned}$$

Now solve for t in each equation.

22. For (b), the distance is $\|\vec{a} - \vec{q}\|$ where $\vec{q}$ is the position vector of the point on the plane closest to A. Show that $\vec{q} = \vec{a} - t\vec{n}$ for some scalar t, and use the fact (from (a)) that $\vec{q} \bullet \vec{n} = k$.

23. Let $\vec{v}$ and $\vec{v}_1$ be the position vectors of points on the two planes, so that $\vec{v} \bullet \vec{n} = k$ and $\vec{v}_1 \bullet \vec{n} = k_1$. Argue (as in Example 12) that the required distance is $\|proj_{\vec{n}}(\vec{v} - \vec{v}_1)\|$.

3.4 The Cross Product

1. For (b), $\vec{v} \times \vec{w} = det \begin{bmatrix} \vec{i} & 3 & 1 \\ \vec{j} & -5 & 1 \\ \vec{k} & 1 & 2 \end{bmatrix} = \begin{bmatrix} (-5)\cdot 2 - 1\cdot 1 \\ -(3\cdot 2 - 1\cdot 1) \\ 3\cdot 1 - (-5)\cdot 1 \end{bmatrix} = \begin{bmatrix} -11 \\ -5 \\ 8 \end{bmatrix}$
[ILAW: Exploration 3.4.1.]

2. The area for a triangle ABC is half the area of the parallelogram determined by the vectors $\overrightarrow{AB}$ and $\overrightarrow{AC}$, hence half the length of their cross product.
[ILAW: Exploration 3.4.2.]

Selected Hints and Solutions 547

For (b),
$$\text{Area of } \triangle ABC = \tfrac{1}{2}\left\|\overrightarrow{AB} \times \overrightarrow{AC}\right\|$$
$$= \tfrac{1}{2}\left\|\begin{bmatrix} 1 \\ 2 \\ -6 \end{bmatrix} \times \begin{bmatrix} -1 \\ 1 \\ 1 \end{bmatrix}\right\|$$
$$= \tfrac{1}{2}\left\|\begin{bmatrix} 8 \\ 5 \\ 3 \end{bmatrix}\right\|$$
$$= \tfrac{1}{2}\sqrt{98}.$$

3. This is a simple matter of calculation, but a very useful fact.

4. The line in (b) must be perpendicular to the direction vectors of both lines, and so it must be parallel to their cross product. Hence $[7\ 0\ 1]^T \times [2\ 3\ 0]^T = [-3\ 2\ 21]^T$ is a direction vector for the new line. The scalar equations are therefore $[x\ y\ z]^T = [2\ -1\ 3]^T + t[-3\ 2\ 21]^T$.
   For (d), the intersection of the two given planes is the line $[x\ y\ z]^T = [\tfrac{3}{7}\ -\tfrac{1}{7}\ 0]^T + t[2\ 1\ 1]^T$.
   [ILAW: Exploration 1.2.3.]

5. For (b), the vector $\overrightarrow{AB} \times \overrightarrow{AC} = [5\ -10\ -35]^T$ can be used for a normal to the plane; the equation becomes $5x - 10y - 35z = 5$, or simply $x - 2y - 7z = 1$.
   For (d), the direction vector of the line can be used as a normal for the plane, the plane equation becomes $8x - 2y + 0z = 10$, or simply $4x - y = 5$.

6. One solution is to find the equation of the plane containing the first three points A, B, and C, and verify whether the point D satisfies the equation. For (b), a normal for the plane through the points A, B, and C is $\vec{n} = \overrightarrow{AB} \times \overrightarrow{AC} = [-4\ 2\ -6]^T$, giving the equation $-4x + 2y - 6z = -2$, or simply $2x - y + 3z = 1$. Since the point D also satisfies this equation, we have shown that the four point do lie on the same plane.
   A related method for three points is covered in Exercise 14.

7. Since the two lines are parallel, you may as well choose any point P on the first line, and find the shortest distance between that point and the second line, right?
   [ILAW: Exploration 3.3.2.]

8. For (c), $\vec{d_0} \times \vec{d_1} = [19\ 1\ -11]^T$, and $\overrightarrow{P_0 P_1} = [-1\ -7\ 4]^T$. Therefore $\vec{v} = \text{proj}_{\vec{d_0} \times \vec{d_1}}(\overrightarrow{P_0 P_1}) = -\tfrac{10}{69}[19\ 1\ -11]^T$, and the shortest distance is $\|\vec{v}\| = \tfrac{10}{69}\sqrt{483}$ as in Example 4. [ILAW: Exploration 3.4.1.]

9. The volume of the parallelepiped determined by $\vec{u}$, $\vec{v}$ and $\vec{w}$ is $|\vec{u} \bullet (\vec{v} \times \vec{w})| = |\det[\vec{u}\ \vec{v}\ \vec{w}]|$ and therefore can be computed in two ways. You can compute $|\vec{u} \bullet (\vec{v} \times \vec{w})|$, or simply compute $|\det[\vec{u}\ \vec{v}\ \vec{w}]|$.
   For (b), we get $|-7| = 7$.
   [ILAW: Explorations 2.1.1, 3.4.1 and 3.2.1.]

10. Both problems rely on Theorem 5.
    [ILAW: Explorations 3.4.1, 3.4.2, and 3.4.3.]

11. Use the properties listed in Theorem 3. We can get
$$(a\vec{v} + b\vec{w}) \times (c\vec{v} + d\vec{w})$$
$$= ((a\vec{v} + b\vec{w}) \times c\vec{v}) + ((a\vec{v} + b\vec{w}) \times d\vec{w})$$
$$= (a\vec{v} \times c\vec{v}) + (b\vec{w} \times c\vec{v}) + (a\vec{v} \times d\vec{w}) + (b\vec{w} \times d\vec{w}).$$

Continue using Theorem 3.

12. If $\vec{u} + \vec{v} + \vec{w} = \vec{0}$, then $\vec{u} \times \vec{v} = (-\vec{v} - \vec{w}) \times \vec{v}$. The other case is similar.

13. The volume of the parallelepiped determined by $\vec{v}$, $\vec{w}$ and $\vec{v} \times \vec{w}$ is the same as the volume of the parallelepiped determined by $\vec{n} = \vec{v} \times \vec{w}$, $\vec{v}$, and $\vec{w}$, and therefore is given by Theorem 6:

$$|\vec{n} \bullet (\vec{v} \times \vec{w})| = \|\vec{n} \times \vec{n}\| = \|\vec{n}\|^2.$$

14. Three points A, B and C all lie on the same line if and only if the vectors $\overrightarrow{AB}$ and $\overrightarrow{AC}$ are parallel.

15. Four points A, B, C, and D all lie on the same plane if and only if the vector $\overrightarrow{AB}$ lies in the plane determined by the three points A, C and D, and thus if and only if $\overrightarrow{AB}$ is perpendicular to the normal to the plane (given by $\overrightarrow{AC} \times \overrightarrow{AD}$). Alternatively, think of the volume of teh parallelopiped determined by $\overrightarrow{AB}$, $\overrightarrow{AC}$ and $\overrightarrow{AD}$.

16. Use Theorem 6.

17. Proceed in a similar fashion to the proof of Theorem 1.

18. The number $\|\overrightarrow{P_0P} \times \vec{d}\|$ is the area of the parallelogram determined by these two vectors, which can also be computed by the base ($\|\vec{d}\|$) times the height (the shortest distance).

19. This really is a problem for the last section. Find a point $P(x, y, z)$ on the plane and proceed as in Example 12 of Section 3.3.

3.5 Matrix Transformations of $\mathbb{R}^2$

1. For (b), T is given by $T \begin{bmatrix} x \\ y \end{bmatrix} = \begin{bmatrix} 3 & -1 \\ 1 & 2 \end{bmatrix} \begin{bmatrix} x \\ y \end{bmatrix}$.

   For (d), $T \begin{bmatrix} x \\ y \end{bmatrix} = \begin{bmatrix} -1 & 1 \\ 1 & 1 \end{bmatrix} \begin{bmatrix} x \\ y \end{bmatrix}$.

2. (d). We have $\begin{bmatrix} x \\ y \end{bmatrix} = x \begin{bmatrix} 1 \\ -1 \end{bmatrix} + (x + y) \begin{bmatrix} 0 \\ 1 \end{bmatrix}$ so the linearity of T gives

$$T \begin{bmatrix} x \\ y \end{bmatrix} = xT \begin{bmatrix} 1 \\ -1 \end{bmatrix} + (x+y)T \begin{bmatrix} 0 \\ 1 \end{bmatrix}$$
$$= x \begin{bmatrix} 1 \\ 0 \end{bmatrix} + (x+y) \begin{bmatrix} 2 \\ -3 \end{bmatrix}$$
$$= \begin{bmatrix} 3x + 2y \\ -3x - 3y \end{bmatrix}.$$

Hence T has matrix $\begin{bmatrix} 3 & 2 \\ -3 & -3 \end{bmatrix}$.

[ILAW: Exploration 3.5.2.]

3. Theorems 1 and 3 provide the matrices. For (b), the matrix is $\frac{1}{5} \begin{bmatrix} -3 & 4 \\ 4 & 3 \end{bmatrix}$.

   For (d), we get $\begin{bmatrix} 0 & 1 \\ -1 & 0 \end{bmatrix}$, and for (f), the matrix is $\frac{1}{2} \begin{bmatrix} 1 & -1 \\ -1 & 1 \end{bmatrix}$.

   [ILAW: Exploration 3.5.1 and 3.5.3.]

Selected Hints and Solutions 549

4. For (b), if T was induced by the matrix $\begin{bmatrix} a & b \\ c & d \end{bmatrix}$, then we would have $cx+dy = y^2$ for all x and y. If $x = 1$ and $y = 0$ we get $c = 0$; hence if $y = 1$ we get $d = 1$, while $y = 2$ gives $d = 2$, a contradiction.

5. For (b), T is given by $T\begin{bmatrix} x \\ y \end{bmatrix} = \begin{bmatrix} p & r \\ q & s \end{bmatrix}\begin{bmatrix} x \\ y \end{bmatrix}$.

6. For (b), the matrix of T is $\frac{1}{5}\begin{bmatrix} -4 & 3 \\ 3 & 4 \end{bmatrix}$, $T\begin{bmatrix} 1 \\ 1 \end{bmatrix} = \frac{1}{5}\begin{bmatrix} -1 \\ 7 \end{bmatrix}$, and $T\begin{bmatrix} 2 \\ -1 \end{bmatrix} = \frac{1}{5}\begin{bmatrix} -11 \\ 2 \end{bmatrix}$.

   For (d), the matrix of T is $\frac{1}{\sqrt{2}}\begin{bmatrix} 1 & 1 \\ -1 & 1 \end{bmatrix}$, $T\begin{bmatrix} 1 \\ 1 \end{bmatrix} = \frac{1}{\sqrt{2}}\begin{bmatrix} 2 \\ 0 \end{bmatrix}$, and $T\begin{bmatrix} 2 \\ -1 \end{bmatrix} = \frac{1}{\sqrt{2}}\begin{bmatrix} 1 \\ -3 \end{bmatrix}$.

   For (f), the matrix of T is $\frac{1}{5}\begin{bmatrix} 4 & -2 \\ -2 & 1 \end{bmatrix}$, $T\begin{bmatrix} 1 \\ 1 \end{bmatrix} = \frac{1}{5}\begin{bmatrix} 2 \\ -1 \end{bmatrix}$, and $T\begin{bmatrix} 2 \\ -1 \end{bmatrix} = \begin{bmatrix} 2 \\ -1 \end{bmatrix}$.

   [ILAW: Exploration 3.5.1 and 3.5.3.]

7. For (b), T is the projection on $y = -x$. For (d), T is the reflection in $y = 2x$, and for (f), T is the rotation through $\pi/3$.
   [ILAW: Exploration 3.5.1 and 3.5.2.]

8. For (b), the matrix of T is kA.

9. The transformation is given by
$$T\begin{bmatrix} x \\ y \end{bmatrix} = \begin{bmatrix} ax \\ ay \end{bmatrix} = \begin{bmatrix} a & 0 \\ 0 & a \end{bmatrix}\begin{bmatrix} x \\ y \end{bmatrix}.$$
   Use Exploration 3.5.3 to visualize the effect on the unit square.

10. For (b),
$$P_L(\vec{v}+\vec{w}) = \frac{(\vec{v}+\vec{w})\cdot\vec{d}}{\|\vec{d}\|^2}\vec{d} = \frac{\vec{v}\cdot\vec{d}+\vec{w}\cdot\vec{d}}{\|\vec{d}\|^2}\vec{d} = \frac{\vec{v}\cdot\vec{d}}{\|\vec{d}\|^2}\vec{d} + \frac{\vec{w}\cdot\vec{d}}{\|\vec{d}\|^2}\vec{d}$$
$$= P_L(\vec{v}) + P_L(\vec{w}).$$

   Similarly $P_L(k\vec{v}) = \frac{(k\vec{v})\cdot\vec{d}}{\|\vec{d}\|^2}\vec{d} = \frac{k(\vec{v}\cdot\vec{d})}{\|\vec{d}\|^2}\vec{d} = kP_L(\vec{v})$.

   Hence P_L is linear.

11. For (b), let A be the matrix of T. Then $T(-\vec{v}) = A(-\vec{v}) = -A\vec{v} = -T(\vec{v})$.

12. If $a \neq 0$, the line L has equation $y = \frac{b}{a}x$; now use Theorem 1. Check the case $a = 0$ separately.

13. For (b), T is the X-shear of 5, so T^{-1} is the X shear of -5. Thus $A^{-1} = \begin{bmatrix} 1 & -5 \\ 0 & 1 \end{bmatrix}$.

    For (d), $T = R_{\pi/4}$ so $T^{-1} = R_{-\pi/4}$. Thus
$$A^{-1} = \begin{bmatrix} \cos(-\pi/4) & -\sin(-\pi/4) \\ \sin(-\pi/4) & \cos(-\pi/4) \end{bmatrix} = \frac{1}{\sqrt{2}}\begin{bmatrix} -1 & 1 \\ -1 & 1 \end{bmatrix}.$$

    [ILAW: Exploration 3.5.5.]

14. For (b) the composition gives a reflection in the Y-axis.
    [ILAW: Exploration 3.5.4.]

15. The required rotation is by the angle π.
    [ILAW: Exploration 3.5.4.]

16. For (b), the matrix of the composite is
    $\begin{bmatrix} 0 & -1 \\ -1 & 0 \end{bmatrix} \begin{bmatrix} 0 & 1 \\ 1 & 0 \end{bmatrix} = \begin{bmatrix} -1 & 0 \\ 0 & -1 \end{bmatrix}$, which is the rotation through π.
    For (d), the matrix of the composite is $\begin{bmatrix} 0 & -1 \\ 1 & 0 \end{bmatrix} \begin{bmatrix} 1 & 0 \\ 0 & -1 \end{bmatrix} = \begin{bmatrix} 0 & 1 \\ 1 & 0 \end{bmatrix}$, which is the reflection in the line $y = x$.
    [ILAW: Exploration 3.5.4.]

17. For (a), $P_m \circ Q_m$ means first reflect in the line L with equation $y = mx$, then project on L — the result is to project on L. Similarly $Q_m \circ P_m$ means first project on L, then reflect in L — i.e. project on L.
    For (b), following Example 2, $P_m \circ Q_m$ has matrix

    $$\frac{1}{1+m^2} \begin{bmatrix} 1 & m \\ m & m^2 \end{bmatrix} \cdot \frac{1}{1+m^2} \begin{bmatrix} 1-m^2 & 2m \\ 2m & m^2-1 \end{bmatrix}$$
    $$= \frac{1}{(1+m^2)^2} \begin{bmatrix} 1+m^2 & m(1+m^2) \\ m(1+m^2) & m^2(1+m^2) \end{bmatrix}$$
    $$= \frac{1}{1+m^2} \begin{bmatrix} 1 & m \\ m & m^2 \end{bmatrix}$$

    Thus $P_m \circ Q_m = P_m$. Similarly $Q_m \circ P_m = P_m$.

18. For (a), the resulting vector $P_{m_1}(\vec{v})$ is orthogonal to the line $y = mx$, so $P_m \circ P_{m_1}(\vec{v}) = \vec{0}$. For (b), use the fact that m_1 must be $-1/m$, and then use Theorem 1.
    [ILAW: Exploration 3.5.4.]

19. For (a), $T(\vec{v})$ is already projected, so doing it again won't do anything. For (b), multiply the matrices from Theorem 1.
    [ILAW: Exploration 3.5.4.]

20. For (a), reflecting twice brings you back to where you started. For (b), multiply the matrix from Theorem 1 with itself to verify that you get the identity matrix.
    [ILAW: Exploration 3.5.4.]

21. (a). If T and T_1 are isometries, then we have

    $$\|T(T_1(\vec{v})) - T(T_1(\vec{w}))\| = \|T_1(\vec{v}) - T_1(\vec{w})\| = \|\vec{v} - \vec{w}\|$$

    and therefore $T \circ T_1$ is an isometry.

    (b). If the two reflections have matrices A and B, compute $det(AB)$ and use Theorem 7.
    [ILAW: Exploration 3.5.4 and 3.5.7.]

22. Given a reflection with matrix Q_m, let A denote the matrix of reflection in the X-axis. Use Exercise 20 and Theorem 7 to show that $Q_m A^{-1}$ is a reflection.

23. For (a), use the proof of Theorem 4.

Selected Hints and Solutions 551

24. For (a), use Theorem 7. For (b), use the fact that $\vec{v} \bullet \vec{w} = \vec{v}^T \vec{w}$ for all columns $\vec{v}$ and $\vec{w}$ to show that $\|A\vec{v}\|^2 = \|\vec{v}\|^2$ for all $\vec{v}$ in $\mathbb{R}^2$.

Chapter 4

THE VECTOR SPACE $\mathbb{R}^n$

4.1 Subspaces and Spanning

1. Recall that vectors in $\mathbb{R}^1$ are merely real numbers and can be written with or without parentheses. For example, $4 = (4)$ in $\mathbb{R}^1$.
   [ILAW: Exploration 4.1.2.]

2. The value of $-n$ is just the length of the vector.
   (b) $\mathbb{R}^6$ Note that zeros at the end of a vector still affect the vector's length.
   (d) $\mathbb{R}^1$ Remember that real numbers are elements of $\mathbb{R}^1$.

3. Recall that for two vectors to be equal, they must first of all have the same dimensions, and then each of the corresponding entries must be equal.
   (c) Not equal. Same numbers used, but the corresponding entries don't match.
   (e) Not equal. Different dimensions.

4. In each case determine expressions for vectors in $null A$ and $im A$. [ILAW: Exploration 4.1.3.]
   (d) $null A = span\{[-\frac{1}{5}\ 0\ 0\ 1]^T\}$ using the Gaussian algorithm, and $im A = span\{[5\ 0\ 0\ 5]^T, [0\ 1\ 0\ 0]^T, [-7\ 4\ 1\ 0]^T, [1\ 0\ 0\ 1]^T\}$ using Example 12. Can you simplify the spanning set for $im A$?

5. For each of these problems, you need to show that each of the three axioms used in the definition of a subspace are satisfied.
   (a) This is not a subspace because the zero vector $[0\ 0\ 0]^T$ is not in U. It is not possible to generate $[0\ 0\ 0]^T$ from the general form $[s\ t\ 1]^T$ since there is always a 1 in the third position.
   (b) Here, all three axioms hold for U :

   S1. Choose $t = 0$, then $[0\ 0\ 0]^T$ is in U.

   S2. Consider $X = [0\ t_1\ 0]^T$ and $Y = [0\ t_2\ 0]^T$, both in U. Then $X + Y = [0\ t_1\ 0]^T + [0\ t_2\ 0]^T = [0\ t_1 + t_2\ 0]^T$ is also in U.

   S3. Let $X = [0\ t\ 0]^T$ and choose any real number r, then $rX = r[0\ t\ 0]^T = [0\ rt\ 0]^T$ which is also in U.

6. If it turns out that X is indeed a member of $span\{Y, Z\}$, find the linear combination. If X is *not* in $span\{Y, Z\}$, you have a little more work to do to show why not (see answer for part (b).) [ILAW: Exploration 4.1.4.]
   (b) The general form for a linear combination of Y and Z is $sY + tZ = [2s + t\ -s-t\ -3t\ 2s+t]^T$. So if X is in $span\{Y, Z\}$, it will have this form, but the first and last entries of X are not equal as they are in $sY + tZ$. Thus X cannot be a linear combination of Y and Z.
   (d) $X = 3Y + 4Z$.

7. One way to attack this problem is to use a system of equations. Demonstrating that the vectors of the set span $\mathbb{R}^4$ is the same as showing that any vector $[a \ b \ c \ d]^T$ is a linear combination of the given vectors, that is there exist $s, t, p,$ and q such that

$$sX_1 + tX_2 + pX_3 + qX_4 = \begin{bmatrix} X_1 & X_2 & X_3 & X_4 \end{bmatrix} \begin{bmatrix} s \\ t \\ p \\ q \end{bmatrix} = \begin{bmatrix} a \\ b \\ c \\ d \end{bmatrix}.$$

If this system always has a solution, then any vector in $\mathbb{R}^4$ is a linear combination of the given vectors. By Theorem 5 § 1.5, the system always has a solution if and only if the coefficient matrix $[X_1 \ X_2 \ X_3 \ X_4]$ is invertible. This is easy to check.

8. (a) This is false. In order to prove this, you need to find an example of a subspace U where $X + Y$ is in U, but at least one of X, Y is not in U.

Try $U = \{[0 \ t \ 0]^T \mid t \text{ in } \mathbb{R}\}$. This was shown in the previous question to be a subspace. You will agree that the vectors $X = [1 \ 1 \ 0]^T$ and $Y = [-1 \ 0 \ 0]^T$ are not in U. However, $X + Y = [0 \ 1 \ 0]^T$, which is in U. The same technique can be applied to any known subspace (try it with the subspace from Example 3).

(b) True: $X = 1X$ is in U.

9. Is $[1 \ -1 \ 0]^T$ in U?

10. To prove that for $A = B$ for two sets A and B, you need to prove that both $A \subseteq B$ and $B \subseteq A$. The fact that $a \neq 0$ is useful to match the coefficient of X while proving the case that $span\{aX\} \subseteq span\{X\}$.

11. Use the definition of $span\{X\}$.

12. A subspace U of $\mathbb{R}^3$ must be *contained* in $\mathbb{R}^3$, that is every vector in U must be in $\mathbb{R}^3$. Is $(0, 0, 0)$ in $\mathbb{R}^2$?

13. It is always the case that $span\{Y, X_2, \cdots, X_k\} \subseteq span\{X_1, X_2, \cdots, X_k\}$ because each of $Y, X_1, X_2, \cdots, X_k$ is in $span\{X_1, X_2, \cdots, X_k\}$. The fact $a_1 \neq 0$ is to prove the case that $span\{X_1, X_2, \cdots, X_k\} \subseteq span\{Y, X_2, \cdots, X_k\}$.

14. Given that Y is in $U = span\{X_1, X_2, \cdots, X_k\}$, how can Y be written in terms of all of the vectors X_i? Use this expression in conjunction with $AY = 0$ as your starting point.

15. For both parts remember that to prove equality of sets, you must prove that each set is a subset of the other.

(a) CASE 1: $null(A) \subseteq (UA)$. This case is straightforward if you use the definitions. Suppose you have a vector X in $null(A)$, then $AX = 0$. But now, $(UA)X = U(AX) = U(0) = 0$, and we are done.

CASE 2: $null(UA) \subseteq (A)$. This case will require the fact that U is invertible. Given X in $null(UA)$, we have that $(UA)X = 0$. Applying U^{-1} gives $AX = U^{-1}(UAX) = U^{-1}0 = 0$.

(b) The solution to this question is similar to part (a). The case where you show $imA \subseteq im(AV)$ uses the fact that V is invertible.

Selected Hints and Solutions 553

16. (a) Apply axiom S3 to aX, and see if you can find the vector X as a scalar multiple of aX.

    (b) In this case, apply S2 to X and $X + Y$. Again, try to find Y in the set of vectors generated from the axiom.

17. If A be a vector in $span\{\mathcal{X}\}$, then $A = s_1 X_1 + s_2 X_2 + \cdots + s_k X_k$ where all s_i are in $\mathbb{R}$, and all X_i are in $\mathcal{X}$. Now use the fact that $\mathcal{X} \subseteq \mathcal{Y}$.

18. Hint for proving S3: Substitute rX for X in AX. Manipulate this and use the equation $AX = BX$ to get $A(rX) = B(rX)$.

19. (a) Hint for proving S2: Substitute $X + Y$ for X in AX. By rearranging this and using the equation $AX = \lambda X$ to get $A(X + Y) = \lambda(X + Y)$.

    (b) Notice that $AX = \lambda X$ is the same as $(A - \lambda I)X = 0$.

20. Every subspace must contain the zero vector.

21. The only two subspaces are $\{0\}$ and $span\{V\}$. It is clear that $\{0\}$ and $span\{V\}$ are subspaces (see Example 2 and Theorem 1). To prove that there are no other subspaces, suppose that $U \subseteq span\{V\}$ such that $U \neq \{0\}$, and assume the U is a subspace. Choose a vector Y in U, and show that $span\{Y\} = span\{V\}$.

22. CASE 1: Show that if U is a subspace of $\mathbb{R}^n$ then S2 and S3 hold. This is obvious.

    CASE 2: Show the other direction, that if S2 and S3 hold for U then U is a subspace of $\mathbb{R}^n$. All that remains to show is that S1 holds. What happens to S3 if you take $r = 0$? Why must U be nonempty?

23. (c) To show uniqueness, the goal is to show that if X has two different representations then these two forms are actually the same. So suppose that $X = Y + Z$ and $X = Y_1 + Z_1$, for Y and Y_1 in U, and Z and Z_1 in W. We want to show that $Y_1 = Y$ and $Z_1 = Z$. It follows that $Y + Z = Y_1 + Z_1$, which can be rearranged as $Y - Y_1 = Z_1 - Z$. However, $Y - Y_1$ is in U and $Z_1 - Z$ is in W, and since the two are equal, they must be in $U \cap W$. Since $U \cap W = \{0\}$, $Y - Y_1 = 0$ and $Z_1 - Z = 0$, and so $Y = Y_1$ and $Z = Z_1$.

4.2 Linear Independence

1. Follow the set up of Example 1 in the text and you solve the system of equations that will arise. Each set of vectors is independent because the only solution to the system is the trivial one.
   [ILAW: Exploration 1.3.1 or 1.2.2, and 4.2.3.]

2. (c) $-2[1 \ -1 \ 0]^T - [3 \ 2 \ -1]^T + [5 \ 0 \ -1]^T = [0 \ 0 \ 0]^T$.

   (f) Independent.
   [ILAW: Exploration 4.2.1.]

3. In Theorem 2, the invertibility of the square matrix A is equivalent to independence of either the rows or the columns. [ILAW: Explorations 4.2.1 and 4.2.2.]

   (b) Invertible. All rows are independent (equivalently, all columns are independent).

   (c) Not invertible. The rows are not independent (and neither are the columns).

4. Form a square matrix from the vectors and if the matrix is invertible the vectors are independent.

   (a) $det \begin{bmatrix} 3 & -2 \\ 1 & 2 \end{bmatrix} = 8$. Thus A is invertible, and so by Theorem 2 its columns, which are the vectors of the set, are independent.

   (c) $det \begin{bmatrix} 5 & -5 & 4 \\ -2 & 0 & -2 \\ 1 & -5 & 0 \end{bmatrix} = 0$. Thus A is not invertible, and by Theorem 2 its columns, which are the vectors in the question, are not independent.

5. Form an $n \times n$ matrix from the vectors and if the matrix is invertible the vectors span $\mathbb{R}^n$.

   (a) $det \begin{bmatrix} 4 & 2 \\ -7 & -5 \end{bmatrix} = -6$. Hence A is invertible and so its columns do span $\mathbb{R}^2$.

   (c) $det \begin{bmatrix} -8 & -5 & 7 & 6 \\ -1 & -5 & 5 & 1 \\ 3 & 4 & 9 & -16 \\ 6 & -9 & 0 & 3 \end{bmatrix} = 0$. Clearly, A is not invertible, and so it follows from Theorem 2 that its columns are dependent.

6. In order to use Theorem 2, the matrix formed by the set of vectors must be square. In the following exercises, such sets of vectors were used and Theorem 2 could be applied:

   - Exercise 1 (a) (c)
   - Exercise 2 (a) (b) (c) (g)

7. (b) True. *Hint:* $aX + bY = 0$ is the same as $aX + bY + 0Z = 0$.

   (d) False. Consider Example 1 from the notes.

   (f) False. Example: If $X = -Y$, $a = b = 1$, and $c = 0$ then $aX + bY + cZ = X - X + 0Z = 0$.

   (h) True. This follows from the definition of linear dependence.

8. If a linear combination vanishes, take the dot product of both sides with $\vec{u}$.

9. Before doing any long calculations, see if you can spot that one vector can be made from a linear combination of the other vectors. If so, you know the set is dependent.

   (b) Independent.

   (d) Dependent: $X + Y = (W + X) - (Z + W) + (Y + Z)$.

10. How do the following theorems relate? Theorem 2 §2.2, Theorem 2 §4.2.

11. We want to show that the only solution to
    $$t_1 X_1 + t_2 X_2 + \cdots + t_k X_k = 0$$
    is the trivial one where $t_i = 0$ for all i. If you multiply this equation on the left by A, how does this relate to the independence of the B_i?

12. If you set the linear combination
    $$a_1 X_1 + a_2(X_1 + X_2) + a_3(X_1 + X_2 + X_3) + \cdots + a_k(X_1 + X_2 + \cdots + X_k)$$
    equal to zero, what are the coefficients of $X_1, X_2, \cdots X_k$, and how can you use the independence of $\{X_1, X_2, \cdots X_k\}$ to show that $a_1 = a_2 = \cdots = a_k = 0$?

Selected Hints and Solutions

13. Approach this question in the same manner as the preceding question.
14. Follow the steps of the Independence Test.
    Step 1: *Set a linear combination of the larger set equal to zero:*
    $$t_0 Y + t_1 X_1 + t_2 X_2 + \cdots + t_k X_k = 0.$$
    Step 2: *Show that the only solution to this is the trivial one where all $t_i = 0$.*
    There are two cases. The first is when $t_0 = 0$. In this case, the above equation becomes
    $$0Y + t_1 X_1 + t_2 X_2 + \cdots + t_k X_k = 0.$$
    Since $\{X_1, X_2, \cdots, X_k\}$ is independent, it follows that $t_1 = t_2 = \cdots = t_k = 0$, which combined with $t_0 = 0$ forms the trivial solution.
    The second case is when $t_0 \neq 0$. Here, we can divide by t_0 giving
    $$Y = \frac{t_1 X_1 + t_2 X_2 + \cdots + t_k X_k}{-t_0},$$
    which is in $span\{X_1, X_2, \cdots, X_k\}$. This case cannot occur because of the condition that Y is not in $span\{X_1, X_2, \cdots, X_k\}$.
15. This is an extension of Exercise 7(b).
16. This is an extension of Exercise 7(c)

4.3 Dimension

1. The dimension of a subspace is given by the number of vectors in a basis of the subspace. So just count the number of vectors in the basis.
   (a) 2   (b) 3   (c) 2   (d) 3   (e) 4
2. In each part, the number of vectors in the set is equal to the dimension of the indicated space. By Theorem 4, it suffices to prove only one of the two conditions of the definition of a basis. That is, you only need to show either that the set is linearly independent or that it spans the space.
3. Since each set spans the given space, all that remains is to verify whether vectors are independent in the space. One short-cut is to apply the Corollary to Theorem 1: if the vectors are in $\mathbb{R}^n$ but there are more than n vectors, they cannot be independent in $\mathbb{R}^n$.
   (a) The set contains 4 vectors in $\mathbb{R}^3$, so they cannot be independent, and hence are not a basis of $\mathbb{R}^3$.
   (b) Not a basis. Too many vectors to be independent.
   (c) This set has 3 vectors in $\mathbb{R}^3$, so they might be independent. But the vectors are dependent, and so cannot form a basis of $\mathbb{R}^3$.
   (d) This is a basis. The vectors are independent.
4. The Invariance Theorem (Theorem 2) says that if two sets are bases of the same subspace, they must have the same number of vectors. This holds only for part (c). So the remaining parts cannot be bases of the same subspace.
5. (b) $\{E_1, E_2, E_3\} = \left\{ \begin{bmatrix} 1 \\ 0 \\ 0 \end{bmatrix}, \begin{bmatrix} 0 \\ 1 \\ 0 \end{bmatrix}, \begin{bmatrix} 0 \\ 0 \\ 1 \end{bmatrix} \right\}.$
   (d) $\{E_1\} = \{1\}$.

6. Each spanning set contains a basis of the subspace by Theorem 3. Listed are the dimensions of each subspace: (a) 2   (b) 2   (c) 2   (d) 3.

   Different bases will be of the same dimension (see Theorem 2). What happens if you use different vectors in the spanning set?
   [ILAW: Exploration 4.3.2.]

7. Can you write U as the span of n vectors? If so, then the question resembles the previous question.
   (a) $[a \ \ a+b \ \ a-b \ \ b]^T = a[1 \ 1 \ 1 \ 0]^T + b[0 \ 1 \ -1 \ 1]^T$. $dimU = 2$.
   (c) $dimU = 3$.
   (e) $dimU = 3$.

8. Find a spanning set as in Example 3 §1.3. How many parameters are there?
   [ILAW: Exploration 4.1.3.]
   (a) One possible basis for $nullA$ is $\{[4 \ 1 \ 2 \ 0 \ 0]^T, [-4 \ 0 \ 0 \ 1 \ 0]^T, [2 \ 1 \ 0 \ 0 \ 2]^T\}$.

9. (a) You can prove that U is a subspace by proving each of the three axioms, however you can save yourself some work by noticing that U is the nullspace of A^T, and we have shown that the nullspace of any matrix is a subspace. One basis for U is $\{[2 \ -2 \ 3 \ 0]^T, [-7 \ -2 \ 0 \ 3]^T\}$.
   (c) U can be rewritten as a nullspace of what matrix? One basis for U is $\{[1 \ 0 \ 0 \ 0]^T, [0 \ -4 \ 2 \ 1]^T\}$.

10. (b) False. Consider the standard basis of $\mathbb{R}^3$.
    (d) False. Verify that the following set is a basis of $\mathbb{R}^5$:
    $\{[1 \ -1 \ 0 \ 0 \ 0]^T, [1 \ 1 \ 0 \ 0 \ 0]^T, [1 \ 1 \ 1 \ 0 \ 0]^T, [1 \ 1 \ 1 \ 1 \ 0]^T, [1 \ 1 \ 1 \ 1 \ 1]^T\}$.
    This counterexample is not unique. Can you use Exploration 4.3.1 to find more?

    (f) False. Verify that
    $$\{[1 \ 0 \ 0 \ 0]^T, [0 \ 1 \ 0 \ 0]^T, [0 \ 0 \ 1 \ 0]^T, [0 \ 0 \ 0 \ 1]^T\} \text{ and}$$
    $$\{[-1 \ 0 \ 0 \ 0]^T, [0 \ -1 \ 0 \ 0]^T, [0 \ 0 \ -1 \ 0]^T, [0 \ 0 \ 0 \ -1]^T\}$$
    are bases of $\mathbb{R}^4$. What happens when you form a new set by adding corresponding vectors from each basis together? Is this new set a basis of $\mathbb{R}^4$?

11. Consider Example 3.

12. Recall Theorem 3 part (2). Is one vector by itself independent?

13. (a) By Theorem 3 part (2), we can enlarge this one vector to a basis. Gradually add independent vectors until you have a basis.
    [ILAW: Exploration 4.3.2.]
    (b) Does Theorem 3 apply to *all* vectors?
    (c) Again, use Theorem 3. Make sure that you prove both directions:

    1. If the condition holds, then there exists a basis containing both X and Y.
    2. If there exists a basis containing both X and Y, then the condition holds.

14. Theorem 3 will be useful here. Again, ensure that you prove both directions:

    1. If the condition holds, then the set contains a basis.

Selected Hints and Solutions   557

   2. If the set contains a basis, then the condition holds.

15. (a) A quick solution is the standard basis of $\mathbb{R}^3$. Find another.

    (b) Equivalently, can you find an invertible 3×3 matrix whose columns add up to 0?

16. Check to see if you can use the time-saving results in Theorem 4 to show linear independence and/or spanning.

17. The key to these questions lies in Theorem 3 §1.4 which shows that $[X\ Y]A = [aX + bY\ cX + dY]$. Use Theorem 2 §4.2.

18. How can you go from $t_1(AX_1) + t_2(AX_2) + \cdots + t_n(AX_n) = 0$ to $t_1 X_1 + t_2 X_2 + \cdots + t_n X_n = 0$?

19. Use Theorem 4 to set upper and lower limits on $dim U$. You will find that $dim U = 0$ or $dim U = 1$. The first case gives $U = \{0\}$, whereas if $dim U = 1 = dim W$ then $U = W$ by Theorem 4 (2).

20. Similar set up as the previous problem.

21. $U \cap W \subseteq U$ is really two cases: $U \cap W = U$ and $U \cap W \subset U$. See where each case leads and apply Theorem 4 if necessary.

22. (a) *Hint*: If $\{X_1, X_2, \cdots, X_k\}$ is a basis of $null A$, show that $\{V^{-1}X_1, V^{-1}X_2, \cdots, V^{-1}X_k\}$ is a basis of $null(AV)$.

    (b) *Hint*: If $\{Y_1, Y_2, \cdots, Y_k\}$ is a basis of $im(UA)$, show that $\{U^{-1}Y_1, U^{-1}Y_2, \cdots, U^{-1}Y_k\}$ is a basis of $im A$.

23. Since $\{X_1, X_2, \cdots, X_k\}$ is the largest independent subset of U, it remains to show that $span\{X_1, X_2, \cdots, X_k\} = U$. The hint is setting up a proof by contradiction. By assuming that $span\{X_1, X_2, \cdots, X_k\} \neq U$ we are looking for a contradiction to prove that this cannot happen. By Lemma 1, the set $\{Y, X_1, X_2, \cdots, X_k\}$ is independent in U, but has $k + 1$ vectors in it. This is the contradiction since this is larger than the maximal independent set given earlier. Hence it must be that $span\{X_1, X_2, \cdots, X_k\} = U$.

24. This problem is similar to the previous one. In order to prove that $\{X_1, X_2, \cdots, X_k\}$ is a basis, all that you need to do is prove that it is independent. Again, the hint describes the set up for proof by contradiction and Theorem 3 § 4.2 provides the contradiction. If $\{X_1, X_2, \cdots, X_k\}$ were dependent one of the vectors is a linear combination of the others (call this one X_i), and so there exists a spanning set of U that is smaller than the minimal spanning set, namely the minimal spanning set with X_i removed.

25. *Hint on showing that* $\{X_1, \cdots, X_d, Y_1, \cdots, Y_k, Z_1, \cdots, Z_m\}$ *is a basis of* $U + W$:
    If the linear combination $\sum r_i X_i + \sum s_i Y_i + \sum t_i Z_i = 0$ vanishes, show that $\sum t_i Z_i \in W \cap (U \cap W) = U \cap W$.

    Hint on the final step: Once you have determined the basis of $U + W$, what is $dim(U + W)$?

4.4 Rank

1. (a) $row A = span\{[4\ -6], [7\ 6]\}$,
    $col A = span\{[4\ 7]^T, [-6\ 6]^T\}$.

    (c) $row A = span\{[0\ -4\ -8\ -4], [8\ 2\ -8\ 0], [-3\ 1\ -1\ -2]\}$,
    $col A = span\{[0\ 8\ -3]^T, [-4\ 2\ 1]^T, [-8\ -8\ -1]^T\}$.

2. The key to this problem is to recognize that each matrix on the left is the result of row (or column) operations performed on a matrix from the right. Then Lemma 1 applies.
   - (d) came from row operations on (a), so their row spaces match.
   - (b) came from column operations on (c), so their column spaces match.
   - (f) is the result of row operations on (e), so their row spaces match.

3. Find a basis for each space and then apply Theorem 1.
   [ILAW: Exploration 4.3.2 or 4.4.2.]

4. These questions are modelled after Example 1. In each part, decide whether $U = \operatorname{row} A$ or $U = \operatorname{col} A$.

5. Follow the method of Example 2.
   [ILAW: Exploration 4.4.2.]

6. Refer to Example 3 for a detailed explanation how to find $\operatorname{null} A$.
   [ILAW: Exploration 4.4.3.]

7. (b) False. Consider the matrix $A = \begin{bmatrix} 0 & 1 & 0 \\ 1 & 0 & 0 \end{bmatrix}$.

   (e) True. A row of zeros indicates that A does not have full rank, since the row of zeros is a linear combination of the other rows each multiplied by zero. So the rank of A is at most $m - 1$, that is $\operatorname{rank} A \leq m - 1 < m$.

8. By the Corollary to Theorem 1 § 4.3, the columns cannot be independent.

9. There are two tools you can use for this question: the Corollary to Theorem 1 § 4.3 (as used above), or use the Rank Theorem to find the maximal number of columns/rows that can be independent.

10. Find a contradiction if you assume that both the rows and columns are independent.

11. Use Theorem 6.

12. Notice that this question is the logical equivalent to Exercise 7.

13. All that you need is Corollary 1 of the Rank Theorem.

14. (a) Write $B = [B_1 \ B_2 \ \cdots \ B_n]$ where B_i are the columns of B. Then $AB = [AB_1 \ AB_2 \ \cdots \ AB_n]$. Now you are ready to apply equation (*).

    (b) Is $\operatorname{col} A$ a subset of $\operatorname{col} AB$, or is it the other way around? Once you determine this, apply Theorem 4 §4.3.

    (c) By Corollary 3 of Theorem 2 and (b) we have
    $$\operatorname{rank}(AB) = \operatorname{rank}(AB)^T = \operatorname{rank}(B^T A^T) \leq \operatorname{rank} B^T = \operatorname{rank} B.$$

15. (a) In order to apply the Theorem in the hint, recognize that if $X \in \operatorname{col} B$ we can write $X = t_1 B_1 + t_2 B_2 + \cdots + t_k B_k$, where B_i are the columns of B.

    (b) After applying the Theorem given in the hint, note that since $\operatorname{col} B \subseteq \operatorname{null} A$, $AB_i = 0$ for each column i of B.

16. (a) If $B \in \operatorname{col} A$, then B can be written as a linear combination of the columns of A.

    (b) What does $\operatorname{rank} A = \operatorname{rank}[A \ B]$ say about B relative to the columns of A?

17. Hint: $[I \ Z] \begin{bmatrix} I \\ Z^T \end{bmatrix} = [I \ Z][I \ Z]^T$. How many leading ones are there in $[I \ Z]$?

Selected Hints and Solutions 559

18. (a) By Theorem 4 (4) we know that AA^T is invertible, so $AA^T(AA^T)^{-1} = I_m$. This is $AB = I_m$ for $B = A^T(AA^T)^{-1}$.

    (b) By the Rank Theorem, $\text{rank} A \leq m$, so given that $\text{rank}(AB) \leq \text{rank} A$ it remains to show that $\text{rank}(AB) \geq m$.

19. (a) It is trivial to show that $\text{null} A \subseteq \text{null}(A^T A)$. To show that $\text{null}(A^T A) \subseteq \text{null} A$, choose $X \in \text{null}(A^T A)$. Then $A^T A X = 0$, and left multiplication by X^T allows you to apply the hint.

    (b) Apply Theorem 6 twice together with (a).

    (c) In order to apply Theorem 4 §4.3, first use Theorem 5 to show that $\dim\{im(A^T A)\}$ and $\dim\{im(A^T)\}$ both equal $\text{rank} A$.

20. (a) Multiply left and right sides of $AA^\# A = A$ by A^{-1}.

    (b) By Theorem 4 (4), we know that AA^T is invertible, so right multiply $AA^\# A = A$ by A^T, giving $AA^\# = I_m$. Now left multiply by A^T and recognize that $(AA^T)^T = A^T A$ is also invertible.

4.5 Orthogonality

1. See the computation preceding Example 2.
    (b) $\sqrt{76}$
    (d) 3.

2. Confirm that the dot product of each pair of vectors is zero.
   [ILAW: Exploration 4.5.1.]

3. First, verify that each set of vectors is an orthogonal basis of $\mathbb{R}^3$ using Exploration 4.5.2. Then when normalizing the orthogonal basis, calculate the length of each vector. [ILAW: Exploration 4.5.1.]
   (a) $\{\frac{1}{\sqrt{5}}[1\ 0\ 2]^T, \frac{1}{\sqrt{30}}[-2\ 5\ 1]^T, \frac{1}{\sqrt{6}}[2\ 1\ -1]^T\}$.

4. Let $X_1 = [1\ -2\ 1\ -1]^T$ and $X_2 = [2\ 1\ -1\ 1]^T$. Determine that $Y = -X_1 + X_2$. Since Z must also be in U, we can write $Z = aX_1 + bX_2$ for some a and b. What conditions does the orthogonality of Y and Z put on the variables a and b?
   [ILAW: Exploration 4.1.4.]

5. For convenience, label the given set $\{X_1, X_2, X_3, Y\}$ where the X_i are the given vectors with known values and $Y = [a\ b\ c\ d]^T$. Orthogonality forces $(X_i)^T Y = 0$ for each i, so we can set up the system

$$\begin{bmatrix} (X_1)^T \\ (X_2)^T \\ (X_3)^T \end{bmatrix} \begin{bmatrix} a \\ b \\ c \\ d \end{bmatrix} = 0.$$

   [ILAW: Explorations 1.4.1 or 1.2.2.]

6. (b) The coefficients are $\frac{-4a+b+5c}{\sqrt{42}}$, $\frac{-a+b-c}{\sqrt{3}}$ and $\frac{2a+3b+c}{\sqrt{14}}$.
   [ILAW: Exploration 4.5.2.]

7. (b) $\{[2\ 1]^T, [-1\ 2]^T\}$
   (d) $\{[0\ 1\ 1]^T, [1\ 0\ 0]^T, [0\ -2\ 2]^T\}$
   [ILAW: Exploration 4.5.3.]

8. (b) $Q = \frac{1}{\sqrt{5}}\begin{bmatrix} 2 & -1 \\ 1 & 2 \end{bmatrix}$, $R = \frac{1}{\sqrt{5}}\begin{bmatrix} 5 & 3 \\ 0 & 1 \end{bmatrix}$

(d) $Q = \frac{1}{\sqrt{3}}\begin{bmatrix} 1 & 1 & 0 \\ -1 & 0 & 1 \\ 0 & 1 & 1 \\ 1 & -1 & 1 \end{bmatrix}$, $R = \frac{1}{\sqrt{3}}\begin{bmatrix} 3 & 0 & -1 \\ 0 & 3 & 1 \\ 0 & 0 & 2 \end{bmatrix}$

[ILAW: Exploration 4.5.4.]

9. (b) False. Consider the counterexample $\{X, Y\} = \{[1\ 0]^T, [0\ 1]^T\}$.

(d) True. In order for $\{X_1, X_2, Y_1, Y_2, Y_3\}$ to be orthogonal, $X_i \bullet X_j = 0$, $Y_i \bullet Y_j = 0$, $X_i \bullet Y_j = 0$ for all appropriate i and j. Each of these hold by the assumptions.

10. Use Theorem 1 part (3) to show that $a_i X_i \bullet a_j X_j = a_i a_j (X_i \bullet X_j)$.

11. (a) Use the calculation in Example 2 to find expressions for $\|X + Y\|$ and $\|X - Y\|$.

(b) Substitute the expressions from part (a) into $\|X - Y\| = \|X + Y\|$ and manipulate this new formula to show that $X \bullet Y = 0$.

12. (a) Again, use the formula from Example 2.

(b) Use the formula from Example 2 to expand $\|X + (Y + Z)\|^2$.

13. (a) Start by using the formulae derived in Exercise 11 to find an expression for $\|X + Y\|^2 - \|X - Y\|^2$.

(b) Similarly to (a), find an expression for $\|X + Y\|^2 + \|X - Y\|^2$.

14. If X is an eigenvector of $A^T A$, then $A^T A X = \lambda X$. Use this to show that $\|AX\|^2 = \lambda \|X\|^2$.

15. Assign $Q = [C_1\ C_2\ \cdots\ C_n]$. Show that $C_i \bullet C_i = 1$ and $C_i \bullet C_j = 0$ for $i \neq j$. How does this relate to the entries of $Q^T Q$?

16. Since X is in $\mathbb{R}^n = \text{span}\{F_1, F_2, \cdots, F_m\}$ write $X = a_1 F_1 + a_2 F_2 + \cdots + a_m F_m$, and use this to show $\|X\| = X \bullet X = 0$.

17. Showing $X = Y$ is equivalent to showing $X - Y = 0$. Rearrange the question so that you can apply Exercise 16.

18. (a) Consider the Orthogonal Lemma

(b) Which multiples of G have length 1?

19. Since $\{X_1, X_2, \cdots, X_m\}$ is already orthogonal, verify that the Gram-Schmidt Algorithm gives $F_i = X_i$ for all i.

20. Let A be the matrix with the X_i^T as rows and let B be the matrix with the Y_i as columns. Does the product AB enable you to use Theorem 2 §4.2?

21. Express X and Y by applying the Expansion Theorem. Then apply Theorem 1 keeping in mind that $G_i \bullet G_j = 0$ for $i \neq j$.

22. Let R be the row-echelon form of A, and use the Gram-Schmidt Algorithm on the nonzero rows of R fom the bottom up. Use Lemma 1 §1.6.

4.6 Projections and Approximations

1. There are two methods for solving these types of problems. One is given in Example 4, and the other involves systems of linear equations. We will do part (a) as in Example 4, and use the second method on part (b).

SELECTED HINTS AND SOLUTIONS

(a) An orthogonal basis of U is $\{[1\ 2\ -1]^T, [11\ -2\ 7]^T\}$. This gives $\text{proj}_U(X) = \frac{1}{29}[85\ -26\ 62]^T$, and so $X - P = \frac{1}{29}[2\ -3\ -4]^T$.

(b) $U^\perp = \text{span}\{[2\ -11\ 1]^T\}$, so we can write X as a sum of a vector $r[-1\ 0\ 2]^T + s[3\ 1\ 5]^T$ from U, and a vector $t[2\ -11\ 1]^T$ from $U^\perp$. Thus we can write

$$r\begin{bmatrix}-1\\0\\2\end{bmatrix} + s\begin{bmatrix}3\\1\\5\end{bmatrix} + t\begin{bmatrix}2\\-11\\1\end{bmatrix} = \begin{bmatrix}1\\1\\3\end{bmatrix}.$$

Solving this system gives $r = \frac{1}{3}, s = \frac{10}{21}, t = \frac{-1}{21}$. Thus $\frac{1}{21}[23\ 10\ 64]^T$ is the vector in U and $\frac{1}{21}[-2\ 11\ -1]^T$ is the vector from $U^\perp$.
[ILAW: Exploration 4.6.2.]

2. Follow the projection calculations in the solution to Example 4, and check your answers.
[ILAW: Exploration 4.6.2.]

3. (b) $[0\ -1\ 1]^T$

   (d) $\frac{1}{6}[25\ -7\ 16\ -6]^T$.
   [ILAW: Exploration 4.6.3.]

4. (b) $\frac{1}{6}[5\ 2]^T$

   (d) $\frac{1}{1339}[722\ -1722\ -1460]^T$.
   [ILAW: Exploration 4.6.4.]

5. (b) $[\frac{74}{55}\ \frac{137}{220}\ -\frac{21}{44}]^T$.
   [ILAW: Exploration 4.6.5.]

6. (b) $[\frac{429}{133}\ -\frac{743}{798}\ -\frac{219}{266}\ \frac{13}{57}]^T$.
   [ILAW: Exploration 4.6.5.]

7.
$$M = \begin{bmatrix} f_1(x_1) & f_2(x_1) & f_3(x_1) \\ f_1(x_2) & f_2(x_2) & f_3(x_2) \\ f_1(x_3) & f_2(x_3) & f_3(x_3) \\ f_1(x_4) & f_2(x_4) & f_3(x_4) \end{bmatrix} = \begin{bmatrix} x_1 & (x_1)^2 & (-1)^{x_1} \\ x_2 & (x_2)^2 & (-1)^{x_2} \\ x_3 & (x_3)^2 & (-1)^{x_3} \\ x_4 & (x_4)^2 & (-1)^{x_4} \end{bmatrix}$$

and

$$Y = \begin{bmatrix} y_1 \\ y_2 \\ y_3 \\ y_4 \end{bmatrix} = \begin{bmatrix} 1 \\ 3 \\ 1 \\ 0 \end{bmatrix}.$$

Using these matrices for M and Y, solving the normal equation gives the best approximation $Z = [a_0\ a_1\ a_2]^T = \frac{1}{36}[-19\ 11\ 13]^T$.

8. (a) To show that $U^\perp \subseteq \{X \text{ in } \mathbb{R}^n \mid AX^T = 0\}$, use the fact that

$$AX^T = [X \bullet Y_1\ X \bullet Y_2\ \cdots\ X \bullet Y_m]^T$$

to show that if a vector X is in $U^\perp$ then $AX^T = 0$.
To show that $\{X \text{ in } \mathbb{R}^n \mid AX^T = 0\} \subseteq U^\perp$, use the same fact given in part (a).

(b) $U^\perp = \text{span}\{[5\ 3\ 1\ 0], [-5\ -2\ 0\ 1]\}$.

9. (b) True. We know from the Projection Theorem that $proj_U(X)$ is in U, so if it is also in $U^\perp$, then $proj_U(X) = 0$. Now, by the other statement of the Projection Theorem, $X = X - 0 = X - proj_U(X) \in U^\perp$.

   (d) True. Using the Projection Theorem, if $proj_U(X) = 0$ the $X - 0 = X$ is in $U^\perp$.

10. Let $X = a_1 F_1 + a_2 F_2 + \cdots + a_m F_m$ for an orthogonal basis $\{F_1, F_2, \cdots, F_m\}$ of U. Use this in the definition of $proj_U(X)$ to show that $proj_U(X) = a_1 F_1 + a_2 F_2 + \cdots + x_m F_m$. The Projection Theorem is used to show that if $proj_U(X) = X$ then X is in U.

11. (a) *First Direction*: Let $U^\perp = \mathbb{R}^n$. If X is be a vector in U, then $X \in U \subseteq \mathbb{R}^n = U^\perp$. Now that X is in $U^\perp$, calculate $X \cdot X$.

    (b) Apply property (2) of Theorem 3 to part (a).

12. *First direction*: Suppose that $proj_U(X) = 0$. Use the Projection Theorem to show that X is in $U^\perp$.

    Second direction: Suppose that X is in $U^\perp$ and $\{F_1, F_2, \cdots, F_m\}$ is an orthogonal basis of U. First confirm that $X \bullet F_i = 0$ for all i, and use this to show that $proj_U(X) = 0$.

13. (a) First show that $\frac{(X+Y)\bullet F_i}{|F_i|^2} F_i = \frac{X\bullet F_i}{|F_i|^2} F_i + \frac{Y\bullet F_i}{|F_i|^2} F_i$ for any i.

    (b) First show that $\frac{(aX)\bullet F_i}{|F_i|^2} F_i = a\left(\frac{X\bullet F_i}{|F_i|^2}\right) F_i$ for any i.

    (c) Look at Exercise 10.

14. If X is in $\mathbb{R}^n$, express it as $X = \frac{X\bullet F_1}{|F_1|^2} F_1 + \cdots + \frac{X\bullet F_k}{|F_k|^2} F_k + \cdots + \frac{X\bullet F_n}{|F_n|^2} F_n$.

15. Use Exercise 14 to show $U^\perp = span\{F_{k+1}, F_{k+2}, \cdots, F_n\}$ where F_i are vectors from an orthogonal basis of U. Apply the result of Exercise 8(a) to show that

    $$U = U^{\perp\perp} = \{X \text{ in } \mathbb{R}^n \mid AX^T = 0\}$$

    for $A = [F_{k+1} \ F_{k+2} \ \cdots \ F_n]^T$.

16. (a) *First direction*: Let $X \in U^\perp \cap W^\perp$. Use the fact that X is in both $U^\perp$ and $W^\perp$ to show that $(U + W) \bullet X = 0$.

    Second direction: Let $X \in (U + W)^\perp$, so $X \bullet Y = 0$ for all Y in $U + W$. Is $U \subseteq U + W$? What does this tell you about $X \bullet Z$ if $Z \in U$? Is this the same for if $Z \in W$?

    (c) Use part (b) twice, comparing $U^\perp$ and $W^\perp$ to $(U \cap W)^\perp$.

    (d) *Hint for proving* $dim(U^\perp + W^\perp) = dim((U \cap W)^\perp)$: From (a) and Exercise 2 §4.3 we get $dim(U^\perp + W^\perp) = dim(U^\perp) + dim(W^\perp) - dim((U + W)^\perp)$. Use Theorem 3 (1) to show that $dim(U^\perp + W^\perp) = n + (dim(U + W) - dim(U) - dim(W))$. Follow this with one more application of Exercise 25 §4.3 to give $dim(U^\perp + W^\perp) = n - dim(U \cap W)$.

17. (c) Recall that for any two matrices A and B, $(A + B)^T = A^T + B^T$ and, if $AB = 0 = BA$, $(A + B)^2 = (A + B)(A + B)$.

4.7 Orthogonal Diagonalization

1. (b) False. Consider the matrix $\begin{bmatrix} \frac{1}{2} & -1 \\ \frac{1}{2} & 1 \end{bmatrix}$.

(d) False. Consider the matrix $\frac{1}{\sqrt{2}}\begin{bmatrix} 1 & 1 \\ -1 & 1 \end{bmatrix}$.

(f) True. Calculate $(P^T A P)^T$ and use the fact that $A^T = A$.

2. In each case simply normalize each column. For (d) the result is

$$\frac{1}{\sqrt{6}}\begin{bmatrix} \sqrt{2} & 0 & 2 \\ -\sqrt{2} & \sqrt{3} & 1 \\ \sqrt{2} & \sqrt{3} & -1 \end{bmatrix}.$$

3. One possibility is $Z = [\frac{2}{3} \ \frac{1}{3} \ \frac{-2}{3}]$. If X and Y denote the existing rows and $U = \text{span}\{X, Y\}$, then $\dim U = 2$, so $\dim U^{\perp} = 1$ by Theorem 3 §4.6. Hence every vector in $U^{\perp}$ is a multiple of Z, and the only ones of length 1 are $\pm Z$.

4. The expansion is $Y = (Y \bullet X_1)X_1 + (Y \bullet X_2)X_2 + \cdots + (Y \bullet X_n)X_n$ so $Y = (Y \bullet X_1)X_1$ because Y is orthogonal to $X_2, \cdots, X_n$. Since $\|Y\| = 1$, show that the only possibilities are $\pm Y$.

5. In each case find the (positive) eigenvalues of the symmetric matrix A, find an orthogonal basis of each eigenspace (using Gram-Schmidt if necessary), and normalize to get an orthonormal basis. Take P to be the matrix with these orthonormal vectors as its columns.

(d) $P = \frac{1}{\sqrt{182}}\begin{bmatrix} 3\sqrt{13} & 2\sqrt{14} & 3 \\ -\sqrt{13} & 0 & 13 \\ 2\sqrt{13} & -3\sqrt{14} & 2 \end{bmatrix}$.

(f) $P = \frac{1}{2}\begin{bmatrix} \sqrt{2} & 0 & 1 & 1 \\ \sqrt{2} & 0 & -1 & -1 \\ 0 & \sqrt{2} & 1 & -1 \\ 0 & \sqrt{2} & -1 & 1 \end{bmatrix}$.

6. $(cP)^T(cP) = c^2 P^T P$.

7. If A is $n \times n$, let $A = \begin{bmatrix} a & X \\ 0 & B \end{bmatrix}$ in block form where X is a row of length $(n-1)$. From $A^T A = I$, deduce that $a^2 = 1$ and $aX = 0$, so $a = \pm 1$ and $A = \begin{bmatrix} a & 0 \\ 0 & B \end{bmatrix}$. Now repeat the argument on B.

8. Use the Hint. Don't forget to prove the converse.

9. (a) The columns of I are orthonormal no matter in which order they are written.

10. The rows of A^T are $C_1^T, C_2^T, \cdots, C_n^T$.

11. For (i)$\Rightarrow$(ii) take $P = \begin{bmatrix} \frac{C_1}{\|C_1\|} & \frac{C_2}{\|C_2\|} & \cdots & \frac{C_n}{\|C_n\|} \end{bmatrix}$; for (ii)$\Rightarrow$(iii) show that $P^T P = I$; for (iii)$\Rightarrow$(i) and use the preceding exercise.

12. Use the Hint.

13. Use the Hint.

14. If $A = B^2$ where B is symmetric, and $AX = \lambda X$ where $X \neq 0$, show that $\|BX\|^2 = \lambda \|X\|^2$. For the converse, use the Hint.

15. As in the proof of Lemma 3, show that $(AX) \bullet Y = X \bullet (A^T Y)$, and proceed as in the proof of Theorem 8.

16. (a) (ii). If $A \overset{\circ}{\sim} B$, say $B = P^T A P$, then $A = PBP^T = (P^T)^T B(P^T)$.
    (d) A and B are similar.

17. If E_j denotes column j of I, show that $E_i^T A E_j$ is the (i,j)-entry of A.

18. (b) If P is orthogonal and symmetric, then $E^2 = \frac{1}{4}(I^2 - IP - PI + P^2) = \frac{1}{4}(2I - 2P) = E$, and $E^T = \frac{1}{2}(I^T - P^T) = E$.

19. Use the Hint.

20. (a) If A is $n \times n$, write $A = \begin{bmatrix} 0 & Y \\ 0 & B \end{bmatrix}$ in block form where B is $(n-1) \times (n-1)$.
    Show that $A^k = \begin{bmatrix} 0 & YB^{k-1} \\ 0 & B^k \end{bmatrix}$ for each $k \geq 1$. By induction on n some power of B is zero.
    (c) By Theorem 9 let $P^T A P = B$ where B is upper triangular and P is orthogonal. Let D be the diagonal matrix with the same diagonal as B, and let $X = B - D$. Then X is nilpotent by (a). Use $S = PBP^T$ and $N = PXP^T$.

21. (a) If $AX = \lambda X$ show that $A^k X = \lambda^k X$.
    (b) By Theorem 9, let $P^T A P = B$ where P is orthogonal and B is upper triangular. But A and B are similar so the diagonal entries of B are the eigenvalues of A, and so are all zero. Now use (a) of the preceding exercise.
    (c) If $P^{-1}AP = D$ is diagonal and $A \neq 0$ is nilpotent, then $D = 0$ by (a). Hence $A = 0$.

22. First apply Theorem 9 to A^T.

23. Let $A = [C_1 \ C_2 \ \cdots \ C_n]$ where C_j is column j of A. Complete the set of nonzero C_j to an orthogonal basis of $\mathbb{R}^n$, and let P denote the orthogonal matrix with these basis vectors as its columns, keeping the normalization of each nonzero C_j in column j of P. Then take $D = diag(d_1, d_2, \cdots, d_n)$ where
$$d_j = \begin{cases} \|C_j\| & \text{if } C_j \neq 0 \\ 0 & \text{if } C_j = 0 \end{cases}. \text{ Verify that } A = PD.$$

4.8 Quadratic Forms

1. (b) False. Consider $\begin{bmatrix} 1 & 1 \\ 0 & 2 \end{bmatrix}$.
   (d) True. See the discussion following Theorem 3.
   (f) False. Consider $\begin{bmatrix} 1 & 2 \\ 0 & 2 \end{bmatrix}$.

2. (b) $A = \begin{bmatrix} 2 & 3 & 3 \\ 3 & 1 & -1 \\ 3 & -1 & 5 \end{bmatrix}$.

3. (b) $y_1 = \frac{1}{\sqrt{5}}(2x_1 + x_2)$, $y_2 = \frac{1}{\sqrt{5}}(x_1 - 2x_2)$, and $q = 3y_1^2 - 2y_2^2$.
   (d) $y_1 = \frac{1}{\sqrt{4+2\sqrt{2}}}(x_1 + (1+\sqrt{2})x_2)$, $y_2 = \frac{1}{\sqrt{4+2\sqrt{2}}}(x_1 + (1-\sqrt{2})x_2)$, and $q = (1+2\sqrt{2})y_1^2 + (1-2\sqrt{2})y_2^2$.
   (f) $y_1 = \frac{1}{\sqrt{5}}(x_1 - 2x_2)$, $y_2 = \frac{1}{\sqrt{5}}(-2x_1 + x_3)$, $y_3 = \frac{1}{3}(2x_1 + x_2 + 2x_3)$, and $q = 9y_1^2 + 9y_2^2$.

4. (b) Ellipse.

Selected Hints and Solutions

5. (b) Maximum $\frac{1}{2}(5\sqrt{2}-1)$, minimum $-\frac{1}{2}(5\sqrt{2}+1)$.
   (d) Maximum $2\sqrt{3}-1$, minimum $-(2\sqrt{3}+1)$.

6. (b) $A = U^T U$ where $U = \frac{\sqrt{3}}{3}\begin{bmatrix} 3 & -2 \\ 0 & \sqrt{2} \end{bmatrix}$.

   (d) $A = U^T U$ where $U = \frac{\sqrt{5}}{5}\begin{bmatrix} 5 & 3 & -2 \\ 0 & 1 & 1 \\ 0 & 0 & \sqrt{15} \end{bmatrix}$.

7. Verify that $E_j^T A E_j$ is exactly the (j,j)-entry of A.

8. (a) Use the Hint.
   (c) For A^2 use (a). If A^m is known to be positive definite, show that the same is true of A^{m+2} by writing $X^T A^{m+2} X = (XA)^T A^m (XA)$.
   (d) If A^m is positive definite for some odd power $m > 1$, show that A^{m-2} is also positive definite using an argument like that in (c).

9. How are the eigenvalues of A^{-1} related to those of A?

10. $X^T(A+B)X = X^T AX + X^T BX$.

11. If A is $n \times n$ and B is $m \times m$ let Z be a column in $\mathbb{R}^{n+m}$. Write $Z = \begin{bmatrix} X \\ Y \end{bmatrix}$ where X is in $\mathbb{R}^n$ and Y is in $\mathbb{R}^m$. Show that $\begin{bmatrix} X \\ Y \end{bmatrix}^T \begin{bmatrix} A & 0 \\ 0 & B \end{bmatrix} \begin{bmatrix} X \\ Y \end{bmatrix} = X^T AX + Y^T BY$.

12. If $X^T(U^T AU)X = 0$, use the fact that A is positive definite to show that $UX = 0$. Now apply Theorem 3 §4.4.

13. Given the Hint, take $B = PD_0 P^T$.

14. Use the Hint.

15. (a) As in the Hint, the matrix $L^{-1} L_1 D_1 = DUU_1^{-1} = C$ is lower triangular (left side) and upper triangular (right side), and so is diagonal. But the diagonal entries of $C = L^{-1} L_1 D_1$ are just those of D_1 because L^{-1} and L_1 are both unit lower triangular (verify). This shows that the diagonal matrix $C = L^{-1} L_1 D_1$ must equal D_1, and so $L = L_1$ and $D_1 = C$. Similarly $U = U_1$ and $D = C$.
    (b) If A is positive definite, let $A = U^T U$ be its Cholesky factorization. Let D be the diagonal matrix with the same (positive) diagonal entries as U. Verify that $U_1 = D^{-1} U$ is unit upper triangular, and that $A = U_1^T D^2 U_1$.

16. As in the Hint, $X = P^T Y$ gives $x_1 = \cos\theta\, y_1 + \sin\theta\, y_2$ and $x_2 = -\sin\theta\, y_1 + \cos\theta\, y_2$. Substitute in the equation $ax_1^2 + bx_1 x_2 + cx_2^2 = 1$, and show that the coefficient of $y_1 y_2$ is $(a-c)\sin(2\theta) + b\cos(2\theta)$.

17. (a) As in the Hint, if $X^T CX = 0$ for all X, then $(X+Y)^T C(X+Y) = 0$ so (since $X^T CY = Y^T CX$—it is 1×1—we get $X^T CY = 0$ for all X and Y. Now use the fact that $E_i^T C E_j$ is the (i,j)-entry of C.

4.9 Linear Transformations

1. (b) True, by Theorem 5.
   (d) False. Consider the transformation in (b), with $X = \begin{bmatrix} 1 & 1 \end{bmatrix}^T$ and $Y = \begin{bmatrix} 1 & -1 \end{bmatrix}^T$.

2. If X is between Y and Z, write $X = (1-f)Y + fZ$ where $0 \leq f \leq 1$. Apply T and use T1 and T2.

3. aI_n.

4. The $m \times n$ matrix $[I_n \ 0]$ in block form.

5. The $m \times n$ matrix $\begin{bmatrix} I_m \\ 0 \end{bmatrix}$ in block form.

6. (a) $T \begin{bmatrix} x_1 \\ x_2 \\ x_3 \end{bmatrix} = \begin{bmatrix} 3x_2 - x_3 \\ -4x_1 + 3x_2 - 2x_3 \\ 3x_2 \end{bmatrix}$.

7. (b) The basis $\mathcal{F}$ is orthogonal so the Expansion Theorem (Theorem 6 §4.5) gives

$$[x_1 \ x_2 \ x_3]^T =$$
$$\frac{x_1 + 2x_2 + 2x_3}{9} \begin{bmatrix} 1 \\ 2 \\ 2 \end{bmatrix} + \frac{2x_1 + x_2 - 2x_3}{9} \begin{bmatrix} 2 \\ 1 \\ -2 \end{bmatrix} + \frac{2x_1 - 2x_2 + x_3}{9} \begin{bmatrix} 2 \\ -2 \\ 1 \end{bmatrix}.$$

Writing $X = [x_1 \ x_2 \ x_3]^T$, Theorem 7 gives $C_{\mathcal{F}}(T(X)) = M_{\mathcal{F}}(T)C_{\mathcal{F}}(X)$, that is

$$C_{\mathcal{F}}(T(X)) = \begin{bmatrix} 3 & -5 & 1 \\ 1 & 1 & 5 \\ -2 & 2 & 7 \end{bmatrix} \tfrac{1}{9} \begin{bmatrix} x_1 + 2x_2 + 2x_3 \\ 2x_1 + x_2 - 2x_3 \\ 2x_1 - 2x_2 + x_3 \end{bmatrix}$$
$$= \tfrac{1}{9} \begin{bmatrix} -5x_1 - x_2 + 17x_3 \\ 13x_1 - 7x_2 + 5x_3 \\ 16x_1 - 16x_2 - x_3 \end{bmatrix}.$$

Hence you can calculate $T(X)$.

8. Since $X \neq 0$ it is part of a basis $\{X, X_2, X_3, \cdots, X_n\}$ of $\mathbb{R}^n$ by Theorem 3 §4.3. Hence, by Theorem 5, there is a linear transformation $T : \mathbb{R}^n \to \mathbb{R}^m$ such that $T(X) = Y$ and (say) $T(X_i) = 0$ for $2 \leq i \leq n$.

9. (b) Let $\{E_1, \cdots, E_n\}$ denote the standard basis of $\mathbb{R}^n$, and write $T(E_i) = a_i$ for each i. Take $X_0 = [a_1 \ a_2 \ \cdots \ a_n]^T$.

10. (b) Take $X_0 = T(1)$.

11. Every vector X in $\mathbb{R}^n$ is a linear combination of the Y_i.

12. Let X and Y be in $kerT$, so $T(X) = 0$ and $T(Y) = 0$. Hence $T(X+Y) = T(X) + T(Y) = 0 + 0 = 0$. This shows that $kerT$ is closed under addition, part of the Subspace Test.

13. If $T(X)$ and $T(Y)$ are in imT then $T(X) + T(Y) = T(X+Y)$ is also in imT, so imT is closed under addition. Complete the Subspace Test.

14. (a) Let $a_1X_1 + a_2X_2 + \cdots + a_nX_n = 0$. Apply T and use Theorem 1 and the independence of $\{T(X_1), T(X_2), \cdots, T(X_n)\}$.

15. If $\mathcal{F} = \{F_1, F_2, \cdots, F_n\}$, we have $P_{\mathcal{G} \leftarrow \mathcal{F}} = [C_{\mathcal{G}}(F_1) \ C_{\mathcal{G}}(F_2) \ \cdots \ C_{\mathcal{G}}(F_n)]^T$. In this case $P_{\mathcal{G} \leftarrow \mathcal{F}} = [C_{\mathcal{G}}([1 \ 1]^T) \ C_{\mathcal{G}}([0 \ 1]^T)] = \begin{bmatrix} 1 & -1 \\ -1 & 2 \end{bmatrix}$ because $[1 \ 1]^T = [2 \ 3]^T - [1 \ 2]^T$ and $[0 \ 1]^T = -[2 \ 3]^T + 2[1 \ 2]^T$.

SELECTED HINTS AND SOLUTIONS

16. Mimic the proof of $T(aX) = aT(X)$ in Theorem 9.

17. If S and T are distance preserving then $\|S(X) - S(Y)\| = \|X - Y\|$ for all X and Y, with a similar condition on T. So $\|(S \circ T)(X) - (S \circ T)(Y)\| = \|S[T(X)] - S[T(Y)]\|$ $= \|T(X) - T(Y)\|$ because S is distance preserving. Continue.

18. We have $(Q_{T(Y)} \circ T)(X) = Q_{T(Y)}[T(X)] = T(X) + T(Y)$. Now use the linearity of T, and continue.

19. If X is in $\mathbb{R}^n$, use the Projection Theorem (Theorem 2 §4.6) to see that $P(X)$ and $P(X) - X$ are orthogonal. Using the Hint, apply Pythagoras' (Theorem 4 §4.5) twice to conclude that $\|T(X)\| = \|X\|$.

20. Let E_1 and E_2 be orthogonal unit vectors in the plane, and let N be a unit normal to the plane. Show $\mathcal{F} = \{E_1, E_2, N\}$ is an orthonormal basis of $\mathbb{R}^3$. If T is a reflection in the plane, show that that $T(E_i) = E_i$ for each i, and that $T(N) = -N$. Hence show that the $\mathcal{F}$-matrix of T is diagonal, and apply Theorem 8.

21. (a) Show that $M_{\mathcal{E}}(T_A) = A$ where $\mathcal{E}$ is the standard basis, and apply Theorem 8.

    (b) Apply Lemma 3 with a basis of eigenvectors of A.

22. If $\mathcal{F} = \{F_1, F_2, \cdots, F_n\}$, show that $C(F_i)$ is column i of the identity matrix. Alternatively, apply Lemmas 1 and 2.

23. Use the Hint.

24. (b) Apply the preceding exercise to (a).

25. These follow directly from the definition of isometries, or from Theorem 11 using the Corollary to Theorem 4 §4.7.

26. (a) Use Theorem 11 and the Corollary to Theorem 4 §4.7.

    (b) If $\mathcal{F}$ is any basis, use Examples 15 and 16 to show that show that $det(M_{\mathcal{F}}(T)) = 1$ for any rotation and $det(M_{\mathcal{F}}(T)) = 1$ for any reflection. Then use the Hint.

27. Use the Hint.

28. Use the Hint.

4.10 Complex Matrices

1. (b) False. $\frac{1}{\sqrt{2}}\begin{bmatrix} 1 & -1 \\ 1 & 1 \end{bmatrix}$

2. (c) $\|Z\| = 2\sqrt{5}$, $\|W\| = 2\sqrt{2}$; not orthogonal as $\langle Z, W \rangle = 4 + 5i$ is nonzero.

3. (a) Normal only.

   (b) Hermitian (and normal), not unitary.

   (c) None.

   (d) Unitary (and normal), not Hermitian.

4. (a) $U = \frac{1}{\sqrt{2}} \begin{bmatrix} 1 & i \\ i & 1 \end{bmatrix}$, $U^*ZU = \begin{bmatrix} 0 & 0 \\ 0 & 2 \end{bmatrix}$.

(b) $U = \frac{1}{\sqrt{14}} \begin{bmatrix} -2 & 3-i \\ 3+i & 2 \end{bmatrix}$, $U^*ZU = \begin{bmatrix} -1 & 0 \\ 0 & 6 \end{bmatrix}$.

(c) $U = \frac{1}{\sqrt{3}} \begin{bmatrix} 1+i & 0 & -1 \\ 0 & \sqrt{3} & 0 \\ 1 & 0 & 1-i \end{bmatrix}$, $U^*ZU = \begin{bmatrix} 2 & 0 & 0 \\ 0 & 2 & 0 \\ 0 & 0 & -1 \end{bmatrix}$.

(d) $U = \frac{1}{\sqrt{3}} \begin{bmatrix} \sqrt{3} & 0 & 0 \\ 0 & 1+i & 1 \\ 0 & -1 & 1-i \end{bmatrix}$, $U^*ZU = \begin{bmatrix} 1 & 0 & 0 \\ 0 & 0 & 0 \\ 0 & 0 & 3 \end{bmatrix}$.

5. Every complex eigenvalue of A must be a root of the characteristic polynomial, and so must be 1. Hence $U^{-1}AU = I$, whence $A = I$, a contradiction.

6. Write $U = \begin{bmatrix} a & b \\ c & d \end{bmatrix}$ and suppose that $U^{-1}AU = T = \begin{bmatrix} p & z \\ 0 & q \end{bmatrix}$. We have $AU = UT$ and comparing entries gives $-c = ap$ and $a = cp$. Hence p is real because a and c are real and they are not both zero (because U is invertible). But $a(1 + p^2) = 0 = c(1 + p^2)$, which leads to a contradiction.

7. If H and H_1 are Hermitian, show that $\overline{HH_1} = (H_1H)^T$, which is $(HH_1)^T$ by hypothesis. The converse is similar.

8. Use the definitions.

9. (b) The diagonal entries must have absolute value 1.

10. (a) Use Theorem 3.

(b) By (a), $H^2 = H^*H = Z^*ZZ^*Z = Z^*IZ = H$. Note that Z need not be square.

11. If $A = \begin{bmatrix} a & b \\ c & d \end{bmatrix}$ is real and normal, we have $AA^T = A^TA$, and equating entries gives $b^2 = c^2$ and $ac + bd = ab + cd$. If $b = c$ then A is symmetric. Otherwise $b = -c$ and you can show that $d = a$.

12. Use the fact that $\langle Z, W \rangle = Z^T \bar{W}$ for all columns Z and W.

13. Use the definitions and Theorem 3.

14. Use the definitions and Theorem 3.

15. (c) Use Exercise 12.

(d) Use (c).

16. Use Theorem 4 and (d) of the preceding exercise.

17. Use the definitions and Theorem 3.

18. Use the definitions and Theorem 3.

19. Apply Schur's Theorem to G^*, and apply $*$ to both sides.

20. (a) If $Z = [z_{ij}]$ write $z_{ij} = a_{ij} + b_{ij}i$ where a_{ij} and b_{ij} are real. Use $A = [a_{ij}]$ and $B = [b_{ij}]$.

(b) Show first that $Z^* = A^T - iB^T$.

21. (c) Mimic the proof of Theorem 4.

(d) If Z is any matrix, show that $\frac{1}{2}(Z + Z^*)$ is Hermitian and $\frac{1}{2}(Z - Z^*)$ is skew Hermitian.

Selected Hints and Solutions 569

22. See Exercises 20 and 21 in Section 4.7.
23. Let $\lambda_1, \cdots, \lambda_n$ be the eigenvalues of the Hermitian matrix H. They are the roots of $c_H(x)$, so $c_H(x) = (x-\lambda_1)(x-\lambda_2)\cdots(x-\lambda_n)$. Thus the coefficients of $c_H(x)$ are sums of products of the λ_i, and Theorem 4 applies.
24. Use the Hints.

4.11 Singular Value Decomposition

1. $|k|\sigma_1 \geq |k|\sigma_2 \geq \cdots \geq |k|\sigma_n$.
2. In the proof of Theorem 2 we saw that $(AA^T)P = P(\Sigma\Sigma^T)$. Use this to deduce that $(AA^T)P_i = \sigma_i^2 P_i$ for each i.
3. $A^{-1} = (Q^{-1})^T \Sigma^{-1} P^{-1}$. Now write the nonzero diagonal entries in Σ^{-1} in the reverse order, and use the Corollary to Theorem 4 §4.7. Note that this (with Theorem 2) shows that the singular values of A^{-1} are the reciprocals of those of A.
4. If $A = P\Sigma Q$ is a singular value decomposition for the square matrix A then $det A = det P \, det \Sigma \, det Q$. Use the Corollary to Theorem 4 §4.7 and the definition of Σ.
5. If $B = PA$ where P is orthogonal, compute $B^T B$ and use the definition of the singular values.
6. By Theorem 5 §1.5, a square matrix is invertible if and only if all its eigenvalues are nonzero. Apply this to the matrix $A^T A$, and use Theorem 3 §4.4.
7. We have
$$A = [P_1 \; P_2 \cdots P_m] \, diag(\sigma_1, \cdots, \sigma_r, 0, \cdots, 0) \begin{bmatrix} Q_1^T \\ Q_1^T \\ \vdots \\ Q_1^T \end{bmatrix}$$
$$= [\sigma_1 P_1 \; \sigma_2 P_2 \cdots \sigma_r P_r \; 0 \cdots 0] \begin{bmatrix} Q_1^T \\ Q_1^T \\ \vdots \\ Q_1^T \end{bmatrix}.$$
Use block multiplication.
8. Each $\lambda_i \geq 0$ by the discussion preceding Lemma 1, so D is a real diagonal matrix with positive diagonal entries and $Q^T(A^T A)Q = D^2$. Verify that $PDQ^T = A$, and show that P is orthogonal.
9. We have $AA^+ = P(\Sigma\Sigma^+)P^T$ and $\Sigma\Sigma^+ = \begin{bmatrix} I_r & 0 \\ 0 & 0 \end{bmatrix}_{m \times m}$ is diagonal. Hence AA^+ is symmetric. The rest is routine.
10. If μ is an eigenvalue of A, show first that μ^2 is an eigenvalue of $A^2 = A^T A$.
11. It need not even be symmetric. Use Example 3 to find examples.
12. If λ is any eigenvalue of G, show first that $\lambda = 1$. Now use the fact that G is diagonalizable (it is symmetric).

13. Since G is invertible, it is positive definite. The eigenvalues of G^{-1} are the reciprocals of those of G.

14. How are the eigenvalues of kG related to those of G?

15. Write $H = U^T GU$. Then H is symmetric since G is; use condition (2) preceding Example 3.

16. You must verify that G is positive semidefinite and P and Q are orthogonal.

17. Take any polar form for A^T, and transpose.

18. (b) You will need the fact that if $\{Y_1, \cdots, Y_n\}$ is orthogonal, then
$$\|Y_1 + \cdots + Y_n\|^2 = \|Y_1\|^2 + \cdots + \|Y_n\|^2.$$

19. Use the Hint.

Chapter 5

VECTOR SPACES

5.1 Examples and Basic Properties

1. (a) True. The zero vector.
   (b) True. Every multiple of any nonzero vector.
   (c) False. 1 is not in $span\{1-x, 2+x-x^2\}$.
   (d) True. It is a standard trigonometric identity that $1 = cos^2 x + sin^2 x$ for all real x.
   (e) False. If $x = a\cos x + b\sin x$, it would have to hold for $x = 0$ and $\frac{\pi}{2}$.
   (f) False. For example, $\frac{1}{2} \cdot 3$ is not in this set.
   (g) False. For example $\sqrt{2} \cdot 3$ is not in $\mathbb{Q}$.
   (h) False. The vector space $\mathbb{R}$ has only the subspaces $\{\mathbf{0}\}$ and $\mathbb{R}$.

2. (a) Not a vector space. Axiom S3 fails (and no others).
   (b) Not a vector space. Axiom S5 fails (and no others).
   (c) Not a vector space. Axiom S5 fails (and no others).
   (d) Not a vector space. Axiom S5 fails (and no others).
   (e) Not a vector space. Axiom S2 fails (and no others).
   (f) Not a vector space. Axioms S4 and S5 fail (and no others).

3. (a) Not a vector space. Axiom A5 fails (and no others).
   (b) This is a vector space. The axioms are rules of complex addition and multiplication.
   (c) This is a vector space because $\mathbf{0} + \mathbf{0} = \mathbf{0}$ and $a\mathbf{0} = \mathbf{0}$ for all a.
   (d) Not a vector space. Axioms A1, A4, A5 and S1 all fail.
   (e) Not a vector space. Axiom A1 fails (and no others).
   (f) Not a vector space. Axiom S3 fails (and no others).
   (g) Not a vector space. Axioms S4 and S5 fail (and no others).
   (h) Not a vector space. Axioms S2 and S3 fail (and no others).

SELECTED HINTS AND SOLUTIONS

4. The zero vector is $[0, -1]$, and the negative of $[x\ y]$ is $[-x\ -2-y]$.
   Axiom S3 is verified as follows:
   $$\begin{aligned} a \cdot [x\ y] \dot{+} b \cdot [x\ y] &= [ax,\ ay+a-1] \dot{+} [bx,\ by+b-1] \\ &= [ax+bx,\ (ay+a-1)+(by+b-1)+1] \\ &= [(a+b)x,\ (a+b)y+(a+b)-1] \\ &= (a+b) \cdot [x\ y]. \end{aligned}$$

5. The zero vector is 1, and the negative of a is $\frac{1}{a}$. If we write $\dot{+}$ for the new addition, then $v \dot{+} w = vw$ is the usual product. Then Axiom S2 is verified as follows:
   $$a \cdot (v \dot{+} w) = a \cdot (vw) = (vw)^a = v^a w^a = v^a \dot{+} w^a = a \cdot v \dot{+} a \cdot w.$$

6. (a) Not a subspace. For example, the zero polynomial is not in U.
   (b) Not a subspace. For example, it is not closed under addition.
   (c) Not a subspace. Here, U is not *contained* in $\mathbb{P}_3$. Note that it is a subspace of $\mathbb{P}_4$.
   (d) Subspace.
   (e) Subspace.
   (f) Subspace.

7. (a) Subspace.
   (b) Subspace.
   (c) Subspace.
   (d) Subspace.
   (e) Not a subspace. Not closed under addition.
   (f) Not a subspace. Not closed under addition or scalar multiplication.
   (g) Subspace.

8. (a) Not a subspace. For example, it does not contain the zero function.
   (b) Subspace.
   (c) Not a subspace. For example, it does not contain the zero function.
   (d) Not a Subspace. For example, it does not contain the negative of a function.
   (e) Subspace.
   (f) Subspace.
   (g) Subspace. The verification requires the following formulas from calculus: $(f+g)' = f' + g'$, and $(rf)' = rf'$ for every real number r.
   (h) Subspace. The verification requires the following formulas from calculus: $\int_0^1 [f(x) + g(x)] dx = \int_0^1 f(x) dx + \int_0^1 g(x) dx$, and $\int_0^1 rf(x) dx = r \int_0^1 f(x) dx$ for all real numbers r.

9. (a) $\mathbf{v}_1$ is not in, $\mathbf{v}_2$ is in.
   (c) $\mathbf{v}_1$ is in, $\mathbf{v}_2$ is not in.
   (e) $\mathbf{v}_1$ is not in, $\mathbf{v}_2$ is in because of the trigonometric identity $cos(2x) = cos^2 x - sin^2 x$.

10. (b) No. 1 is not in the span of these vectors.

11. The Gaussian algorithm works just as for the numerical systems in Section 1.2. For (b) we obtain: $\mathbf{x} = 2\mathbf{t}$, $\mathbf{y} = \mathbf{0}$, and $\mathbf{z} = \mathbf{t}$ where $\mathbf{t}$ is an arbitrary vector.

12. (a) 0 is in U (take $A = 0$), and sums and scalar products of vectors of the form AY_0 are again of this form.

13. This is because the zero function is continuous, and sums and scalar multiples of continuous functions are again continuous (a theorem of calculus). So the subspace test applies.

14. (a) If $\mathbf{0}$ and $\mathbf{z}$ are both zero vectors then (using Axiom A4) $\mathbf{0} + \mathbf{z} = \mathbf{z}$ because $\mathbf{0}$ is a zero vector, and $\mathbf{0} + \mathbf{z} = \mathbf{0}$ because $\mathbf{z}$ is a zero vector.

15. If x is any solution then $\mathbf{x} + \mathbf{v} = \mathbf{w}$. Using Axiom A5, add $-\mathbf{v}$ to both sides to get $(\mathbf{x} + \mathbf{v}) + (-\mathbf{v}) = \mathbf{w} + (-\mathbf{v}) = \mathbf{w} - \mathbf{v}$ (using the definition of $\mathbf{w} - \mathbf{v}$). Now Axioms A3, A4 and A5 apply to the left side to give (verify) $\mathbf{x} = \mathbf{w} - \mathbf{v}$.

16. Using several axioms, we get $a(-\mathbf{v}) = a[(-1)\mathbf{v}] = [a(-1)]\mathbf{v} = (-a)\mathbf{v}$, and this equals $-(a\mathbf{v})$ by what is already proved in Theorem 2 (5).

17. By Theorem 2 we have $-\mathbf{0} = (-1)\mathbf{0} = \mathbf{0}$.

18. Let U be a subspace of $\mathbb{R}\mathbf{v}$ containing a nonzero vector $\mathbf{u}$. If $\mathbf{u} = a\mathbf{v}$ where $a \neq 0$, show that every vector in $\mathbb{R}\mathbf{v}$ is a scalar multiple of $\mathbf{u}$.

19. $\text{span}\{a\mathbf{v}\}$ consists of all scalar multiples of $a\mathbf{v}$, and hence of all scalar multiples of $\mathbf{v}$ because $a \neq 0$.

20. (b) Compute $(af_1)(x)$ for any real number a, and any x in D.

21. (a) First compute $(a-b)\mathbf{v}$.

22. (a) First compute $a(\mathbf{v} - \mathbf{w})$.

23. U satisfies the conditions in the Subspace Test.

24. (a) You can directly show that every vector is a linear combination of the spanning vectors, or you can use Theorem 4 as in Example 14.

25. Use the fact that transpose respects linear combinations: $(r_1 A_1 + \cdots + r_n A_n)^T = r_1 A_1^T + \cdots + r_n A_n^T$.

26. Can $BY = 0$ for all B in $\mathbb{M}_{m,n}$?

27. This follows directly from the definitions.

28. If they did, show that they would all be in $\mathbb{P}_n$ for some n.

29. If $\mathbf{u}$ is in U, then $\mathbf{0} = 0\mathbf{u}$ is also in U.

30. Just do it.

31. (a) It clearly holds for $n = 1$. If it holds for some value of $n \geq 1$, consider $n+1$ vectors $\mathbf{v}_1, \mathbf{v}_2, \cdots, \mathbf{v}_n, \mathbf{v}_{n+1}$, we begin by using Axiom S2:

$$\begin{aligned} a(\mathbf{v}_1 + \mathbf{v}_2 + \cdots + \mathbf{v}_n + \mathbf{v}_{n+1}) &= a((\mathbf{v}_1 + \mathbf{v}_2 + \cdots + \mathbf{v}_n) + \mathbf{v}_{n+1}) \\ &= a(\mathbf{v}_1 + \mathbf{v}_2 + \cdots + \mathbf{v}_n) + a\mathbf{v}_{n+1} \\ &= (a\mathbf{v}_1 + a\mathbf{v}_2 + \cdots + a\mathbf{v}_n) + a\mathbf{v}_{n+1} \\ &= a\mathbf{v}_1 + a\mathbf{v}_2 + \cdots + a\mathbf{v}_n + a\mathbf{v}_{n+1} \end{aligned}$$

where the induction assumption was used at the second last line.

5.2 Independence and Dimension

1. (a) False. (b) True. (c) True. (d) False. (e) True. (f) False.
   (g) True. (h) False. (i) False. (j) True. (k) False. (l) False.
   (m) True.

Selected Hints and Solutions

2. (a) Not independent. (b) Independent. (c) Not independent.
   (d) Not independent. (e) Not independent. (f) Independent.
   (g) Not independent. (h) Independent.

3. In each case the dimension d is given below; any d independent vectors is a basis. Theorem 3 and 4 may be useful.

   (a) Dimension is 2. (b) Dimension is 2. (c) Dimension is 2.
   (d) Dimension is 3. (e) Dimension is 2. (f) Dimension is 2.
   (g) Dimension is 1. (h) Dimension is 3. (i) Dimension is 3.
   (j) Dimension is 2. (k) Dimension is 3. (l) Dimension is 3.

4. If the coefficients of each polynomial in a set sum to zero, the same is true for the coefficients of any linear combination of these polynomials.

5. (b) A nontrivial linear combination of the vectors in the dependent set can be used to get a nontrivial linear combination of the vectors in the larger set.

6. (a) Suppose that $az + bz^2 = 0$ where a and b are real. Then $z(a + bz) = 0$, so $a + bz = 0$ because $z \neq 0$ (it is not real). Now argue that $a = 0$ and $b = 0$.

7. $\mathcal{B}$ is independent by Example 2. Use Theorem 4.

8. Since V is finite dimensional, let $\{\mathbf{b}_1, \mathbf{b}_2, \cdots, \mathbf{b}_m\}$ be a (finite) spanning set for V. If $\mathbb{S} = \{\mathbf{s}_1, \mathbf{s}_2, \cdots\}$ then each $\mathbf{b}_i$ is a linear combination of a finite number of the $\mathbf{s}_j$. Show that V is spanned by all these $\mathbf{s}_j$, and then use Theorem 3.

9. If $aI + bA^2 + cA^2 = 0$ where a, b and c are real numbers, multiply by A^2 to conclude $a = 0$. Continue in this way. The generalization: If $A^{n+1} = 0$ but $A^n \neq 0$, then $\{I, A, A^2, \cdots, A^n\}$ is independent. Write out a proof.

10. If $aA + bB = 0$ where a and b are real numbers, take the transpose and apply the hypotheses.

11. Remember that is f and g are functions and a linear combination vanishes, $af + bg = 0$, it means that $(af + bg)(x) = 0(x)$ for all x in the domain, that is $af(x) + bg(x) = 0$ for all x.

12. This is routine matrix algebra. It is important that P and Q are invertible.

13. If $a_1 X_1 + a_2 X_2 + \cdots + a_k X_k = 0$, multiply by A.

14. Think about how $deg\{p(x)q(x)\}$ compares with $deg\, p(x)$ and $deg\, q(x)$.

15. Remember the definition of scalar multiplication in $\mathbb{F}[a, b]$.

16. Show first that $r(a + bx) + s(a_1 + b_1 x) = 0$ if and only if $\begin{bmatrix} a & a_1 \\ b & b_1 \end{bmatrix} \begin{bmatrix} r \\ s \end{bmatrix} = \begin{bmatrix} 0 \\ 0 \end{bmatrix}$. Use Theorem 4.

17. Remember the trigonometric identity for $cos(2\theta)$. The dimension is 2.

18. Two such matrices are $\begin{bmatrix} 1 & a \\ 0 & 0 \end{bmatrix}$ and $\begin{bmatrix} 1 & 0 \\ a & 0 \end{bmatrix}$ for any real number a.

19. Use the Hint.

20. Note that all even powers of x are even polynomials, and odd powers are odd. The dimensions depend on whether n is even or odd.

21. See Exercise 19, Section 5.1.

22. Use part (1) of Theorem 5 and the preceding exercise. In $\mathbb{R}^2$, show that the dimension 1 subspaces are the lines through the origin.

23. Use Exercise 22. Show that $\mathbb{R}\mathbf{v} = \mathbb{R}(a\mathbf{v})$ for any $a \neq 0$ in $\mathbb{R}$.

24. (b) and (d) are independent, the other two are not.

25. Consider the proof of Theorem 3 §4.2.

26. Use Theorem 3.

27. Use the definition of independence.

28. Use the definition of independence.

29. If it is not independent, show that some $\mathbf{v}_i$ is a linear combination of the others, and hence that the rest of the $\mathbf{v}_i$ span V.

30. If some nontrivial linear combination vanishes, show that V can be spanned by all the $\mathbf{v}_i$ except one.

31. It contains $\mathbb{P}$.

32. By Theorem 3 choose a basis $\{\mathbf{v}_1, \cdots, \mathbf{v}_k, \mathbf{v}_{k+1}, \cdots, \mathbf{v}_m\}$ of W such that $\{\mathbf{v}_1, \cdots, \mathbf{v}_k\}$ is a basis of U. Use this to construct X.

33. (a) show that $\mathbb{S}$ contains an infinite independent set.

34. Note that there is a sequence in U beginning with *any* choice of x_0 and x_1. Use the Hint.

35. (b) If a linear combination vanishes, use the fact that $U \cap W = \{\mathbf{0}\}$.
    (c) and (d) Use the Hints.

5.3 Linear Transformations

1. (a) False (b) False (c) False (d) True (e) False (f) False
   (g) True (h) True (i) False (j) True (k) False (l) False

2. (a) True (b) False (c) True (d) True
   (e) True (f) False (g) True (h) False

3. (a) Showing that $T(rz) = rT(z)$ requires the fact that $\bar{r} = r$ for all real r.
   (d) This uses Theorem 1 §4.5.
   (i) For this you must be careful about the definition of a constant function.

4. (b) $rank(A+B)$ need not equal $rankA + rankB$. (However, $rank(rA) = r\,rankA$ for all r and A – can you prove it?)

5. (b) $T \begin{bmatrix} a & b \\ c & d \end{bmatrix} = \begin{bmatrix} a+c \\ 2a+b \end{bmatrix}$.

6. Use the fact that T is linear to show first that $T(\mathbf{e}) = 5\mathbf{e} - \mathbf{f}$ and $T(\mathbf{f}) = 3\mathbf{e}$.

7. Use Theorem 1.

8. Begin with the definition of subtraction: $\mathbf{v} - \mathbf{v}_1 = \mathbf{v} + (-\mathbf{v}_1)$.

9. We have $T(1) = \mathbf{v}_0$ for some $\mathbf{v}_0$ in V. Use the Hint.

10. (a) If $\mathcal{B} = \{\mathbf{e}_1, \cdots, \mathbf{e}_n\}$ let $T(\mathbf{e}_i) = y_i$ for each i.

11. Use Theorem 3 §5.2 and Theorem 3.

Selected Hints and Solutions 575

12. (b)⇒(c). We have $imT = \{T(\mathbf{v}) \mid \mathbf{v} \text{ in } V\}$. So if $imT = \{\mathbf{0}\}$ then $T(\mathbf{v}) = \mathbf{0}$ for all $\mathbf{v}$ in V.

13. (c) Given $\mathbf{w}$ in W, write $\mathbf{w}$ as a linear combination of the vectors in $\mathcal{C}$, and use the linearity of T.

14. (a) Use Theorem 4.

15. (b) Let $\{\mathbf{u}_1, \cdots, \mathbf{u}_m\}$ be a basis of U and extend it (by Theorem 3§5.2) to a basis
    $\{\mathbf{u}_1, \cdots, \mathbf{u}_m, \mathbf{u}_{m+1}, \cdots, \mathbf{u}_n\}$ of V. Then use Theorem 3.

16. Use the Dimension Theorem.

17. If you can think of a linear transformation $T : \mathbb{M}_{m,n} \to \mathbb{M}_{m,n}$ with $kerT = U$ and $imT = W$, then the Dimension Theorem does most of the work for you.

18. Exhibit the set of solutions as the kernel of a linear transformation.

19. Use the Hint.

20. Use the Hint.

21. Use the Hint.

22. Find a linear transformation T such that imT is the set of all matrices of the desired form.

23. Adapt the suggestion in the preceding exercise. Show that $q(x+1) = q(x)$ if and only if $q(x)$ is constant. (If $deg\, q(x) = n \geq 1$, look at the coefficient of x^{n-1}.)

24. $\{1, x, x^2, \cdots, x^n\}$ is a basis of $\mathbb{P}_n$. You will need Theorem 2.

25. (b) Verify first that a complex number w is pure imaginary (that is $w = bi$ for some real number b) if and only if $\bar{w} = -w$. Then use the Hint and the linearity of T.

26. (b) Use the Hint, the linearity of T and Theorem 2.

27. By the Hint and our hypothesis, $T(E_{ii}) = T(E_{i1}E_{1i}) = T(E_{1i}E_{i1}) = T(E_{11})$ for each i. Now look at $T(E_{ij})$ where $i \neq j$.

28. (a) Write $dimV = n$ and $dimW = m$ for simplicity, and suppose that $\{\mathbf{v}_1, \cdots, \mathbf{v}_n\}$ and $\{\mathbf{w}_1, \cdots, \mathbf{w}_m\}$ are bases of V and W respectively. If $n \leq m$ define $T : V \to W$ by $T(\mathbf{v}_i) = \mathbf{w}_i$ for $i = 1, 2, \cdots, n$. For the converse, use the Dimension Theorem.

5.4 Isomorphisms and Matrices

1. (a) False   (b) True   (c) False   (d) False
   (e) False   (f) True   (g) True   (h) False

2. (b) $T^{-1}(z) = \bar{z}$.
   (d) $T^{-1}\{[r\ s]^T\} = r + (s-r)x$.
   (f) $T^{-1}(\mathbf{v}) = \frac{1}{k}\mathbf{v}$.

3. It may help to look at the standard matrix of T.

4. If r is any real number then $2^r > 0$, so 2^r is in V.

5. Use Theorems 4 and 5.

6. (b) $C_{\mathcal{B}}(\mathbf{v}) = [1\ 0\ 1\ -2]^T$.

7. (b) $M_{\mathcal{DB}}(T) = \begin{bmatrix} 1 & 0 & 0 & 0 \\ 0 & 0 & 1 & 0 \\ 0 & 1 & 0 & 0 \\ 0 & 0 & 0 & 1 \end{bmatrix}$.

8. (b) $T \begin{bmatrix} x \\ y \\ z \end{bmatrix} = \begin{bmatrix} x & -x+2y+z \\ y+2z & x \end{bmatrix}$.

9. $M_{\mathcal{DB}}(T) = \begin{bmatrix} 0 & 1 & 0 & 0 \\ 0 & 0 & 2 & 0 \\ 0 & 0 & 0 & 3 \end{bmatrix}$.

10. First confirm that $C_{\mathcal{E}}(Y) = Y$ for all Y in $\mathbb{R}^m$.

11. If $\mathcal{D} = \{\mathbf{d}_1, \cdots, \mathbf{d}_n\}$, what is $C_{\mathcal{D}}(\mathbf{d}_j)$?

12. As in Exercise 10, $C_{\mathcal{E}}(Y) = Y$ for all Y in $\mathbb{R}^n$.

13. (b) If $(S \circ T)(\mathbf{v}) = \mathbf{0}$ then $T(\mathbf{v})$ is in $\ker S$; use Theorem 4 §5.3.

14. (b) Show first that $S(T(\mathbf{v})) = S(T_1(\mathbf{v}))$ for all $\mathbf{v}$ in V.

15. Use the definitions of "one-to-one" and "onto", and Theorem 4 §5.3.

16. Show first that T^{-1} exists (the Corollary to Theorem 5 §5.3 is useful), then use Theorems 4, 5 and 6.

17. The verifications are routine. This is in contrast to the finite dimensional situation in the preceding exercise, and also to the Corollary to Theorem 5 §5.3.

18. Given the Hint, show that W has a basis of the form $\{T(\mathbf{b}_1), T(\mathbf{b}_2), \cdots, T(\mathbf{b}_n), \mathbf{d}_1, \mathbf{d}_2, \cdots, \mathbf{d}_k\}$. Then define S by $S(T(\mathbf{b}_i)) = \mathbf{b}_i$ for $1 \leq i \leq n$, and $S(\mathbf{d}_j) = \mathbf{0}$ for $1 \leq i \leq k$ and use Theorem 2 §5.3.

19. Given the Hint, define $S : W \to V$ by $S(T(\mathbf{b}_i)) = \mathbf{b}_i$ for $1 \leq i \leq m$. Use Theorem 2 §5.3.

20. (c) No, not even if $n = 2$.

21. (b) Use (a) in the case $m = n$.
    (c) If E_j is column j of the identity matrix, observe that AE_j is column j of A.
    (d) Use the Hint. This also follows from Theorem 10 and part (e) of this exercise.
    (e) You will need the fact that $M_{\mathcal{F}}(Y) = Y$ for all Y.

22. Compare columns in $M_{\mathcal{DB}}(T) = M_{\mathcal{DB}}(T_1)$, use the fact that the coordinate map $C_{\mathcal{D}}$ is one-to-one, and use Theorem 2 §5.3.

23. Use the Hint.

24. Use the Hint.

25. To show that $S(\mathbf{w}+\mathbf{w}_1) = S(\mathbf{w})+S(\mathbf{w}_1)$, begin by computing $T\{S(\mathbf{w}+\mathbf{w}_1)\}$ and $T\{S(\mathbf{w}) + S(\mathbf{w}_1)\}$.

26. (b) Use the Hint.

5.5 Linear Operators and Similarity

1. (a) False. (b) True. (c) True. (d) True by Theorems 3 and 4. (e) False. (f) False. (g) True. (h) False.

2. (a) $M_\mathcal{B}(T) = \begin{bmatrix} 2 & -1 & 0 \\ 1 & 1 & 1 \\ -1 & 0 & 1 \end{bmatrix}$.

3. (a) $M_\mathcal{B}(T) = \begin{bmatrix} 1 & 1 & 1 & 0 \\ -1 & 0 & 2 & 0 \\ 0 & 1 & 0 & 1 \\ 0 & 0 & 1 & -1 \end{bmatrix}$.

4. $T^{-1}(2x - 3) = 5 + 4x$.

5. $T^{-1}\begin{bmatrix} 2 & 3 \\ -1 & 1 \end{bmatrix} = \begin{bmatrix} 0 & 4 \\ 1 & 1 \end{bmatrix}$.

6. The standard basis of matrix units.

7. (c) $\{1 - x, 2 + 3x\}$.

8. (b) No, 1 is in the kernel.

9. (b) $P_{\mathcal{B} \leftarrow \mathcal{D}} = \begin{bmatrix} 2 & 0 & 1 \\ -1 & 1 & 1 \\ 0 & -1 & 1 \end{bmatrix}$.

10. (b) $T^{-1}(\mathbf{b}_2) = \mathbf{b}_1 - \mathbf{b}_3$.

11. $-(15\mathbf{b}_1 + 18\mathbf{b}_2 + 7\mathbf{b}_3)$.

12. If $\mathcal{B} = \{\mathbf{b}_1, \cdots, \mathbf{b}_n\}$, define T so that $T(\mathbf{b}_j)$ has column j of A as its $\mathcal{B}$-coordinate vector.

13. T is a scalar operator–see Example 5 §5.3.

14. (a) $\{\mathbf{v}, \mathbf{v} - 2\mathbf{w}\}$. Lemma 2 allows you to pass back and forth between eigenvectors of T and those of $M_\mathcal{B}(T)$.

15. $M_\mathcal{B}(T) = diag(1, 2, 3, \cdots, n, n+1)$, which is invertible. Use Theorem 1.

16. $M_\mathcal{B}(T) = diag(1 + k, 1 + k, 1 + k, 1 - k)$. Use Theorem 1.

17. (b) If $\vec{n}$ is any nonzero vector perpendicular to the line, argue geometrically that $T(\vec{n}) = -\vec{n}$.

18. Use the definition of $det T$, together with part (2) of Theorem 1.

19. Use Lemma 4 §4.7.

20. (b) Show that R is an isomorphism by Theorem 2 §5.4, and that $R \circ S = T$ by Theorem 2 §5.3, applied to the basis $\mathcal{B}$.

5.6 Invariant Subspaces

1. (a) True (b) True (c) False (d) True
   (e) True (f) False (g) True (h) True

2. Show that $T^2 = T$ by showing that $T^2(\mathbf{b}_i) = T(\mathbf{b}_i)$ for each i. $\mathcal{B}_0 = \{\mathbf{b}_1, \mathbf{b}_2 + 2\mathbf{b}_4\}$.

3. (a) By Lemma 1 is suffices to show that $T(1)$ and $T(x + x^2)$ are both in U.
   (b) Use Theorem 1.
   (c) Here Theorem 3 §2.2 is very useful.

4. (a) Use the definition of T-invariant subspace.

5. Use the definition of a subspace.

6. In each case you must show that $U \cap W = \{\mathbf{0}\}$ (that is, the only vector in both U and W is the zero vector) and $U + W = V$ (that is, every vector in V is the sum of a vector in U and one in W).

7. $M_\mathcal{B}(T) = \begin{bmatrix} P & 0 \\ 0 & Q \end{bmatrix}$ where $P = \begin{bmatrix} 2 & 1 \\ -5 & -2 \end{bmatrix}$ and $Q = \begin{bmatrix} 1 & 2 \\ -1 & -1 \end{bmatrix}$.

8. (a) $\mathbb{R}\mathbf{v}$ is a subspace for every vector $\mathbf{v}$ in V.
   (b) If $\{\mathbf{v}_1, \cdots, \mathbf{v}_n\}$ is a basis of V and $T(\mathbf{v}_1) = \lambda \mathbf{v}_1$ and $T(\mathbf{v}_i) = \mu \mathbf{v}_i$, consider $T(\mathbf{v}_1 + \mathbf{v}_i)$.

9. If $U \neq \{\mathbf{0}\}$ is T-invariant for every T, choose $\mathbf{0} \neq \mathbf{u} \in U$. Given any vector $\mathbf{v}$ in V, show there exists an operator T such that $T(\mathbf{u}) = \mathbf{v}$. Theorem 3 §5.3 is useful here.

10. Use Theorem 4.

11. If $a\mathbf{u} + b\mathbf{w} = \mathbf{0}$ with $a \neq 0$, show that $\mathbf{u} \in U \cap W$.

12. Apply Theorem 4 to the subspace $U \oplus W$.

13. Use Theorems 3 and 5.

14. (b) If $dim(E_\lambda(T)) + dim(E_\mu(T)) = n$, use the preceding exercise, Theorem 4 and Example 2.
    (c) Use Theorem 4 to show that $dim(E_\lambda(T) + E_\mu(T)) \geq 2$, and conclude that $E_\lambda(T) + E_\mu(T) = V$. Use Theorem 5.
    (d) Let $T : \mathbb{R}^3 \to \mathbb{R}^3$ be multiplication by any non-diagonalizable 3×3 matrix.

15. Use the Hint and the definition of a T-simple space.

16. (b) If $\mathbf{v}$ is in imS, say $\mathbf{v} = S(\mathbf{w})$, then $T(\mathbf{v}) = (T \circ S)(\mathbf{w}) = (S \circ T)(\mathbf{w})$ is in imS. Hence imS is T-invariant.

17. We have $U = span\{\mathbf{v}, T(\mathbf{v}), T^2(\mathbf{v}), \cdots, T^{m-1}(\mathbf{v})\}$. If $\mathbf{w}$ is one of these generators of U, it is clear that $T(\mathbf{w})$ is in U, except if $\mathbf{w} = T^{m-1}(\mathbf{v})$. For this, use equation (**) preceding Theorem 6.

18. (a) False. Construct an example where U (say) is two-dimensional and the restriction of T to U is not diagonalizable.
    (b) True.

19. Verify that V has no subspaces *at all* except $\{\mathbf{0}\}$ and V.

20. This requires only the definition of eigenvectors, and of T-invariant subspaces.

21. (a) Use the preceding exercise.

22. If $\{\mathbf{v}, \mathbf{w}\}$ is a basis of eigenvectors of T, show that $\mathbb{R}\mathbf{v} \oplus \mathbb{R}\mathbf{w} = V$. For the converse, if $V = U \oplus W$ where U and W are T-invariant subspaces other than $\{\mathbf{0}\}$ and V, what are $dimU$ and $dimW$?

23. What is the action of the restriction of T to the subspace $E_\lambda(T)$?

24. Just use the definition of T-invariant subspaces.

25. (a) You must first check that T is **well defined**. (If $\mathbf{v} = \mathbf{u} + \mathbf{w}$ where $\mathbf{u}$ is in U and $\mathbf{w}$ is in W, and also $\mathbf{v} = \mathbf{u}_1 + \mathbf{w}_1$ where $\mathbf{u}_1$ is in U and $\mathbf{w}_1$ is in W, then $T(\mathbf{v}) = \mathbf{u}$ and also $T(\mathbf{v}) = \mathbf{u}_1$, so T makes no sense unless $\mathbf{u} = \mathbf{u}_1$.) Use Theorem 3.
    (b) Use the Hint.

Selected Hints and Solutions

26. Mimic the solution of Example 12.
27. (b) Let $\mathcal{B} = \{\mathbf{b}_1, \cdots, \mathbf{b}_n\}$. Then column j of $M_\mathcal{B}(S+T)$ is $C_\mathcal{B}\{(S+T)(\mathbf{b}_j)\} = C_\mathcal{B}\{S(\mathbf{b}_j) + T(\mathbf{b}_j)\}$. Use the fact that $C_\mathcal{B}$ is linear.

5.7 General Inner Products

1.

	(a)	(b)	(c)	(d)	(e)	(f)
P1	True	True	True	True	True	True
P2	False	True	True	True	False	True
P3	False	True	True	True	False	True
P4	False	False	True	False	False	False

2. Use the fact that $\mathbf{0} = 0\mathbf{v}$ for every vector $\mathbf{v}$.
3. Use the fact that $\|\mathbf{z}\|^2 = \langle \mathbf{z}, \mathbf{z} \rangle$ for all vectors $\mathbf{z}$. Use the linearity of the inner product.
4. (a) See the comment in the preceding exercise.
   (b) If $\langle , \rangle$ and $\langle , \rangle_1$ are inner products, write the corresponding norms as $\|\mathbf{v}\| = \sqrt{\langle \mathbf{v}, \mathbf{v} \rangle}$ and $\|\mathbf{v}\|_1 = \sqrt{\langle \mathbf{v}, \mathbf{v} \rangle_1}$. Then having the same norm function means that $\|\mathbf{v}\| = \|\mathbf{v}\|_1$ for every vector $\mathbf{v}$.
5. First reduce it to showing that if $\langle \mathbf{v}, \mathbf{v}_i \rangle = 0$ holds for each i then $\mathbf{v} = \mathbf{0}$. For this, write $\mathbf{v}$ as a linear combination of the $\mathbf{v}_i$.
6. (a) Since the domain is $\{1, 2, \cdots, n\}$ here, we have $f = 0$ if and only if $f(1) = f(2) = \cdots = f(n) = 0$.
   (c) Use the Expansion Theorem (Theorem 5).
7. Combine the Hint and Theorem 2.
8. The dot product is an inner product on $\mathbb{R}^n$. Use the fact that $C_\mathcal{B}$ is one-to-one and linear.
9. (b) Use the Hint and the fact that $X^T A Y$ is symmetric (it is 1×1) for all X and Y and all $n \times n$ matrices A.
10. (a) If the orthogonal basis is p_0, p_1 and p_2, we have $p_1(x) = x - 1$.
    (b) If the orthogonal basis is p_0, p_1 and p_2, we have $p_1(x) = x - \frac{1}{2}$.
11. By Gram Schmidt, $U = span\{I, K\}$ where $K = \begin{bmatrix} 0 & 1 \\ 1 & 0 \end{bmatrix}$. Answer: $\frac{1}{2}\begin{bmatrix} 4 & 1 \\ 1 & 4 \end{bmatrix}$.
12. Look at the argument leading to Theorem 2, and at the definition of $C_\mathcal{B}(\mathbf{v})$.
13. Use the Hint.
14. $\|\mathbf{v}\|^2 = \langle \mathbf{v}, \mathbf{v} \rangle$.
15. Use the Hint and the linearity of the inner product.
16. You will need the fact that $U \subseteq U + W$ and $W \subseteq U + W$. Use the definition of $U^\perp$ and $W^\perp$.
17. (a) Use the fact that $U \cap W$ is contained in both U and W.
    (c) Exercise 25 §4.3 asserts that $dim(P + Q) = dim P + dim Q - dim(P \cap Q)$ for any subspaces P and Q of a finite dimensional space. Apply this twice, beginning with $dim(U^\perp + W^\perp)$. Then apply the Hint.

18. Use Theorem 10 with $U = U_2$.

19. To show that $\{\mathbf{b}_1, \mathbf{b}_2, \cdots, \mathbf{b}_n\}$ is independent for each n, use induction on n together with Lemma 1 §5.2.

20. Use the definition of $\langle , \rangle$.

21. Use the Hints.

22. (a) For (3), calculate $\|r\mathbf{v}\|^2 = \langle r\mathbf{v}, r\mathbf{v} \rangle$.
    (b) If $\| \ \|$ came from an inner product, there would (by Theorem 2) be a matrix P such that $[x \ y] P \begin{bmatrix} x \\ y \end{bmatrix} = |x| + |y|$ for all x and y. Show that this is impossible by examining various choices for x and y.

23. Using the Hint, this is a routine exercise in integration.

24. $p_6(x) = \frac{1}{16}(231x^6 - 315x^4 + 105x^2 - 5)$.

25. Recall that a function f is even if $f(-x) = f(x)$, and f is odd if $f(-x) = -f(x)$. Use Theorem 12 and induction.

26. Use Theorem 12 and induction on m.

27. (b) Use (a) and induction on m.

Index

absolute value, 136
altitude, 199
angle, 505
    acute, 506
    cosine, 506
    radian measure, 505
    right, 506
    sine, 506
    standard position, 505

back-substitution, 23
basis, 231, 269, 415
    change matrix, 446
    coordinate vector, 446
    standard, 231

Cauchy inequality, 293
Cauchy-Schwarz inequality, 488
Cholesky
    algorithm, 344
    factorization, 343
cofactor, 99
column space, 280
complex column
    set
        orthogonal, 381
        orthonormal, 381
complex matrix
    characteristic polynomial, 380
    conjugate, 378
    eigenvalue, 380
    eigenvector, 380
    orthogonal, 381
complex number, 147
    absolute value, 149
    argument, 154
    conjugate, 148
    conjugate matrix, 158
    Euler's formula, 154
    imaginary part, 147
    imaginary unit, 147
    length, 377
    modulus, 149
    polar form, 154
    pure imaginary, 147
    real part, 147
    roots of unity, 157
    unit circle, 155
complex plane, 148
    imaginary axis, 148
    real axis, 148
conic, 30
coordinate, 360
    change matrix, 361
    vector, 360
coordinate vector, 218
cosine, 506

determinant, 99
    cofactor, 99
    Cramer's rule, 114
    elementary operation, 101
    Laplace expansion, 100
    sign, 99
dimension, 270, 415
direction, 180
direction vector, 206
discriminant, 150
dynamical system, 135
    matrix recurrence, 135
    trajectory, 142

equivalence relation, 129, 442
Euclidean n-space, 250
    basis, 269
    column space, 280, 286
    dimension, 270
    eigenspace, 253
    image, 252, 286
    null space, 252, 286

row space, 280
spanning set, 256
standard basis, 270
Steinitz Exchange Lemma, 276
subspace, 251
  orthogonal complement, 307
  orthogonal projection, 309
  proper, 251
  zero, 251
vector, 250
  linear combination, 254
evaluation, 130, 426

Fibonacci sequence, 165
Fourier
  approximation, 497
  coefficients, 497
  series, 498
function, 402
  constant, 411
  differentiable, 173
  domain, 402
  even, 473
  feasible region, 345
  objective, 345
  odd, 473
  pointwise operation, 173, 402
  real valued, 173

Gaussian quadrature, 490
geometry
  analytic, 179
  axis, 179
  coordinates, 179
  origin, 179
  synthetic, 179
Gram-Schmidt Algorithm, 299
graph, 46
  adjacency matrix, 46
  directed, 46
  edge, 46
  path, 46
  vertex, 46

image, 374
independent, 260, 413, 417
  lemma, 271, 417
induction, 509
inner product, 377, 485
  space, 485

norm, 488
standard, 377
interpolating polynomial, 169

kernel, 374

Lagrange interpolation polynomial, 490
Lagrange polynomials, 489
least squares approximation, 317
Legendre polynomials, 500
  differential equation, 501
  generating function, 501
  Rodrigue's formula, 501
line, 206
  direction vector, 206
  scalar equations, 207
  vector equation, 207
linear combination, 32, 412
  trivial, 260, 412
  vanishing, 261, 412
linear equation, 14
  coefficient, 14
  constant term, 14
  solution, 15
    nontrivial, 30
    trivial, 30
  system
    forward substitution, 79
  system of, 15
    associated homogeneous, 42
    augmented matrix, 16
    back substitution, 79
    basic solution, 32
    best approximation, 313
    coefficient matrix, 16
    consistent, 15, 25
    Cramer's rule, 114
    elementary row operation, 17
    Gaussian Elimination, 22
    general solution, 22
    homogeneous, 30
    inconsistent, 15, 25
    leading 1, 20
    leading variables, 22
    normal equations, 313
    parameters, 16
    reduced row-echelon, 20
    row-echelon, 20
    solution to, 15
  variable, 14

Index

linear functional, 374, 437
linear operator, 425
    identity, 426
    scalar, 426
    zero, 426
linear programming, 348
    simplex algorithm, 348
linear transformation, 425
    codomain, 425
    composite, 442
    domain, 425
    equal, 428
    evaluation, 426
    fundamental identities, 444
    image, 425, 430
    inverse, 444
    kernel, 430
    one-to-one, 432
    onto, 432
LU-Algorithm, 81

magnitude, 180
Markov chain, 89
    graph, 90
    probabibilty vector, 89
    regular, 90
    stage, 89
    state, 89
    state vector, 89
    steady state, 89
    stochastic matrix, 89
    transition matrix, 89
    transition probability, 89
matrix, 2
    adjacency, 46
    adjoint, 55, 111
    adjoint formula, 112
    adjugate, 111
    associative law, 39
    block, 43
    block form, 43
    block lower triangular, 110
    block partitioning, 43
    block upper triangular, 110
    characteristic polynomial, 121
    coefficient matrix, 42
    column, 2
    column matrix, 2
    commute, 38
    companion, 481
    compatible for multiplication, 38, 45
    conjugate, 158
    defective, 128
    determinant, 55
    diagonal, 13, 50, 124
    diagonalizable, 125
    diagonalize, 120
    difference, 4
    distributive law, 40
    dot product, 36
    eigenvalue, 121, 253
        dominant, 136
        multiplicity, 127
    eigenvector, 121, 253
        basic, 127
    elementary, 69
    entries, 2
    equal, 3
    identity, 39
    inverse, 54
    invertible, 54
    leading column, 80
    lower reduced, 80
    lower triangular, 62, 78
    LU-factorization, 80
    main diagonal, 8
    negative, 5
    normal, 384
    of constants, 42
    of variables, 15, 42
    orthogonal, 326
    orthogonally diagonalizable, 328
    permutation, 83
    polar decomposition, 394
    positive semidefinite, 393
    powers, 38
    principal axes, 329
    product, 37
    rank, 24, 281
    recurrence, 135
    row, 2
    row matrix, 2
    scalar product, 6
    similar, 128, 366
    similarity class, 367
    skew-symmetric, 431
    square matrix, 2
    stochastic, 89
    sum, 4

symmetric, 10
trace, 332
transpose, 9
triangular, 62, 78
unit, 407
upper triangular, 62
Vandermonde, 170
Vandermonde determinant, 170
zero matrix, 5
Moore-Penrose inverse, 393

normal, 211
numerical integration, 490

operator
    characteristic polynomial, 461
    determinant, 461
    diagonalizable, 462
    eigenspace, 462
    eigenvector, 462
    idempotent, 477
    inverse, 375
    invertible, 375
    rank, 461
    reducible, 476
    restriction, 470
    trace, 461
orthogonal, 198, 489
orthonormal, 489

Parseval's Formula, 502
period, 508
    periodic functions, 508
plane, 211
    normal, 211
    scalar equation, 211
    vector equation, 211
polynomial, 130, 401
    characteristic, 121
    coefficients, 401
    companion matrix, 481
    constant coefficient, 401
    degree, 401
    evaluation, 130, 426
    even, 423
    indeterminant, 401
    leading coefficient, 401
    odd, 423
    root, 121
    zero, 401

polynomials, 511
principal submatrix, 342
probability, 86
pseudo-inverse, 393
Pythagoras' Theorem, 183, 190, 296, 489

quadratic
    irreducible, 150
quadratic form, 337
    diagonalized, 338
quadratic formula, 150
    discriminant, 150

random variable, 348
    expectation, 349
    mean, 349
    principal components, 350
    standard deviation, 349
    total variance, 350
    variance, 349
Rayleigh quotient, 397
recurrence relation, 162
    Fibonacci sequence, 165
    linear, 162
reflection, 375
rhombus, 198
Rodrigue's formula, 501
root, 121
row space, 280

scalar, 180
similar triangle, 506
sine, 506
singular value decomposition, 388
    condition number, 391
    fundamental subspaces, 392
    singular values, 388
Spectral Theorem, 384
standard position, 505
statistics
    principal components, 350
    total variance, 350
    trend analysis, 494
subspace, 251, 405
    complement, 473
    direct sum, 473
    intersection, 260, 424, 472
    invariant, 469
    proper, 251

Index

SVD
  summation notation, 47
  zero, 251
sum, 424, 472

singular value decomposition, 391

transformation, 228, 354
  B-matrix, 456
  DB-matrix, 449
  action, 354, 358
  composite, 238, 357, 442
  compression, 236
  distance preserving, 368
  expansion, 235
  identity, 229, 354
  image, 235, 354
  inverse, 240
  invertible, 240
  isometry, 242, 368
  isomorphism, 439
  linear, 231, 354, 425
  standard matrix, 356
  linear operator, 354
  matrix, 229, 355
  projection, 229
  reflection, 228, 229
  rotation, 233
  shear, 236
  standard matrix, 356
  translation, 231
  zero, 229, 354
triangle, 506
triangle inequality, 294, 488
unit ball, 347
unit circle, 155, 505

Vandermonde, 170
vector, 180, 250, 399
  addition, 184, 400
  parallelogram law, 184
  tip-to-tail law, 185
  angle, 196, 293
  between, 373
  components, 181
  coordinate vector, 218
  difference, 186
  direction, 180, 206
  distance, 294

dot product, 194, 291
equality, 183
geometric, 180
geometry, 182
length, 180, 291
magnitude, 180
matrix form, 181
negative, 182, 184, 400
normalizing, 295
orthogonal, 198, 295
orthogonal set, 295, 489
orthonormal set, 295
parallelogram, 184
determined, 184
position, 205
position vector, 181
projection, 200
scalar multiplication, 188
geometric description, 188
scalar product, 182
standard position, 181
subtraction, 186
sum, 182
tail, 180
tip, 180
unit, 293
zero, 182, 183, 400
vector space, 399
axioms, 399
basis, 415
dimension, 415
finite dimensional, 417
infinite dimensional, 416
inner product, 485
intersection, 424
isomorphic, 439
proper subspace, 405
scalar, 399
scalar multiplication, 400
standard basis, 415
subspace, 251, 405
complement, 473
direct sum, 473
intersection, 472
invariant, 469
orthogonal complement, 492
proper decomposition, 478
sum, 472
sum, 424
T-irreducible, 478

T-simple, 478
zero space, 403
vectors
   dependent, 264, 417
   independent, 260, 413, 417, 495
   linear combination, 406
   linear dependent, 417
   linear independent, 413
   linearly independent, 260, 495
   normalize, 489
   orthogonal, 488
   orthogonal set, 495
   orthonormal set, 489
   span, 406